Viability Theory

Jean-Pierre Aubin • Alexandre M. Bayen
Patrick Saint-Pierre

Viability Theory

New Directions

Second Edition

Springer

Prof. Jean-Pierre Aubin
VIMADES
14 rue Domat
75005 Paris
France
aubin.jp@gmail.com

Prof. Dr.-Ing. Alexandre M. Bayen
University of California at Berkeley
Electrical Engineering and Computer Sciences
Civil and Environmental Engineering
Sutardja Dai Hall 642
Berkeley CA 94720
USA
bayen@berkeley.edu

Prof. Patrick Saint-Pierre
Université Paris Dauphine
LEDA-SDFi
12 rue de la Roue
92140 Clamart
France
patrick.saint.pierre@gmail.com

ISBN 978-3-642-16683-9 e-ISBN 978-3-642-16684-6
DOI 10.1007/978-3-642-16684-6
Springer Heidelberg Dordrecht London New York

Library of Congress Control Number: 2011932723

First edition was published with Birkhäuser Boston.

Cover design: eStudio Calamar, Berlin/Figueres

Printed on acid-free paper

Springer is part of Springer Science+Business Media (www.springer.com)

À Pierre-Cyril Aubin-Frankowska et à sa maman, Hélène Frankowska,

À la mémoire de François Bayen,

À Catherine, François, Claire, Jean-Benoît, Guillaume Saint-Pierre et à Odile Perrier-Saint-Pierre.

Acknowledgments

The authors would like to thank Hélène Frankowska for her implicit and explicit contributions to the writing of the book, as well as the members of the "viability group" who participated to its elaboration: Anne Bru-Gorre, Pierre Cardaliaguet, Luxi Chen, Christian Claudel, Éva Crück, Anya Désilles, Gaël Désilles, Olivier Dordan, Halim Doss, luc Doyen, Marie-Hélène Durand, Alexandra Fronville, Thomas Lorenz, Sophie Martin, Juliette Mattioli, Frank Maurin, Alberto Murillo, Laurent Najman, Dominique Pujal, Marc Quincampoix and Nicolas Seube without forgetting Alain Haurie, Michèle Breton and Georges Zaccour.

We also would like to thank André Giroux for his carcful reading of the first part of this book.

The authors want to thank Claire Tomlin, Ian Mitchell, Meeko Oishi, Shankar Sastry, George Pappas, John Lygeros and Craig Evans for fruitful discussions leading to the initial formulation of the landing problem and reachability analysis. The chapter on the traffic and conservations laws greatly benefited from fruitful conversations with Carlos Daganzo, Miroslav Krstic, Michael Cassidy, Benedetto Piccoli, Christian Claudel, Sebastien Blandin and Xavier Litrico.

The simulations using the software of the viability algorithms are due to Patrick Saint-Pierre and the figures to Alexandre Bayen and Patrick Saint-Pierre.

The experiment on the pioneer has been made by Éva Crück, Gaël Désilles, Benjamin Frenais de Coutard, Aurélien Gosselin, Fabrice Leroux at Délégation Générale à l'Armement (DGA). The design and implementation of the viability algorithm used to compute the optimal trajectories of the submarine was done by Andrew Tinka. The submarine used for the experiments presented in this book was manufactured by the University of Porto, by the team of Joao Sousa, who is gratefully acknowledged for enabling field experimental tests in the United States.

The authors would like to thank Werner Müller for his 6-years patience as well as Catriona Byrne and Martina Bihn for their help for publishing this book.

Contents

Part I Viability Kernels and Examples

Part III First-Order Partial Differential Equations

Part IV Appendices

List of Figures

Chapter 1
Overview and Organization

Viability theory designs and develops mathematical and algorithmic methods for investigating the *adaptation to viability constraints of evolutions governed by complex systems under uncertainty* that are found in many domains involving living beings, from biological evolution to economics, from environmental sciences to financial markets, from control theory and robotics to cognitive sciences. It involves interdisciplinary investigations spanning fields that have traditionally developed in isolation.

The purpose of this book is to present an initiation to applications of viability theory, explaining and motivating the main concepts and illustrating them with numerous numerical examples taken from various fields.

Viability Theory. New Directions plays the role of a second edition of *Viability Theory*, [18, Aubin] (1991), presenting advances occurred in set-valued analysis and viability theory during the two decades following the publication of the series of monographs: *Differential Inclusions. Set-Valued Maps and Viability Theory*, [25, Aubin & Cellina] (1984), *Set-valued Analysis*, [27, Aubin & Frankowska] (1990), *Analyse qualitative*, [85, Dordan] (1995), *Neural Networks and Qualitative Physics: A Viability Approach*, [21, Aubin] (1996), *Dynamic Economic Theory: A Viability Approach*, [22, Aubin] (1997), *Mutational, Morphological Analysis: Tools for Shape Regulation and Morphogenesis*, [23, Aubin] (2000), *Mutational Analysis*, [150, Lorenz] (2010) and *Sustainable Management of Natural Resources*, [77, De Lara & Doyen] (2008).

The monograph *La mort du devin, l'émergence du démiurge. Essai sur la contingence et la viabilité des systèmes*, [24, Aubin] (2010), divulges vernacularly the motivations, concepts, theorems and applications found in this book. Its English version, *The Demise of the Seer, the Rise of the Demiurge. Essay on contingency, viability and inertia of systems*, is under preparation.

However, several issues presented in the first edition of *Viability Theory*, [18, Aubin] are not covered in this second edition for lack of room. They concern Haddad's viability theorems for functional differential inclusions where both the dynamics and the constraints depend on the history (or path) of

J.-P. Aubin et al., *Viability Theory*, DOI 10.1007/978-3-642-16684-6_1,

the evolution and the Shi Shuzhong viability theorems dealing with partial differential evolution equation (of parabolic type) in Sobolev spaces, as well as fuzzy control systems and constraints, and, above all, differential (or dynamic) games. A sizable monograph on tychastic and stochastic viability and, for instance, their applications to finance, would be needed to deal with uncertainty issues where the actor has no power on the choice of the uncertain parameters, taking over the problems treated in this book in the worst case (tychastic approach) or in average (stochastic approach).

We have chosen an outline, which is increasing with respect to mathematical technical difficulty, relegating to the end the proofs of the main Viability and Invariance Theorems (see Chap. 19, p.769).

The proofs of the theorems presented in *Set-valued analysis* [27, Aubin & Frankowska] (1990) and in convex analysis (see *Optima and Equilibria*, [19, Aubin]), are not duplicated but referred to. An appendix, *Set-Valued Analysis at a Glance* (18, p. 713) provides without proofs the statements of the main results of set-valued analysis used in these monographs. The notations used in this book are summarised in its Sect. 18.1, p. 713.

1.1 Motivations

1.1.1 Chance and Necessity

The purpose of viability "theory" (in the sense of a sequence [*theôria*, procession] of mathematical tools sharing a common background, and not necessarily an attempt to explain something [*theôrein*, to observe]) is to attempt to answer directly the question of dynamic adaptation of uncertain evolutionary systems to environments defined by constraints, that we called viability constraints for obvious reasons. Hence the name of this body of mathematical results developed since the end of the 1970s that needed to forge a differential calculus of set-valued maps (set-valued analysis), differential inclusions and differential calculus in metric spaces (mutational analysis). These results, how imperfect they might be to answer this challenge, have at least been motivated by social and biological sciences, even though constrained and shaped by the mathematical training of their authors.

It is by now a consensus that the evolution of many variables describing systems, organizations, networks arising in biology and human and social sciences do not evolve in a deterministic way, not even always in a stochastic way as it is usually understood, but evolve with a Darwinian flavor.

Viability theory started in 1976 by translating mathematically the title

Chance	and	*Necessity*
$\Updownarrow$		$\Updownarrow$
$x'(t) \in F(x(t))$	&	$x(t) \in K$

of the famous 1973 book by *Jacques Monod*, *Chance and Necessity* (see [163, Monod]), taken from an (apocryphical?) quotation of Democritus who held that *"the whole universe is but the fruit of two qualities, chance and necessity"*.

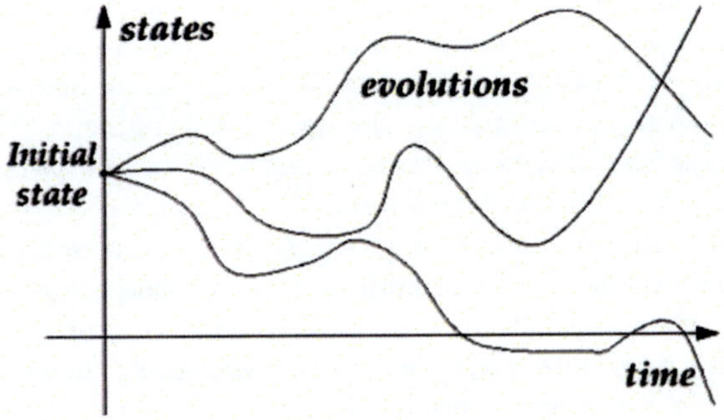

Fig. 1.1 The mathematical translation of "chance".
*The mathematical translation of "**chance**" is the differential inclusion $x'(t) \in F(x(t))$, which is a type of evolutionary engine (called an* evolutionary system*) associating with any initial state x the subset $\mathcal{S}(x)$ of evolutions starting at x and governed by the differential inclusion above. The figure displays evolutions starting from a give initial state, which are functions from time (in abscissas) to the state space (ordinates).*

The system is said to be *deterministic* if for any initial state x, $\mathcal{S}(x)$ is made of one and only one evolution, whereas "contingent uncertainty" happens when the subset $\mathcal{S}(x)$ of evolutions contains more than one evolution for at least one initial state. "Contingence is a non-necessity, it is a characteristic attribute of freedom", wrote *Gottfried Leibniz*.

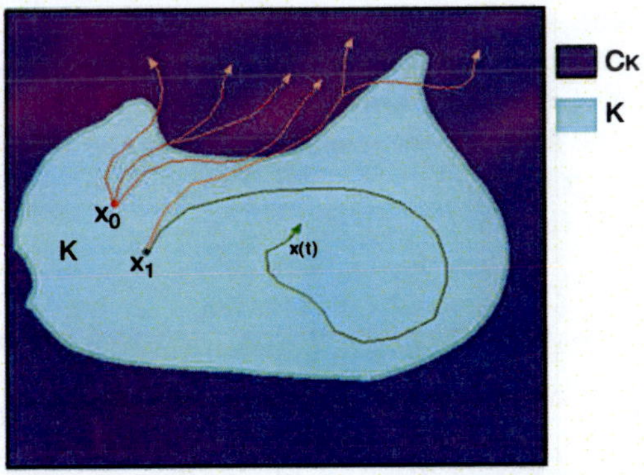

Fig. 1.2 The mathematical translation of "necessity".
*The mathematical translation of "**necessity**" is the requirement that for all $t \geq 0$, $x(t) \in K$, meaning that at each instant, "viability constraints" are*

satisfied by the state of the system. The figure represents the state space as the plane, and the environment defined as a subset. It shows two initial states, one, x_0 from which all evolutions violate the constraints in finite time, the other one x_1, from which starts one viable evolution and another one which is not viable.

One purpose of viability theory is to attempt to answer directly the question that some economists, biologists or engineers ask: *"Complex organizations, systems and networks, yes, but for what purpose?"* The answer we suggest: *"to adapt to the environment."*

This is the case in economics when we have to adapt to scarcity constraints, balances between supply and demand, and many other constraints.

This is also the case in biology, since Claude Bernard's "constance du milieu intérieur" and Walter Cannon's "homeostasis". This is naturally the case in ecology and environmental studies.

This is equally the case in control theory and, in particular, in robotics, when the state of the system must evolve while avoiding obstacles forever or until they reach a target.

In summary, *the environment is described by viability constraints* of various types, a word encompassing polysemous concepts as *stability, confinement, homeostasis, adaptation*, etc., expressing the idea that some variables must obey some constraints (representing physical, social, biological and economic constraints, etc.) that can never be violated. So, viability theory started as *the confrontation of evolutionary systems governing evolutions and viability constraints* that such evolutions must obey.

At the same time, controls, subsets of controls, in engineering, regulons (regulatory controls) such as prices, messages, coalitions of actors, connectionnist operators in biological and social sciences, which parameterize evolutionary systems, do evolve: *Their evolution must be consistent with the constraints, and the targets or objectives they must reach in finite or prescribed time.* The aim of viability theory is to provide the "regulation maps" associating with any state the (possibly empty) subset of controls or regulons governing viable evolutions.

Together with the selection of evolutions governed by teleological objectives, mathematically translated by intertemporal optimality criteria as in optimal control, viability theory offers other selection mechanisms by requiring evolutions to obey several forms of *viability* requirements. In social and biological sciences, intertemporal optimization can be replaced by *myopic, opportunistic, conservative and lazy* selection mechanisms of viable evolutions that involve present knowledge, sometimes the knowledge of the history (or the path) of the evolution, instead of anticipations or knowledge of the future (whenever the evolution of these systems cannot be reproduced experimentally). Other forms of uncertainty do not obey statistical laws, but also take into account unforeseeable rare events (tyches, or perturbations, disturbances) that must be avoided at all costs (precautionary principle[1]). These systems can be regulated by using regulation (or cybernetical) controls that have to be chosen as feedbacks for guaranteeing the viability of constraints and/or the capturability of targets and objectives, possibly against perturbations played by "Nature", which we call *tyches.*

However, there is no reason why collective constraints are satisfied at each instant by evolutions under uncertainty governed by evolutionary systems. This leads us to the study of *how to correct either the dynamics, and/or the constraints* in order to restore viability. This may allow us to provide an explanation of the formation and the evolution of controls and regulons through regulation or adjustment laws that can be designed (and computed) to insure viability, as well as other procedures, such as using *impulses* (evolutions with infinite velocity) governed by other systems, or by regulating the evolution of the environment.

Presented in such an evolutionary perspective, this approach of (complex) evolutionary systems departs from main stream modelling by a direct approach:

1 [Direct Approach.] It consists in studying properties of evolutions governed by an evolutionary system: gather the larger number of properties of evolutions starting from each initial state. It may be an information both costly and useless, since our human brains cannot handle simultaneously too many observations and concepts.

Moreover, it may happen that evolutions starting from a given initial state satisfy properties which are lost by evolutions starting from another initial state, even close to it (sensitivity analysis) or governed by (stability analysis).

Viability theory rather uses instead an *inverse approach*:

2 [Inverse Approach.] A set of prescribed properties of evolutions being given, study the (possibly empty) subsets of initial states from which

1. starts at least one evolution governed by the evolutionary system satisfying the prescribed properties,
2. all evolutions starting from it satisfy these prescribed properties.

These two subsets coincide whenever the evolutionary system is deterministic.

[1] Stating that one should limit, bound or even forbid potential dangerous actions, without waiting for a scientific proof of their hazardous consequences, whatever the economic cost.

Stationarity, periodicity and asymptotic behavior are examples of classical properties motivated by physical sciences which have been extensively studied.

We thus have to add to this list of classical properties other ones, such as concepts of *viability* of an environment, of *capturability* of a target in finite time, and of other concepts combining properties of this type.

1.1.2 Motivating Applications

For dealing with these issues, one needs *"dedicated"* concepts and formal tools, algorithms and mathematical techniques motivated by complex systems evolving under uncertainty. For instance, and without going into details, we can mention systems sharing common features:

1. ***Systems designed by human brains*** in the sense that agents, actors, decision-makers act on the evolutionary system, as in engineering. ***Control theory and differential games***, conveniently revisited, provide numerous metaphors and tools for grasping viability questions. Problems in *control design, stability, reachability, intertemporal optimality, tracking of evolutions, observability, identification and set-valued estimation*, etc., can be formulated in terms of viability and capturability concepts investigated in this book.
   Some technological systems such as robots of all types, from drones, unmanned underwater vehicles, etc., to animats (artificial animals, a contraction of anima-materials) need "embedded systems" implementations *autonomous* enough to regulate viability/capturability problems by adequate regulation (feedback) control laws. Viability theory provides algorithms for computing the feedback laws by modular and portable software flexible enough for integrating new problems when they appear (hybrid systems, dynamical games, etc.).
2. ***Systems observed by human brains***, are more difficult to understand since human beings did not design or construct them. Human beings live, think, are involved in socio-economic interactions, but struggle for grasping why and how they do it, at least, why. This happens for instance in the following fields:

   - ***economics***, where the viability constraints are the scarcity constraints among many other ones. We can replace the fundamental Walrasian model of resource allocations by decentralized dynamical model in which the role of the controls is played by the prices or other economic decentralizing messages (as well as coalitions of consumers, interest rates, and so forth). The regulation law can be interpreted as the behavior of Adam Smith's invisible hand choosing the prices as a function of allocations of commodities,

- ***finance***, where shares of assets of a portfolio play the role of controls for guaranteing that the values of the portfolio remains above a given time/price dependent function at each instant until the exercise time (horizon), whatever the prices and their growth rates taken above evolving bounds,
- ***dynamical connectionnist networks and/or dynamical cooperative games***, where coalitions of players may play the role of controls: each coalition acts on the environment by changing it through dynamical systems. The viability constraints are given by the architecture of the network allowed to evolve,
- ***Population genetics***, where the viability constraints are the ecological constraints, the state describes the phenotype and the controls are genotypes or fitness matrices.
- ***sociological sciences***, where a society can be interpreted as a set of individuals subjected to viability constraints. Such constraints correspond to what is necessary for the survival of the social organization. Laws and other cultural codes are then devised to provide each individual with psychological and economical means of survival as well as guidelines for avoiding conflicts. Subsets of cultural codes (regarded as cultures) play the role of regulation parameters.
- ***cognitive sciences***, in which, at least at one level of investigation, the variables describe the sensory-motor activities of the cognitive system, while the controls translate into what could be called conceptual controls (which are the synaptic matrices in neural networks.)

Theoretical results about the ways of thinking described above are useful for the understanding of non teleological evolutions, of the inertia principle, of the emergence of new regulons when viability is at stakes, of the role of different types of uncertainties (contingent, tychastic or stochastic), of the (re)designing of regulatory institutions (regulated markets when political convention must exist for global purpose, mediation or metamediation of all types, including law, social conflicts, institutions for sustainable development, etc.). And progressively, when more data gathered by these institutions will be available, qualitative (and sometimes quantitative) prescriptions of viability theory may be useful.

1.1.3 Motivations of Viability Theory from Living Systems

Are social and biological systems sufficiently similar to systems currently studied in mathematics, physics, computer sciences or engineering? Eugene Wigner's considerations on *the unreasonable effectiveness of mathematics in the natural sciences* [215, Wigner] are even more relevant in life sciences.

For many centuries, human minds used their potential "mathematical capabilities" to describe and share their "mathematical perceptions" of the world. This mathematical capability of human brains is assumed to be analogous to the language capability. Each child coming to this world uses this specific capability in social interaction with other people to join at each instant an (evolving) consensus on the perception of their world by learning their mother tongue (and few others before this capability fades away with age). We suggest the same phenomenon happens with mathematics. They play the "mathematical role" of *metaphors* that language uses for allowing us to understand a new phenomenon by metaphors comparing it with previously "understood phenomena". Before it exploded recently in a Babel skyscraper, this "mathematical father tongue" was quite consensual and perceived as universal. This is this very universality which makes mathematics so fascinating, deriving mathematical theories or tools motivated by one field to apply them to several other ones. However, apparently, because up to now, the mathematical "father tongue" was mainly shaped by "simple" physical problems of the inert part of the environment, letting aside, with few exceptions, the living world. For good reasons. Fundamental simple principles, such as the *Pierre de Fermat*'s "variational principle", including *Isaac Newton*'s law thanks to *Maupertuis*'s least action principle, derived explanations of complex phenomena from simple principles, as *Ockham*'s razor prescribes: This "law of parsimony" states that an explanation of any phenomenon should make as few assumptions as possible, and to choose among competing theories the one that postulates the fewest concepts. This is the result of an "abstraction process", which is the (poor) capability of human brains that select among the perceptions of the world the few ones from which they may derive logically or mathematically many other ones. *Simplifying complexity* should be the purpose of an emerging science of complexity, if such a science will emerge beyond its present fashionable status.

So physics, which could be defined as the part of the cultural and physical environment which is understandable by mathematical metaphors, has not yet, in our opinion, encapsulated the mathematical metaphors of living systems, from organic molecules to social systems, made of human brains controlling social activities. The reason seems to be that the adequate mathematical tongue does not yet exist. And the challenge is that before creating it, the present one has to be forgotten, de-constructed. This is quite impossible because mathematicians have been educated *in the same way* all over the world, depriving mathematics from the Darwinian evolution which has operated on languages. This uniformity is the strength and the weakness of present day mathematics: its universality is partial. The only possibility to mathematically perceive living systems would remain a dream: to gather in secluded convents young children with good mathematical capability, but little training in the present mathematics, under the supervision or guidance of economists or biologists without mathematical training. They possibly could come up with new mathematical languages unknown to us providing the long expected *unreasonable effectiveness of mathematics in the social and biological sciences.*

Even the concept of natural number is oversimplifying, by putting in a same equivalence class so several different sets, erasing their *qualitative* properties or hiding them behind their *quantitative* ones. Numbers, then next measurements, and then, statistics, how helpful they are for understanding and controlling the physical part of the environment, may be a drawback to address the qualitative aspects of our world, left to plain language for the quest of elucidation. We may have to return to the origins and explore new "qualitative" routes, without overusing the mathematics that our ancestors accumulated so far and bequeathed to us.

Meanwhile, we are left with this paradox: "simple" physical phenomena are explained by more and more sophisticated and abstract mathematics, whereas "complex" phenomena of living systems use, most of the time, relatively rudimentary mathematical tools. For instance, the mathematical tools used so far did not answer the facts that, for instance:

1. economic evolution is *never* at equilibrium (stationary state),
2. and thus, there were no need that this evolution converges to it, in a stable or unstable way,
3. elementary cognitive sciences cannot accept the rationality assumption of human brains,
4. and even more that they can be reduced to utility functions, the existence of which was already questioned by *Henri Poincaré* when he wrote to *Léon Walras* that "*Satisfaction is thus a magnitude, but not a measurable magnitude*" (numbers are not sufficient to grasp satisfaction),
5. uncertainty can be mathematically captured only by probabilities (numbers, again),
6. chaos, defined as a property of deterministic systems, is not fit to represent a nondeterministic behavior of living systems which struggle to remain as stable (and thus, "non chaotic") as possible,
7. intertemporal optimality, a creation of the human brain to explain some physical phenomena, is not the only creation of Nature, (in the sense that "Nature" created it only through human brains!),
8. those human brains should complement it by another and more recent principle, *adaptation of transient evolutions to environments,*
9. and so on.

These epistemological considerations are developed in *La mort du devin, l'émergence du démiurge. Essai sur la contingence et la viabilité des systèmes* (*The Demise of the Seer, the Rise of the Demiurge. Essay on contingency, viability and inertia of systems*).

1.1.4 Applications of Viability Theory to... Mathematics

It is time to cross the interdisciplinary gap and to confront and hopefully to merge points of view rooted in different disciplines.

After years of study of various problems of different types, motivated by robotics (and animat theory), game theory, economics, neuro-sciences, biological evolution and, unexpectedly, by financial mathematics, these few relevant features common to all these problems were uncovered, after noticing the common features of the proofs and algorithms.

This history is a kind of *mathematical striptease*, the modern version of what Parmenides and the pre-Socratic Greeks called *a-letheia*, the discovering, un-veiling of the world that surrounds us. This is exactly the drive to "abstraction", isolating, in a given perspective, the relevant information in each concept and investigating the interplay between them. Indeed, one by one, slowly and very shyly, the required properties of the control system were taken away (see Sect. 18.9, p. 765 for a brief illustration of the Graal of the Ultimate Derivative).

Mathematics, thanks to its abstraction power by isolating only few key features of a class of problems, can help to bridge these barriers as long as it proposes new methods motivated by these new problems instead of applying the classical ones only motivated until now by physical sciences. Paradoxically, the very fact that the mathematical tools useful for social and biological sciences are and have to be quite sophisticated impairs their acceptance by many social scientists, economists and biologists, and the gap threatens to widen.

Hence, viability theory designs and devises a mathematical tool-box universal enough to be efficient in many apparently different problems. Furthermore, using methods that are rooted neither in linear system theory nor in differential geometry, the results:

1. hold true for *nonlinear systems*,
2. are *global* instead of being local,
3. and allow an *algorithmic treatment* without loss of information due to the treatment of classical equivalent problems (systems of first-order partial differential equations for instance).

Although viability theory has been designed and developed for studying the evolutions of uncertain systems confronted to viability constraints arising in socioeconomic and biological sciences, as well as in control theory, it had also been used as a mathematical tool for deriving purely mathematical results. These tools enrich the panoply of those diverse and ingenious techniques born out of the pioneering works of Lyapunov and Poincaré more than one century ago. Most of them were motivated by physics and mechanics, not necessarily designed to adaptation problems to environmental or viability constraints. These concepts and theorems provide deep insights in the behavior of complex evolutions that even simple deterministic dynamical systems provide.

Not only the viability/capturability problems are important in themselves in the fields we have mentioned, but it happens that many other important concepts of control theory and mathematics can be formulated in terms of viability kernels or capture basins under auxiliary systems. Although they were designed to replace the static concept of equilibrium by the dynamical concept of viability kernel, to offer an alternative to optimal control problems by introducing the concepts of inertia functions, heavy evolutions and punctuated equilibrium, it happens that nevertheless, the viability results:

1. provide new insights and results in the study of Julia sets and Fatou dusts and fractals, all concepts closely related to viability kernels;
2. are useful for providing the final version of the inverse function theorem for set-valued maps;
3. offer new theorems on the existence of equilibria and the study of their asymptotic properties, even for determinist systems, where the concepts of attractors (as the Lorenz attractors) are closely related to the concept of viability kernels;
4. Provide tools to study several types of first-order partial differential equations, conservation laws and Hamilton–Jacobi–Bellman equations, following a long series of articles by Hélène Frankowska who pioneered the use of viability techniques, and thus to solve many intertemporal optimization problems.

Actually, the role played by the Viability Theorem in dynamical and evolutionary systems is analogous to the one played by the Brouwer Fixed Theorems in static nonlinear analysis. Numerous static problems solved by the equivalent statements and consequences of this Brouwer Theorem in nonlinear analysis can be reformulated in an evolutionary framework and solved by using viability theory.

The miraculous universality of mathematics is once again illustrated by the fact that some viability tools, inspired by the evolutionary questions raised in life sciences, also became, in a totally unpredicted way, tools to be added to the existing panoply for solving engineering and purely mathematical problems well outside their initial motivation.

1.2 Main Concepts of Viability Theory

Viability theory incorporates some mathematical features of uncertainty without statistical regularity, deals not only with optimality but also with viability and *decisions taken at the appropriate time.* Viability techniques are also *geometric* in nature, but they do not require smoothness properties usually assumed in differential geometry. They not only deal with asymptotic behavior, but also and mainly with *transient* evolutions and capturability of targets in finite or prescribed time. They are *global* instead of local, and truly *nonlinear* since they bypass linearization techniques of the dynamics around equilibria, for instance. They bring other insights to the decipherability of complex, paradoxical and strange dynamical behaviors by providing other types of mathematical results and algorithms. And above all, they have been motivated by dynamical systems arising in issues involving living beings, as well as networks of systems (or organizations, organisms).

In a nutshell, viability theory investigates evolutions:

1. in *continuous time, discrete time, or a "hybrid"* of the two when *impulses* are involved,
2. constrained to *adapt* to an environment,
3. evolving under *contingent and/or tychastic uncertainty*,
4. using for this purpose *controls*, *regulons* (regulation controls), subsets of regulons, and in the case of networks, connectionnist matrices,
5. *regulated* by *feedback laws* (static or dynamic) that are then "computed" according to given principles, such as the *inertia principle*, intertemporal optimization, etc.,
6. co-evolving with their environment (*mutational and morphological viability*),
7. and corrected by introducing adequate controls (*viability multipliers*) when viability or capturability is at stakes.

1.2.1 Viability Kernels and Capture Basins Under Regulated Systems

We begin by defining set-valued maps (see Definition 18.3.1, p. 719):

> **3** ***[Set-Valued Map]*** *A* set-valued *map $F : X \rightsquigarrow Y$ associates with any $x \in X$ a subset $F(x) \subset Y$ (which may be the empty set $\emptyset$). It is a (single-valued) map $f := F : X \mapsto Y$ if for any x, $F(x) := \{y\}$ is reduced to a single element y. The symbol "$\rightsquigarrow$" denotes set-valued maps whereas the classical symbol "$\mapsto$" denotes single-valued maps.*
> *The* graph $\text{Graph}(F)$ *of a set-valued map F is the set of pairs $(x, y) \in X \times Y$ satisfying $y \in F(x)$. If $f := F : X \mapsto Y$ is a single-valued map, it coincides with the usual concept of graph. The* inverse F^{-1} *of F is the set-valued map from Y to X defined by*
>
> $$x \in F^{-1}(y) \iff y \in F(x) \iff (x, y) \in \text{Graph}(F)$$

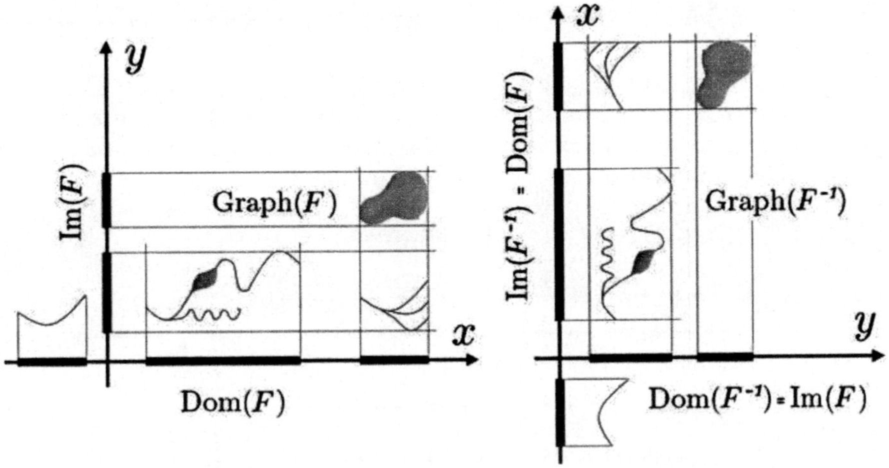

Fig. 1.3 Graphs of a set-valued map and of the inverse map.

The main examples of evolutionary systems, those "engines" providing evolutions, are associated with differential inclusions, which are multi-valued or set-valued differential equations:

> 4 ***[Evolutionary Systems]*** *Let* $X := \mathbb{R}^d$ *be a finite dimensional space, regarded as the* state space, *and* $F : X \rightsquigarrow X$ *be a set-valued map associating with any state* $x \in X$ *the set* $F(x) \subset X$ *of* velocities *available at state* x. *It defines the* differential inclusion $x'(t) \in F(x(t))$ *(boiling down to an ordinary differential equation whenever* F *is single-valued). An* evolution $x(\cdot) : t \in [0, \infty[\mapsto x(t) \in \mathbb{R}^d$ *is a function of time taking its values in a vector space* $\mathbb{R}^d$. *Let* $\mathcal{C}(0, +\infty; X)$ *denote the space of* continuous evolutions *in the state space* X. *The* evolutionary system $\mathcal{S} : X \rightsquigarrow \mathcal{C}(0, +\infty; X)$ *maps any initial state* $x \in X$ *to the set* $\mathcal{S}(x)$ *of evolutions* $x(\cdot)$ *starting from* x *and governed by differential inclusion* $x'(t) \in F(x(t))$. *It is* deterministic *if the evolutionary system* $\mathcal{S}$ *is single-valued and* non deterministic *if it is set-valued.*

The main examples of differential inclusions are provided by

> 5 ***[Parameterized Systems]*** *Let* $\mathcal{U} := \mathbb{R}^c$ *be a space of parameters. A parameterized system is made of two "boxes":*
> *1 - The* "input–output box" *associating with any evolution* $u(\cdot)$ *of the parameter* (input) *the evolution governed by the differential equation* $x'(t) = f(x(t), u(t))$ *starting from an initial state* (open loop),

2 - The non deterministic "output–input box", *associating with any state a subset $U(x)$ of parameters (*output*).*
It defines the set-valued map F associating with any x the subset $F(x) := \{f(x,u)\}_{u \in U(x)}$ of velocities parameterized by $u \in U(x)$. The associated evolutionary system $\mathcal{S}$ maps any initial state x to the set $\mathcal{S}(x)$ of evolutions $x(\cdot)$ starting from x ($x(0) = x$) and governed by

$$x'(t) = f(x(t), u(t)) \text{ where } u(t) \in U(x(t)) \tag{1.1}$$

or, equivalently, to differential inclusion $x'(t) \in F(x(t))$.

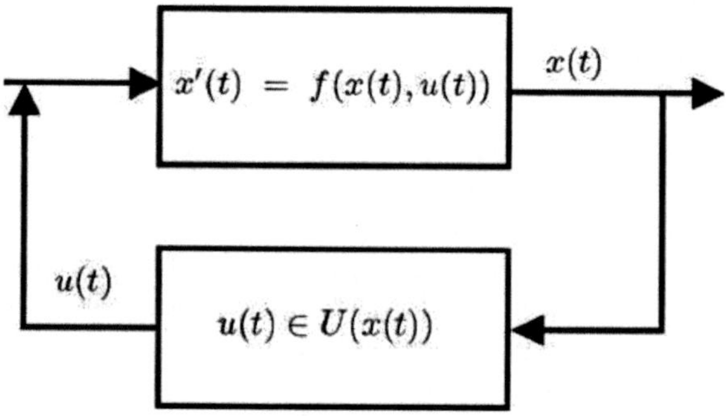

Fig. 1.4 Parameterized Systems.
The input–output and output–input boxes of a parameterized systems are depicted in this diagram: the controlled dynamical system at the input–output level, the "cybernetical" map imposing state-dependent constraints on the control at the output–input level.

The parameters range over a state-dependent *"cybernetic" map* $U : x \rightsquigarrow U(x)$, providing the system opportunities to adapt at each state to viability constraints (often, as slowly as possible) and/or to regulate intertemporal optimal evolutions.

The nature of the parameters differs according to the problems and to questions asked: They can be:

- "controls", whenever a controller or a decision maker "pilots" the system by choosing the controls, as in engineering,
- "regulons" or regulatory parameters in those natural systems where no identified or consensual agent acts on the system,
- "tyches" or disturbances, perturbations under which nobody has any control.

We postpone to Chap. 2, p. 43 more details on these issues.

To be more explicit, we have to introduce formal definitions describing the concepts of viability kernel of an environment and capturability of a target under an evolutionary system.

6 ***[Viability and Capturability]*** *If a subset* $K \subset \mathbb{R}^d$ *is regarded as an* environment *(defined by* viability constraints*), an evolution* $x(\cdot)$ *is said to be* viable in the environment $K \subset \mathbb{R}^d$ on an interval $[0, T[$ *(where* $T \leq +\infty$*) if for every time* $t \in [0, T[$, $x(t)$ *belongs to* K.
If a subset $C \subset K$ *is regarded as a target, an evolution* $x(\cdot)$ captures C *if there exists a finite time* T *such that the evolution is viable in* K *on the interval* $[0, T[$ *until it reaches the target at* $x(T) \in C$ *at time* T. *See Definition 2.2.3, p. 49.*

Viability and capturability are the main properties of evolutions that are investigated in this book.

1.2.1.1 Viability Kernels

We begin by introducing the *viability kernel of an environment* under an evolutionary system associated with a nonlinear parameterized system.

7 ***[Viability Kernel]*** *Let* K *be an environment and* $\mathcal{S}$ *an evolutionary system. The* viability kernel of K under the evolutionary system $\mathcal{S}$ *is the set* $\mathrm{Viab}_{\mathcal{S}}(K)$ *of initial states* $x \in K$ *from which starts at least one evolution* $x(\cdot) \in \mathcal{S}(x)$ *viable in* K *for all times* $t \geq 0$:

$$\mathrm{Viab}_{\mathcal{S}}(K) := \{x_0 \in K \mid \exists x(\cdot) \in \mathcal{S}(x_0) \ \ \textit{such that}\ \forall t \geq 0,\ x(t) \in K\}$$

Two extreme situations deserve to be singled out: The environment is said to be

1. viable *under* $\mathcal{S}$ *if it is equal to its viability kernel:* $\mathrm{Viab}_{\mathcal{S}}(K) = K$,
2. a repeller *under* $\mathcal{S}$ *if its viability kernel is empty:* $\mathrm{Viab}_{\mathcal{S}}(K) = \emptyset$.

It is equivalent to say that all evolutions starting from a state belonging to the complement of the viability kernel in K leave the environment *in finite time*.

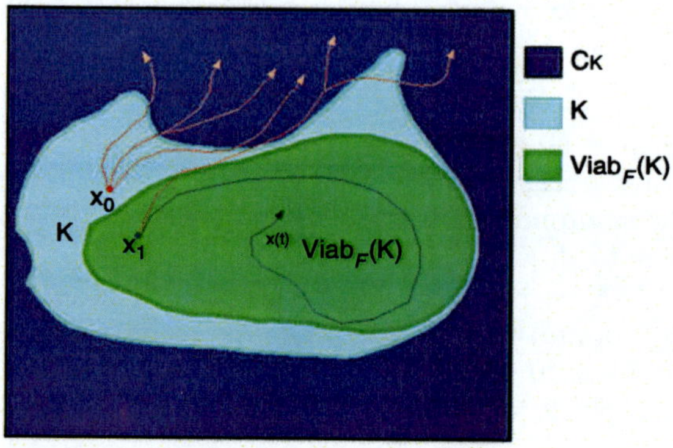

Fig. 1.5 Illustration of a viability kernel.
From a point x_1 in the viability kernel of the environment K starts **at least** *one evolution viable in K forever.* **All** *evolutions starting from $x_0 \in K$ outside the viability kernel leave K in finite time.*

Hence, the viability kernel plays the role of a *viabilimeter*, the *"size"* of which measuring the degree of viability of an environment, so to speak.

1.2.1.2 Capture Basins

8 ***[Capture Basin Viable in an Environment]*** *Let K be an environment, $C \subset K$ be a target and $\mathcal{S}$ an evolutionary system. The* capture basin of C (viable in K) under the evolutionary system $\mathcal{S}$ *is the set* $\text{Capt}_{\mathcal{S}}(K,C)$ *of initial states $x \in K$ from which starts at least one evolution $x(\cdot) \in \mathcal{S}(x)$ viable in K on $[0,T[$ until the finite time T when the evolution reaches the target at $x(T) \in C$.*

For simplicity, we shall speak of capture basin without explicit mention of the environment whenever there is no risk of confusion.

It is equivalent to say that, starting from a state belonging to the complement in K of the capture basin, **all** evolutions remain outside the target until they leave the environment K.

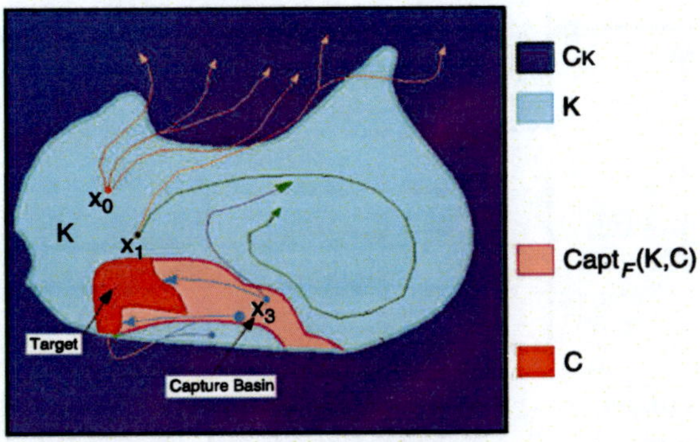

Fig. 1.6 Illustration of a capture basin.
From a state x_3 in the capture basin of the target C viable in the environment K starts **at least** *one evolution viable in K until it reaches C in finite time.* ***All*** *evolutions starting from $x_1 \in K$ outside the capture basin remain outside the target C forever or until they leave K.*

The concept of capture basin of a target requires that at least one evolution reaches the target in finite time, and *not only asymptotically*, as it is usually studied with concepts of *attractors* since the pioneering works of *Alexandr Lyapunov* going back to 1892.

1.2.1.3 Designing Feedbacks

The theorems characterizing the viability kernel of an environment of the capture basin of a target also provide a regulation map $x \rightsquigarrow R(x) \subset U(x)$ regulating evolutions viable in K:

9 *[**Regulation Map**] Let us consider control system*

$$\begin{cases} (i) \;\; x'(t) = f(x(t), u(t)) \\ (ii) \; u(t) \in U(x(t)) \end{cases}$$

and an environment K. A set-valued map $x \rightsquigarrow R(x) \subset U(x)$ is called a regulation map governing viable evolutions *if the viability kernel of K is viable under the control system*

$$\begin{cases} (i) \;\; x'(t) = f(x(t), u(t)) \\ (ii) \; u(t) \in R(x(t)) \end{cases}$$

Actually, we are looking for single-valued regulation maps governing viable evolutions, which are usually called feedbacks.

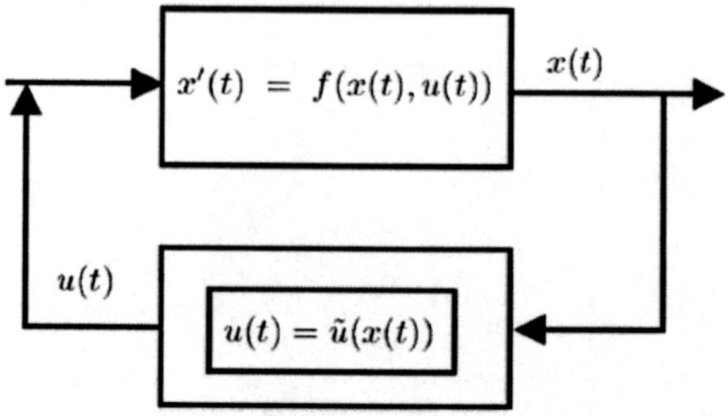

Fig. 1.7 Feedbacks.
The single-valued maps $x \mapsto \widetilde{u}(x)$ *are called the* feedbacks *(or servomechanisms, closed loop controls, etc.) allowing one to pilot evolutions by using controls of the form* $u(t) := \widetilde{u}(x(t))$ *in system* $\mathcal{S}$ *defined by (1.1), p. 14:* $x'(t) = f(x(t), u(t))$ *where* $u(t) \in U(x(t))$*. Knowing such a feedback, the evolution is governed by ordinary differential equation* $x'(t) = f(x(t), \widetilde{u}(x(t)))$.

Hence, knowing the regulation map R, *viable feedbacks* are single-valued regulation maps.

Building viable feedbacks, combining prescribed feedbacks to govern viable evolutions, etc., are issues investigated in Chap. 11, p. 437.

1.2.2 Viability Kernel and Capture Basin Algorithms

These kernels and basins can be approximated by discrete analog methods and computed numerically thanks to *Viability Kernel and Capture Basin Algorithms*. Indeed, all examples shown in this introductory chapter have been computed using software packages based on these algorithms, developed by LASTRE (Laboratoire d'Applications des Systèmes Tychastiques Régulés). These algorithms compute not only the viability kernel of an environment or the capture basin of a target, but also the regulation map and viable feedbacks regulating evolutions viable in the environment (until reaching the target if any). By opposition to *"shooting methods"* using classical solvers of differential equations, these *"viability algorithms"* allow evolutions governed by these feedbacks to remain viable. Shooting methods do not provide the corrections needed to keep the solutions viable in given sets. This is one of the reasons why shooting methods cannot compute evolutions viable in attractors, or in stable manifolds, in fractals of Julia sets, even in the case of discrete and deterministic systems, although the theory states that they should remain viable in such sets. Viability algorithms allow us to govern evolutions satisfying these viability requirements: They are designed to do so.

1.2.3 Restoring Viability

There is no reason why an arbitrary subset K should be viable under a control system. The introduction of the concept of viability kernel does not exhaust the problem of restoring viability. One can imagine several other methods for this purpose:

1. Keep the constraints and change the initial dynamics by introducing regulons that are "viability multipliers";
2. Keep the constraints and change the initial conditions by introducing a *reset map* Φ mapping any state of K to a (possibly empty) set $\Phi(x) \subset X$ of new "initialized states" (*impulse control*);
3. Keep the same dynamics and let the set of constraints evolve according to *mutational equations* (see *Mutational and morphological analysis: tools for shape regulation and morphogenesis*, [23, Aubin] and *Mutational Analysis*, [150, Lorenz]).

Introductions to the two first approaches, viability multipliers and impulse systems, are presented in Chap. 12, p.485.

1.3 Examples of Simple Viability Problems

In order to support our claim that viability problems are more or less hidden in a variety of disciplines, we have selected a short list of very simple examples which can be solved in terms of viability kernels and capture basins developed in this monograph.

10 ***Heritage.*** *Whenever the solution to a given problem can be formulated in terms of viability kernel, capture basin, or any of the other combinations of them, this solution inherits the properties of kernels and basins gathered in this book. In particular, the solution can be computed by the Viability Kernel and Capture Basin Algorithms.*

Among these problems, some of them can be formulated directly in terms of viability kernels or capture basins. We begin our gallery by these ones.

However, many other problems are viability problems in disguise. They require some mathematical treatment to uncover them, by introducing auxiliary environments and auxiliary targets under auxiliary systems. After some accidental discoveries and further development, unexpected and unsuspected examples of viability kernels and capture basins have emerged in optimal control theory. Optimal control problems of various types appear in engineering, social sciences, economics and other fields. They have been investigated with methods born out of the study of calculus of variations by Hamilton and Jacobi, adapted to differential games and control problems by Isaacs and Bellman (under the name *"dynamical programming"*). The ultimate aim is to provide regulation laws allowing us to pilot optimal evolutions. They are derived from the derivatives of the *"value function"*, which is classically a solution to the Hamilton–Jacobi–Bellman partial differential equations in the absence of viability constraints on the states. It turns out that these functions can be characterized by means of viability kernels of auxiliary environments or of capture basins of auxiliary targets under auxiliary dynamical systems.

The same happy events also happened in the field of systems of first-order partial differential equations. These examples are investigated in depth in Chaps. 4, p. 125, 6, p. 199, 16, p. 631 and 17, p. 681. They are graphs of solutions of systems of partial differential equations of first order or of Hamilton–Jacobi–Bellman partial differential equations of various types arising in optimal control theory. The feature common to these problems is that environments and targets are either graphs of maps from a vector space to another or epigraphs of extended functions.

All examples which are presented have been computed using the Viability Kernel or Capture Basin Algorithms, and represent benchmark examples which can now be solved in a standard manner using viability algorithms.

1.3.1 Engineering Applications

A fundamental problem of engineering is "obstacle avoidance", which appears in numerous application fields.

1.3.1.1 Rallying a Target While Avoiding Obstacles

Chapter 5, p. 179, *Avoiding Skylla and Charybdis*, illustrates several concepts on the same Zermelo navigation problem of ships aiming to rally a harbor while avoiding obstacles called Skylla and Charybdis in our illustration, in reference to the Homer *Odyssey*. The environment is the sea, the target the harbor. The chapter presents the computations of.

1. the domain of the *minimal length function* (see Definition 4.4.1, p.140) which is contained in the viability kernel,
2. the complement of the viability kernel which is the domain of the *exit function*, measuring the largest survival time before the ship sinks (see Definition 4.3.3, p.135);
3. the *capture basin of the harbor*, which is the domain of the *minimal time function*;
4. the *attraction basin*, which is the domain of the *Lyapunov function*;

5. the *controllability basin*, which is the domain of the value function of the optimal control problem minimizing the cumulated (square of the) velocity.

For each of these examples, the graphs of the feedback maps associating with each state the velocity and the steering direction of the ship are computed, producing evolutions (viable with minimal length, non viable *persistent evolutions*, *minimal time evolution* reaching the harbor in finite time, *Lyapunov evolutions* asymptotically converging to the harbor and *optimal evolutions* minimizing the intertemporal criterion). In each case, examples of trajectories of such evolutions are displayed.

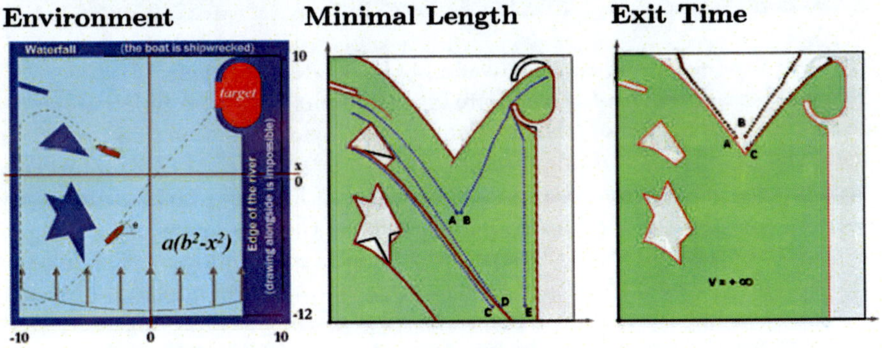

Fig. 1.8 Minimal length and exit time functions.
The left figure describes the non smooth environment (the complement of the two obstacles and the dyke in the square) and the harbor, regarded as the target. The wind blows stronger in the center than near the banks, where it is weak enough to allow the boat to sail south. The figure in the center displays the viability kernel of the environment, which is the domain of the minimal length functions. The viability kernel is empty north of the waterfall and has holes south of the two mains obstacles and in the harbor on the east. The feedback map governs the evolution of minimal length evolutions which converge to equilibria. The presence of obstacles implies three south–east to north–west discontinuities. For instance, trajectories starting from C and D in the middle figure go north and south of the obstacle. The set of equilibria is not connected: some of them are located in a vertical line near the west bank, the other ones on a vertical line in the harbor. The trajectory of the evolution starting from A sails to an equilibrium near the west bank and from B sails to an equilibrium in the harbor. The figure on the right provides trajectories of non viable evolutions which are persistent in the sense that they maximize their exit time from the environment. The ones starting from A and B exit through the waterfall, the one starting from C exits through the harbor. The complement of the viability kernel is the domain of the exit time function.

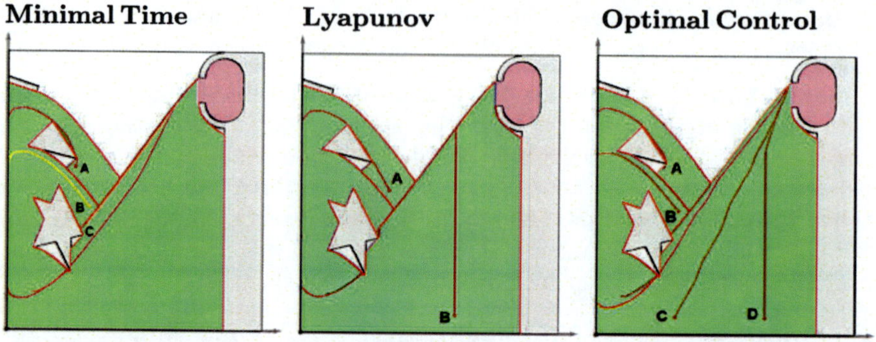

Fig. 1.9 Minimal time, Lyapunov and optimal value functions.
These three figures display the domains of the minimal time function, which is the capture basin of the harbor, of the Lyapunov function, which is the attraction basin and of the value function of an optimal control problem (minimizing the cumulated squares of the velocity). These domains are empty north of the waterfall and have "holes" south of the two mains obstacles and of the harbor on the east. The presence of obstacles induces three lines of discontinuity, delimitating evolutions sailing north of the northern obstacle, between the two obstacles and south or east of the southern obstacle, as it is shown for the evolutions starting from A, B and C in the capture basin, domain of the minimal time function. The trajectories of the evolutions starting from A and B are the same after they leave the left bank for reaching the harbor.

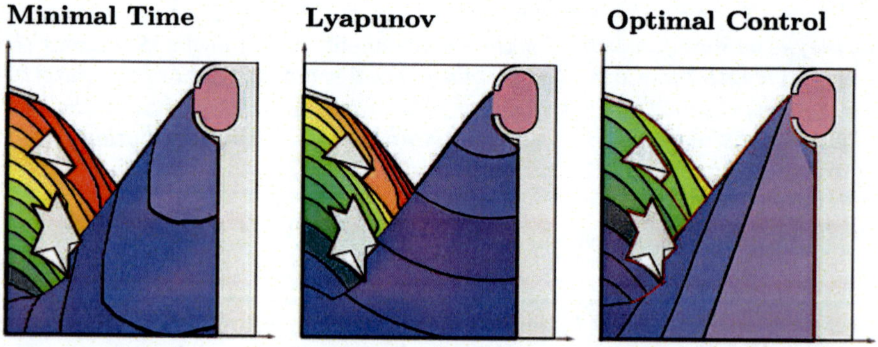

Fig. 1.10 Isolines of the minimal time, Lyapunov and value functions.
These figures display the isolines of these functions on their respective domains, indicating in the color scale the value of these functions. For the minimal time function, the level curves provide the (minimum) time needed to reach the target.

1.3.1.2 Autonomous Navigation in an Urban Network

Viability concepts have not only been simulated for a large variety of problems, but been implemented in field experiments. This is the case of obstacle avoidance on which an experimental robot (Pioneer 3AT of *activmedia robotics*) has been programmed to reach a target from any point of its capture basin (see Sect. 3.2, p.105). The environment is a road network, on which a target to be reached in minimal time has been assigned. Knowing the dynamics of the pioneer robot, both the capture basin and, above all, the feedback map, have been computed. The graph of the feedback (the command card) has been embedded in the navigation unit of the robot. Two sensors (odometers and GPS) tell the robot were it is, but the robot does not use sensors locating the obstacles nor the target.

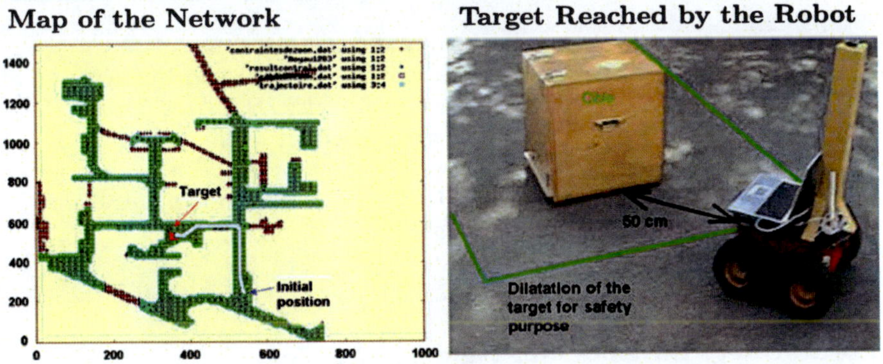

Fig. 1.11 Experiment of the viability algorithm for autonomous navigation. *The* left figure *displays the trajectory of the robot in the map of the urban network. The right figure is a close up photograph of the robot near the target. Despite the lack of precision of the GPS, the trajectory of the robot followed exactly the one which was simulated.*

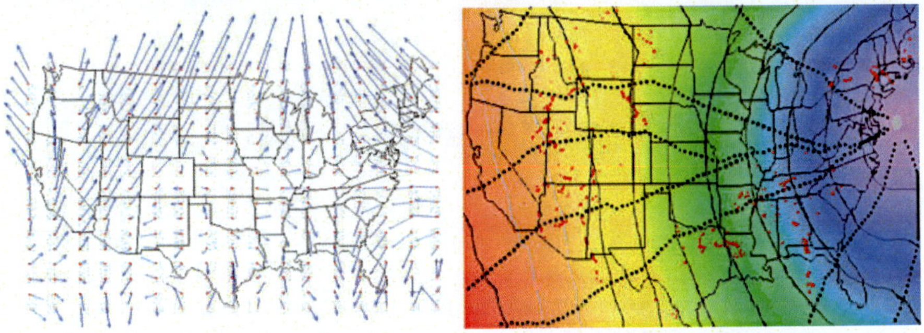

Fig. 1.12 Wind-optimal high altitude routes.
The left figure *shows the wind-field over the continental US (source: NOAA). The right figure displays the wind-optimal routes avoiding turbulence to reach the New York airport in minimal time.*

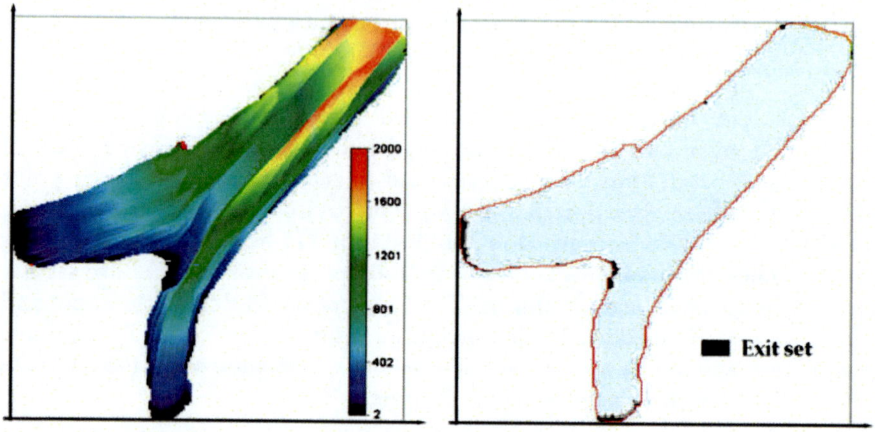

Fig. 1.13 Persistent evolutions of drifters and their exit sets.
The figure displays the exit time function (in seconds) and the exit sets of drifters in the Georgianna Slough of the San Francisco Bay Area (see Fig. 1.12, p.23). The exit sets (right *subfigure) are the places of the boundary of the domain of interest where these drifters can be recovered. The level-curves of the exit time function are indicated in the left subfigure. In this example, drifters evolve according to currents (see Sect. 3.4, p.119).*

1.3.1.3 Safety Envelopes for Landing of Plane

The state variables are the velocity V, the flight path angle γ and the altitude z of a plane. We refer to Sect. 3.3, p. 108 for the details concerning the flight dynamics. The environment is the "flight envelope" taking into account the fight constraints and the target is the "zero altitude" subset of the flight envelope, called the *"touch down envelope"*.

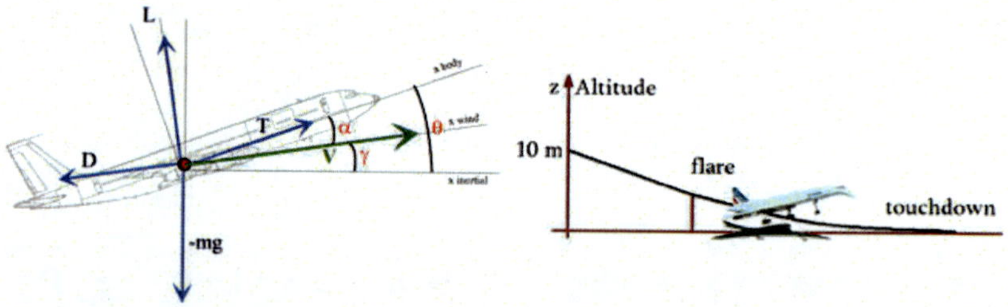

Fig. 1.14 Model of the touch down manoeuver of an aircraft.
In order to ensure safety in the last phase of landing, the airplane must stay within a "flight envelope" which plays the role of environment.

The *Touch Down Envelope* is the set of flight parameters at which it is safe to touch the ground. The *touch down safety envelope* is the set of altitude, velocities and path angles from which one can reach the touch down envelope. The touch down safety envelope can then be computed with the viability algorithms (for the technical characteristics of a DC9-30) since it is a capture basin:

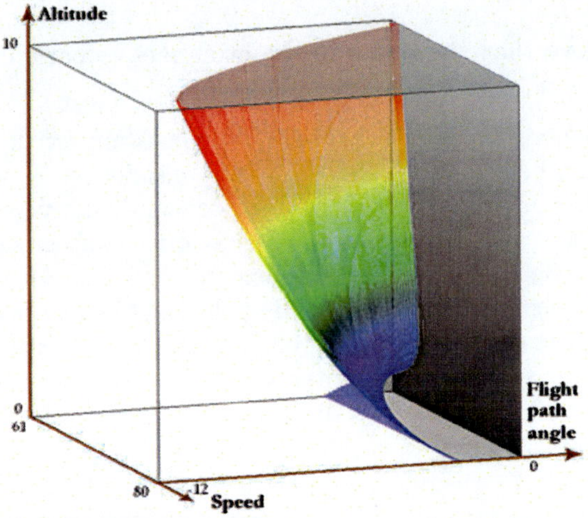

Fig. 1.15 The "Touch down safety envelope".
The 3-d boundary of the safety envelope is displayed. The colorimetric scale indicates the altitude.

This topic is developed in Sect. 3.3, p. 108.

1.3.2 Environmental Illustrations

1.3.2.1 Management of Renewable Resources

This example is fully developed in Sects. 9.2, p. 321 and 7.3, p. 262.

The management of renewable resources requires both dynamics governing the renewable resource (fishes) and the economics (fisheries), and biological and economic viability constraints.

The following classical example (called Verhulst-Schaeffer example) assumes that the evolution of the population of a fish species $x(t)$ is driven by the logistic Verhlust differential equation $x'(t) = r(t, x(t))x(t) := rx(t)\left(1 - \frac{x(t)}{b}\right)$ (see Sect. 7.2, p. 248) and that its growth rate is depleted proportionally to the fishing activity $y(t)$: $y'(t) = (r(t, x(t)) - v(t))y(t)$ (see Sect. 7.3, p. 262). The controls are the velocities $v(t)$ of the fishing activity (see (7.6), p. 263).

The ecological[2] environment consists of the number of fishes above a certain threshold in order to survive (on the right of the black area on Fig. 7.8). The economic environment is the subset of the ecological environment, since, without renewable biological resources there is no economic activity whatsoever. In this illustration, the economic environment is the subset above the hyperbola: fishing activity is not viable (grey area) outside of it.

We shall prove that the economic environment is partitioned into three zones:

- *Zone* (1), where economic activity is consistent with the ecological viability: it is the ecological and economic paradise;
- *Zone* (2), an economic purgatory, where economic activity will eventually disappear, but can revive later since enough fishes survive (see the trajectory of the evolution starting form A);
- *Zone* (3), the ecological and economic hell, where economic activity leads both to the eventual bankruptcy of fisheries and eventual extinction of fishes, like the sardines in Monterey after an intensive fishery activity during the 1920–1940s before disappearing in the 1950s together with Cannery Row.

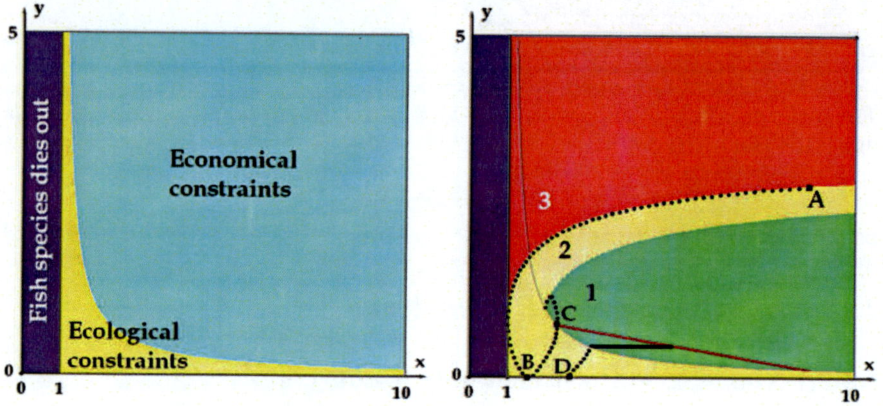

Fig. 1.16 Permanence Basin of a Sub-environment.
Zone (1) *is the* viability kernel *of the economic environment and* Zone (2) *is the* permanence kernel *(see Definition 9.2.3, p.323) of the economic environment, which is viable subset of the ecological environment. In* Zone (3), *economic activity leads both to the bankruptcy of fisheries and extinction of fishes.*

[2] The root *"eco"* comes from the classical Greek *"oiko"*, meaning house.

We shall see in Sect. 7.3.3, p. 269 that the concept of crisis function measures the shortest time spent outside the target, equal to zero in the viability kernel, finite in the permanence kernel and infinite outside of the permanence kernel: See Fig. 7.8, p. 270.

1.3.2.2 Climate-Economic Coupling and the Transition Cost Function

This example is a metaphor of Green House Gas emissions. For stylizing the problems and providing a three-dimensional illustration, we begin by isolating three variables defining the state:

- the concentration $x(t)$ of greenhouse gases,
- the short-term pollution rate $y(t) \in \mathbb{R}_+$,
- generated by economic activity summarized by a macro-economic variable $z(t) \in \mathbb{R}_+$ representing the overall economic activity producing emissions of pollutants.

The control represents an economic policy taken here as the velocity of the economic activity.

The ultimate constraint is the simple limit b on the concentration of the greenhouse gases: $x(t) \in [0, b]$.

We assume known the dynamic governing the evolution of the concentration of greenhouse gases, the emission of pollutants and of economic activity, depending, in the last analysis, on the economic policy, describing how one can slow or increase the economic activity. The question asked is how to compute the transitions cost (measured, for instance, by the intensity of the economic policy). Namely, transition cost is defined as the largest intensity of the economic activities consistent with the bound b on concentration of greenhouse gases. The transition cost function is the smallest transaction cost over the intensities of economic policies. This is a crucial information to know what would be the price to pay for keeping pollution under a given threshold. Knowing this function, we can derive, for each amount of transition cost, what will be the three-dimensional subset of triples (concentration-emission-economic activity) the transition costs of which are smaller than or equal to this amount.

We provide these subsets (called level-sets) for four amounts c, including the most interesting one, $c = 0$, for an example of dynamical system:

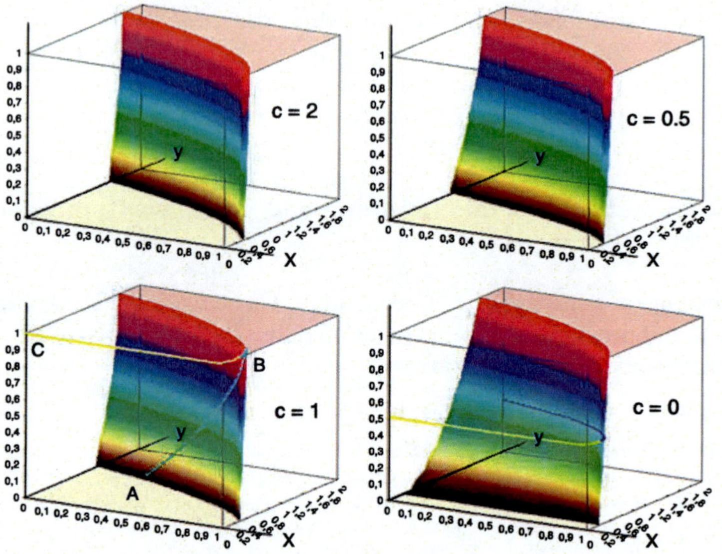

Fig. 1.17 Level-sets of the transition function.
We provide the level-sets of the transition function for the amounts $c =$ 0, 0.5, 1 & 2. *The trajectories of some evolutions are displayed (see The Coupling of Climate and Economic Dynamics: Essays on Integrated Assessment, [113, Haurie & Viguier]).*

Since this analysis involves macroeconomic variables which are not really defined and the dynamics of which are not known, this illustration has only a qualitative and metaphoric value, yet, instructive enough.

1.3.3 Strange Attractors and Fractals

1.3.3.1 Lorenz Attractors

We consider the celebrated Lorenz system (see Sect. 9.3, p. 344), known for the "chaotic" behavior of some of its evolutions and the "strange" properties of its attractor. *The classical approximations of the Lorenz attractor by commonly used "shooting methods" are not contained in the mathematical attractor*, because the attractor is approximated by the union of the trajectories computed for a large, but finite, number of iterations (computers do not experiment infinity). Theorem 9.3.12, p.352 states that the Lorenz attractor is contained in the backward viability kernel (the viability kernel under the Lorenz system with the opposite sign). If the solution starts outside the backward viability theorem, then it cannot reach the attractor in finite time, but only approaches it asymptotically. Hence, the first task is to study and compute this backward viability kernel containing the attractor.

However, the attractor being invariant, evolutions starting from the attractor are viable in it. Unfortunately, the usual *shooting methods* do not provide a good approximation of the Lorenz attractor. Even when one element the attractor is known, the extreme sensitivity of the Lorenz system on initial conditions forbids the usual *shooting methods* to govern evolutions viable in the attractor.

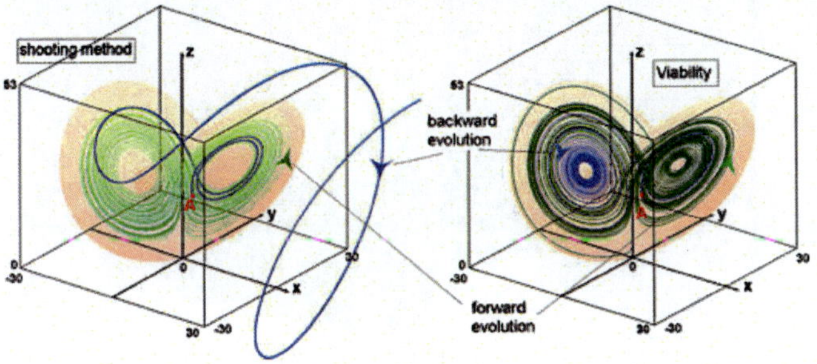

Fig. 1.18 Viable evolutions in the backward viability kernel.
Knowing that the attractor is contained in the backward viability kernel computed by the viability algorithm and is forward invariant, the viability algorithm provides evolutions viable in this attractor, such as the one displayed in the figure on the right. It "corrects" the plain shooting method using a finite-difference approximation of the Lorenz systems (such as the Runge–Kutta method used displayed in the figure on the left) which are too sensitive to the round-up errors. The viability algorithm "tames" viable evolutions (which loose their wild behavior on the attractor) by providing a feedback governing viable evolution, as can be shown in the figure on the right, displaying a viable evolution starting at A. Contrary to the solution starting at A the shooting method provided in the figure on the left, this "viable" evolution is viable.

The Lorenz system has fascinated mathematicians since the discovery of its properties.

1.3.3.2 Fluctuations Under the Lorenz System

Another feature of the wild behavior of evolutions governed by the Lorenz system is their "fluctuating property": most of them flip back and forth from one area to another. This is not the case of all of them, though, since the evolutions starting from equilibria remain in this state, and therefore, do not fluctuate. We shall introduce the concept of *fluctuation basin* (see Definition 9.2.1, p.322), which is the set of initial states from which starts a fluctuating evolution. It can be formulated in terms of viability kernels and capture basins, and thus computed with a viability algorithm.

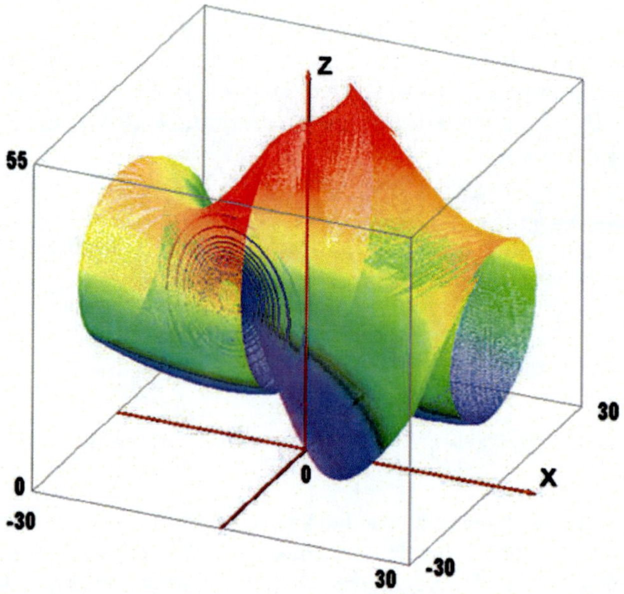

Fig. 1.19 Fluctuation basin of the Lorenz system.
The figure displays the (very small) complement *of the fluctuation: outside of it, evolutions flip back and forth from one half-cube to the other. The complement of the fluctuation basin is viable.*

1.3.3.3 Fractals Properties of Some Viability Kernels

Viability kernels of compact sets under a class of discrete set-valued dynamical systems are Cantor sets having fractal dimensions, as it is explained in Sect. 2.9.4, p. 79. Hence they can be computed with the Viability Kernel Algorithm, which provides an exact computation up to the pixel, instead of approximations as it is usually obtained by "shooting methods".

This allows us to revisit some classical examples, among which we quote here the *Julia sets*, studied in depth for more than a century. They are (the boundaries of) the subsets of initial complex numbers z such that their iterates $z_{n+1} = z_n^2 + u$ (or, equivalently, setting $z = x + iy$ and $u = a + ib$, $(x_{n+1}, y_{n+1}) = (x_{n+1}^2 - y_{n+1}^2 + a, 2x_{n+1}y_{n+1} + b)$), remain in a given ball. As is clear from this definition, Julia sets are (boundaries) of viability kernels, depending on the parameter $u = a + ib$.

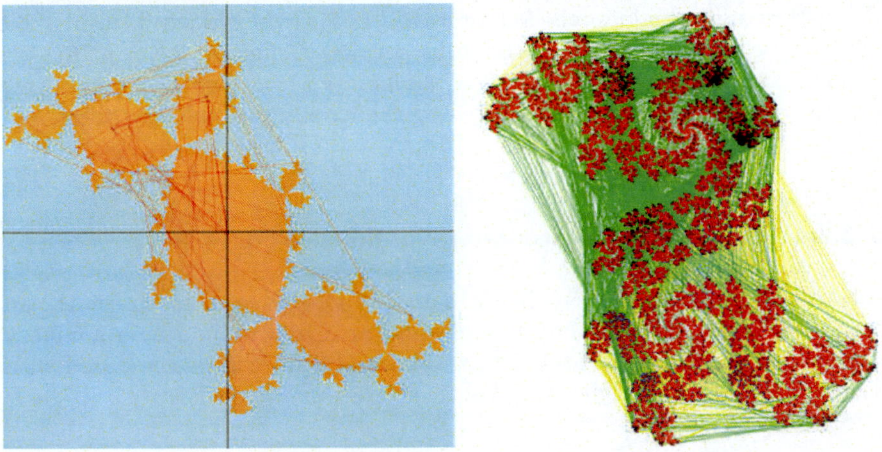

Fig. 1.20 Julia sets and fractals.
Unlike "shooting methods", the viability kernel algorithm provides the exact viability kernels and the viable iterates for two values of the parameter u. On the left, the viability kernel as a non-empty interior and its boundary, the Julia set, *has fractal properties. The second example, called* Fatou dust, *having an empty interior, coincides with its boundary, and thus is a Julia set. The discrete evolutions are not governed by the standard dynamics, which are sensitive to round-up errors, but by the corrected one, which allows the discrete evolutions to be viable on the Julia sets.*

Actually, viability kernel algorithms also provide the Julia set, which is the boundary of the viability kernel, equal to another boundary kernel thanks to Theorem 9.2.18, p. 339:

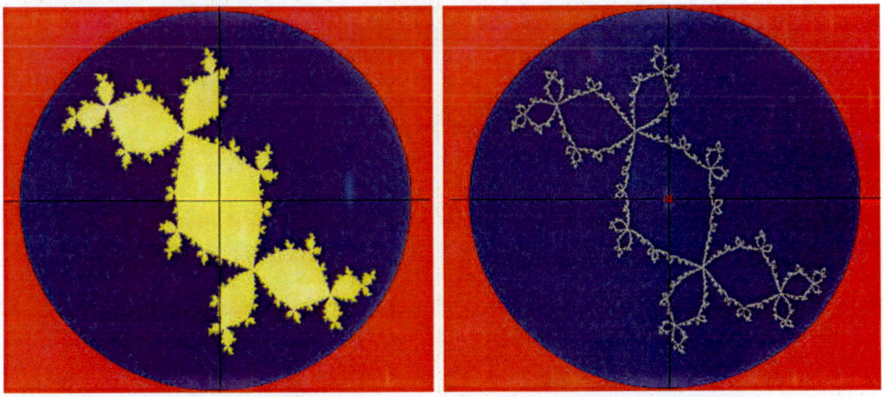

Fig. 1.21 Julia sets and filled-in Julia sets.

The left figure *reproduces the* left figure *of Fig. 2.4, which is the* filled-in Julia set, *the boundary of which is the* Julia set *(see Definition 2.9.6, p.76). We shall prove that this boundary is itself a viability kernel (see Theorem 9.2.18, p. 339), which can then be computed by the viability kernel algorithm.*

1.3.3.4 Computation of Roots of Nonlinear Equations

The set of all roots of a nonlinear equation located in a given set A can be formulated as the viability kernel of an auxiliary set under an auxiliary differential equation (see Sect. 4.4, p. 139), and thus, can be computed with the Viability Kernel Algorithm. We illustrate this for the map $f : \mathbb{R}^3 \mapsto \mathbb{R}$ defined by

$$f(x, y, z) = (|x - \sin z| + |y - \cos z|) \cdot (|x + \sin z| + |y + \cos z|)$$

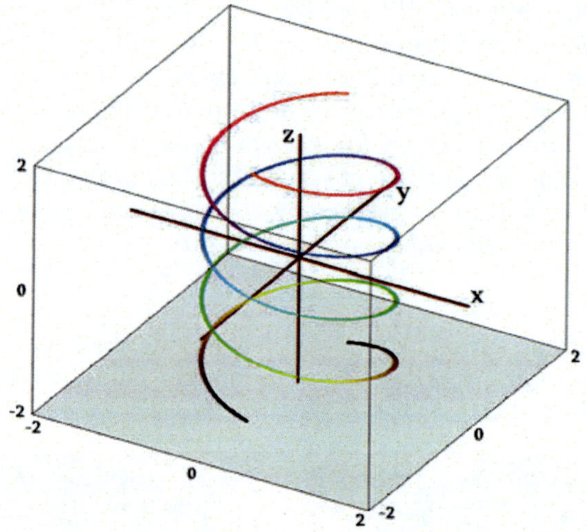

Fig. 1.22 The set of roots of nonlinear equations.
The set of roots of the equation

$$(|x - \sin z| + |y - \cos z|) \cdot (|x + \sin z| + |y + \cos z|) \ = \ 0$$

is a double helix. It is recovered in this example by using the viability kernel algorithm.

1.4 Organization of the Book

This book attempts to bridge conflicting criteria: logical and pedagogical. The latter one requires to start with easy examples and complexify them, one after the other. The first approach demands to start from the general to the particular (top–down), not so much for logical reasons than for avoiding to duplicate proofs, and thus space in this book and time for some readers. We have chosen after many discussions and hesitations some compromise, bad as compromises can be.

When needed, the reader can work out the proofs of the theorems scattered in the rest of the book and their consequences. We have tried to develop a rich enough system of internal references to make this task not too difficult.

The *heart of the book* is made of the three first Chaps. 2, p. 43, 4, p. 125 and 6, p. 199. They are devoted to the exposition of the main concepts and principal results of viability, dynamic intertemporal optimization (optimal control) and heavy evolutions (inertia functions). Most of the proofs are not provided, only the simple ones which are instructive. However, "survival kits" summarizing the main results are provided to be applied without waiting for the proofs and the underlying mathematical machinery behind.

These three chapters should allow the reader to master the concepts of those three domains and use them for solving a manifold of problems: engineering, requiring both viability and intertemporal optimization problems, life sciences, requiring both viability and inertia concepts and... mathematics, where more viability concepts and theorems are provided either for their intrinsic interest or for mathematical applications (detection maps, connection and collision kernels, chaos à la Saari, etc.) to connected problems (chaos and attractors, local stability and instability, Newton methods for finding equilibria, inverse mapping theorem, etc.).

1.4.1 Overall Organization

The contents of viability theory may be divided in two parts:

- *Qualitative concepts*, exploring the concepts and providing the results at the level of subsets, for instance, the "viability kernel", presented in Chaps. 2, p. 43, *Viability and Capturability*, 8, p. 273, *Connection Basins*, 9, p. 319 *Local and Asymptotic Properties of Equilibria*, 10, p. 375 *Viability and Capturability Properties of Evolutionary Systems* and 11, p. 437 *Regulation of Control Systems*, the two last ones providing the proofs. Chapter 2, p. 43 and its *Viability Survival Kit*, Sect. 2.15, p. 98, is designed to allow the non mathematician to grasp the applications.
- *Quantitative concepts* in the form of dynamical indicators. For instance, the "exit function" not only provides the viability kernel, over which

this exit function takes infinite values, but, outside the viability kernel, *measures* the exit time, a kind of survival time measuring the largest time spent in the environment before leaving it. The main results are presented in Chaps. 4, p. 125 *Viability and Dynamic Intertemporal Optimality.* Sect. 4.11, p. 168 provides the *Optimal Control Survival Kit* allowing the reader to study the illustrations and applications without having to confront the mathematical proofs presented in Chaps. 13, p.523, and 17, p. 681.

- *Inertial concepts* using quantitative on the velocities of the regulons. For instance, Chaps. 6, p. 199 *Inertia Functions, Viability Oscillators and Hysteresis.*

On the other hand, we propose three types of applications: to social sciences, to engineering and to... mathematics, since viability theory provides tools for solving first-order systems of partial differential equations, opening the gate for many other applications.

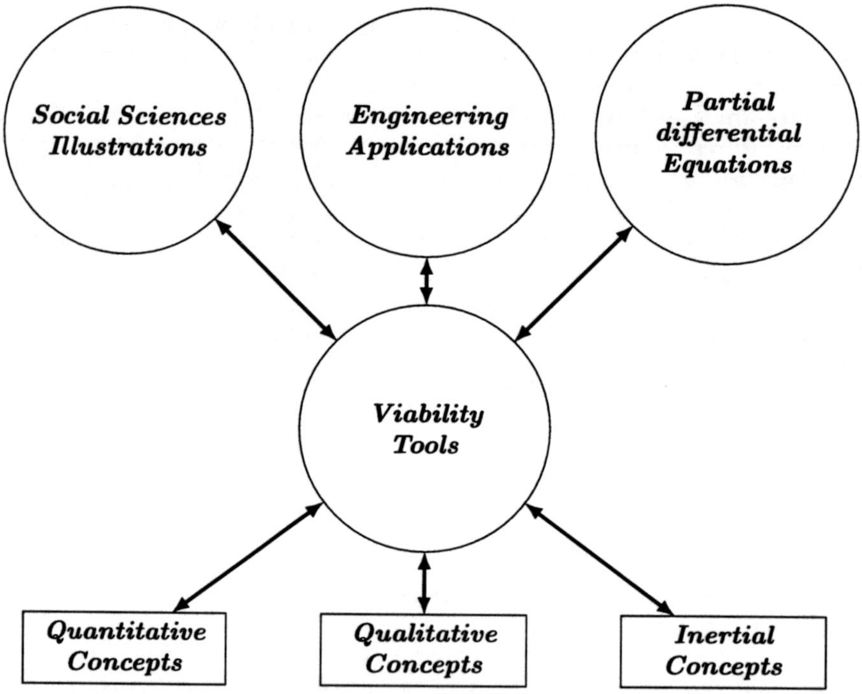

1.4.2 Qualitative Concepts

Qualitative concepts of viability theory constitute the main body of this book. We have chosen to regroup the main concepts and to state the main results in Chap. 2, p. 43. It has been written to start slowly following a *bottom–up* approach presenting the concepts from the simplest ones to more general ones, instead of starting with the more economical and logical *top–down* approach starting with the definition of the *evolutionary systems*. This chapter ends with a *viability survival kit* summarizing the most important results used in the applications. Hence, in a first reading, the reader can skip the four other chapters of this book. They can be omitted upon first reading.

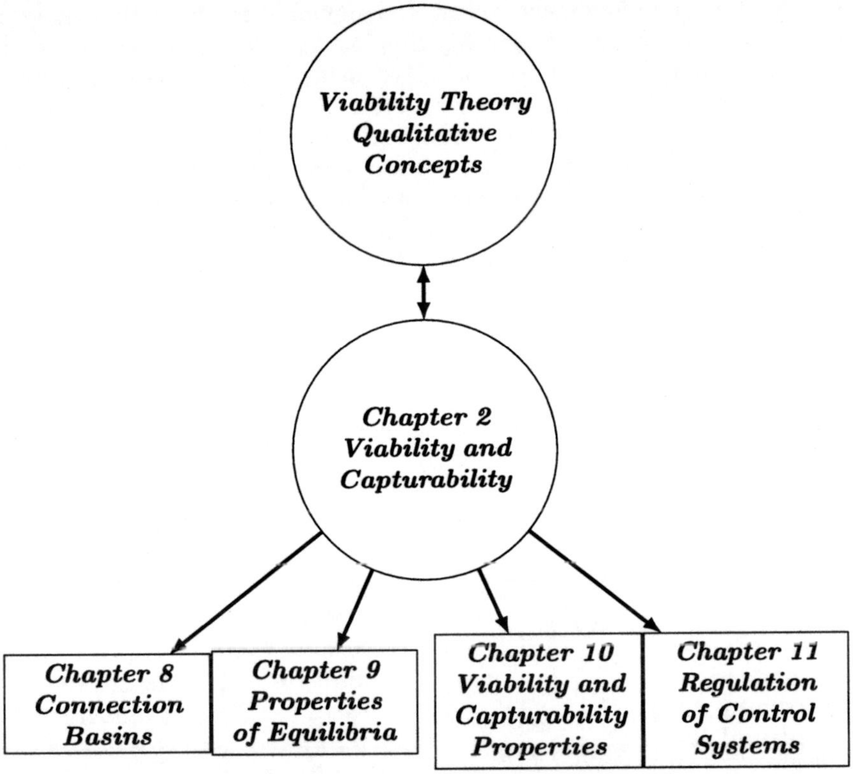

Chapter 8, p. 273 presents time dependent evolutionary systems, constraints and targets. These concepts allow us to study connection and collision basins. Chapter 9, p. 319 deals with the concepts of permanence kernels and fluctuation basins that we use to studies more classical issues of dynamical systems such as local and asymptotic stability properties of viable subsets and equilibria. Chapters 10, p. 375 and 11, p. 437 provide the proofs of the results stated in the viability survival kit of Chap. 2, p. 43, in the framework of general evolutionary systems, first, and next, of parameterized systems, where the tangential regulation maps governing the evolution of viable evolutions ares studied.

1.4.3 Quantitative Concepts

We chose to interrupt the logical course of studying the *qualitative* properties first before tackling the *quantitative* ones. The reason is that they provide the most interesting application to different issues of intertemporal optimization. Chapter 4, p. 125 is devoted to optimization of "averaging or integral criteria" on evolutions governed by controlled and regulated systems, among them the exit time function of an environment, the minimal time to reach a target in a viable way, the minimal length function, occupational costs and measures, attracting and Lyapunov functions, crisis, safety and restoration functions, Eupalinian and collision functions, etc. As for Chap. 2, p. 43, we do not, at this stage, provide all the proofs, in particular, the proofs concerning Hamilton–Jacobi–Bellman partial differential equations and their rigorous role in their use for building the feedbacks regulating viable and/or optimal evolutions. They are relegated in Chap. 17, p. 681 at the end of the book, since it involves the most technical statements and proofs.

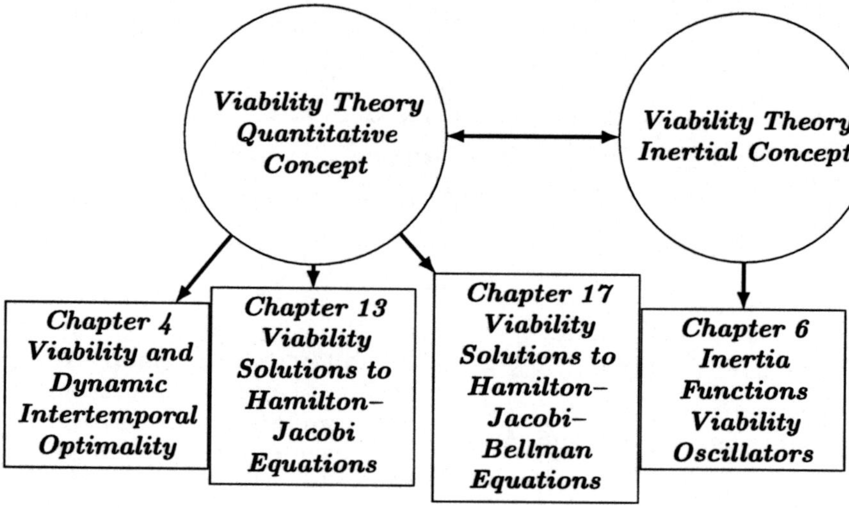

Chapter 6, p. 199 proposes the study of the main qualitative property of regulated systems which motivated viability theory from the very beginning: the concept of *inertia functions*, which, on top of their intrinsic interest, furnish the keys to the definition of the *inertia principle*, satisfied by *heavy evolutions*, as well as *hysterons* (hysteresis loops) and other interesting features.

1.4.4 First-Order Partial Differential Equations

Although first-order partial differential equations were not part of the original motivations of viability theory, it happened that some of its results were efficient for studying some of them: Chap. 16, p. 631 *A Viability Approach to Conservation Laws* is devoted to the prototype of *conservations laws*, the Burgers partial differential equation, and Chap. 17, p. 681 *A Viability Approach to Hamilton–Jacobi Equations* to Hamilton–Jacobi–Bellman equations motivated by optimal control problems and more generally, by intertemporal optimization problems.

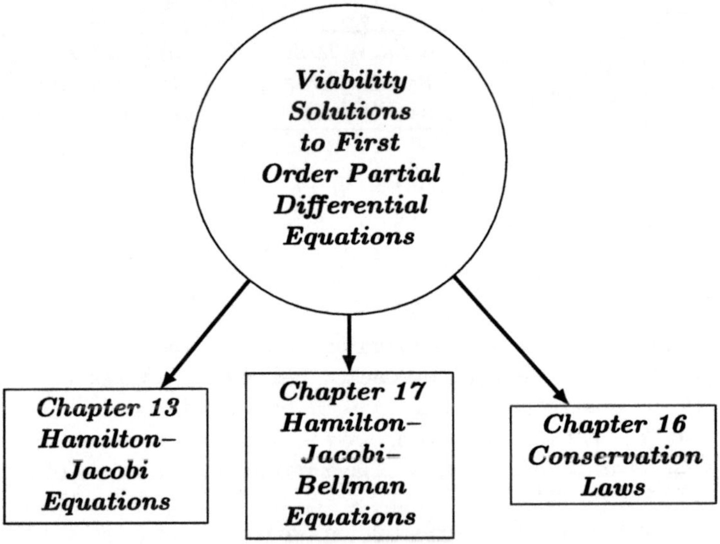

Chapters 4, p. 125, 6, p. 199, 15, p. 603, and 14, p. 563 provide other examples of Hamilton–Jacobi–Bellman equations.

1.4.5 Social Sciences Illustrations

Two chapters are devoted to some questions pertaining to this domain to social sciences issues.

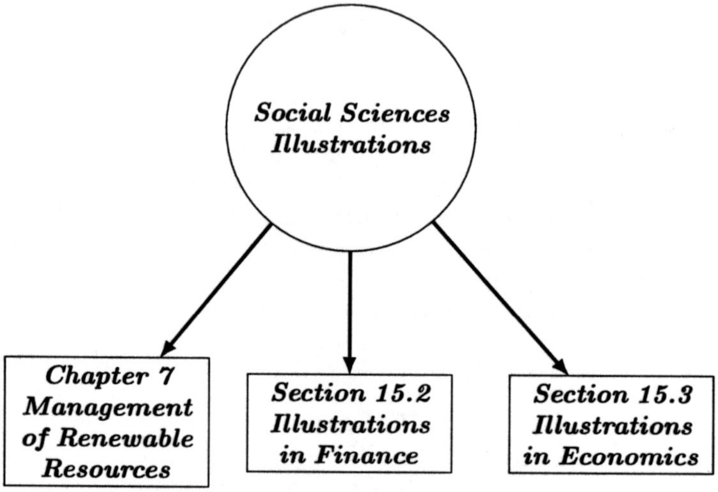

Chapter 7, p. 247 *Management of Renewable Resources* presents in the simplest framework of one dimensional model a study of the management of ecological renewable resources, such as fishes, coupled with economic constraints.

Section 15.2, p. 605 *Illustrations in Finance* deals with one very specific, yet, important, problem in mathematical finance, known under the name of *implicit volatility.* Direct approaches (see Box) assumes the volatility (stochastic or tychastic) to be known for deriving the value of portfolios. But volatility is not known, and an inverse approach is required to deduce the *implicit volatility* from empirical laws obeyed by the values of portfolios.

Section 15.3, p. 620 *Illustrations in Economics* deals with links between micro-economic and macro-economic approaches of the same dynamical economic problem. We use duality theory of convex analysis to prove that the derivation of macroeconomic value from microeconomic behavior of economic agents is equivalent to the derivation of the behavior of these agents from the laws governing the macroeconomic value.

1.4.6 Engineering Applications

The two chapters placed under this label deal directly with engineering issues.

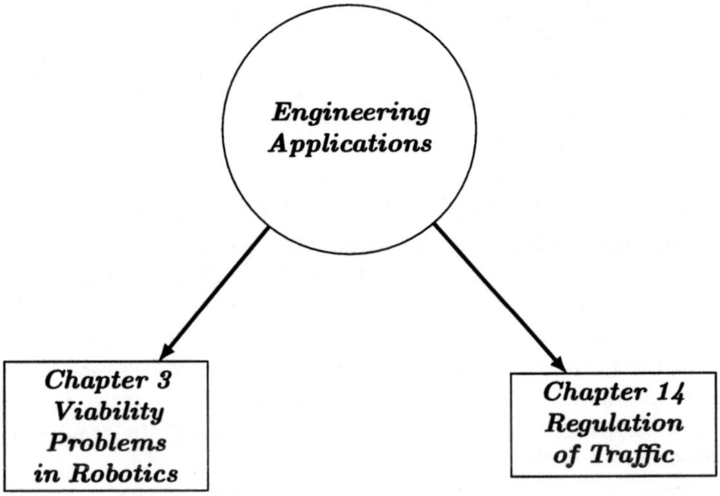

Chapter 3, p.105, *Viability Problems in Robotics*, studies two applications to robotics, one concerning field experiment of the viability feedback allowing a robot to rally a target in a urban environment while avoiding obstacles, the second one studying the safety envelope of the landing of a plane as well as the regulation law governing the safe landing evolutions viable in this envelope.

Chapter 14, p. 563 *Regulation of Traffic* revisits the standard macroscopic model of *Lighthill – Whitham – Richards*, providing the Moskowitz traffic functions governed by an example of Hamilton–Jacobi partial differential equation. It deals with new ways of measuring traffic. The first one uses "Lagrangian" data by monitoring some probe vehicles instead of static sensors measuring of traffic density, which amounts to replacing standard boundary conditions. The second one shows that the fundamental relation linking density to flows of traffic, determining the shape of the Hamilton–Jacobi–Bellman equation, can be replaced by a "dual" relation linking macroscopic traffic velocity and flows. This relation could be obtained from satellite measurements in the future.

1.4.7 The Four Parts of the Monograph

The book is organized according to increasing mathematical difficulty. The first part, *Viability Kernels and Examples*, provides the main definitions (Chaps. 2, p.43, for the qualitative concepts, 4, p.125 for the qualitative ones, and 6, p.199 for the inertial ones). The most useful statements are not proved at this stage, but summarized in the *Viability Survival Kit*, Sect. 2.15, p. 98, and the *Optimal Control Survival Kit*, Sect. 4.11, p. 168. We provide illustrative applications at the end of these three chapters respectively: Chaps. 3, p.105, 5, p.179, and 7, p.247.

The second part, *Mathematical Properties of Viability Kernels*, is devoted to some other concepts and issues (Chaps. 8, p.273, and 9, p.319) and the proofs of the viability properties under general evolutionary systems (Chap. 10, p.375) and under regulated systems Chap. 11, p.437). The main Viability and Invariance Theorems are proved in Chap. 19, p.769, at the very end of the book, due to the technical difficulties.

The third part, *First-Order Partial Differential Equations*, presents the viability approach for solving first-order partial differential equations: Hamilton–Jacobi equations (Chap. 13, p.523), Hamilton–Jacobi–Bellman equations (Chap. 17, p.681), and Burgers equations (Chap. 16, p.631) as a prototype of conservation laws. Chapter 14, p. 563, deals with regulation of traffic and Chap. 15, p. 603 with financial and economic issues.

Besides the main Viability and Invariance Theorems, proved in Chap. 19, p.769, and Chaps. 16, p. 631, 17, p. 681, the proofs of the results presented in this book are technically affordable.

What may be difficult is the novelty of the mathematical approach based on set-valued analysis instead of analysis, dealing with subsets rather than with functions. Unfortunately, at the beginning of the Second World War, a turn has been taken, mainly by *Nicolas Bourbaki*, to replace set-valued maps studied since the very beginning of the century, after the pioneering views by *Georg Cantor*, *Paul Painlevé*, *Maurice Fréchet*, *Felix Hausdorff*, *Kazimierz Kuratowski*, *René Baire* and *Georges Bouligand*. The idea was that a set-valued map $F : X \rightsquigarrow Y$ was a single-valued map (again denoted by) $F : X \mapsto \mathcal{P}(Y)$, so that, it was enough to study single-valued map. One could have said that single-value maps being particular cases of set-valued maps, it would be sufficient to study set-valued maps. The opposite decision was taken and adopted everywhere after the 1950s. This is unfortunate, just because the $\mathcal{P}(Y)$ does not inherit the structures and the properties of the state Y. So, the main difficulty is to cure the single-valued brainwashing that all mathematicians have been subjected to. Adapting to a new "style" and another vision is more difficult than mastering new techniques, but in the same "style".

This is the reason why we summarized the main results of set-valued map analysis in Chap. 18, p. 713 and we postponed at the end of the book Chap. 19, p.769 gathering the proofs of the viability theorems. They constitute the fourth part of the book.

Many results of this book are unpublished. The first edition of *Viability theory*, [18, Aubin] (1991), contained 542 entries. It soon appeared that the number of fields of motivation and applications having increased, it would be impossible to make a relevant bibliography due to the lack of space. On the other hand, Internet allows to get the bibliographical information easily. So, we present only a list of monographs to conclude this book.

Part I
Viability Kernels and Examples

Chapter 2
Viability and Capturability

2.1 Introduction

This rather long chapter is the central one. It is aimed at allowing the reader to grasp enough concepts and statements of the principal results proved later on in the book to read directly and independently most of the chapters of the book: Chaps. 4, p. 125 and 6, p. 199 for qualitative applications, Chap. 8, p. 273 and 9, p. 319 for other quantitative concepts, Chaps. 7, p. 247 and 15, p. 603 for social science illustrations, Chaps. 3, p. 105, 14, p. 563 and 16, p. 631 in engineering, even though, here and there, some results require statements proved in the mathematical Chaps. 10, p. 375 and 11, p. 437.

This chapter defines and reviews the basic concepts: evolutions and their properties, in Sect. 2.2, p. 45, and next, several sections providing examples of evolutionary systems. We begin by the simple single-valued discrete systems and differential equations.

We next introduce parameters in the dynamics of the systems, and, among these parameters, distinguish constant coefficients, controls, regulons and tyches (Sect. 2.5, p. 58) according to the roles they will play. They motivate the introduction of controlled discrete and continuous time systems. All these systems generate *evolutionary systems* defined in Sect. 2.8, p. 68, the abstract level where it is convenient to study the viability and capturability properties of the evolutions they govern, presented in detail in Chap. 10, p. 375.

Next, we review the concepts of viability kernels and capture basins, under discrete time controlled systems in Sect. 2.9, p. 71. In this framework, the computation of the regulation map is easy and straightforward. We present the viability kernel algorithm in Sect. 2.9.2, p. 74 and use it to compute the Julia sets and Fatou dusts in Sect. 2.9.3, p. 75 and show in Sect. 2.9.4, p. 79 how the celebrated fractals are related to viability kernels under the class of discrete disconnected systems.

J.-P. Aubin et al., *Viability Theory*, DOI 10.1007/978-3-642-16684-6_2,

Viability kernels and capture basins under continuous time controlled systems are the topics of Sect. 2.10, p. 85. We illustrate the concepts of viability kernels by computing the viability kernel under the (backward) Lorenz system, since we shall prove in Chap. 9, p. 319 that it contains the famous Lorenz attractor.

Viability kernels and capture basins single out initial states from which **at least** one discrete evolution is viable forever or until it captures a target. We are also interested to the invariance kernels and absorption basins made of initial states from which **all** evolutions are viable forever or until they capture a target. One mathematical reason is that these concepts are, in some sense, "dual" to the concepts of viability kernels and capture basins respectively, and that they will play a fundamental role for characterizing them. The other motivation is the study of "tychastic" systems where the parameters are tyches (perturbations, disturbances, etc.) which translate one kind of uncertainty without statistical regularity, since tyches are neither under the control of an actor nor chosen to regulate the system.

We also address computational issues:

- viability kernel and capture basin algorithms for computing viability kernels and capture basins under discrete system in Sect. 2.9.2, p. 74,
- and next, discretization issues in Sect. 2.14, p. 96.

For computing viability kernels and capture basins under continuous time controlled systems, we proceed in two steps. First, approximate the continuous time controlled systems by discretized time controlled systems, so that viability kernels and capture basins under discretized systems converge to the viability kernels and capture basins under the continuous time evolutionary system, and next, compute the viability kernels and capture basins under the discretized time controlled systems by the viability kernel and capture basin algorithms.

This section just mentions these problems and shows how the (trivial) characterization of viability kernels and capture basins under discretized controlled systems gives rise to the tangential conditions characterizing them under the continuous time controlled system, studied in Chap. 11, p. 437.

The chapter ends with a "viability survival kit" in Sect. 2.15, p. 98, which summarizes the most important statements necessary to apply viability theory without proving them. They are classified in three categories:

- At the most general level, where simple and important results and are valid without any assumption,
- At the level of evolutionary systems, where viability kernels and capture basins are characterized in terms of local viability properties and backward invariance,
- At the level of control systems, where viability kernels and capture basins are characterized in terms of "tangential characterization" which allows us to define and study the regulation map.

2.2 Evolutions

Let X denote the *state space* of the system. Evolutions describe the behavior of the state of the system as a function of time.

Definition 2.2.1 ***[Evolutions and their Trajectories]*** *The time t ranges over a set $\mathbb{T}$ that is in most cases,*

1. *either the* **discrete time set** *of times $j \in \mathbb{T} := \mathbb{N} := \{0,\ldots,+\infty\{$ ranging over the set of positive integers $j \in \mathbb{N}$,*
2. *or the* **continuous time set** *of times $t \in \mathbb{T} := \mathbb{R}_+ := [0,\ldots,+\infty[$ ranging over the set of positive real numbers or scalars $t \in \mathbb{R}_+$.*

Therefore, evolutions are functions $x(\cdot) : t \in \mathbb{T} \mapsto x(t) \in X$ *describing the evolution of the state $x(t)$. The* trajectory (or orbit) *of an evolution $x(\cdot)$ is the subset $\{x(t)\}_{t\in\mathbb{T}} \subset X$ of states $x(t)$ when t ranges over $\mathbb{T}$.*

Warning: The terminology "trajectory" is often used as a synonym of evolution, but inadequately: *a trajectory is the range of an evolution.*

Unfortunately, for discrete time evolutions, tradition imposes upon us to regard discrete evolutions as *sequences* and to use the notation $\overrightarrow{x} : j \in \mathbb{N} \mapsto x_j := x(j) \in X$. We shall use this notation when we deal explicitly with discrete time. We use the notation $x(\cdot) : t \in \mathbb{R}_+ \mapsto x(t) \in X$ for continuous time evolutions and whenever the results we mention are valid for both continuous and discrete times. It should be obvious from the context whether $x(\cdot)$ denotes an evolution when time ranges over either discrete $\mathbb{T} := \mathbb{N}_+$ time set or continuous $\mathbb{T} := \mathbb{R}_+$ time set.

The choice between these two representations of time is not easy. The "natural" one, which appears the simplest for non mathematicians, is the choice of the set $\mathbb{T} := \mathbb{N}_+$ of discrete times. It has drawbacks, though. On one hand, it may be difficult to find a common *time scale* for the different components of the state variables of the state space of a given type of models. On the other hand, by doing so, we deprive ourselves from the concepts of velocity, acceleration and other dynamical concepts introduced by Isaac Newton (1642–1727), that are not well taken into account by discrete time systems as well as of the many results of the differential and integral calculus gathered for more than four centuries since the invention of the infinitesimal calculus by Gottfried Leibniz (1646–1716). Therefore, the choice between these two representations of times being impossible, we have to investigate both discrete and continuous time systems. Actually, the results dealing with viability kernels and capture basins use the same proofs, as we shall see in Chap. 10, p. 375. Only the viability characterization becomes dramatically simpler, not to say trivial, in the case of discrete systems.

Note however that for computational purposes, we shall approximate continuous time systems by discrete time ones where the time scale becomes infinitesimal.

Warning: *Viability properties of the discrete analogues of continuous-time systems can be drastically different*: we shall see on the simple example of the *Verhulst* logistic equation that the interval $[0, 1]$ is invariant under the continuous system

$$x'(t) = rx(t)(1 - x(t))$$

for all $r \geq 0$ whereas the viability kernel of $[0, 1]$ under its discrete counterpart

$$x_{n+1} = rx_n(1 - x_n)$$

is a Cantor subset of $[0, 1]$ when $r > 4$. Properties of discrete counterparts of continuous time dynamical systems can be different from their discretizations. These discretizations, under the assumptions of adequate convergence theorems, share the same properties than the continuous time systems.

We shall assume most of the time that:

1. the state space is a finite dimensional vector space $X := \mathbb{R}^n$,
2. evolutions are *continuous* functions $x(\cdot) : t \in \mathbb{R}_+ \mapsto x(t) \in X$ describing the evolution of the state $x(t)$.

We denote the space of continuous evolutions $x(\cdot)$ by $\mathcal{C}(0, \infty; X)$.

Some evolutions, mainly motivated by physics, are classical: *equilibria and periodic evolutions*. But these properties are not necessarily adequate for problems arising in economics, biology, cognitive sciences and other domains involving living beings. Hence we add the concept of evolutions *viable in a environment or capturing a target in finite time* to the list of properties satisfied by evolutions.

2.2.1 Stationary and Periodic Evolutions

We focus our attention to specific properties of evolutions, denoting by $\mathcal{H} \subset \mathcal{C}(0, \infty; X)$ the subset of evolutions satisfying these properties. For instance, the most common are stationary ones and periodic ones:

Definition 2.2.2 *[Stationary and Periodic Evolutions]*

1. *The subset* $\mathcal{X} \subset \mathcal{C}(0, \infty; X)$ *of* stationary evolutions *is the subset of evolutions* $x : t \mapsto x$ *when* x *ranges over the state space* X.

2. *The subset* $\mathcal{P}_T(X)$ *of* T-periodic evolutions *is the subset of evolutions* $x(\cdot) \in \mathcal{C}(0, \infty; X)$ *such that,* $\forall t \geq 0,\ x(t+T) = x(t)$.

Stationary evolutions are periodic evolutions for all periods T.

Stationary and periodic evolutions have been a central topic of investigation in dynamical systems motivated by physical sciences. Indeed, the brain, maybe because it uses periodic evolutions of neurotransmitters through subsets of synapses, has evolved to recognize periodic evolutions, in particular those surrounding us in daily life (circadian clocks associated with the light of the sun). Their extensive study is perfectly legitimate in physical sciences, as well as their new developments (bifurcations, catastrophes, dealing with the dependence of equilibria in terms of a parameter, and chaos, investigating the absence of continuous dependence of evolution(s) with respect to the initial states, for instance). However, even though we shall study evolutions regulated by constant parameters, bifurcations are quite difficult to observe, as it was pointed out in Sect. 3.3 of the book *Introduction to nonlinear systems and chaos* by *Stephen Wiggins*:

11 ***[On the Interpretation and Application of Bifurcation Diagrams: A Word of Caution]*** *At this point, we have seen enough examples so that it should be clear that the term bifurcation refers to the phenomenon of a system exhibiting qualitatively new dynamical behavior as parameters are varied. However, the phrase "as parameters are varied" deserves careful consideration [...] In all of our analysis thus far the parameters have been constant. The point is that we cannot think of the parameter as varying in time, even though this is what happens in practice. Dynamical systems having parameters that change in time (no matter how slowly!) and that pass through bifurcation values often exhibit behavior that is very different from the analogous situation where the parameters are constant.*

The situation in which *coefficients* are kept constant is familiar in physics, but, in engineering as well as in economic and biological sciences, they may have to vary with time, playing the roles of controls in engineering, of *regulons* in social and biological sciences, or *tyches*, when they play the role of random variables when uncertainty does not obey statistical regularity, as we shall see in Sect. 2.5, p. 58.

Insofar as physical sciences privilege the study of stability or chaotic behavior around *attractors* (see Definition 9.3.8, p. 349) and their *attraction basins* (see Definition 9.3.3, p. 347), the thorough study of *transient evolutions* have been neglected, although they pervade economic, social and biological sciences.

2.2.2 Transient Evolutions

Theory of Games and Economic Behavior. John von Neumann (1903–1957) and Oskar Morgenstern (1902–1976) concluded the first chapter of their monograph *"Theory of Games and Economic Behavior"* (1944) by these words: *Our theory is thoroughly static. A dynamic theory would unquestionably be more complete and therefore, preferable. But there is ample evidence from other branches of science that it is futile to try to build one as long as the static side is not thoroughly understood. [...] Finally, let us note a point at which the theory of social phenomena will presumably take a very definite turn away from the existing patterns of mathematical physics. This is, of course, only a surmise on a subject where much uncertainty and obscurity prevail [...] A dynamic theory, when one is found, will probably describe the changes in terms of simpler concepts.*

Unfortunately, the concept of equilibrium is polysemous. The mathematical one, which we adopt here, expresses stationary – time independent – evolution, that is, no evolution. The concept of equilibrium used by von Neumann and Morgenstern is indeed this static concept, derived from the concept of general equilibrium introduced by Léon Walras (1834–1910) in his book *Éléments d'économie politique pure* (1873) as an equilibrium (stationary point) of his tâtonnement process.

Another meaning results from the articulation between dynamics and *viability constraints*: This means here that, starting from any initial state satisfying these constraints, **at least** one evolution satisfies these constraints at each instant (such an evolution is called *viable*). An equilibrium can be viable or not, these two issues are independent of each other.

The fact that many scarcity constraints in economics are presented in terms of "balance", such as the balance of payments, may contribute to the misunderstanding. Indeed, the image of a balance conveys both the concept of equalization of opposite forces, hence of constraints, and the resulting stationarity – often called "stability", again, an ambivalent word connoting too many different meanings.

This is also the case in biology, since Claude Bernard (1813-1878) introduced the notion of constancy of inner milieu (constance du milieu intérieur). In 1898 he wrote: *Life results form the encounter of organisms and milieu, [...] we cannot understand it with organisms only, or with milieu only.* This idea was taken up under the name of "homeostasis" by Walter Cannon (1871-1945) in his book *Bodily changes in pain, hunger, fear and rage* (1915). This is again the case in ecology and environmental studies, as well as in many domains of social and human sciences when organisms adapt or not to several forms of viability constraints.

2.2.3 Viable and Capturing Evolutions

Investigating evolutionary problems, in particular those involving living beings, should start with identifying the constraints bearing on the variables which cannot – or should not – be violated. Therefore, if a subset $K \subset X$ represents or describes an environment, we mainly consider evolutions $x(\cdot)$ *viable in the environment* $K \subset X$ in the sense that

$$\forall t \geq 0, \;\; x(t) \in K \tag{2.1}$$

or *capturing the target* C in the sense that they are viable in K until they reach the target C *in finite time*:

$$\exists\, T \geq 0 \text{ such that } \begin{cases} x(T) \in C \\ \forall t \in [0,T], \; x(t) \;\in\; K \end{cases} \tag{2.2}$$

We complement Definition 6, p. 15 with the following notations:

Definition 2.2.3 ***[Viable and Capturing Evolutions]*** *The subset of evolutions* viable *in* K *is denoted by*

$$\mathcal{V}(K) := \{x(\cdot) \mid \forall t \geq 0, \;\; x(t) \in K\} \tag{2.3}$$

and the subset of evolutions capturing *the target* C *by*

$$\mathcal{K}(K,C) := \{x(\cdot) \mid \exists\, T \geq 0 \text{ such that } x(T) \in C \;\&\; \forall t \in [0,T], \; x(t) \;\in\; K\} \tag{2.4}$$

We also denote by

$$\mathcal{V}(K,C) := \mathcal{V}(K) \cup \mathcal{K}(K,C) \tag{2.5}$$

the set of evolutions viable in K outside C, *i.e. that are viable in* K *forever or until they reach the target* C *in finite time.*

Example: The first examples of such environments used in control theory were vector (affine) subsets because, historically, analytical formulas could be obtained. Nonlinear control theory used first geometrical methods, which required *smooth equality constraints*, yielding environments of the form

$$K := \{x \mid g(x) = 0\} \text{ where } g : X := \mathbb{R}^c \mapsto Y := \mathbb{R}^b \; (b < c) \text{ is smooth}$$

Subsets such as smooth manifolds (Klein bottle, for instance), having no boundaries, the viability and invariance problems were evacuated. This is no longer the case when the environment is defined by inequality constraints, even smooth ones, yielding subsets of the form

$$K := \{x \mid g(x) \leq 0\}$$

the boundary of which is a proper subset. Subsets of the form

$$K := \{x \in L \mid g(x) \in M\}$$

where $L \subset X$, $g : X \mapsto Y$ and where $M \subset Y$ are typical environments encountered in mathematical economics. It is for such cases that mathematical difficulties appeared, triggering viability theory.

Constrained subsets in economics and biology are generally not smooth. The question arose to build a theory and forge new tools that did require neither the smoothness nor the convexity of the environments. *Set-valued analysis*, motivated in part by these viability and capturability issues, provided such tools.

Remark. These constraints can depend on time (time-dependent constraints), as we shall see in Chap. 8, p. 273, upon the state, the history (or the path) of the evolution of the state. *Morphological equations* are kind of differential equations governing the evolution of the constrained state $K(t)$ and can be paired with evolutions of the state. The issues are dealt in [23, Aubin]. □

Remark. We shall introduce other families of evolutions, such as the viable evolutions *permanent* in a cell $C \subset K$ of *fluctuating* around C which are introduced in bio-mathematics (see Definition 9.2.1, p. 322). In "qualitative physics", a sequence of tasks or objectives is described by a family of subsets regarded as qualitative cells. We shall investigate the problem of finding evolutions visiting these cells in a prescribed order (see Sect. 8.8, p. 302). □

These constraints have to be confronted with evolutions. It is now time to describe how these evolutions are produced and to design mathematical translations of several evolutionary mechanisms.

2.3 Discrete Systems

Discrete evolutionary systems can be defined on any metric state space X.

12 *Examples of State Spaces for Discrete Systems:*

1. *When $X := \mathbb{R}^d$, we take any of the equivalent vector space metrics for which the addition and the multiplication by scalars is continuous.*
2. *When $X_\rho := \rho\mathbb{Z}^d$ is a grid with step size ρ, we take the discrete topology, defined by $d(x,x) := 0$ and $d(x,y) := 1$ whenever $x \neq y$. A sequence of elements $x_n \in X$ converges to x if there exists an integer N such that for any $n \geq N$, $x_n = x$, any subset is both closed and open, the compacts are finite subsets. Any single-valued map from $X := \mathbb{Z}^d$ to some space E is continuous.*

Deterministic discrete systems, defined by

$$\forall j \geq 0,\ x_{j+1} = \varphi(x_j) \text{ where } \varphi : x \in X \mapsto \varphi(x) \in X$$

are the simplest ones to formulate, but not necessarily the easiest ones to investigate.

Definition 2.3.1 ***[Evolutionary Systems associated with Discrete Systems]*** *Let X be any metric space and $\varphi : X \mapsto X$ be the single-valued map associating with any state $x \in X$ its successor $\varphi(x) \in X$.*
The space of discrete evolutions $\overrightarrow{x} := \{x_j\}_{j\in\mathbb{N}}$ is denoted by $X^{\mathbb{N}}$. The evolutionary system *$S_\varphi : X \mapsto X^{\mathbb{N}}$ associated with the map $\varphi : x \in X \mapsto \varphi(x) \in X$ associates with any $x \in X$ the set $S_\varphi(x)$ of discrete evolutions $\overrightarrow{x}$ starting at $x_0 = x$ and governed by the discrete system*

$$\forall j \geq 0,\ x_{j+1} = \varphi(x_j)$$

An equilibrium of a discrete dynamical system is a stationary evolution governed by this system.

An equilibrium $\overline{x} \in X$ (stationary point) of an evolution $\overrightarrow{x}$ governed by the discrete system $x_{j+1} = \varphi(x_j)$ is a *fixed point* of the map φ, i.e., a solution to the equation $\varphi(\overline{x}) = \overline{x}$. There are two families of Fixed Point Theorems based:

1. either on the simple Banach–Cacciopoli–Picard Contraction Mapping Theorem in complete metric spaces,
2. or on the very deep and difficult 1910 Brouwer Fixed Point Theorem on convex compact subsets, the cornerstone of nonlinear analysis.

Example: The Quadratic Map The quadratic map φ associates with $x \in [0,1]$ the element $\varphi(x) = rx(1-x) \in \mathbb{R}$, governing the celebrated *discrete logistic system* $x_{j+1} = rx_j(1-x_j)$. The fixed points of φ are 0 and $c := \frac{r-1}{r}$, which is smaller than 1. We also observe that $\varphi(0) = \varphi(1) = 0$ so that the successor of 1 is the equilibrium 0.

For $K := [0,1] \subset \mathbb{R}$ to be a state space under this discrete logistic system, we need φ to map $K := [0,1]$ to itself, i.e., that $r \leq 4$. Otherwise, for $r > 4$, the roots of the equation $\varphi(x) = 1$ are equal to $a := \frac{1}{2} - \frac{\sqrt{r^2-4r}}{2r}$ and $b := \frac{1}{2} + \frac{\sqrt{r^2-4r}}{2r}$, where $b < c$. We denote by $d \in [0,a]$ the other root of the equation $\varphi(d) = c$. Therefore, for any $x \in]a,b[$, $\varphi(x) > 1$.

A way to overcome this difficulty is to *associate with the single-valued* $\varphi : [0,1] \mapsto \mathbb{R}$ *the set-valued map* $\Phi : [0,1] \rightsquigarrow [0,1]$ *defined by* $\Phi(x) := \varphi(x)$ when $x \in [0,a]$ and $x \in [b,1]$ and $\Phi(x) := \emptyset$ when $x \in]a,b[$. Let us set

$$\omega^\flat(y) := \frac{1}{2} - \frac{\sqrt{r^2-4ry}}{2r} \text{ and } \omega^\sharp(y) := \frac{1}{2} + \frac{\sqrt{r^2-4ry}}{2r}$$

The inverse Φ^{-1} is defined by

$$\Phi^{-1}(y) := \left(\omega^\flat(y), \omega^\sharp(y)\right)$$

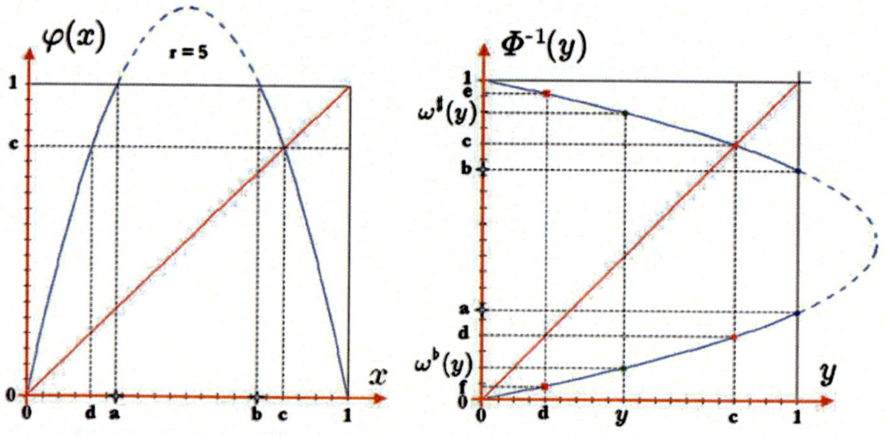

Fig. 2.1 Discrete Logistic System.
The graph of the function $x \mapsto \varphi(x) := rx(1-x)$ *for* $r = 5$ *is displayed as a function* $\varphi : [0,1] \mapsto \mathbb{R}$ *as a set-valued map* $\Phi : [0,1] \rightsquigarrow [0,1]$ *associating with any* $x \in [a,b]$ *the empty set. Equilibria are the abscissas of points of the intersection of the graph* Graph(φ) *of* φ *and of the bisectrix. We observe that* 0 *and the point* c *(to the right of* b*) are equilibria. On the right, the graph of the inverse is displayed, with its two branches.*

The predecessors $\Phi^{-1}(0)$ and $\Phi^{-1}(c)$ of equilibria 0 and c are initial states of viable discrete evolutions because, starting from them, the equilibria are their successors, from which the evolution remains in the interval forever. They are made of $\omega^\sharp(0) = 1$ and of $c_1 := \omega^\flat(c)$. In the same way, the four predecessors $\Phi^{-2}(0) := \Phi^{-1}(\Phi^{-1}(0)) = \{\omega^\flat(1) = a, \omega^\sharp(1) = b\}$ and $\Phi^{-2}(c)$ are initial states of viable evolutions, since, after two iterations, we obtain the two equilibria from which the evolution remains in the interval forever. And so on: The subsets $\Phi^{-p}(0)$ and $\Phi^{-p}(c)$ are made of initial states from which start evolutions which reach the two equilibria after p iterations, and thus, which are viable in K. They belong to the *viability kernel* of K (see Definition 2.9.1, p. 71 below). This study will resume in Sect. 2.9.4, p. 79.

2.4 Differential Equations

2.4.1 Determinism and Predictability

We begin by the simplest class of continuous time evolutionary systems, which are associated with differential equations

$$x'(t) = f(x(t))$$

where $f : X \mapsto X$ is the single-valued map associating with any state $x \in X$ its velocity $f(x) \in X$.

Definition 2.4.1 *[**Evolutionary Systems associated with Differential Equations**] Let $f : X \mapsto X$ be the single-valued map associating with any state $x \in X$ its velocity $f(x) \in X$.*
The evolutionary system *$\mathcal{S}_f : X \rightsquigarrow \mathcal{C}(0, +\infty; X)$ defined by $f : X \mapsto X$ is the set-valued map associating with any $x \in X$ the set $\mathcal{S}_f(x)$ of evolutions $x(\cdot)$ starting at x and governed by differential equation*

$$x'(t) = f(x(t))$$

The evolutionary system is said to be deterministic *if $\mathcal{S}_f : X \rightsquigarrow \mathcal{C}(0, +\infty; X)$ is single-valued. An equilibrium of a differential equation is a stationary solution of this equation.*

An equilibrium $\overline{x}$ (stationary point) of a differential equation $x'(t) = f(x(t))$ being a constant evolution, its velocity is equal to 0, so that it *is characterized as a solution to the equation $f(\overline{x}) = 0$.*

The evolutionary system $\mathcal{S}_f$ associated with the single-valued map f is a priori a set-valued map, taking:

1. nonempty values $\mathcal{S}_f(x)$ whenever there exists a solution to the differential equation starting at x, guaranteed by (local) existence theorems (the Peano Theorem, when f is continuous),
2. at most one value $\mathcal{S}_f(x)$ whenever uniqueness of the solution starting at x is guaranteed. There are many sufficient conditions guaranteeing the uniqueness: f is Lipschitz, by the Cauchy–Lipschitz Theorem, or f is monotone in the sense that there exists a constant $\lambda \in \mathbb{R}$ such that

$$\forall x, y \in X, \ \langle f(x) - f(y), x - y \rangle \ \leq \ \lambda \|x - y\|^2$$

(we shall not review other uniqueness conditions here.)

Since the study of such equations, linear and nonlinear, has, for a long time, been a favorite topic among mathematicians, the study of dynamical systems has for a long time focussed on equilibria: existence, uniqueness, stability, which are investigated in Chap. 9, p. 319.

Existence and uniqueness of solutions to a differential equation was identified with the mathematical description of determinism by many scientists after the 1796 book *L'Exposition du système du monde* and the 1814 book *Essai philosophique sur les probabilités* by *Pierre Simon de Laplace* (1749–1827):

Determinism and predictability. *"We must regard the present state of the universe as the effect of its anterior state and not as the cause of the state which follows. An intelligence which, at a given instant, would know all the forces of which the nature is animated and the respective situation of the beings of which it is made of, if by the way it was wide enough to subject these data to analysis, wouldembrace in a unique formula the movements of the largest bodies of the universe and those of the lightest atom: Nothing would be uncertain for it, and the future, as for the past, would present at its eyes." Does it imply what is meant by "predictability"?*

Although a differential equation assigns a unique velocity to each state, this does not imply that the associated evolutionary system $\mathcal{S} : X \rightsquigarrow \mathcal{C}(0, +\infty; X)$ is *deterministic*, in the sense that it is univoque (single-valued). It may happen that several evolutions governed by a differential equation start from a same initial state. *Valentin-Joseph Boussinesq* (1842–1929) used this lack of uniqueness property of solutions to a differential equation starting at some initial state (that he called "bifurcation", with a different meaning that this word has now, as in Box 11, p. 47) to propose that the multivalued character of evolutions governed by a univoque mechanism describes the evolution of living beings.

The lack of uniqueness of some differential equations does not allows us to regard differential equations as a model for a deterministic evolution. Determinism can be translated by evolutionary systems which associate with any initial state one and only one evolution.

But even when a differential equation generates a deterministic evolutionary system, Laplace's enthusiasm was questioned by *Henri Poincaré* in his study of the evolution of the three-body problem, a simplified version of the evolution of the solar system in his famous 1887 essay. He observed in his 1908 book "*La science et l'hypothèse*" that tiny differences of initial conditions implied widely divergent positions after some time:

Predictions. *"If we knew exactly the laws of Nature and the situation of the universe at the initial moment, we could predict exactly the situation of that same universe at a succeeding moment. But even if it was the case that the natural laws had no longer any secret for us, we could still know the situation approximately. If that enabled us to predict the succeeding situation with the same approximation, that is all we require, and we should say that the phenomenon has been predicted, that it is governed by the laws. But it is not always so; it may happen that small differences in the initial conditions produce very great ones in the final phenomena. A small error in the former will produce an enormous error in the latter.* Prediction becomes impossible and we obtain a fortuitous phenomenon."

The sensitive dependence on initial conditions is one prerequisite of "chaotic" behavior of evolutions, resurrected, because, two centuries earlier, even before Laplace, *Paul Henri Thiry, Baron d'Holbach* wrote in one of his wonderful books, the 1770 *Système de la nature*:

Holbach. *"Finally, if everything in nature is linked to everything, if all motions are born from each other although they communicate secretely to each other unseen from us, we must hold for certain that there is no cause small enough or remote enough which sometimes does not bring about the largest and the closest effects on us. The first elements of athunderstorm may gather in the arid plains of Lybia, then will come to us with the winds, make our weather heavier, alter the moods and the passions of a man of influence, deciding the fate of several nations".*

Some nonlinear *differential equations* produce *chaotic* behavior, quite *unstable*, sensitive to initial conditions and producing fluctuating evolutions (see Definition 9.2.1, p. 322). However, for many problems arising in biological, cognitive, social and economic sciences, we face a completely *orthogonal situation*, governed by *differential inclusions, regulated or controlled systems, tychastic or stochastic systems*, but producing evolutions as *regular or stable* (in a very loose sense) as possible for the sake of adaptation and viability required for life.

2.4.2 Example of Differential Equations: The Lorenz System

Since uncertainty is the underlying theme of this book, we propose to investigate the Lorenz system of differential equations, which is deterministic, but unpredictable in practice, as a simple example to test results presented in this book.

Studying a simplified meteorological model made of a system of three differential equations, the meteorologist *Edward Lorenz* discovered by chance (the famous *serendipity*) in the beginning of the 1960s that for certain parameters for which the system has three non stable equilibria, the "limit set" was quite strange, "chaotic" in the sense that many evolutions governed by this system "fluctuate", approach one equilibrium while circling around it, then suddenly leave away toward another equilibrium around which it turns again, and so on. In other words, this behavior is strange in the sense that the limit set of an evolution is not a trajectory of a periodic solution.[1]

Predictability: Does the flap of a butterfly's wing in Brazil set off a tornado in Texas? *After* Henri Poincaré *who discovered the lack of predictability of evolutions of the three-body problem, Lorenz presented in 1979 a lecture to the American Association for the Advancement of Sciences with the above famous title.*

Lorenz introduced the following variables:

1. x, proportional to the intensity of convective motion,

[1] As in the case of two-dimensional systems, thanks to the Poincaré–Bendixon Theorem.

2. y, proportional to the temperature difference between ascending and descending currents,
3. z, proportional to the distortion (from linearity) of the vertical temperature profile.

Their evolution is governed by the following system of differential equations:

$$\begin{cases} (i) \quad x'(t) = \sigma y(t) - \sigma x(t) \\ (ii) \quad y'(t) = rx(t) - y(t) - x(t)z(t) \\ (iii) \quad z'(t) = \ x(t)y(t) - bz(t) \end{cases} \tag{2.6}$$

where the positive parameters σ and b satisfy $\sigma > b+1$ and r is the normalized Rayleigh number.

We observe that the vertical axis $(0,0,z)_{z\in\mathbb{R}}$ is a symmetry axis, which is also the viability kernel of the hyperplane $(0,y,z)$ under the Lorenz system, from which the solutions boil down to the exponentials $(0,0,ze^{-bt})$.

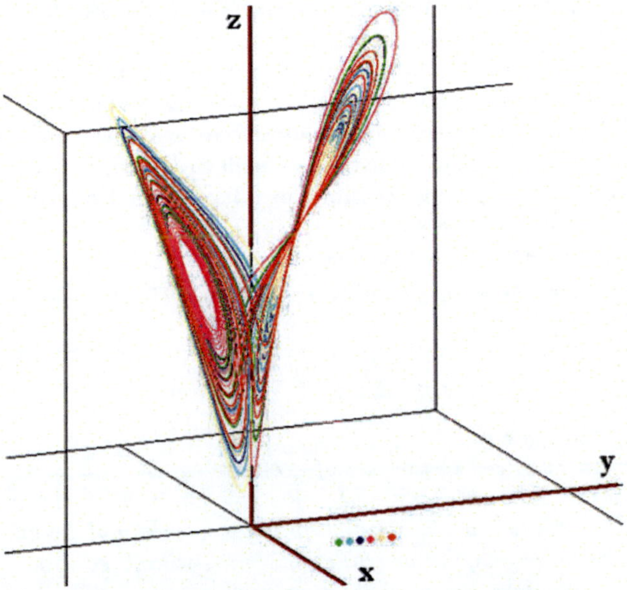

Fig. 2.2 Trajectories of six evolutions of the Lorenz system. *starting from initial conditions* $(i, 50, 0)$, $i = 0, \ldots, 5$. *Only the part of the trajectories from step times ranging between* 190 *and* 200 *are shown for clarity.*

If $r \in]0,1[$, then 0 is an asymptotically stable equilibrium. If $r = 1$, the equilibrium 0 is "neutrally stable". When $r > 1$, the equilibrium 0 becomes unstable and two more equilibria appear:

$$e_1 := \left(\sqrt{b(r-1)}, \sqrt{b(r-1)}, r-1\right) \ \& \ e_2 := \left(-\sqrt{b(r-1)}, -\sqrt{b(r-1)}, r-1\right)$$

Setting $r^\star := \dfrac{\sigma(\sigma+b+3)}{\sigma-b-1}$, these equilibria are stable when $r^\star > 1$ and unstable when $r > r^\star$. We take $\sigma = 10$, $b = \frac{8}{3}$ and $r = 28$ in the numerical experiments.

2.5 Regulons and Tyches

For physical and engineering systems controlled in an optimal way by optimal controls, agents are decision makers or identified actors having access to the controls of the system.

For systems involving living beings, agents interfering with the evolutionary mechanisms are often *myopic, conservative, lazy and opportunistic*, from molecules to (wo)men, exhibiting some *contingent* freedom to choose among some *regulons* (regulatory parameters) to govern evolutions.

In both cases, controls and regulons may have to protect themselves against *tychastic* uncertainty, obeying no statistical regularity.

These features are translated by adding to state variables other ones, parameters, among which (constant) coefficients, controls, regulons and tyches. These different names describe the different questions concerning their role in the dynamics of the system.

In other words, the *state* of the system evolves according to evolutionary laws involving *parameters*, which may in their turn depend on *observation variables* of the states:

Definition 2.5.1 ***[Classification of Variables]***

1. states *of the system;*
2. parameters, *involved in the law of evolution of the states;*
3. values, indicators *which provide some information on the system, such as exit functions, minimal time functions, minimal length functions, Lyapunov functions, value functions in optimal control, value of a portfolio, monetary mass, congestion traffic, etc.;*
4. observations *on the states, such as measurements, information, predictions, etc., given or built.*

We distinguish several categories of parameters, according to the existence or the absence of an actor (controller, agent, decision-maker, etc.) acting on them on one hand, or the degree of knowledge or control on the other hand, and to explain their role:

Definition 2.5.2 *[Classification of Parameters] Parameters can be classified in the following way:*

1. constant coefficients *which may vary, but not in time (see Box 11, p. 47),*
2. *parameters under the command of an identified actor, called* controls *(or decision parameters),*
3. *parameters evolving independently of an identified actor, which can themselves be divided in two classes:*
   a. Regulons or regulation controls,
   b. Tyches, perturbations, disturbances, random events.

These parameters participate in different ways to the general concept of uncertainty. *A given system can depend only on controls, and is called a* controlled system, *or on regulons, and is called a* regulated system *or on tyches, and is called a* tychastic system. *It also can involve two or three of these parameters: for instance, if it involves controls and tyches, it is called a* tychastic controlled system, *and, in the case of regulons and tyches, a* tychastic regulated system.

The study of parameterized systems thus depends on the interpretation of the parameters, either regarded as controls and regulons on one hand, or as tyches or random variables on the other.

1. In *control theory*, it is common for parameters to evolve in order to solve some specific requirements (optimality, viability, reachability) by **at least one evolution** governed by an identified actor (agent, decision-maker, etc.). Control theory is also known under the names of *automatics*, and, when dealing with mechanical systems, *robotics*. Another word, *Cybernetics*, from the Greek *kubernesis*, "control", "govern", as it was suggested first by André Ampère (1775–1836), and later, by *Norbert Wiener* (1894-1964) in his famous book *Cybernetics or Control and Communication in the Animal and the Machine* published in 1948, is by now unfortunately no longer currently used by American specialists of control theory. In physics and engineering, the actors are well identified and their purpose clearly defined, so that only state, control and observation variables matter.
2. In biological, cognitive, social and economic sciences, these parameters *are not under the control* of an identified and consensual agent involved in the evolutionary mechanism governing the evolutions of the state of the system. In those so called "soft sciences" involving uncertain evolutions of systems (organizations, organisms, organs, etc.) of living beings, the situation is more complex, because the identification of actors governing the evolution of parameters is more questionable, so that we regard in this case these parameters as regulons (for regulatory parameters).

13 *Examples of States and Regulons*

Field	*State*	*Regulon*	*Viability*	*Actors*
Economics	*Physical goods*	*Fiduciary goods*	*Economic scarcity*	*Economic agents*
Genetics	*Phenotype*	*Genotype*	*Viability or homeostasis*	*Bio-mechanical metabolism*
Sociology	*Psychological state*	*Cultural codes*	*Sociability*	*Individual actors*
Cognitive sciences	*Sensorimotor states*	*Conceptual codes*	*Adaptiveness*	*Organims*

The main question raised about controlled systems is to find "optimal controls" optimizing an intertemporal criteria, or other objectives, to which we devote Chap. 4, p. 125. The questions raised about regulated system deal with "inertia principle" (keep the regulon constant as long as viability is not at stakes), inertia functions, heavy evolutions, etc., which are exposed in Chap. 6, p. 199.

3. However, even control theory has to take into account some uncertainty (disturbances, perturbations, etc.) that we summarize under the name of *tyches*. Tyches describe uncertainties played by an indifferent, maybe hostile, Nature.

Tyche. *Uncertainty without statistical regularity can be translated mathematically by parameters on which actors, agents, decision makers, etc. have no controls. These parameters are often perturbations, disturbances (as in "robust control" or "differential games against nature") or more generally, tyches (meaning "chance" in classical Greek, from the Goddess Tyche) ranging over a state-dependent* tychastic map. *They could have be called "random variables" if this terminology were not already preempted by probabilists.*

This is why we borrow the term of *tychastic evolution* to *Charles Peirce* who introduced it in 1893 under the title *evolutionary love*:

Tychastic evolution. *"Three modes of evolution have thus been brought before us: evolution by fortuitous variation, evolution by mechanical necessity, and evolution by creative love. We may term them tychastic evolution, or tychasm, anancastic evolution, or anancasm, and agapastic evolution, or agapasm." In this paper, Peirce associates the concept of* anancastic evolution *with the Greek concept of necessity,* ananke, *anticipating the "chance and necessity" framework that motivated viability theory.*

When parameters represent tyches (disturbances, perturbations, etc.), we are interested in "robust" control of the system in the sense that **all evolutions** of the evolutionary system starting from a given initial state satisfy a given evolutionary property.

Fortune *fortuitously* left its role to randomness, originating in the French *"randon"*, from the verb *"randir"*, sharing the same root than the English "to run" and the German *rennen*. When running too fast, one looses the control of himself, the race becomes a poor "random walk", bumping over *scandala* (stones scattered on the way) and falling down, *cadere* in Latin, a matter of *chance* since it is the etymology of this word. Hazard was imported by William of Tyre from the crusades from Palestine castle named after a dice game, *az zahr*. Now dice, in Latin, is *alea*, famed after Julius Caesar's *alea jacta est*, which was actually thrown out the English language: chance and hazard took in this language the meaning of danger, itself from Latin *dominarium*. Being used in probability, the word *random* had to be complemented by *tyche* for describing evolutions without statistical regularity prone to extreme events.

Zhu Xi (1130–1200), 朱熹 one of the most important unorthodox neo-Confucian of the Song dynasty, suggested that *"if you want to treat everything, and as changes are infinite, it is difficult to predict, it must, according to circumstances, react to changes (literally,* "follow, opportunity, reaction, change*)*, instead of a priori action.

The four ideograms *follow*, *opportunity*, *reaction*, *change*:

随机应变

are combined to express in Chinese:

1. by the first half, *"follow, opportunity"*, 随机, the concept of randomness or stochasticity,
2. while Shi Shuzhong has proposed that the second half, *"reaction, change"*, 应变; translate the concept of tychasticity,
3. and *"no, necessary"*, 未定, translates contingent.

2.6 Discrete Nondeterministic Systems

Here, the time set is $\mathbb{N}$, the state space is any metric set X and the evolutionary space is the space $X^{\mathbb{N}}$ of sequences $\overrightarrow{x} := \{x_j\}_{j\in\mathbb{N}}$ of elements $x_j \in X$. The space of parameters (controls, regulons or tyches) is another set denoted by $\mathcal{U}$. The evolutionary system is defined by the discrete parameterized system (φ, U) where:

1. $\varphi : X \times \mathcal{U} \mapsto X$ is a map associating with any state-parameter pair (x,u) the successor $\varphi(x,u)$,
2. $U : X \rightsquigarrow \mathcal{U}$ is a set-valued map associating with any state x a set $U(x)$ of parameters *feeding back* on the state x.

Definition 2.6.1 *[Discrete Systems with State-Dependent Parameters] A discrete parameterized system $\Phi := \varphi(\cdot, U(\cdot))$ defines the evolutionary system $\mathcal{S}_\Phi : X \rightsquigarrow X^{\mathbb{N}}$ in the following way: for any $x \in X$, $\mathcal{S}_\Phi(x)$ is the set of sequences $\overrightarrow{x}$ governed by*

$$\begin{cases} (i) \;\; x_{j+1} = \varphi(x_j, u_j) \\ (ii) \; u_j \in U(x_j) \end{cases} \tag{2.7}$$

starting from x.

When the parameter space is reduced to a singleton, we find discrete equations $x_{j+1} = \varphi(x_j)$ as a particular case. They generate deterministic evolutionary systems $\mathcal{S}_\varphi : X \mapsto X^{\mathbb{N}}$.

Setting

$$\Phi(x) := \varphi(x, U(x)) = \{\varphi(x,u)\}_{u\in U(x)}$$

the subset of all available successors $\varphi(x,u)$ at x when u ranges over the set of parameters allows us to treat these dynamical systems as difference inclusions:

Definition 2.6.2 *[Difference Inclusions] Let $\Phi(x) := \varphi(x, U(x))$ denote the set of velocities of the parameterized system. The evolutions $\overrightarrow{x}$ governed by the parameterized system*

$$\begin{cases} (i) \;\; x_{j+1} = \varphi(x_j, u_j) \\ (ii) \; u_j \in U(x_j) \end{cases} \tag{2.8}$$

are governed by the difference inclusion

$$x_{j+1} \in \Phi(x_j) \tag{2.9}$$

and conversely.

An equilibrium of a difference inclusion is a stationary solution of this inclusion.

Actually, any difference inclusion $x_{j+1} \in \Phi(x_j)$ can be regarded as a parameterized system (φ, U) by taking $\varphi(x,u) := u$ and $U(x) := \Phi(x)$.

Selections of the set-valued map U are *retroactions* (see Definition 1.7, p. 18) governing specific evolutions. Among them, we single out the following

- **The Slow Retroaction.** It is associated with a given fixed element $a \in X$ (for instance, the origin in the case of a finite dimensional vector space). We set
$$u^{\circ}(x) := \left\{ y \in U(x) \mid d(a,y) = \inf_{z \in U(x)} d(a,z) \right\}$$
The evolutions governed by the dynamical system
$$\forall n \geq 0, \; x_{n+1} \in \varphi(x_n, u^{\circ}(x_n))$$
are called *slow evolutions*, i.e., evolutions associated with parameters remaining as close as possible to the given element a. In the case of a finite dimensional vector space, *slow evolutions are evolutions associated with controls with minimal norm.*
- **The Heavy Retroaction.** We denote by $\mathcal{P}(\mathcal{U})$ the hyperspace (see Definition 18.3.3, p. 720) of all subsets of U. Consider the set-valued map $\mathbb{S} : \mathcal{P}(\mathcal{U}) \times \mathcal{U} \rightsquigarrow \mathcal{U}$ associating with any pair (A,u) the subset $\mathbb{S}(A,u) := \{v \in A \mid d(u,v) = \inf_{w \in A} d(u,w)\}$ of "best approximations of u by elements of A". The evolutions governed by the dynamical system
$$\forall n \geq 0, \; x_{n+1} \in \varphi(x_n, \mathbf{s}(U(x_n), u_{n-1}))$$
are called *heavy evolutions.*
This amounts to taking at time n a regulon $u_n \in s(U(x_n), u_{n-1})$ as close as possible to the regulon u_{n-1} chosen at the preceding step. If such a regulon u_{n-1} belongs to $U(x_n)$, it can be kept at the present step n. This is in this sense that the selection $\mathbf{s}(U(x), u)$ provides a heavy solution, since the regulons are kept constant during the evolution as long as the viability is not at stakes.
For instance, when the state space is a finite dimensional vector space X supplied with a scalar product and when the subsets $U(x)$ are closed and convex, the projection theorem implies that the map $\mathbf{s}(U(x), u)$ is single-valued.

2.7 Retroactions of Parameterized Dynamical Systems

2.7.1 *Parameterized Dynamical Systems*

The space of parameters (controls, regulons or tyches) is another finite dimensional vector space $\mathcal{U} := \mathbb{R}^c$.

Definition 2.7.1 ***[Evolutionary Systems associated with Control Systems]*** *We introduce the following notation*

1. $f : X \times \mathcal{U} \mapsto X$ *is a map associating the* velocity $f(x,u)$ *of the state* x *with any state-control pair* (x,u)*,*
2. $U : X \rightsquigarrow \mathcal{U}$ *is a set-valued map associating a set* $U(x)$ *of controls* feeding back *on the state* x*.*

The evolutionary system $\mathcal{S} : X \rightsquigarrow \mathcal{C}(0,+\infty;X)$ *defined by the control system* (f,U) *is the set-valued map associating with any* $x \in X$ *the set* $\mathcal{S}(x)$ *of evolutions* $x(\cdot)$ *governed by the control (or regulated) system*

$$\begin{cases} (i) \ x'(t) = f(x(t),u(t)) \\ (ii) \ u(t) \in U(x(t)) \end{cases} \tag{2.10}$$

starting from x*.*

Remark. Differential equation (2.10)(i) is an "input-output map" associating an output-state with an input-control. Inclusion (2.10)(ii) associates input-controls with output-states, *"feeds back"* the system (the a priori feedback relation is set-valued, otherwise, we just obtain a differential equation). See Figure 1.4, p. 14. □

Remark. We have to give a meaning to the differential equation $x'(t) = f(x(t),u(t))$ and inclusion $u(t) \in U(x(t))$ in system (2.10). Since the parameters are not specified, this system is not valid for any $t \geq 0$, but only for *"almost all"* $t \geq 0$ (see Theorems 19.2.3, p.771 and 19.4.3, p.783). We delay the consequences of the Viability Theorem with such mathematical property. The technical explanations are relegated to Chap. 19, p.769 because they are not really used in the rest of the book. By using graphical derivatives $Dx(t)(1)$ instead of the usual derivatives, the "differential" equation $Dx(t)(1) \ni f(x(t),u(t))$ providing the same evolutions holds true for any $t \geq 0$ (see Proposition 19.4.5, p.787). □

2.7.2 Retroactions

In control theory, open and closed loop controls, feedbacks or retroactions provide the central concepts of cybernetics and general systems theory:

Definition 2.7.2 *[**Retroactions**] Retroactions are single-valued maps $\widetilde{u} : (t,x) \in \mathbb{R}_+ \times X \mapsto \widetilde{u}(t,x) \in \mathcal{U}$ that are plugged as inputs in the differential equation*

$$x'(t) = f(x(t), \widetilde{u}(t, x(t))) \tag{2.11}$$

In control theory, state-independent retroactions $t \mapsto \widetilde{u}(t,x) := u(t)$ are called open loop controls *whereas time-independent retroactions $x \mapsto \widetilde{u}(t,x) := \widetilde{u}(x)$ are called* closed loop controls *or* feedbacks. *See Figure 1.7, p. 18.*

The class $\widetilde{\mathcal{U}}$ in which retroactions are taken must be consistent with the properties of the parameterized system so that

- the differential equations $x'(t) = f(x(t), \widetilde{u}(t, x(t)))$ have solutions[2],
- for every $t \geq 0$, $\widetilde{u}(t,x) \in U(x)$.

When no state-dependent constraints bear on the controls, i.e., when $U(x) = U$ does not depend on the state x, then open loop controls can be used to parameterize the evolutions $S(x, u(\cdot))(\cdot)$ governed by differential equations (2.10)(i).

This is no longer the case when the constraints on the controls depend on the state. In this case, we parameterize the evolutions of control system (2.10) by closed loop controls or retroactions.

Inclusion (2.10)(ii), which associates input-controls with output-states, "feeds back" the system in a set-valued way. Retroactions can be used to parameterize the evolutionary system spanned by the parameterized system (f, U): with any retroaction $\widetilde{u}$ we associate the evolutionary system $S(\cdot, \widetilde{u})$ generated by the differential equation

$$x'(t) = f(x(t), \widetilde{u}(t, x(t)))$$

Whenever a class $\widetilde{\mathcal{U}}$ has been chosen, we observe the following

$$\{S(x, \widetilde{u})\}_{\widetilde{u} \in \widetilde{\mathcal{U}}} \subset \mathcal{S}(x)$$

[2] Open loop controls can be only measurable. When the map $f : X \times \mathcal{U} \mapsto X$ is continuous, the Carathéodory Theorem states that differential equation (2.10)(i) has solutions even when the open loop control is just measurable, (and thus, can be discontinuous as a function of time). It is the case whenever they take their values in a finite set, in which case they are usually called "bang-bang controls" in the control terminology.

The evolutionary system can be parameterized by a feedback class $\widetilde{\mathcal{U}}$ if equality holds true.

The choice of an adequate class $\widetilde{\mathcal{U}}$ of feedbacks regulating specific evolutions satisfying required properties is often an important issue. Finding them may be a difficult problem to solve. Even though one could solve this problem, computing or using a feedback in a class too large may not be desirable whenever feedbacks are required to belong to a class of specific maps (constant maps, time-dependent polynomials, etc.). Another issue concerns the use of a prescribed class of retroactions and to "combine" them to construct new feedbacks for answering some questions, viability or capturability, for instance. This issue is dealt with in Chap. 11, p. 437.

Remark. For one-dimensional systems, retroactions are classified in positive retroactions, when the phenomenon is "amplified", and negative ones in the opposite case. They were introduced in 1885 by French physiologist *Charles-Edouard Brown-Séquard* under the nicer names *"dynamogenic" and "inhibitive" retroactions respectively.* □

The concepts of retroaction and feedback play a central role in control theory, for building servomechanisms, and then, later, in all versions of the "theory of systems" born from the influence of the mathematics of their time on biology, as the Austrian biologist *Ludwig von Bertalanffy* (1901-1972) in his book *Das biologische Weltbild* published in 1950, and after *Jan Smuts* (1870-1950) in his 1926 *Holism and evolution*. The fact that not only effects resulted from causes, but that also effects retroacted on causes, "closing" a system, has had a great influence in many fields.

2.7.3 Differential Inclusions

In the early times of (linear) control theory, the set-valued map was assumed to be constant ($U(\cdot) = U$) and even the parameter set U was taken to be equal to the entire vector space $\mathcal{U} := \mathbb{R}^c$. In this case, the parameterized system is a system of parameterized differential equations, so that the theory of (linear) differential equations could be used.

The questions arose to consider the case of state-dependent constraints bearing on the controls. For example, set-valued maps of the form $U(x) := \prod_{j=1}^{m}[a_j(x), b_j(x)]$ summarize state-dependent constraints of the form:

$$\forall t \geq 0,\ \forall j = 1, \dots, m,\ a_j(x(t)) \leq u_j(t) \leq b_j(x(t))$$

When the constraints bearing on the parameters (controls, regulons, tyches) are state dependent, we can no longer use differential equations. We must appeal to the theory of differential inclusions, initiated in the early

1930's by *André Marchaud* and *Sanislas Zaremba*, and next, by the Polish and Russian schools, around *Tadeusz Ważewski* and *Alexei Filippov*, who laid the foundations of the mathematical theory of differential inclusions after the 1950's.

Indeed, denoting by

$$F(x) := f(x, U(x)) = \{f(x,u)\}_{u \in U(x)}$$

the subset of all available velocities $f(x,u)$ at x when u ranges over the set of parameters, we observe the following:

Lemma 2.7.3 *[Differential Inclusions] Let $F(x) := f(x, U(x))$ denote the set of velocities of the parameterized system. The evolutions $x(\cdot)$ governed by the parameterized system*

$$\begin{cases} (i) \ x'(t) = f(x(t), u(t)) \\ (ii) \ u(t) \in U(x(t)) \end{cases} \tag{2.12}$$

are governed by the differential inclusion

$$x'(t) \in F(x(t)) \tag{2.13}$$

and conversely.

An equilibrium of a differential inclusion is a stationary solution of this inclusion.

By taking $f(x,u) := u$ and $U(x) := F(x)$, any differential inclusion $x'(t) \in F(x(t))$ appears as a parameterized system (f, U) parameterized by its velocities. *Whenever we do not need to write the controls explicitely, it is simpler to consider a parameterized system as a differential inclusion.* Most theorems on differential equations can be adapted to differential inclusions (some of them, the basic ones, are indeed more difficult to prove), but they are by now available.

However, there are examples of differential inclusions without solutions, such as the simplest one:

Example of Differential Inclusion Without Solution: The constrained set is $K := [a, b]$ and the subsets of velocities are singletons except at one point $c \in]a, b[$, where $F(x) := \{-1, 1\}$:

$$F(x) := \begin{cases} +1 & \text{if } x \in [a, c[\\ -1 \text{ or } +1 & \text{if } x = c \\ -1 & \text{if } x \in]c, b] \end{cases}$$

No evolution can start from c. Observe that this is no longer a counter-example when $F(c) := [-1, +1]$, since in this case c is an equilibrium, since its velocity 0 belongs to $F(c)$.

Remark. Although a differential inclusion assigns several velocities to a same states, this does not imply that the associated evolutionary system is non deterministic. It may happen for certain classes of differential inclusions. This is the case for instance when there exists a constant $\lambda \in \mathbb{R}$ such that

$$\forall x, y \in X,\ \forall u \in F(x),\ \forall v \in F(y),\ \ \langle u - v, x - y \rangle \leq \lambda \|x - y\|^2$$

because in this case evolutions starting from each initial state, if any, are unique. □

For discrete dynamical systems, the single-valuedness of the dynamics $\varphi : X \mapsto X$ is equivalent to the single-valuedness of the associated evolutionary system $\mathcal{S}_\varphi : X \mapsto X^{\mathbb{N}}$. This is no longer the case for continuous time dynamical systems:

Warning: *The deterministic character of an evolutionary system generated by a parameterized system is a concept different from the set-valued character of the map F. What matters is that the evolutionary system $\mathcal{S}$ associated with the parameterized system is single-valued (deterministic evolutionary systems) or set-valued (nondeterministic evolutionary systems).*

It is the case, for instance, for set-valued maps F which are *monotone set-valued maps* in the sense that

$$\forall\ y \in \mathrm{Dom}(F),\ \forall\, u\ \in\ U(x),\ y\ \in\ V(y),\ \ \langle u - v, x - y \rangle\ \leq\ 0$$

2.8 Evolutionary Systems

Therefore, we shall study general evolutionary systems defined as set-valued maps $X \rightsquigarrow \mathcal{C}(0, +\infty; X)$ satisfying given requirements listed below. For continuous time evolutionary systems, the state space X is a finite dimensional vector space for most examples. However, besides the characterization of *regulation maps*, which are specific for control systems, many theorems are true even in cases when the evolutionary system is not generated by control systems or differential inclusions, and for infinite dimensional vector spaces X.

Other Examples of State Spaces:

1. When X := $\mathcal{C}(-\infty, 0; X)$ is the space of evolution histories (see Chap. 12 of the first edition of [18, Aubin]), we supply it with the metrizable compact convergence topology,
2. When X is a space of spatial functions when one deals with partial differential inclusions or distributed control systems, we endow it with its natural topology for which it is a complete metrizable spaces,
3. When $X := \mathcal{K}(\mathbb{R}^d)$ is the set of nonempty compact subsets of the vector space $\mathbb{R}^d$, we use the Pompeiu–Hausdorff topology (morphological and mutational equations, presented in [23, Aubin]).

The algebraic structures of the state space appear to be much less relevant in the study of evolutionary systems. Only the following algebraic operations on the evolutionary spaces $\mathcal{C}(0, +\infty; X)$ are used in the properties of viability kernels and capture basins:

Definition 2.8.1 *[Translations and Concatenations]*

1. ***Translation*** *Let $x(\cdot) : \mathbb{R}_+ \mapsto X$ be an evolution. For all $T \geq 0$, the translation (to the left) $\kappa(-T)(x(\cdot))$ of the evolution $x(\cdot)$ is defined by $\kappa(-T)(x(\cdot))(t) := x(t+T)$ and t the translation (to the right) $\kappa(+T)(x(\cdot))(t) := x(t-T)$,*
2. ***Concatenation*** *Let $x(\cdot) : \mathbb{R}_+ \mapsto X$ and $y(\cdot) : \mathbb{R}_+ \mapsto X$ be two evolutions. For all $T \geq 0$, the concatenation $(x(\cdot) \diamond_T y(\cdot))(\cdot)$ of the evolutions $x(\cdot)$ and $y(\cdot)$ at time T is defined by*

$$(x(\cdot) \diamond_T y(\cdot))(t) := \begin{cases} x(t) & \text{if } t \in [0, T] \\ \kappa(+T)(y(\cdot))(t) := y(t-T) & \text{if } t \geq T \end{cases} \tag{2.14}$$

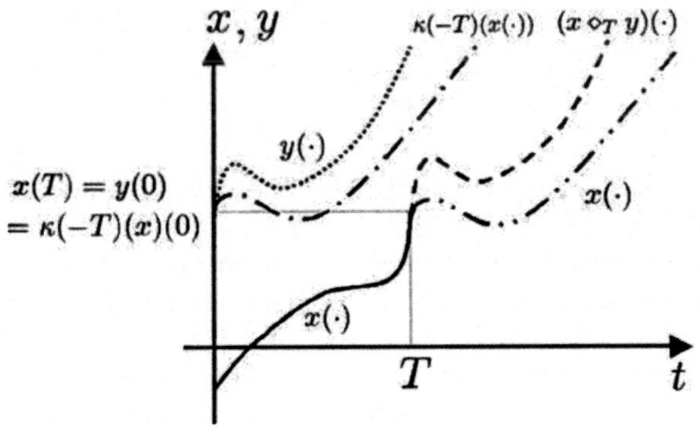

Fig. 2.3 Translations and Concatenations.
• *plain (—): $x(\cdot)$ for $t \in [0, T]$;*
• *dash dot dot $(- \cdot \cdot)$: $x(\cdot)$ for $t \geq T$;*

- *dot* ($\cdots$): $y(\cdot)$;
- *dash dot* ($-\cdot$): $\kappa(-T)(x(\cdot))$;
- *dashed* ($-\ -$): $(x \diamond_T y)(\cdot)$.

$x(\cdot)$ is thus the union of the plain and the dash dot dot. The concatenation $x(\cdot) \diamond_T y(\cdot)$ of x and y is the union of the plain and the dashed.

The concatenation $(x(\cdot) \diamond_T y(\cdot))(\cdot)$ of two continuous evolutions at time T is continuous if $x(T) = y(0)$. We also observe that $(x(\cdot) \diamond_0 y(\cdot))(\cdot) = y(\cdot)$, that $\forall T \geq S \geq 0,\ (\kappa(-S)(x(\cdot) \diamond_T y(\cdot))) = (\kappa(-S)x(\cdot)) \diamond_{T-S} y(\cdot)$ and thus, that

$$\forall T \geq 0,\ \kappa(-T)(x(\cdot) \diamond_T y(\cdot)) = y(\cdot)$$

The adaptation of these definitions to discrete time evolutions is obvious:

$$\begin{cases} (i)\ \kappa(-N)(\overrightarrow{x})_j := x_{j+N} \\ (ii)\ (\overrightarrow{x} \diamond_N \overrightarrow{y})_j := \begin{cases} x_j & \text{if } 0 \leq j < N \\ y_{j-N} & \text{if } j \geq N \end{cases} \end{cases} \tag{2.15}$$

We shall use only the following properties of evolutionary systems:

Definition 2.8.2 ***[Evolutionary Systems]*** *Let us consider a set-valued map $\mathcal{S} : X \rightsquigarrow \mathcal{C}(0,+\infty;X)$ associating with each initial state $x \in X$ a (possibly empty) subset of* evolutions *$x(\cdot) \in \mathcal{S}(x)$ starting from x in the sense that $x(0) = x$. It is said to be an* evolutionary system *if it satisfies*

1. *the* translation property*: Let $x(\cdot) \in \mathcal{S}(x)$. Then for all $T \geq 0$, the translation $\kappa(-T)(x(\cdot))$ of the evolution $x(\cdot)$ belongs to $\mathcal{S}(x(T))$,*
2. *the* concatenation property*: Let $x(\cdot) \in \mathcal{S}(x)$. Then for every $T \geq 0$ and $y(\cdot) \in \mathcal{S}(x(T))$, the concatenation $(x(\cdot) \diamond_T y(\cdot))(\cdot)$ belongs to $\mathcal{S}(x)$.*

The evolutionary system is said to be deterministic *if $\mathcal{S} : X \rightsquigarrow \mathcal{C}(0,+\infty;X)$ is single-valued.*

There are several ways for describing continuity of the evolutionary system $x \rightsquigarrow \mathcal{S}(x)$ with respect to the initial state, regarded as stability property: Stability means generally that the solution of a problem depends continuously upon its data or parameters. Here, for differential inclusions, the data are usually and principally the initial states, but can also be other parameters involved in the right hand side of the differential inclusion. We shall introduce them later, when we shall study the topological properties of the viability kernels and capture basins (See Sect. 10.3.2, p. 387 of Chap. 10, p. 375).

2.9 Viability Kernels and Capture Basins for Discrete Time Systems

2.9.1 Definitions

Definition 6, p. 15 can be adapted to discrete evolution $\overrightarrow{x}$: it is *viable in a subset* $K \subset X$ (an environment) if:

$$\forall n \geq 0, \quad x_n \in K \tag{2.16}$$

and *capture* a target C if it is viable in K until it reaches the target C in *finite time*:

$$\exists\, N \geq 0 \text{ such that} \begin{cases} x_N \in C \\ \forall n \leq N, \ x_N \ \in \ K \end{cases} \tag{2.17}$$

Consider a set-valued map $\Phi : X \rightsquigarrow X$ from a metric space X to itself, governing the evolution $\overrightarrow{x} : n \mapsto x_n$ defined by

$$\forall j \geq 0, \ x_{j+1} \in \Phi(x_j)$$

and the associated evolutionary system $\mathcal{S}_\Phi : X \rightsquigarrow X^{\mathbb{N}}$ associating with any $x \in X$ the set of evolutions $\overrightarrow{x}$ of solutions to the above discrete system starting at x. Replacing the space $\mathcal{C}(0, +\infty; X)$ of continuous time-dependent functions by the space $X^{\mathbb{N}}$ of discrete-time dependent functions (sequences) and making the necessary adjustments in definitions, we can still regard $\mathcal{S}_\Phi$ as an evolutionary system from X to $X^{\mathbb{N}}$.

The viability kernels $\text{Viab}_{(\varphi,U)}(K,C) := \text{Viab}_\Phi(K,C) := \text{Viab}_{\mathcal{S}_\Phi}(K,C)$ and the invariance kernels $\text{Inv}_\Phi(K,C) := \text{Inv}_{\mathcal{S}_\Phi}(K,C)$ are defined in the very same way:

Definition 2.9.1 ***[Viability Kernel under a Discrete System]*** *Let $K \subset X$ be an environment and $C \subset K$ a target.*

The subset $\text{Viab}_\Phi(K,C)$ *of initial states $x_0 \in K$ such that* ***at least one evolution*** $\overrightarrow{x} \in \mathcal{S}_\Phi(x_0)$ *starting at x_0 is viable in K for all $n \geq 1$ or viable in K until it reaches C in finite time is called the* viability kernel of K with target C *under* $\mathcal{S}$.

When the target $C = \emptyset$ is the empty set, we say that $\text{Viab}_\Phi(K) = \text{Viab}_\Phi(K, \emptyset)$ *is the* viability kernel *of K.*

The subset $\text{Capt}_\Phi(K,C)$ *of initial states $x_0 \in K$ such that* ***at least one evolution*** $\overrightarrow{x} \in \mathcal{S}_\Phi(x_0)$ *starting at x_0 is viable in K until it reaches C in finite time is called the* capture basin of C viable in K *under* $\mathcal{S}_\Phi$.

We say that

1. *a subset* K *is* viable outside the target $C \subset K$ *under the discrete system* S_Φ *if* $K = \text{Viab}_\Phi(K, C)$ *and that* K *is* viable *under* S_Φ *if* $K = \text{Viab}_\Phi(K)$,
2. *that* C *is* isolated *in* K *if* $C = \text{Viab}_\Phi(K, C)$,
3. *that* K *is a* repeller *if* $\text{Viab}_\Phi(K) = \emptyset$, *i.e. if the empty set is isolated in* K.

We introduce the discrete invariance kernels and absorption basins:

Definition 2.9.2 ***[Invariance Kernel under a Discrete System]*** *Let* $K \subset X$ *be a environment and* $C \subset K$ *a target.*

The subset $\text{Inv}_\Phi(K, C) := \text{Inv}_{S_\Phi}(K, C)$ *of initial states* $x_0 \in K$ *such that* ***all evolutions*** $\overrightarrow{x} \in S_\Phi(x_0)$ *starting at* x_0 *are viable in* K *for all* $n \geq 1$ *or viable in* K *until they reach* C *in finite time is called the* discrete invariance kernel of K with target C *under* S_Φ.

When the target $C = \emptyset$ *is the empty set, we say that* $\text{Inv}_\Phi(K) := \text{Inv}_\Phi(K, \emptyset)$ *is the* discrete invariance kernel *of* K.

The subset $\text{Abs}_\Phi(K, C)$ *of initial states* $x_0 \in K$ *such that* ***all evolutions*** $\overrightarrow{x} \in S_\Phi(x_0)$ *starting at* x_0 *are viable in* K *until they reach* C *in finite time is called the* absorption basin of C invariant in K *under* S_Φ.

We say that

1. *a subset* K *is* invariant outside a target $C \subset K$ *under the discrete system* S_Φ *if* $K := \text{Inv}_\Phi(K, C)$ *and that* K *is* invariant *under* S_Φ *if* $K = \text{Inv}_\Phi(K)$,
2. *that* C *is* separated *in* K *if* $C = \text{Inv}_\Phi(K, C)$.

In the discrete-time case, the following characterization of viability and invariance of K with a target $C \subset K$ is a tautology:

Theorem 2.9.3 ***[The Discrete Viability and Invariance Characterization]*** *Let* $K \subset X$ *and* $C \subset K$ *be two subsets and* $\Phi : K \rightsquigarrow X$ *govern the evolution of the discrete system. Then the two following statements are equivalent*

1. *K is viable outside C under Φ if and only if*

$$\forall x \in K \backslash C,\ \Phi(x) \cap K \neq \emptyset \tag{2.18}$$

2. *K is invariant outside C under Φ if and only if*

$$\forall x \in K \backslash C,\ \Phi(x) \subset K \tag{2.19}$$

Unfortunately, the analogous characterization is much more difficult in the case of continuous time control systems, where the proofs of the statements require almost all fundamental theorems of functional analysis to be proved (see Chap. 19, p.769).

Remark. The fact that the above characterizations of viability and invariance in terms of (2.18) and (2.19) are trivial does not imply that using them is necessarily an easy task: Proving that $\Phi(x) \cap K$ is not empty or that $\Phi(x) \subset K$ can be difficult and requires some sophisticated theorems of nonlinear analysis mentioned in Chap. 9, p. 319. We shall meet the same obstacles – but compounded – when using the Viability Theorem 11.3.4, p. 455 and Invariance Theorem 11.3.7, p. 457 for continuous time systems. □

For discrete systems $x_{j+1} \in \Phi(x_j) := \varphi(x_j, U(x_j))$, it is also easy to construct the regulation map R_K governing viable evolutions in the viability kernel:

Definition 2.9.4 *[**Regulation Maps**] Let (φ, U) be a discrete parameterized system, K be an environment and $C \subset K$ be a target. The* regulation map R_K *is defined on the viability kernel of K by $\forall x \in \mathrm{Viab}_{(\varphi,U)}(K,C)\backslash C$,*

$$R_K(x) := \{u \in U(x) \text{ such that } \varphi(x,u) \in \mathrm{Viab}_{(\varphi,U)}(K,C)\} \tag{2.20}$$

The regulation map is computed from the discrete parameterized system (φ, U), the environment K and the target $C \subset K$.

For discrete-time parameterized systems (φ, U), all evolutions governed by the discrete parameterized subsystem (φ, R_K) are viable in the viability kernel of K with target C. Unfortunately, this important property is no longer necessarily true for continuous-time systems.

Theorem 2.9.5 *[**Invariance Property of Regulation Maps**] The regulation map R_K satisfies*

$$\mathrm{Viab}_{(\varphi,U)}(K,C) = \mathrm{Inv}_{(\varphi,R_K)}(K,C)$$

All other submaps $P \subset R_K$ also satisfy

$$\mathrm{Viab}_{(\varphi,U)}(K,C) = \mathrm{Inv}_{(\varphi,P)}(K,C) \tag{2.21}$$

The regulation map is the largest map satisfying this property.

Proof. Theorem 2.9.3, p. 72 and Definition 2.9.4, p. 73 imply that the regulation map R_K satisfy

$$\mathrm{Viab}_{(\varphi,U)}(K,C) = \mathrm{Inv}_{(\varphi,R_K)}(K,C)$$

1. If $Q \subset R_K \subset U$ is a set-valued map defined on $\mathrm{Viab}_{(\varphi,U)}(K,C)$, then inclusions

$$\begin{cases} \mathrm{Viab}_{(\varphi,U)}(K,C) = \mathrm{Inv}_{(\varphi,R_K)}(K,C) \subset \mathrm{Inv}_{(\varphi,Q)}(K,C) \\ \subset \mathrm{Viab}_{(\varphi,Q)}(K,C) \subset \mathrm{Viab}_{(\varphi,R_K)}(K,C) \subset \mathrm{Viab}_{(\varphi,U)}(K,C) \end{cases}$$

imply that all the subsets coincide, and in particular, that $\mathrm{Inv}_{(\varphi,Q)}(K,C) = \mathrm{Viab}_{(\varphi,U)}(K,C)$.
2. The regulation map R_K is the largest one by construction satisfying (2.21), p. 73, because if a set-valued map $P \supset R_K$ is strictly larger than R_K, then there would exist an element $(x_0,u_0) \in \mathrm{Graph}(P) \setminus \mathrm{Graph}(R_K)$, i.e., such that $\varphi(x_0,u_0) \notin \mathrm{Viab}_{(\varphi,U)}(K,C)$. But since $\mathrm{Inv}_{(\varphi,P)}(K,C) \subset \mathrm{Viab}_{(\varphi,U)}(K,C)$, all elements $\varphi(x,u)$ when $u \in P(x_0)$ belong to $\mathrm{Viab}_{(\varphi,U)}(K,C)$, a contradiction. □

2.9.2 Viability Kernel Algorithms

For evolutionary systems associated with discrete dynamical inclusions and control systems, the *Viability Kernel Algorithm* and the *Capture Basin Algorithm* devised by Patrick Saint-Pierre allow us to

1. compute the viability kernel of an environment or the capture basin of a target under a control system,
2. compute the graph of the regulation map governing the evolutions viable in the environment, forever or until they reach the target in finite time.

This algorithm *manipulates subsets instead of functions*, and is part of the emerging field of *"set-valued numerical analysis"*.

The viability kernel algorithm provides the exact subset of initial states of the state space from which at least one evolution of the discrete system remains in the constrained set, forever or until it reaches the target in finite time, *without computing these evolutions.*

However, viable evolutions can be obtained from any state in the viability kernel or the capture basin by using the regulation map. The viability kernel algorithms provide the regulation map by computing their graphs, which are also subsets.

This regulation map allows us to "tame" evolutions to maintain them in the viability kernel or the capture basin. Otherwise, using the initial dynamical system instead of the regulation map, evolutions may quickly leave the environment, above all for systems which are sensitive to initial states, such as the Lorenz system.

Consequently, viability kernel and capture basin algorithms face the same "dimensionality curse" than algorithms for solving partial differential equations or other "grid" algorithms. They manipulate indeed "tables" of points in the state space, which become very large when the dimension of the state space is larger than 4 or 5. At the end of the process, the graph of the regulation map can be recovered and stored in the state-control space, which requires higher dimensions. Once the graph of the regulation map is stored, it is then easy to pilot evolutions which are viable forever or until they reach their target.

Despite these shortcomings, the viability kernel algorithms present some advantage over the simulation methods, known under the name of *shooting methods*. These methods compute the evolutions starting at each point and check whether or not at least one evolution satisfies the required properties. They need much less memory space, but demand a considerable amount of time, because, the number of initial states of the environment is high, and second, in the case of controlled systems, the set of evolutions starting from a given initial state becomes huge.

On the other hand, viability properties and other properties of this type, such as asymptotic properties, *cannot be checked on computers*. For instance, one cannot verify whether an evolution is viable forever, since computers provide evolutions defined on a finite number of time steps.

Nothing guarantees that the *finite* time chosen to stop the computation of the solution is large enough to check whether a property bearing on the whole evolution is valid. Such property can be satisfied for a given number of times, without implying that it still holds true later on, above all for systems, like the Lorenz one, which are sensitive to initial conditions.

Finally, starting from an initial state in the viability kernel or the capture basin, shooting methods use solvers which do not take into consideration the corrections *for imposing the viability of the solution*, for instance. Since the initial state is only an approximation of the viability kernel, the absence of these corrections does not allow us to "tame" evolutions which then may leave the environment, and very quickly for systems which are sensitive to initial states, such as the Lorenz system or the discrete time dynamics related to Julia sets.

2.9.3 Julia and Mandelbrot Sets

Studies of dynamical systems (bifurcations, chaos, catastrophe) focus on the dependence on some properties of specific classes of dynamical systems of constant parameters u (which, in contrary to the control case, are not allowed to evolve with time): The idea is to study a given property in terms of the parameter u of a discrete dynamical system $x_{j+1} = \varphi(x_j, u)$ where u is a parameter ranging over a subset $\mathcal{U}$.

Benoît Mandelbrot introduced in the late 1970's the Mandelbrot sets and functions in his investigation of the fractal dimension of subsets:

Definition 2.9.6 *[The Mandelbrot Function] For a discrete dynamical system $x_{j+1} = \varphi(x_j, u)$ where u is a parameter ranging over a subset $\mathcal{U}$, the* Mandelbrot function *$\mu : X \times \mathcal{U} \mapsto \mathbb{R}_+ \cup \{+\infty\}$ associates with any pair (x, u) the scalar*

$$\mu(x, u) := \sup_{j \geq 0} \|x_j\|$$

where $x_{j+1} = \varphi(x_j, u)$ and $x_0 = x$.
The subset $K_u := \text{Viab}_\varphi(B(0,1))$ is the filled-in Julia set *and its boundary $J_u := \partial K_u$ the* Julia set.

The Mandelbrot function μ is characterized through the viability kernel of an auxiliary system:

Lemma 2.9.7 *[Viability Characterization of the Mandelbrot Function] Let us associate with the map φ the following map $\Phi : X \times \mathcal{U} \times \mathbb{R} \mapsto X \times \mathcal{U} \times \mathbb{R}$ defined by $\Phi(x, u, y) = (\varphi(x, u), u, y)$. Consider the subset*

$$K := \{(x, u, y) \in X \times \mathcal{U} \times \mathbb{R} \mid \|x\| \leq y\}$$

Then the Mandelbrot function is characterized by the formula

$$\mu(x, u) = \inf_{(x,u,y) \in \text{Viab}_\Phi(K)} y$$

or, equivalently,

$$\mu(x, u) \leq y \text{ if and only if } x \in \text{Viab}_{\varphi(\cdot,u)}(B(0, y))$$

Proof. Indeed, to say that (x, u, y) belongs to the viability kernel of $K := \{(x, u, y) \mid \|x\| \leq y\}$ means that the solution (x_j, u, y) to the auxiliary system satisfies

$$\forall j \geq 0, \ \|x_j\| \leq y$$

i.e., that $\mu(x, u) \leq y$. □

This story was initiated by *Pierre Fatou* and *Gaston Julia*:

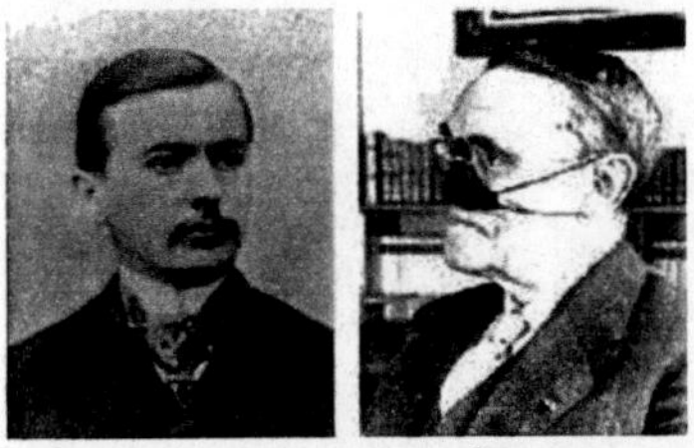

Pierre Fatou and Gaston Julia. *Pierre Fatou [1878-1929] and Gaston Julia [1893-1978] studied in depth the iterates of complex function*
$z \mapsto z^2 + u,$
or, equivalently, of the map
$(x, y) \mapsto \varphi(x, y) := (x^2 - y^2 + a, 2xy + b)$
when $z = x + iy$ *and* $u = a + ib$.
The subset $K_u := \mathrm{Viab}_\varphi(B(0,1))$ *is the* filled-in Julia set *for this specific map* φ *and its boundary* $J_u := \partial K_u$ *the* Julia set. *The subsets whose filled-in Julia sets have empty interior are called* Fatou dust.

Therefore, the viability kernel algorithm allows us to compute the Julia sets, offering an alternative to "shooting methods". These shooting methods compute solutions of the discrete system starting from various initial states and check whether a given property is satisfied or not. Here, this property is the viability of the evolution *for a finite number of times.* Instead, the Viability Kernel Algorithm provides the set of initial states from which at least one evolution is viable forever, *without computing all evolutions to check whether one of them satisfies it.*

Furthermore, the regulation map built by the Viability Kernel Algorithm *provides evolutions which are effectively viable in the viability kernel.* This is a property that shooting methods cannot provide:

- first, because the viability kernel is not known precisely, but only approximatively,
- and second, because even if we know that the initial state belongs to the viability kernel, the evolution governed by such a program is not necessarily viable

The reason why this happens is that programs computing evolutions are independent of the viability problem. They do not make the corrections at each step guaranteeing that the new state is (approximatively) viable in K, contrary to the one computed with Viability Kernel Algorithm.

The more so in this case, since this system is sensitive to initial data.

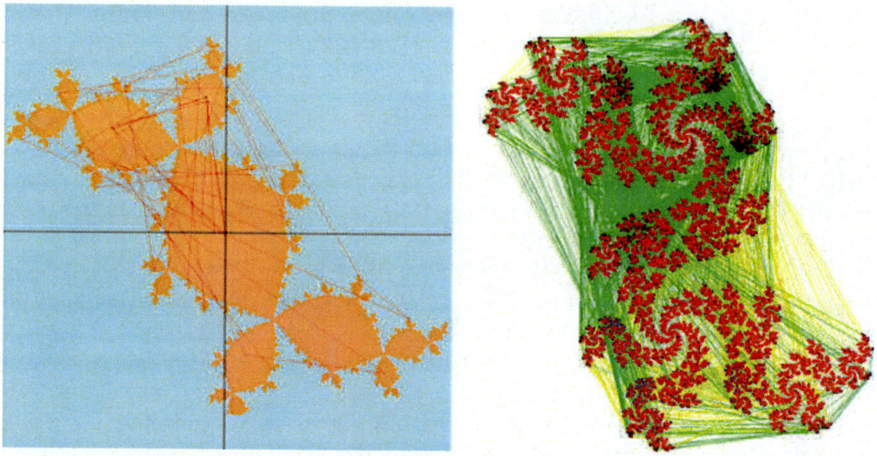

Fig. 2.4 Julia sets: Douady Rabbit and Fatou Dust.
We refer to Figure 2.5 for the computation of its boundary, the Julia set, thanks to Theorem 9.2.18. Even for discrete systems, the round-off errors do not allow the discrete evolution to remain in filled-in Julia set, which is the viability kernel of the ball, whereas the viability kernel algorithm provides both the filled-in Julia set, its boundary and evolutions which remain in the Julia set.

In 1982, a deep theorem by *Adrien Douady* and *Hubbard* states that K_u is connected if and only if $\mu(0, u)$ is finite.

Theorem 9.2.18, p. 339 states that the Julia set is the viability kernel of $K \setminus C$ if and only if it is viable and C absorbs the interior of K_u. In this case, the Viability Kernel Algorithm also provides the Julia set J_u by computing the viability kernel of $K \setminus C$.

We illustrate this fact by computing the viability kernel of the complement of a ball $B(0, \alpha) \subset K$ in K whenever the interior of K_u is not empty (we took $u = -0.202 - 0.787i$). We compute the viability kernel for $\alpha :=$ 0.10, 0.12, 0.14 and 0.16, and we observed that this viability kernel is equal to the boundary for $\alpha = 0.16$. In this case, the ball $B(0, 0.16)$ is absorbing the interior of the filled-in Julia set. The resulting computations can be seen in Figures 2.4 and 2.5.

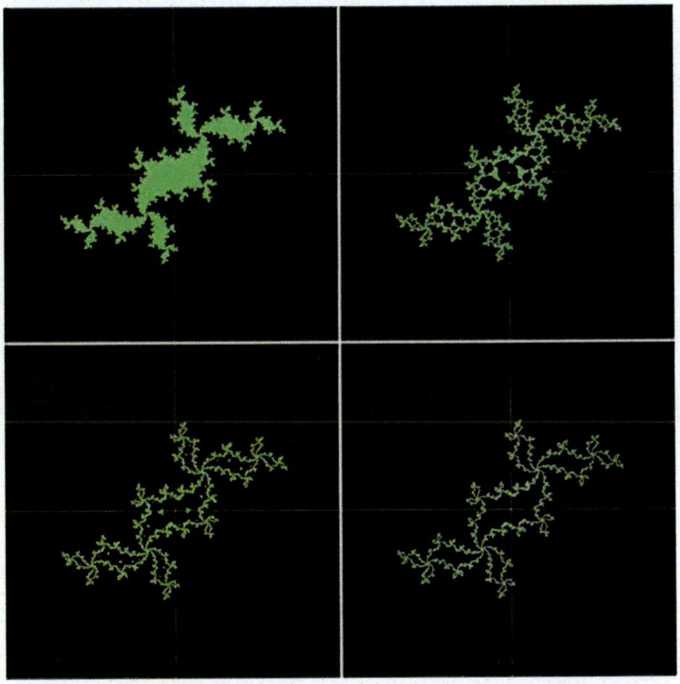

Fig. 2.5 Computation of the Julia set.
The figure on the left is the filled-in Julia set K_u with $u = -0.202 - 0.787i$, which is the viability kernel of the unit ball. The other figures display the viability kernels of $K \setminus B(0,\alpha)$ for $\alpha := 0.10,\ 0.12,\ 0.14$ and 0.16. We obtain the Julia set for $\alpha = 0.16$. Theorem 9.2.18, p. 339 states that the ball $B(0, 0.16)$ absorbs the interior of the filled-in Julia set.

2.9.4 Viability Kernels under Disconnected Discrete Systems and Fractals

If Φ is disconnecting, the viability kernel is a Cantor set, with further properties (self similarity, fractal dimension). Recall that Φ^{-1} denotes the inverse of Φ.

Definition 2.9.8 *[Hutchinson Maps] A set-valued map Φ is said to be* disconnecting *on a subset K if there exists a finite number p of functions $\alpha_i : K \mapsto X$ such that*

$$\forall x \in K,\ \Phi^{-1}(x) := \bigcup_{i=1}^{p} \alpha_i(x)$$

and such that there exist constants $\lambda_i \in]0,1[$ *satisfying: for each subset* $C \subset K$*,*

$$\begin{cases} (i) \quad \forall i = 1, \ldots, p,\ \alpha_i(C) \ \subset \ C \ (\alpha_i \text{ is } \textit{antiextensive}) \\ (ii) \ \ \forall i \neq j,\ \alpha_i(C) \cap \alpha_j(C) = \emptyset \\ (iii) \ \forall i = 1, \ldots, p,\ \mathrm{diam}(\alpha_i(C)) \leq \lambda_i \mathrm{diam}(C) \end{cases}$$

If the functions $\alpha_i : K \mapsto K$ *are contractions with Lipschitz constants* $\lambda_i \in]0,1[$*, then* Φ^{-1} *is called an* Hutchinson map *(introduced in 1981 by* John Hutchinson *and also called an* iterated function system *by* Michael Barnsley.*)*

We now define Cantor sets:

Definition 2.9.9 *[Cantor Sets] A subset K is said to be*

1. perfect *if it is closed and if each of its elements is a limit of other elements of K,*
2. totally disconnected *if it contains no nonempty open subset,*
3. *a* Cantor set *if it is non-empty compact, totally disconnected and perfect.*

The famous Cantor Theorem states:

Theorem 2.9.10 *[The Cantor Theorem] The viability kernel of a compact set under a disconnecting map is an uncountable Cantor set.*

The Cantor set is a viability kernel and the Viability Kernel Algorithm is the celebrated construction procedure of the Cantor set.

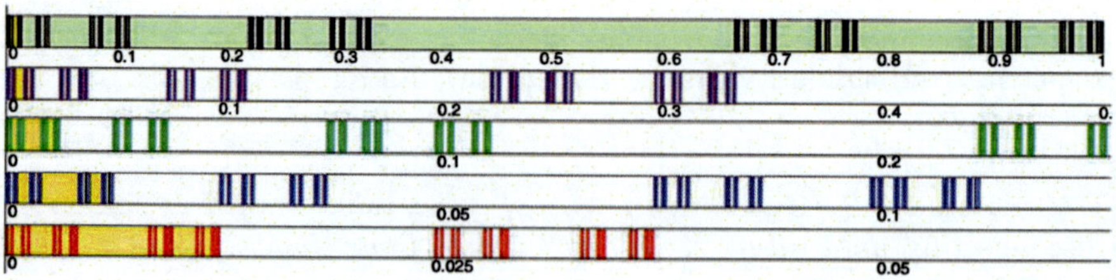

Fig. 2.6 Example: Cantor Ternary Map.

Corollary 2.9.11 *[The Cantor Ternary Set] The Cantor ternary set* $\mathbf{C}$ *is the viability kernel of the interval* $[0,1]$ *under the* Cantor Ternary Map Φ *defined on* $K := [0,1] \subset \mathbb{R}$ *by*

$$\Phi(x) := (3x, 3(1-x))$$

The Cantor Ternary Set is a self similar (see Definition 2.9.14, p. 84), symmetric, uncountable Cantor set with fractal dimension $\frac{\log 2}{\log 3}$ *(see Definition 2.9.13, p. 83) and satisfies* $\mathbf{C} = \alpha_1(\mathbf{C}) \cup \alpha_2(\mathbf{C})$ *and* $\alpha_1(\mathbf{C}) \cap \alpha_2(\mathbf{C}) = \emptyset$.

Proof. The *Cantor Ternary Map* is disconnecting because

$$\Phi^{-1}(x) := \left(\alpha_1(x) := \frac{x}{3}, \alpha_2(x) := 1 - \frac{x}{3}\right)$$

so that $\alpha_1(K) = \left[0, \frac{1}{3}\right]$ and $\alpha_2(K) = \left[\frac{2}{3}, 1\right]$ and that the α_i's are antiextensive contractions of constant $\frac{1}{3}$. □

Example: Quadratic Map In Sect. 2.3, p. 50, we associated with the quadratic map $\varphi(x) := 5x(1-x)$ the set-valued map $\Phi : [0,1] \rightsquigarrow [0,1]$ defined by $\Phi(x) := \varphi(x)$ when $x \in [0,a]$ and $x \in [b,1]$ and $\varphi(x) := \emptyset$ when $x \in]a,b[$, where $a := \frac{1}{2} - \frac{\sqrt{5}}{10}$ and $b := \frac{1}{2} + \frac{\sqrt{5}}{10}$ are the roots of the equation $\varphi(x) = 1$.

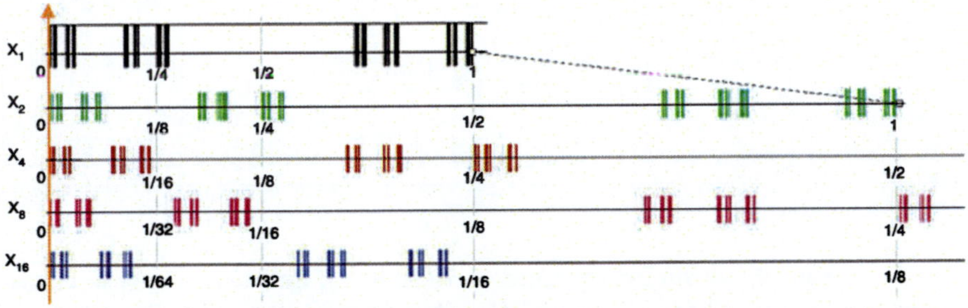

Fig. 2.7 Viability Kernel under the Quadratic Map.
The viability kernel of the interval $[0,1]$ *under the quadratic map* Φ *associated with the map* $\varphi(x) := \{5x(1-x)\}$ *is an uncountable, symmetric Cantor set.*

The interval $[0,1]$ *is viable under the Verhulst logistic differential equation* $x'(t) = rx(t)(1 - x(t))$ *whereas its viability kernel is a Cantor set for its discrete counterpart* $x_{n+1} = rx_{n+1}(1 - x_{n+1})$ *when* $r > 4$.

Proof. Indeed, in this case, the inverse Φ^{-1} is defined by

$$\Phi^{-1}(y) := \left(\omega^{\flat}(y), \omega^{\sharp}(y)\right)$$

where we set

$$\omega^{\flat}(y) := \frac{1}{2} - \frac{\sqrt{r^2 - 4ry}}{2r} \text{ and } \omega^{\sharp}(y) := \frac{1}{2} + \frac{\sqrt{r^2 - 4ry}}{2r}$$

(see Sect. 2.3.1, p. 51). □

The intervals $\omega^{\flat}(K) = \left[0, \left(\frac{1}{2} - \frac{\sqrt{r^2-4r}}{2r}\right)\right]$ and $\omega^{\sharp}(K) = \left[\left(\frac{1}{2} + \frac{\sqrt{r^2-4r}}{2r}\right), 1\right]$ are disjoint intervals which do not cover $[0,1]$. The maps $\omega^{\flat}$ and $\omega^{\sharp}$ are antiextensive and contractions.

We know that the interval $[0,1]$ is viable under the *Verhulst* logistic equation, whereas for $r > 4$, we saw that the discrete viability kernel is a Cantor subset of $[0,1]$. But $[0,1]$ is still viable under the discretizations of the *Verhulst* logistic equation:

Proposition 2.9.12 ***[Discretization of the Verhulst Logistic Equation]*** *The interval* $[0,1]$ *is viable under the explicit discretization* Φ_h *of the* Verhulst *logistic equation, defined by*

$$\Phi_h(x) := rhx\left(\frac{1+rh}{rh} - x\right)$$

Proof. Indeed, Φ_h is surjective from $[0,1]$ to $[0,1]$, and thus, $[0,1]$ is viable under Φ_h: Starting from $x_0 \in [0,1]$, the discrete evolution $\overrightarrow{x}$ defined by

$$x_{n+1} = rhx_n\left(\frac{1+rh}{rh} - x_n\right)$$

remains in K. □

This is an example illustrating the danger of using "discrete analogues" of continuous time differential equations instead of their discretizations. The latter share the same properties than the differential equation (under adequate assumptions), whereas discrete analogues may not share them. This is the case for the quadratic map, the prototype of maps producing chaos, analogues of the Verhulst logistic equation.

Example: Sierpinski Gasket

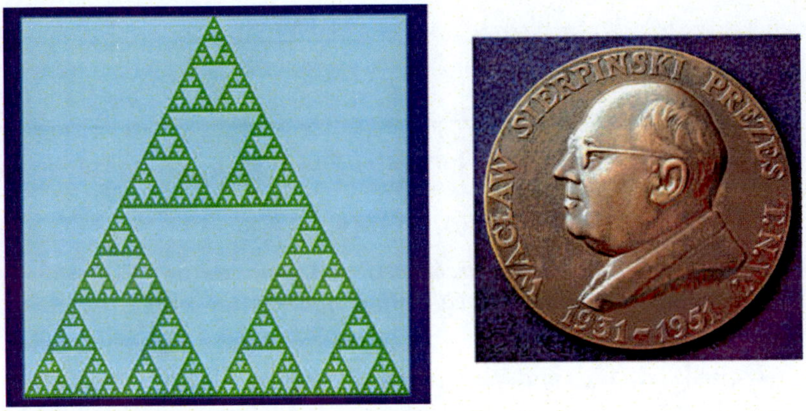

Fig. 2.8 The Sierpinski Gasket.
The Sierpinski Gasket is the viability kernel of the square $[0,1]^2$ *under the discrete map associating with each pair* (x,y) *the subset* $\Phi(x,y) := \left\{(2x,2y),(2x-1,2y),(2x-\frac{1}{2},2y-1)\right\}$ *of 3 elements. Since this map is disconnecting, the Sierpinski Gasket is a self similar, uncountable Cantor set with fractal dimension* $\frac{\log 3}{\log 2}$ *(*left figure*), named from* Waclaw Sierpinski *(1882–1969) (*right figure*).*

2.9.4.1 Fractal Dimension of Self-Similar Sets

Some viability kernels under discrete disconnecting maps have a fractal dimension that we now define:

Definition 2.9.13 ***[Fractal Dimension]*** *Let* $K \subset \mathbb{R}^d$ *be a subset of* $\mathbb{R}^d$ *and* $\nu_K(\varepsilon)$ *the smallest number of* ε*-cubes* $\varepsilon[-1,+1]^d$ *needed to cover the subset* K*. If the limit*

$$\dim(K) := \lim_{\varepsilon \mapsto 0+} \frac{\log(\nu_K(\varepsilon))}{\log\left(\frac{1}{\varepsilon}\right)}$$

exists and is not an integer, it is called the fractal dimension *of* K*.*

To say that K has a fractal dimension $\delta := \dim(K)$ means that the smallest number $\nu_K(\varepsilon)$ of ε-cubes needed to cover K behaves like $\frac{a}{\varepsilon^\delta}$ for some constant $a > 0$.

Actually, it is enough to take subsequences $\varepsilon_n := \lambda^n$ where $0 < \lambda < 1$ converging to 0 when $n \to +\infty$, so that

$$\dim(K) := \lim_{n \to +\infty} \frac{\log(\nu_K(\lambda^n))}{n \log\left(\frac{1}{\lambda}\right)}$$

Definition 2.9.14 *[Self-Similar Sets] Functions α_i are called* similarities *if*

$$\forall x,\ y \in K,\ d(\alpha_i(x), \alpha_i(y)) = \lambda_i d(x, y)$$

Let Φ be a disconnecting map associated with p similarities α_i.

A subset $K_\propto$ is said to be self-similar under Φ if

$$K_\propto = \bigcup_{i=1}^{p} \alpha_i(K_\propto) \text{ and the subsets } \alpha_i(K_\propto) \text{ are pairwise disjoint.}$$

For example,

1. the Cantor set is self-similar:

$$\mathbf{C} = \alpha_1(\mathbf{C}) \cup \alpha_2(\mathbf{C})$$

It is the union of two similarities of constant $\frac{1}{3}$,

2. the Sierpinski gasket is self-similar[3]:

$$\mathbf{S} = \Phi^{-1}(\mathbf{S}) = \bigcup_{i=1}^{3} \alpha_i(\mathbf{S})$$

It is the union of three similarities of constant $\frac{1}{2}$.

[3] Actually, the subsets are not pairwise disjoint, but the above results hold true when the intersections $\alpha_i(\mathbf{C}) \cap \alpha_j(\mathbf{C})$ are manifolds of dimension strictly smaller than the dimension of the vector space.

Lemma 2.9.15 *[Fractal Dimension of Self-Similar Sets] If the p similarities α_i have the same contraction rate $\lambda < 1$, then the fractal dimension of a self-similar set $K_\propto = \bigcup_{i=1}^{p} \alpha_i(K_\propto)$ is equal to*

$$\dim(K_\propto) = \frac{\log(p)}{\log\left(\frac{1}{\lambda}\right)}$$

Consequently,

1. The fractal dimension of the Cantor set is equal to $\frac{\log 2}{\log 3}$: $p = 2$ and $\lambda = \frac{1}{3}$,
2. The fractal dimension of the Sierpinski gasket is equal to $\frac{\log 3}{\log 2}$: $p = 3$ and $\lambda = \frac{1}{2}$.

2.10 Viability Kernels and Capture Basins for Continuous Time Systems

Let $\mathcal{S} : X \rightsquigarrow \mathcal{C}(0,\infty;X)$ denote the evolutionary system associated with parameterized dynamical system (2.10) and $\mathcal{H} \subset \mathcal{C}(0,\infty;X)$ be a subset of evolutions sharing a given set of properties.

2.10.1 Definitions

When the parameterized system is regarded as a control system, we single out the *inverse image* (see Definition 18.3.3, p. 720) of $\mathcal{H}$ under the evolutionary system:

Definition 2.10.1 *[Inverse Image under an Evolutionary System] Let $\mathcal{S} : X \rightsquigarrow \mathcal{C}(0,\infty;X)$ denote an evolutionary system and $\mathcal{H} \subset \mathcal{C}(0,\infty;X)$ a subset of evolutions sharing a given set of properties. The set*

$$\mathcal{S}^{-1}(\mathcal{H}) := \{x \in X \mid \mathcal{S}(x) \cap \mathcal{H} \neq \emptyset\} \tag{2.22}$$

of initial states $x \in X$ from which starts ***at least one evolution*** *$x(\cdot) \in \mathcal{S}(x)$ satisfying the property $\mathcal{H}$ is the* inverse image *of $\mathcal{H}$ under $\mathcal{S}$.*

For instance, taking for set $\mathcal{H} := \mathcal{X}$ defined as the set of stationary evolutions, we obtain the set of all equilibria x of the evolutionary system: at least one evolution $x(\cdot) \in \mathcal{S}(x)$ remains constant and equal to x. In the same way, taking for set $\mathcal{H} := \mathcal{P}_T(X)$ the set of T-periodic evolutions, we obtain the set of points through which passes at least one T-periodic evolution of the evolutionary system.

When we take $\mathcal{H} := \mathcal{V}(K,C)$ to be the set of evolutions viable in a constrained subset $K \subset X$ outside a target $C \subset K$ (see 2.5, p. 49), we obtain the *viability kernel* $\text{Viab}_{\mathcal{S}}(K,C)$ of K with target C:

Definition 2.10.2 ***[Viability Kernel and Capture Basin]*** *Let $K \subset X$ be a environment and $C \subset K$ be a target.*

1. *The subset* $\text{Viab}_{\mathcal{S}}(K,C)$ *of initial states $x_0 \in K$ such that* ***at least one evolution*** *$x(\cdot) \in \mathcal{S}(x_0)$ starting at x_0 is viable in K for all $t \geq 0$ or viable in K until it reaches C in finite time is called the* viability kernel of K with target C *under $\mathcal{S}$.*
   When the target $C = \emptyset$ is the empty set, we say that $\text{Viab}_{\mathcal{S}}(K) := \text{Viab}_{\mathcal{S}}(K,\emptyset)$ *is the* viability kernel *of K. We set* $\text{Capt}_{\mathcal{S}}(K,\emptyset) = \emptyset$.
2. *The subset* $\text{Capt}_{\mathcal{S}}(K,C)$ *of initial states $x_0 \in K$ such that* ***at least one evolution*** *$x(\cdot) \in \mathcal{S}(x_0)$ starting at x_0 is viable in K until it reaches C in finite time is called the* capture basin of C viable in K *under $\mathcal{S}$. When $K = X$ is the whole space, we say that* $\text{Capt}_{\mathcal{S}}(C) := \text{Capt}_{\mathcal{S}}(X,C)$ *is the* capture basin *of C. (see Figure 5.2, p. 182)*

We say that

1. *a subset K is* viable *under $\mathcal{S}$ if $K = \text{Viab}_{\mathcal{S}}(K)$,*
2. *K is* viable outside the target $C \subset K$ *under the evolutionary system $\mathcal{S}$ if $K = \text{Viab}_{\mathcal{S}}(K,C)$,*
3. *C is* isolated *in K if $C = \text{Viab}_{\mathcal{S}}(K,C)$,*
4. *K is a* repeller *if $\text{Viab}_{\mathcal{S}}(K) = \emptyset$, i.e., if the empty set is isolated in K.*

Remark: Trapping Set. A connected closed viable subset is sometimes called a *trapping set*. In the framework of differential equations, *Henri Poincaré* introduced the concept of *shadow* (in French, *ombre*) of K, which is the set of initial points of K from which (all) evolutions leave K in finite time. It is thus equal to the complement $K \backslash \text{Viab}_{\mathcal{S}}(K)$ of the viability kernel of K in K. □

Remark. Theorem 9.3.13, p. 353 provides sufficient conditions (the environment K is compact and backward viable, the evolutionary system is upper semicompact) for the viability kernel to be nonempty.

Another interesting case is the one when the viability kernel $\text{Viab}_{\mathcal{S}}(K) \subset \text{Int}(K)$ of K is contained in the interior of K (in this case, $\text{Viab}_{\mathcal{S}}(K)$) is said to be *source* of K (see Definition 9.2.3, p. 323). □

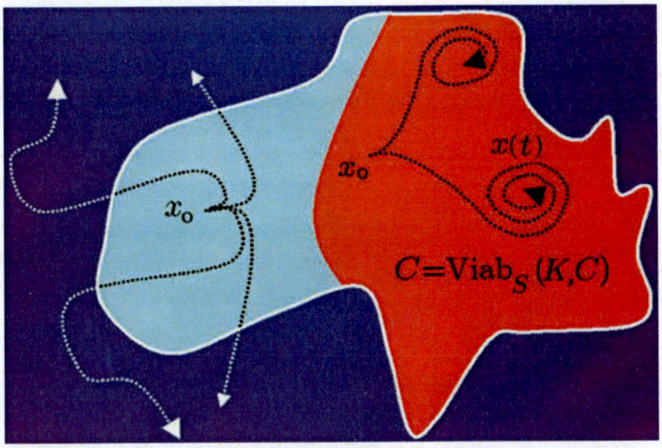

Fig. 2.9 Viability Outside a Target and Isolated Target.
If C is isolated, all evolutions starting in K outside of C are viable outside C before leaving K in finite time.

2.10.2 Viability Kernels under the Lorenz System

We resume our study of the Lorenz system (2.6), p. 57 initiated in Sect. 2.4.2, p. 56.

We provide the viability kernel of the cube $[-\alpha, +\alpha] \times [-\beta, +\beta] \times [-\gamma, +\gamma]$ under the Lorenz system (2.6), p. 57 and the backward Lorenz system

$$\begin{cases} (i) & x'(t) = -\sigma y(t) + \sigma x(t) \\ (ii) & y'(t) = -rx(t) + y(t) + x(t)z(t) \\ (iii) & z'(t) = -x(t)y(t) + bz(t) \end{cases}$$

We call "backward viability kernel" the viability kernel under the backward system.

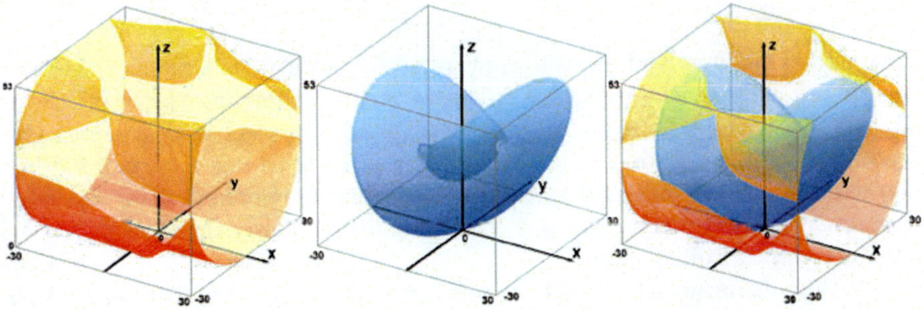

Fig. 2.10 Viability Kernels of a Cube K under Forward and Backward Lorenz Systems.
The figure displays the forward viability kernel of the cube K (left), the backward viability kernel (center) and the superposition of the two (right). We take $\sigma > b + 1$, Proposition 8.3.3, p. 282 implies that whenever the viability kernel of the backward system is contained in the interior of K, the backward viability kernel is contained in the forward viability kernel. Proposition 9.3.11, p. 351 implies that the famous Lorenz attractors (see Definition 9.3.8, p. 349) is contained in the backward viability kernel.

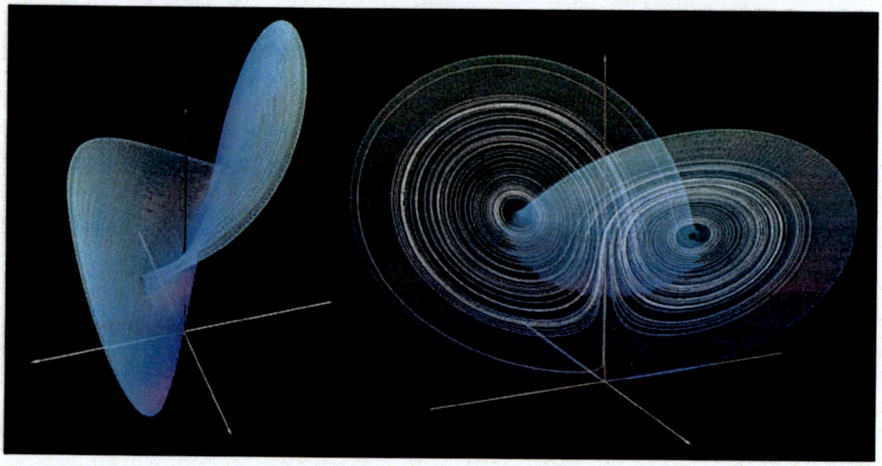

Fig. 2.11 Backward Viability Kernel and Viable Evolution.
This figure displays another view of the backward viability kernel and a viable evolution. They are computed with the viability kernel algorithm.

2.11 Invariance Kernel under a Tychastic System

The questions involved in the concepts of viability kernels and capture basins ask only of the existence of an evolution satisfying the viability or the viability/capturability issue. In the case of parameterized systems, this lead to the interpretation of the parameter as a control or a regulon. When the parameters are regarded as tyches, disturbances, perturbations, etc., the questions are dual: they require that all evolutions satisfy the viability or the viability/capturability issue.

We then introduce the "dual" concept of *invariance kernel and absorption basin*:

Here, we regard the parameterized system

$$x'(t) \;=\; f(x(t), v(t)) \text{ where } v(t) \;\in\; V(x(t)) \tag{2.23}$$

where $v(t)$ is no longer a control or a regulon, but a tyche, where the set of tyches is $\mathcal{V}$ and where $V : X \rightsquigarrow \mathcal{V}$ is a *tychastic map* as a *tychastic system.* Although this system is formally the same that control system (1.1), p. 14

$$x'(t) = f(x(t), u(t)) \text{ where } u(t) \;\in\; U(x(t))$$

the questions asked are different: We no longer check whether a given property is satisfied by **at least** one evolution governed by the control or regulated system, but by **all** evolutions governed by the tychastic system.

When the parameterized system is regarded as a tychastic system, it is natural to consider the *core* (see Definition 18.3.3, p. 720) of a set of evolutions under a tychastic system:

Definition 2.11.1 *[Core under an Evolutionary System] Let $\mathcal{S} : X \rightsquigarrow \mathcal{C}(0,\infty;X)$ denote an evolutionary system and $\mathcal{H} \subset \mathcal{C}(0,\infty;X)$ a subset of evolutions sharing a given property. The set*

$$\mathcal{S}^{\ominus 1}(\mathcal{H}) := \{x \in X \mid \mathcal{S}(x) \;\subset\; \mathcal{H}\} \tag{2.24}$$

of initial states $x \in X$ from which ***all evolutions*** *$x(\cdot) \in \mathcal{S}(x)$ satisfy the property $\mathcal{H}$ is called the* core *of $\mathcal{H}$ under $\mathcal{S}$.*

Taking $\mathcal{H} := \mathcal{V}(K, C)$, we obtain the *invariance kernel* $\mathrm{Inv}_{\mathcal{S}}(K, C)$ of K with target C:

Definition 2.11.2 *[Invariance Kernel and Absorption Basin] Let $K \subset X$ be a environment and $C \subset K$ be a target.*

1. *The subset* $\mathrm{Inv}_{\mathcal{S}}(K, C)$ *of initial states $x_0 \in K$ such that* ***all evolutions*** *$x(\cdot) \in \mathcal{S}(x_0)$ starting at x_0 are viable in K for all $t \geq 0$ or viable in K until they reach C in finite time is called the* invariance kernel of K with target C *under $\mathcal{S}$.*
   When the target $C = \emptyset$ is the empty set, we say that $\mathrm{Inv}_{\mathcal{S}}(K) := \mathrm{Inv}_{\mathcal{S}}(K, \emptyset)$ *is the* invariance kernel *of K.*
2. *The subset* $\mathrm{Abs}_{\mathcal{S}}(K, C)$ *of initial states $x_0 \in K$ such that* ***all evolutions*** *$x(\cdot) \in \mathcal{S}(x_0)$ starting at x_0 are viable in K until they reach C in finite time is called the* absorption basin of C invariant in K *under $\mathcal{S}$.*
   When $K = X$ is the whole space, we say that $\mathrm{Abs}_{\mathcal{S}}(X, C)$ *is the* absorption basin *of C.*

We say that

1. *a subset* K *is* invariant *under* $\mathcal{S}$ *if* $K = \mathrm{Inv}_{\mathcal{S}}(K)$,
2. K *is* invariant outside a target $C \subset K$ *under the evolutionary system* $\mathcal{S}$ *if* $K = \mathrm{Inv}_{\mathcal{S}}(K, C)$,
3. C *is* separated *in* K *if* $C = \mathrm{Inv}_{\mathcal{S}}(K, C)$.

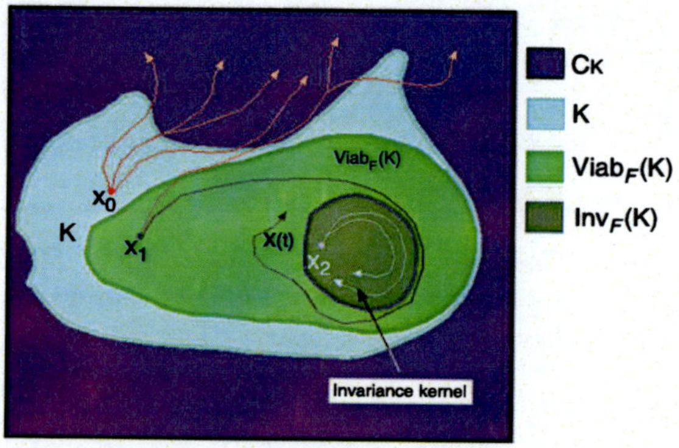

Fig. 2.12 Figure of an Invariance Kernel.
A state x_2 *belongs to the invariance kernel of the environment* K *under an evolutionary system if* **all** *the evolutions starting from it are viable in* K *forever. Starting from a state* $x_1 \in K$ *outside the invariance kernel,* **at least** *one evolution leaves the environment in finite time.*

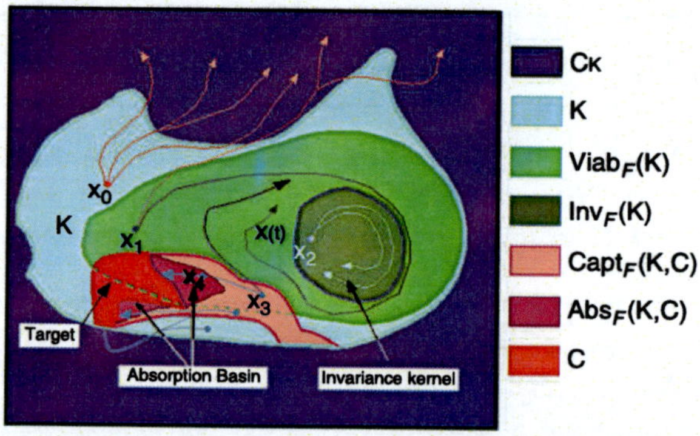

Fig. 2.13 Figure of an Absorption Basin.
All *evolutions starting from a state* x_4 *in the absorption basin of the target* C *invariant in the environment* K *are viable in* K *until they reach* C *in finite time.* **At least** *one evolution starting from* $x_3 \in K$ *outside the absorption basin remains viable outside the target* C *forever or until it leaves* K.

These are four of the main concepts used by viability theory. Other definitions, motivations and comments are given in Chap. 2, p. 43, their general properties in Chap. 10, p. 375, whereas their characterization in terms of tangential conditions are presented in Chap. 11, p. 437. Many other subsets of interest of initial conditions from which at least one or all evolution(s) satisfies(y) more and more complicated interesting properties will be introduced all along the book. They all are combinations in various ways of these basic kernels and basins.

For instance, *tychastic control systems* (or *dynamical games*) involve both regulons and tyches in the dynamics. Tyches describe uncertainties played by an indifferent, maybe hostile, Nature. Regulons are chosen among the available ones by the system in order *to adapt its evolutions regardless of the tyches*. We introduce the concept of *tychastic (or guaranteed) viability kernel*, which is the subset of initial states from which *there exists a regulon such that, for all tyches*, the associated evolutions are *viable* in the environment forever.

The set of initial states from which *there exists a regulon such that, for all tyches*, the associated evolutions reach the target *in finite time* before possibly violating the constraints is called the *tychastic (or guaranteed) absorption basin* of the target invariant in the environment.

Remark: Semi-permeability. We deduce from the definitions that from any $x \in \mathrm{Viab}_S(K,C) \setminus \mathrm{Inv}_S(K,C)$,

1. *there exists at least one evolution which is viable in* $\mathrm{Viab}_S(K,C)$ *until it may reach the target* C,
2. *there exists at least one evolution which leaves* $\mathrm{Viab}_S(K,C)$ *in finite time, and is viable in* $K \setminus C$ *until it leaves* K *in finite time.*

The latter property is a *semi-permeability property*:

1. the boundary of the invariance kernel separates the set of initial states from which all evolutions are viable in K until they may reach the target from the set of initial states satisfying the above property,
2. the boundary of the viability kernel separates the set of initial states from which there exists at least two different evolutions satisfying the above property from the set of initial states from which all evolutions are viable in $K \setminus C$ as long as it is viable in K.

Therefore $x \in \mathrm{Viab}_S(K,C) \setminus \mathrm{Inv}_S(K,C)$ is the set where some uncertainty about viability prevails, Outside the viability kernel, only one property is shared by **all** evolutions starting from an initial state: either they are viable in K until they may reach the target, or they leave C in finite time and are viable in $K \setminus C$ until they leave K in finite time.

The Quincampoix Barrier Theorems 10.5.19, p. 409 and 10.6.4, p. 413 provide precise statements of the properties of the boundaries of the viability and invariance kernels. □

2.12 Links between Kernels and Basins

Viability kernels and absorption basins are linked to each other by complementarity, as well as invariance kernels and capture basins:

Definition 2.12.1 ***[Complement of a Subset]*** *The complement of the subset $C \subset K$ in K is the set $K \backslash C := K \cap \complement C$ of elements $x \in K$ not belonging to C. When $K := X$ is the whole space, we set $\complement\, C := X \backslash C$. Observe that*

$$K \backslash C = \complement C \backslash \complement K \text{ and } \complement(K \setminus C) = C \cup \complement K$$

The following useful consequences relating the kernels and basins follow readily from the definitions:

Lemma 2.12.2 ***[Complements of Kernels and Basins]*** *Kernels and Basins are exchanged by complementarity:*

$$\begin{cases} (i) \ \ \complement\mathrm{Viab}_{\mathcal{S}}(K,C) = \mathrm{Abs}_{\mathcal{S}}(\complement C, \complement K) \\ (ii) \ \complement\mathrm{Capt}_{\mathcal{S}}(K,C) = \mathrm{Inv}_{\mathcal{S}}(\complement C, \complement K) \end{cases} \tag{2.25}$$

Remark. This would suggest that only two of these four concepts would suffice. However, we would like these kernels and basins to be closed under adequate assumption, and for that purpose, we need the four concepts, since the complement of a closed subset is open. But every statement related to the closedness property of these kernels and basins provide corresponding results on openness properties of their complements, as we shall see in Sect. 10.3.2 p. 387. □

The next result concerns the a priori futile or subtle differences between viability kernels with targets (concept proposed by Marc Quincampoix) and capture basins:

Lemma 2.12.3 ***[Comparison between Viability Kernels with Targets and Capture Basins]*** *The viability kernel of K with target C and the capture basin of C viable in K are related by formulas*

$$\mathrm{Viab}_{\mathcal{S}}(K,C) = \mathrm{Viab}_{\mathcal{S}}(K \setminus C) \cup \mathrm{Capt}_{\mathcal{S}}(K,C) \tag{2.26}$$

Hence the viability kernel with target C coincides with the capture basin of C viable in K if $\mathrm{Viab}_{\mathcal{S}}(K \setminus C) = \emptyset$, i.e., if $K \setminus C$ is a repeller. This is particularly the case when the viability kernel $\mathrm{Viab}_{\mathcal{S}}(K)$ of K is contained in the target C, and more so, when K itself is a repeller.

Proof. Actually, we shall prove that

$$\begin{cases} (i)\ \mathrm{Viab}_{\mathcal{S}}(K,C) \setminus \mathrm{Capt}_{\mathcal{S}}(K,C) \subset \mathrm{Viab}_{\mathcal{S}}(K \setminus C) \\ (ii)\ \mathrm{Viab}_{\mathcal{S}}(K,C) \setminus \mathrm{Viab}_{\mathcal{S}}(K \setminus C) \subset \mathrm{Capt}_{\mathcal{S}}(K,C) \end{cases}$$

Indeed, inclusion $\mathrm{Viab}_{\mathcal{S}}(K \setminus C) \cup \mathrm{Capt}_{\mathcal{S}}(K,C) \subset \mathrm{Viab}_{\mathcal{S}}(K,C)$ being obvious, the opposite inclusion is implied by, for instance,

$$\mathrm{Viab}_{\mathcal{S}}(K,C) \setminus \mathrm{Capt}_{\mathcal{S}}(K,C) \subset \mathrm{Viab}_{\mathcal{S}}(K \setminus C) \tag{2.27}$$

because

$$\begin{cases} \mathrm{Viab}_{\mathcal{S}}(K,C) = \mathrm{Capt}_{\mathcal{S}}(K,C) \cup (\mathrm{Viab}_{\mathcal{S}}(K,C) \setminus \mathrm{Capt}_{\mathcal{S}}(K,C)) \\ \subset \mathrm{Viab}_{\mathcal{S}}(K \setminus C) \cup \mathrm{Capt}_{\mathcal{S}}(K,C) \end{cases}$$

For proving the first formula

$$\mathrm{Viab}_{\mathcal{S}}(K,C) \setminus \mathrm{Capt}_{\mathcal{S}}(K,C) \subset \mathrm{Viab}_{\mathcal{S}}(K \setminus C) \tag{2.28}$$

we observe that Lemma 2.12.2, p. 92 implies that $\mathrm{Viab}_{\mathcal{S}}(K,C) \setminus \mathrm{Capt}_{\mathcal{S}}(K,C) = \mathrm{Viab}_{\mathcal{S}}(K,C) \cap \mathrm{Inv}_{\mathcal{S}}(\complement C, \complement K)$ by formula (2.25)(i). Take any $x \in \mathrm{Viab}_{\mathcal{S}}(K,C) \cap \mathrm{Inv}_{\mathcal{S}}(\complement C, \complement K)$. Since $x \in \mathrm{Viab}_{\mathcal{S}}(K,C)$, there exists at least one evolution $x(\cdot) \in \mathcal{S}(x)$ either viable in K forever or reaching C in finite time. But since $x \in \mathrm{Inv}_{\mathcal{S}}(\complement C, \complement K)$, all evolutions starting from x are viable in $\complement C$ forever or until they leave K in finite time. Hence the evolution $x(\cdot)$ cannot reach C in finite time, and thus, is viable in K forever, hence cannot leave K in finite time, and thus is viable in $\complement C$, and consequently, in $K \setminus C$.

Next, let us prove inclusion $\mathrm{Viab}_{\mathcal{S}}(K,C) \setminus \mathrm{Viab}_{\mathcal{S}}(K \setminus C) \subset \mathrm{Capt}_{\mathcal{S}}(K,C)$. Lemma 2.12.2, p. 92 implies that $\complement \mathrm{Viab}_{\mathcal{S}}(K \setminus C) = \mathrm{Abs}_{\mathcal{S}}(X, C \cup \complement K)$. Therefore, for any $x \in \mathrm{Viab}_{\mathcal{S}}(K,C) \setminus \mathrm{Viab}_{\mathcal{S}}(K \setminus C) = \mathrm{Viab}_{\mathcal{S}}(K,C) \cap \mathrm{Abs}_{\mathcal{S}}(X, C \cup \complement K)$, there exists an evolution $x(\cdot) \in \mathcal{S}(x)$ viable in K forever or until a time $t^\star < +\infty$ when $x(t^\star) \in C$, and all evolutions starting at x

either leave K in finite time or reach C in finite time. Hence, $x(\cdot)$ being forbidden to leave K in finite time, must reach the target in finite time. □

Lemma 2.12.4 ***[Partition of the Viability Kernel with Targets]*** *The following equalities hold true:*

$$\mathrm{Capt}_{\mathcal{S}}(K,C) \cap \mathrm{Inv}_{\mathcal{S}}(K \setminus C) \;=\; \mathrm{Abs}_{\mathcal{S}}(K,C) \cap \mathrm{Viab}_{\mathcal{S}}(K \setminus C) \;=\; \emptyset$$

Therefore, equality $\mathrm{Viab}_{\mathcal{S}}(K\setminus C) = \mathrm{Inv}_{\mathcal{S}}(K\setminus C)$ *implies that* $\mathrm{Viab}_{\mathcal{S}}(K \setminus C)$ *and* $\mathrm{Capt}_{\mathcal{S}}(K,C)$ *form a partition of* $\mathrm{Viab}_{\mathcal{S}}(K,C)$.

For invariance kernels, we obtain:

Lemma 2.12.5 ***[Comparison between Invariance Kernels with Targets and Absorption Basins]*** *The invariance kernel of K with target C and the absorption basin of C viable in K coincide whenever $K \setminus C$ is a repeller.*

Proof. We still observe that the invariance kernel $\mathrm{Inv}_{\mathcal{S}}(K,C)$ of K with target C coincides with the absorption basin $\mathrm{Abs}_{\mathcal{S}}(K,C)$ of C invariant in K whenever the viability kernel $\mathrm{Viab}_{\mathcal{S}}(K\setminus C)$ is empty. □

Therefore, *the concepts of viability and of invariance kernels with a target allow us to study both the viability and invariance kernels of a closed subset and the capture and absorption basins of a target.*

Remark: Stochastic and Tychastic Properties. There are natural and deeper mathematical links between viability and capturability properties under stochastic and tychastic systems. A whole book could be devoted to this topic. We just develop in this one few remarks in Sect. 10.10, p. 433. □

2.13 Local Viability and Invariance

We introduce the weaker concepts of *local* viability and invariance:

Definition 2.13.1 ***[Local Viability and Invariance]***
Let $\mathcal{S} : X \rightsquigarrow \mathcal{C}(0,\infty;X)$ be an evolutionary system and a subset $K \subset X$.

1. *A subset K is said to be* locally viable *under $\mathcal{S}$ if from any initial state $x \in K$* there exists at least one evolution $x(\cdot) \in \mathcal{S}(x)$ *and a strictly positive time $T_{x(\cdot)} > 0$ such that $x(\cdot)$ is viable in K on the nonempty interval $[0, T_{x(\cdot)}[$ (it is thus* viable *if $T_{x(\cdot)} = +\infty$),*
2. *A subset K is said to be* locally invariant *under $\mathcal{S}$ if from any initial state $x \in K$ and* for any evolution $x(\cdot) \in \mathcal{S}(x)$, *there exists a strictly positive $T_{x(\cdot)} > 0$ such that $x(\cdot)$ is viable in K on the nonempty interval $[0, T_{x(\cdot)}[$ (it is thus* invariant *if $T_{x(\cdot)} = +\infty$).*

The (local) viability property of viability kernels and invariance property of invariance kernels are particular cases of viability property of inverse images of sets of evolutions and invariance property of their cores when the sets of evolutions are (locally) stable under translations. Local viability kernels are studied in Sect. 10.4.3, p. 396. For the time, we provide a family of examples of subsets (locally) viable and invariant subsets built from subsets of evolutions stable (or invariant) under translation.

Definition 2.13.2 ***[Stability Under Translation]*** *A subset $\mathcal{H} \subset \mathcal{C}(0, \infty; X)$ of evolutions is* locally stable under translation *if for every $x(\cdot) \in \mathcal{H}$, there exists $T_{x(\cdot)} > 0$ such that for every $t \in [0, T_{x(\cdot)}[$, the translation $\kappa(-t)(x(\cdot))(\cdot)$ belongs to $\mathcal{H}$. It is said to be* stable under translation *if we can always take $T_{x(\cdot)} = +\infty$.*

Inverse images (resp. cores) of subsets of evolutions stable under translation (resp. concatenation) are viable (resp. invariant) subsets:

Proposition 2.13.3 ***[Viability of Inverse Images and Invariance of Cores]*** *Let $\mathcal{S} : X \rightsquigarrow \mathcal{C}(0, \infty; X)$ be an evolutionary system and $\mathcal{H} \subset \mathcal{C}(0, \infty; X)$ be a subset of evolutions. If $\mathcal{H}$ is (locally) stable under translation, then*

1. *its inverse image $\mathcal{S}^{-1}(\mathcal{H}) := \{x \in X \mid \mathcal{S}(x) \cap \mathcal{H}\}$ under $\mathcal{S}$ is (locally) viable,*
2. *its core $\mathcal{S}^{\ominus 1}(\mathcal{H}) := \{x \in X \mid \mathcal{S}(x) \subset \mathcal{H}\}$ under $\mathcal{S}$ is (locally) invariant.*

(See Definition 18.3.3, p. 720).

Proof. 1. The (local) translation property of $\mathcal{S}$ implies the (local) viability of the inverse image $\mathcal{S}^{-1}(\mathcal{H})$. Take $x_0 \in \mathcal{S}^{-1}(\mathcal{H})$ and prove that there exists an evolution $x(\cdot) \in \mathcal{S}(x_0)$ starting at x_0 viable in $\mathcal{S}^{-1}(\mathcal{H})$ on some interval $[0, T_{x(\cdot)}]$. Indeed, there exists an evolution $x(\cdot) \in \mathcal{S}(x_0) \cap \mathcal{H}$ and

$T_{x(\cdot)} > 0$ such that for every $t \in [0, T_{x(\cdot)}[$, the translation $\kappa(-t)(x(\cdot))(\cdot)$ belongs to $\mathcal{H}$. It also belongs to $\mathcal{S}(x(t))$ thanks to the translation property of evolutionary systems. Therefore $x(t)$ does belong to $\mathcal{S}^{-1}(\mathcal{H})$ for every $t \in [0, T_{x(\cdot)}[$.

2. The concatenation property of $\mathcal{S}$ implies the local invariance of the core $\mathcal{S}^{\ominus 1}(\mathcal{H})$. Take $x_0 \in \mathcal{S}^{\ominus 1}(\mathcal{H})$ and prove that for all evolutions $x(\cdot) \in \mathcal{S}(x_0)$ starting at x_0, there exists $T_{x(\cdot)}$ such that $x(\cdot)$ is viable in $\mathcal{S}^{\ominus 1}(\mathcal{H})$ on the interval $[0, T_{x(\cdot)}]$. Indeed, take any such evolution $x(\cdot) \in \mathcal{S}(x_0)$ which belongs to $\mathcal{H}$ by definition. Thus there exists $T_{x(\cdot)} > 0$ such that for every $t \in [0, T_{x(\cdot)}]$, the translation $\kappa(-t)(x(\cdot))(\cdot)$ belongs to $\mathcal{H}$. Take any $t \in [0, T_{x(\cdot)}[$ and any evolution $y(\cdot) \in \mathcal{S}(x(t))$. Hence the t-concatenation $(x \diamond_t y)(\cdot)$ belongs to $\mathcal{S}(x_0)$ by definition of evolutionary systems, and thus to $\mathcal{H}$ because $x_0 \in \mathcal{S}^{\ominus 1}(\mathcal{H})$. Since $\mathcal{H}$ is locally stable under translation, we deduce that $y(\cdot) = (\kappa(-t)((x \diamond_t y(\cdot))))(\cdot)$ also belongs to $\mathcal{H}$. Since this holds true for every any evolution $y(\cdot) \in \mathcal{S}(x(t))$, we infer that $x(t) \in \mathcal{S}^{\ominus 1}(\mathcal{H})$. □

The study of local viability is continued in Sect. 10.4.3, p. 396.

2.14 Discretization Issues

The task for achieving this objective is divided in two different problems:

1. Approximate the continuous problem by discretized problem (in time) and digitalized on a grid (in state) by *difference inclusions* on *digitalized sets*. Most of the time, the real mathematical difficulties come from the proof of the convergence theorems stating that the limits of the solutions to the approximate discretized/digitalized problems converge (in an adequate sense) to solutions to the original continuous-time problem.
2. Compute the viability kernel or the capture basin of the discretized/digitalized problem with a specific algorithm, also providing the viable evolutions, as mentioned in Sect. 2.9.2, p. 74.

Let h denote the time discretization step. There are many more or less sophisticated ways to discretize a continuous parameterized system (f, U) by a discrete one (ϕ_h, U). The simplest way is to choose the explicit scheme $\phi_h(x, u) := x + hf(x, u)$. Indeed, the discretized system can be written as

$$\frac{x_{j+1} - x_j}{h} = f(x_j, u_j) \text{ where } u_j \in U(x_j)$$

The simplest way to digitalize a vector space $X := \mathbb{R}^d$ is to embed a (regular) *grid*[4] $X_\rho := \rho\mathbb{Z}^d$ in X. Points of the grid are of the form $x :=$

[4] supplied with the metric $d(x, y)$ equal to 0 if $x = y$ and to 1 if $x \neq y$.

$(\rho n_i)_{i=1,\ldots,n}$ where for all $i = 1, \ldots, n$, n_i ranges over the set $\mathbb{Z}$ of positive or negative integers.

We cannot define the above discrete system on the grid X_ρ, because there is no reason why for any $x \in X_\rho$, $\phi_h(x,u)$ would belong to the grid X_ρ. Let us denote by $B := [-1,+1]^d$ the unit square ball of X^d. One way to overcome this difficulty is to "add" the set $\rho B = [-\rho,+\rho]^d$ to $\phi_h(x,u)$. Setting $\lambda A + \mu B := \{\lambda x + \mu y\}_{x \in A,\, y \in B}$ when $A \subset X$ and $B \subset X$ are nonempty subsets of a vector space X, we obtain the following example:

Definition 2.14.1 *[Explicit Discrete/Digital Approximation] Parameterized control systems*

$$x'(t) = f(x(t), u(t)) \text{ where } u(t) \in U(x(t))$$

can be approximated by discrete/digital parameterized systems

$$\frac{x_{j+1} - x_j}{h} \in f(x_j, u_j) + \rho B \text{ where } u_j \in U(x_j)$$

which is a discrete system $x_{j+1} \in \Phi_{h,\rho}(x_j)$ on X_ρ where

$$\Phi_{h,\rho}(x) := x + hf(x, U(x)) + \rho h B$$

We can also use implicit difference schemes:

Definition 2.14.2 *[Implicit Discrete/Digital Approximation] Parameterized control systems*

$$x'(t) = f(x(t), u(t)) \text{ where } u(t) \in U(x(t))$$

can be approximated by discrete/digital parameterized systems

$$\frac{x_{j+1} - x_j}{h} \in f(x_{j+1}, u_{j+1}) \text{ where } u_{j+1} \in U(x_{j+1})$$

which is a discrete system $x_{j+1} \in \Psi_{h,\rho}(x_j)$ on X_ρ where

$$\Psi_{h,\rho}(x) := (\mathbf{I} - hf(\cdot, U(\cdot)))^{-1}(x) + \rho h B$$

Characterization Theorem 2.9.3, p. 72 of viability and invariance under discrete systems, applied to the explicit discretization of control systems, indicates how tangential conditions for characterizing viability and invariance under control systems did emerge:

Lemma 2.14.3 *[**Discretized Regulation Map**] Let us introduce the discretized* regulation map $R_{K_{h,\rho}}$ *defined by*

$$\forall x \in K, \ R_{K_{h,\rho}}(x) := \left\{ u \in U(x) \text{ such that } f(x,u) \in \frac{K-x}{h} + \rho B \right\} \tag{2.29}$$

Then K is viable (resp. invariant) under the discretized system if and only if $\forall\, x \in K$, $R_{K_{h,\rho}}(x) \neq \emptyset$ (resp. $\forall\, x \in K$, $R_{K_{h,\rho}}(x) = U(x)$).

For proving viability and invariance theorems in Chap. 11, p. 437, we shall take the limit in the results of the above lemma, and in particular, for the kind of "difference quotient" $\dfrac{K-x}{h}$, if one is allowed to say so.

14 *[**How Tangential Conditions Emerge**] The Bouligand-Severi tangent cone $T_K(x)$ (see Definition 11.2.1, p. 442) to K at $x \in K$ is the upper limit in the sense of Painlevé-Kuratowski of $\frac{K-x}{h}$ when $h \to 0$: $f(x,u)$ is the limit of elements v_n such that $x + h_n v_n \in K$ for some $h_n \to 0+$, i.e., of velocities $v_n \in \dfrac{K-x}{h_n}$.*
Consequently, "taking the limit", formally (for the time), we obtain the emergence of the (continuous-time) tangential condition

$$\forall x \in K, \ R_K(x) := \{u \in U(x) \text{ such that } f(x,u) \in T_K(x)\} \tag{2.30}$$

where $T_K(x)$ is the Bouligand-Severi tangent cone to K at $x \in K$ (see Definition 11.2.1, p.442).

This tangential condition will play a crucial role for characterizing viability and invariance properties for continuous-time systems in Chap. 11, p. 437.

2.15 A Viability Survival Kit

The mathematical properties of viability and invariance kernels and capture and absorption basins are presented in detail in Chap. 10 p. 375 for evolutionary systems and in Chap. 11, p. 437 for differential inclusions and control systems, where we can take advantage of tangential conditions involving tangent cones to the environments. This section presents few selected statements that are most often used, restricted to viability kernels and capture basins only. Three categories of statements are presented:

- The first one provides characterizations of viability kernels and capture bilateral fixed points, which are simple, important and are valid without any assumption.
- The second one provides characterizations in terms of local viability properties and backward invariance, involving topological assumptions on the evolutionary systems.
- The third one characterizes viability kernels and capture basins under differential inclusions in terms of tangential conditions, which furnishes the regulation map allowing to pilot viable evolutions (and optimal evolutions in the case of optimal control problems).

2.15.1 Bilateral Fixed Point Characterization

We consider the maps $(K, C) \mapsto \text{Viab}(K, C)$ and $(K, C) \mapsto \text{Capt}(K, C)$. The properties of these maps provide fixed point characterizations of viability kernels of the maps $K \mapsto \text{Viab}(K, C)$ and $C \mapsto \text{Viab}(K, C)$ and fixed point characterizations of capture basins of the maps $K \mapsto \text{Capt}(K, C)$ and $C \mapsto \text{Capt}(K, C)$. We refer to Definition 2.10.2, p.86 for the definitions of viable and isolated subsets.

Theorem 2.15.1 *[The Fundamental Characterization of Viability Kernels] Let $\mathcal{S} : X \rightsquigarrow \mathcal{C}(0, +\infty; X)$ be an evolutionary system and $K \subset X$ be a environment. The viability kernel $\text{Viab}_{\mathcal{S}}(K) := \text{Viab}_{\mathcal{S}}(K, \emptyset)$ of K (see Definition 2.10.2, p. 86) is the* ***unique*** *subset D contained in K that is both*

1. ***viable*** *in K (and is the largest viable subset $D \subset K$ contained in K),*
2. ***isolated*** *in K (and is the smallest subset $D \subset K$ isolated in K):*

i.e., the bilateral fixed point

$$\text{Viab}_{\mathcal{S}}(\text{Viab}_{\mathcal{S}}(K)) = \text{Viab}_{\mathcal{S}}(K) = \text{Viab}_{\mathcal{S}}(K, \text{Viab}_{\mathcal{S}}(K)) \tag{2.31}$$

For capture basins, we shall prove

Theorem 2.15.2 *[The Fundamental Characterization of Capture Basins] Let $\mathcal{S} : X \rightsquigarrow \mathcal{C}(0, \infty; X)$ be an evolutionary system, $K \subset X$ be an environment and $C \subset K$ be a nonempty target. The capture basin*

$\mathrm{Capt}_{\mathcal{S}}(K,C)$ *of* C *viable in* K *(see Definition 2.10.2, p. 86) is the* **unique** *subset* D *between* C *and* K *that is both*

1. ***viable outside*** C *(and is the largest subset* $D \subset K$ *viable outside* C*),*
2. *satisfying* $\mathrm{Capt}_{\mathcal{S}}(K,C) = \mathrm{Capt}_{\mathcal{S}}(K, \mathrm{Capt}_{\mathcal{S}}(K,C))$ *(and is the smallest subset* $D \supset C$ *to do so):*

i.e., the bilateral fixed point

$$\mathrm{Capt}_{\mathcal{S}}(\mathrm{Capt}_{\mathcal{S}}(K,C),C) = \mathrm{Capt}_{\mathcal{S}}(K,C) = \mathrm{Capt}_{\mathcal{S}}(K,\mathrm{Capt}_{\mathcal{S}}(K,C)) \tag{2.32}$$

2.15.2 Viability Characterization

However important Theorems 2.15.1, p. 99 and 2.15.2, p. 99 are, isolated subsets are difficult to characterize, in contrast to viable or locally viable subsets (see Definition 2.13.1, p. 94). It happens that isolated subsets are, under adequate assumptions, backward invariant (see Sect. 10.5.2, p. 401). Hence we shall introduce the concept of *backward evolutionary system* (see Definition 8.2.1, p. 276) and the concept of *backward invariance*, i.e., of invariance with respect to the backward evolutionary system (see Definition 8.2.4, p. 278). Characterizing viability kernels and capture basins in terms of forward viability and backward invariance allows us to use the results on viability and invariance.

Definition 2.15.3 *[**Backward Relative Invariance**] A subset* $C \subset K$ *is* backward invariant relatively to K *under* $\mathcal{S}$ *if for every* $x \in C$*, for every* $t_0 \in]0,+\infty[$*, for all evolutions* $x(\cdot)$ *arriving at* x *at time* t_0 *such that there exists* $s \in [0,t_0[$ *such that* $x(\cdot)$ *is viable in* K *on the interval* $[s,t_0]$*, then* $x(\cdot)$ *is viable in* C *on the same interval.*

If K is itself backward invariant, any subset backward invariant relatively to K is actually backward invariant.

Viability results hold true whenever the evolutionary system is upper semicontinuous (see Definitions 18.4.3, p. 729).

Using the concept of backward invariance, we provide a further characterization of viability kernels and capture basins:

Theorem 2.15.4 ***[Characterization of Viability Kernels]*** *Let us assume that $\mathcal{S}$ is upper semicompact and that the subset K is closed. The viability kernel* $\text{Viab}_{\mathcal{S}}(K)$ *of a subset K under $\mathcal{S}$ is the* **unique** *closed subset $D \subset K$ satisfying*

$$\begin{cases} (i) & D \text{ is } \textit{viable} \text{ under } \mathcal{S} \\ (ii) & D \textit{ is backward invariant} \text{ under } \mathcal{S} \\ (iii) & K \backslash D \text{ is a } \textit{repeller} \text{ under } \mathcal{S}. \end{cases} \tag{2.33}$$

For capture basins, we obtain

Theorem 2.15.5 ***[Characterization of Capture Basins]*** *Let us assume that $\mathcal{S}$ is upper semicompact, that the environment $K \subset X$ and the target $C \subset K$ are closed subsets satisfying*

1. *K is* backward invariant
2. *$K \setminus C$ is a* repeller *($\text{Viab}_{\mathcal{S}}(K \setminus C) = \emptyset$)*

Then the viable capture basin $\text{Capt}_{\mathcal{S}}(K, C)$ *is the* **unique** *closed subset D satisfying $C \subset D \subset K$ and*

$$\begin{cases} (i) & D \backslash C \text{ is } \textit{locally viable} \text{ under } \mathcal{S} \\ (ii) & D \textit{ is relatively backward invariant with respect to } K \text{ under } \mathcal{S}. \end{cases} \tag{2.34}$$

2.15.3 Tangential Characterization

These theorems, which are valid for any evolutionary system, paved the way to go one step further when the evolutionary system is associated with a differential inclusion (and control systems, as we shall see in Sect. 11.3.1, p. 453). We mentioned, in the case of discrete systems, how tangential conditions (2.30), p. 98 did emerge when we characterized viable and invariance (see Box 14, p. 98). Actually, we shall use the closed convex hull $T_K^{\star\star}(x)$ of the tangent cone $T_K(x)$ (see Definition 11.2.1, p. 442) for this purpose.

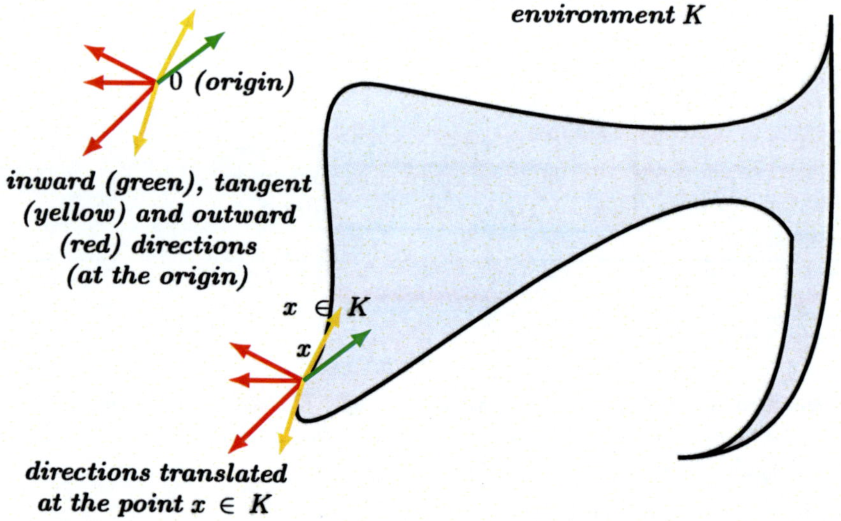

Fig. 2.14 Schematic Representation of Tangent Cones.
We represent the environment K, an element $x \in K$ and the origin. Six vectors v are depicted: one which points inward K, and thus tangent to K, two tangent vectors which are not inward and three outward vectors. Their translations at x belong to K for the inward vector, "almost" belong to K for the two tangent and not inward vectors (see Definition 11.2.1, p. 442) and belong to the complement of K for the three outward vectors.

Not only Viability and Invariance Theorems provide characterizations of viability kernels and capture basins, but also the *regulation map* $R_D \subset F$ which governs viable evolutions:

Definition 2.15.6 *[**Regulation Map**] Let us consider three subsets $C \subset D \subset K$ (where the target C may be empty) and a set-valued map $F : X \rightsquigarrow X$.*

The set-valued map $R_D : x \in D \rightsquigarrow F(x) \cap T_D^{\star\star}(x) \subset X$ is called the regulation map of F on $D \setminus C$ if

$$\forall\, x \in D \setminus C, \;\; R_D(x) \; := \; F(x) \cap T_D^{\star\star}(x) \; \neq \; \emptyset \qquad (2.35)$$

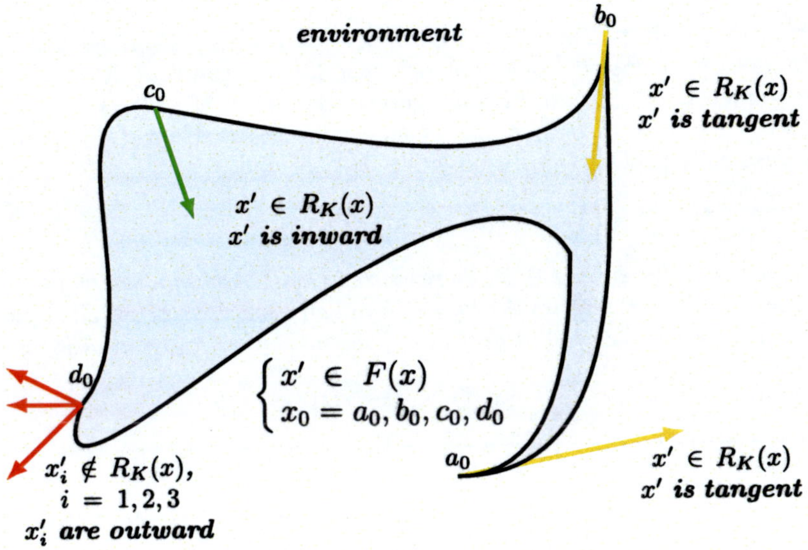

Fig. 2.15 Schematic Illustration of the Regulation Map.
In this scheme, we describe four situations at elements a_0, b_0, c_0 and $d_0 \in K$. At a_0 and b_0, the right hand side of the differential inclusions contains tangent velocities to K, so that we can expect an evolution to be viable. At c_0, this hope is even more justified because the velocity points in the interior of K. Finally, at d_0, all velocities point outward K, and it is intuitive that all evolutions leave K instantaneously. The viability theorem states that these intuition and hopes are correct for any closed subset K and for Marchaud maps.

The Viability and Invariance Theorems imply that

Theorem 2.15.7 ***[Tangential Characterization of Viability Kernels]*** *Let us assume that F is Marchaud (see Definition 10.3.2, p. 384) and that the subset K is closed. The viability kernel* $\text{Viab}_S(K)$ *of a subset K under S is the* ***largest*** *closed subset $D \subset K$ satisfying*

$$\forall x \in D, \;\; R_D(x) := F(x) \cap T_D^{\star\star}(x) \neq \emptyset \tag{2.36}$$

Furthermore, for every $x \in D$, there exists at least one evolution $x(\cdot) \in S(x)$ viable in D and all evolutions $x(\cdot) \in S(x)$ viable in D are governed by the differential inclusion

$$x'(t) \in R_D(x(t))$$

For capture basins, we obtain

Theorem 2.15.8 *[Tangential Characterization of Capture Basins] Let us assume that F is Marchaud, that the environment $K \subset X$ and the target $C \subset K$ are closed subsets such that $K \setminus C$ is a repeller ($\text{Viab}_F(K \backslash C) = \emptyset$). Then the viable-capture basin $\text{Capt}_{\mathcal{S}}(K, C)$ is the **largest** closed subset D satisfying $C \subset D \subset K$ and*

$$\forall x \in D \backslash C, \;\; F(x) \cap T_D^{\star\star}(x) \neq \emptyset$$

Furthermore, for every $x \in D$, there exists at least one evolution $x(\cdot) \in \mathcal{S}(x)$ viable in D until it reaches the target C and all evolutions $x(\cdot) \in \mathcal{S}(x)$ viable in D until they reach the target C are governed by the differential inclusion

$$x'(t) \in R_D(x(t))$$

Further important properties hold true when the set-valued map F is Lipschitz (see Definition 10.3.5, p. 385).

Theorem 2.15.9 *[Characterization of Viability Kernels] Let us assume that (f, U) is both Marchaud and Lipschitz and that the subset K is closed. The viability kernel $\text{Viab}_F(K)$ of a subset K under $\mathcal{S}$ is the **unique** closed subset $D \subset K$ satisfying*

- *$K \setminus D$ is a repeller;*
- *and the* Frankowska property*:*

$$\begin{cases} (i) \;\forall x \in D, \;\; F(x) \cap T_D^{\star\star}(x) \neq \emptyset \\ (ii) \;\forall x \in D \cap \text{Int}(K), \;\; -F(x) \subset T_D^{\star\star}(x) \\ (ii) \;\forall x \in D \cap \partial K, \;\; -F(x) \cap T_K^{\star\star}(x) = -F(x) \cap T_D^{\star\star}(x) \end{cases} \quad (2.37)$$

For capture basins, we obtain

Theorem 2.15.10 *[Characterization of Capture Basins] Let us assume that (f, U) is Marchaud and Lipschitz and that the environment $K \subset X$ and the target $C \subset K$ are closed subsets such that $K \setminus C$ is a repeller ($\text{Viab}_F(K \backslash C) = \emptyset$). Then the viable-capture basin $\text{Capt}_F(K, C)$ is the **unique** closed subset D satisfying*

- *$C \subset D \subset K$,*
- *and the* Frankowska property *(2.37), p. 104.*

Chapter 3
Viability Problems in Robotics

3.1 Introduction

This chapter studies three applications to robotics, one focussing on field experiments of the viability feedback allowing a robot to rally a target in an urban environment while avoiding obstacles, the second one dealing with the safety envelope of the landing of a plane as well as the regulation law governing the safe landing evolutions viable in this envelope, and the third one focused on navigation of submarines in rivers.

3.2 Fields Experiment with the Pioneer

Viability theory not only computes the capture basin of the target viable in a given environment, but also provides the dedicated feedback piloting the state variables towards the target from any state of the capture basin. For this purpose, the time is discretized in time steps, the environment and the target are stored in a grid, and the viability software uses the capture basin algorithm to compute the graph of the feedback represented as a subset of a grid of state-control pairs. The graph of this feedback is integrated in the embedded software of the robot. At each time step, sensors locate the position of the robot and its direction. The feedback provides the controls.

The experimental robot presented is a Pioneer 3AT of *activmedia robotics* and the environment is a road network, on which a target to be reached in minimal time has been assigned.

Two sensors (odometers and GPS) are used for its localization, but the robot does not use the other sensors locating the obstacles and the target.

J.-P. Aubin et al., *Viability Theory*, DOI 10.1007/978-3-642-16684-6_3,

Fig. 3.1 Pioneer Robot used for the field experiment.
View of the Pionner, its "localization" sensors (GPS, left *and odometer,* right*) and the computer containing the tabulated graph of the viable feedback map (see Fig. 3.2, p.107) connected to the navigation system of the robot.*

The variables, controls, dynamics, constraints and target are as follow:

1. **State variables:** x_i, $i = 1, 2$, positions of the vehicle, θ, heading (direction),
2. **Controls:** u, total velocity, ω, angular velocity
3. **Control System**

$$\begin{cases} (i)\ x_1'(t) &= u(t)\cos(\theta(t)) \\ (ii)\ x_2'(t) &= u(t)\sin(\theta(t)) \\ (iii)\ \theta'(t) &= \omega(t) \end{cases} \tag{3.1}$$

4. **State Constraints and Target:**
The position constraints and targets have been discretized on a grid (see the left figure of Fig. 1.11, p.23) and the direction is bounded: $\theta \in [\theta^\flat, \theta^\sharp]$.
5. **Constraints on controls:** Constraints are added which encode position dependent bounds on velocity and angular rotation velocity:

$$u \in [-\mu^\flat(x_1, x_2, \theta), \mu^\sharp(x_1, x_2, \theta)], \ \& \ \omega \in [-\omega^\flat(x_1, x_2, \theta), \omega^\sharp(x_1, x_2, \theta)]$$

Knowing the graph of feedback map computed by the viability algorithm, which is a five-dimensional set of vectors (x, y, θ, u_1, u_2), at each time step,

- The sensors measure the position and the heading (x, y, θ);
- The software takes into account this state of the robot and provides the controls (u_1, u_2) such that (x, y, θ, u_1, u_2) belongs to the graph of the feedback;
- These controls are transmitted to the navigation software of the Pioneer robot which waits for the next time step to update the velocity controls for moving to the next position

and repeats these steps.

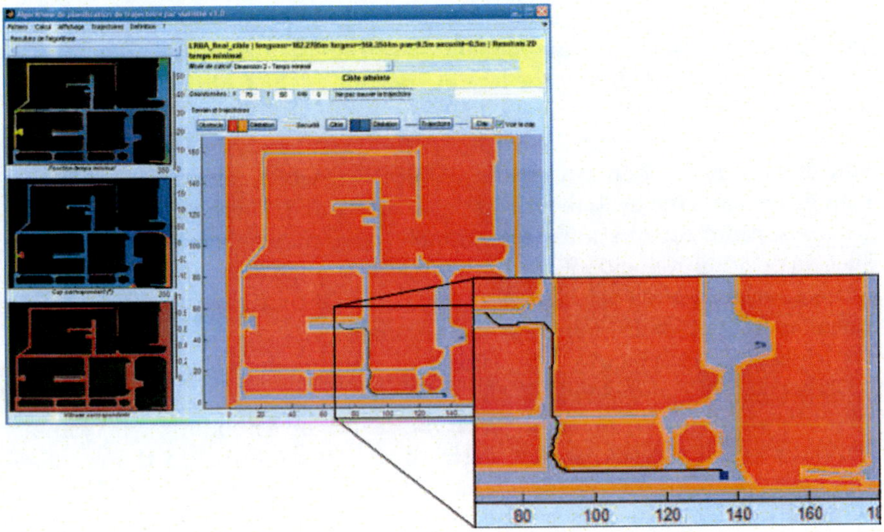

Fig. 3.2 Computation of the feedback for the Pioneer Robot.
The left column of the computer screen displays three windows providing in color scales the values of the minimal time at each position in the road (first window from the top), the direction (second window) and the velocities (provided by the feedback map). The main window displays the map of the road network and the trajectory of the robot, which is enlarged in the second figure.

This viability software was used for field implementation.

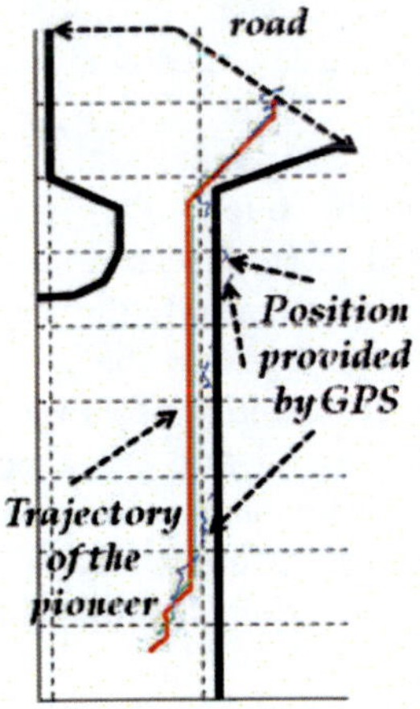

The robot must remain on the road, represented by thick lines. The actual trajectory of the robot rallying the target in minimum time while remaining in the road is indicated in grey. This actual trajectory is subjected to position errors estimated from the odometer and the GPS (providing sometimes positions outside the road). The dedicated feedback computed by the viability algorithm taking into account the environment allows the software to correct these errors.

3.3 Safety Envelopes in Aeronautics

In this section, we present an application of viability theory to *safety analysis in aeronautics.* We focus on the landing portion of the flight of a large civilian airliner (DC9-30).

One of the key technologies for design and analysis of safety critical and human-in-the-loop systems is *verification*, which allows for heightened confidence that the system will perform as desired. Verification means that from an initial set of states (for example, aircraft configurations, or states, such as position, velocity, flight path angle and angle of attack), a system can reach another desired set of states (*target*) while remaining in an acceptable range of states, called *envelope* in aeronautics, (i.e., an environment, in the viability terminology). The subset of states from which the target can be reached while remaining in the envelope is sometimes called the set of *controllable states* or *maximal controllable set* in aeronautics: in viability terminology, this is the *capture basin* of the target viable in the envelope. For example, if an aircraft is landing, the initial set of states is the set of acceptable (viable) aircraft configurations of the aircraft a few hundred feet before landing, the target is the set of acceptable aircraft states at touch down, and the envelope is the range of states in which it is safe to operate the aircraft. A *safe landing evolution* is one which starts from the envelope until it reaches the target in finite time.

Viability theory provides a framework for computing a capture basin of a given target, and thus, the maximal controllable set.

The benefit of this approach, is that it provides a *verification* (for the mathematical models used) that the system will remain inside the envelope and reach target.

This is an "inverse approach" (see Box 2, p. 5), to be contrasted with "direct approaches" (see Box 1, p. 5), using simulation methods, such as Monte-Carlo methods. They do not provide any guarantee that evolutions starting from configurations that are not part of the testing set of the simulation will land safely. Monte Carlo methods have historically been used to explore the possible evolutions a system might follow. The more finely gridded the state-space, the more information the Monte Carlo simulations will provide. However, this class of methods is fundamentally limited in that it provides no information about initial conditions outside the grid points. This provides a good motivation for the use of viability techniques: they provide both capture basin (maximal controllable set) and the a posteriori feedback governing safe evolutions, i.e., the *certificate guaranteeing* that from this *set*, an evolution will land safely if piloted using this feedback. Monte-Carlo methods may provide such initial states and evolutions piloted with a priori feedbacks if one is... lucky.

The first subsection of this section presents the model of the longitudinal dynamics of the aircraft, as well as the definition of the safety envelopes in the descent (flare) mode of the aircraft. The following subsection presents the computation of both the capture basin of the touch down target and the a posteriori *viability feedback* dedicated to safe touch down, in the sense that is its not given a priori, but computed from the target and the environment.

3.3.1 Landing Aircraft: Application of the Capture Basin Algorithm

3.3.1.1 Physical Model, Equations of Motion

This section presents the equations of motion used to model the aircraft's behavior.

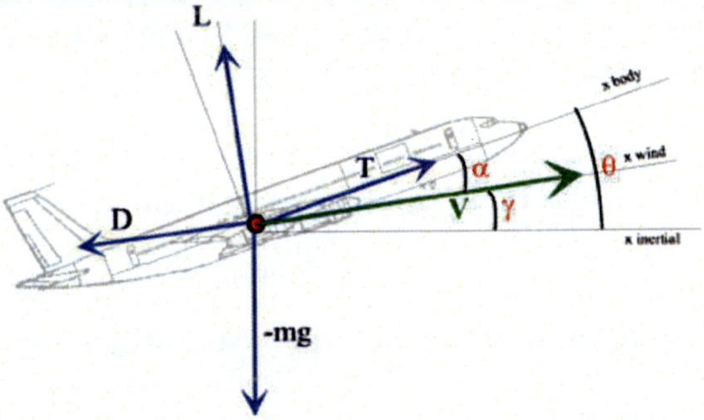

Fig. 3.3 Aircraft dynamic model.
Point mass force diagram for the longitudinal dynamics of the aircraft. Equation (3.2) is given in the inertial frame. V, γ, α, L, D *and* T *are used in (3.2) as displayed here.*

The aerodynamic properties of the aircraft, which are used for this model are derived from empirical data as well as fundamental principles. We model the longitudinal dynamics of an aircraft (see Fig. 3.3).

15 States and Controls. The state variables are:

1. the velocity V,
2. the flight path angle γ,
3. The altitude z.

We denote $x = (V, \gamma, z)$ the state of the system.
The controls (inputs) are:

1. the thrust T
2. the angle of attack α.

(to be controlled by the pilot or the autopilot). We denote $u = (T, \alpha)$ the control of the system.

We consider a point mass model where the aircraft is subjected to the gravity force mg, thrust T, lift L and drag D. We call m the mass of the aircraft. The equations of motion for this system read:

$$\frac{d}{dt}\begin{bmatrix} V \\ \gamma \\ z \end{bmatrix} = \begin{bmatrix} \frac{1}{m}[T\cos\alpha - D(\alpha, V) - mg\sin\gamma] \\ \frac{1}{mV}[T\sin\alpha + L(\alpha, V) - mg\cos\gamma] \\ V\sin\gamma \end{bmatrix} \tag{3.2}$$

Typically, landing is operated at $T_{\text{idle}} = 0.2 \cdot T_{\max}$ where $T_{\max}$ is the maximal thrust. This value enables the aircraft to counteract the drag due to flaps, slats and landing gear. Most of the parameters for the DC9-30 can be found in the literature. The values of the numerical parameters used for the DC9-30 in this flight configuration are $m = 60,000\,\text{kg}$, $T_{\max} = 160,000\text{N}$ and $g = 9.8\,\text{ms}^{-2}$. The lift and drag forces are thus

$$\begin{aligned} L(\alpha, V) &= 68.6\ (1.25 + 4.2\alpha)V^2 \\ D(\alpha, V) &= [2.7 + 3.08\ (1.25 + 4.2\alpha)^2]V^2 \end{aligned} \tag{3.3}$$

(they are expressed in Newtons if V is taken in m/s). We refer to the specialized literature for the methods leading to these formulas.

3.3.1.2 Landing Maneuvers

In a typical autoland maneuver, the aircraft begins its approach approximately 10 nautical miles from the touchdown point. The aircraft descends towards the *glideslope*, an inertial beam which the aircraft can track. The landing gear is down, and the pilot sets the flaps at the first high-lift configuration in the landing sequence.

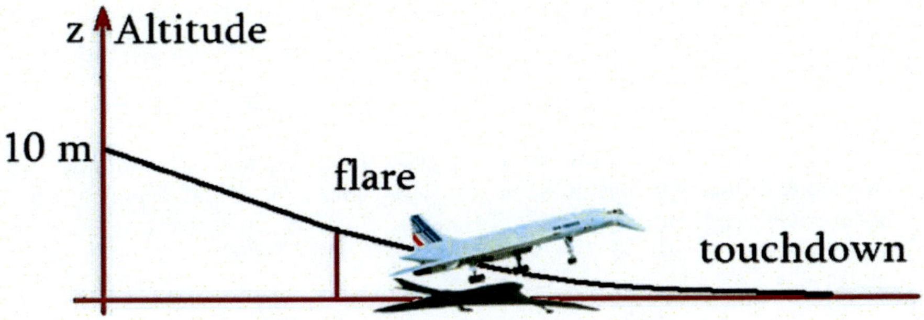

Fig. 3.4 Autoland Profile.
Typical Autoland Profile and flare section of the flight before touch down studied in this example.

The autopilot captures the glideslope signal around five nautical miles from the touch down point. The pilot increases flap deflection to effect a descent without increasing velocity. The pilot steps the flaps through the different flap settings, reaching the highest deflection when the aircraft reaches 1,000 feet altitude. At approximately 50 feet, the aircraft leaves the glideslope and begins the flare maneuver, which allows the aircraft to touchdown smoothly on the runway with an appropriate descent rate. The deflection of the slats is correlated with the deflection of the flaps in an automated way.

3.3.1.3 Safety Envelopes

Flight operating conditions are defined by the limits of aircraft performance, as well as by airport or the Federal Aviation Administration (FAA) regulations. During descent and flare, the aircraft proceeds through successive flap and slat settings. In each of these settings, the safe set is defined by bounds on the state variables. The maximal allowed velocity $V_{\max}$ is dictated by regulations. The minimal velocity is related to the stall velocity by $V_{\min} = 1.3 \cdot V_{\text{stall}}$. The minimal velocity is an FAA safety recommendation, the aircraft might become uncontrollable below V_{stall}.

During descent, the aircraft tracks the glideslope and must remain within $\pm 0.7°$ of the glideslope angle γ_{GS}. By regulation, as the aircraft reduces its descent rate to land smoothly (in the last 50 feet before touch down), the flight path angle γ in "flare mode" can range from $\gamma_{\min} = -3.7°$ to 0.

During descent and flare, thrust T should be at idle, but the pilot can use the full range of angle of attack α. Other landing procedures in which T varies are used as well.

16 Flight and Touch down envelopes: The environment and the target. The environment

$$K_{\text{flare}} := [V_{\min}, V_{\max}] \times [\gamma_{\min}, 0] \times \mathbb{R}_+ \tag{3.4}$$

is the flight envelope in flare mode, or *flare envelope*, when the velocity lies between minimal velocity $V_{\min} = 1.3 \cdot V_{\text{stall}}$ and a maximal velocity, when the negative path angle is larger than $\gamma_{\min} = -3.7°$ and the altitude is positive.
The target

$$\begin{cases} C_{\text{touch down}} := \\ \{(V, \gamma, z) \in [V_{\min}, V_{\max}] \times [\gamma_{\min}, 0] \times \{0\} \text{ such that } V \sin\gamma \geq \dot{z}_0\} \end{cases} \tag{3.5}$$

is the *touch down envelope*, when the velocity and the negative path angle are constrained by $V \sin\gamma \geq \dot{z}_0$ and the altitude is equal to 0.
The constraints on the controls are $U(x) := [T_{\min}, T_{\max}] \times [\alpha_{\min}, \alpha_{\max}]$

At touch down ($z = 0$, but not $V = 0$ which describes the stopping time of the aircraft), the restrictions are the same as flight parameters of the flight envelope, except for the descent velocity. This last requirement becomes $\dot{z}(t) > \dot{z}_0$, where $\dot{z}_0$ represent the maximal touch down velocity (in order not to damage the landing gear). This condition thus reads $V \sin\gamma \geq \dot{z}_0$.

The physical constraints of this problem are thus defined in terms of the safety envelope and the target, mathematically expressed by (3.4) and (3.5) respectively.

We now illustrate the difficulty of keeping the system inside the constraint set K_{flare} defined by (3.4) and of bringing it to the target $C_{\text{touch down}}$ defined by (3.5).

The environment (the flight envelope in flare mode) K_{flare} is a box in the (V, γ, z) space. Similarly, the target $C_{\text{touch down}}$ is a subset of the $z = 0$ face of K_{flare}, defined by $0 \geq V \sin\gamma \geq \dot{z}_0$ and $V \in [V_{\min}, V_{\max}]$.

Note that it is common for landing maneuvers to operate at constant thrust, which does not mean that, in the current model, the controls are constant, it only means that one of the control variables is constant for the duration of the maneuver (or some portion of it).

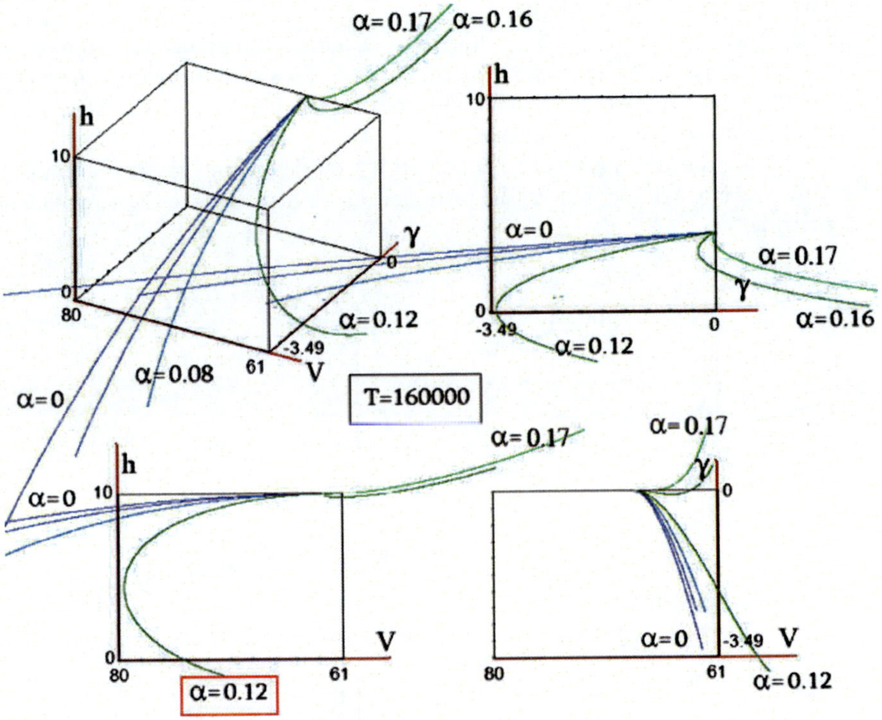

Fig. 3.5 Trajectories of evolutions with constant thrust.
Example of trajectories of evolutions emanating from a point at the $\gamma = 0$ *face of* K_{flare}*, with maximal thrust and different angles of attack* α*. Some of these evolutions are not viable in* K_{flare} *and do not reach* C *in finite time.*

Figure 3.5 displays several trajectories of evolutions obtained by applying $T = T_{\text{max}}$ with different values of α. As can be seen from this figure, the trajectories obtained in the (V, γ, z) space exit K_{flare} in finite time. This means that the control (α, T_{max}) does not enable one to reach the target C starting from the corresponding initial condition.

Similarly, Fig. 3.6 shows trajectories of several evolutions in the (V, γ, z) space obtained with a fixed α and different values for T. As can be seen from this figure, some of these trajectories are not viable in K and do not reach the target $C_{\text{touch down}}$ before they hit the ground of exit K_{flare}.

This can be seen for the $T = 0$ and $T = 20,000$ trajectories, which start from the point B in Fig. 3.6, and exit K at B_1. Note that other trajectories stay in K, for example the one obtained for $T = 160,000$, which manages to reach the ground safely. Maneuvers can include several sequences for which some of the parameters are fixed, in the present case, using a constant thrust for descent is a common practice in aviation. The figure enables one to see the evolution of the flight parameters, V and z as the altitude decreases (see in particular top right subfigure).

This figure also displays the trajectory of one viable evolution emanating from B which reaches the target C in finite time. This evolution is governed by the a posteriori viability feedback computed by the Capture Basin Algorithm providing minimal time evolutions.

These two figures illustrate the difficulty of designing viable evolutions for this system, which thus underlines the usefulness of the Capture Basin Algorithm.

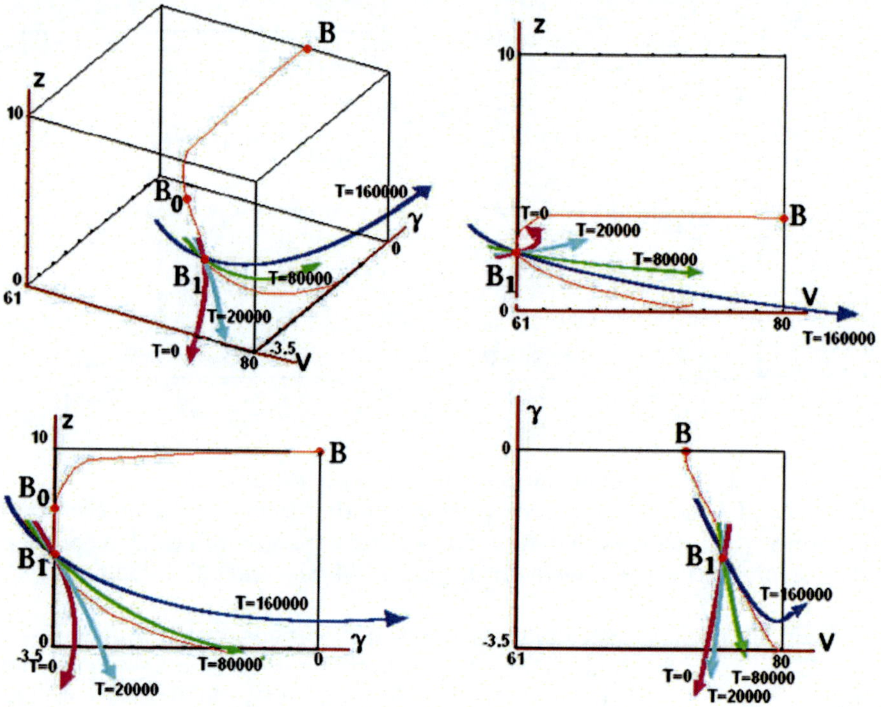

Fig. 3.6 Trajectories of evolutions with constant thrust.
Example of trajectories of evolutions emanating from a point at the $\gamma = 0$ face of K_{flare}, with different thrusts and a fixed angle of attack α. Some of these trajectories are not viable in K_{flare} and do not reach $C_{\text{touch down}}$ in finite time. However, the trajectory of the evolution starting at B and governed by the viable feedback synthesized by the Capture Basin Algorithm is viable.

3.3.2 Computation of the Safe Flight Envelope and the Viability Feedback

In the process of landing described in the previous sections, the following question is now of interest: starting from a given position in space (altitude z), with given flight conditions (velocity V and flight path angle γ), with fixed thrust, is there a viable feedback map (i.e., a switching policy made of a set of successive flap deflections/retractions), for which there exists a control (angle of attack α) able to bring the aircraft safely to the ground?

The *safe touch down envelope* is the set of states (V, γ, z) where $z > 0$ (aircraft in the air), from which it is possible to touch down the ground safely, (i.e. to reach $C_{\text{touch down}}$) while doing a safe flare (i.e. while staying in K_{flare}). It is therefore the capture basin

$$\text{Capt}_{(3.2)}(K_{\text{flare}}, C_{\text{touch down}}) \tag{3.6}$$

of the target viable in K_{flare}.

We now provide both the capture basin and the graph of the a posteriori viability feedback computed in terms of the target and the environment by the *Capture Basin Algorithm*. Knowing this a posteriori viability feedback dedicated to this task, one knows from the theorems and still can "*verify*" that any evolution starting from the capture basin (the safe flight envelope) and governed by this dedicated a posteriori feedback is viable until it reaches the target.

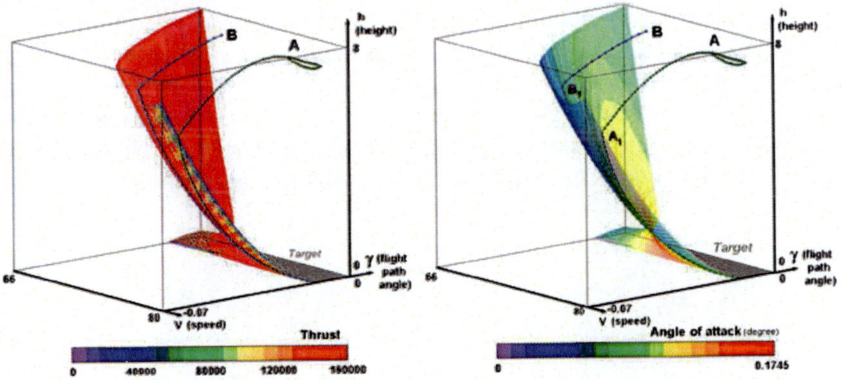

Fig. 3.7 Capture basin, thrust and angle of attack feedback.
Left: The thrust component of the feedback to apply at the boundary of $\text{Capt}_{(3.2)}(K_{\text{flare}}, C_{\text{touch down}})$ *to maintain the aircraft in the safe flight envelope. The thrust is indicated by the colorimetric scale on the boundary of the capture basin* $\text{Capt}_{(3.2)}(K_{\text{flare}}, C_{\text{touch down}})$. *Right: The angle of attack component is samely indicated by the colorimetric scale on the boundary of the capture basin with the trajectories of two viable evolutions of the landing aircraft.*

Figure 3.7 (left and right) shows the west boundary of $\text{Capt}_{(3.2)}(K_{\text{flare}}, C_{\text{touch down}})$, Fig. 3.7 (left) the thrust component and Fig. 3.7 (right) the angle of attack of the feedback map $(V, \gamma, z) \mapsto (T, \alpha) \in R(V, \gamma, z)$ needed on the boundary of $\text{Capt}_{(3.2)}(K_{\text{flare}}, C_{\text{touch down}})$ to govern a safe evolution of the aircraft in K_{flare} until landing at $C_{\text{touch down}}$.

The target $C_{\text{touch down}}$ is not represented explicitly in Fig. 3.7, but one can visually infer it from

$$C_{\text{touch down}} = \text{Capt}_{(3.2)}(K_{\text{flare}}, C_{\text{touch down}}) \cap ([V_{\min}, V_{\max}] \times [\gamma_{\min}, 0] \times \{0\})$$

It is the set of values of (V, γ) such that $\dot{z}_0 \leq V \sin \gamma \leq 0$, which is the almost rectangular slab at the top of the grey bottom of the cube. Reaching any (V, γ) value in this slab represents a safe touch down. Several trajectories are represented in this figure, one starting from the point A, evolving inside the capture basin until it reaches the boundary, at point A_1. At that point, it has to follow the feedback prescribed by the capture basin algorithm, to stay inside K, until it lands safely at the end of the trajectory. The evolution between B and B_1 is of the same type, it reaches the boundary at B_1, and then needs to apply the feedback to reach the target safely.

As can be seen, the thrust needs to be maximal almost everywhere far from the ground, which is intuitive: close to $\gamma = \gamma_{\min}$, high thrust is needed to prevent the aircraft from descending along a too steep flight path angle, or reaching $z = 0$ with a too fast descent velocity. Note that close to the $V = V_{\max}$ boundary of K_{flare}, lower values of T are required, which again is intuitive: in order not to exceed the maximal velocity required by K_{flare}, a low value of thrust must be applied. Finally, close to the ground (low values of z and V), low thrust is required to touch down, since the aircraft is almost on the ground.

The evolutions of the flight parameters $(V(t), \gamma(t), z(t))$, as well as the evolution of the viable feedback $(T(t), \alpha(t))$ are displayed in Figs. 3.8 and 3.9 for the two evolutions emanating from B and A respectively.

1. **Trajectory emanating from** A.

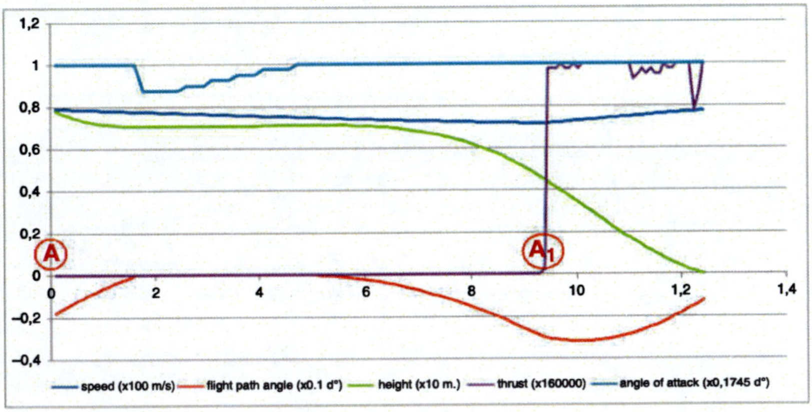

Fig. 3.8 Graph of the evolution starting from A.
Evolutions of the flight parameters and the control variables for the landing starting from state A.

We first describe the trajectory emanating from the point A. This corresponds to a descent at idle (no thrust) until thrust is turned on.
The trajectory evolves in the capture basin $\text{Capt}_{(3.2)}(K_{\text{flare}}, C_{\text{touch down}})$ until a point A_1 where it hits the boundary of that set. The corresponding history of the flight parameters can be seen in Fig. 3.7. As can be seen from this figure, until the trajectory reaches the point A_1, the aircraft is in idle mode (zero thrust), until thrust is turned back on to avoid leaving $\text{Capt}_{(3.2)}(K_{\text{flare}}, C_{\text{touch down}})$ (which would lead to an unsafe landing, i.e. the impossibility of reaching the target while staying in the constraint set). As can be seen from the other flight parameters, speed is almost constant during the landing, altitude decreases with a plateau (corresponding to a zero flight path angle). The sharpest descent happens at the end around A_1. As can be seen from Fig. 3.7 right, as soon as the trajectory hits A_1, it stays along the boundary of the capture basin, i.e. viability is ensured by applying the viability feedback (shown in Fig. 3.7).

2. **Trajectory emanating from B.**

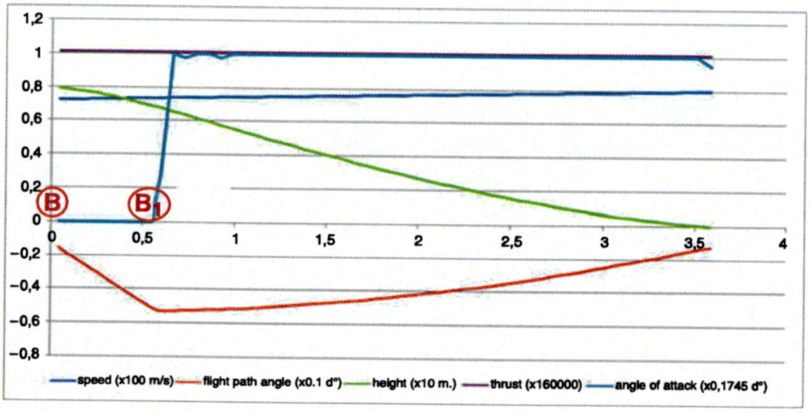

Fig. 3.9 Graph of the evolution starting from B.
Evolutions of the flight parameters and the control variables for the landing starting from state B.

We then describe the trajectory emanating from the point B, visible in Fig. 3.7 right. This corresponds to a descent at constant thrust but zero angle of attack until the angle of attack is raised (flare) to achieve a safe landing. Starting from point B, the trajectory reaches the boundary of $\text{Capt}_{(3.2)}(K_{\text{flare}}, C_{\text{touch down}})$ at point B_1, where it is forced to increase the angle of attack (which will in turn increase lift), to be able to reach the target safely. As can be seen from the temporal histories of the flight parameters (see Fig. 3.9, at B_1, the angle of attack increases, leading to a reduction of the decrease of the flight path angle. This is relatively intuitive, the increased lift due to a higher angle of attack decreases the rate at which the flight path angle decreases (hence after $t = 0.5$, one can see that the flight path angle starts to increase again. The trajectory reaches the target along the boundary of the capture basin, as can be seen from the temporal histories, with a slight increase in speed, and with a monotonic decay in altitude.

These two examples illustrate the power of the viability method used, which is able to adapt the feedback (by varying the thrust and the angle of attack) based on the scenario, in order to preserve viability while descending in the constraint set.

3.4 Path Planing for a Light Autonomous Underwater Vehicle

This section presents the results of the application of the capture basin algorithm to a minimal time problem (see Sect. 4.3, p.132). We are interested in finding the fastest way for a Light Autonomous Underwater Vehicle (LAUV) to reach a target location.

3.4.1 Dynamics, Environment and Target

We treat the submarine as a three degree of freedom planar vehicle. The vehicle has a propeller for longitudinal actuation and fins for lateral and vertical actuation. However, we decouple the actuators and use the vertical actuation only to maintain a constant depth.

We characterize the state of the system with three variables: x, y for earth-fixed East and North coordinates, and ψ for earth-fixed heading. For this application it is acceptable to assume that the effect of currents can be captured by superposition; in other words, the velocity of the surrounding water can simply be added to the velocity of the vehicle due to actuation.

The LAUV is modelled by the tree-dimensional system

$$\begin{cases} (i) \quad x'(t) &= u(t)\cos(\psi(t)) + v_{\mathrm{cx}}(x(t), y(t)) \\ (ii) \quad y'(t) &= u(t)\sin(\psi(t)) + v_{\mathrm{cy}}(x(t), y(t)) \\ (iii) \; \psi'(t) &\in [-r_{\mathbf{max}}, r_{\mathbf{max}}] \end{cases} \tag{3.7}$$

where $v_{\mathrm{cx}}(x, y)$ and $v_{\mathrm{cy}}(x, y)$ are the components of the water velocity.

The deployment area was a 400×300 m rectangle containing the junction of the Sacramento River and the Georgianna Slough in California, USA. Under normal conditions, water flows from North to South, at speeds ranging from 0.5 to 1.5 m/s.

Bathymetric data for the region is available. The channel depth drops steeply from the bank, and is deeper than 2 m at all points away from the shore, so operations can be safely conducted as long as the LAUV does not come within 5 m of the shore.

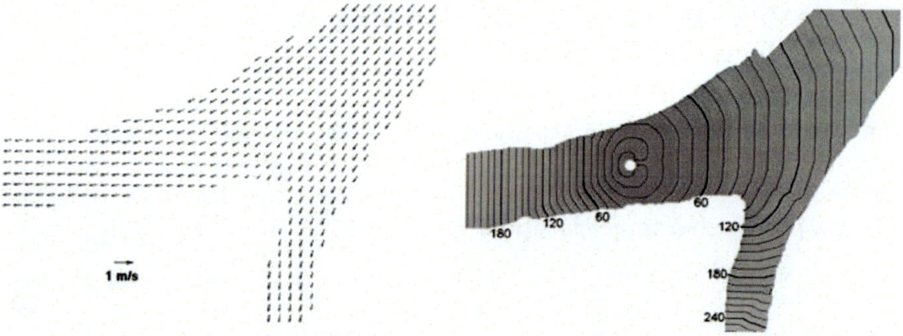

Fig. 3.10 Current Map and Isochrones of the Minimal Time Function.
The left subfigure *displays the currents in the Sacramento river and the Georgianna Slough where the LAUV was deployed for the test. The current is used in the computation of minimum time function, which itself serves to compute the optimal trajectory. Isochrones of the minimal time functions for various starting positions and initial direction East are displayed in the* right subfigure. *Numbered contours give time to target in seconds.*

Figure 3.10, p.120 shows the results of the minimal time computation using the capture basin algorithm for the LAUV model. The effect of the anisotropic current on the minimal time function is clearly visible in the figure showing the isochrones, and the presence of discontinuities in the feedback map is apparent in the eight subfigures showing the retroaction below (see Fig. 3.12, p.122).

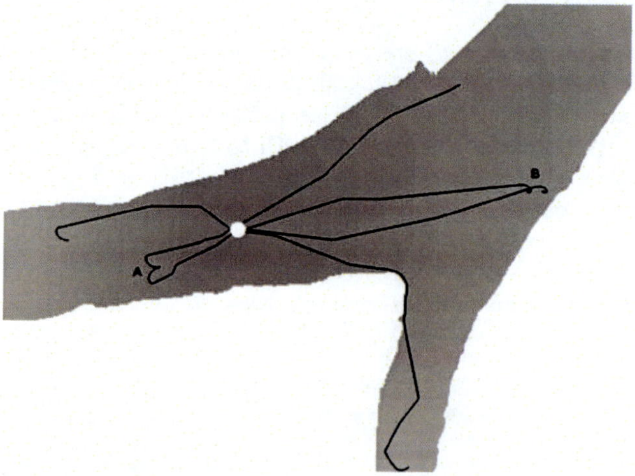

Fig. 3.11 Sample of generated trajectories.
Sample of generated trajectories. Pair "A" have the same starting position, directions differ by 15°. Pair "B" have the same starting direction, positions separated by 10 m.

In the figure above, the trajectories labelled "A" show how a change in initial direction (270° versus 255°) can result in dramatically different actions. The trajectories labelled "B" show how two starting points, separated by 10m, with the same initial direction 90°, can take very different paths to the target.

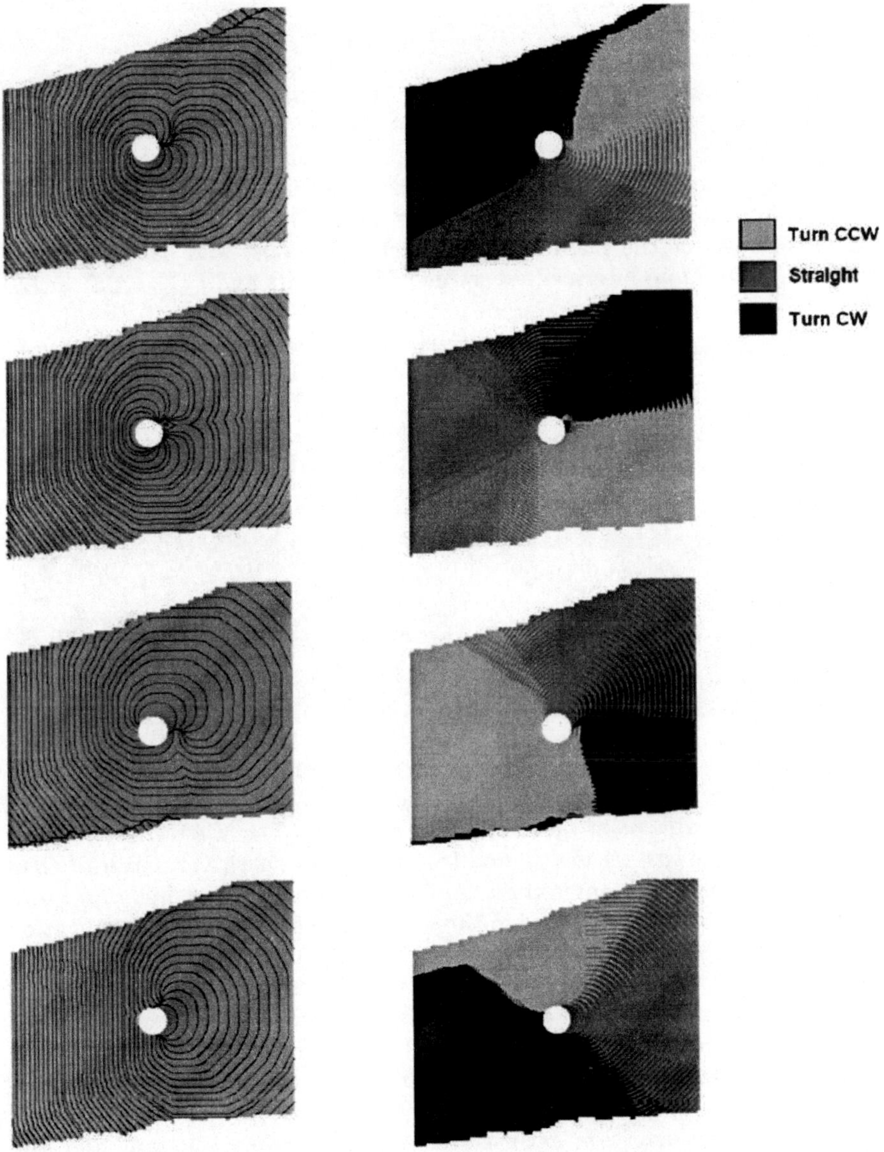

Fig. 3.12 Minimal time function (*left*) and feedback map (*right*).
*Minimal time function (*left*) and feedback map (*right*) for a 100 × 100 m region around the target, and initial direction (from* top*) North, East, South,*

West. The direction feedback map indicates directions of heading change (CW stands for clockwise *and CCW for* counterclockwise*).*

3.4.2 Experimental Results

For the experiment, we use the following values which result from our tests. The maximum speed of the submarine in still water was experimentally determined to be 2 m/s, and at that speed, the maximum turn rate was 0.5 rad/s. By fitting a hydrodynamic model to these values, the turn rate/speed relationship could be estimated for lower speeds. The value 2 m/s was judged to be too fast for the planned experiments, for safety and other logistical reasons. We used an intermediate value of $V = 1\,\mathrm{m/s}$ and $r_{\max} = 0.2\,\mathrm{rad/s}$.

Fig. 3.13 Light Autonomous Underwater Vehicle (LAUV).
The Light Autonomous Underwater Vehicle (LAUV) from Porto University used for the implementation of the algorithm. The LAUV is a small 110 × 16 cm low-cost submarine with a maximum operating depth of 50 m for oceanographic and environmental surveys designed and built at Porto University. It has one propeller and three control fins. The onboard navigation suite includes a Microstrain low-cost inertial measurement unit, a Honeywell depth sensor, a GPS unit and a transponder for acoustic positioning (GPS does not work underwater). This transponder is also used for receiving basic commands from the operator. The LAUV has a miniaturized computer system running modular controllers on a real-time Linux kernel.

The viability algorithm returns the minimum time function and the feedback map for the chosen target. Due to the way the minimum time function is built, it is efficient to calculate the feedback map at the same time. The feedback map is a function that returns the direction command necessary to stay on the optimal trajectory at each position/direction in the viable set. Ideally, the LAUV would use this feedback map directly. In our experiment, we instead use this feedback map to pre-calculate the optimal trajectory. This is done by selecting a starting point and direction, then finding the optimal trajectory from that point by using the feedback map to find successive points. In other words, there is no need to perform a dynamic programming optimization on the minimum time function; the feedback map made the trajectory calculation straightforward.

We ran several experiments with the trajectory data sets provided by the optimal control algorithm. This took place in second week of November 2007. The qualitative behavior of the LAUV did not change significantly across several experiments, and showed good agreement with predicted results.

Figure 3.14, p.123 displays the results for the experiment involving the optimal trajectory planning. In this experiment the LAUV was deployed at the location labelled *start*, where it drifted with the current while waiting for the *startmission* command.

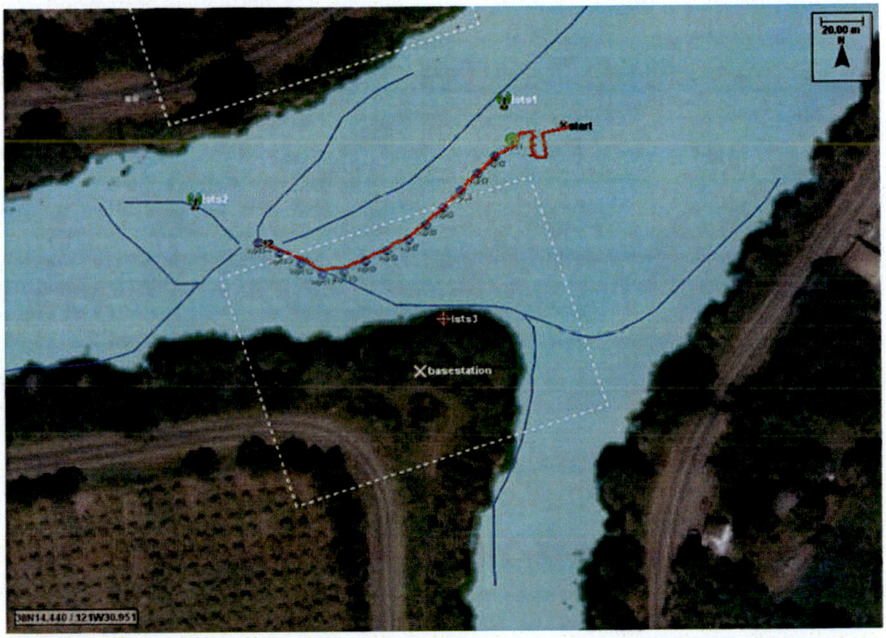

Fig. 3.14 Minimal time evolution of the submarine.
The minimal time evolutions of the submarine are computed by the capture basin algorithm viable in the Georgianna Slough. The submarine starts from the start point and rallies the target following the optimal trajectory obtained using the minimum time function.

Figure 3.15, p.124 shows the deviation between the planned and actual trajectory.

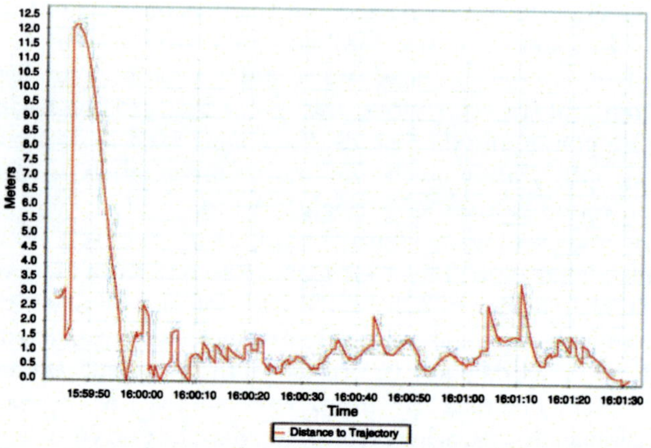

Fig. 3.15 Deviation between planned and actual trajectory.
The deviation between planned and actual trajectory is obtained by comparing the estimated position of the submarine during the operation (by acoustic methods) with the planned position obtained from the viability algorithm.

Chapter 4
Viability and Dynamic Intertemporal Optimality

4.1 Introduction

We consider throughout this chapter the parameterized system (2.10), p.64:

$$x'(t) = f(x(t), u(t)) \text{ where } u(t) \in U(x(t))$$

In this chapter, we denote by $\mathcal{P}$ the evolutionary system associating with any x the set $\mathcal{P}(x)$ of *state-control pairs* $(x(\cdot), u(\cdot))$ governed by the control system and starting from x. This evolutionary system is generated by the parameterized system (2.10), but, in contrast to the evolutionary system $\mathcal{S}$, it associates evolutions of state-control pairs instead of states.

We also introduce a "transient state" cost function (often called a Lagrangian) $\mathbf{l} : X \times \mathcal{U} \mapsto \mathbb{R}_+$ associating with any state-controlled evolution $(x(\cdot), u(\cdot)) \in \mathcal{P}(x)$ and any $t \geq 0$ a positive cost $\mathbf{l}(x(t), u(t)) \geq 0$. Given a discount rate $m \geq 0$, the "discounted cumulated" cost along the evolution is then measured by the integral $\int_0^{+\infty} e^{-m\tau}\mathbf{l}(x(\tau), u(\tau))d\tau$. The *value function*

$$\overset{\infty}{\mathbf{V}}_{\mathbf{l}}(x) := \inf_{(x(\cdot),u(\cdot))\in\mathcal{P}(x)} \int_0^{+\infty} e^{-m\tau}\mathbf{l}(x(\tau), u(\tau))d\tau$$

describes this infinite horizon intertemporal optimization problem.

The objective of this chapter is to find a retroaction governing optimal evolutions which minimize this intertemporal criterion (see Definition 2.7.2, p. 65), and, for that purpose, two miracles happened:

1. the breakthrough uncovered by Hamilton and Jacobi is that the computation of the retroaction involves the value function and its partial derivatives, obtained as a solution to a partial differential equation, called the *Hamilton–Jacobi–Bellman partial differential equation*. This was discovered two centuries ago by *William Hamilton* and *Carl Jacobi* in the

J.-P. Aubin et al., *Viability Theory*, DOI 10.1007/978-3-642-16684-6_4,

framework of the calculus of variations, and adapted half a century ago to control systems by *Richard Bellman* and *Rufus Isaacs*, who extended these concepts to differential games.
Intertemporal optimization is treated in depth in Chap. 17, p. 681. Being the one of the most technical chapter, it is relegated to the end of the book.

Fig. 4.1 Hamilton–Jacobi–Bellman–Isaacs.
Hamilton (Sir William Rowan) [1805–1865], Jacobi (Carl Gustav Jakob) [1804–1851], Bellman (Richard Ernest) [1920–1984] and Isaacs (Rufus Philip) [1914–1981]

2. another breakthrough is that it is possible to bypass this approach by characterizing this value function by a viability kernel and thus, to *derive the retroaction map without solving this partial differential equation.* The way to do this is to associate with the initial control system (2.10):

$$x'(t) \;=\; f(x(t),u(t)) \text{ where } u(t) \;\in\; U(x(t))$$

governing the evolution $x(\cdot)$ of the state, another system governing an auxiliary variable $y(\cdot)$, for most examples, a scalar. In this case, it would be

$$y'(t) \;=\; -\mathbf{l}(x(t),u(t))$$

It would be convenient to regard the initial controlled system (2.10) as a *microsystem*, regulating the evolution of a state variable x, whereas the *macrosystem* regulates the evolution of $y(\cdot)$, regarded as a macro variable. The distinction between "micro" and "macro" is that *the dynamics governing the evolution of the state does not depend on the macro variable,* whereas the evolution of the macro variable depends upon the state variable either through its control law or trough constraints linking micro and macro variables. These systems are also called *structured systems* (see Chap. 13, p.523). The question arises whether these two systems, micro and macro, are *consistent*, in the sense, for example, that the evolutions of the micro and macro variables satisfy imposed links. This is a viability problem.
In economics, for instance, the micro system governs the evolution of the transactions among many economic agents, in a decentralized way, whereas

the macro-variables are monetary ones, under the control of bankers. The agents, whose behaviors are studied in micro-economics, are constrained by fiduciary variables without acting on them, whereas bankers or the abstract concept of "Market", regulating these variables taking into account the behavior of agents.
In engineering, the macro variables are observations, which summarize the behavior of the state variables. Micro–macro systems are prevalent in physics.

We shall "embed" the optimization problem in the framework of a viability/capturability problem of micro–macro targets viable in a micro–macro environment under the auxiliary micro–macro system and prove that the viable evolutions of the auxiliary micro–macro systems generates optimal evolutions of the state variable. Hence *the regulation map governing the evolution of micro–macro variables regulates the evolution of optimal evolutions*, and thus, the solution to our intertemporal optimization problem.

We shall attempt in Sect. 4.12, p. 171 to *clarify* without proofs the links between viability theory on one hand, intertemporal optimization, micro–macro systems and Hamilton–Jacobi–Bellman partial differential equations on the other hand. These links are *established* and proved later in the adequate sections of the book (see Chaps. 13, p.523 and 18.8, p.734).

This is why and how viability theory can bring a new – and unified – light on these intertemporal optimization problems.

With this panoramic landscape described at a high level, we begin our presentation in Sect. 4.2, p. 129 by a review of some examples of intertemporal criteria we shall minimize over evolutions governed by control systems, as well as *extended functions* which "integrate" constraints in the intertemporal criteria, allowing them to take infinite values outside the environment where they were initially defined.

We then proceed in Sect. 4.3, p. 132 with the main example of such functions, the *exit function*, which could have been called a "survival function" since it measures the time spent by an evolution in the environment before leaving it, and the *minimal time function*, measuring the time needed for an evolution to reach a target. We proceed with the exposition of *minimal length functions* measuring the smallest length of an evolution starting from a given initial state viable in the environment. This is the "spatial" counterpart of the concept of exit function, the domain of which is contained in the viability kernel. Minimal length evolutions converge to equilibria of the system, which range over the states where the minimal length function vanishes.

These notions are useful to "quantify" the qualitative concepts of viability kernels and capture basins. We shall characterize them in terms of viability kernels and capture basins, which allows them to enjoy their properties and to be computed.

Section 4.5, p. 142 provides functions which allows us to quantify some of the numerous concepts of *asymptotic stability*, following the road opened to us in 1899 by *Alexandr Lyapunov*. We shall introduce attracting and exponential Lyapunov functions and characterize them in terms of viability kernels.

The next example we present in Sect. 4.6, p. 152 is the one dealing with safety and transgression functions. Starting from inside the environment, the safety function associates with the initial state the worst distance of the states of the evolution to the boundary of the environment, which measures a kind of safety. In a dual way, starting outside the environment, the transgression function measures the worst distance of the states of the evolution to the environment, measuring how far from the environment the evolution will be. We shall address the same theme in Sect. 9.2.2, p. 326 where space is replaced by the time spent by an evolution outside the environment measured by the *crisis function*, which quantifies the concept of *permanence kernel* (see Definition 9.2.3, p. 323).

These are examples of infinite horizon intertemporal optimization problems we present in Sect. 4.7, p. 155, with a summary of their main properties derived from their characterization in terms of viability kernels.

We provide the example of intergenerational optimization in Sect. 4.8, p. 158, one of the useful concepts underlying the notion of sustainable development. At initial time, one minimizes an intertemporal criterion from 0 to $+\infty$, in such a way that for each later time t, the value of the intertemporal criterion from 0 to $+\infty$ must be bounded by a function depending on the state at time t of the evolution.

Finite-horizon intertemporal optimization is paradoxically somewhere more complicated because it involves time as an explicit extra variable. However, the approach is the same, and described in Sect. 4.9, p. 162.

We provide another application to the concept of *occupational measure*, an important tool introduced by physicists for "describing" through a measure the behavior of an evolution governed by a determinist system: knowing any criterion function, its integral under the occupational measure automatically provides the value of this criterion over the path of the evolution. We shall extend this concept to the case when the evolution is governed by a control system. In this case, we associate with any state a closed convex subset of measures, called the *generalized occupational measure*, which allows us to recover directly without solving them the optimal value of the criterion. As can be expected of a topic treated in this book, the concept of occupational measure is related to capture basins.

These are examples of a general theory linking intertemporal optimization theory with viability and capturability concepts under an underlying auxiliary *micro–macro* controlled system on one hand, and with Hamilton–Jacobi–Bellman partial differential equations on the other one. The later links are known for a long time. The novelty consists in the links with viability theory which allows us to add new tools for studying these problems, including the case when the evolution must face viability (or state) constraints.

An "optimal control survival kit" summarizing the main results of the Hamilton–Jacobi–Bellman strategy to study intertemporal optimization is provided in Sect. 4.11, p. 168. These results, quite technical, are presented and proved in full details at the end of the book, in Chap. 17, p. 681.

4.2 Intertemporal Criteria

4.2.1 Examples of Functionals

Naturally, there are as many examples of functionals, and as many intertemporal optimization problems over evolutions governed by controlled systems. In other words, a zoo. So that we face the same difficulties as our zoologist colleagues, to classify those examples.

We shall cover many criteria, some classical, some more original. *At no extra cost*, because all these examples of a whole class of systems share the fundamental pair of properties.

We provide here a list of examples of functionals that we shall optimize over the subset of evolutions governed by an evolutionary system and starting from an initial state. They can be classified under several criteria:

1. **Integral and Non Integral Functionals**
   These functionals are based on evaluation functions $\mathbf{l} : X \mapsto \mathbb{R}_+$ (for example, $\mathbf{l}(x) := \|x\|$). We associate with any evolution $x(\cdot) \in \mathcal{C}(0,\infty;X)$ the associated functional

$$\begin{cases} (i) \;\; \mathbf{J}_{\mathbf{l}}(T;x(\cdot)) := \;\; \mathbf{l}(x(T)) \\ (ii) \;\; \mathbf{J}_{\mathbf{l}}(x(\cdot)) \;\; := \left(\int_0^T \mathbf{l}(x(t))^p dt\right)^{\frac{1}{p}} \\ (iii) \;\; \mathbf{J}_{\mathbf{l}}(x(\cdot)) \;\; := \;\; \sup_{t\in[0,T[} \mathbf{l}(x(t)) \end{cases}$$

   with the favorite $p=1$ (integral functional) or $p=2$ (quadratic functional). The case $p = 2$ is chosen whenever we need to differentiate the functional, because Hilbertian techniques are very well adapted, above all to linear problems.
   The first and second cases are the most natural ones to be formulated, the second one is the simplest one mathematically in a linear world, and the third one, that we can regard as associated with $p = +\infty$, provides non differentiable functionals. The third case has been relatively neglected in the extent that tools of smooth differential and integral analysis and functional analysis are required.
   However, the viability approach does not need to differentiate the value function. Hence viability techniques treat integral functionals and non

integral ones in the same way (our favorite and innovating case in this book being $p = \infty$).
In control theory, the case (i) corresponds to "terminal costs", the case (ii) corresponds to "running costs" and the case (iii) corresponds to a maximal cost, which sometimes appears in the domain of "robust control".
In life sciences, the case (iii) is prevalent, as we shall explain in Chap. 6, p. 199, which is devoted to it.

2. **Combining Integral and Non Integral Functionals Depending on Controls**
In the case of control problems, the functions $\mathbf{l}(x,u)$ can depend upon the controls and can be regarded as a Lagrangian, interpreted as a *transient cost function*, which may involve explicitly the controls $u(\cdot)$, as well as a *spot cost function* function $\mathbf{c}$ and a *environmental cost function* $\mathbf{k}$:

$$\begin{cases} \mathbf{J}_{\mathbf{l}}(x(\cdot)) & := \int_0^T \mathbf{l}(x(t),u(t))dt \\ \mathbf{J}_{\mathbf{l},\mathbf{c}}(T,x(\cdot)) & := \mathbf{c}(x(T)) + \int_0^T \mathbf{l}(x(t),u(t))dt \\ \mathbf{I}_{\mathbf{l},\mathbf{k}}(T;x(\cdot)) & := \sup_{t\in[0,T]} \left(\mathbf{k}(x(t),u(t)) + \int_0^t \mathbf{l}(x(\tau),u(\tau))d\tau\right) \\ \mathbf{L}_{\mathbf{l},\mathbf{k},\mathbf{c}}(T,x(\cdot)) & := \max\left(\mathbf{I}_{\mathbf{l},\mathbf{k}}(T;x(\cdot)), \mathbf{J}_{\mathbf{l},\mathbf{c}}(T,x(\cdot))\right) \end{cases}$$

and so on... See Definition 17.3.2, p. 687 for the largest class of functionals that we shall study in depth in this book. However, the larger the class, the more technical and less intuitive the formulas... This is the reason why we postpone this general study at the end of the book.

3. **Versatility and volatility**
They are functionals on derivatives of (differentiable) evolutions. We define the *versatility* of the evolution with respect to $\mathbf{l}$ on the interval $[0,T[$ by

$$\mathrm{Vers}_{\mathbf{l}}(x(\cdot)) := \sup_{t\in[0,T[} \mathbf{l}(x'(t)) \tag{4.1}$$

and the *volatility* of the evolution with respect to $\mathbf{l}$ by

$$\mathrm{Vol}_{\mathbf{l}}(x(\cdot)) := \left(\frac{1}{T}\int_0^T \mathbf{l}(x'(t))^2 dt\right)^{\frac{1}{2}}$$

For example, taking $\mathbf{l}(x) := \|x\|$, the versatility and volatility are

$$\begin{cases} (i)\ \mathrm{Vers}_{\mathbf{l}}(x(\cdot)) := \sup_{t\in[0,T[} \|x'(t)\| \\ (ii)\ \mathrm{Vol}_{\mathbf{l}}(x(\cdot)) := \left(\frac{1}{T}\int_0^T \|x'(t)-m\|^2 dt\right)^{\frac{1}{2}} \end{cases}$$

For a control system, the versatility of the control variable is called the *inertia*, and the minimal inertia over viable evolutions is called the *inertia function*. For a tychastic system, the versatility of the tychastic variable is called the *palikinesia* and the minimal palikinesia over viable evolutions is called the *palikinesia function.*

4.2.2 Extended Functions and Hidden Constraints

The above examples of intertemporal criteria functions do not take into account the environment K over which they are defined. The trick for involving constraints into the cost functions $\mathbf{k}$ and $\mathbf{c}$, for instance, is, in a minimization framework, to assign the value $+\infty$ (infinite cost) outside the environment, since no "minimizer" will choose a state with an infinite cost. Hence if we consider an environment $K \subset X$ and a function $\mathbf{v}_K : K \mapsto \mathbb{R}$ (with finite values), we can extend it to the function $\mathbf{v} : X \mapsto \mathbb{R} \cup \{+\infty\}$ defined on the whole space, but taking finite and infinite values: $\mathbf{v}(x) := \mathbf{v}_K(x)$ whenever $x \in K$, $\mathbf{v}(x) := +\infty$ whenever $x \notin K$. These numerical functions taking infinite values, which will be extensively used in this book, are called *extended functions.*

For that purpose, we use the convention $\inf\{\emptyset\} := +\infty$ and $\sup\{\emptyset\} := -\infty$.

Definition 4.2.1 *[Extended Functions] A function* $\mathbf{v} : X \mapsto \overline{\mathbb{R}} := \mathbb{R} \cup \{-\infty\} \cup \{+\infty\}$ *is said to be an* extended function. *Its* domain $\mathrm{Dom}(\mathbf{v})$ *defined by*

$$\mathrm{Dom}(\mathbf{v}) := \{x \in X \mid -\infty < \mathbf{v}(x) < +\infty\}$$

is the set of elements on which the function is finite.

The domain of an extended function incorporates implicitly state constraints hidden in the extended character of the function $\mathbf{v}$.

We refer to Sect. 18.6, p. 742 for further details on epigraphical analysis. It was discovered in the 1960s with the development of convex analysis founded by *Moritz Fenchel, Jean Jacques Moreau* and *Terry Rockafellar* that numerous properties relevant to the optimization of general functionals, involving the order relation of $\mathbb{R}$ and inequalities, are read through their epigraphs or hypographs:

Definition 4.2.2 *[Epigraph of a Function] Let* $\mathbf{v} : X \mapsto \overline{\mathbb{R}}$ *be an extended function.*

1. *Its* epigraph $\mathcal{E}p(\mathbf{v})$ *is the set of pairs* $(x, y) \in X \times \mathbb{R}$ *satisfying* $\mathbf{v}(x) \leq y$.

2. *Its* hypograph $\mathcal{H}yp(\mathbf{v})$ *is the set of pairs* $(x, y) \in X \times \mathbb{R}$ *satisfying* $\mathbf{v}(x) \geq y$.
3. *Its* graph $\text{Graph}(\mathbf{v})$ *is the set of pairs* $(x, y) \in \text{Dom}(\mathbf{v}) \times \mathbb{R}$ *satisfying* $\mathbf{v}(x) = y$.

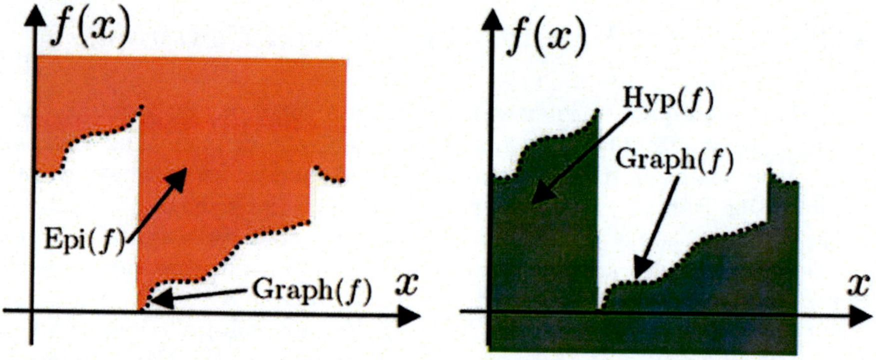

Fig. 4.2 Epigraph and hypograph.
Illustrations of the epigraph (left) *and the hypograph* (right) *of an Extended Function. The epigraph is the set of points above the graph. The hypograph is the set below the graph.*

For more details on tubes and their level sets, see Sect. 10.9, p. 427.

4.3 Exit and Minimal Time Functions

4.3.1 Viability and Capturability Tubes

Until now, we studied viability without time horizon. However, it is reasonable to also study the case in which the concepts of viability and capturability depend on time. They provide examples of tubes.

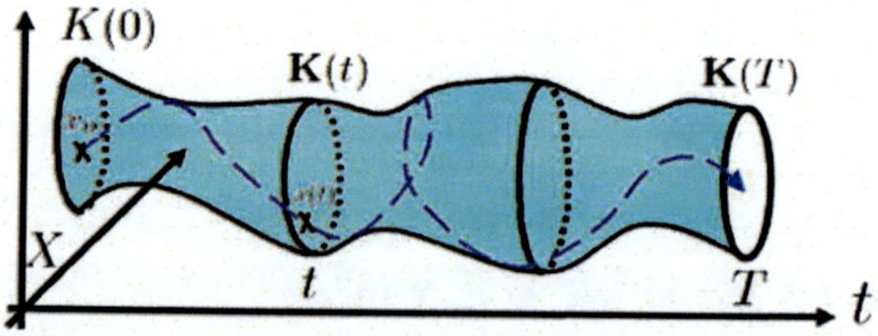

Fig. 4.3 Illustration of a tube.
Tubes are *nicknames for "set-valued evolutions"* $\mathbf{K} : t \in \mathbb{R} \rightsquigarrow \mathbf{K}(t) \subset X$.

We begin here the study of time-dependent viability kernels and capture basins, so to speak. We shall pursue the thorough study of tubes in Chap. 8, p. 273, and in particular, the study of reachable maps (see Proposition 8.4.3, p. 286) and detectors (see Theorem 8.10.6, p. 314).

Definition 4.3.1 ***[Viability, Capturability and Exact Capturability Tubes]*** *Let $\mathcal{S} : X \rightsquigarrow \mathcal{C}(0, +\infty; X)$ be an evolutionary system and C and K be two closed subsets such that $C \subset K$. The T-viability kernels, the T-capture basins and the T-exact capture basin are defined in the following way:*

1. *the T-viability kernel $\mathrm{Viab}_{\mathcal{S}}(K)(T)$ of K under $\mathcal{S}$ is the set of elements $x \in K$ from which starts at least one evolution $x(\cdot) \in \mathcal{S}(x)$ viable in K on the interval $[0, T]$,*
2. *the T-capture basin $\mathrm{Capt}_{\mathcal{S}}(K, C)(T)$ is the set of elements $x \in K$ from which starts at least one evolution $x(\cdot) \in \mathcal{S}(x)$ viable in K until it reaches the target C before time T.*
3. *the T-exact capture basin $\widehat{\mathrm{Capt}}_{\mathcal{S}}(K, C)(T)$ is the set of elements $x \in K$ from which starts at least one evolution $x(\cdot) \in \mathcal{S}(x)$ viable in K until it reaches the target C at exactly time T.*

We shall say that the set-valued maps $T \rightsquigarrow \mathrm{Viab}_{\mathcal{S}}(K)(T)$, $T \rightsquigarrow \mathrm{Capt}_{\mathcal{S}}(K, C)(T)$ and $T \rightsquigarrow \widehat{\mathrm{Capt}}_{\mathcal{S}}(K, C)(T)$ are respectively the viability tube, *the* capturability tube *and the* exact capturability tube.

We can characterize the graphs of these viability and capturability tubes as viability kernels (actually, capture basin):

Theorem 4.3.2 ***[Graphs of Viability and Capturability Tubes]*** *Let us consider*

$$\begin{cases} (i) \ \ \tau'(t) = -1 \\ (ii) \ x'(t) = f(x(t), u(t)) \\ \qquad \text{where } u(t) \in U(x(t)) \end{cases} \tag{4.2}$$

1. *The graph of the viability tube $\mathrm{Viab}_{\mathcal{S}}(K)(\cdot)$ is the capture basin of $\{0\} \times K$ viable in $\mathbb{R}_+ \times K$ under the system (4.2):*

$$\mathrm{Graph}(\mathrm{Viab}_{\mathcal{S}}(K)(\cdot)) = \mathrm{Capt}_{(4.2)}(\mathbb{R}_+ \times K, \{0\} \times K)$$

2. *The graph of the capturability tube $\mathrm{Capt}_{\mathcal{S}}(K, C)(\cdot)$ is the capture basin of $\mathbb{R}_+ \times C$ viable in $\mathbb{R}_+ \times K$ under the system (4.2):*

$$\mathrm{Graph}(\mathrm{Capt}_{\mathcal{S}}(K,C)(\cdot)) = \mathrm{Capt}_{(4.2)}(\mathbb{R}_{+}\times K, \mathbb{R}_{+}\times C)$$

3. *The graph of the exact capturability tube* $\widehat{\mathrm{Capt}}_{\mathcal{S}}(K,C)(\cdot)$ *is the capture basin of* $\{0\}\times C$ *viable in* $\mathbb{R}_{+}\times K$ *under the system (4.2):*

$$\mathrm{Graph}(\widehat{\mathrm{Capt}}_{\mathcal{S}}(K,C)(\cdot)) = \mathrm{Capt}_{(4.2)}(\mathbb{R}_{+}\times K, \{0\}\times C)$$

Proof.

1. to say that (T,x) belongs to the capture basin of $\mathbb{R}_{+}\times K$ with target $\{0\}\times K$ under evolutionary system (4.2) amounts to saying that there exists an evolution $x(\cdot)\in\mathcal{S}(x)$ starting at x such that $(T-t,x(t))$ is viable in $\mathbb{R}_{+}\times K$ forever or until it reaches $\{0\}\times K$ at some time $t^{\star}$. But $T-t$ leaves $\mathbb{R}_{+}$ at time T and the solution reaches the target $\{0\}\times K$ at time $t=T$. This means that $x(\cdot)\in\mathcal{S}(x)$ is a solution to the evolutionary system viable in K on the interval $[0,T]$, i.e., that x belongs to $\mathrm{Viab}_{\mathcal{S}}(K)(T)$
2. to say that (T,x) belongs to the capture basin of $\mathbb{R}_{+}\times K$ with target $\mathbb{R}_{+}\times C$ under auxiliary system (4.2) amounts to saying that there exists an evolution $x(\cdot)\in\mathcal{S}(x)$ starting at x such that $(T-t,x(t))$ is viable in $\mathbb{R}_{+}\times K$ until it reaches $(T-s,x(s))\in\mathbb{R}_{+}\times C$ at time s. Since $T-s\geq 0$, this means that $x(\cdot)$ is an evolution to the evolutionary system $\mathcal{S}(x)$ viable in K on the interval $[0,s]$ and that $x(s)\in C$, i.e., that x belongs to $\mathrm{Capt}_{\mathcal{S}}(K,C)(T)$
3. to say that (T,x) belongs to the viability kernel of $\mathbb{R}_{+}\times K$ with target $\{0\}\times C$ under auxiliary system (4.2) amounts to saying that there exists an evolution $x(\cdot)\in\mathcal{S}(x)$ starting at x such that $(T-t,x(t))$ is viable in $\mathbb{R}_{+}\times K$ forever or until it reaches $(T-s,x(s))\in\{0\}\times C$ at time s. Since $T-s=0$, this means that $x(\cdot)$ is an evolution in $\mathcal{S}(x)$ viable in K on the interval $[0,T]$ and that $x(T)\in C$, i.e., that x belongs to $\widehat{\mathrm{Capt}}_{\mathcal{S}}(K,C)(T)$ □

Hence the graphs of the viability, capturability and exact capturability tubes inherit the general properties of capture basins.

We refer to Sects. 4.3, p. 132 and 10.4, p. 392 for other characterizations of these tubes.

4.3.2 Exit and Minimal Time Functions

The inverse of the tubes $T\rightsquigarrow \mathrm{Viab}_{\mathcal{S}}(K)(T)$ and $T\rightsquigarrow \mathrm{Capt}_{\mathcal{S}}(K,C)(T)$ are respectively the hypograph of the exit function and the epigraph of the minimal time function:

Definition 4.3.3 *[Exit and Minimal Time Functionals] Let $K \subset X$ be a subset.*

1. *The functional $\tau_K : \mathcal{C}(0,+\infty;X) \mapsto \mathbb{R}_+ \cup \{+\infty\}$ associating with $x(\cdot)$ its* exit time $\tau_K(x(\cdot))$ *defined by*

$$\tau_K(x(\cdot)) \ := \ \inf\{t \in [0,\infty[\ | \ x(t) \notin K\}$$

is called the exit functional.

2. *Let $C \subset K$ be a target. We introduce the* (constrained) minimal time functional *(or* minimal time*)* $\varpi_{(K,C)}$ *defined by*

$$\varpi_{(K,C)}(x(\cdot)) \ := \ \inf\{t \geq 0 \ | \ x(t) \in C \ \& \ \forall s \in [0,t], \ x(s) \in K\}$$

associating with $x(\cdot)$ its minimal time. *If $K = X$ is the entire space, we set $\varpi_C(x(\cdot)) := \varpi_{(X,C)}(x(\cdot))$*

We observe that

$$\tau_K(x(\cdot)) = \varpi_{\complement K}(x(\cdot))$$

These being defined, we apply these functionals to evolutions provided by an evolutionary system:

Definition 4.3.4 *[Exit and Minimal Time Functions] Consider an evolutionary system $\mathcal{S} : X \rightsquigarrow \mathcal{C}(0,+\infty;X)$.*

Let $K \subset X$ and $C \subset K$ be two subsets.

1. *The (extended) function $\tau_K^\sharp : K \mapsto \mathbb{R}_+ \cup \{+\infty\}$ defined by*

$$\tau_K^\sharp(x) := \sup_{x(\cdot)\in\mathcal{S}(x)} \tau_K(x(\cdot))$$

is called the (upper) exit function *(instead of functional): see Fig. 5.11, p. 190.*

2. *The (extended) function $\varpi^\flat_{(K,C)} : K \mapsto \mathbb{R}_+ \cup \{+\infty\}$ defined by*

$$\varpi^\flat_{(K,C)}(x) := \inf_{x(\cdot)\in\mathcal{S}(x)} \varpi_{(K,C)}(x(\cdot))$$

is called the (lower constrained) minimal time function *(see Fig. 5.12, p. 192).*

These links can be interpreted by saying that the *exit functions* $\tau_K^\sharp$ and the *minimal time functions* $\varpi_{(K,C)}^\flat$ "quantify" the concepts of viability kernels and capture basins. Proposition 10.4.5, p. 394 states that they are actually equalities (10.11), p. 395 whenever the evolutionary system is upper semicompact and that they are respectively upper semicontinuous and lower semicontinuous without further assumptions. *Concerning the capture basins* $\mathrm{Capt}_S(K,C)$, *we need to assume that* $K \setminus C$ *is a repeller to guarantee that the capture basin is closed (see Theorem 10.3.10, p. 388) although it is* always *the domain of an upper semicontinuous function* (whenever the evolutionary system is upper semicompact). This is a weaker property than closedness, but still, carries some useful information. □

These links between these concepts are illustrated in the two following figures:

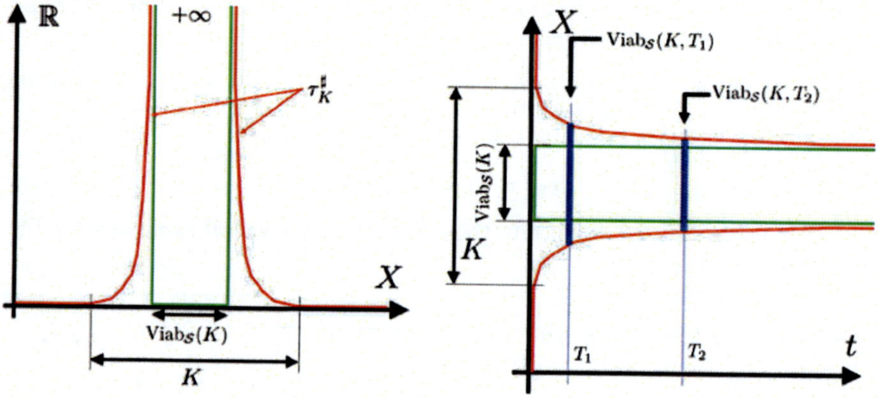

Fig. 4.4 Hypograph of an upper exit function and graph of the viability tube.

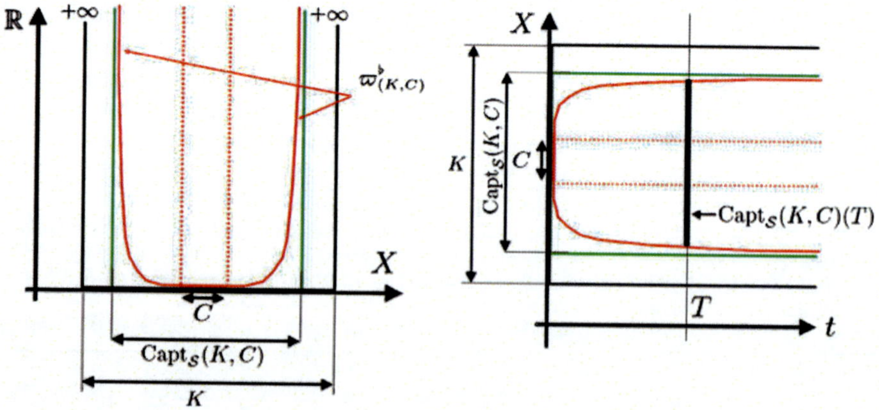

Fig. 4.5 Epigraph of a lower minimal time function and graph of the capturability tube.

Proposition 4.3.5 *[Elementary Properties of Exit and Minimal Time functions]*

1. *Behavior under translation: Setting* $(\kappa(-s)x(\cdot))(t) := x(t+s)$,

$$\forall s \in [0, \tau_K(x(\cdot))],\ \tau_K((\kappa(-s)x(\cdot))(\cdot)) = \tau_K(x(\cdot)) - s \qquad (4.3)$$

2. *Monotonicity Properties: If* $K_1 \subset K_2$, *then* $\tau_{K_1}(x(\cdot)) \leq \tau_{K_2}(x(\cdot))$ *and if furthermore,* $C_1 \supset C_2$, *then* $\varpi_{(K_1,C_1)}(x(\cdot)) \leq \varpi_{(K_2,C_2)}(x(\cdot))$
3. *Behavior under union and intersection:*

$$\tau_{\bigcap_{i=1}^n K_i}(x(\cdot)) = \min_{i=1,\ldots,n} \tau_{K_i}(x(\cdot)) \ \&\ \varpi_{\bigcup_{i=1}^n C_i}(x(\cdot)) = \min_{i=1,\ldots,n} \varpi_{C_i}(x(\cdot))$$

4. *Exit time of the complement of a target in the environment:*

$$\forall x \in K \backslash C,\ \tau_{K\backslash C}(x(\cdot)) = \min(\varpi_C(x(\cdot), \tau_K(x(\cdot))))$$

5. *Behavior under product:*

$$\tau_{\prod_{i=1}^n K_i}(x_1(\cdot), \ldots, x_n(\cdot)) = \min_{i=1,\ldots,n} \tau_{K_i}(x_i(\cdot))$$

Proof. The first two properties being obvious, we note that the third holds true since the infimum on a finite union of subsets is the minimum of the infima on each subsets by Lemma 18.2.7, p.718. Therefore, the fourth one follows from

$$\begin{cases} \tau_{K\backslash C}(x(\cdot)) = \varpi_{\complement(K\backslash C)}(x(\cdot)) = \varpi_{C \cup \complement K}(x(\cdot)) = \\ \min(\varpi_C(x(\cdot)), \varpi_{\complement K}(x(\cdot))) = \min(\varpi_C(x(\cdot), \tau_K(x(\cdot)))) \end{cases}$$

Observing that when $K := K_1 \times \cdots \times K_n := \prod_{i=1}^n K_i$ where the environments $K_i \subset X_i$ are subsets of vector spaces X_i,

$$\complement\left(\prod_{j=1}^n K_j\right) = \bigcup_{j=1}^n \left(\prod_{i=1}^{j-1} X_i \times \complement K_j \times \prod_{l=j+1}^n X_l\right)$$

the last formula follows from

$$\begin{aligned}\tau_{\prod_{i=1}^n K_i}(x_1(\cdot),\ldots,x_n(\cdot)) &:= \inf\{t\geq 0 \mid x(t)\in \complement K\}\\ &= \min_{j=1,\ldots,n}\left(\inf\{t|x_j(t)\in\complement K_j\}\right)\\ &= \min_{j=1,\ldots,n}\tau_{K_j}(x_j(\cdot))\end{aligned}$$ □

We recall that $\mathcal{S}$ is the evolutionary system generated by control system $x'(t) = f(x(t),u(t))$ where $u(t)\in U(x(t))$ and we introduce the auxiliary micro–macro system

$$\begin{cases} (i)\ x'(t) = f(x(t),u(t)) \text{ where } u(t)\in U(x(t))\\ (ii)\ \tau'(t) = -1\end{cases} \tag{4.4}$$

We now prove that exit and minimal time functions are related to viability and capturability tube:

Theorem 4.3.6 *[Viability Characterization of Minimal Time and Exit Functions] The exit and minimal time functions are respectively equal to*

1. *The exit function $\tau_K^\sharp(\cdot)$, related to the viability kernel by the following formula*

$$\tau_K^\sharp(x) = \sup_{(x,T)\in\mathrm{Capt}_{(4.4)}(K\times\mathbb{R}_+,K\times\{0\})} T = \sup_{(T,x)\in\mathrm{Graph}(\mathrm{Viab}_{\mathcal{S}}(K)(\cdot))} T \tag{4.5}$$

2. *The minimal time function $\varpi_{K,C}^\flat(\cdot)$, related to the capture basin by the following formula*

$$\begin{cases}\varpi_{(K,C)}^\flat(x) = \inf_{(x,T)\in\mathrm{Capt}_{(4.4)}(K\times\mathbb{R}_+,C\times\mathbb{R}_+)} T\\ = \inf_{(T,x)\in\mathrm{Graph}(\mathrm{Capt}_{\mathcal{S}}(K,C))(\cdot)} T\end{cases} \tag{4.6}$$

Proof. This is a consequence of Theorem 4.9.2, p. 163 with $\mathbf{l}(x,u) := 1$. □

Remark. Inclusions

$$\begin{cases}\mathrm{Viab}_{\mathcal{S}}(K)(T) \subset \left\{x\in K \mid \tau_K^\sharp(x)\geq T\right\}\\ \\ \mathrm{Capt}_{\mathcal{S}}(K,C)(T)\subset\left\{x\in X\mid \varpi_{(K,C)}^\flat(x)\leq T\right\}\end{cases}$$

are obviously always true, so that inequalities below follow

$$\begin{cases} (i) \ \tau_K^\sharp(x) \ \geq \ \sup_{(T,x)\in\text{Graph}(\text{Viab}_S(K)(\cdot))} T \\ (ii) \ \varpi_{(K,C)}^\flat(x) \ \leq \ \inf_{(T,x)\in\text{Graph}(\text{Capt}_S(K,C)(\cdot))} T \end{cases} \tag{4.7}$$

Let us prove that for any $S < \tau_K^\sharp(x)$, $S \leq \sup_{(T,x)\in\text{Graph}(\text{Viab}_S(K))(\cdot)} T$, so that equality (4.7)(i) will be satisfied. By the very definition of the supremum, there exists $x(\cdot) \in \mathcal{S}(x)$ such that $S < \tau_K(x(\cdot))$, and thus, such that $x(\cdot)$ is viable in K on the interval $[0, S]$, i.e., such that $x \in \text{Viab}_S(K)(S)$. This implies that $S \leq \sup_{(T,x)\in\text{Graph}(\text{Viab}_S(K))(\cdot)} T$.

The proof of inequality (4.7)(ii) is analogous. □

Remark: The associated Hamilton–Jacobi–Bellman Equation. — Denote by $f_i(x,u)$ the ith component of $f(x,u)$. The exit and minimal time functions $\tau_K^\sharp$ and $\varpi_{(K,C)}^\flat$ are the largest positive solutions to the Hamilton–Jacobi–Bellman partial differential equation

$$\inf_{u\in U(x)} \sum_{i=1}^{n} \frac{\partial \mathbf{v}(x)}{\partial x_i} f_i(x,u) + 1 = 0$$

satisfying respectively the conditions $\mathbf{v} = 0$ on $\complement K$ for the exit function and that $\mathbf{v} = 0$ on C and $\mathbf{v} = +\infty$ on $\complement K$ for the minimal time function. □

The study of exit and minimal time functions, which inherit the properties of capture basins, continues in Sect. 10.4, p. 392. Among them, we shall prove the fact that the supremum in the definition of the exit function $\tau_K^\sharp$ and the infimum in the definition of the minimal time function $\varpi_{(K,C)}^\flat$ are achieved by evolutions, respectively called *persistent evolutions* and *minimal time evolutions* (see Definitions 10.4.2, p. 393 and 10.4.3, p. 393 and Theorem 10.4.4, p. 394 as well as Sect. 10.4.3, p. 396 on exit sets $\text{Exit}_S(K) := \left\{x \in K \text{ such that } \tau_K^\sharp(x) = 0\right\}$).

4.4 Minimal Length Function

A subsequent question related to viability is to find a viable evolution minimizing its length

$$\mathbf{J}(x(\cdot)) \ := \ \int_0^\infty \|x'(\tau)\| d\tau$$

among the evolutions starting from an initial state $x \in K$ and viable in K. Viable geodesics connecting $y \in K$ to $z \in K$ are the viable evolutions connecting them with minimal length (see Sect. 8.5.1.1, p.292).

Definition 4.4.1 ***[Minimal Length Function]*** *The* minimal length function $\gamma_K(x) : X \mapsto \mathbb{R}_+ \cup \{+\infty\}$ *associates with* $x(\cdot)$ *its* minimal length *defined by*

$$\gamma_K(x) := \inf_{x(\cdot)\in\mathcal{S}^K(x)} \int_0^{\infty} \|x'(\tau)\| d\tau$$

where $\mathcal{S}^K(x)$ *denotes the set of evolutions starting from* $x \in K$ *and viable in* K*. The domain of the minimal length function can be regarded as the "minimal length viability kernel".*

The lower level sets $\{x \in K \mid \gamma_K(x) \leq \lambda\}$ of the minimal length function is the set of initial states x from which starts a viable evolution of prescribed length smaller than or equal to λ.

The minimal length function is characterized in terms of the viability kernel of an auxiliary micro–macro system:

$$\begin{cases} (i) \;\; x'(t) = f(x(t), u(t)) \\ (ii) \; y'(t) = -\|f(x(t), u(t))\| \\ \quad\;\; \text{where } u(t) \in U(x(t)) \end{cases} \tag{4.8}$$

subject to the constraint

$$\forall t \geq 0, \;\; (x(t), y(t)) \in K \times \mathbb{R}_+$$

Theorem 4.4.2 ***[Viability Characterization of the Minimal Length Function]*** *The minimal length function is related to the viability kernel of* $K \times \mathbb{R}_+$ *under the auxiliary micro–macro system (4.8) by the following formula*

$$\gamma_K(x) = \inf_{(x,y)\in \text{Viab}_{(4.8)}(K\times\mathbb{R}_+)} y$$

Its domain is the viability kernel $\text{Viab}_{\mathcal{S}}(K)$ *of the subset* K *and the equilibrium set coincides with evolutions of length equal to zero (see Fig. 5.13, p. 194).*

Proof. This is a consequence of Theorem 4.7.2, p. 156 with $\mathbf{l}(x, u) := \|f(x(t), u(t))\|$. □

We observe that equilibria are evolutions of length equal to 0, and thus, are located in the finite length viability kernel. But there is no reason why should an equilibrium exists in this set. However, we shall prove in Sect. 9.5, p.360 that this is the case when the control system is Marchaud.

Theorem 4.4.3 ***[Existence of an Equilibrium in the Finite Length Viability Kernel]*** *Assume that the control system (f, U) is Marchaud, that the environment K is closed and that the finite length viability kernel is not empty. Then the control system has an equilibrium $(\overline{x}, \overline{u}) \in \mathrm{Graph}(U)$ such that $f(\overline{x}, \overline{u}) = 0$.*

Proof. Theorem 4.4.2, p.140 states that the epigraph of the minimum length function is a viability kernel. Since the control system is Marchaud, the viability kernel is closed and thus, the minimum length function is lower semicontinuous (see Theorem 17.3.6, p.692). Furthermore, Theorem 19.4.1, p.781 on necessary condition implies that

$$\begin{cases} \forall\,(x, \gamma_K(x)) \in \mathcal{E}p(\gamma_K) \;=\; \mathrm{Viab}_{(4.8)}(K \times \mathbb{R}_+),\ \exists\, u \,\in\, U(x) \ \text{ such that} \\ (f(x,u), -\|f(x,u)\|) \,\in\, T_{\mathcal{E}p(\gamma_K)}(x, \gamma_K(x)) \;= \mathcal{E}p(D_\uparrow \gamma_K(x)) \end{cases}$$

This implies that

$$\forall\, x \in \mathrm{Dom}(\gamma_K),\ \exists\ u \,\in\, U(x) \ \text{ such that } D_\uparrow \gamma_K(x)(f(x,u)) \;\le\; -\|f(x,u)\| \tag{4.9}$$

Since the minimal length function γ_K is lower semicontinuous and positive, bounded on the closed subset K, Theorem 18.6.15, p.751 (Ekeland's variational principe) states that there exists $x_\varepsilon \,\in\, X$ satisfying property (18.20)(ii), p.751:

$$\forall\, v \,\in\, X,\ 0 \,\le\, D_\uparrow V(x_\varepsilon)(v) + \varepsilon\|v\| \tag{4.10}$$

Take now $\varepsilon \in]0,1[$. Hence, choosing x_ε in inequality (4.9), p.141, taking $u_\varepsilon \in U(x_\varepsilon)$ in inequality (4.10) and setting $v_\varepsilon := f(x_\varepsilon, u_\varepsilon)$, we infer that

$$0 \,\le\, D_\uparrow V(x_\varepsilon)(v_\varepsilon) + \varepsilon\|v_\varepsilon\| \,\le\, (\varepsilon - 1)\|v_\varepsilon\|$$

Consequently, $v_\varepsilon := f(x_\varepsilon, u_\varepsilon) = 0$, so that x_ε is an equilibrium. □

Remark: The *"equilibrium alternative"*. Theorem 4.4.3, p.141 can be restated by saying that whenever the system is Marchaud and K is viable,

1. either there exists a viable equilibrium in K;
2. or the length of all evolutions is infinite □

Proposition 4.4.4 ***[Asymptotic Properties of Minimum Length Evolutions.]*** *Let $x(\cdot) \,\in\, \mathcal{S}_{(f,U)}(x)$ be a minimum length evolution. Then $\lim_{t \mapsto +\infty} x(t)$ exists and is an equilibrium.*

Proof. Viable evolutions $(x(\cdot), \gamma_K(x(\cdot)))$ regulated by the auxiliary system (4.8), p.140 and starting from $(x_0, \gamma_K(x_0))$ satisfy

$$\forall\, t \geq s \geq 0, \;\; \gamma_K(x(t)) \leq \gamma_K(x(s)) - \int_s^t \|x'(\tau)\| d\tau \;\leq\; \gamma_K(x(s))$$

Hence $t \mapsto \gamma_K(x(t))$ is decreasing and bounded below by 0, so that $\gamma_K(x(t))$ converges to some $\alpha \geq 0$ when $t \to +\infty$. Consequently,

$$\|x(t) - x(s)\| \;=\; \left\| \int_s^t x'(\tau) d\tau \right\| \;\leq\; \int_s^t \|x'(\tau)\| d\tau \;\leq\; \gamma_K(x(s)) - \gamma_K(x(t))$$

converges to $\alpha - \alpha = 0$ when $t \geq s$ goes to $+\infty$. Hence the Cauchy criterion implies that the sequence $x(t)$ converges to some $\overline{x}$ when t goes to $+\infty$. The limit set of $x(\cdot)$ being a singleton, Theorem 9.3.11, p.351, implies that it is viable, and thus, an equilibrium. □

Remark: The associated Hamilton–Jacobi–Bellman Equation. The **minimal length function** γ_K is the smallest positive solution to the Hamilton–Jacobi–Bellman partial differential equation

$$\forall x \in K, \;\; \inf_{u \in U(x)} \left(\sum_{i=1}^{n} \frac{\partial \mathbf{v}(x)}{\partial x_i} f_i(x,u) + \|f(x,u)\| \right) = 0 \;\; \square$$

4.5 Attracting and Exponential Lyapunov for Asymptotic Stability

4.5.1 Attracting Functions

Let us consider a target $C \subset K$. We denote by $d_{(K,C)}$ the *constrained distance* to the subset C in K defined by

$$d_{(K,C)}(x) \;:=\; \begin{cases} \inf_{y \in C} \|x - y\| & \text{if } x \in K \\ +\infty & \text{if } x \notin K \end{cases}$$

We set $d_C(x) := d_{(X,C)}(x)$ when the environment is the whole space.

We say that a subset C is clustering an evolution $x(\cdot)$ in K if there exists a subsequence $d_{(K,C)}(x(t_n))$ converging to 0, i.e., such that

$$\liminf_{t \to +\infty} d_{(K,C)}(x(t)) = 0$$

A sufficient condition implying this clustering property is provided by the following elementary lemma:

Lemma 4.5.1 *[Integral of a Positive Lagrangian] Let us consider a positive continuous function* $\mathbf{l} : X \mapsto \mathbb{R} \cup \{+\infty\}$ *independent of the control and an evolution* $x(\cdot)$ *such that* $\int_0^{+\infty} \mathbf{l}(x(t))dt < +\infty$. *Hence*

$$\forall t \geq 0, \ \mathbf{l}(x(t)) < +\infty \text{ and } \liminf_{t \to +\infty} \mathbf{l}(x(t)) = 0$$

Proof. Assume that $\int_0^{\infty} \mathbf{l}(x(\tau))d\tau < +\infty$ is finite and that

$$\liminf_{t \to +\infty} \mathbf{l}(x(t)) \geq 2c > 0$$

The latter condition means that there would exist $T > 0$ such that

$$\forall t \geq T, \ \mathbf{l}(x(t)) \geq c > 0$$

Hence $\int_T^S \mathbf{l}(x(\tau))d\tau \geq c(S - T)$ and converges to $+\infty$ with S, a contradiction. □

Definition 4.5.2 *[Attracting Function to a Target] Let us consider a target* $C \subset K$ *and the distance function* $d_{(K,C)}$ *to a subset* C. *Let us consider the functional*

$$\mathbf{J}(x(\cdot)) := \int_0^{\infty} d_{(K,C)}(x(\tau))d\tau$$

called the attraction functional. *The extended* attracting function $\zeta_{(K,C)}(\cdot)$ *is defined by*

$$\forall x \in K, \ \zeta_{(K,C)}(x) := \inf_{x(\cdot) \in \mathcal{S}(x)} \int_0^{\infty} d_{(K,C)}(x(\tau))d\tau$$

Lemma 4.5.1, p. 143 and Definition 4.5.2, p. 143 imply that for any x such that $\zeta_{(K,C)}(x) < +\infty$, there exists an evolution $x(\cdot)$ viable in K and such that

$$\liminf_{t \to +\infty} d_{(K,C)}(x(t)) = 0$$

This property justifies the name of attracting function.

Proposition 4.5.3 *[Viability Characterization of the Attracting Function] Let us consider the auxiliary control system*

$$\begin{cases} (i) \ x'(t) = f(x(t), u(t)) \\ (ii) \ y'(t) = -d_{(K,C)}(x(t)) \\ \quad \text{where } u(t) \in U(x(t)) \end{cases} \tag{4.11}$$

Then the attracting function is equal to

$$\zeta_{(K,C)}(x) = \inf_{(x,y) \in \mathrm{Capt}_{(4.11)}(K \times \mathbb{R}_+)} y$$

Proof. This is a consequence of Theorem 4.7.2, p. 156 with $\mathbf{l}(x, u) := d_C(x)$. □

Example We consider

- the differential equation $x'(t) = -x(t)((x(t) - 1)^2 + 0.1)$,
- the distance function $\mathbf{d}(x) := |x|$ to the equilibrium $\{0\}$,
- the environment $[-10, +10]$

The Viability Kernel Algorithm provides the graphs of the attracting and the exponential Lyapunov functions used to compute the viability kernel of the environment of $\mathcal{E}p(|\cdot|)$ under the auxiliary micro–macro systems $x'(t) = -x(t)((x(t)-1)^2+0.1)$ and $y'(t) = -|x(t)|$ and the environment $[-10, +10] \times \mathbb{R}_+$ under the auxiliary micro–macro system $x'(t) = -x(t)((x(t) - 1)^2 + 0.1)$ and $y'(t) = -my$ respectively.

They are respectively the epigraphs of the attracting function $\zeta_{([-10,+10],\{0\})}$ and the exponential Lyapunov function of the equilibrium $\{0\}$ under the differential equation $x'(t) = -x(t)((x(t) - 1)^2 + 0.1)$ for the distance function $\mathbf{d}(x) := |x|$. □

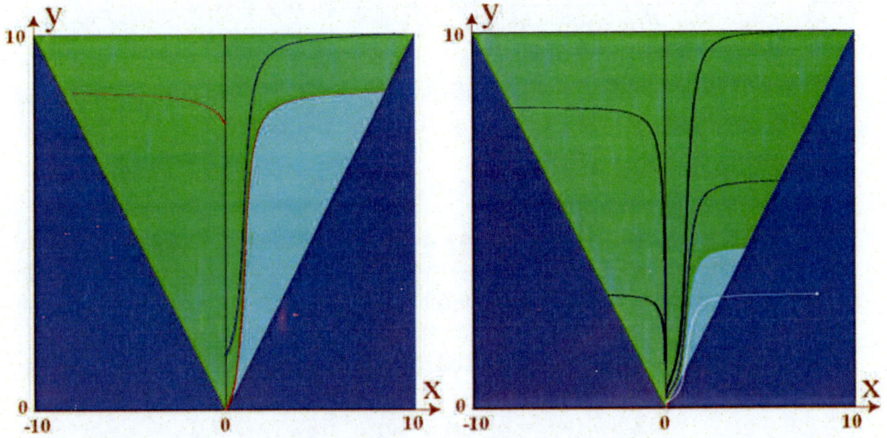

Fig. 4.6 Examples of Attracting and Lyapunov Functions.
*The x-axis is the axis of the variable, the y-axis the axis of the attracting function (*left*) and the exponential Lyapunov function (*right*). The grey area is the epigraph of the function.*

4.5.2 Exponential Lyapunov Functions

We can accelerate the convergence of the distance $d_{(K,C)}(x(t))$ of an evolution $x(\cdot)$ to 0 when $t \mapsto +\infty$ whenever one can establish an estimate of the form

$$\forall t \geq 0, \;\; d_C(x(t)) \leq ye^{-mt}$$

This idea goes back to *Alexandr Lyapunov.*

Alexandr Mikhalovitch Lyapunov (1857–1918). *Many methods for studying stability and asymptotic stability of an equilibrium, of the trajectory of a periodic evolution (limit cycles) or of any viable subset C have been designed by* Alexandr Mikhalovitch Lyapunov *and* Henri Poincaré *in 1892.*

4.5.2.1 Examples of Lyapunov Functions

- **Asymptotic Lyapunov Stability**
  Let us consider a target $C \subset K$. Recall that $d_{(K,C)}$ denotes the function:

$$d_{(K,C)}(x) := \begin{cases} \inf_{y \in C} \|x - y\| & \text{if } x \in K \\ +\infty & \text{if } x \notin K \end{cases}$$

We ask if one can find an evolution $x(\cdot) \in \mathcal{S}(x)$ viable in K and some positive constant y satisfying inequalities of the type

$$\forall t \geq 0, \ d_{(K,C)}(x(t)) \leq ye^{-mt}$$

If so, and if m is strictly positive, we deduce that the evolution $x(\cdot)$ is viable in K and converges to C when $t \to +\infty$. Then, if the above property holds true for any initial state x in a neighborhood of C, we are answering the question of *exponential asymptotic stability* in an exponential way.
This property can be reformulated by stating that the function $\lambda_{d_{(K,C)}}$ defined by

$$\lambda_{d_{(K,C)}}(x) := \inf_{x(\cdot) \in \mathcal{S}(x)} \sup_{t \geq 0} e^{mt} d_{(K,C)}(x(t))$$

is finite at x. We shall say the function $\lambda_{d_{(K,C)}}$ (providing the smallest constant y in the above inequality) is the m-Lyapunov function associated with the distance function $d_{(K,C)}$ and that its domain $\text{Dom}(\lambda_{d_{(K,C)}})$ is the exponential attraction basin.

- **a posteriori Estimates**
We are looking whether one can find an evolution $x(\cdot) \in \mathcal{S}(x)$ and some positive constant y satisfying inequalities of the type

$$\forall t \geq 0, \ \|x(t)\| \leq ye^{-mt}$$

This property can be reformulated by stating that the 0-order inertia function μ_m defined by

$$\mu_m(x) := \inf_{x(\cdot) \in \mathcal{S}(x)} \sup_{t \geq 0} e^{mt} \|x(t)\|$$

is finite at x.
In this case, we deduce that there exists at least one evolution satisfying a priori estimates on the growth of the solution, where $\mu_m(x)$ provides the smallest constant y in the above inequality.
In particular, when $m = 0$, we can determine whether or not there exists at least one *bounded evolution* starting from x.
Instead of looking for a priori estimates on the evolutions starting from a given initial state by providing an estimate $\mu_m(x)$ of the norms of the states of the evolution, we may obtain a posteriori estimates singling out the initial states satisfying, for a given constant c, the set

$$\mathbf{L}^{\leq}_{\mu_0}(c) := \{x \in \text{Dom}(U) \text{ such that } \mu_0(x) \leq c\}$$

(see Comments 1, p. 5 and 2, p. 5).

$$\mathbf{L}^{\leq}_{\alpha_0}(c) \ := \ \{x \in \mathrm{Dom}(U) \text{ such that } \alpha_0(x) \ \leq \ c\}$$

provides the subset of initial states from which at least one evolution is governed by open-loop controls bounded by the given constant c. Instead of looking for a priori estimates on the regulons, we are looking for a posteriori estimates singling out what are the initial states satisfying this property (see Comments 1, p. 5 and 2, p. 5).

- **Mandelbrot Functions**
  The Mandelbrot function is involved in the study of Julia and Mandelbrot subsets presented in Chap. 2, p. 43. Its continuous analogue is a Lyapunov function that we define in the general case of a parameterized differential equation $x'(t) = f(x(t), u)$ parameterized by *constant coefficients* u. We are looking for the subset of pairs (x, u) such that the solution to the above differential equation starting from x is bounded. This can be reformulated by checking whether the Mandelbrot function

$$\mu(x, u) := \sup_{t \geq 0} \|x(t)\|$$

is finite. Therefore, $\mu(x, u) = \lambda_{\|(\cdot)\|}(x, u)$, where $\lambda_{\|(\cdot)\|}$ is the function associated to the auxiliary micro–macro system

$$\begin{cases} (i) \ \ x'(t) = f(x(t), u(t)) \\ (ii) \ u'(t) = 0 \ \square \end{cases}$$

4.5.2.2 Definition of Lyapunov Functions

The problem is to find functions $\lambda_{\mathbf{v}}$ *associated with functions* $\mathbf{v}$ such as the distance function $d_{(K,C)}$ or the norm of the vector space:

Definition 4.5.4 *[**Exponential Lyapunov Functions**] The* exponential Lyapunov function $\lambda_{\mathbf{v}}(m; \cdot) : X \mapsto \mathbb{R}_+ \cup \{+\infty\}$ *associated with* $\mathbf{v}$ *is defined by*

$$\lambda_{\mathbf{v}}(m; x) := \inf_{x(\cdot) \in \mathcal{S}(x)} \sup_{t \geq 0} e^{mt} \mathbf{v}(x(t))$$

Therefore, for any $x \in \mathrm{Dom}(\lambda_{\mathbf{v}}(m; \cdot))$, *regarded as the* exponential attraction basin *of* $\mathbf{v}$, *and for any* $\varepsilon > 0$, *there exists at least one evolution* $x(\cdot) \in \mathcal{S}(x)$ *such that*

$$\forall t \geq 0, \ \mathbf{v}(x(t)) \leq e^{-mt}(\lambda_{\mathbf{v}}(m; x) + \varepsilon)$$

If $m > 0$, *this implies that* $\mathbf{v}(x(t))$ *converges to* 0 *exponentially when* $t \to +\infty$ *and if* $m = 0$, *that* $\mathbf{v}(x(t))$ *remains bounded.*

The exponential Lyapunov function can be characterized in terms of the viability kernel of the epigraph of the function $\mathbf{v}$ under the auxiliary micro–macro system:

$$\begin{cases} (i) \; x'(t) = f(x(t), u(t)) \\ (ii) \; y'(t) = -my(t) \\ \quad\quad \text{where } u(t) \in U(x(t)) \end{cases} \tag{4.12}$$

subject to the constraint

$$\forall t \geq 0, \;\; (x(t), y(t)) \in \mathcal{E}p(\mathbf{v})$$

Proposition 4.5.5 *[Viability Characterization of the Exponential Lyapunov Function] The exponential Lyapunov function $\lambda_{\mathbf{v}}$ is related to the viability kernel of the epigraph $\mathcal{E}p(\mathbf{v})$ of the function $\mathbf{v}$ under auxiliary micro–macro system (4.12) by the following formula*

$$\lambda_{\mathbf{v}}(m; x) = \inf_{(x,y) \in \text{Viab}_{(4.12)}(\mathcal{E}p(\mathbf{v}))} y$$

Proof. Indeed, to say that (x, y) belongs to the viability kernel of $\mathcal{E}p(\mathbf{v})$ under auxiliary micro–macro system (4.12) amounts to saying that there exists an evolution $t \mapsto (x(t), y(t))$ starting at (x, y) and governed by the auxiliary micro–macro system such that, for all $t \geq 0$, $u(t) \in U(x(t))$. By definition of (4.12), we know that for all $t \geq 0$, this evolution also satisfies for all $t \geq 0$,

$$\mathbf{v}(x(t)) \leq y(t) = e^{-mt} y$$

Therefore

$$\sup_{t \geq 0} e^{mt} \mathbf{v}(x(t)) \leq y$$

and thus, $\lambda_{\mathbf{v}}(m; x) \leq \inf_{(x,y) \in \text{Viab}_{(4.12)}(\mathcal{E}p(\mathbf{v}))} y$. □

We refer to Fig. 4.6, p. 145 where we compute the exponential function for the standard example $x'(t) = -x(t)((x(t) - 1)^2 + 0.1)$.

Remark: The Separation Function. The same type of approach applies for the *separation function* $\rho_{\mathbf{w}}(x) : X \mapsto \mathbb{R}_+ \cup \{+\infty\}$ associated with $\mathbf{w}$ defined by

$$\rho_{\mathbf{w}}(x) := \sup_{x(\cdot) \in \mathcal{S}(x)} \inf_{t \geq 0} e^{mt} \mathbf{w}(x(t))$$

Therefore, for any $x \in \text{Dom}(\rho_{\mathbf{w}})$, regarded as the *separation domain* of $\mathbf{w}$, and for any $\varepsilon > 0$, there exists an evolution $x(\cdot) \in \mathcal{S}(x)$ such that

$$\forall t \geq 0, \ \mathbf{w}(x(t)) \ \geq \ e^{-mt}(\rho_{\mathbf{w}}(x) + \varepsilon)$$

If $m < 0$, this implies that $\mathbf{w}(x(t)) \to +\infty$ when $t \to +\infty$ and if $m = 0$, that $\mathbf{w}(x)(t) \geq \rho(x)$.

They can be characterized in terms of viability kernels by introducing the auxiliary micro–macro system:

$$\begin{cases} (i) \ \ x'(t) = f(x(t), u(t)) \\ (ii) \ y'(t) = -my(t) \\ \quad\quad \text{where } u(t) \in U(x(t)) \end{cases} \tag{4.13}$$

subject to the constraint

$$\forall t \geq 0, \ \ (x(t), y(t)) \ \in \ \mathcal{H}yp(\mathbf{w})$$

Then the same type of proof implies the formula

$$\rho_{\mathbf{w}}(x) \ = \ \sup_{(x,y) \in \mathrm{Viab}_{(4.13)}(\mathcal{H}yp(\mathbf{w}))} y \ \square$$

Remark: The associated Hamilton–Jacobi–Bellman Equation. The *Lyapunov function* $\lambda_{\mathbf{v}}$ is the smallest solution to the Hamilton–Jacobi–Bellman partial differential equation

$$\inf_{u \in U(x)} \sum_{i=1}^{n} \frac{\partial \mathbf{v}(x)}{\partial x_i} f_i(x, u) + m\mathbf{v}(x) = 0$$

larger than the function $\mathbf{v}$. $\square$

4.5.3 The Montagnes Russes Algorithm for Global Optimization

Consider a positive extended function $\mathbf{v} : X \mapsto \mathbb{R}_+ \cup \{+\infty\}$ satisfying

$$\inf_{x \in X} \mathbf{v}(x) = 0$$

and the set $\mathbf{v}^{-1}(0)$ of its global minima.

A way to introduce the "Gradient Descent Method" is to use the simple differential inclusion

$$\forall t \geq 0, \ x'(t) \in B$$

where B denotes the unit ball of the finite dimensional vector space X, leaving open the direction to be chosen by the algorithm, as in the method of simulated annealing or some versions of *Newton type Algorithms* for finding equilibria (see Sect. 9.6.2, p. 363). However, instead of choosing the velocities at random and being satisfied by convergence in probability, we shall apply it not to the original function $\mathbf{v}$, but to its Lyapunov function $\lambda_{\mathbf{v}}$ under the differential equation $x'(t) \in B$ (see *Mathematical Morphology*, [166, Najman]). In this case, we know that starting from an initial state $x \in \mathrm{Dom}(\mathbf{v})$, and for any $\varepsilon > 0$, there exists an evolution $x(\cdot)$ satisfying $x(0) = x$ and $\sup_{t \geq 0} \|x'(t)\| \leq 1$ such that

$$\mathbf{v}(x(t)) \leq \lambda_{\mathbf{v}}(x(t)) \leq e^{-mt}(\lambda_{\mathbf{v}}(x) + \varepsilon)$$

Therefore, any cluster point $x^\star$ of the evolution $t \mapsto x(t)$ when $t \to +\infty$ is a global minimum of $\mathbf{v}$.

Hence, the epigraph of the function $\lambda_{\mathbf{v}}$ is the viability kernel of the epigraph $\mathcal{E}p(v)$ of the function $\mathbf{v}$ under the auxiliary micro–macro system $x'(t) \in B$ and $y'(t) = -my(t)$.

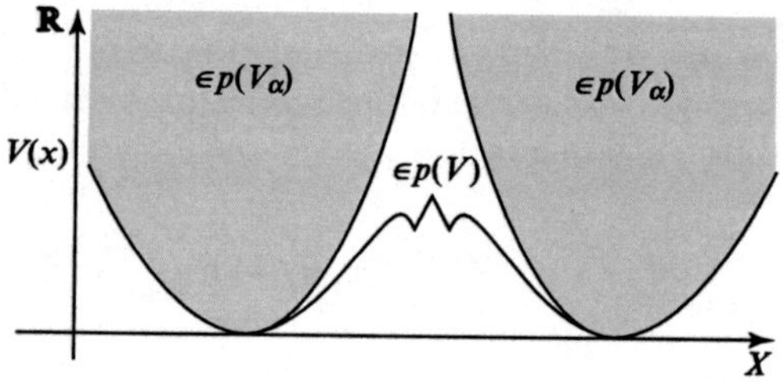

Fig. 4.7 The Montagnes Russes Algorithm.
Epigraph of the auxiliary function involved in the Montagnes Russes Algorithm.

17 ***The Montagnes Russes Algorithm.*** *Although we know that* $\mathbf{v}(x(t))$ *converges to* v_0 *when* $t \to +\infty$*, the function* $t \mapsto \mathbf{v}(x(t))$ *is not necessarily decreasing. Along such a solution, the function* $\mathbf{v}$ *jumps above local maxima, leaves local minima, play "Montagnes Russes" (called "American Mountains" in Russian and "Big Dipper" in American English!), but, ultimately, converges to its infimum.*

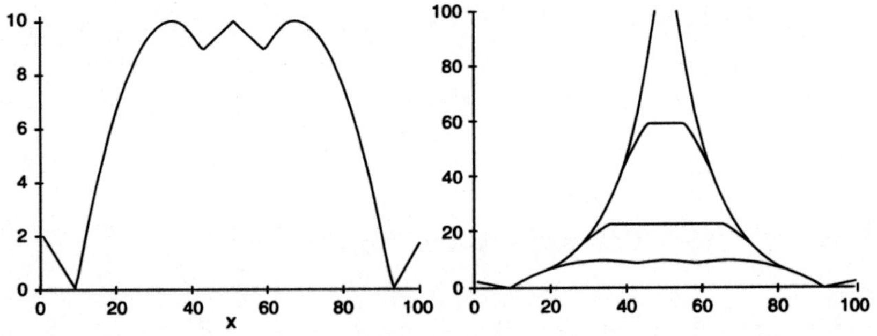

Fig. 4.8 Example 1.
*The epigraph of the original function (see Fig. 4.7) is displayed (*left*) and several steps of the Montagnes Russes Algorithm are shown on a different scale (*right*).*

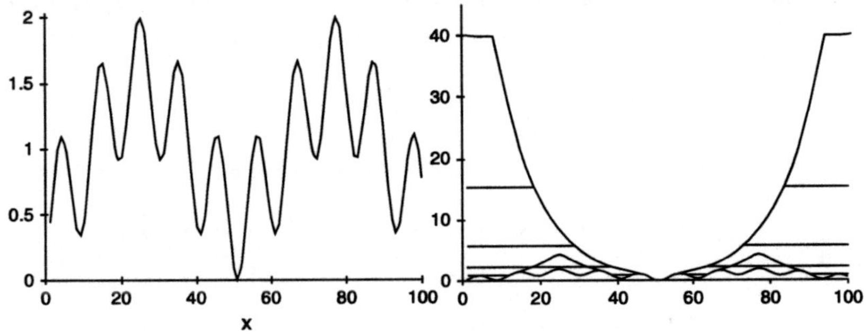

Fig. 4.9 Example 2.
*The epigraph of the original function $x \mapsto \mathbf{v}(x) := 1 - \cos(2x)\cos(3x)$ having many local minima and only one global minimum is displayed (*left*) and several steps of the Montagnes Russes Algorithm are shown on a different scale (*right*).*

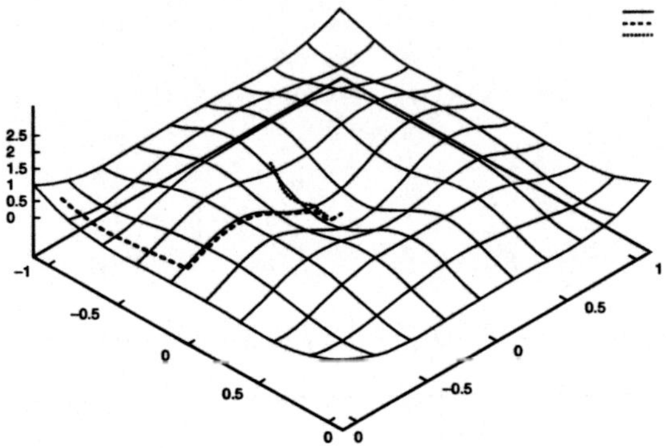

Fig. 4.10 Example 3.
The descent algorithm stops at local minima. The Montagnes Russes algorithm is the descent algorithm applied to the optimal exponential

Lyapunov function, the minima of which are the global minima of the function $x \mapsto \mathbf{v}(x) := 1 - \cos(2\|x\|)\cos(3\|x\|)$ *defined in* $\mathbb{R}^2$. *Source: Laurent Najman.*

4.6 Safety and Transgression Functions

Consider an environment K with a nonempty interior $\overset{\circ}{K}$. Starting from an initial state $x \in \overset{\circ}{K}$, i.e., a state such that the distance $d(x, \complement K) > 0$, we would like to know the smallest distance $\inf_{t \geq 0} d(x(t), \complement K)$ of the evolution to the boundary of K: the further away from the boundary the state is, the more secure the state of the evolution is.

In a dual way, starting from an initial state outside the environment, we would like to measure the largest distance $\sup_{t \geq 0} d(x(t), K)$ of the evolution to K. This is a measure to viability transgression, which tells how far from the environment the state of the evolution will remain during the course of time. This idea is parallel to the concept of *crisis function* investigated later (see Sect. 9.2.2, p. 326) which measures the time spent by an evolution outside of the environment.

Consider an evolutionary system $\mathcal{S} : X \rightsquigarrow \mathcal{C}(0, +\infty; X)$ generated by a control problem

$$x'(t) = f(x(t), u(t)) \text{ where } u(t) \in U(x(t)) \tag{4.14}$$

We thus introduce the safety and transgression functionals and functions:

Definition 4.6.1 *[Safety and Transgression Functions] We associate with a subset K* safety functional σ_K *and* transgression functional ξ_K *defined by:*

$$\begin{cases} (i) \ \sigma_K(x(\cdot)) := \inf_{t \geq 0} d(x(t), \complement K) \\ (ii) \ \xi_K(x(\cdot)) := \sup_{t \geq 0} d(x(t), K) \end{cases} \tag{4.15}$$

The extended functions

$$\begin{cases} (i) \ \sigma_K^\sharp(x) := \sup_{x(\cdot) \in \mathcal{S}(x)} \sigma_K(x(\cdot)) \\ (ii) \ \xi_K^\flat(x) := \inf_{x(\cdot) \in \mathcal{S}(x)} \xi_K(x(\cdot)) \end{cases} \tag{4.16}$$

are called safety function $\sigma_K^\sharp$ *and* transgression function $\xi_K^\flat$ *respectively.*

We can characterize these two functions as the viability kernel of auxiliary environments $\mathcal{H}yp(d(\cdot,\complement K))$ and $\mathcal{E}p(d(\cdot,K))$ under auxiliary micro–macro systems of the form

$$\begin{cases} (i)\ x'(t) &= f(x(t),u(t)) \text{ where } u(t) \in U(x(t)) \\ (ii)\ y'(t) &= 0 \end{cases} \tag{4.17}$$

Theorem 4.6.2 ***[Viability Characterization of the Safety and Transgression Functions.]*** *The* safety function *and the* transgression function *are equal to:*

$$\forall\, x \in K, \begin{cases} (i)\ \ \sigma_K^\sharp(x) &= \sup_{(x,y)\in \text{Viab}_{(4.17)}(\mathcal{H}yp(d(\cdot,\complement K)))} y \\ (ii)\ \ \xi_K^\flat(x) &= \inf_{(x,y)\in \text{Viab}_{(4.17)}(\mathcal{E}p(d(\cdot,K)))} y \end{cases} \tag{4.18}$$

Proof. 1. If $(x,y) \in \text{Viab}_{(4.17)}(\mathcal{H}yp(d(\cdot,\complement K)))$, then there exists an evolution $x(\cdot) \in \mathcal{S}(x)$ such that $(x(t),y) \in \mathcal{H}yp(d(\cdot,\complement K))$ for all $t \geq 0$, i.e., such that, for all $t \geq 0$, $y \leq d(x(t),\complement K)$. By taking the infimum of $d(x(\cdot),\complement K)$ over $t \geq 0$, we infer that

$$y^\sharp := \sup_{(x,y)\in \text{Viab}_{(4.17)}(\mathcal{H}yp(d(\cdot,\complement K)))} y \leq \sigma_K^\sharp(x)$$

For proving the opposite inequality, take $\mu < \sigma_K^\sharp(x)$. By definition of the supremum, there exists an evolution $x(\cdot) \in \mathcal{S}(x)$ such that $\mu \leq \sigma_K(x(\cdot))$, and thus, such that for all $t \geq 0$, $\mu \leq d(x(t),\complement K)$. This implies that $(x(\cdot),\mu)$ is viable in $\mathcal{H}yp(d(\cdot,\complement K))$, and therefore, that $\mu \leq y^\sharp$. Letting μ converge to $\sigma_K^\sharp(x)$ implies the opposite inequality we were looking for.

2. If $(x,y) \in \text{Viab}_{(4.17)}(\mathcal{E}p(d(\cdot,K)))$, then there exists an evolution $x(\cdot) \in \mathcal{S}(x)$ such that $(x(t),y) \in \mathcal{E}p(d(\cdot,K))$ for all $t \geq 0$, i.e., such that, for all $t \geq 0$, $y \geq d(x(t),K)$. By taking the supremum over $t \geq 0$, we infer that

$$y^\flat := \inf_{(x,y)\in \text{Viab}_{(4.17)}(\mathcal{E}p(d(\cdot,K)))} y \geq \xi_K^\flat(x)$$

For proving the opposite inequality, take $\mu > \xi_K^\flat(x)$. By definition of the infimum, there exists an evolution $x(\cdot) \in \mathcal{S}(x)$ such that $\mu \geq \xi_K(x(\cdot))$, and thus, such that for all $t \geq 0$, $\mu \geq d(x(t),K)$. This implies that $(x(\cdot),\mu)$ is viable in $\mathcal{E}p(d(\cdot,K))$, and therefore, that $\mu \geq y^\flat$. Letting μ converge to $\xi_K^\flat(x)$ implies the opposite inequality we were looking for. □

We observe that whenever K is closed,

$$\begin{cases} (i)\ K \ominus y \overset{\circ}{B} = \{x \text{ such that } y \le d(x, \complement K)\} \\ (ii)\ K - yB = \{x \text{ such that } d(x,K) \le y\} \end{cases} \tag{4.19}$$

where $\overset{\circ}{B}$ is the open unit ball and where the Minkowski sums and differences are defined by:

Definition 4.6.3 *[Minkowski Sums and Differences of Sets] Let $A \subset X$ and $B \subset X$ two subsets of a vector space X. The subset $A - B := \bigcup_{b \in B} (A - b)$ is the* Minkowski sum *of A and $-B$ and $A \ominus B := \bigcap_{b \in B} (A - b)$ is the* Minkowski difference *between A and B.*

In mathematical morphology, a convex compact subset B containing the origin (such as a ball) is regarded as a "*structuring element*", and these subsets are called *dilations* and *erosions* of K by the "*structuring element*" B respectively (see *Morphologie Mathématique*, [187, Schmitt M. & Mattioli], and *Mathematical Morphology*, [166, Najman]). To say that $x \in A - B$ means that $(x + B) \cap A \neq \emptyset$ and to say that $x \in A \ominus B$ amounts to saying that $x + B \subset A$. We note that *$A \ominus B$ is closed whenever A is closed* and that *$A - B$ is closed whenever A is closed and B is compact.* We observe that

$$\complement(A - B) = \complement(A) \ominus B$$

We also observe that whenever $A_1 \subset A_2$, $B \ominus A_2 \subset B \ominus A_1$ and whenever P is a convex cone,

$$B \ominus (A + P) \subset B \ominus A \subset (B + P) \ominus (A + P) \tag{4.20}$$

so that whenever $B = B + P$, then $B \ominus A = B \ominus (A + P)$. Indeed, since $0 \in P$, $A \subset A + P$ so that $B \ominus (A + P) \subset B \ominus A$. For any $x \in B \ominus A$, then $x + A \subset B$ and thus, $x + A + P \subset B + P$, so that x belongs to $(B + P) \ominus (A + P)$.

We regard $y \ge 0$ as a degree of safety of an evolution if $x(\cdot)$ viable in $K \ominus y \overset{\circ}{B}$ as well as a degree of transgression if $x(\cdot)$ is viable in $K - yB$. This suggests to study the viability kernels of the erosions $K \ominus y \overset{\circ}{B}$ and the dilations $K - yB$ respectively. We denote by $\text{Graph}(K - (\cdot)B)$ and $\text{Graph}(K \ominus (\cdot) \overset{\circ}{B})$ the graphs of the set-valued maps $y \rightsquigarrow K - yB$ and $y \rightsquigarrow K \ominus y \overset{\circ}{B}$.

$$\begin{cases} (i)\ y'(t) = 0 \\ (ii)\ x'(t) = f(x(t), u(t)) \text{ where } u(t) \in U(x(t)) \end{cases} \tag{4.21}$$

Theorem 4.6.4 ***[Minkowski Characterizations of the Safety and Transgression Functions.]*** *Let us consider the Minkowski sum and difference $K - yB$ and $K \ominus y \overset{\circ}{B}$. For all $x \in K$,*

$$\begin{cases} (i) \ \sigma_K^\sharp(x) \ = \ \sup_{(y,x)\in \mathrm{Viab}_{(4.21)}(\mathrm{Graph}(K\ominus(\cdot)\overset{\circ}{B}))} y \\ (ii) \ \xi_K^\flat(x) \ = \ \inf_{(y,x)\in \mathrm{Viab}_{(4.21)}(\mathrm{Graph}(K-(\cdot)B))} y \end{cases} \tag{4.22}$$

This example is the particular case of a more general situation when we have to inverse the set-valued map associating with a parameter y the viability kernel of an environment depending upon y with a target depending upon this parameter under an evolutionary system depending also on this parameter, investigated in Sect. 10.9, p. 427. These crucial questions are related to the important class of *parameter identification*.

Remark: The associated Hamilton–Jacobi–Bellman Equation. The safety function $\sigma_K^\sharp$ is the largest positive solution to the Hamilton–Jacobi–Bellman partial differential equation

$$\inf_{u \subset U(x)} \sum_{i=1}^{n} \frac{\partial \mathbf{v}(x)}{\partial x_i} f_i(x,u) + d(x, \complement K) = 0$$

on the subset K, and the transgression function $\xi_K^\sharp$ is the smallest positive solution to the Hamilton–Jacobi–Bellman partial differential equation

$$\inf_{u \in U(x)} \sum_{i=1}^{n} \frac{\partial \mathbf{v}(x)}{\partial x_i} f_i(x,u) + d(x, K) = 0$$

on the subset $\complement \overset{\circ}{K}$. □

4.7 Infinite Horizon Optimal Control and Viability Kernels

Minimal time and exit functions, attracting and Lyapunov functions, safety and transgression functions, crisis functions, etc., are all examples of selection of "optimal evolutions" in $\mathcal{S}(x)$ starting from a given initial state x according to a given intertemporal optimality criterion.

The usual shape of a criterion on the space $\mathcal{C}(0, +\infty; X)$ of continuous functions involves:

1. a "final state" cost function $\mathbf{c} : X \mapsto \mathbb{R}_+ \cup \{+\infty\}$,
2. a "transient state" cost function (often called a Lagrangian) $\mathbf{l} : X \times \mathcal{U} \mapsto \mathbb{R}_+$.

Namely, consider parameterized system (2.10):

$$x'(t) = f(x(t), u(t)) \text{ where } u(t) \in U(x(t))$$

generating the evolutionary system $\mathcal{P}$ associating with any x the set of pairs $(x(\cdot), u(\cdot))$ governed by (2.10) and starting from x.

Definition 4.7.1 ***[Infinite Horizon Optimization Problem]*** *Given the system (2.10) and the cost function* $\mathbf{l} : X \times \mathcal{U} \mapsto \mathbb{R}_+$*, the function* $\overset{\infty}{\mathbf{V}}$ *defined by*

$$\overset{\infty}{\mathbf{V}}(x) := \inf_{(x(\cdot),u(\cdot))\in\mathcal{P}(x)} \sup_{t\geq 0} \left(e^{-mt}\mathbf{c}(x(t)) + \int_0^t e^{-m\tau}\mathbf{l}(x(\tau), u(\tau))d\tau \right) \tag{4.23}$$

is called the valuation function *of the infinite horizon control problem: find an evolution* $(\overline{x}(\cdot), \overline{u}(\cdot)) \in \mathcal{P}(x)$ *minimizing the intertemporal cost functional*

$$\overset{\infty}{\mathbf{V}}(x) = \sup_{t\geq 0} \left(e^{-mt}\mathbf{c}(\overline{x}(t)) + \int_0^t e^{-m\tau}\mathbf{l}(\overline{x}(\tau), \overline{u}(\tau))d\tau \right)$$

See Fig. 5.14, p. 196.

We relate the valuation functions of an infinite horizon control problem to viability kernels of epigraphs of functions under the auxiliary micro–macro system

$$\begin{cases} (i)\ x'(t) = f(x(t), u(t)) \\ (ii)\ y'(t) = my(t) - \mathbf{l}(x(t), u(t)) \\ \quad \text{where } u(t) \in U(x(t)) \end{cases} \tag{4.24}$$

Theorem 4.7.2 ***[Viability Characterization of the Valuation Function]*** *The valuation function*

$$\overset{\infty}{\mathbf{V}}(x) := \inf_{(x(\cdot),u(\cdot))\in\mathcal{P}(x)} \sup_{t\geq 0} \left(e^{-mt}\mathbf{c}(x(t)) + \int_0^t e^{-m\tau}\mathbf{l}(x(\tau), u(\tau))d\tau \right)$$

is related to the viability kernel $\text{Viab}_{(4.24)}(\mathcal{E}p(\mathbf{c}))$ *of the epigraph of the function* $\mathbf{c}$ *under the auxiliary micro–macro system* (4.24) *by the formula*

$$\overset{\infty}{\mathbf{V}}(x) = \inf_{(x,y)\in \text{Viab}_{(4.24)}(\mathcal{E}p(\mathbf{c}))} y$$

Proof. Indeed, to say that (x, y) belongs to $\text{Viab}_{(4.24)}(\mathcal{E}p(\mathbf{c}))$ means that there exists an evolution $t \mapsto (x(t), y(t))$ starting at (x, y) governed by the auxiliary evolutionary system (4.24) such that

$$\forall t \geq 0,\ (x(t), y(t)) \in \mathcal{E}p(\mathbf{c})$$

By the very definition of the epigraph, these inequalities can be written in the form

$$\forall t \geq 0,\ \ y(t)\ \geq\ \mathbf{c}(x(t))$$

We now recall that the evolution $(x(\cdot), u(\cdot))$ belongs to $\mathcal{P}(x)$ so that

$$y(t) = e^{mt}\left(y - \int_0^t e^{-m\tau}\mathbf{l}(x(\tau), u(\tau))d\tau\right)$$

Hence the above inequalities become

$$\forall t \geq 0,\ \ y\ \geq\ e^{-mt}\mathbf{c}(x(t)) + \int_0^t e^{-m\tau}\mathbf{l}(x(\tau), u(\tau))d\tau$$

Therefore,

$$\overset{\infty}{\mathbf{V}}(x) \leq \inf\{y \text{ such that } (x, y) \in \text{Viab}_{(4.24)}(\mathcal{E}p(\mathbf{c}))\}$$

Conversely, it is enough to prove that for any $\varepsilon > 0$,

$$\inf\{y \text{ such that } (x, y) \in \text{Viab}_{(4.24)}(\mathcal{E}p(\mathbf{c}))\} \leq \overset{\infty}{\mathbf{V}}(x) + \varepsilon$$

and to let ε converge to 0.

By definition of the infimum, there exists an evolution $(x(\cdot), u(\cdot)) \in \mathcal{P}(x)$ such that

$$\overset{\infty}{\mathbf{V}}(x) = \sup_{t\geq 0}\left(e^{-mt}\mathbf{c}(x(t)) + \int_0^t e^{-m\tau}l(x(\tau), u(\tau))d\tau\right) \leq \overset{\infty}{\mathbf{V}}(x) + \varepsilon$$

Setting

$$y_\varepsilon(t) = e^{mt}\left(\overset{\infty}{\mathbf{V}}(x) + \varepsilon - \int_0^t e^{-m\tau}\mathbf{l}(x(\tau), u(\tau))d\tau\right)$$

we observe that $(x(\cdot), y_\varepsilon(\cdot))$ is a solution to the auxiliary micro–macro system (4.24) starting at $(x, \overset{\infty}{\mathbf{V}}(x)+\varepsilon)$ and viable in the epigraph of the function $\mathbf{c}$. Hence the pair $(x, \overset{\infty}{\mathbf{V}}(x)+\varepsilon)$ belongs to its viability kernel, so that

$$\inf\{y \text{ such that } (x,y) \in \text{Viab}_{(4.24)}(\mathcal{E}p(\mathbf{c}))\} \leq \overset{\infty}{\mathbf{V}}(x)+\varepsilon$$

Therefore,

$$\overset{\infty}{\mathbf{V}}(x) = \inf\{y \text{ such that } (x,y) \in \text{Viab}_{(4.24)}(\mathcal{E}p(\mathbf{c}))\} \quad \square$$

Remark: The associated Hamilton–Jacobi–Bellman Equation. The **valuation function** $\overset{\infty}{\mathbf{V}}$ of an infinite horizon control problem is the smallest solution to the Hamilton–Jacobi–Bellman partial differential equation

$$\inf_{u\in U(x)}\left(\sum_{i=1}^{n}\frac{\partial \mathbf{v}(x)}{\partial x_i}f_i(x,u)+l(x,u)\right)-m\mathbf{v}(x) = 0$$

larger than or equal to the function $\mathbf{c}$.

These results have been obtained in collaboration with Hélène Frankowska (see also *Infinite Horizon Optimal Control: Theory and Applications*, [55, Carlson & Haurie] on this topic). $\square$

4.8 Intergenerational Optimization

We introduce not only a positive Lagrangian $\mathbf{l}$, but also a "generational cost function" $\mathbf{d}: X \mapsto \mathbb{R}_+$ and denote by $\mathcal{D}(x)$ the subset of evolutions $(x(\cdot), u(\cdot)) \in \mathcal{P}(x)$ satisfying the *intergenerational constraints*

$$\forall t \geq 0, \int_t^{\infty} e^{-m(\tau-t)}\mathbf{l}(x(\tau), u(\tau))d\tau \leq \mathbf{d}(x(t)) \tag{4.25}$$

This expresses the fact that at each instant t, the *future cumulated cost* $\int_t^{\infty} e^{-m(\tau-t)}\mathbf{l}(x(\tau), u(\tau))d\tau$ should be below a given cost $\mathbf{d}(x(t))$ of the value of the state at time t.

Definition 4.8.1 *[The Intergenerational Valuation Function] The intergenerational valuation function* $\overset{\infty}{\mathbf{V}}_{\mathbf{d}}$ *of the infinite horizon intertemporal minimization problem*

$$\overset{\infty}{\mathbf{V}}_{\mathbf{d}}(x) := \inf_{(x(\cdot),u(\cdot))\in\mathcal{D}(x)} \int_0^{\infty} e^{-m\tau}\mathbf{l}(x(\tau),u(\tau))d\tau$$

over the subset $\mathcal{D}(x)$ of all the evolutions satisfying the intergenerational constraints (4.25)

$$\forall t \geq 0, \int_t^{\infty} e^{-m(\tau-t)}\mathbf{l}(x(\tau),u(\tau))d\tau \leq \mathbf{d}(x(t))$$

is called the intergenerational valuation function. *They express that at each future time $t \geq 0$, the intertemporal value generation $\int_t^{\infty} e^{-m(\tau-t)}$ $\mathbf{l}(x(\tau),u(\tau))d\tau$ of the state $x(t)$ of the future "generation" starting at time t up to $+\infty$ is smaller than or equal to a predetermined "generational" upper cost $\mathbf{d}(x(t))$ that the present generation leaves or bequests to the future generations at time t.*

Let us observe that if $x(\cdot)$ satisfies the intergenerational constraints, then

$$0 \leq \int_0^{\infty} e^{-m\tau}\mathbf{l}(x(\tau),u(\tau))d\tau \leq \mathbf{d}(x)$$

so that

$$0 \leq \overset{\infty}{\mathbf{V}}_{\mathbf{d}}(x) \leq \mathbf{d}(x)$$

Hence, whenever $\mathbf{d}(x)$ is finite, inequalities (4.25), p. 158 imply that

$$\begin{cases} \forall t \geq 0, \int_0^{\infty} e^{-m\tau}\mathbf{l}(x(\tau),u(\tau))d\tau \\ \leq e^{-mt}\mathbf{d}(x(t)) + \int_0^{t} e^{-m\tau}\mathbf{l}(x(\tau),u(\tau))d\tau \end{cases}$$

We introduce the auxiliary control system

$$\begin{cases} (i)\ x'(t) = f(x(t),u(t)) \\ (ii)\ y'(t) = my(t) - \mathbf{l}(x(t),u(t)) \\ \quad\text{where } u(t) \in U(x(t)) \end{cases} \tag{4.26}$$

Theorem 4.8.2 ***[Viability Characterization of the Intergenerational Function]*** *Assume that the extended function $\mathbf{d}$ is nontrivial and positive and that the Lagrangian $\mathbf{l}$ is positive. Consider the viability kernel* $\text{Viab}_{(4.26)}(\mathcal{K})$ *of the subset*

$$\mathcal{K} := \{(x,y) \in X \times \mathbb{R}_+ \mid y \leq \mathbf{d}(x)\}$$

under auxiliary the set-valued evolutionary system (4.26). Then

$$\forall x \in \mathcal{K},\ \overset{\infty}{\mathbf{V}}_{\mathbf{d}}(x) = \inf\{y|(x,y) \in \text{Viab}_{(4.26)}(\mathcal{K})\}$$

Proof. We set

$$\mathbf{v}(x) := \inf\{y|(x,y) \in \text{Viab}_{(4.26)}(\mathcal{K})\}$$

For proving inequality $\mathbf{v}(x) \leq \overset{\infty}{\mathbf{V}}_{\mathbf{d}}(x)$, let us take any evolution $(x(\cdot), u(\cdot)) \in \mathcal{D}(x)$ satisfying the intergenerational constraints (4.25).

We set

$$z(t) := \int_t^{+\infty} e^{-m(\tau-t)}\mathbf{l}(x(\tau), u(\tau))d\tau$$

and observe that the pair $(x(\cdot), z(\cdot))$ is an evolution to auxiliary micro–macro system (4.26) starting at

$$\left(x, \int_0^{+\infty} e^{-m\tau}\mathbf{l}(x(\tau), u(\tau))d\tau\right)$$

Since $(x(\cdot), u(\cdot)) \in \mathcal{D}(x)$ satisfies the intergenerational constraints (4.25), the means that

$$\forall t \geq 0,\ z(t) \leq \mathbf{d}(x(t))$$

Since this can also be written in the form

$$\forall t \geq 0,\ (x(t), z(t)) \in \mathcal{K}$$

we deduce that the pair $\left(x, \int_0^{+\infty} e^{-m\tau}\mathbf{l}(x(\tau), u(\tau))d\tau\right)$ belongs to the viability kernel $\mathcal{W} := \text{Viab}_{(4.26)}(\mathcal{K})$. Therefore

$$\mathbf{v}(x) \leq \int_0^{+\infty} e^{-m\tau}\mathbf{l}(x(\tau), u(\tau))d\tau \tag{4.27}$$

and thus, we infer that $\mathbf{v}(x) \leq \overset{\infty}{\mathbf{V}}_{\mathbf{d}}(x)$.

For proving the opposite inequality, take any pair (x, y) in the viability kernel

$$\mathcal{W} := \text{Viab}_{(4.26)}(\mathcal{K}) \subset \mathcal{K}$$

This means that there exists an evolution $(\widetilde{x}(\cdot), y(\cdot))$ to the auxiliary micro–macro system starting at (x, y) such that

$$\forall t \geq 0, \ (\widetilde{x}(t), y(t)) \in \mathcal{W} \subset \mathcal{K}$$

where

$$y(t) = e^{mt}\left(y - \int_0^t e^{-m\tau}\mathbf{l}(\widetilde{x}(\tau), \widetilde{u}(\tau))d\tau\right)$$

By the very definition of $\mathcal{W}$, this is equivalent to say that

$$\forall t \geq 0, \ 0 \leq y(t) \leq \mathbf{d}(\widetilde{x}(t))$$

i.e.,

$$\forall t \geq 0, \ 0 \leq e^{mt}y - \int_0^t e^{-m(\tau-t)}\mathbf{l}(\widetilde{x}(\tau), \widetilde{u}(\tau))d\tau \leq \mathbf{d}(\widetilde{x}(t)) \tag{4.28}$$

The first inequality $0 \leq e^{mt}y - \int_0^t e^{-m(\tau-t)}\mathbf{l}(\widetilde{x}(\tau), \widetilde{u}(\tau))d\tau$ implies that

$$\forall t \geq 0, \ \int_0^t e^{-m\tau}\mathbf{l}(\widetilde{x}(\tau), \widetilde{u}(\tau))d\tau \ \leq \ y$$

so that, letting t converge to ∞,

$$\int_0^\infty e^{-m\tau}\mathbf{l}(\widetilde{x}(\tau), \widetilde{u}(\tau))d\tau \ \leq \ y \tag{4.29}$$

The second inequality $e^{mt}y - \int_0^t e^{-m(\tau-t)}\mathbf{l}(\widetilde{x}(\tau), \widetilde{u}(\tau))d\tau \leq \mathbf{d}(\widetilde{x}(t))$ implies that

$$\begin{cases} \int_t^\infty e^{-m(\tau-t)}\mathbf{l}(\widetilde{x}(\tau), \widetilde{u}(\tau))d\tau \\ = e^{mt}\int_0^\infty e^{-m\tau}\mathbf{l}(\widetilde{x}(\tau), \widetilde{u}(\tau))d\tau - \int_0^t e^{-m(\tau-t)}\mathbf{l}(\widetilde{x}(\tau), \widetilde{u}(\tau))d\tau \\ \leq e^{mt}\int_0^\infty e^{-m\tau}\mathbf{l}(\widetilde{x}(\tau), \widetilde{u}(\tau))d\tau - e^{mt}y + \mathbf{d}(\widetilde{x}(t)) \ \leq \ \mathbf{d}(\widetilde{x}(t)) \end{cases}$$

so that the evolution satisfies the intergenerational constraints (4.25). Hence

$$\overset{\infty}{\mathbf{V}}_{\mathbf{d}}(x) \leq \int_0^\infty e^{-m\tau}\mathbf{l}(\widetilde{x}(\tau), \widetilde{u}(\tau))d\tau \leq y \leq \mathbf{d}(x) \tag{4.30}$$

Properties (4.30) can be rewritten in the form

$$\overset{\infty}{\mathbf{V}}_{\mathbf{d}}(x) \leq \mathbf{v}(x) := \inf_{(x,y)\in\mathcal{W}} y \leq \mathbf{d}(x) \tag{4.31}$$

and thus, we have proved that the valuation function $\overset{\infty}{\mathbf{V}}_{\mathbf{d}}(x)$ coincides with the function $\mathbf{v}$ associated with the viability kernel $\mathcal{W}$ of $\mathcal{K}$ under (4.26). □

4.9 Finite Horizon Optimal Control and Capture Basins

Consider parameterized system (2.10), p. 64:

$$x'(t) = f(x(t), u(t)) \text{ where } u(t) \in U(x(t))$$

generating the evolutionary system denoted $\mathcal{P}$ and

1. a "final state" cost function $\mathbf{c} : X \mapsto \mathbb{R}_+ \cup \{+\infty\}$,
2. a "transient state" cost function (often called a Lagrangian) $\mathbf{l} : X \times \mathcal{U} \mapsto \mathbb{R}_+$.

We introduce a finite horizon T that is used as a parameter in the argument of the valuation function:

Definition 4.9.1 ***[Intertemporal Optimization Problem]*** *Given an evolutionary system* $\mathcal{S} : X \mapsto \mathcal{C}(0,\infty;X)$ *and cost functions* $\mathbf{c}{:}X \mapsto \mathbb{R}_+ \cup \{+\infty\}$ *and* $\mathbf{l}{:}X \mapsto \mathbb{R}_+$, *the function* $\mathbf{V}$ *defined by*

$$\mathbf{V}(T,x) := \inf_{(x(\cdot),u(\cdot))\in\mathcal{P}(x)} \left(\mathbf{c}(x(T)) + \int_0^T \mathbf{l}(x(\tau),u(\tau))d\tau \right) \tag{4.32}$$

is called the valuation function *of the finite horizon control problem: find an evolution* $(\overline{x}(\cdot), \overline{u}(\cdot)) \in \mathcal{P}(x)$ *minimizing the intertemporal cost functional*

$$\mathbf{V}(T,x) = \mathbf{c}(\overline{x}(T)) + \int_0^T \mathbf{l}(\overline{x}(\tau),\overline{u}(\tau))d\tau$$

This is known as the Bolza problem *in control theory and, when* $\mathbf{l}(x,u) \equiv 0$, *the* Mayer problem.

We introduce the auxiliary micro–macro system

$$\begin{cases} (i) \ \tau'(t) = -1 \\ (ii) \ x'(t) = f(x(t),u(t)) \\ (iii) \ y'(t) = -\mathbf{l}(x(t),u(t)) \\ \qquad \text{where } u(t) \in U(x(t)) \end{cases} \tag{4.33}$$

We set $\mathbf{0}$ defined by

$$\mathbf{0}(t,x) = \begin{cases} 0 & \text{if } \quad t \geq 0, \\ +\infty & \text{if not} \end{cases} \tag{4.34}$$

and extend the function $\mathbf{c} : X \mapsto \mathbb{R} \cup \{+\infty\}$ to the function $\mathbf{c}_\infty : \mathbb{R}_+ \times X$ defined by

$$\mathbf{c}_\infty(t,x) := \begin{cases} \mathbf{c}(x) & \text{if } \quad t=0 \\ +\infty & \text{if not} \end{cases} \tag{4.35}$$

Theorem 4.9.2 ***[Viability Characterization of the Valuation Function]*** *The valuation function*

$$\mathbf{V}(T,x) := \inf_{x(\cdot)\in\mathcal{S}(x)} \left(\mathbf{c}(x(T)) + \int_0^T \mathbf{l}(x(\tau),u(\tau))d\tau \right)$$

is related to capture basin $\mathrm{Capt}_{(4.33)}(\mathcal{E}p(\mathbf{0}),\mathcal{E}p(\mathbf{c}_\infty))$ *of the epigraph of the function* $\mathbf{0}$ *with the target equal to the epigraph of the function* $\mathbf{c}_\infty$ *under auxiliary micro–macro system (4.33) by the formula*

$$\mathbf{V}(T,x) = \inf_{(T,x,y)\in\mathrm{Capt}_{(4.33)}(\mathcal{E}p(\mathbf{0}),\mathcal{E}p(\mathbf{c}_\infty))} y$$

Proof. (Theorem 4.7.2) Indeed, to say that (T,x,y) belongs to $\mathrm{Capt}_{(4.33)}(\mathcal{E}p(\mathbf{0}),\mathcal{E}p(\mathbf{c}_\infty))$ means that there exist an evolution

$$t \mapsto (T-t, x(t), y(t))$$

governed by auxiliary evolutionary system (4.33), p.162 starting at (T,x,y) and some time $\overline{t}\in[0,T]$ such that

$$\begin{cases} (i) \;\; (T-\overline{t}, x(\overline{t}), y(\overline{t})) \in \mathcal{E}p(\mathbf{c}_\infty) \\ (ii) \; \forall t\in[0,\overline{t}], \;\; (T-t,x(t),y(t)) \in \; \mathcal{E}p(\mathbf{0}) \end{cases}$$

because $(T-t,x(t),y(t)) \;\notin\; \mathcal{E}p(\mathbf{0})$ whenever $t>T$.

By the very definition of the epigraph, these inequalities can be written in the form

$$\begin{cases} (i) \;\; y(\overline{t}) \;\geq\; \mathbf{c}_\infty(T-\overline{t}, x(\overline{t})) \\ (ii) \; \forall t\in[0,\overline{t}], \;\; y(t) \;\geq\; \mathbf{0}(T-t,x(t)) \end{cases}$$

By definition of the function $\mathbf{c}_\infty$, the first inequality implies that $\overline{t} = T$ (otherwise, we would have $\infty > y(t) \geq \mathbf{c}_\infty(T-t,x(t)) \;=\; +\infty$) so that the above system of inequalities is equivalent to

$$\begin{cases} (i) \;\; y(T) \;\geq\; \mathbf{c}(x(T)) \\ (ii) \; \forall t\in[0,T], \;\; y(t) \;\geq\; 0 \end{cases}$$

We now recall that the evolution $x(\cdot)$ belongs to $\mathcal{S}(x)$ and that $y(t) = y - \int_0^t \mathbf{l}(x(\tau),u(\tau))d\tau$ by (4.33)(iii). Hence the above inequalities become

$$\begin{cases} (i)\ \ y \ \geq \ \mathbf{c}(x(T)) + \displaystyle\int_0^T \mathbf{l}(x(\tau), u(\tau))d\tau \\ (ii)\ \forall t \in [0, T],\ \ y \ \geq \ \displaystyle\int_0^t \mathbf{l}(x(\tau), u(\tau))d\tau \end{cases}$$

Since the first inequality holds true whenever (T, x, y) belongs to the capture basin $\text{Capt}_{(4.33)}(\mathcal{E}p(\mathbf{0}), \mathcal{E}p(\mathbf{c}_\infty))$, it implies also that

$$\mathbf{V}(T, x) \leq \mathbf{c}(x(T)) + \int_0^T \mathbf{l}(x(\tau), u(\tau))d\tau \leq \inf_{(T,x,y)\in\text{Capt}_{(4.33)}(\mathcal{E}p(\mathbf{0}),\mathcal{E}p(\mathbf{c}_\infty))} y$$

Conversely, we know that for any $\varepsilon > 0$, there exists an evolution $x_\varepsilon(\cdot) \in \mathcal{S}(x)$ such that

$$\mathbf{c}(x_\varepsilon(T)) + \int_0^T \mathbf{l}(x_\varepsilon(\tau), u_\varepsilon(\tau))d\tau \leq \mathbf{V}(T, x) + \varepsilon$$

Setting $y_\varepsilon(t) := \mathbf{V}(T, x) + \varepsilon - \int_0^t \mathbf{l}(x_\varepsilon(\tau), u_\varepsilon(\tau))d\tau$, we infer that $t \mapsto (T - t, x_\varepsilon(t), y_\varepsilon(t))$ is a solution to the auxiliary evolutionary system starting at $(T, x, \mathbf{V}(T, x) + \varepsilon)$ reaching the target $\mathcal{E}p(\mathbf{c}_\infty)$ in finite time since

$$\left(0, x_\varepsilon(T), \mathbf{V}(T, x) + \varepsilon - \int_0^T \mathbf{l}(x_\varepsilon(\tau), u_\varepsilon(\tau))d\tau\right) \in \mathcal{E}p(\mathbf{c}_\infty)$$

Hence $(T, x, \mathbf{V}(T, x) + \varepsilon)$ belongs to $\text{Capt}_{(4.33)}(\mathcal{E}p(\mathbf{0}), \mathcal{E}p(\mathbf{c}_\infty))$, so that $\inf_{(T,x,y)\in\text{Capt}_{(4.33)}(\mathcal{E}p(\mathbf{0}),\mathcal{E}p(\mathbf{c}_\infty))} y \leq \mathbf{V}(T, x) + \varepsilon$. Letting ε converge to 0, we obtain the converse inequality

$$\inf_{(T,x,y)\in\text{Capt}_{(4.33)}(\mathcal{E}p(\mathbf{0}),\mathcal{E}p(\mathbf{c}_\infty))} y \leq \inf_{x(\cdot)\in\mathcal{S}(x)} \left(\mathbf{c}(x(T)) + \int_0^T \mathbf{l}(x(\tau), u(\tau))d\tau\right)$$

In summary, we proved that

$$\mathbf{V}(T, x) = \inf\{y \text{ such that } (T, x, y) \in \text{Capt}_{(4.33)}(\mathcal{E}p(\mathbf{0}), \mathcal{E}p(\mathbf{c}_\infty))\} \ \square$$

4.10 Occupational Costs and Measures

Instead of measuring occupational costs, some physicists and other scientists use to measure "*occupational measures*" $\frac{1}{T}\int_0^T \varphi(x(t), u(t))dt$ over state-control evolutions governed by control system (2.10):

$$x'(t) = f(x(t), u(t)) \text{ where } u(t) \in U(x(t))$$

generating the evolutionary system $\mathcal{P}$ associating with any x the set of state-control pairs $(x(\cdot), u(\cdot))$ governed by (2.10) and starting from x. We denote by $\mathcal{P}^K(x)$ the set of pairs $(x(\cdot), u(\cdot)) \in \mathcal{P}(x)$ such that $x(\cdot)$ is viable in K.

As in the infinite horizon case, we define:

Definition 4.10.1 *[Occupational Costs] We associate with any positive lower semicontinuous function* $\varphi : \text{Graph}(U) \mapsto \mathbb{R}_+$ *with linear growth the* occupational measure $N_K(x, \varphi)$ *defined by*

$$N_K(x, \varphi) := \inf_{(x(\cdot),u(\cdot)) \in \mathcal{P}^K(x)} \frac{1}{T} \int_0^T \varphi(x(\tau), u(\tau)) d\tau \tag{4.36}$$

As for occupational cost, occupational measures can be couched in terms of capture basins:

Proposition 4.10.2 *[Viability Characterization of Occupational Costs] Let us associate with control system (2.10) and the function* φ *the auxiliary differential inclusion*

$$\begin{cases} (i) \ \tau'(t) = -1 \\ (ii) \ x'(t) = f(x(t), u(t)) \\ (iii) \ y'(t) = -\varphi(x(t), u(t)) \\ \quad \text{where } u(t) \in U(x(t)) \end{cases} \tag{4.37}$$

Then the occupational cost $N_K(x, \varphi)$ *is equal to*

$$N_K(x, \varphi) = \frac{1}{T} \inf_{(T,x,y) \in \text{Viab}_{(4.37)}(\mathbb{R}_+ \times K \times \mathbb{R}_+)} y$$

Proof. It is an immediate consequence of Theorem 4.7.2, p. 156. □

When the control system boils down to a differential equation $x'(t) = f(x(t))$ generating a deterministic evolutionary system $x(\cdot) := \mathcal{S}_f(x)$ (see Definition 2.4.1, p. 53), the function

$$d\mu_K(x) : \varphi \mapsto \frac{1}{T} \int_0^T \varphi(x(t)) dt := \frac{1}{T} \int_0^T \varphi((\mathcal{S}_f(x))(t)) dt$$

where φ ranges over the space $\mathcal{C}_b(K)$ of bounded continuous functions on K is a continuous linear functional. Consequently, it is a measure $d\mu_K(x)$, called the *occupational measure*, since the (topological) dual $\mathcal{M}(K) := (\mathcal{C}_b(K))^\star$ is the space of Radon measures with compact support on the locally compact subset K. We summarize studies by *Zvi Arstein, Valdimir Gaitsgory* and *Marc Quincampoix.*

Naturally, the linearity of this function is lost when the evolutionary system is no longer deterministic. Therefore, the functions $\varphi \in \mathcal{C}_b(\text{Graph}(U)) \mapsto N_K(x,\varphi) \in \mathbb{R}_+ \cup \{+\infty\}$ are no longer linear, and thus, are no longer measures in the sense of measure theory.

However, being the infimum of continuous linear functions with respect to test functions φ, they are upper semicontinuous positively homogeneous functions on the test function space $\mathcal{C}_b(\text{Graph}(U))$.

Since any upper semicontinuous positively homogenous function is the support function of a closed convex subset of its dual, we can associate with the function $\varphi \mapsto N_K(x,\varphi)$ the closed convex $dM_K(x) \subset \mathcal{M}(\text{Graph}(U))$.

Naturally, when the system is deterministic, this closed convex subset is reduced to a singleton $dM_K(x) = d\mu_K(x)$, which is the usual occupational Radon measure (with compact support) because the function $\varphi \mapsto N_K(x,\cdot)$ is linear and continuous on $\mathcal{C}_b(\text{Graph}(U))$.

Definition 4.10.3 ***[Generalized Occupational Measures under Control Systems]*** *Let us consider control system (2.10):*

$$x'(t) = f(x(t),u(t)) \text{ where } u(t) \in U(x(t))$$

and a closed environment K. The closed convex subset

$$\begin{cases} dM_K(x) := \Big\{ \mu \in \mathcal{M}(\text{Graph}(U)) \mid \forall \varphi \in \mathcal{C}_b(\text{Graph}(U)), \\ \displaystyle\int_{\text{Graph}(U)} \varphi(y,u)d\mu(y,u) \geq N_K(x,\varphi) \Big\} \end{cases} \tag{4.38}$$

is the generalized occupational measure *at x.*

Therefore, summarizing, we obtain the following formulas:

Theorem 4.10.4 ***[Viability Kernels and Occupational Measures]*** *Knowing either the viability kernel* $\text{Viab}_{(4.37)}(\mathbb{R}_+ \times K \times \mathbb{R}_+)$ *or the generalized occupational measure $dM_K(x)$, the occupational cost $N_K(x,\varphi)$ can* *be recovered thanks to the following formulas:*

$$\begin{cases} N_K(x,\varphi) & = \frac{1}{T} \inf_{(T,x,y)\in \mathrm{Viab}_{(4.37)}(\mathbb{R}_+\times K\times\mathbb{R}_+)} y \\ & := \inf_{d\mu\in dM_K(x)} \int_{\mathrm{Graph}(U)} \varphi(y,u)d\mu(y,u) \end{cases} \tag{4.39}$$

Remark: Infinite Horizon Occupational Measures. When the horizon is infinite, several approaches can be used, among which we suggest a simple one: use probability measures of the form

$$N_K^m(x,\varphi) := \inf_{(x(\cdot),u(\cdot))\in\mathcal{P}(x)} m\int_0^{+\infty} e^{-mt}\varphi(x(t),u(t))dt$$

Introducing the auxiliary micro–macro system

$$\begin{cases} (i)\ x'(t) & = f(x(t),u(t)) \\ (ii)\ y'(t) & = my(t)-\varphi(x(t),u(t)) \\ \quad \text{where } u(t) & \in U(x(t)) \end{cases} \tag{4.40}$$

Theorem 4.7.2, p. 156 implies that $x \mapsto N_K^m(x,\varphi)$ can be characterized by formula

$$N_K^m(x,\varphi) = \inf_{(x,y)\in \mathrm{Viab}_{(4.40)}(K\times\mathbb{R}_+)} y$$

Also, since the function $\varphi \mapsto N_K^m(x,\varphi)$ being upper semicontinuous, positively homogeneous and concave, we can characterize it by the support function of a closed convex subset $dM_K^m(x) \subset \mathcal{C}_b(\mathrm{Graph}(U))$ (see (4.38), p.166), so that

$$N_K^m(x,\varphi) = \inf_{d\mu\in dM_K^m(x)} \int_{\mathrm{Graph}(U)} \varphi(y,u)d\mu(y,u)$$

The drawback with this approach is that the concept of generalized occupational measure depends upon the discount rate m. Another strategy defines limiting occupational costs

$$N_K^\infty(x,\varphi) := \inf_{(x(\cdot),u(\cdot))\in\mathcal{P}(x)} \sup_{T\geq 0} \frac{1}{T}\int_0^T \varphi(x(\tau),u(\tau))d\tau$$

the study of which being outside the scope of this book. □

4.11 Optimal Control Survival Kit

We summarize the questions that are answered by using the properties of the capture basins in the framework of finite-horizon control problems (4.32)

$$\mathbf{V}(T,x) := \inf_{(x(\cdot),u(\cdot))\in\mathcal{P}(x)} \left(\mathbf{c}(x(T)) + \int_0^T \mathbf{l}(x(\tau),u(\tau))d\tau \right)$$

that we shall state and prove rigorously in Chap. 17, p. 681. The adaptation of these statements to infinite horizon optimization is left to the reader. Let the horizon T and the initial state x be fixed.

We shall assume once and for all in this section that:

1. the control system $x'(t) = f(x(t),u(t))$ where $u(t) \in U(x(t))$ is Marchaud,
2. the function $x \mapsto \mathbf{c}(x)$ is lower semicontinuous,
3. the function $u \mapsto \mathbf{l}(x,u)$ is positive and convex,
4. the function $(x,u) \mapsto \mathbf{l}(x,u)$ is lower semicontinuous,
5. $l^\sharp := \sup_{(x,u)\in \mathrm{Graph}(U)} \mathbf{l}(x,u)$ is finite.

We begin by the existence of at least one optimal evolution: Recall that the associated micro–macro system (4.33), p. 162 is defined by

$$\begin{cases} (i) \;\; \tau'(t) = -1 \\ (ii) \;\; x'(t) = f(x(t),u(t)) \\ (iii) \; y'(t) = -\mathbf{l}(x(t),u(t)) \\ \qquad \text{where } u(t) \in U(x(t)) \end{cases}$$

Theorem 17.3.6, p. 692 implies

Theorem 4.11.1 *[Existence of Optimal Evolutions] The valuation function* $\mathbf{V}$ *is lower semicontinuous and its epigraph*

$$\mathcal{E}p(\mathbf{V}) \;=\; \mathrm{Capt}_{(4.33)}(\mathcal{E}p(\mathbf{0}), \mathcal{E}p(\mathbf{c}_\infty))$$

(where $\mathbf{0}$ *is the function defined by (4.34), p.162) is equal to the capture basin above. Furthermore,*

1. *there exists at least one optimal evolution minimizing the valuation functional* $\mathbf{V}$*,*
2. *optimal evolutions* $x(\cdot)$ *are the components of the evolutions* $t \mapsto (T - t, x(t), y(t))$ *governed by auxiliary micro–macro system and viable on* $[0,T]$*.*

Theorem 17.3.9, p. 696 states that optimal evolutions do satisfy the Isaac–Bellman's dynamical optimality principle:

Theorem 4.11.2 ***[Dynamic Programming Equation]*** *Optimal evolutions $x(\cdot)$ satisfy* dynamic programming equation

$$\mathbf{V}(T-s,x(s))+\int_0^s \mathbf{l}(x(\tau),u(\tau))d\tau \;=\; \mathbf{V}(T,x) \qquad (4.41)$$

for all $s \in [0,T]$.

When the evolutionary system is the solution map of control system (2.10), p. 64:

$$x'(t) = f(x(t),u(t)) \text{ where } u(t) \in U(x(t))$$

the tangential characterization of the viability kernels provided by the Theorem 11.4.6, p. 463 implies the following statement:

Theorem 4.11.3 ***[Hamilton–Jacobi–Bellman Equation for the Valuation Function]*** *The valuation function is the unique solution (defined in an adequate "generalized" sense) to Hamilton–Jacobi–Bellman partial differential equation*

$$-\frac{\partial \mathbf{V}(t,x)}{\partial t}+\inf_{u\in U(x)}\left(\sum_{i=1}^{n}\frac{\partial \mathbf{V}(t,x)}{\partial x_i}f_i(x,u)+\mathbf{l}(x,u)\right)=0$$

satisfying the initial condition $\mathbf{V}(0,x)=\mathbf{c}(x)$.

The sign "minus" in front of $\frac{\partial \mathbf{V}(t,x)}{\partial t}$ appears here since we study the valuation function $\mathbf{V}(t,x)$ parameterized by the horizon here denoted by t instead of the standard value function where the horizon is fixed and the current time is denoted by t, providing the Hamilton–Jacobi–Bellman equation in the traditional form

$$+\frac{\partial \mathbf{V}(t,x)}{\partial t}+\inf_{u\in U(x)}\left(\sum_{i=1}^{n}\frac{\partial \mathbf{V}(t,x)}{\partial x_i}f_i(x,u)+\mathbf{l}(x,u)\right)=0$$

The valuation function is useful in the extent that it allows us to build the *regulation map* that provides the controls or regulons governing the evolution of optimal solutions:

Definition 4.11.4 ***[Regulation Map]*** *The regulation map for the intertemporal optimization problem is defined by*

$$\mathbb{R}(t,x) := \left\{ u \in U(x) \mid -\frac{\partial \mathbf{V}(t,x)}{\partial t} + \sum_{i=1}^{n} \frac{\partial \mathbf{V}(t,x)}{\partial x_i} f_i(x,u) + \mathbf{l}(x,u) = 0 \right\}$$

We shall prove the following regulation mechanism:

Theorem 4.11.5 ***[Regulating Optimal Evolutions]*** *Optimal solutions are governed by the control system*

$$\forall t \in [0,T], \quad \begin{cases} (i) \ x'(t) = f(x(t),u(t)) \\ (ii) \ u(t) \in \mathbb{R}(T-t,x(t)) \end{cases} \tag{4.42}$$

For simplicity of the exposition, we restricted in this chapter the concepts of intertemporal optimization to problems without state constraints. We can add state constraints, or even, functional viability constraints, which are associated with any positive extended function $\mathbf{k} \leq \mathbf{c}$ by formula

$$\text{Capt}_{(4.33)}(\mathcal{E}p(\mathbf{k}), \mathcal{E}p(\mathbf{c}_\infty))$$

Theorem 17.3.4, p. 689 implies that the capture basin of the epigraph of the function $\mathbf{c}$ viable in the epigraph of the function $\mathbf{k}$ is the epigraph of valuation function of an intertemporal optimization problem described in Definition 17.3.2, p. 687.

Remark. The viability solution to these intertemporal optimization problems is a "concrete" one, which can be computed by the viability algorithm. We just prove that for Marchaud control systems, the valuation function is lower semicontinuous. Next, regularity issues arise: under which conditions is the valuation function continuous, Lipschitz, more or less differentiable, as a function of the state or as a function of time along an (optimal) evolution, etc. These issues are not dealt with in this book, but an immense literature is available. The questions related to necessary conditions, an adaptation to intertemporal optimization problems of the *Fermat rule*, known under the concept of *Pontriagin principle*, are not addressed. For a survey, we refer to [103, 104, Frankowska]. □

4.12 Links Between Viability Theory and Other Problems

We already have seen many examples of intertemporal optimization problems over evolutions of state-control pairs governed by micro–macro systems of the form

$$\begin{cases} (i) \ x'(t) &= f(x(t), u(t)) \\ (ii) \ y'(t) &= my(t) - \mathbf{l}(x(t), u(t)) \text{ where } u(t) \in U(x(t)) \end{cases}$$

and to environmental and epigraphical environments $\mathcal{E}p(\mathbf{k})$ and epigraphical targets $\mathcal{E}p(\mathbf{c})$.

We also mentioned that the value functions of these intertemporal optimization problems are solutions to Hamilton–Jacobi–Bellman equations, as is proved rigourously in Theorem 17.4.3, p. 701.

We devote this section to these three types or problems and try to uncover in a more systematic way the links between each of these three points of view and viability kernels and their properties, and, consequently, between each of these three problems.

4.12.1 Micro–Macro Dynamic Consistency

As we saw in the introduction,

Definition 4.12.1 *[Micro–Macro Systems] A system*

$$\begin{cases} (i) \ x'(t) &= f(x(t), u(t)) \text{ where } u(t) \in U(x(t)) \\ (ii) \ y'(t) &= g(x(t), y(t), u(t)) \text{ where } u(t) \in \mathbf{V}(x(t), y(t)) \end{cases} \tag{4.43}$$

governing the evolution of pairs $(x(t), y(t)) \in X \times Y$ is called a micro–macro system *whenever the dynamics governing the evolution $x(\cdot)$ of the state does not depend on the macro variable $y(\cdot)$, whereas the evolution of the macro variable depends upon the state variable either through its control law or trough constraints linking micro and macro variables.*

Naturally, we assume that whenever $u \in \mathbf{V}(x, y)$, then $u \in U(x)$. Since the micro–macro states range over the product $X \times Y$, an environment (denoted by) $\mathcal{K} \subset X \times Y$ is a subset of $X \times Y$, as well as a target $\mathcal{C} \subset \mathcal{K} \subset X \times Y$, so that the viability kernel $\text{Viab}_{(4.43)}(\mathcal{K}, \mathcal{C}) \subset X \times Y$ with target $\mathcal{C}$ viable in $\mathcal{K}$ is a subset of $X \times Y$. In the graphical framework we advocate in this book, any subset $\mathcal{V} \subset X \times Y$ can be regarded as the graph

$$\mathcal{V} =: \text{Graph}(\mathbf{V})$$

of the set-valued map defined by

$$\forall x \in X, \ \mathbf{V}(x) := \{y \in Y \text{ such that } (x,y) \in \mathcal{V}\}$$

(see Definition 18.3.1, p. 719). Therefore, we can associate with the subsets $\mathcal{C}$ and $\mathcal{K}$ three set-valued maps

$$\begin{cases} (i) & \text{Graph}(\mathbf{C}) := \mathcal{C} \\ (ii) & \text{Graph}(\mathbf{K}) := \mathcal{K} \\ (iii) & \text{Graph}(\mathbf{V}_{(\mathbf{K},\mathbf{C})}) := \text{Viab}_{(4.43)}(\mathcal{K},\mathcal{C}) \end{cases}$$

satisfying

$$\forall x \in X, \ \mathbf{C}(x) \subset \mathbf{V}_{(\mathbf{K},\mathbf{C})}(x) \subset \mathbf{K}(x)$$

Theorem 2.15.1, p. 99 states that the set-valued map $\mathbf{V}_{(\mathbf{K},\mathbf{C})}$ is the *unique* set-valued map from X to Y satisfying the above inclusions and

$$\mathbf{V}_{(\mathbf{V}_{(\mathbf{K},\mathbf{C})},\mathbf{C})} = \mathbf{V}_{(\mathbf{K},\mathbf{C})} = \mathbf{V}_{(\mathbf{K},\mathbf{V}_{(\mathbf{K},\mathbf{C})})}$$

Consequently, whenever set-valued map $\mathbf{K}$ and $\mathbf{C} \subset \mathbf{K}$ are given, the set-valued map $\mathbf{V}_{(\mathbf{K},\mathbf{C})}$ can be regarded as a *viability graphical solution* of a micro–macro problem associated with the data $\mathbf{K}$ and $\mathbf{C}$.

The set-valued map $\mathbf{V}_{(\mathbf{K},\mathbf{C})}$ associates with any x the subset of elements $y \in \mathbf{K}(x)$ such that there exists a micro–macro $(x(\cdot), y(\cdot))$ evolution governed by (4.43), p. 171 and a time $t^\star$ such that $y(t) \in \mathbf{K}(x(t))$ for all $t \in [0,T]$ and $y(t^\star) \in \mathbf{C}(x(t^\star))$. We shall see in Sect. 4.12.3, p. 174 that this set-valued map $\mathbf{V}_{(\mathbf{K},\mathbf{C})}$ is the *unique solution* of a partial differential equation or partial differential inclusion, among which we shall find Burgers type partial differential equation (see Chap. 16, p. 631).

4.12.2 The Epigraphical Miracle

We now turn our attention to the special but very important case when $Y := \mathbb{R}$, i.e., when the macro variable is a scalar. In this case, we recall (Definition 18.6.5, p. 745) that a subset $\mathcal{V} \subset X \times \mathbb{R}$ is an *epigraph* if

$$\mathcal{V} + \{0\} \times \mathbb{R}_+ = \mathcal{V}$$

that epigraphs of functions are epigraph and that, conversely, if the epigraphs are closed, they are the epigraphs of (lower semicontinuous) extended functions.

Theorem 4.12.2 *[Viability Kernels of Epigraphs] Consider control systems of the form*

$$\begin{cases} (i) \ x'(t) = f(x(t),u(t)) \text{ where } u(t) \in U(x(t)) \\ (ii) \ y'(t) = g(x(t),y(t),u(t)) \text{ where } u(t) \in \mathbf{V}(x(t),y(t)) \end{cases} \quad (4.44)$$

where $y(t) \in \mathbb{R}$ is a real variable.

Assume that the micro–macro environment $\mathcal{K}$ and targets $\mathcal{C}$ are epigraphs. So is the viability kernel $\text{Viab}_{(4.43)}(K,C)$.

Proof. Let us consider a pair $(x,y) \in \text{Viab}_{(4.44)}(\mathcal{K},\mathcal{C})$ and $\widetilde{y} \geq y$. Then, there exist an evolution $(x(\cdot),y(\cdot))$ starting at (x,y) and a finite time $t^\star$ such that $(x(t),y(t)) \in \mathcal{K}$ forever or until the finite time $t^\star$ when $(x(t^\star),y(t^\star)) \in \mathcal{C}$. If $\widetilde{y} > y$, there exists an evolution $\widetilde{y}(\cdot)$ governed by $\widetilde{y}'(t) = g(x(t),\widetilde{y}(t),u(t))$ larger than or equal to $y(\cdot)$. Indeed, since $\widetilde{y}(0) > y(0)$ and since the functions $\widetilde{y}(\cdot)$ and $y(\cdot)$ are continuous, either $\widetilde{y}(t) > y(t)$ for all positive $t \geq 0$ or there exists a smallest finite time $\widehat{t}$ when $\widetilde{y}(\widehat{t}) = y(\widehat{t})$. It is then enough to concatenate the evolution $\widetilde{y}(\cdot)$ with the solution $y(\cdot)$ at time $\widehat{t}$ (see Definition 2.8.1, p. 69) to observe that the concatenation is still an evolution starting at $\widetilde{y}$ and larger or equal to the evolution $y(\cdot)$. Therefore, $(x(t),\widetilde{y}(t)) = (x(t),y(t)+p(t))$ where $p(t) \geq 0$, so that $(x(t),\widetilde{y}(t)) \in \mathcal{K}$ forever or until the finite time $t^\star$ where $(x(t^\star),\widetilde{y}(t^\star)) \in \mathcal{C}$. This means that $(x,\widetilde{y})$ belongs to the viability kernel $\text{Viab}_{(4.44)}(\mathcal{K},\mathcal{C})$, which is, then, an epigraph. □

Therefore, if the environment $\mathcal{K} := \mathcal{E}p(\mathbf{k})$ and the target $\mathcal{C} := \mathcal{E}p(\mathbf{c})$ are epigraphs of extended functions $\mathbf{k} \leq \mathbf{c}$, and whenever the viability kernel is closed, it is the epigraph of the function denoted by $\mathbf{v}_{(\mathbf{k},\mathbf{c})}$, satisfying inequalities

$$\mathbf{k}(x) \leq \mathbf{v}_{(\mathbf{k},\mathbf{c})}(x) \leq \mathbf{c}(x)$$

Theorem 2.15.1, p. 99 states that the extended function $\mathbf{v}_{(\mathbf{k},\mathbf{c})}$ is the *unique* extended function from X to $\mathbb{R} \cup \{+\infty\}$ satisfying the above inequalities and

$$\mathbf{v}_{(\mathbf{v}_{(\mathbf{k},\mathbf{c})},\mathbf{c})} = \mathbf{v}_{(\mathbf{k},\mathbf{c})} = \mathbf{v}_{(\mathbf{k},\mathbf{v}_{(\mathbf{k},\mathbf{c})})}$$

We shall see in Sect. 4.12.3, p. 174 that this extended function $\mathbf{v}_{(\mathbf{k},\mathbf{c})}$ is the *unique solution* of a Hamilton–Jacobi–Bellman partial differential equation, as we have seen in these examples.

4.12.3 First-Order Systems of Partial Differential Equations

Since we have related viability kernels under auxiliary micro–macro systems as graphs of set-valued maps or epigraphs of extended functions (when $Y := \mathbb{R}$), the question arises to take advantage of these graphical or epigraphical representations for formulating the statements of Viability Theorem 11.3.4, p. 455 and Invariance Theorem 11.3.7, p. 457.

The basic idea goes back to *Fermat* in 1637, when, translated in modern terms, he defined the tangent to the graph of a function (from $\mathbb{R}$ to $\mathbb{R}$) is the graph of its derivative. This was the starting point of the differential calculus of set-valued maps: the graph of the derivative of a set-valued map is the tangent cone to its graph (see Definition 18.5.3, p. 739 and Theorem 18.5.4, p. 739). This is how the tangential conditions characterizing the graph of the set-valued map $\mathbf{V}_{(\mathbf{K},\mathbf{C})}$ can be translated by saying that $\mathbf{V}_{(\mathbf{K},\mathbf{C})}$ is a solution to a first-order "partial differential inclusion". Needless to say, in the case we recover usual systems of first-order with smooth solution, the above set-valued solution coincides with the regular one.

When $Y := \mathbb{R}$ and since we related viability kernels under auxiliary micro–macro systems as epigraphs of extended functions, we follow the same approach, defining "epiderivatives" of extended functions through their epigraphs: the epigraph of an epiderivative is the tangent cone to the epigraph of the function (see Definition 18.6.9, p. 747 and Theorem 18.6.10, p. 748).

It is then enough to "translate" the tangential conditions on the epigraph of the extended function $\mathbf{v}_{(\mathbf{k},\mathbf{c})}$, which is a viability kernel under a micro–macro system, in terms of epiderivatives for stating that it is a generalized (extended) solution of Hamilton–Jacobi–Bellman partial differential equations of the type

$$\mathbf{h}\left(x, \mathbf{v}(x), \frac{\partial \mathbf{v}(x)}{\partial x}\right) := \inf_{u \in \mathbf{V}(x,\mathbf{v}(x))} \left(\left\langle \frac{\partial \mathbf{v}(x)}{\partial x}, f(x,u)\right\rangle - g(x,\mathbf{v}(x),u)\right) = 0$$

with adequate conditions depending upon the auxiliary environment $\mathcal{E}p(\mathbf{k})$ and target $\mathcal{E}p(\mathbf{b})$, where the function $\mathbf{h}:(x,y,p) \in X \times \mathbb{R} \times X^\star \mapsto \mathbf{h}(x,y,p)$ defined by

$$\mathbf{h}(x,y,p) := \inf_{u \in \mathbf{V}(x,y)} (\langle p, f(x,u)\rangle - g(x,y,u))$$

is called an *Hamiltonian*.

The Hamilton–Jacobi–Bellman partial differential equation depends only the data f, g and $\mathbf{V}$ describing the micro–macro system, whereas inequalities

$$\mathbf{k}(x) \le \mathbf{v}_{(\mathbf{k},\mathbf{c})}(x) \le \mathbf{c}(x)$$

depend on the environment and target functions $\mathbf{k}$ and $\mathbf{c}$. Hence, there are as many solutions to the Hamilton–Jacobi–Bellman partial differential equation as pairs $(\mathbf{k}, \mathbf{c})$ of functions such that $\mathbf{k} \leq \mathbf{c}$.

In this way, the Viability and Invariance Theorems provide existence and uniqueness of a solution to a Hamilton–Jacobi–Bellman equation. Naturally, since the solutions are not necessarily differentiable in the classical sense, the derivatives involved are epiderivatives, which coincide with usual derivatives whenever the solution is differentiable. These conditions state that the value function $\mathbf{v}_{(\mathbf{k},\mathbf{c})}$ is a "generalized solution" to the above Hamilton–Jacobi–Bellman equation. They bear as many names as concepts of tangents and normal provided by set-valued analysis (for instance, viscosity solutions, Barron–Jensen/Frankowska solutions, etc. See Definition 17.4.2, p. 701). This is at the level of the translation of tangential properties of graphs and epigraphs that technical difficulties arise, and not at the level of existence, uniqueness and stability properties of the solution to the partial differential equation, which derives directly from the existence, uniqueness and stability properties of viability kernels. Above all, *this is no longer useful*, since Viability Kernel Algorithms provide directly both the value function and the optimal regulation map. We shall postpone the thorough study to the end of this book (see Chap. 17, p. 681), since the translation from tangential properties to partial differential equation is quite technical.

The inverse question arises: what are the Hamilton–Jacobi–Bellman partial differential equations

$$\mathbf{h}\left(x, \mathbf{v}(x), \frac{\partial \mathbf{v}(x)}{\partial x}\right) = 0$$

which can be derived from the tangential characterization under an unknown micro–macro system? Not all, naturally, since a necessary condition is that for any $(x, y) \in X \times \mathbb{R}$, the Hamiltonian $p \mapsto \mathbf{h}(x, y, p)$ is concave and upper semicontinuous. In this case, we can always associate with the Hamiltonian $\mathbf{h}$ a micro–macro controlled system

$$\begin{cases} (i) \ x'(t) &= u \\ (ii) \ y'(t) &= -\mathbf{l}(x(t), y(t), u(t)) \end{cases}$$

where the (artificial) controls u are chosen in X and where $\mathbf{l}$ is a Lagrangian associated with the Hamiltonian in an adequate way. In this case, controls and partial derivatives of the value functions are in duality, a situation common in mechanics and physics, which provided the motivations for *Jean Jacques Moreau* to be a co-founder of convex analysis.

Once related to a micro–macro system where the macro-system is affine with respect to y, their solutions are value functions of intertemporal optimization of criteria concealed in the partial differential equation.

We shall illustrate this important method relying in convex analysis (Sect. 18.7, p. 755) in Sect. 14, p. 563 for nonlinear Hamilton–Jacobi–Bellman partial differential equations arising in finance (Sect. 15.2, p. 605), in micro–macro economics (Sect. 15.3, p. 620) and a macroscopic approach to traffic theory (Chap. 14, p. 563).

4.12.4 Intertemporal Optimization Under State Constraints

We already have related several intertemporal optimization problems over evolutions of state-control pairs governed by a control system

$$\begin{cases} (i)\ x'(t) &= f(x(t),u(t)) \\ (ii)\ y'(t) &= my(t) - \mathbf{l}(x(t),u(t)) \text{ where } u(t) \in U(x(t)) \end{cases}$$

and to environmental and epigraphical environments $\mathcal{E}p(\mathbf{k})$ and epigraphical targets $\mathcal{E}p(\mathbf{c})$.

In this chapter, we started from functionals involving the transient costs $\mathbf{l}$ and spot costs $\mathbf{c}$ involved in the criterion, and the constraints where plain environmental constraints $K \subset X$, with which we associate the *indicator function* $\mathbf{k} := \psi_K$ of K (see Definition 18.6.1, p. 743).

Naturally, if the links between value function of some classical control problems (*Bolza problems*) and viability kernel of epigraphs with epigraphical targets under adequate micro–macro systems are straightforward, it is not always the case: the discovery of some micro–macro systems may require some ingenuity.

However, we can proceed differently, and compute the viability kernels of the epigraphical environments with epigraphical targets and discover the associated intertemporal criterion (see Sect. 17.3.1, p. 685). This provides a factory of intertemporal optimization problems which will be reviewed in Chap. 17, p. 681.

Once this micro–macro system is found, we use the links between micro–macro systems and first-order partial differential equations to derive the Hamilton–Jacobi–Bellman equation providing the value function for each problem associated with a discount factor $\mathbf{m}$ (which may depend upon the state-control pair), the transient cost $\mathbf{l}$, satisfying the conditions

$$\mathbf{k}(x) \leq \mathbf{v}_{(\mathbf{k},\mathbf{c})}(x) \leq \mathbf{c}(x)$$

associated with the constraint $\mathbf{k}$ and spot costs $\mathbf{c}$.

Therefore, value functions inherit the properties of viability kernels and capture basins, and *optimal evolutions can be regulated by the regulation maps* provided by Viability Theorem 11.3.4, p. 455 and Invariance Theorem 11.3.7, p. 457. Furthermore, they can be computed by the Viability Kernel Algorithms.

4.12.5 The Viability Strategy for Dynamic Optimization

The list of intertemporal optimization problems we provided in this Section is far to be exhaustive, and we shall review other later in the book. For instance, by taking invariance kernels instead of viability kernels, we obtain optimization problems of the same nature where the infimum over the set of evolutions is replaced by the supremum.

On the other hand, we shall see that other problems in dynamical economics and social sciences are naturally formulated in terms of viability and/or invariance kernels of epigraphs of functions with targets which are themselves epigraphs of functions.

According to the choice of these functions and of the auxiliary problems, we can cover a wide spectrum of dynamical optimization problems.

Being related to viability kernels, the Viability Kernel Algorithm allows us to compute these valuation functions. We can derive from the characterization of the viability kernel in terms of tangential conditions that these valuation functions are solutions to Hamilton–Jacobi–Bellman equations in a adequate generalized sense, since these functions are not necessarily differential, not even continuous.

Therefore, we suggest:

1. to relate a whole class of problems to much simpler corresponding viability/capturability problems,
2. to "solve" these viability/capturability problems at this general level and gather as much characterization theorems and properties of various kinds as possible,
3. to approximate and compute these valuation functions by using viability/capturability algorithms and software,
4. when the environments and the targets are epigraphs of functions or graphs of maps, to use set-valued analysis and nonsmooth analysis for translating the general results on viability kernels and for giving a precise meaning to the concept of generalized solutions to systems of first-order Hamilton–Jacobi–Bellman differential equations, as we shall do in Chaps. 13, p.523, and 17, p. 681.

Chapter 5
Avoiding Skylla and Charybdis

5.1 Introduction: The Zermelo Navigation Problem

This chapter provides a similar viability framework with two-dimensional nonsmooth environments, targets and a nonlinear control system for which we illustrate and compare basic concepts, such as minimal length and exit time functions, minimal time, Lyapunov and value function of an optimal control problem.

Ernst Zermelo [1871–1953]. *The German mathematician* Ernst Zermelo *is mainly known for his major contributions to the foundations of mathematics when* David Hilbert *challenged his colleagues with the first of his 23 celebrated problem during the 1900 conference of the International Congress of Mathematicians in Paris, dealing with the* continuum hypothesis *introduced by* Georg Cantor. *Before that, he started his mathematical investigations on the calculus of variations and studied hydrodynamics under the guidance of* Max Planck. *As early as 1935, Zermelo resigned his chair to protest Hitler's regime, to which he was reinstated at the end of World War II.*

18 ***The Zermelo Navigation Problem.*** In his 1935 book [54, *Calculus of Variations and partial differential equations of the first order*], *Constantin Carathéodory* mentions that Zermelo "completely solved by an extraordinary ingenious method" the "Zermelo Navigation Problem" stated as follows: *In an unbounded plane where the wind distribution is given by a*

J.-P. Aubin et al., *Viability Theory*, DOI 10.1007/978-3-642-16684-6_5,

vector field as a function of position and time, a ship moves with constant velocity relative to the surrounding air mass. How much the ship be steered in order to come from a starting point to a given goal in the shortest time?

The state variables x and y denote the coordinates of the moving ship, $f(x, y)$ and $g(x, y)$ the components of the wind velocity. There are two control variables for governing the evolution of the ship: the *velocity* v, the norm $\|v\|$ of which is bounded: $\|v\| \leq c$, and the *steering direction* u, that is, the angle which the vector of the relative velocity forms with the x-direction. Indeed, the components of the absolute velocity are $f(x(t), y(t)) + v(t) \cos u(t)$ and $g(x(t), y(t)) + v(t) \sin u(t)$. We can also incorporate state-dependent constraints on velocity by a bound $c(x, y)$ and constraints on the steering direction described by the bounds $\alpha_{\min}(x, y)$ and $\alpha_{\max}(x, y)$.

The evolution of the ship is thus governed by the following control system

$$\begin{cases} (i) \ \ x'(t) = f(x(t), y(t)) + v(t) \cos u(t) \\ (ii) \ y'(t) = g(x(t), y(t)) + v(t) \sin u(t) \\ \quad \text{where } u(t) \in [\alpha_{\min}(x(t), y(t)), \alpha_{\max}(x(t), y(t))], \|v(t)\| \leq c(x(t), y(t)) \\ \quad 0 \leq \alpha_{\min}(x, y) \leq \alpha_{\max}(x, y) \leq 2\pi \text{ and } 0 \leq c(x(t), y(t)) \end{cases} \tag{5.1}$$

We shall investigate Zermelo's problem, where we replace the "unbounded plane" by an arbitrary closed *environment* $K \subset \mathbb{R}^2$ *with obstacles*, the "goal" being the *harbor*, playing the role of the "target" $C \subset K$ in our terminology. This is of course is a more challenging problem because of these added constraints.

We instantiate these results numerically when the evolution of the ship is governed by Zermelo's equation

$$\begin{cases} (i) \ \ x'(t) = v(t) \cos u(t) \\ (ii) \ y'(t) = a\left(b^2 - x^2\right) + v(t) \sin u(t) \\ \quad \text{where } u(t) \in [0, 2\pi], \ \|v(t)\| \ \leq c \end{cases} \tag{5.2}$$

with $a = \dfrac{1}{80}$, $b = 12$ and $c = 1$.

The environment and target shown next provide the basis for five two-dimensional examples for the sake of comparison. We chose a compromise between enough complexity for underlying interesting features and simplicity to have the same figures to adapt to.

5.2 The Odyssey

Odyssey. *Odysseus (or Ulysses) was warned by Circe that if he sailed too close to Scylla, she would attack and swallow his crew. If he sailed near by Charybdis, she would suck the ship down the sea.*

So was Odysseus' dilemma: fight Scylla, and thus, be attracted to her down below the sea by Charybdis, or sail too close to Scylla, and loose six sailors. But Odysseus survived 10 years to reach Ithaca and the arms of Penelope. Had Odysseus known viability theory, that even his *hubris* could not replace, he could have saved his crew by finding a safe route to Ithaca avoiding the monsters from any point of the capture basin, and in particular from the point B of Figure 5.6, p. 185 (left), where he would have had the provocative pleasure to sail safely between them. Either to the secure part of the harbor where the winds would not blow him away, i.e., the viability kernel of the harbor, which he could have done starting from the permanence kernel, since he is still living in our minds, or in the insecure part, where the ship wrecks, also for eternity.

Fig. 5.1 Charybdis and Scylla.
Since the dispute goes on to guess where these monsters were, either in the Strait of Messina, or Cape Skilla, we chose to locate it in Yellowstone Park (Left). The environment K is schematized by a square sea (Right), with a waterfall on the north, cliffs the west and the east where the ship cannot berth, the two monsters Scylla Charybdis south of a dyke. The target C is the harbor. The winds blow with a stronger force in the center, so that ships cannot sail south in the center, but weak enough along the banks to allow the ships to sail back south.

This light reminder illustrates on this example how to escape Odysseus dilemma. Not only starting from the capture basin, but by reaching the harbor in finite (and shortest) time, or in an asymptotic way, or with the shortest route, or by minimizing the effort, or for reaching the harbor of Ithaca, and stay there forever in the arms of Penelope, the real happy end of Homer's Odyssey. Or, if by a bad trick of Tyche, he left, even for a second, the capture basin, he will survive as long as possible until he fought Scylla, or be drowned by Charybdis or die in the waterfall.

5.3 Examples of Concepts and Numerical Computations

5.3.1 *Dynamical Morphological Structure of the Environment and its Behavior*

We first illustrate some of the basic concepts presented in this book (we do not explicitly mention control system (5.2), p. 180 because it is the only one used in this chapter).

- *Kernels and Basins.* Figure 5.2 (**Left**), p. 182 displays, the *viability kernel* Viab(K) of the environment K, including naturally the harbor $C \subset K$ (left) (see Definition 2.10.2, p.86). Fig. 5.2 (**Right**), p. 182 describes the *capture basin* Capt(K, C) of the harbor, in dark, superposed to the viability kernel Viab(K). It displays the subset Viab$(K) \setminus$ Capt(K, C) of initial states viable in the environment but not reaching the harbor in finite time.

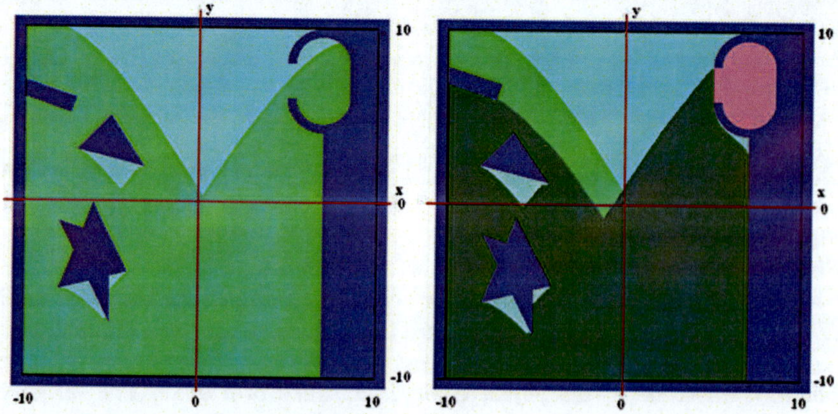

Fig. 5.2 Viability Kernel and Capture Basin.

The subset $\text{Viab}(K \setminus C)$ of evolutions viable in $K \setminus C$ is depicted in Fig. 5.3 (**Left**), p. 183. This subset being not empty, $K \setminus C$ is not a *repeller*. Figure 5.3 (**Right**), p. 183 displays the viability kernel $\text{Viab}(C) \subset C$ of the harbor, which is not empty.

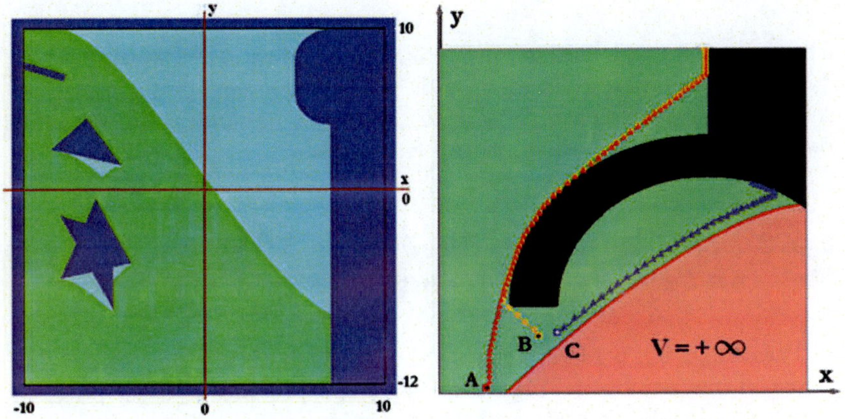

Fig. 5.3 Viability Kernel outside the Harbor and close up view of the Viability Kernel of the Harbor.
The trajectories of three evolutions are displayed: starting from A, B and C, outside of the viability kernel of the harbor, they leave the environment in finite time, the two first ones outside and the third one inside the harbor.

- *Equilibrium and Exit Sets.*

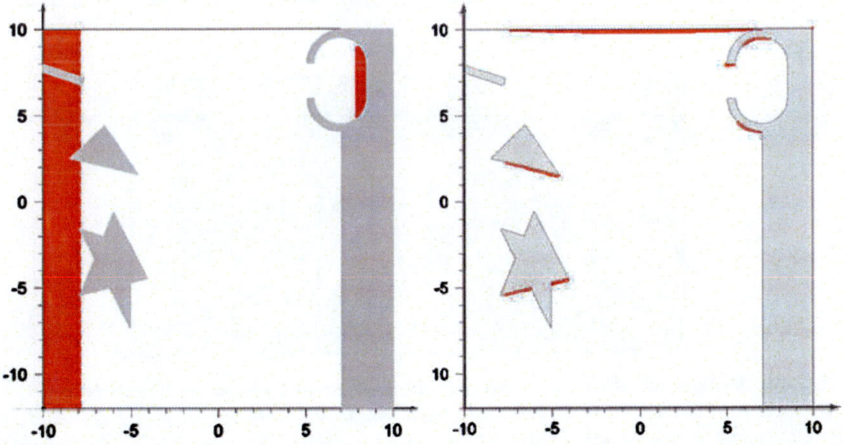

Fig. 5.4 Equilibrium and Exit Sets.
(Left). The equilibrium subset *(see Definition 9.2.13, p.332) is the subset of states of the environment which remain still, either because the velocity*

is equal to zero or because the directions of the winds and the ship are "balanced". There are equilibria close to the left bank, and, inside the harbor, close to the right bank. (Right). The figure displays the exit set (see Definition 10.4.7, p.396), contained $\partial K \setminus \mathrm{Viab}(K)$, where all evolutions leave the environment immediately.

These kernels and basins, providing a "qualitative" description of the dynamical behavior consistent with the environment and the target, are the domains of indicator functions, adding a "quantitative" provided by *indicators.* We classify them in two categories.

5.3.2 Behavior in the Environment: Minimal Length and Persistent Evolutions

- Theorem 4.4.2, p.140 states that the finite length viability kernel contains the viable equilibria. Figure 5.10, p. 188 provides a description of the minimal length function and its properties.
- Definition 4.3.4, p.135 implies that the complement of the viability kernel of the environment is the domain of the exit time function, indicating the maximal time a persistent evolution can remain in the environment before leaving it. Figure 5.11, p. 190 provide a description of the exit time function and its properties.

We reproduce below trajectories of minimal length and persistent evolutions for the sake of comparison of behavior of evolutions starting from the finite length viability kernel and the complement of the viability kernel.

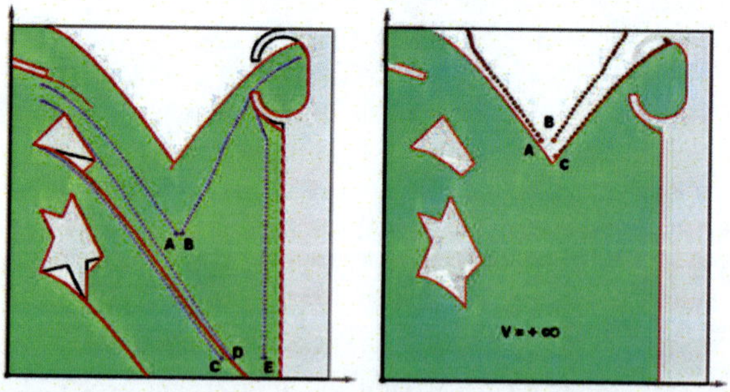

Fig. 5.5 Viable Minimal Length and Non-Viable Persistent Evolutions. *Since the equilibria are contained in the* finite length viability kernel *(see Definition 4.4.1, p.140), some finite length evolution converges to them. Figure 5.4, p. 184 shows the trajectories of some evolutions reaching equilibria. In the same way, Fig. 5.4 (Right, p. 184) displays* persistent evolutions *reaching in finite time the exit set through which it leaves K.*

5.3.3 *Behavior In and Out the Harbor: Minimal Time, Lyapunov and Optimal Evolutions*

- *Minimal Time Functions.* Definition 4.3.4, p.135 implies that the capture basin is the domain of the minimal time function, indicating the minimal time to reach the target, p. 192).
- *Lyapunov Functions.* Among all indicators describing the asymptotic behavior of the evolutions, we chose the exponential *Lyapunov function* which converges exponentially to the target. Its domain is the exponential attraction basin. Evolutions converging asymptotically to the harbor are depicted in Fig. 5.13, p. 194.
- We describe the behavior of the solutions of an intertemporal control problem, minimizing a given intertemporal criterion. The value function (see Definition 4.9.1, p.162) of this optimal problem with viability constraints (also referred as a problem with state constraints) is computed in Fig. 5.14, p. 196.

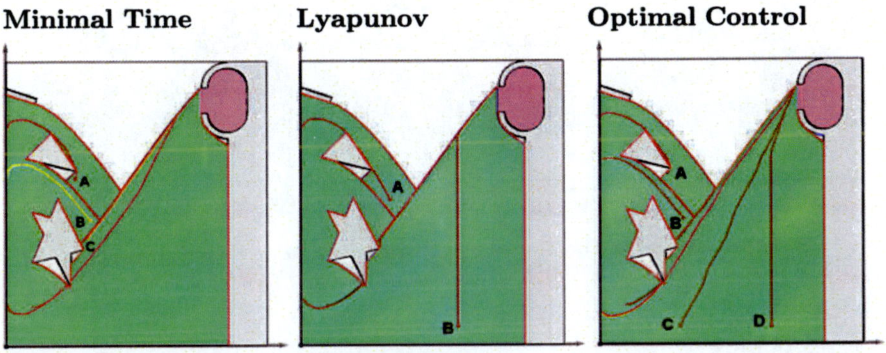

Fig. 5.6 Comparison of the Trajectories.
These figures compare trajectories of the minimal time evolutions (Left), of Lyapunov evolutions (Middle) and evolutions minimizing the cumulated squares of the velocity. (Right). Comments are provided in the snapshots below.

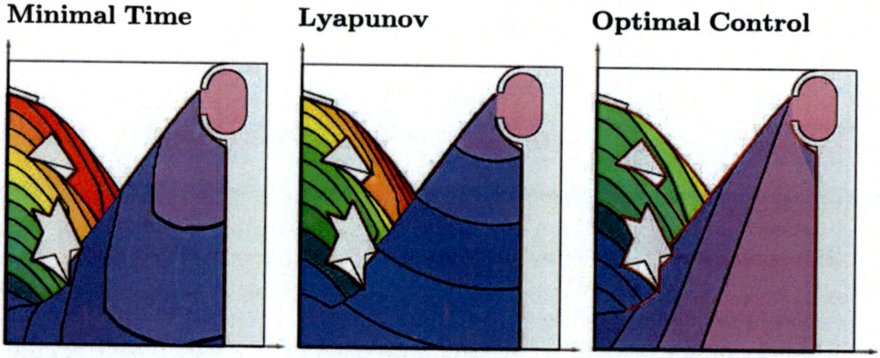

Fig. 5.7 Comparison of Projections of the Indicator.

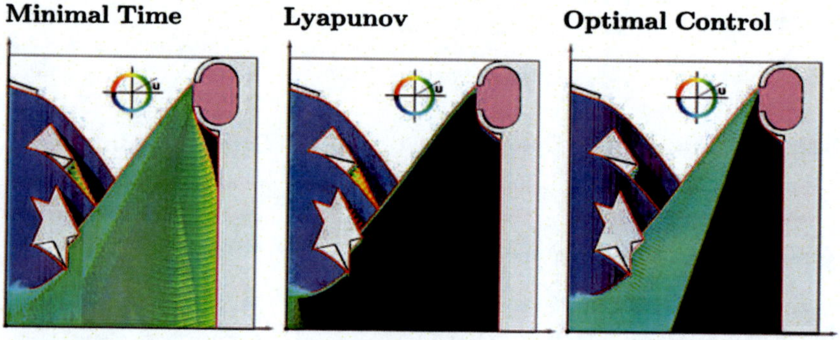

Fig. 5.8 Comparison of the Projections of the Angular Regulation Map.

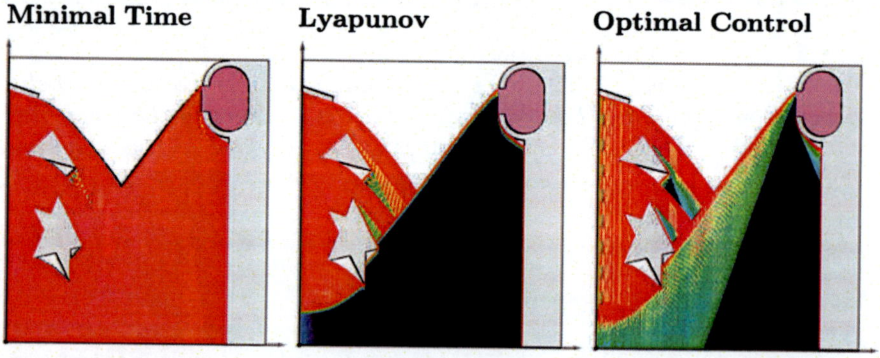

Fig. 5.9 Comparison of the Projections of the Velocity Regulation Map.

5.4 Viewgraphs Describing Five Value Functions

We present the snapshots of the minimal length function, the exit time function, the minimal time function, the Lyapunov function and value function of an optimal control problem according to the same scheme: each result is displayed on two pages facing each other.

1. *The even page.* The three dimensional graph of the indicators (minimal length, exit time, minimal time, Lyapunov and optimal functions) are displayed and commented on left,
2. *The odd page.* provides six two dimensional figures.

- **Left in the top row**. This figure provides the projection of the value function on its domain, with color scale helping to visualize the shape of the indicator, which takes infinite values.
- **Right in the top row**. The presence of obstacles destroys the continuity of the indicator, which also takes infinite values. This figure displays the projections of its discontinuities, separating zones with different qualitative behaviors. It also shows some trajectories of the evolutions optimizing the indicator (minimal length, persistent, minimal time, Lyapunov and optimal evolutions).
- **Left in the middle row**. The angular feedback map $\widetilde{u}(\cdot)$ associates with any state x the angular control $\widetilde{u}(x)$ to be used at this state for governing the optimal evolution through the open loop control $u(t) := \widetilde{u}(x(t))$. This figure provides the projection of the angular feedback map on the domain of the value function, with a color code providing the value of the *steering direction* on each state of the environment.
- **Right in the middle row**. The graph of the evolution $t \mapsto u(t) := \widetilde{u}(x(t))$ of the *steering direction* $u(t)$ along some evolutions are displayed in this figure to provide an easy and instructive way, showing the time behavior of each of the prototypical evolutions.
- **Left in the bottom row**. The velocity feedback map $\widetilde{v}(\cdot)$ associates with any state x the velocity control $\widetilde{v}(x)$ to be used at this state for governing the optimal evolution through the open loop control $v(t) := \widetilde{v}(x(t))$.
- **Right in the bottom row**. This figure displays the graphs of the velocity open loop control $t \mapsto v(t) := \widetilde{v}(x(t))$ along the prototypical evolutions.

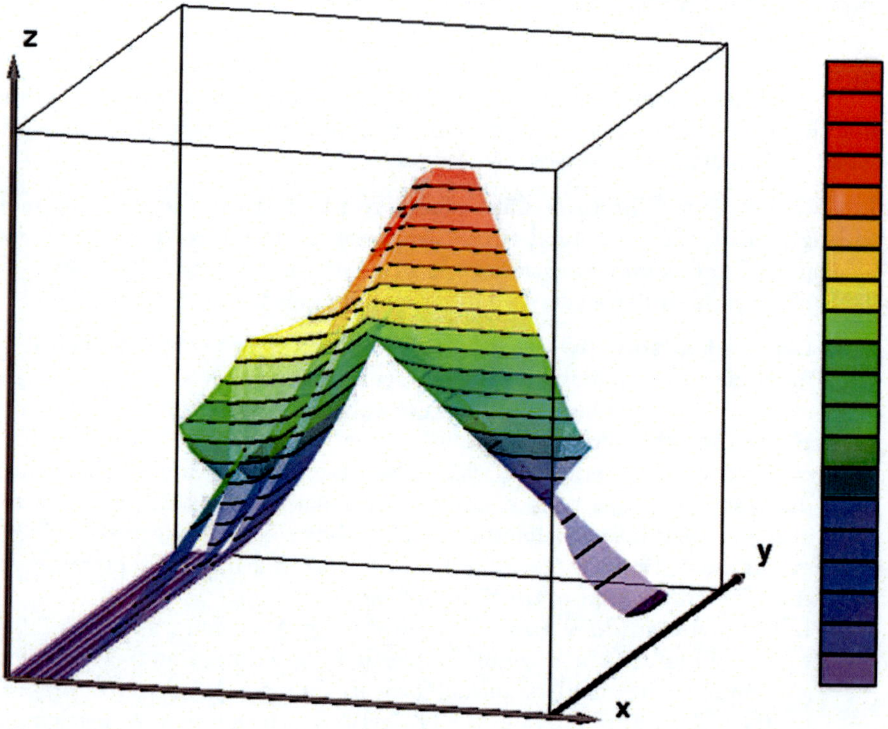

Fig. 5.10 The Minimal Length Function.
The viability kernel contains the domain of the minimal length function, which is the colored part of the projection of its graph. Obstacles can cause discontinuities. In the figure at the right of the first row, we chose to represent four trajectories of evolutions. The evolutions starting from initial states A and B in both sides of the western discontinuity sale west, south and north of the southern obstacle. The evolutions starting from C and D, the one starting from C goes west and the one starting form D heads east, berthing inside the harbor. These minimal length evolutions converge to equilibria, some of them along the left bank, the other ones on the right bank of the harbor. Observe that behind the two obstacles and in the south corner of the harbor, the minimal length is infinite.

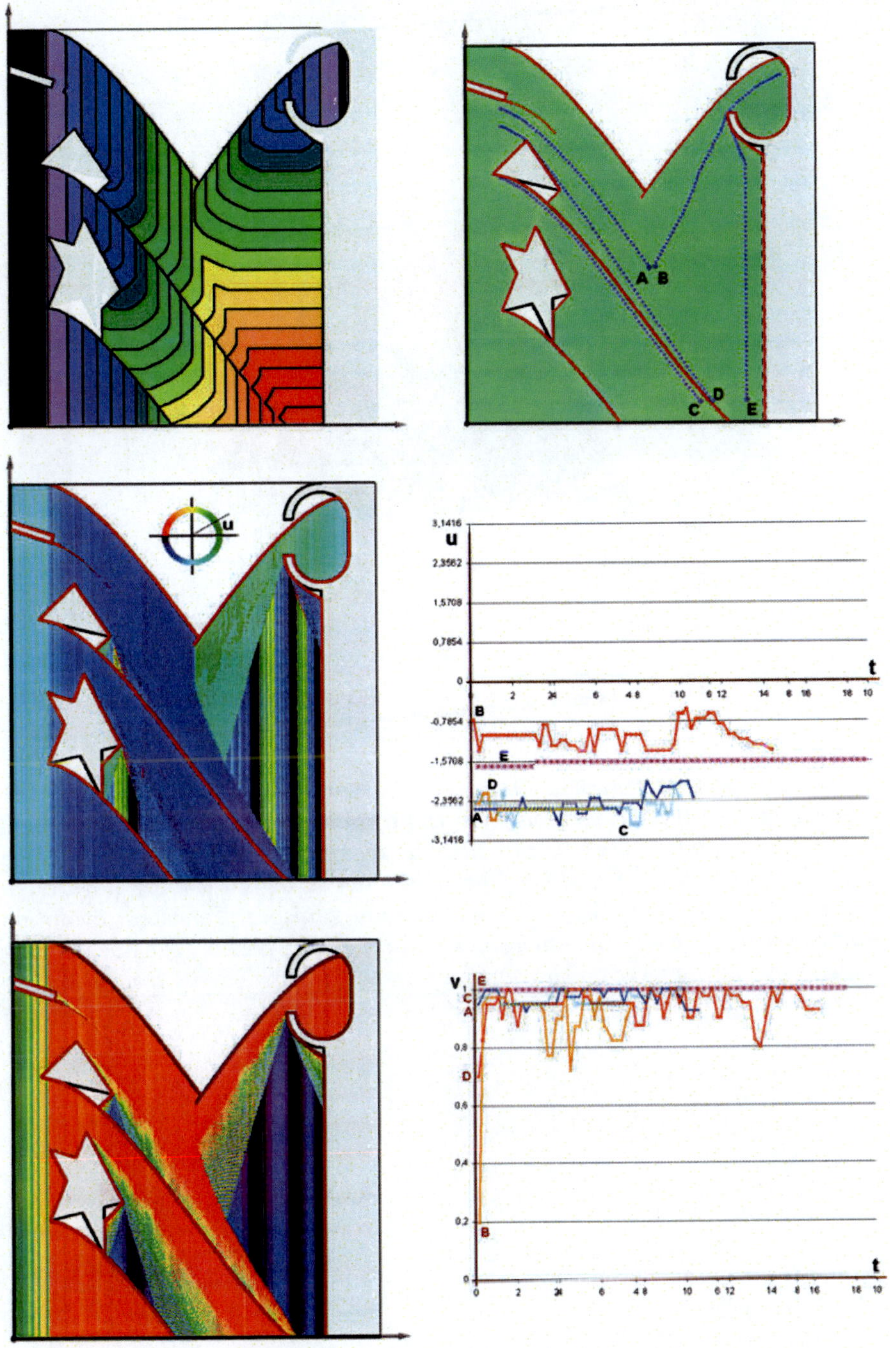
A B
C D E
u
u
3,1416
2,3562
1,5708
0,7854
0
-0,7854
-1,5708
-2,3562
-3,1416
t
B
E
D
A
C
V
E
C
A
D
B
1
0,8
0,6
0,4
0,2
0
t

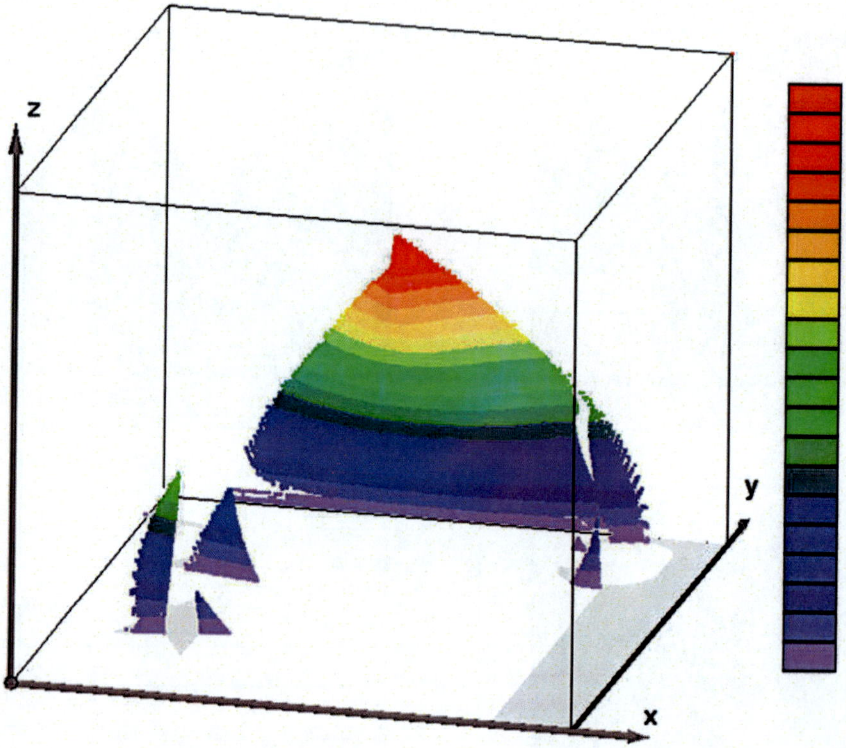

Fig. 5.11 The Exit Time Function.
The exit time function is defined on the complement of the viability kernel (where it is equal to $+\infty$). It provides the maximal "survival time" in which persistent evolutions remain in the environment before leaving it through the exit set, on which the exit time function vanishes. Persistent evolution starting from C leaves the environment through the harbor, the one starting from A leaves the river on the north–east and the one leaving from B on the north–west.

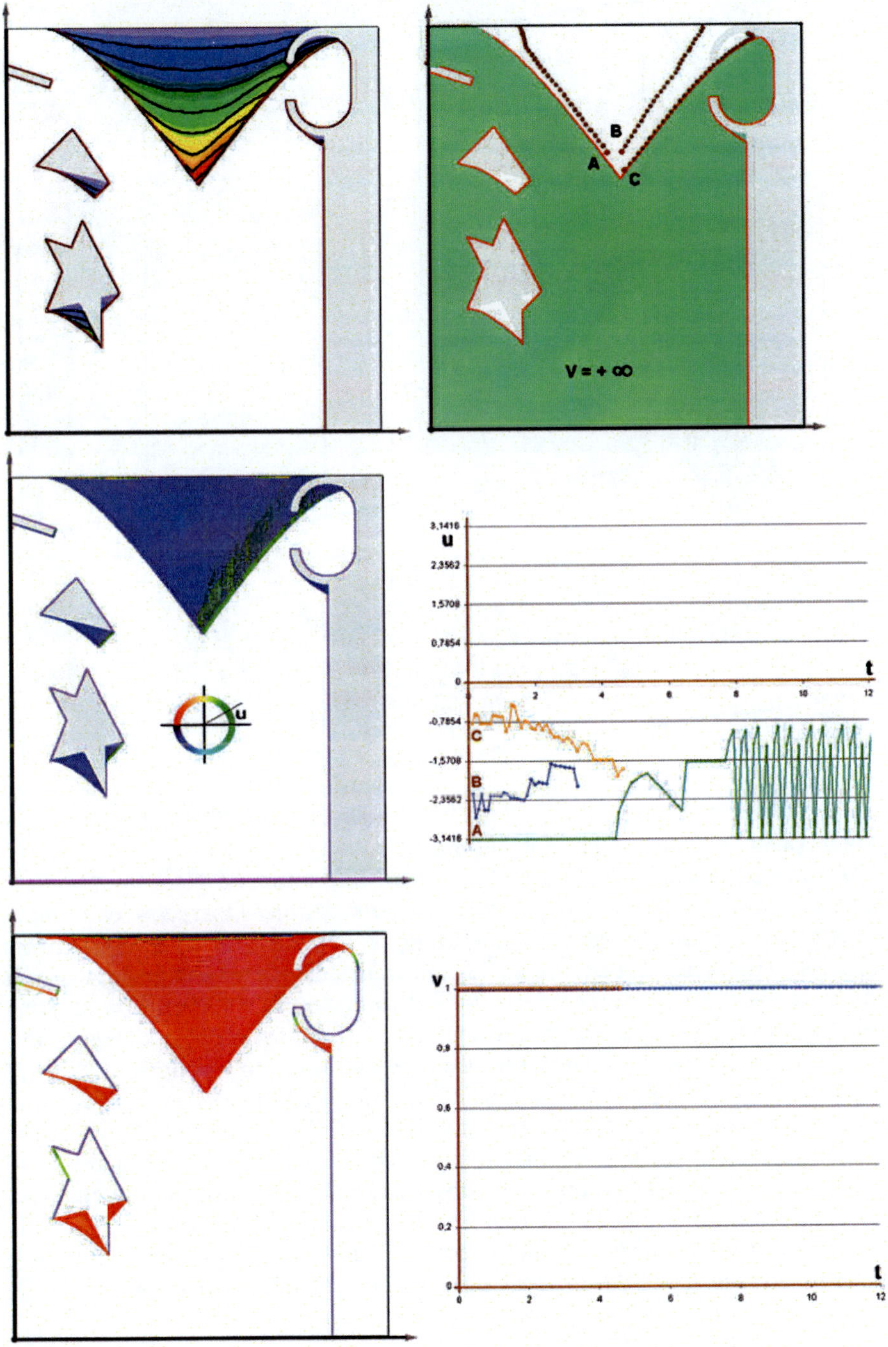
B
A
C
V = + ∞
u
3,1416
u
2,3562
1,5708
0,7854
0
t
2
4
6
8
10
12
-0,7854
C
-1,5708
B
-2,3562
A
-3,1416
V
1
0,8
0,6
0,4
0,2
0
t
0
2
4
6
8
10
12

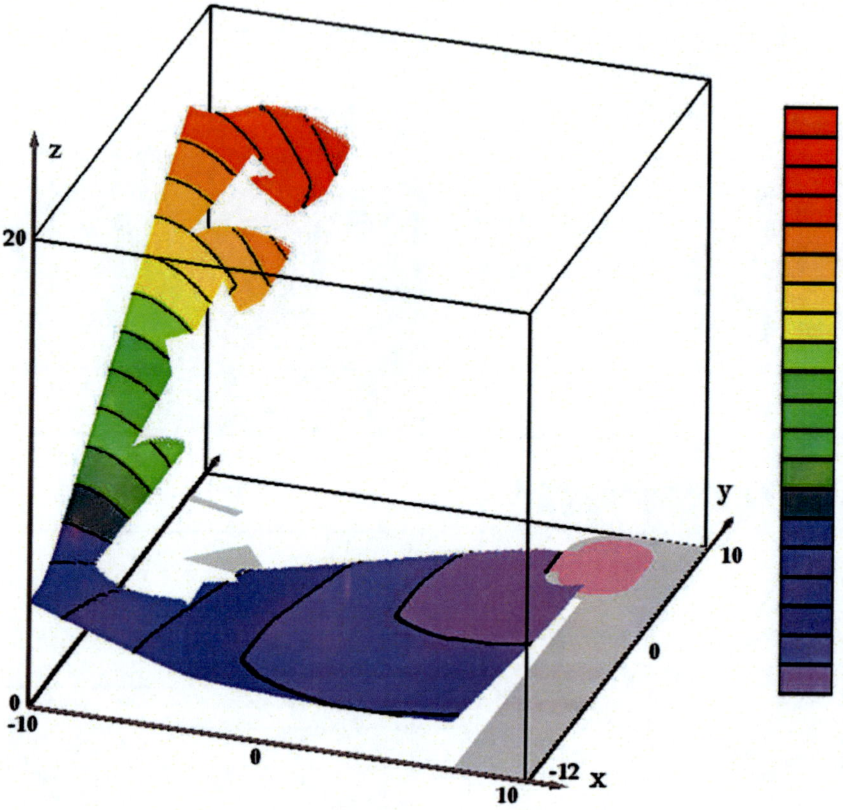

Fig. 5.12 The Minimal Time Function.
The capture basin being the domain of the minimal time function, it is the colored part of the projection of its graph. Obstacles can imply discontinuities. In the figure at the right of the first row of next page, we chose to represent three trajectories of evolutions starting from inial states A, B and C. Being north of the first discontinuity, the evolution starting from A turns around the northern obstacle towards the left bank, sales south far enough to bypass the southern obstacle and then, to the harbor. Since B is located between the two discontinuities, the evolution sales west between the two obstacles and then follows the same trajectory than A. From C, the evolution flies directly to the harbor. The first figure of the third row shows that the norm of the velocity remains maximal and constant and that the evolution is governed only through the steering control. Observe that behind the two obstacles and in the south corner of the harbor, the minimal time is infinite, so that these white areas do not belong to the capture basin.

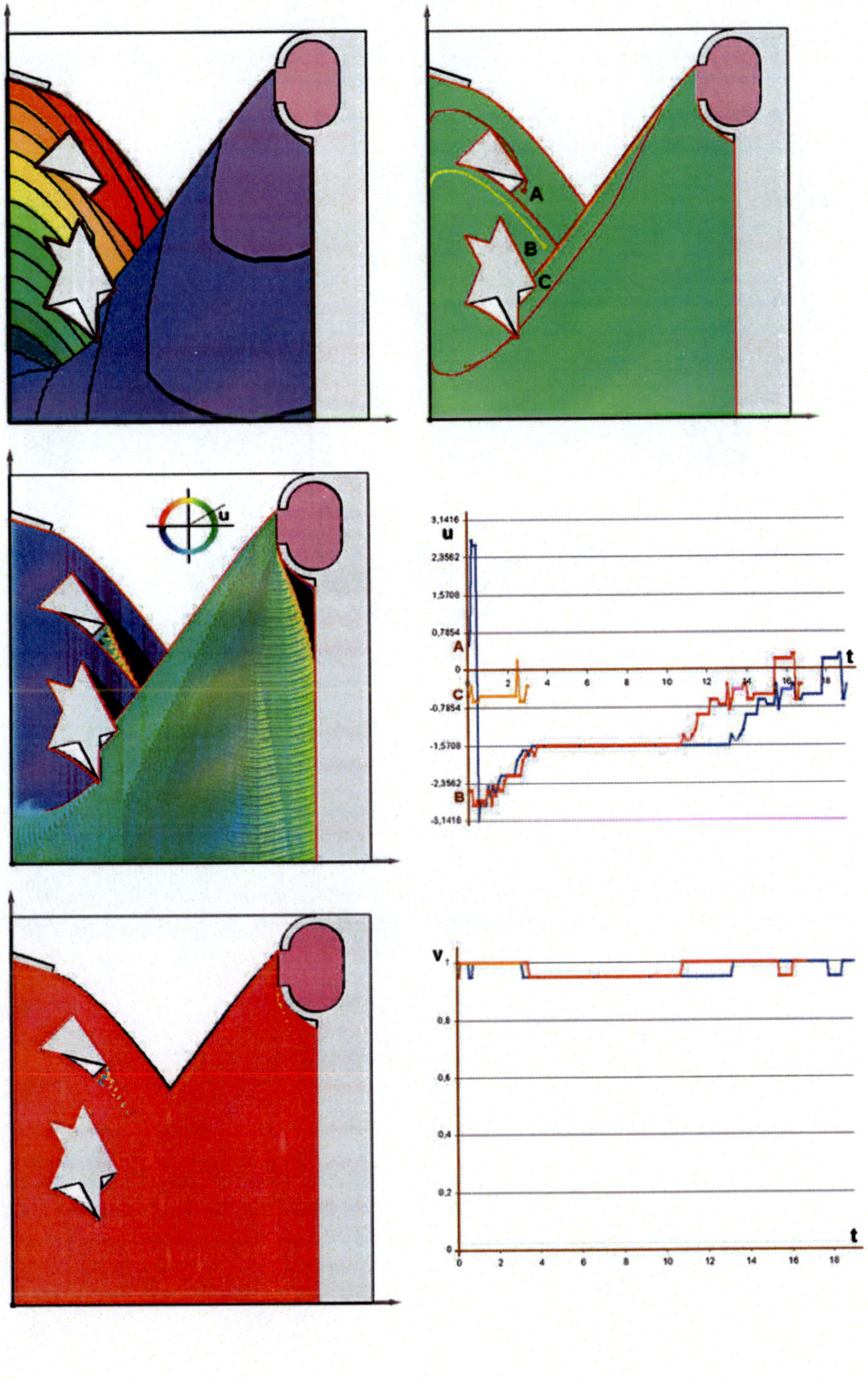
A
B
C
u
3,1416
2,3562
1,5708
0,7854
0
-0,7854
-1,5708
-2,3562
-3,1416
t
v
0,8
0,6
0,4
0,2
0
2
4
6
8
10
12
14
16
18

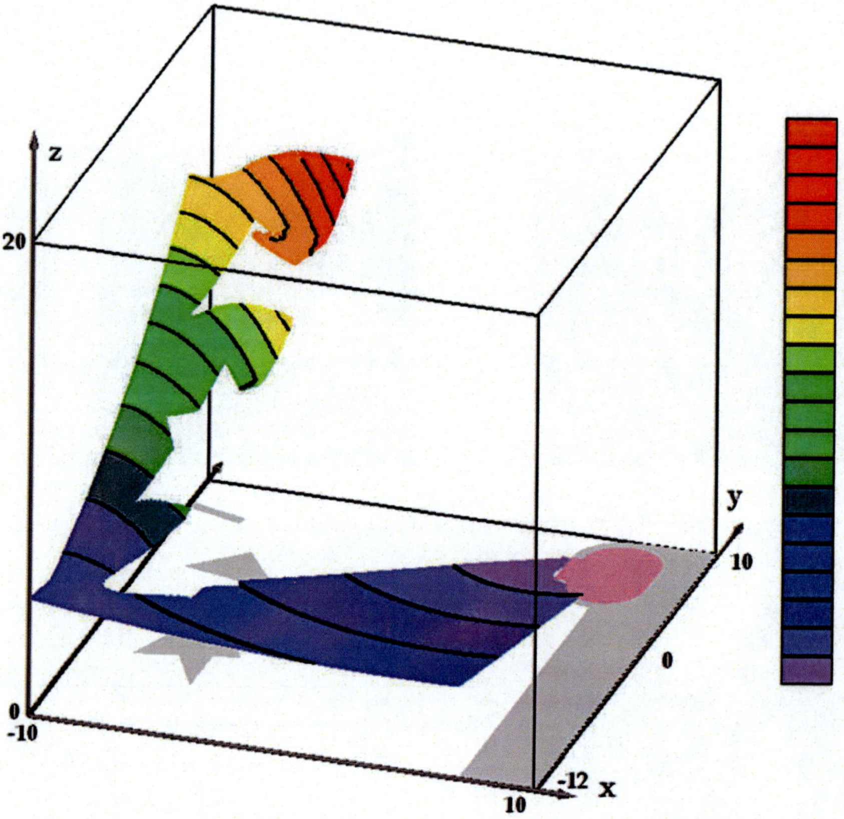

Fig. 5.13 The Lyapunov Function.
The exponential growth of an evolution is measured by $\sup_{t\geq 0} e^{mt} d_{(K,C)}(x(t))$ *(see Definition 4.5.4, p.147). The Lyapunov function minimizes the exponential graph of evolutions starting from a given state. Its graph is displayed. The trajectories of evolutions starting from A (going around the two obstacles) and from B are governed by the feedbacks in the two last rows.*

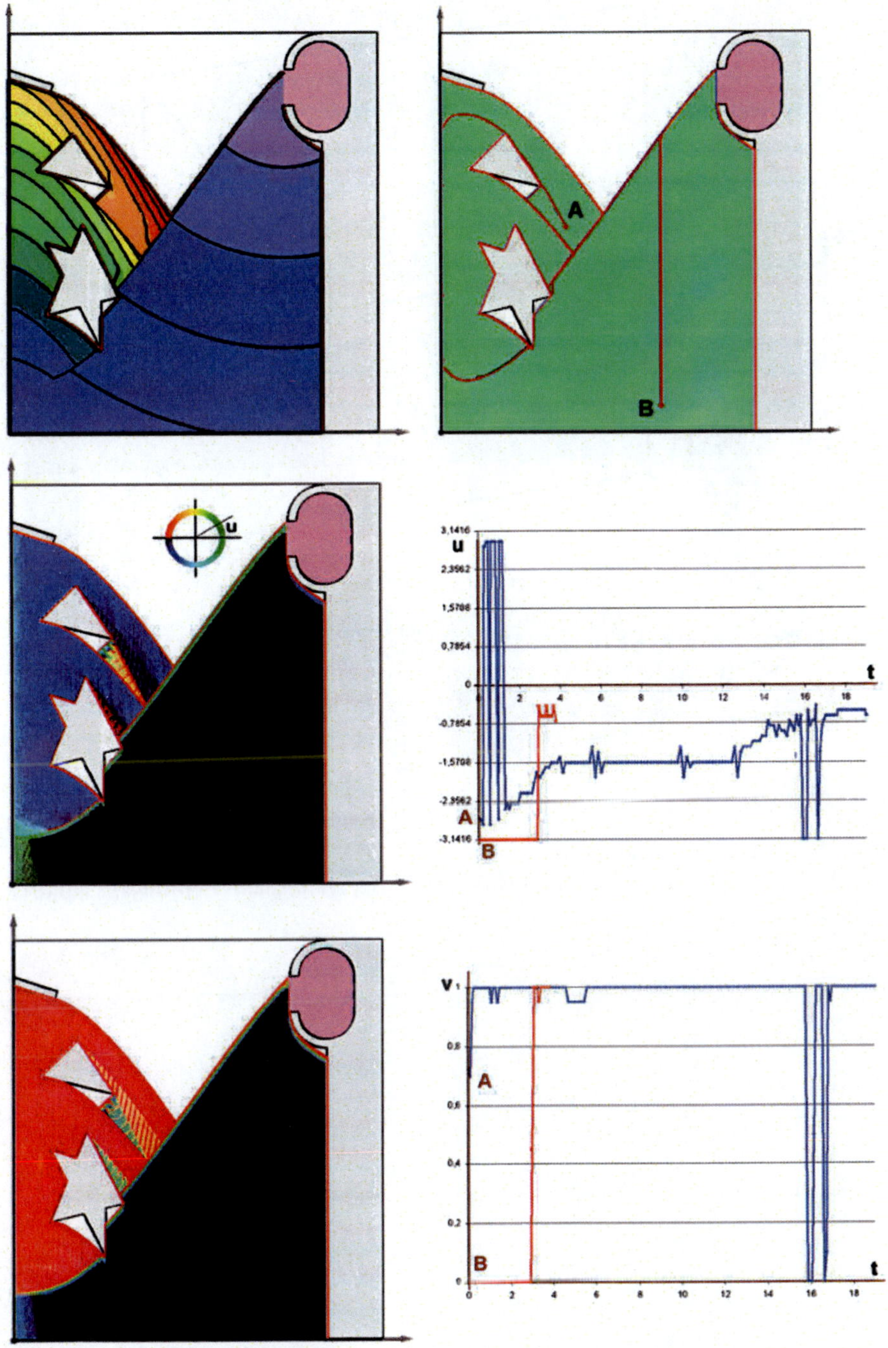
A
B
u
3,1416
u
2,3562
1,5708
0,7854
0
t
2
4
6
8
10
12
14
16
18
-0,7854
-1,5708
-2,3562
A
-3,1416
B
V 1
0,8
A
0,6
0,4
0,2
B
t
0
0
2
4
6
8
10
12
14
16
18

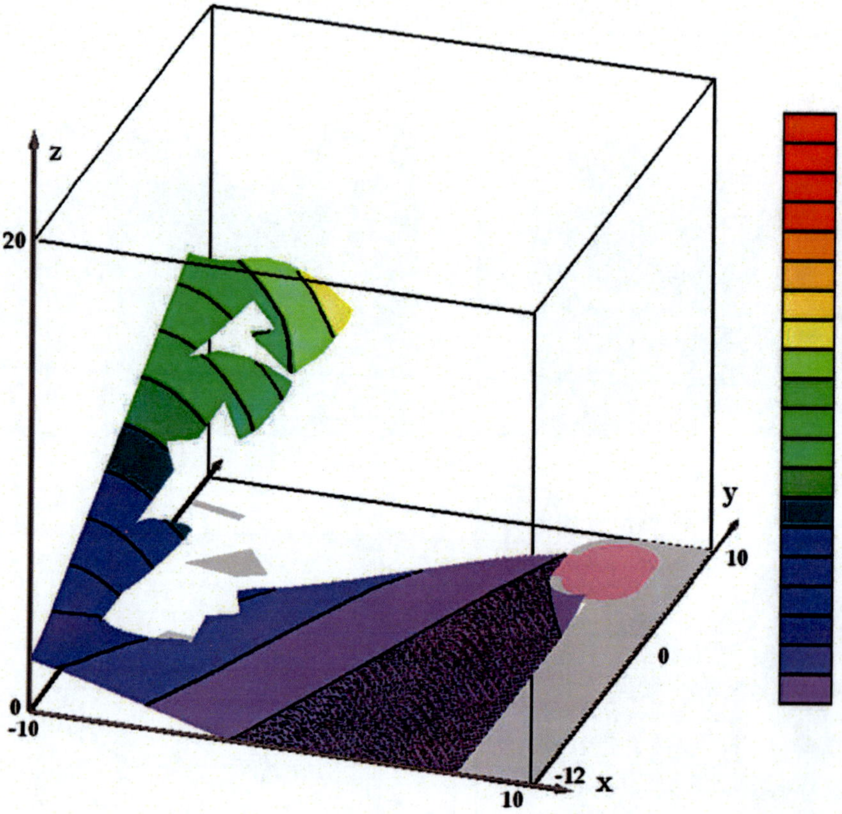

Fig. 5.14 The Value Function.
In this example, we chose the evolutions minimizing the cumulated squares of velocities.

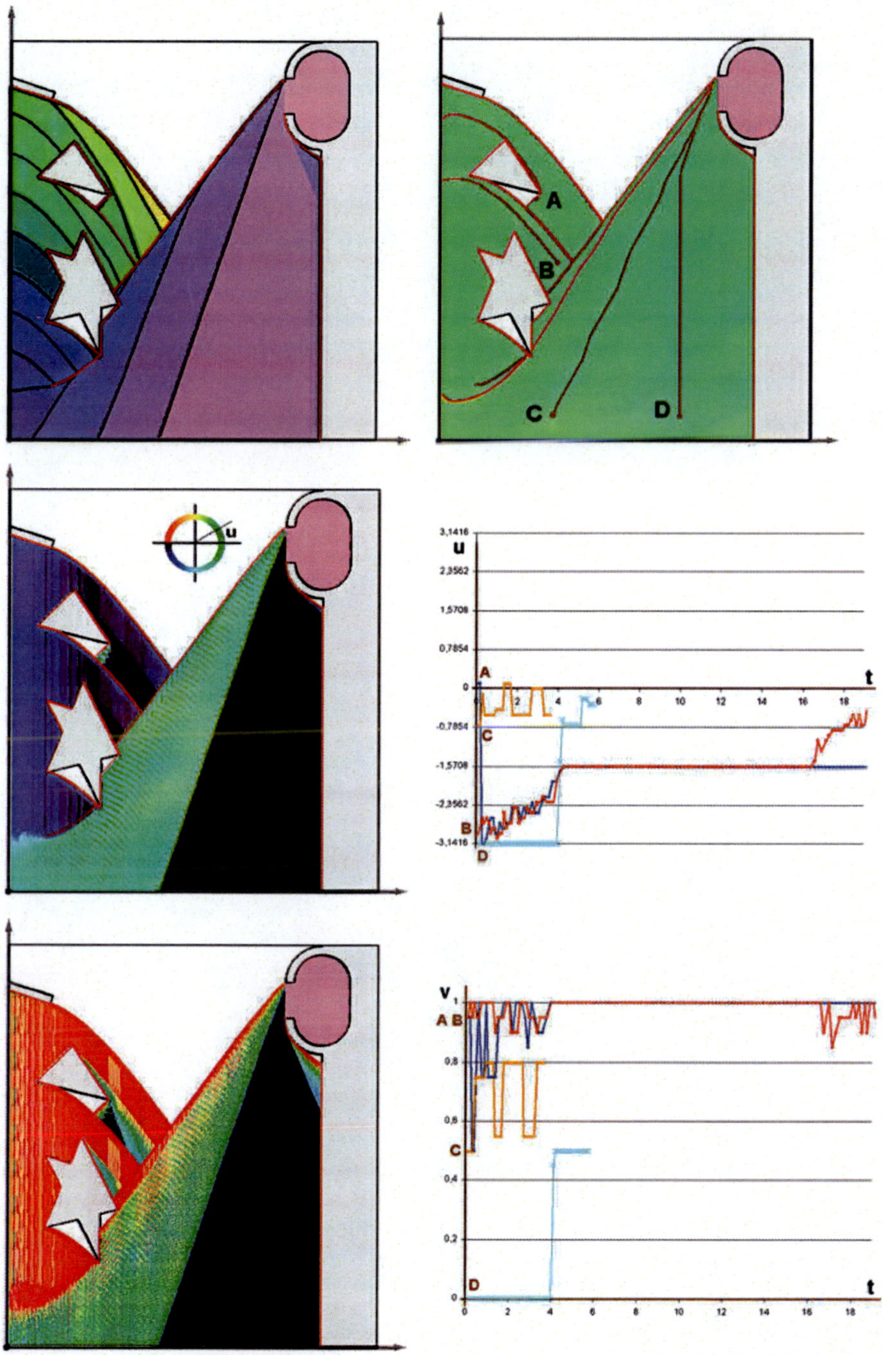
A
B
C
D
u
3,1416
2,3562
1,5708
0,7854
0
-0,7854
-1,5708
-2,3562
-3,1416
t
2
4
6
8
10
12
14
16
18
v
1
0,8
0,6
0,4
0,2
0

Chapter 6
Inertia Functions, Viability Oscillators and Hysteresis

6.1 Introduction

This chapter is devoted to the original motivation of viability theory, which was (and still is) an attempt to mathematically capture some central ideas of evolution of species of Darwinian type, turning around the concept of "punctuated equilibrium" of Eldredge and Gould.

For that purpose, we investigate first in Sect. 6.2, p. 202 whether intertemporal optimization studied in Chap. 4, p. 125 can help us in this undertaking. We observe that this approach requires the existence of a decision-maker acting on the controls of the system, using an intertemporal optimization criterion, which involves some knowledge of the future, for choosing the "best" decision once and for all at the initial time.

We may consider another point of view, in which, in an evolutionary context, it may be wiser to *choose a decision at the right time rather than an intertemporal optimal one taken at the wrong time*. As mentioned in Sect. 2.5, p. 58, many systems involving living organisms are regulated by regulatory controls, called "regulons". They evolve, but there is no consensus on the identity of "who" is acting on them. So, we assume that if such an actor does exist, "he" is *lazy, myopic, conservative and opportunistic*, who, whenever it is possible, will keep the regulons constant as long as viability is not at stakes, by ***minimizing at each instant***, but not in an intertemporal, the ***velocity of the regulons*** (*heavy evolutions*). Indeed these are different behaviors than the ones of an "intertemporal optimizer", where an identified agent acts on the controls for optimizing a given criterion. We thus review in Sect. 6.3, p. 203 the concept of "punctuated equilibrium" for which we shall propose a rigorous mathematical definition (Definition 6.4.9, p. 216) in Sect. 6.4, p. 207.

Actually, the rest of this chapter is not only devoted to issues involving inertia functions for defining the inertia principle, punctuated equilibria and heavy evolutions, but also, in Sect. 6.5, p. 233, to viability oscillators and hysterons, useful for several biological and physical problems, and finally, in

J.-P. Aubin et al., *Viability Theory*, DOI 10.1007/978-3-642-16684-6_6,

Sect. 6.6, p. 242, to the regulation of systems where the controls are involved not only in the dynamics, but also in the environments and the targets.

These issues share the fact that they can be approached by *metasystems*, defined as follows: a metasystem is made of the adjunction to the controlled system

$$x'(t) = f(x(t), u(t)) \text{ where } u(t) \in U(x(t)) \tag{6.1}$$

of the differential equation

$$u'(t) = v(t)$$

which are coupled through the set-valued map U. In other words, a metasystem governs the state-control pairs $(x(\cdot), u(\cdot))$ governed by the system

$$\begin{cases} (i) \ x'(t) = f(x(t), u(t)) \\ (ii) \ u'(t) = v(t) \end{cases}$$

viable in the graph Graph(U) *of the set-valued map* U. Or, equivalently the states of the metasystem are the state-control pairs $(x(\cdot), u(\cdot))$ and its controls, the "metacontrols", are the velocities of the original system (6.1), p. 200.

Metasystems govern evolutions by *continuous open loop controls instead of measurable ones.* They can be investigated using viability tools.

- We begin the study of *inertia functions in Sect. 6.4, p.207.* The objective is to derive *heavy* evolutions of regulated system (6.1), p. 200: *at each instant, the norm of the velocity of the viable controls is minimized.* This is no longer an intertemporal minimization problem studied in Chap. 4, p. 125, but a *spot minimization* problem. This behavior is consistent with the required *myopic, exploratory but lazy, opportunistic but conservative* behavior we are about to describe:
  - *Myopic*, since the knowledge or the future is not required.
  - *Exploratory* thanks to the set of available controls.
  - *Lazy*, since at each instant, he minimizes the norm of the velocity of the regulon.
  - *Opportunistic*, since the controls have to be modified whenever viability is at stakes.
  - *Conservative*, since the controls are kept constant whenever the viability constraints are satisfied.

  Heavy evolutions are the ones which satisfy the principle stating that, given an inertia threshold, heavy evolutions keep their regulon constant until the *warning time* (called *kairos*, see Box 24, p. 218) **when** the state of the heavy evolution system reaches a *critical zone* (see Definition 6.4.9, p. 216), **where** the regulons **have** to evolve (the decision taken at the right moment) in order to remain viable.
  If by any chance, an equilibrium lies in the critical zone, a heavy evolution which reaches it remains there forever. Otherwise, defining the *viability*

niche of a (constant) regulon u as the viability kernel of the environment under the dynamics $x'(t) = f(x(t), u)$, a heavy evolution which reaches the (nonempty) viability niche of a regulon remains there. This is a "lock in" property explaining punctuated evolution (under a constant environment). We adapt these results for high-order systems where we study the behavior of high-order derivatives of the controls. We also extend this idea to control problems involving the derivatives of the controls, useful in economics and ecology for evaluating the ***transition costs*** in the sense that instead of the norm of the velocity of the regulon, one minimizes a Lagrangian $\mathbf{l}(x, u, v)$ *taking into account the cost of the velocity v of the regulon*, measuring what is understood as a transition cost.

In order to explain these uncommon mathematical concepts, we illustrate them by the simplest one-dimensional example we can think of: regulated system $x'(t) = u(t)$ (called *simple integrator* in control theory), where the open loop control $u(t)$ translates our ignorance of the dynamical system governing the evolution of the state. A situation familiar in life sciences, where "models", in the physicist sense of the word, are difficult to design, contrary to engineering, where models of man made products are easier to uncover. Hence, imposing viability constraints $x(t) \in K$ and an inertia threshold $\|u'(t)\| \leq c$ are enough to produce heavy evolutions which can be computed both analytically and by the Viability Kernel Algorithm.
Note that while system $x'(t) = u(t)$ looks like a trivial system to study, the introduction of viability constraints $x(t) \in K$ and inertia threshold $|x'(t)| \leq c$ for producing heavy and cyclic evolutions introduce significant difficulties that can be solved using viability tools.

- We examine in Sect. 6.5, p. 233 how this system provides not only heavy evolutions, but also "cyclic evolutions", in the framework of this simple example for the sake of clarity. Since the above system is not periodic, we cannot call them periodic evolutions. It turns out that viability and inertia threshold are sufficient to provide regulons governing cyclic evolutions. This approach may be more relevant for explaining the functioning of biological clocks, for example. They at least produce *hysteresis loops*, evolutions going back and forth between two equilibria by using velocities the norm of which are equal to the inertia threshold.
- Section 6.6, p. 242 briefly deals with systems in which the controls or the regulons are involved in the environment and/or the target. This is the case whenever we "control a target", instead of the dynamics, for instance, for solving a given set of properties. The use of metasystems allows us to handle this question, since we do not have to assume that the set-valued map U is Marchaud, but only closed. The relation $u(t) \in U(x(t))$ can be inverted and becomes the relation $x(t) \in U^{-1}(u(t))$, where $K(u) := U^{-1}(u)$ can be regarded as a *control dependent environment*.

6.2 Prerequisites for Intertemporal Optimization

Intertemporal minimization problem implicit requirements that deserve to be made explicit are as follows:

19 ***[Implicit Prerequisites for Intertemporal Optimization]***
Intertemporal optimization requires:

1. *the existence of an actor (agent, decision-maker, controller, etc.),*
2. *an optimality criterion,*
3. *that decisions are taken once and for all at the initial time,*
4. *a knowledge of the future (or of its anticipation).*

We already mentioned that in systems involving living beings, there is no consensus on the necessity of an actor governing the evolution of regulons.

The choice of criteria **c** and **l** in intertemporal optimization problems (see Definition 4.9.1, p. 162) is open to question even in static models, even when multicriteria or several decision makers are involved in the model.

Furthermore, the choice (even conditional) of the controls is made *once and for all* at some initial time, and thus *cannot be changed at each instant so as to take into account possible modifications of the environment of the system*, thus forbidding ***adaptation*** to viability constraints.

Finally, the intertemporal cost involving the knowledge of the state $x(t)$ at future times $t \in [0, T]$ requires some knowledge of the future. Most systems we investigate involve, at least, implicitly, a *myopic* behavior. Therefore, they are unable to take into account the future, whereas their evolutions are certainly constrained by their history. The knowledge of the future needs to assume some regularity (for instance, periodicity, cyclicity) of the phenomena (as in mechanics), or to make anticipations, or to *demand experimentation.* Experimentation, by assuming that the evolution of the state of the system starting from a given initial state for a same period of time will be the same whatever the initial time, allows one to translate the time interval back and forth, and, thus, to "know" the future evolution of the system. But in life sciences, and, in particular, in economics, the systems are irreversible, *their dynamics may disappear and cannot be recreated*, forbidding any insight into the future.

Hence, we are left to make forecasting, prediction, *anticipations* of the future, which are extrapolations of past evolutions, constraining in the last analysis the evolution of the system to be a function of its history (See Chap. 11 of the first edition of [18, Aubin]). However, we should note the quotation:

20 ***Paul Valéry (1871–1945)*** *Forecasting is a dream from which event wakes us up.* (La prévision est un rêve duquel l'événement nous tire.)

After all, in biological evolution, intertemporal optimization can be traced back to Sumerian mythology which is at the origin of Genesis: one Decision-Maker, deciding what is good and bad and choosing the best (fortunately, on an intertemporal basis with infinite horizon, thus wisely postponing to eternity the verification of optimality), knowing the future, and having taken the optimal decisions, well, during one week...

We had to wait for *Alfred Wallace* (1823–1913) to question this view in 1858 in his essay *On the Tendency of Varieties to Depart Indefinitely from Original Type* which he sent to *Darwin* (1809–1882) who had been working on his celebrated *Origin of Species* (1859) since 1844. Selection by viability and not by intertemporal optimization motivated viability theory presented in this book.

This is the reason why we shall assume in Sect. 6.3, p. 203 that, in living system, there is no identified actor governing the evolution of the regulons, or that, if such an actor exists, he is *myopic, lazy, opportunistic and conservative*, and that the evolution of the controls, regarded in this case as regulons, obey the *inertia principle* (see Comment 21, p. 204).

However, history has its own perversions, since Hélène Frankowska has shown that these very techniques of viability theory designed to replace optimal control theory are also very useless to solve optimal control problems and other intertemporal optimization problems in the Hamilton–Jacobi–Bellman tradition.

6.3 Punctuated Equilibria and the Inertia Principle

More generally, in regulated systems, agents acting on state variables are well identified, but no identified actor governs the evolution of the regulons. In the absence of such an actor piloting the regulons, or by assuming that this actor is *myopic, lazy, opportunistic and conservative*, we cannot assume any longer that the controls are chosen to minimize an intertemporal criterion, due to the prerequisites for optimal control stated in Comment 19, p. 202 (the existence of an actor, of an optimality criterion, that decisions are taken once and for all at the initial time, a knowledge of the future) are not met.

We may assume instead that regulons evolve as "slowly" as possible because the change of regulons (or controls in engineering) is costly, even very costly. Hence we are led to assume that the regulons are constrained by some *inertia threshold* that can be estimated through some measure of their *velocities* (see definition (4.1), p. 130 of the *versatility* of an evolution). We may even look for heavy evolutions when the regulons evolve as slowly as possible.

The situation in which controls are kept constant (they are called *coefficients* in this case) is familiar in physics, because these physical coefficients are assumed to *remain constant.* Important properties (set of equilibria, stability or instability of equilibria) are then studied in terms of such parameters, as in bifurcation theory, catastrophe theory, chaotic behavior, etc. (see for instance *Stephen Wiggins*'s Quotation 11, p. 47 and [26, Aubin & Ekeland]).

However, evolutions under constant coefficients may not satisfy required properties, such as viability, capturability or optimality. Then the question arises to study *when, where and how* coefficients must cease to be constant and start to "evolve" in order to guarantee the viability property, for instance. In this case, their status as "coefficients" is modified, and they become controls or regulons, according to the context (engineering or life sciences).

Whenever the viability property is concerned, we shall give a name to this phenomenon which seems to be shared by so many systems dealing with living beings:

> **21** ***[The Inertia Principle]*** *In a loose way, the* inertia principle *states that the* "regulons" of the system are kept constant as long as possible and changed only when viability or inertia is at stake.

The inertia principle provides a mathematical explanation of the emergence of the concept of *punctuated equilibrium* introduced in paleontology by *Nils Eldredge* and *Stephen J. Gould* in 1972 (see Comment 23, p. 216, which we link another ecological concept, the ecological niche). We shall translate these biological concepts into rigorous mathematical definitions (see Definition 6.4.9, p. 216) and prove the existence of "heavy evolutions locking in" the viability niche of a punctuated equilibrium whenever the evolution of the regulon $u(\cdot)$ reaches such a punctuated equilibrium on its way. It runs against the teleological trend assigning aims to be achieved (in even an optimal way) by the state of the system and the belief that actors control the system for such purposes (see Sect. 6.2, p. 202). The concept of "locking-in" had been introduced in different fields of economics (of innovation), and with other mathematical techniques, for explaining why, once adopted, some technologies, which may look non-optimal in regard of some criterion, are still adopted, whereas some solutions, maybe optimal with respect to such or such criterion, are not adopted. The same phenomenon appears in biological evolution, and may be explained mathematically by the inertia principle in the paradigm of adaptation to an environment. These are *the very considerations which triggered the investigations of what became viability theory at the end of the 1970s.*

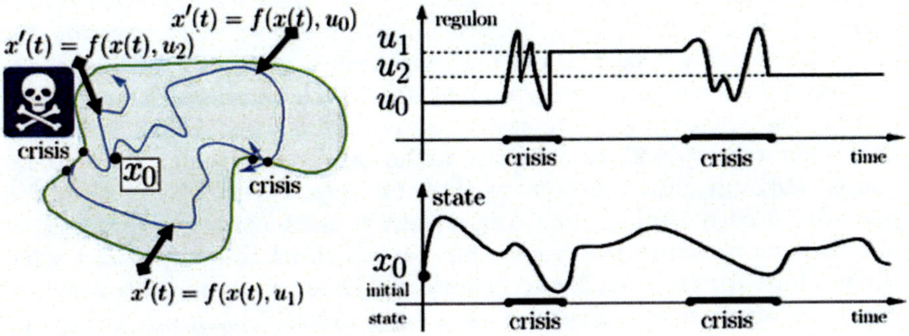

Fig. 6.1 Punctuated Evolution.
Starting from x_0 with the constant regulon u_0, the solution evolves in K until time t_1, (first punctuated equilibrium phase*) when the state $x(t_1)$ is about to leave the environment K and when the constant regulon u_0* must *start to evolve. Then a* critical phase *happens during which velocities also evolve (as slowly as possible) to maintain viability, until time t_1 when the control $u_1 := u(t_1)$ can remain constant during a nonempty time interval:* second punctuated equilibrium phase, *after which a second critical phase.*

However, they were anticipated by *Darwin* himself, who added in the sixth edition of is celebrated book the following underlined sentence:

Addition to Chapter XI of the sixth edition of
***Origin of Species* by Charles Darwin**

"Summary of the preceding and present chapters"

Darwin (1809–1882). *I have attempted to show that the geological record is extremely imperfect; that only a small portion of the globe has been geologically explored with care; that only certain classes of organic beings have been largely preserved in a fossil state; that the number both of specimens and of species, preserved in our museums, is absolutely as nothing compared with the numberof generations which must have passed away even during a single formation; that, owing to subsidence being almost necessary for the accumulation of deposits rich in a fossil species of many kinds, and thick enough to outlast future degradation, great intervals of time must have elapsed between most of our successive formations; that there has probably been more extinction during the periods of subsidence, and more variation during the periods of elevation, and during the latter the record*

will have been least perfectly kept; that each single formation has not been continuously deposited; that the duration of each formation is, probably short compared with the average duration of specific forms; that migration has played an important part in the first appearance of new forms in any one area and formation; that widely ranging species are those which have varied most frequently, and have oftenest given rise to new species; that varieties have at first been local; **and lastly, although each species must have passed through numerous transitional stages, it is probable that the periods, during which each underwent modification, though many and long as measured by years, have been short in comparison with the periods during which each remained in an unchanged condition.** *These causes, taken conjointly, will to a large extent explain why though we do find many links- we do not find interminable varieties, connecting together all extinct and existing forms by the finest graduated steps. It should also be constantly borne in mind that any linking variety between two forms, which might be found, would be ranked, unless the whole chain could be perfectly restored, as a new and distinct species; for it is not pretended that we have any sure criterion by which species and varieties can be discriminated".*

For instance, this has been documented in paleontology:

> **22 *Excavations at Lake Turkana.*** *Excavations at Kenya's Lake Turkana have provided clear evidence of evolution from one species to another. The rock strata there contain a series of fossils that show every small step of an evolution journey that seems to have proceeded in fits and starts. Examination of more than 3,000 fossils by* Peter Williamson *showed how 13 species evolved. He indicated in a 1981 article in* Nature *that the animals stayed much the same for immensely long stretches of time. But twice, about two million years ago and then, 700,000 years ago, the pool of life seemed to explode – triggered, apparently, by a drop in the lake's water level. Intermediate forms appeared very quickly, new species evolving in 5,000–50,000 years, after millions of years of constancy, leading paleontologists to challenge the accepted idea of continuous evolution by proposing the concept of punctuated equilibrium.*

The question arises how to change constant regulons when the viability condition is about to be violated. This can be done:

1. either brutally, as in Sect. 10.8, p. 422, by changing *persistent evolutions* (see Definition 10.4.2, p. 393 and Theorem 10.4.4, p. 394), remaining viable as long as possible until the moment when the reach their *exit set* (see Sect. 10.8, p. 422) [*hard* version of the inertia principle or *infinite inertia threshold*]

2. or in a smoother manner, by setting a *finite inertia threshold* and minimizing *at each instant* the velocity of the control [*soft* version of the inertia principle].

Naturally, there are many other approaches between these two extremes allowing the inertial principle to be satisfied. The *hard version* of the inertia principle is the topic of Sect. 10.8, p. 422 and, more generally, of impulse control and/or hybrid systems in Sect. 12.3, p. 503. The *soft version* of the inertia principle is the topic of this chapter.

6.4 Inertia Functions, Metasystems and Viability Niches

We consider the parameterized system (6.1), p. 200:

$$x'(t) \;=\; f(x(t),u(t)) \text{ where } u(t) \;\in\; U(x(t))$$

where the set-valued map $U : X \rightsquigarrow \mathcal{U}$ implicitly involves the viability constraints

$$\forall t \geq 0,\; x(t) \in K := \mathrm{Dom}(U)$$

Remark. Conversely, the viability constraint described by an environment K under the parameterized system (6.1), p. 200, can be taken into account by introducing the restriction $U|_K$ of the set-valued map U to K defined by

$$U|_K(x) \;:=\; \begin{cases} U(x) & \text{if } x \in K \\ \emptyset & \text{if } x \notin K \end{cases}$$

This amounts to studying the system

$$x'(t) \;=\; f(x(t),u(t)) \text{ where } u(t) \;\in\; U|_K(x(t))$$

which is the above system (6.1), p. 200, when U is replaced by $U|_K$. This is possible whenever the set-valued map U intervenes through its graph and not as a set-valued map $x \rightsquigarrow U(x)$ which would need to satisfy the assumptions of the Viability Theorem 11.3.4, p. 455. Since the domain of a set-valued map with closed graph is not necessarily closed, using this approach allows us to also deal with more general viability problems where the environment K is just the domain of a closed set-valued map. □

6.4.1 The Inertia Cascade

We classify specific evolutions in increasing inertia order. The most "inert" evolutions are *equilibria* $(x^\star, u^\star)$ of the control system, solutions to

$$f(x^\star, u^\star) = 0 \text{ where } u^\star \in U(x^\star)$$

since both the state and the controls do not evolve (see Sect. 9.5, p. 360).

We next distinguish evolution governed under constant controls:

Definition 6.4.1 *[Evolutions under Constant Controls] Evolutions governed by systems*

$$\forall t \in [t_0, t_1], \ x'(t) = f(x(t), u)$$

subjected to viability constraints

$$\forall t \in [t_0, t_1], \ x(t) \in U^{-1}(u)$$

with constant regulon are called evolutions under constant controls *on the time interval* $[t_0, t_1]$. *They are said* persistent evolutions *(see Definition 10.4.2, p. 393) if they remain in K as long as possible, actually, on the interval* $[0, \tau_K^\sharp(x)]$, *where* $\tau_K^\sharp(x)$ *is its (upper)* exit time *(see Definition 4.3.4, p. 135).*

Proposition 10.5.1, p. 399 states that a persistent evolution reaches the boundary of K and leaves K through its *exit subset* $\text{Exit}_u(K)$ (see Definition 10.4.7, p. 396).

Naturally, viable evolutions under constant controls may not exist, or exist only on a finite time interval, or the class of evolutions under constant controls is too small to contain solutions to given problems (viability, capturability, optimal controls, etc.).

For climbing the next step of the inertia cascade, we can look for evolutions regulated by affine open-loop controls of the form $u(t) := u + u_1 t$, nicknamed *ramp controls* in control theory, regulating what we shall call "ramp evolutions":

Definition 6.4.2 *[Ramp Evolutions] An evolution* $(x(\cdot), u(\cdot))$ *is said to be a* ramp evolution *on a time interval* $[t_0, t_1]$ *if it is regulated by an affine open-loop controls of the form* $u(t) := u + u_1 t$, *called* ramp controls, *the velocities of which are constant.*

Inert evolutions *are evolutions controlled by regulons with velocities the* norm of which is constant *and equal to a finite inertia threshold c:*

$$\forall t \in [t_0, t_1],\ x'(t) = f(x(t), u(t)) \text{ where } \|u'(t)\| \ = c$$

Although we shall concentrate our study on inert evolutions, we shall provide some properties common to evolutions governed by open-loop controls $u(t) := u + u_1 t + \cdots + u_{m-1}\frac{t^{m-1}}{(m-1)!}$ which are $(m-1)$-degree polynomials in time.

More generally, we are interested in evolutions governed by open-loop controls $t \mapsto u(t)$ with bounded derivative $u^{(m)}(t)$ or satisfying $\|u^m(t)\| = c$ for some $m \geq 1$ (see Sect. 6.4.6, p. 226).

In this example of inert evolutions of a given degree, we look for the regulation by polynomial open loop controls of fixed degree m, ranging over the space $\mathbb{R}^{m+1}$ of coefficients. This amounts to regulating the system with combinations of a finite number of controls, the coefficients of polynomials of degree m. This answers the important issue of *quantized controls* (see Comment 33, p. 422). In Sect. 10.8, p. 422, we shall regulate viable evolutions by concatenations of systems governed by a combination (called *amalgam*) of finite number of feedbacks (see Definition 10.8.2, p. 424).

An adequate way to handle concepts of evolutions governed by open-loop polynomial controls and differentiable open-loop controls is by means of inertia functions. These inertia functions measure the versatility of the regulons (0-*order inertia function*), the versatility of their first derivatives and/or of their higher order derivatives.

Definition 6.4.3 *[0-Order Inertia Function] Recall that $\mathcal{S}^K(x)$ denotes the set of state evolutions $x(\cdot)$ governed by the control system (6.1), p. 200:*

$$x'(t) \ = \ f(x(t), u(t)) \text{ where } u(t) \ \in \ U(x(t))$$

satisfying the initial condition $x(0) = x \in K$ and viable in K.

The 0-order inertia function $\alpha_0 : X \mapsto \mathbb{R} \cup \{+\infty\}$ *defined by*

$$\alpha_0(x) := \inf_{x(\cdot) \in \mathcal{S}^K(x)} \sup_{t \geq 0} \|u(t)\| \in [0, +\infty]$$

measures the minimal worst intertemporal norm of the open-loop controls.

Its domain is the subset of initial states from which at least one evolution is governed by bounded open-loop controls and its *c-lower level set* (see Definition 10.9.5, p. 429):

$$\mathbf{L}_{\alpha_0}^{\leq}(c) \ := \ \{x \in \mathrm{Dom}(U) \text{ such that } \alpha_0(x) \ \leq \ c\}$$

provide the subset of initial states from which at least one evolution is governed by open-loop controls bounded by the given constant c. Instead of looking for a priori estimates on the regulons, we are looking for a posteriori estimates singling out what are the initial states satisfying this property (see Comments 1, p. 5 and 2, p. 5).

Lemma 6.4.4 *[Viability Characterization of 0-th order inertia function] Introduce the auxiliary micro–macro system:*

$$\begin{cases} (i) \ x'(t) \ = \ f(x(t), u(t)) \\ (ii) \ y'(t) \ = \ 0 \\ \quad \text{where } \|u(t)\| \ \leq \ y(t) \end{cases} \tag{6.2}$$

The 0-th order inertia function is related to the viability kernel of $K \times \mathbb{R}_+$ under system (6.2) by formula

$$\alpha_0(x) \ = \inf_{(x,y)\in \mathrm{Viab}_{(6.2)}(K\times\mathbb{R}_+)} y$$

6.4.2 First-Order Inertia Functions

We denote by $\mathcal{P}(x, u)$ the set of state-control solutions $(x(\cdot), u(\cdot))$ to the control system (6.1), p. 200:

$$x'(t) \ = \ f(x(t), u(t)) \text{ where } u(t) \ \in \ U(x(t))$$

starting at $(x, u) \in \mathrm{Graph}(U)$ and viable in this graph.

Definition 6.4.5 *[Inertia Functions] The* inertia function α *of the system (6.1), p. 200 is defined by*

$$\alpha(x, u) \ := \inf_{(x(\cdot),u(\cdot))\in\mathcal{P}(x,u)} \sup_{t\geq 0} \|u'(t)\| \ \in \ [0, +\infty]$$

on $\mathrm{Graph}(U)$. *It associates with any state-control pair $(x, u) \in \mathrm{Graph}(U)$ the minimal worst intertemporal inertia $\alpha(x, u)$ of the evolutions $(x(\cdot), u(\cdot)) \in \mathcal{P}(x, u)$ starting from $x \in \mathrm{Dom}(U)$ and $u \in U(x)$.*

Therefore, viable evolutions starting from an initial state-control pair $(x,u) \in \mathrm{Dom}(\alpha)$ can be regulated by *(absolutely) continuous* open-loop controls (i.e., whose derivatives are measurable) instead of "wilder" measurable controls. Hence, the first information provided by the inertia function is to localize the set of initial state-control pairs from which the system can be regulated by smoother controls.

We observe that

$$\alpha(x,u) \;=\; 0 \text{ if and only if } \forall t \geq 0,\ \alpha_0(x(t)) \;=\; \|u\|$$

We shall characterize the inertia function in terms of the viability kernel of $\mathrm{Graph}(U) \times \mathbb{R}_+$ under the specific auxiliary system:

Definition 6.4.6 *[Metasystems] The* metasystem *associated with initial control system (6.1), p. 200 is the auxiliary system*

$$\begin{cases} (i) & x'(t) \;=\; f(x(t),u(t)) \\ (ii) & u'(t) \;=\; v(t) \\ (iii) & y'(t) \;=\; 0 \\ & \quad \text{where } \|v(t)\| \;\leq\; y(t) \end{cases} \tag{6.3}$$

It is regulated by the velocities $v(t) = u'(t)$ of the controls of initial system (6.1), called metacontrols.

Metasystems are used to characterize inertia functions:

Theorem 6.4.7 *[Viability Characterization of Inertia Functions] Recall that $\mathcal{P}(x,u)$ denotes the set of state-control evolutions $(x(\cdot),u(\cdot))$ governed by the regulated system (6.1), p. 200:*

$$x'(t) \;=\; f(x(t),u(t)) \text{ where } u(t) \;\in\; U(x(t))$$

such that $x(0) = x$ and $u(0) = u$. The inertia function is related to the viability kernel of $\mathrm{Graph}(U) \times \mathbb{R}_+$ *under metasystem (6.3) by formula*

$$\alpha(x,u) \;=\; \inf_{(x,u,y) \in \mathrm{Viab}_{(6.3)}(\mathrm{Graph}(U) \times \mathbb{R}_+)} y$$

Proof. Indeed, to saythat (x,u,y) belongs to $\mathrm{Viab}_{(6.3)}(\mathrm{Graph}(U) \times \mathbb{R}_+)$ amounts to saying that there exists an evolution $t \mapsto (x(t),u(t))$ governed by (6.3) where $t \mapsto (x(t),u(t),y(t))$ is governed by control system (6.1), p. 200 and where $y(t) \equiv y$. In other words, the solution $(x(\cdot),u(\cdot)) \in \mathcal{P}(x,u)$ satisfies

$$\forall t \geq 0, \ \|u'(t)\| \leq y$$

so that $\alpha(x,u) \leq \sup_{t\geq 0} \|u'(t)\| \leq y$.

Conversely, if $\alpha(x,u) < +\infty$, we can associate with any $\varepsilon > 0$ an evolution $(x_\varepsilon(\cdot), u_\varepsilon(\cdot)) \in \mathcal{P}(x,u)$ such that

$$\forall t \geq 0, \ \|u_\varepsilon'(t)\| \leq \alpha(x,u) + \varepsilon =: \ y_\varepsilon$$

Therefore, setting $v_{\varepsilon_1}(t) := u_\varepsilon'(t)$ and $y_\varepsilon(t) = y_\varepsilon$, we observe that $t \mapsto (x_\varepsilon(t), u_\varepsilon(t), y_\varepsilon)$ is a solution to the auxiliary system (6.3) viable in $\mathrm{Graph}(U) \times \mathbb{R}_+$, and thus, that (x,u,y_ε) belongs to $\mathrm{Viab}_{(6.3)}(\mathrm{Graph}(U) \times \mathbb{R}_+)$. Hence

$$\inf_{(x,u,y)\in \mathrm{Viab}_{(6.3)}(\mathrm{Graph}(U)\times\mathbb{R}_+)} y \leq y_\varepsilon := \alpha(x,u) + \varepsilon$$

and it is enough to let ε converge to 0. □

The metasystem (6.3) is Marchaud (see Definition 10.3.2, p. 384) whenever the single-valued map f is continuous, Lipschitz (see Definition 10.3.5, p. 385) whenever the single-valued map f is Lipschitz and the "metaenvironment" is closed whenever the graph of U is closed and convex if f is affine and is the graph of U is convex. Hence it inherits of the properties of Marchaud, Lipschitz and convex systems respectively.

Illustration: Newtonian Inertia Function. We begin with a one dimensional example, the simplest one. We have to model a complex system where the dynamics is unknown to us. Hence, we describe our ignorance by an "open" right hand side

$$\forall\, t \geq 0, \ \ x'(t) \ = \ u(t) \text{ where } u(t) \ \in \ \mathbb{R} \tag{6.4}$$

It is tempting to try guessing right hand sides, by selecting feedbacks $x \mapsto \widetilde{u}(x)$, solving

$$x'(t) \ := \ \widetilde{u}(x(t))$$

and checking what are the properties of the evolutions governed by this specific differential equation, for instance, whether they are viable in a given subset K, for instance.

Just for illustrating this point of view, consider simple example of *affine feedbacks* defined by

$$\widetilde{u}(x) \ := \ r(b - x)$$

The solution starting from $x \in [0, b]$ is equal to

$$x(t) = e^{-rt}x + b\left(1 - e^{-rt}\right) = b - (b - x)e^{-rt}$$

It remains confined in the interval $[x, b]$, converges to the equilibrium b when $t \mapsto +\infty$, whereas its derivative decreases to 0 when $t \to +\infty$.

Although numerous other illustrations (a term better adapted than "models" for economic or biological systems) have been proposed because, contrary to the physical and engineering sciences, we really do not know these feedbacks, and, in particular, feedbacks producing "heavy evolutions", or feedbacks governing "cyclic evolutions".

The question remains to know whether this is the only feedback satisfying these properties, and whether one can find the class of all possible feedbacks and choose the one which satisfies further properties. Furthermore, if the open-loop control system and the constraints are complicated (in higher dimensions), guessing feedbacks for governing evolutions satisfying such and such property.

Instead, we can begin with the only knowledge we may have at our disposal, for instance,

$$\begin{cases} (i) & \textbf{viability constraints: } x(t) \in K := [a,b], \;\; 0 < a < b < +\infty \\ (ii) & \textbf{inertia threshholds } c \in \mathbb{R}_+\text{: } \|x''(t)\| \;\leq\; c \end{cases} \tag{6.5}$$

The above dynamical inequality can be written in the form of the regulated "metasystem"

$$\begin{cases} (i) \; x'(t) \;=\; u(t) \\ (ii) \; u'(t) \;=\; v(t) \\ \quad\;\; \text{where } \|v(t)\| \leq c \end{cases} \tag{6.6}$$

(where the controls are the accelerations), the equilibria of which are of the form $(x, 0)$ where $x \in [a, b]$.

These simple constraints are enough to deduce many information over the evolutions governed by the regulated system (6.4), p. 212.

Viability and inertia constraints (6.5), p. 213 are sufficient to deduce information over the evolutions governed by the regulated system (6.4), p. 212 by studying the *Newtonian inertia function* defined by

$$\alpha(x,u) \;:=\; \inf_{(x(\cdot),u(\cdot))\in\mathcal{P}(x,u)} \sup_{t\geq 0} \|u'(t)\| \;=\; \inf_{(x(\cdot),u(\cdot))\in\mathcal{P}(x,u)} \sup_{t\geq 0} \|x''(t)\|$$

where $\mathcal{P}(x, u)$ denote the set of state-control solutions $(x(\cdot), u(\cdot))$ to differential equation (6.4) viable in the interval $[a, b]$ such that $x(0) = x$ and $u(0) = x'(0) = u$.

In Newtonian mechanics, the evolution of the second derivative $x''(t)$, i.e., the acceleration, is inversely proportional to the mass: the smaller the evolution of acceleration, the larger the evolution inertia.

Let us introduce (6.7) the auxiliary system

$$\begin{cases} (i) \quad x'(t) = u(t) \\ (ii) \quad u'(t) = v(t) \\ (iii) \quad y'(t) = 0 \\ \qquad \text{where } \|v(t)\| \leq y(t) \end{cases} \tag{6.7}$$

Theorem 6.4.7, p. 211 states that the inertia function is characterized by formula

$$\alpha(x, u) = \inf_{(x,u,y) \in \mathrm{Viab}_{(6.7)}(K \times \mathbb{R} \times \mathbb{R}_+)} y$$

Therefore, the inertia function inherits all the properties of viability kernel and can be computed thanks to the viability kernel algorithm. In this simple example, we can even provide explicit formulas. One can check for instance the analytical formula of the inertia function:

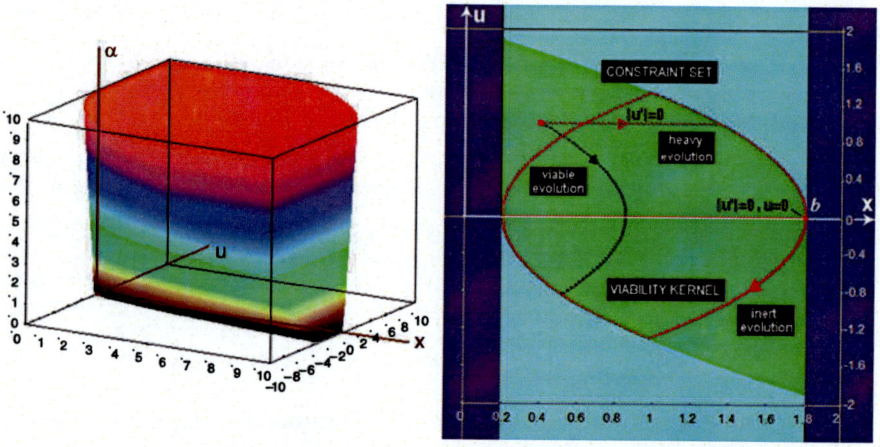

Fig. 6.2 Newtonian Inertia Function.
Left*: The Viability Kernel Algorithm computes the graph of the inertia function.* Right*: A lower lever set (or section) of the inertia function: Its boundary is the critical zone, over which the state evolves with constant acceleration. The trajectory of a heavy evolution minimizing the velocity of the controls and which stops at equilibrium b is shown.*

Lemma 6.4.8 ***[Newtonian Inertia Function]*** *The Newtonian inertia function α defined on $]a, b[\times\mathbb{R}$ is equal to:*

$$\alpha(x, u) := \frac{u^2}{2(b - x)} \ \textit{if } u \geq 0 \text{ and } \frac{u^2}{2(x - a)} \ \textit{if } u \leq 0 \tag{6.8}$$

Its domain is $(\{a\} \times \mathbb{R}_+) \cup (]a, b[\times\mathbb{R}) \cup (\{b\} \times \mathbb{R}_-)$. Hence, from each state-control pair $(x, u) \in \mathrm{Dom}(\alpha)$ starts at least one evolution with acceleration bounded *by $\alpha(x, u)$.*

This allows us to check the precision of the solution provided by the viability kernel algorithm and the analytical formulas, which are not distinguishable at the pixel level.

Remark. We can also check that the inertia function is a solution to the Hamilton–Jacobi partial differential equation

$$\begin{cases} \frac{\partial \alpha(x,u)}{\partial x} u - \alpha(x,u)\frac{\partial \alpha(x,u)}{\partial u} \text{ if } u \geq 0 \\ \frac{\partial \alpha(x,u)}{\partial x} u + \alpha(x,u)\frac{\partial \alpha(x,u)}{\partial u} \text{ if } u \leq 0 \end{cases}$$

Indeed, the partial derivatives of this inertia function is equal to

$$\frac{\partial \alpha(x,u)}{\partial x} := \begin{cases} \frac{u^3}{2\,(b-x)^2} & \text{if } u \geq 0 \\ -\frac{u^3}{2\,(x-a)^2} & \text{if } u \leq 0 \end{cases} \quad \& \quad \frac{\partial \alpha(x,u)}{\partial u} := \begin{cases} \frac{u}{(b-x)} & \text{if } u \geq 0 \\ \frac{u}{(a-x)} & \text{if } u \leq 0 \end{cases}$$

Observe that $\frac{\partial \alpha(x,u)}{\partial u}$ is positive when $u > 0$ and negative when $u < 0$. □

Remark: Finite Horizon Inertia functions. One can be interested by finite horizon inertia function

$$\alpha(T,x,u) := \inf_{(x(\cdot),u(\cdot))\in\mathcal{P}(x,u)} \sup_{t\in[0,T]} \|u'(t)\| \in [0,+\infty]$$

Their study being very similar, we skip it in this book. Let us mention however that introducing the auxiliary system

$$\begin{cases} (i) \;\; \tau'(t) = -1 \\ (ii) \;\; x'(t) = f(x(t),u(t)) \\ (iii) \; u'(t) = v(t) \\ (iv) \;\; y'(t) = 0 \\ \qquad \text{where } \|v(t)\|; \leq y(t) \end{cases} \tag{6.9}$$

the auxiliary environment $\mathcal{K} := \mathbb{R}_+ \times \text{Graph}(U) \times \mathbb{R}_+$ and the target $\mathcal{C} := \{0\} \times \text{Graph}(U) \times \mathbb{R}_+$, then the finite horizon inertia function is related to the capture basin of target $\mathcal{C}$ viable in the environment $\mathcal{K}$ under auxiliary metasystem (6.9) by formula

$$\alpha(T,x,u) = \inf_{(T,x,u,y)\in \text{Capt}_{(6.9)}(\mathbb{R}_+\times\text{Graph}(U)\times\mathbb{R}_+,\{0\}\times\text{Graph}(U)\times\mathbb{R}_+)} y$$

They inherit the properties of capture basins. □

6.4.3 Inert Regulation and Critical Maps

We associate with the inertia function the useful concepts of inert retroaction map and critical map:

Definition 6.4.9 *[Inert Regulation and Critical Maps] We associate with the inertia function α the following set valued maps:*

1. inert regulation map $R_c : x \rightsquigarrow R_c(x) := \{u \in \mathbb{R} \text{ such that } \alpha(x,u) \leq c\}$
2. critical map $\Xi_c : u \rightsquigarrow \Xi_c(u) := \{x \in [a,b] \text{ such that } \alpha(x,u) = c\}$

When $c > 0$, the subset $\Xi_c(u)$ is called the critical zone *of the control u bounded by inertia threshold $c > 0$. When $c = 0$, the subset $\Xi_0(u)$ is called the* viability niche *of the control u. A regulon is a* punctuated equilibrium *(or, actually,* punctuated regulon*) if its viability niche $\Xi_0(u)$ is not empty.*

Therefore, starting from $(x,u) \in \mathrm{Dom}(\alpha)$, then there exists an evolution $x(\cdot)$ regulated by a control $u(\cdot)$ satisfying

$$\forall t \geq 0, \;\; u(t) \in R_{\alpha(x,u)}(x(t))$$

23 *[Mathematical Implementation of Punctuated Equilibria] The concepts of punctuated regulon and their viability niches offer a mathematical translation of the Eldredge and Gould concept of punctuated equilibrium. If a regulon u is a punctuated equilibrium, its viability niche $\Xi_0(u)$ is the viability kernel of $U^{-1}(u)$ under the system $x'(t) = f(x(t),u)$ regulated by the constant punctuated regulon U. Starting from $x \in \Xi_0(u)$, there exists an evolution regulated by this punctuated regulon remaining in its viability niche forever. In other words, the viability niche of a punctuated equilibrium can be regarded as a kind of "punctuated equilibrium": It is no longer reduced to a point, but is a* subset *viable under the system governed by the punctuated equilibrium. If u is the regulon associated to an equilibrium $\overline{x} \in U^{-1}(u)$, solution to equation $f(\overline{x},u) = 0$, then u is naturally a punctuated equilibrium and $\overline{x} \in \Xi_0(u)$.*

Illustration (continuation). We compute now the *inert regulation map* R_c and the *critical map* Ξ_c (see Definition 6.4.9, p. 216):

Lemma 6.4.10 *[Inert Regulation and Critical Maps] We set*

$$r^{\sharp}(x) := \sqrt{2(b-x)},\ r^{\flat}(x) := \sqrt{2\,(x-a)} \text{ and } R(x) := \left[-r^{\flat}(x), +r^{\sharp}(x)\right] \tag{6.10}$$

The regulation map is equal to

$$R_c(x) := \sqrt{c}\left[-r^{\flat}(x), +r^{\sharp}(x)\right] = \sqrt{c}\,R(x)$$

The critical map Ξ_c $(c > 0)$ is defined by

$$\Xi_c(u) = b - \frac{u^2}{2c} \text{ if } u > 0 \text{ and } \Xi_c(u) = a + \frac{u^2}{2c} \text{ if } u < 0$$

The viability niche $\Xi_0(u)$ of the regulon u is empty when $u \neq 0$ and equal to $\Xi_0(0) = [a, b]$ when $u = 0$, which is the set of equilibria.

The graph of the regulation map R_c associated to the inertia function is limited by the two graphs of $-\sqrt{c}\,r^{\flat}$ below and $\sqrt{c}\,r^{\sharp}$ above. The derivatives of these two maps $r^{\sharp}$ and $r^{\flat}$ are given by

$$\frac{dr^{\sharp}(x)}{dx} = \frac{-1}{\sqrt{2\,(b-x)}} \text{ and } \frac{dr^{\flat}(x)}{dx} = \frac{1}{\sqrt{2(x-a)}}$$

The map $r^{\sharp}$ is decreasing and the map $r^{\flat}$ is increasing. □

Remark: Infinite Inertia Thresholds. When $c := +\infty$, Theorem 10.8.3, p. 424 provides a sufficient condition for persistent evolutions associated with a constant control to remain viable until its exit time when it reaches its exit set (playing the role of the critical set for $c := +\infty$) and switches for another constant control (see Sects. 10.8, p. 422 and 10.8, p. 422). □

6.4.4 Heavy Evolutions

How can we implement the inertia principle?

There are numerous methods for regulating evolutions satisfying the inertia principle. We introduce the simplest way to achieve this objective, by selecting at each instant the regulons providing viable evolutions with *minimal velocity*. They obeys this inertia principle because, whenever the control can be constant on some time interval, its velocity is equal to 0, and the evolutions governed by controls with minimal velocity necessarily catches velocities equal to 0. Evolutions obtained in this way are called *"heavy" viable evolutions* in the sense of heavy trends in economics or in *Newtonian mechanics.*

We assume here for simplicity that the solutions to the differential equations $x'(t) = f(x(t), u)$ with constant regulons are unique. This is the case when the maps $x \mapsto f(x, u)$ are Lipschitz or monotone.

For defining heavy solutions, *we still fix a bound c on the norms of the velocities of the regulons* and take any initial state-control pair (x, u) such that $\alpha(x, u) < c$. We then fix the regulon u and consider the evolution $(x_u(t), u)$ under constant control where $x_u(\cdot)$ is the solution to differential equation $x'(t) = f(x(t), u)$ (evolving with velocity $u'(t) = 0$).

As long as $\alpha(x_u(t), u)$ is smaller than the velocity bound c, the regulon u inherited from the past can be maintained, allowing the system to be regulated by this constant control u. Since the state $x_u(\cdot)$ of the system evolves while the regulon remains constant and equal to u, the inertia function $\alpha(x_u(t), u)$ evaluated on such an evolution may increase and eventually overrun the bound c measuring the maximal velocity of the regulons at some time $\sigma_c(x, u)$ at the state $x_u(\sigma_c(x, u)) \in \Xi_c(u)$ providing *warning time* defined as follows:

Definition 6.4.11 *[**Warning Times**] Assume that* $c > \alpha(x, u)$. *Then the* warning time $\sigma_c(x, u) \in \mathbb{R} \cup \{+\infty\}$ *is the first instant when the evolution* $x_u(\cdot)$ *starting from* x *when* $x_u(\sigma_c(x, u)) \in \Xi_c(u)$ *reaches the* critical zone

$$\Xi_c(u) := \{x \in \mathrm{Dom}(U) \text{ such that } \alpha(x, u) = c\}$$

of the regulon u.

This warning time is nothing other than the minimal time function to reach the critical zone $\Xi_c(u)$ being viable in $R_c(u)$. Therefore, it can be computed by the Viability Kernel Algorithm.

Warning times and *critical zones* tell us **when** $(\sigma_c(x, u))$, **where** $(\Xi_c(u))$ and **how** (*minimizing the velocity at each instant*) the regulons must evolve, defining a *viability critical period: To survive*, other regulons must emerge when the state reaches the critical zone of the regulon, in such a way that the new velocities of the regulons are bounded by the *inertia threshold* c until the regulon can again remain constant for a new period of time.

24 *[**Warning Time or "Kairos"**] The concept of* warning time *is a mathematical translation of the anglo-saxon concept of* timing, *or the Italian concept of* tempismo, *modernizing the concept of* kairos *of classical Greece, meaning* propitious *or* opportune *moment. The ancient Greeks used this* qualitative *concept of time by opposition to* chronos, *the* quantitative... *chronological time, which can be measured by clocks.*

Lysippos sculptured a wonderful *concrete* representation of this very *abstract* concept, which was depicted by *Posidippos* who gave his definition of the kairos of *Lysippos*'s bas-relief (in the museum of Torino):

25 ***[Epigram of Posidippos]***

Who and whence was the sculptor? From Sikyon. And his name? Lysippos. And who are you? Time who subdues all things. Why do you stand on tip-toe? I am ever running. And why you have a pair of wings on your feet? I fly with the wind. And why do you hold a razor in your right hand? As a sign to men that I am sharper than any sharp edge. And why does your hair hang over your face? For him who meets me to take me by the forelock. And why, in Heaven's name, is the back of your head bald? Because none whom I have once raced by on my winged feet will now, though he wishes it sore, take hold of me from behind. Why did the artist fashion you? For your sake, stranger, and he set me up in the porch as a lesson.

The razor is reminiscent of *Ockham*'s razor:

26 ***[Ockham's Razor]***

Ockham's Razor *is the principle proposed by* William of Ockham *[1285–1347]:* "Pluralitas non est ponenda sine neccesitate", *which translates as* "entities should not be multiplied unnecessarily". *This "law of parsimony" states that an explanation of any phenomenon should make as few assumptions as possible, and to choose among competing theories the one that postulates the fewest concepts.*

The concept of inertia principle and heavy evolution is also closely connected with the concept of *emergence* in physics (phase transition) and in biological and social sciences.

The concepts of viability niche and heavy evolution are closely related through a concept of "locking-it" introduced in economics of innovation:

Proposition 6.4.12 *[Locking-in Viability Niches] If at some time t_f, $u(t_f)$ is a* punctuated regulon, *then the heavy viable solution enters its viability niche and* may *remain in this viability niche forever whereas the regulon remains equal to this punctuated regulon. In other words,* any heavy evolution arriving at the viability niche of a regulon is "locked-in" this viability niche.

This property explains the survival of species in ecological niches or the socio-economic of selection of commodities, which, far from optimal, become viable once they have been adopted by a sufficient number of organisms.

Remark: The associated Hamilton–Jacobi–Bellman Equation. We supply the finite dimensional vector space X with a norm $x \mapsto \|x\|$ with which we associate its dual norm $\|p\|_\star := \sup_{\|x\|\leq 1} \langle p, x\rangle$. When $X := \mathbb{R}^n$, the dual norm to the norm $\|x\|_\infty := \max_{i=1,\ldots,n} |x_i|$ is the norm $\|p\|_1 := \sum_{i=1}^{n} |p_i|$ and the dual norm to the norm $\|x\|_\alpha := \left(\sum_{i=1}^{n} |x_i|^\alpha\right)^{\frac{1}{\alpha}}$ is the norm $\|x\|_\beta := \left(\sum_{i=1}^{n} |x_i|^\beta\right)^{\frac{1}{\beta}}$ with $\frac{1}{\alpha} + \frac{1}{\beta} = 1$.

One can prove that the inertia function α is the smallest positive lower semicontinuous solution to the Hamilton–Jacobi–Bellman partial differential equation: $\forall (x,u) \in \text{Graph}(U)$,

$$\begin{cases} \left\langle \frac{\partial \mathbf{v}(x,u)}{\partial x}, f(x,u) \right\rangle - \mathbf{v}(x,u) \left\| \frac{\partial \mathbf{v}(x,u)}{\partial u} \right\|_\star \\ = \sum_{i=1}^{n} \frac{\partial \mathbf{v}(x,u)}{\partial x_i} f_i(x,u) - \mathbf{v}(x,u) \left\| \frac{\partial \mathbf{v}(x,u)}{\partial u} \right\|_\star = 0 \end{cases}$$

on the graph of U. Naturally, this statement needs to define rigourously in which sense derivatives are understood in order to give a meaning for solutions which are only lower semicontinuous, and which are not even continuous, even less differentiable. □

Illustration (continuation). We can check that the inertia function is the (smallest positive lower semicontinuous) solution to the Hamilton–Jacobi partial differential equation

$$\frac{\partial \alpha(x,u)}{\partial x} u - \alpha(x,u) \left| \frac{\partial \alpha(x,u)}{\partial u} \right| = 0$$

on $]a,b[\times\mathbb{R}$. It thus splits in two parts:

$$\begin{cases} \frac{\partial\alpha(x,u)}{\partial x}u - \alpha(x,u)\frac{\partial\alpha(x,u)}{\partial u} & = 0 \text{ if } u \geq 0 \\ \frac{\partial\alpha(x,u)}{\partial x}u + \alpha(x,u)\frac{\partial\alpha(x,u)}{\partial u} & = 0 \text{ if } u \leq 0 \end{cases}$$

Indeed, the partial derivatives of this inertia function are equal to

$$\frac{\partial\alpha(x,u)}{\partial x} := \begin{cases} \frac{u^2}{2\,(b-x)^2} & \text{if } u \geq 0 \\ -\frac{u^2}{2\,(x-a)^2} & \text{if } u \leq 0 \end{cases} \quad \& \quad \frac{\partial\alpha(x,u)}{\partial u} := \begin{cases} \frac{u}{(b-x)} & \text{if } u \geq 0 \\ \frac{u}{(a-x)} & \text{if } u \leq 0 \end{cases}$$

Observe that $\frac{\partial\alpha(x,u)}{\partial u}$ is positive when $u > 0$ and negative when $u < 0$. □

6.4.5 The Adjustment Map and Heavy Evolutions

Hamilton–Jacobi–Bellman equations are useful only in the extent where they help us to possibly give analytical formulas of the regulation (see Definition 2.15.6, p. 102) which we shall compute informally:

Definition 6.4.13 *[Adjustment Map] Denote by $\mathcal{U}$ the control space. The adjustment map $G : \mathcal{E}p(\alpha) \rightsquigarrow \mathcal{U}$ is the regulation map of metasystem (6.3): It associates with any $(x,u,c) \in \mathrm{Graph}(U)\times\mathbb{R}_+$ the subset $G(x,u,c)$ of metacontrols v such that*

$$G(x,u,c) := \left\{\|v\| \leq \alpha(x,u) \text{ and } (f(x,u),v,0) \in T^{\star\star}_{\mathcal{E}p(\alpha)}(x,u,c)\right\}$$

where $T^{\star\star}_K(x)$ denotes the convexified tangent cone (see Definition 18.4.8, p. 732).

We derive from Theorem 4.11.5, p. 170 the existence of the regulation map associated with metasystem, which we call an *adjustment map*:

Proposition 6.4.14 *[Characterization of the Adjustment Map] Let us assume that the inertia function α is continuous on its domain. We refer to Definition 18.5.5, p. 740 of the definition of the (convexified) derivative $D^{\star\star}_{\uparrow}\alpha(x,u)$ of the inertia function and set*

$$\begin{cases} (i) \ \Gamma(x,u) := \left\{v \in \alpha(x,u)B \text{ such that } D_{\uparrow}^{\star\star}\alpha(x,u)(f(x,u),v) \le 0\right\} \\ (ii) \ \gamma(x,u) := \inf_{v\in\Gamma(x,u)} \|v\| \end{cases}$$

Then the adjustment map can be written in the following form

$$G(x,u,c) = \begin{cases} \alpha(x,u)B, & \text{if } \alpha(x,u) < c \\ \Gamma(x,u) & \text{if } \alpha(x,u) = c \end{cases}$$

Therefore, all evolutions of metasystem (6.3) starting from (x,u,c) *viable in* $\mathrm{Graph}(U) \times \mathbb{R}_+$ *are regulated by the system of differential inclusions*

$$\begin{cases} (i) \ x'(t) = f(x(t),u(t)) \\ (ii) \ u'(t) \in G(x(t),u(t)) \end{cases} \tag{6.11}$$

We deduce the method for building the adjustment law governing the evolution of heavy evolutions:

Theorem 6.4.15 ***[Regulating Heavy Evolutions]*** *Assume that the inertia function* α *is continuous on its domain and that the function* γ *is upper semicontinuous. Then from any* $(x,u,c) \in \mathcal{E}p(\alpha)$*:*

1. *starts* **at least** *one evolution governed by the system of differential inclusions (6.11), p. 222 viable in the epigraph of the inertia function,*
2. *all evolutions viable in the epigraph of the inertia function are governed by the system of differential inclusions (6.11)*

Heavy evolution *are regulated by the system*

$$\begin{cases} (i) \ x'(t) = f(x(t),u(t)) \\ (ii) \ u'(t) \in g^0(x(t),u(t),y(t)) \end{cases} \tag{6.12}$$

where the heavy adjustment map g^0 *is defined by*

$$g^0(x,u,y) = \begin{cases} \{0\} & \text{if } \alpha(x,u) < y \\ \gamma(x,u)S & \text{if } \alpha(x,u) = y \end{cases}$$

where $S := \{u \text{ such that } \|u\| = 1\}$ *denotes the unit sphere.*

Proof. Observe that the heavy adjustment map $g^0(x,u,y)$ is upper semicontinuous, so that the auxiliary system

$$\begin{cases} (i) \ x'(t) = f(x(t),u(t)) \\ (ii) \ u'(t) \in g^0(x(t),u(t),y(t)) \end{cases}$$

is Marchaud. By construction, the epigraph of the inertia function is viable under this auxiliary system, so that, from any initial (x, u, c) starts an evolution viable in $\mathcal{E}p(\alpha)$ satisfying

$$\|u'(t)\| \leq \|g^0(x(t), u(t), c)\|$$

Being viable in $\mathcal{E}p(\alpha)$, it satisfies

$$u'(t) \in G(x(t), u(t), c)$$

This implies that for almost all $t \geq 0$,

$$u'(t) = g^0(x(t), u(t), c)$$

This concludes the proof. □

We now complete the description of the behavior of heavy evolutions for an inertia threshold c.

Assume that $\alpha(x, u) < c$. Since α is assumed to be continuous, we deduce that whenever $\alpha(x(t), u(t)) < c$, the velocity $u'(t) = 0$ so that the control remains constant. Hence $x(\cdot)$ is regulated by a constant control as long as $\alpha(x(t), u) < c$. Let $t^\star$ be the first instant (kairos) when $\alpha(x(t^\star), u) = c$, i.e., when $x(t^\star) \in \Xi_c(u)$ belongs to the critical zone $\Xi_c(u)$ of the regulon u for the inertia threshold c. Then we know that $u'(t^\star) = \gamma(x(t^\star), u(t^\star))$, so that $\|u'(t^\star)\| = c$.

If the map f is Lipschitz, then the Quincampoix Barrier Theorem (see Theorem 10.5.18, p. 409) implies that $\alpha(x(t), u(t)) = \alpha(x(t^\star), u(t^\star)) = c$ as long as $(x(t), u(t)) \in \text{Int}(\text{Graph}(U))$, since the boundary of the epigraph of α, which is equal to the graph of α at each state-control pair where it is continuous, exhibits the barrier property: The evolution remains in the boundary as long as $(x(t), u(t), c)$ belongs to the interior of $\text{Graph}(U) \times \mathbb{R}_+$. It may change only when the state control pair $(x(t), u(t))$ hits the boundary of $\text{Graph}(U)$.

If $c = 0$, then the viability niche $\Xi(0, u)$ is viable under differential condition $x'(t) = f(x(t), u)$ with constant control.

Illustration (continuation). We have seen that evolutions with constant velocities are not viable in the interval $K := [a, b]$ whereas inert evolutions with strictly positive constant acceleration $c > 0$ are viable.

What happens when we start with controls with inertia $\alpha(x, u) < c$ strictly smaller than an imposed inertia threshold c? We can choose the heavy evolution which is governed by the control constant equal to u until it reaches the critical zone $\Xi_c(u)$ and, next, the inert evolution until it reaches the equilibrium b or a with a regulon equal to 0.

Lemma 6.4.10, p. 216 implies that solutions $x \pm |u|t$ regulated by constant controls $\pm|u|$ reach the critical zone at *warning time* equal to

$$\sigma_c(x,u) \ := \ \frac{(b-x)}{|u|} - \frac{|u|}{2c} \text{ if } u > 0 \text{ and } \sigma_c(x,u) \ := \ \frac{(x-a)}{|u|} - \frac{|u|}{2c} \text{ if } u < 0 \qquad (6.13)$$

because $x + u\sigma_c(x,u) = \Xi_c(u) = b - \frac{u^2}{2c}$. We recall that $\tau_c(\Xi_c(\pm|u|), \pm u) = \frac{|u|}{c}$.

The inertia function α increases over the evolutions under constant regulon u according

$$\forall t \in \left[0, \frac{(b-x)}{u}\right], \ \alpha(x_c(t), u) \ = \ \frac{u^2}{2\left((b-x) - ut\right)}$$

The derivative of the inertia function over the inert evolutions is equal to

$$\frac{d\alpha(x_c(t), u)}{dt} = \frac{u^2}{2\left((b-x) - ut\right)^2}$$

Hence, once a bound c is fixed, the heavy solution evolves with constant regulon u until the last instant $\sigma_c(x,u)$ when the state reaches $\Xi_c(u)$ and the velocity of the regulon $\alpha_c(\Xi_c(u), u) = c$:

$$\forall\, t \in \left[0, \frac{(b-x)}{u} - \frac{|u|}{2c}\right] \begin{cases} u_c(t) & = \ u \\ x_c(t) & = \ x + tu \end{cases}$$

until the *warning time* or *kairos* (see Definition 24, p. 218)

$$\sigma_c(x,u) \ := \ \frac{(b-x)}{|u|} - \frac{|u|}{2c}$$

This is the last moment when we have to change the regulon. After this warning time, the evolution starting at $(\Xi_c(u), u)$ follows the graph of $\sqrt{c}\, r^\sharp$ and is thus equal to

$$\begin{cases} \forall\, t \ \in \ \left[\frac{(b-x)}{u}, \frac{(b-x)}{u} + \frac{|u|}{c}\right], \\ u_c(t) \ := \ u - c(t - \sigma_c(x,u)) = u - c\left(t - \frac{(b-x)}{u} + \frac{|u|}{2c}\right) \\ x_c(t) \ := \ b - \frac{u^2}{2c} + u\left(t - \frac{(b-x)}{u} + \frac{|u|}{2c}\right) - \frac{c\left(t - \frac{(b-x)}{u} + \frac{|u|}{2c}\right)^2}{2} \end{cases}$$

until it reaches equilibrium $(b, 0)$ at time

$$t^\star \ := \ \sigma_c(x,u) + \frac{u}{c} \ = \ \frac{(b-x)}{u} + \frac{|u|}{2c}$$

and thus, remains at equilibrium $(b, 0)$ forever, because the heavy evolution chooses the evolution with minimal norm.

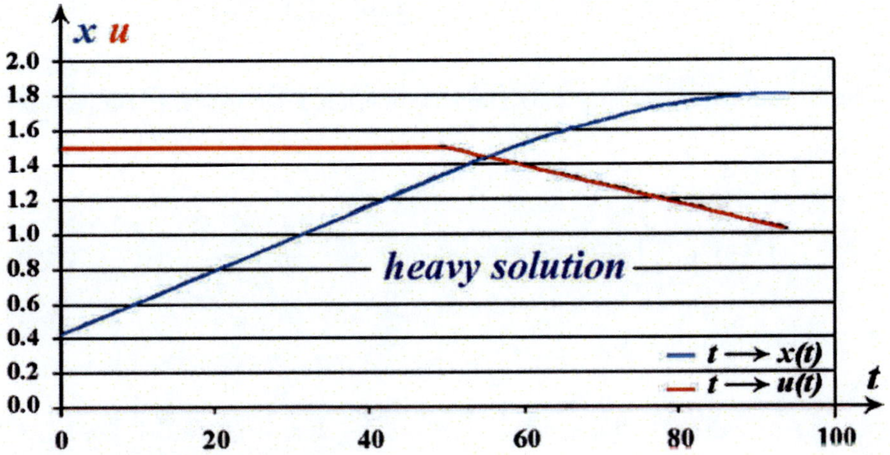

Fig. 6.3 Graph of the Heavy Evolution.
Both the graphs of the heavy evolution (in blue*) and of its control (in* red*) are plotted. They are not computed from the analytical formulas given below, but extracted from the Viability Kernel Algorithm. The* control remains constant *until the trajectory of the (affine) solution hits the boundary of the viability kernel and then slows down when it is controlled with a decreasing linear time dependent controls with velocity equal to* $-c$*. It reaches in finite time the boundary of the constrained interval with a velocity equal to* 0 *and may remain at this equilibrium.*

The *warning signal* $\sigma_c(x, u)$ tells us:

1. *when* the system must start moving its regulon in order to remain viable with an inertia threshold.
2. *where*, at the critical zone $\Xi_c(u)$, on which both the controls $u(t) := u - c(t - \tau_c(x, u))$ and the state ranges over the critical zone $x(t) \in \Xi_c(u - c(t - \tau_c(x, u)))$ until they reach b.

Heavy evolutions are regulated by the following maps:

$$r_c^{\natural}(x) := \begin{cases} \min(+u, +\sqrt{c}\, r^{\sharp}(x)) & \text{if } u > 0 \\ \max(-|u|, -\sqrt{c} r^{\flat}(x)) & \text{if } u < 0 \end{cases}$$

Remark. This study of one dimensional systems can be extended to the case of n decentralized controlled systems $x_i'(t) = u_i(t)$ confronted to decentralized viability constraints

$$\forall\, t \geq 0,\ x(t) \in K := \prod_{i=1}^{n} [a_i, b_i]$$

and to collective inertia constraint

$$\forall t \geq 0, \ \|x'(t)\| := \max_{i=1,\ldots,n} |x_i'(t)| \leq c$$

The multidimensional Newtonian inertia function α defined on $\prod_{i=1}^{n}([a_i, b_i]\times\mathbb{R})$ is equal to:

$$\forall (x,u) \in \prod_{i=1}^{n}([a_i, b_i] \times \mathbb{R}), \ \alpha(x,u) = \sup_{i=1,\ldots,n} \alpha(x_i, u_i) \tag{6.14}$$

where, by (6.8), p. 214,

$$\alpha(x_i, u_i) := \frac{u_i^2}{2(b_i - x_i)} \text{ if } u_i \geq 0 \text{ and } \frac{u_i^2}{2(x_i - a_i)} \text{ if } u_i \leq 0$$

Hence, from each state-control pair $(x,u) \in \mathrm{Dom}(\alpha)$ starts at least one evolution $(x(\cdot), u(\cdot)) \in \mathcal{P}(x,u)$ with *bounded acceleration*, actually, bounded by $\alpha(x,u)$. □

6.4.6 High-Order Metasystems

We continue our investigation of the regulation by twice-differentiable open-loop controls, and, among them, affine controls of the form $u(t) := u + u_1 t$ (called *ramp controls*), where we replaced the former notation $v(t) := u'(t)$ by $u_1(t) := v(t) := u'(t)$.

For that purpose, we denote by $\mathcal{P}_2(x, u, u_1)$ the set of solutions $(x(\cdot), u(\cdot))$ to the control system (6.1), p. 200:

$$x'(t) = f(x(t), u(t)) \text{ where } u(t) \in U(x(t))$$

satisfying the initial conditions

$$x(0) = x, \ u(0) = u, \ u'(0) = u_1$$

Definition 6.4.16 *[Second-Order Inertia Function] The second-order inertia function α of parameterized system (6.1), p. 200 associates the minimal worst intertemporal inertia $\alpha_2(x, u, u_1)$ defined by*

$$\alpha_2(x, u, u_1) := \inf_{(x(\cdot),u(\cdot))\in\mathcal{P}_2(x,u,u_1)} \sup_{t\geq 0} \|u''(t)\| \in [0, +\infty]$$

In the case of differential inclusions $x'(t) \in F(x(t))$ where the controls are the velocities of the state the second-order inertia function becomes the *jerk* function

$$\alpha_2(x,v,\gamma) := \inf_{(x(\cdot),u(\cdot))\in\mathcal{P}_2(x,v,\gamma)} \sup_{t\geq 0} \|u^{(2)}(t)\| \in [0,+\infty]$$

We observe that

$$\alpha_2(x,u,u_1) = 0 \text{ if and only if } \forall t \geq 0,\ \alpha_1(x(t), u+u_1 t) = \|u_1\|.$$

Generally, we associate with any integer $m \geq 0$ the set $\mathcal{P}_m(x,u,\ldots,u_{m-1})$ of solutions $(x(\cdot),u(\cdot))$ to the control system (6.1), p. 200:

$$x'(t) \ = \ f(x(t),u(t)) \text{ where } u(t) \ \in \ U(x(t))$$

satisfying the initial conditions

$$x(0) = x,\ u(0) = u,\ u'(0) = u_1,\ \ldots,\ u^{m-1}(0) = u_{m-1}$$

Definition 6.4.17 *[High-Order Inertia Function] The* m-order inertia function α_1 *of parameterized system(6.1), p. 200 associates the minimal worst intertemporal inertia* $\alpha_m(x,u,\ldots,u_{m-1})$ *of the evolutions starting from* $(x,u,u_1,\ldots,u_{m-1}) \in \mathrm{Graph}(U) \times \mathcal{U}^{m-1}$ *defined by*

$$\alpha_m(x,u,\ldots,u_{m-1}) := \inf_{(x(\cdot),u(\cdot))\in\mathcal{P}_m(x,u,\ldots,u_{m-1})} \sup_{t\geq 0} \|u^{(m)}(t)\| \in [0,+\infty]$$

We observe that $\alpha(x,u,u_1,\ldots,u_{m-1}) = 0$ if and only if

$$\forall t \geq 0,\ \alpha_{m-1}\left(x(t), \sum_{j=0}^{m-1} u_j \frac{t^j}{j!}, \ldots, u_{m-2}+u_{m-1}t\right) = \|u_{m-1}\|.$$

For characterizing high-order inertia functions in terms of viability kernels, we introduce the following auxiliary system, called the m-metasystem:

Definition 6.4.18 *[m-Metasystem] The "metavariables" of the m-metasystem associated with parameterized system (6.1), p. 200:*

$$x'(t) \ = \ f(x(t),u(t)) \text{ where } u(t) \ \in \ U(x(t))$$

are made of sequences $(x, u, u_1, \ldots, u_{m-1}, y)$ *ranging over the "metaenvironment"* $\text{Graph}(U) \times \mathcal{U}^{(m-1)} \times \mathbb{R}_+$, *regulated by "metaregulons"* $u_m \in \mathcal{U}$ *and governed by "metasystem"*

$$\begin{cases} (i) & x'(t) = f(x(t), u(t)) \\ (ii) & u'(t) = u_1(t) \\ \cdots & \cdots \\ (m+1) & u'_{m-1}(t) = u_m(t) \\ (m+2) & y'(t) = 0 \\ & \text{where } \|u_m(t)\| \le y(t) \end{cases} \tag{6.15}$$

We begin by providing the viability characterization of m-order inertia functions:

Theorem 6.4.19 *[Viability Characterization of High-Order Inertia Functions] For any* $m \ge 1$, *the* m*-order inertia function is related to the viability kernel of* $\text{Graph}(U) \times \mathcal{U}^{(m-1)} \times \mathbb{R}_+$ *under metasystem (6.15) by formula*

$$\alpha_m(x, u, \ldots, u_{m-1}) = \inf_{(x,u,\ldots,u_{m-1},y) \in \text{Viab}_{(6.15)}(\text{Graph}(U) \times \mathcal{U}^{(m-1)} \times \mathbb{R}_+)} y$$

Proof. Indeed, to say that $(x, u, \ldots, u_{m-1}, y)$ belongs to $\text{Viab}_{(6.15)}(\text{Graph}(U) \times \mathcal{U}^{(m-1)} \times \mathbb{R}_+)$ amounts to saying that there exists an evolution $t \mapsto (x(t), u(t), u_1(t), \ldots, u_m(t), y(t))$ governed by control system (6.15), p. 228 such that $y(t) = y$ and $u_j(t) = u^{(j)}(t)$. In other words, the solution $(x(\cdot), u(\cdot)) \in \mathcal{P}_m(x, u, \ldots, u_{m-1})$ satisfies

$$\forall t \ge 0, \ \|u^{(m)}(t)\| \le y$$

so that $\alpha_m(x, u, \ldots, u_{m-1}) \le \sup_{t \ge 0} \|u^{(m)}(t)\| \le y$.

Conversely, if $\alpha_m(x, u, \ldots, u_{m-1}) < +\infty$, we can associate with any $\varepsilon > 0$ an evolution $(x_\varepsilon(\cdot), u_\varepsilon(\cdot)) \in \mathcal{P}_m(x, u, \ldots, u_{m-1})$ such that

$$\forall t \ge 0, \ \|u_\varepsilon^m(t)\| \le \alpha(x, u, \ldots, u_{m-1}) + \varepsilon =: y_\varepsilon$$

Therefore, setting $u_{\varepsilon_j}(t) := u_\varepsilon^j(t)$ and $y_\varepsilon(t) = y_\varepsilon$, we observe that $t \mapsto (x_\varepsilon(t), u_\varepsilon(t), u_{\varepsilon_1}(t), u_{\varepsilon_{m-1}}(t), y_\varepsilon)(t)$ is a solution to the auxiliary system (6.15) viable in $\text{Graph}(U) \times \mathcal{U}^{(m-1)} \times \mathbb{R}_+$, and thus, that $(x, u, \ldots, u_{m-1}, y_\varepsilon)$ belongs to $\text{Viab}_{(6.15)}(\text{Graph}(U) \times \mathcal{U}^{(m-1)} \times \mathbb{R}_+)$. Therefore

$$\inf_{(x,u,\ldots,u_{m-1},y)\in \mathrm{Viab}_{(6.15)}(\mathrm{Graph}(U)\times\mathcal{U}^{(m-1)}\times\mathbb{R}_+)} y \le y_\varepsilon := \alpha(x,u,\ldots,u_{m-1})+\varepsilon$$

and it is enough to let ε converge to 0. □

The metasystem (6.15) is Marchaud whenever the single-valued map f is continuous, Lipschitz whenever the single-valued map f is Lipschitz and the "metaenvironment" is closed whenever the graph of U is closed. Hence it inherits of the properties of Marchaud and Lipschitz systems.

Theorem 6.4.20 ***[Existence of Evolutions with Minimal Inertia]*** *If f is continuous and the graph of U is closed, the epigraph of the mth-order inertia function α_m is closed. Furthermore, from any $(x,u,\ldots,u_{m-1}) \in \mathrm{Dom}(\alpha_m)$ starts at least one evolution $(x(\cdot),u(\cdot)) \in \mathcal{P}_m(x,u,\ldots,u_{m-1})$ such that*

$$\alpha_m(x,u,\ldots,u_{m-1}) := \sup_{t\ge 0} \|u^{(m)}(t)\|$$

Proof. Since the auxiliary system (6.15) is Marchaud whenever $m \ge 1$ and f continuous and since the auxiliary environment $\mathrm{Graph}(U)\times\mathcal{U}^{(m-1)}\times\mathbb{R}_+$ is closed by assumption, then the viability kernel $\mathrm{Viab}_{(6.15)}(\mathrm{Graph}(U)\times\mathcal{U}^{(m-1)}\times\mathbb{R}_+)$ is also closed and the upper semi-compactness of the associated evolutionary system implies that a subsequence (again denoted by) of $(x_\varepsilon(\cdot), u_{\varepsilon_0}(\cdot), \cdot, u_{\varepsilon_{m-1}}(\cdot), \alpha_m(x,u,\ldots,u_{m-1})+\varepsilon)$ converges to a solution $(x(\cdot),u(\cdot),\ldots,u_{m-1}(\cdot),\alpha(x,u,\ldots,u_{m-1}))$ satisfying

$$\forall t\ge 0,\ \|u^m(t)\| \le \alpha(x,u,\ldots,u_{m-1})$$

Therefore, the infimum of the m-order inertia function is achieved. □

Remark: Hamilton–Jacobi–Bellman Equations. One can prove that the **inertia function** α_m is the smallest positive solution to the Hamilton–Jacobi–Bellman partial differential equation: $\forall (x,u)\in\mathrm{Graph}(U)$,

$$\begin{cases}\left\langle \dfrac{\partial \mathbf{v}(x,u)}{\partial x}, f(x,u)\right\rangle - \mathbf{v}(x,u)\left\|\dfrac{\partial \mathbf{v}(x,u)}{\partial u}\right\| \\ \displaystyle\sum_{i=1}^n \dfrac{\partial \mathbf{v}(x,u)}{\partial x_i} f_i(x,u) - \mathbf{v}(x,u)\left\|\dfrac{\partial \mathbf{v}(x,u)}{\partial u}\right\| = 0\end{cases}$$

for $m=1$ and, for $m\ge 2$,

$$\begin{cases}\left\langle \dfrac{\partial \mathbf{v}(x,u,u_1,\ldots,u_{m-1})}{\partial x}, f(x,u)\right\rangle + \displaystyle\sum_{j=0}^{m-2}\left\langle \dfrac{\partial \mathbf{v}(x,u,u_1,\ldots,u_{m-1})}{\partial u_j}, u_{j+1}\right\rangle \\ -\mathbf{v}(x,u,u_1,\ldots,u_{m-1})\left\|\dfrac{\partial \mathbf{v}(x,u,u_1,\ldots,u_{m-1})}{\partial u_{m-1}}\right\| = 0\end{cases}$$ □

We can regard the domain and the sections of the inertia functions as graphs of regulation maps expressing the evolution of the derivative $u^{(m-1)}(t)$ of the regulon in terms of the state, the regulon and the derivatives of lower order:

Definition 6.4.21 *[Inert Regulation Maps and Viability Niches] We associate with the mth-order inertia function α_m of system (6.1), p. 200 the inert regulation map $R^m_\infty : \text{Graph}(U) \times \mathcal{U}^{(m-2)} \rightsquigarrow \mathcal{U}$ defined by*

$$\text{Graph}(R^m_\infty) := \{(x, u, \ldots, u_{m-1}) \text{ such that } \alpha_m(x, u, \ldots, u_{m-1}) < +\infty\}$$

and the inert regulation map $R^m_c : \text{Graph}(U) \times \mathcal{U}^{(m-2)} \rightsquigarrow \mathcal{U}$ with finite threshold $c \geq 0$ defined by

$$\text{Graph}(R^m_c) := \{(x, u, \ldots, u_{m-1}) \text{ such that } \alpha_m(x, u, \ldots, u_{m-1}) \leq c\}$$

We deduce the following regulation property of smooth evolutions.

Theorem 6.4.22 *[Regulation of Smooth Evolutions of Parameterized systems] Let us assume that f is continuous and that the graph of U is closed. Then for any initial state $(x, u, \ldots, u_{m-1})$ such that $\alpha_m(x, u, \ldots, u_{m-1}) < +\infty$, there exists a smooth evolution to system (6.1), p. 200:*

$$x'(t) = f(x(t), u(t)) \text{ where } u(t) \in U(x(t))$$

satisfying initial conditions $x(0) = x$, $u(0) = u, \ldots, u^{(m-1)} = u_{m-1}$ and regulated by

$$\begin{cases} u^{(m-1)}(t) \\ \in R^m_{\alpha_m(x,u,\ldots,u_{m-1})}(x(t), u(t), \ldots, u^{(m-2)}(t)) \end{cases}$$

Proof. Since $\alpha_m(x, u, \ldots, u_{m-1}) < +\infty$, we deduce that the inertia function $\alpha_m(x, u, \ldots, u_{m-1}, \alpha_m(x, u, \ldots, u_{m-1}))$ is increasing

$$\begin{cases} \forall t \geq 0,\ \alpha(x(t), u(t), \ldots, u_{m-1}(t)) = \sup_{s \geq t} \|u^m(s)\| \\ \leq \sup_{t \geq 0} \|u^m(t)\| = \alpha(x, u, \ldots, u_{m-1}) \end{cases}$$

This means that

$$u^{(m-1)}(t) \in R^m_{\alpha_m(x,u,\ldots,u_{m-1})}(x(t), u(t), \ldots, u^{(m-2)}(t)) \quad \square$$

The specific case when the threshold c is equal to 0 is worth of attention.

Definition 6.4.23 *[High-Order Viability Niches] We regard the graph*

$$\text{Graph}(R_0^m) := \{(x, u, \ldots, u_{m-1}) \text{ such that } \alpha_m(x, u, \ldots, u_{m-1}) = 0\}$$

as the m-order inertia set and the set $N_m(u, u_1, \ldots, u_{m-1})$ of elements $x \in \text{Dom}(U)$ such that $\alpha_m(x, u, u_1, \ldots, u_{m-1}) = 0$ as the m-order viability niche of $(u, u_1, \ldots, u_{m-1})$.

The m-order viability niches lock-in heavy evolutions:

Proposition 6.4.24 *[Locking-in Viability Niches] From any x belonging to a m-order viability niche $N_m(u, u_1, \ldots, u_{m-1})$ starts at least an evolution*

$$x(t) \in N_m\left(u + u_1 t + \cdots + u_{m-1}\frac{t^{m-1}}{(m-1)!}\right)$$

governed by the open-loop control $u(t) = u + u_1 t + \cdots + u_{m-1}\frac{t^{m-1}}{(m-1)!}$ which is a $(m-1)$-degree polynomial in time.

If at some time t_f, $x(t_f)$ belongs to the m-order viability niche $N_m(u(t_f), u'(t_f), \ldots, u^{(m-1)}(t_f))$, then for $t \geq t_f$, the evolution $x(\cdot) \in \mathcal{S}(x)$ may be regulated by the open-loop polynomial $u(t) = u(t_f) + u'(t_f)(t - t_f) + \cdots + u^{(m-1)}(t_f)\frac{(t-t_f)^{m-1}}{(m-1)!}$ and satisfy

$$x(t) \in N_m\left(u + u_1(t - t_f) + \cdots + u_{m-1}\frac{(t - t_f)^{m-1}}{(m-1)!}\right)$$

for $t \geq t_f$.

6.4.7 The Transition Cost Function

The inertia function defined in Definition 6.4.5 offers the simplest example of cost incurred by changing regulons. The general form of costs incurred by changing regulons is given by:

Definition 6.4.25 *[The Transition Function] Consider a positive extended cost function $\mathbf{c} : X \times \mathcal{U} \mapsto \mathbb{R}_+$, a cumulated transition cost function* *of the regulons* $\mathbf{l} : X \times \mathcal{U} \times \mathcal{U} \rightsquigarrow \mathbb{R}_+ \cup \{+\infty\}$ *and a discount factor* $m(x,u)$ *(which may depend upon both of the states and the controls).* The transition function $\alpha_{(\mathbf{c},\mathbf{l})}$ *is defined by*

$$\begin{cases} \alpha_{(\mathbf{c},\mathbf{l})}(x,u) := \inf_{(x(\cdot),u(\cdot))\in\mathcal{P}(x,u)} \sup_{t\geq 0} \\ \left(e^{-\int_0^t m(x(s),u(s))ds}\mathbf{c}(x(t),u(t))\right. \\ \left. + \int_0^t e^{-\int_0^\tau m(x(s),u(s))ds}\mathbf{l}(x(\tau),u(\tau),u'(\tau))d\tau\right) \end{cases} \quad (6.16)$$

Starting from an initial stat x, it will be advantageous to look for an initial regulon $u \in U(x)$ that minimizes the worst transition cost of regulons.

For characterizing the transition cost function in terms of viability kernels, we introduce the set-valued map

$$V_{\mathbf{c}} : (x,u;y) \rightsquigarrow \{v \in \mathcal{U} \mid \mathbf{c}(x,u) \leq y\}$$

and the following auxiliary metasystem of differential inclusions

$$\begin{cases} (i) \;\; x'(t) = f(x(t),u(t)) \\ (ii) \;\; u'(t) = v(t) \\ (iii) \; y'(t) = m(x(t),u(t))y(t) - \mathbf{l}(x(t),u(t),v(t)) \\ \quad\quad \text{where } v(t) \in V_{\mathbf{c}}(x(t),u(t);y(t)) \end{cases} \quad (6.17)$$

subject to the constraint

$$\forall t \geq 0, \;\; (x(t),u(t),y(t)) \in \mathrm{Graph}(U) \times \mathbb{R}_+$$

Theorem 6.4.26 *[**Viability Characterization of the Transition Cost Function**] The transition cost function is related to the viability kernel of the graph of U under the auxiliary metasystem (6.17) by the following formula*

$$\alpha_{(\mathbf{c},\mathbf{l})}(x,u) = \inf_{(x,u,y)\in \mathrm{Viab}_{(6.17)}(\mathrm{Graph}(U)\times\mathbb{R}_+)} y$$

Proof. Indeed, to say that (x,u,y) belongs to the viability kernel of the graph of U under the auxiliary system (6.17) amounts to saying that there exists an evolution $t \mapsto (x(t),u(t),y(t))$ governed by the auxiliary metasystem such that, for all $t \geq 0$, $u(t) \in U(x(t))$. By definition of (6.17), we know that for all $t \geq 0$, this evolution satisfies also for all $t \geq 0$,

$$\begin{array}{l}\mathbf{c}(x(t),u(t)) \\ \leq y(t) \;=\; e^{\int_0^t m(x(s),u(s))ds}\left(y - \int_0^t e^{-\int_0^\tau m(x(s),u(s))ds}\mathbf{l}(x(\tau),u(\tau),v(\tau))d\tau\right)\end{array}$$

Therefore

$$\begin{cases}\sup_{t\geq 0}\left(e^{-\int_0^t m(x(s),u(s))ds}\mathbf{c}(x(t),u(t))\right. \\ \left. + \int_0^t e^{-\int_0^\tau m(x(s),u(s))ds}\mathbf{l}(x(\tau),u(\tau),u'(\tau))d\tau\right) \leq y\end{cases}$$

and thus, $\alpha_{(\mathbf{c},\mathbf{l})}(x,u) \leq \inf_{(x,u,y)\in \mathrm{Viab}_{(6.17)}(\mathrm{Graph}(U)\times\mathbb{R}_+)} y$.

Conversely, we know that for any $\varepsilon > 0$, there exists an evolution $(x(\cdot),u(\cdot)) \in \mathcal{P}(x,u)$ such that

$$\begin{array}{l}\sup_{t\geq 0}\left(e^{-\int_0^t m(x(s),u(s))ds}\mathbf{c}(x(t),u(t))\right. \\ \left. + \int_0^t e^{-\int_0^\tau m(x(s),u(s))ds}\mathbf{l}(x(\tau),u(\tau),u'(\tau)d\tau)\right) \leq \alpha_{(\mathbf{c},\mathbf{l})}(x,u)+\varepsilon\end{array}$$

Setting

$$\begin{cases} y_\varepsilon(t) \;:=\; e^{\int_0^t m(x(s),u(s))ds}\left(\alpha_{(\mathbf{c},\mathbf{l})}(x,u)+\varepsilon\right. \\ \left. - \int_0^t e^{-\int_0^\tau m(x(s),u(s))ds}\mathbf{l}(x(\tau),u(\tau),u'(\tau))d\tau\right)\end{cases}$$

we infer that $\mathbf{c}(x(t),u(t),v(t)) \leq y_\varepsilon(t)$ and thus, that $t \mapsto (x(t),u(t),y_\varepsilon(t))$ is a solution to the solution to auxiliary evolutionary system (6.17) starting at $(x,u,\alpha_{(\mathbf{c},\mathbf{l})}(x,u)+\varepsilon)$. This evolution is viable in $\mathrm{Graph}(U)\times\mathbb{R}_+$ since $(x(\cdot),u(\cdot)) \in \mathcal{P}(x,u)$, and thus, since $x(t) \in U(x(t))$, or, equivalently, since

$$\forall t \geq 0,\; (x(t),u(t),y_\varepsilon(t)) \in \mathrm{Graph}(U)\times\mathbb{R}_+$$

Hence $(x,u,\alpha_{(\mathbf{c},\mathbf{l})}(x,u)+\varepsilon)$ belongs to the viability kernel $\mathrm{Viab}_{(6.17)}$ $(\mathrm{Graph}(U)\times\mathbb{R}_+)$, so that

$$\inf_{(x,u,y)\in \mathrm{Viab}_{(6.17)}(\mathrm{Graph}(U)\times\mathbb{R}_+)} y \leq \alpha_{(\mathbf{c},\mathbf{l})}(x,u)+\varepsilon$$

Letting ε converge to 0, we obtain the converse inequality. □

6.5 Viability Oscillators and Hysterons

Biology offers to our investigations myriads of biological clocks or oscillators, producing *periodic evolutions*, or, rather, *cyclic evolutions*. This modification of the terminology is justified by the fact that nowadays, periodic evolutions are understood as produced by a system of periodic differential equations. The search of these equations is a very difficult undertaking, and may not be realistic at all. Hence the question arises to look for *other ways to produce periodic solutions*, that we suggest to call *cyclic* to underlie the fact that *they are not solutions of a given system of periodic differential equations.* As often in biology or in life sciences, we face the following dilemma:

27 ***[Simple Dynamics and Complex Constraints]*** *Are the examples of biological clocks produced by complex systems of differential equations or produced by very* simple dynamics, *confronted to a* complex maze of constraints*?*

The authors, being mathematicians, have no competence to answer this question. They can only suggest the possibility of producing mathematically such complex cyclic evolutions from very simple dynamics. The first approach, looking for a priori systems of differential equations pertains to a *direct approach* (see Comment 1, p. 5) whereas the second follows the *inverse approach* (see Comment 2, p. 5).

Remark. For example, the history of the mathematical modelling of the propagation of the nervous influx started in 1903 with very simple *impulse models* by *Louis Lapicque.* At the time, a mathematical theory of impulse systems did not exist yet. It started as such two decades ago, and is briefly presented in Sect. 12.3, p. 503. This triggered the search of systems of differential equations reproducing evolutions "looking like" or reminiscent of the propagation or the nervous influx by the ingenious works of *Alan Hodgkin* and *Andrew Huxley* (nephew of the writer *Aldous Huxley* and grandson of *Thomas Huxley*). These equations reproduce evolutions looking like nervous influxes, but without producing explanations, whereas the Lapicque model, very simple indeed, but involving constraints and impulse evolutions when they reach those constraints, provides both a simple explanation and evolutions looking like the propagation of the nervous influx. □

Therefore, the underlying assumptions of our engine producing cyclic evolutions is

28 ***[Viability Oscillator]*** *A* viability oscillator *is made of:*

1. **viability constraints** *on the state of the system in a given environment,*

2. **inertia thresholds** *imposing a* speed limit *on each component of the evolutions of the regulons.*

When the inertia thresholds are infinite, we shall say that the viability oscillator is impulsive *(see Box 34 and Theorem 10.8.3, p. 424, p. 426).*

The case of impulse cyclic systems can be dealt with the concept of *amalgamated system* (see Definition 10.8.2, p. 424) using alternatively *persistent evolutions* (see Definitions 10.4.2, p. 393 and Theorem 10.4.4, p. 394), remaining viable as long as possible until the moment when they reach their *exit set* (see Sect. 10.4.3, p. 396).

In this section, we shall study the more realistic case when the inertia thresholds are finite.

From now on, we shall restrict our analysis to the simple illustration (6.4), p. 212

$$\forall\, t \geq 0,\;\; x'(t) \;=\; u(t) \text{ where } u(t) \;\in\; \mathbb{R}$$

for the sake of clarity.

6.5.1 *Persistent Oscillator and Preisach Hysteron*

Let us consider the case in which the inertia threshold $c = 0$ is equal to 0. Hence the velocities u are constant and the evolutions $x(t) = x + ut$ are affine. Therefore, the *viability niches* of the regulons (see Definition 6.4.9, p. 216) are empty if the regulon $u \neq 0$ is different from 0 and equal to $K := [a, b]$ when $u = 0$. *Persistent evolutions* $x(t) = x + ut$ (see Definition 6.4.1, p. 208) are regulated by constant velocity (control) u. Their exist times are equal to

$$\tau_K^\sharp(x) \;:=\; \frac{(b-x)}{|u|} \text{ if } u > 0 \text{ and } \tau_K^\sharp(x) \;:=\; \frac{(x-a)}{|u|} \text{ if } u < 0$$

and their exit sets (see Definition 10.4.7, p. 396) are equal to

$$\mathrm{Exit}_u(K) \;:=\; \{b\} \times \mathbb{R}_+ \text{ if } u > 0 \text{ and } \mathrm{Exit}_u(K) \;:=\; \{a\} \times \mathbb{R}_- \text{ if } u < 0$$

They are governed by the metasystem associated with the 0-inertia threshold:

$$\begin{cases} (i) \;\; x'(t) \;=\; u(t) \\ (ii) \; u'(t) \;=\; 0 \end{cases} \tag{6.18}$$

Therefore, starting from a with control $+u$ and b with control $-u$ respectively, the persistent cyclic evolution (with the inertia bound 0) is given by the formulas

$$\begin{cases} \forall t \in [0, \frac{(b-a)}{|u|}], \\ u_a(t) = +u \ \& \ x_a(t) = a + ut \text{ if } u > 0 \\ u_b(t) = -|u| \ \& \ x_b(t) = b - |u|t \text{ if } u < 0 \end{cases} \tag{6.19}$$

Concatenating these evolutions alternatively, we obtain, starting from (a, u) the evolution $u(t) = +u$ & $x(t) = a + ut$ which arrives at the exit set $\{b\} \times \mathbb{R}_+$ at exit time $\frac{(b-a)}{|u|}$ with constant velocity equal to $+u$. In order to pursue the evolution, we have to introduce a *reset map* describing an *impulse* (infinite negative velocity) forcing the velocity to jump from $+u$ to $-u$ when the state reaches b. Next, at exit time $\frac{(b-a)}{|u|}$, the state-control pair starts from (b, u) and is equal to $u(t) = + - u$ and $x(t) = b - ut$ until it arrives at the exit set $\{b\} \times \mathbb{R}_+$ at exit time $2\frac{(b-a)}{|u|}$. In order to pursue the evolution, the *reset map* describing an *impulse* (infinite positive velocity) forcing the velocity to jump from $-u$ to $+u$ when the state reaches a.

We obtained a cyclic evolution with cycle $2\frac{(b-a)}{|u|}$ which oscillates linearly and *continuously* back and forth between a and b with 0 acceleration, but with *discontinuous* controls*jumping alternatively from $+u$ to $-u$ with infinite velocities (infinite acceleration of the state) in order to maintain the viability of the interval K under metasystem* (6.18), p. 235.

This is nothing other than the 1784 *Coulomb* approximation law of friction, an adequate representation of friction for the analysis of many physical systems, appeared in his *Recherches théoriques et expérimentales sur la force de torsion et sur l'élasticité des fils de metal.* It provides a *threshold value* for a net force tending to cause the beginning of a motion, rather than providing an estimate of the actual frictional force. This is the model of the rectangular hysteresis loop, known under the name of Preisach hysteris cycle: Actually, this simplest hysteron was proposed in 1938 by *Preisach* to subsume the behavior of magnetic hysteresis:

Definition 6.5.1 *[The Preisach Hysteron] The* Preisach hysteron *is the set-valued map $\Phi_{Preisach}$ defined by*

$$\begin{cases} \Phi_{Preisach}(a) := +u & \text{if } x = a \\ \Phi_{Preisach}(x) := \{-u, +u\} & \text{if } a \le x < b \\ \Phi_{Preisach}(x) := -u & \text{if } x = b \end{cases}$$

(where u is a given value). Its graph is the Preisach hysteresis loop *contained in the phase space (actually, the state-control space).*

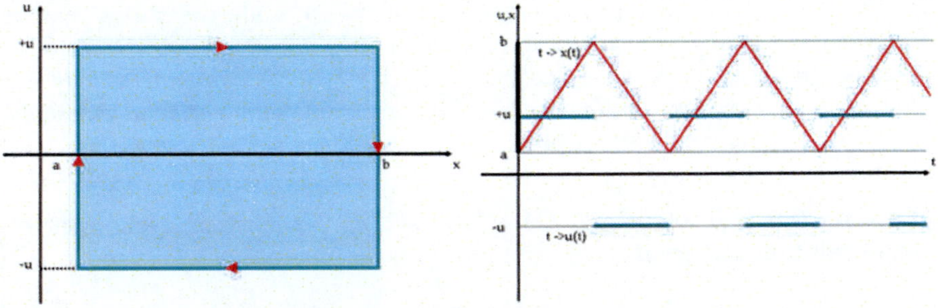

Fig. 6.4 Preisach's Hysteron and Evolution.
Left: *The Preisach hysteris cycle, which is the graph of the Preisach hysteron (see Definition 6.5.1, p. 236).* Right: *Graphs of the evolutions $t \mapsto u(t)$ and $t \mapsto x(t)$ of the persistent cyclic evolution with 0-inertia threshold (see Lemma 6.5.2, p. 238). The acceleration is equal to 0 when $a < x < b$ and equal to $+\infty$ (impulse) when $x = a$ and to $-\infty$ when $x = b$.*

This is the simplest example of a *hysteresis loop*:

29 ***[Hysteresis Loops and Hysterons]*** James Ewin, *a Scottish physicist discovered and coined the word* hysteresis *meaning* lagging behind *in classical Greek. This polysemous word is used in many different fields, including in mathematics, where several mathematical translations have been observed (among which* Visintin *'one, [212, Visintin], related to subsets invariant with respect to affine time scaling). We use it here as an engine, called* hysteron, *producing* hysteresis loops *or* hysteresis cycles*: when the evolution of the state of a system with hysteresis is plotted on a graph against the applied force (in our case, the control), the resulting curve is in the form of a loop.*

These hysteresis issues are closely related with "quantized" controls (in control theory), governed by an *amalgam* of a finite number (here, 2) of controls only (see Comment 33, p. 422).

Starting from $(a, +u)$, the persistent cyclic evolution follows the northern boundary of the graph of the Preisach hysteron with 0 acceleration and positive velocity up to the exit time $\frac{(b-a)}{|u|}$ when the control jumps to the value $-u$ with infinite negative velocity (impulse) while the states reaches the upper boundary b of the interval. Then, starting from $(b, -|u|)$, the persistent cyclic evolution follows the southern boundary of the graph of the Preisach hysteron up to the warning time $2\frac{(b-a)}{|u|}$, when the control jumps with infinite positive velocity (impulse) to the value $+u$ while the states reaches the lower boundary a of the interval, and so on.

Hence *Preisach's classical hysteron subsuming the ferromagnetic loop can thus be built from two principles: viability constraint and* 0*-inertia threshold.* It generates persistent oscillators:

Lemma 6.5.2 ***[Persistent Oscillators]*** *The* 0-persistent cyclic control and evolution *are equal to:*

$$\forall n \geq 0, \begin{cases} \forall t \in \left[2n\dfrac{(b-x)}{u}, (2n+1)\dfrac{(b-x)}{u}\right[, \\ u(t) = u \;\&\; x(t) = a + ut \\ \textit{and} \\ \forall t \in \left[(2n+1)\dfrac{(b-x)}{u}, (2n+2)\dfrac{(b-x)}{u}\right[, \\ u(t) = -u \;\&\; x(t) = b - ut \end{cases}$$

In summary, this cyclic evolution is regulated by accelerations being equal to 0 in the interior of the interval and infinite values at the boundary. The question arises whether we can regulate cyclic evolutions with continuous controls instead of discontinuous ones.

6.5.2 Viability Oscillators and Hysterons

Let us choose an inertia threshold c and assume that $|u| \leq \sqrt{c}\sqrt{b-a}$.

Instead of choosing the heavy evolution which remains at one of the equilibria $(a,0)$ or $(b,0)$ forever by switching the acceleration (velocity of the regulon) to 0, we continue the evolution by keeping the acceleration $-c$ or $+c$ as long as possible, and then, switch again to 0, again, as long as it is possible to obey viability and inertia constraints.

Recall that the critical map is equal to

$$\Xi_c(u) = b - \frac{u^2}{2c} \text{ if } u > 0 \text{ and } \Xi_c(-|u|) = a + \frac{u^2}{2c} \text{ if } u < 0$$

We set

$$x^\star := \frac{b-a}{2} \text{ and } u^\star := \sqrt{b-a}$$

The two values are equal if $u = \sqrt{c}\, u^\star$ where we set $u^\star := \sqrt{b-a}$. We observe that $x^\star := \Xi_c(\sqrt{c}\, u^\star) = \dfrac{b-a}{2}$ does not depend on c.

Hence, by (6.13), p. 224, we deduce that the *warning time* is equal to

$$\begin{cases} (i)\ \sigma_c(a,u) = \sigma_c(b,-|u|) = \dfrac{(b-a)}{|u|} - \dfrac{|u|}{2c} \\ (ii)\ \sigma_c\left(a+\dfrac{u^2}{2c},u\right) = \sigma_c\left(b-\dfrac{u^2}{2c},-|u|\right) = \dfrac{(b-a)}{|u|} - \dfrac{|u|}{c} \end{cases} \tag{6.20}$$

Recall that $\frac{|u|}{c}$ is the time needed to reach $(b,0)$ from $(\Xi_c(u),u)$ (or $(a,0)$ from $(\Xi_c(-|u|),-|u|)$). If $u := \sqrt{c}u^\star$, then $\frac{\sqrt{c}|u^\star|}{c} = \frac{\sqrt{b-a}}{\sqrt{c}}$.

The inert evolutions ranging over the boundary of the regulation map satisfy the *Titanic effect* described by the following property:

Definition 6.5.3 ***[Titanic Effect]*** *A state-control pair* $(x(\cdot),u(\cdot)) \in \mathbb{R}^2$ *satisfies the* Titanic effect *on the interval* I *if*

$$\forall\, t \in I,\ \ x'(t)u'(t) < 0$$

The Titanic effect means that a decrease of the control corresponds to an increase of the state and vice-versa. In our example, starting at a and decelerating with acceleration equal to $-c$, the state $x_a(t)$ increases from a to b whereas the regulon decreases from u_a to 0 at time $\frac{\sqrt{2}}{\sqrt{c}}\tau^\star$.

We define the *smooth heavy hysteresis cycle* $(x_h(\cdot),u_h(\cdot))$ of cycle $2\left(\frac{b-a}{|u|}+\frac{|u|}{c}\right)$ starting at $(\Xi_c(-|u|),u)$ with $u>0$ where $\Xi_c(-|u|) := a+\frac{u^2}{2c}$ in the following way:

1. The state-regulon pair $(x_h(\cdot),u_h(\cdot))$ starts from $(\Xi_c(-|u|),u)$ by taking the velocity of the regulon equal to 0. It remains viable on the time interval $\left[0,\frac{b-a}{|u|}-\frac{|u|}{c}\right]$ until it reaches the state-regulon pair $(\Xi_c(u),u)$ where $\Xi_c(u) := b-\frac{u^2}{2c}$ because $\sigma_c\left(a+\frac{u^2}{2c}\right) = \frac{(b-a)}{|u|} - \frac{|u|}{c}$.
2. The state-regulon pair $(x_h(\cdot),u_h(\cdot))$ starts from $(\Xi_c(u),u)$ at time $\frac{b-a}{|u|}-\frac{|u|}{c}$ by taking the velocity of the regulon (acceleration) equal to $-c$.
   (a) It ranges over the graph of $\sqrt{c}\ \ r^\sharp(x)$ on the time interval $\left[\frac{b-a}{|u|}-\frac{|u|}{c},\frac{b-a}{|u|}\right]$ until it reaches the state-regulon pair $(b,0)$. The heavy evolution would remain at this equilibrium forever with an acceleration equal to 0.

Note that during this time interval, the state-control evolution satisfies the *Titanic effect* (see Definition 6.5.3, p. 239).

(b) However, for defining the heavy hysteresis cycle, we assume that we keep the acceleration equal to $-c$. Hence the state-regulon pair $(x_h(\cdot), u_h(\cdot))$ ranges over the graph of $-\sqrt{c}\ r^\sharp(x)$ on the time interval $\left[\frac{b-a}{|u|}, \frac{b-a}{|u|} + \frac{|u|}{c}\right]$ until it reaches the state-regulon pair $(\Xi_c(u), -|u|)$.

3. The state-regulon pair $(x_h(\cdot), u_h(\cdot))$ starts from $(\Xi_c(u), -|u|)$ at time $\frac{b-a}{|u|} + \frac{|u|}{c}$ by taking the velocity of the regulon equal to 0. It remains viable on the time interval $\left[\frac{b-a}{|u|} + \frac{|u|}{c}, 2\frac{b-a}{|u|}\right]$ until it reaches the state-regulon pair $(\Xi_c(-|u|), -u)$.
4. The state-regulon pair $(x_h(\cdot), u_h(\cdot))$ starts from $(\Xi_c(-|u|), -u)$ at time $2\frac{b-a}{|u|}$ by taking the velocity of the regulon equal to $+c$.

(a) It ranges over the graph of $-\sqrt{c}\ r^\flat(x)$ on the time interval $\left[2\frac{b-a}{|u|}, 2\frac{b-a}{|u|} + \frac{|u|}{c}\right]$ until it reaches the state-regulon pair $(a, 0)$, satisfying the Titanic effect. The heavy evolution would remain at this equilibrium forever with an acceleration equal to 0.

(b) However, for defining the heavy hysteresis cycle, we assume that we keep the acceleration equal to $+c$. Hence the state-regulon pair $(x_h(\cdot), u_h(\cdot))$ ranges over the graph of $+\sqrt{c}\ r^\flat(x)$ on the time interval $\left[2\frac{b-a}{|u|}c + \frac{|u|}{c}, 2\left(\frac{b-a}{|u|} + \frac{|u|}{c}\right)\right]$ until it reaches the state-regulon pair $(\Xi_c(-|u|), u)$.

This ends the smooth heavy hysteresis cycle: When regulated by strictly positive regulons, the state goes from a to b and the state-regulon pair ranges over the graph of $x \mapsto \min(u, \sqrt{c}r^\flat(x), \sqrt{c}\ r^\sharp(x))$ whereas, when regulated by negative regulons, the state goes from b to a, and the state-regulon pair ranges over the graph of $x \mapsto -\min(u, \sqrt{c}r^\flat(x), \sqrt{c}\ r^\sharp(x))$. The evolution $t \mapsto (x_h(t), u_h(t))$ is cyclic of cycle $2\left(\frac{b-a}{|u|} + \frac{|u|}{c}\right)$.

Heavy hysteresis cycles are governed by three velocities of regulons only, $-c$, 0 and c, which can be regarded as "meta-regulons". They provide an example of "quantized" evolutions, governed by a combination (called amalgam), of a finite number of (meta) regulons only.

Quantization is a recent issue in control theory, where, instead of computing the complicated feedback regulating viable evolutions, the question arises to achieve the regulation of viable evolutions with a finite number of controls or feedbacks.

This number can be reduced to two (meta) regulons in the limiting case $u := \sqrt{c}\, u^\star$ and when $\Xi_c(\sqrt{c}\, u^\star) = \Xi_c(-\sqrt{c}\, u^\star) =: x^\star$. In this case, heavy hysteresis cycles are called *smooth inert hysteresis cycle* $x_h(\cdot)$ (of cycle $\frac{4u^\star}{\sqrt{c}}$) in the following way:

1. The state-regulon pair $(x_h(\cdot), u_h(\cdot))$ starts from $(x^\star, \sqrt{c}\, u^\star)$ at time 0 by taking the velocity of the regulon equal to $-c$. It ranges over the graph of $\sqrt{c}\, r^\sharp$ on the time interval $[0, \frac{u^\star}{\sqrt{c}}]$ until it reaches the equilibrium $(b, 0)$ and next, keeping the velocity of the regulon equal to $-c$, it ranges over the graph of $-\sqrt{c}\, r^\sharp$ on the time interval $\left[\frac{u^\star}{\sqrt{c}}, \frac{2u^\star}{\sqrt{c}}\right]$ until it reaches the state-regulon pair $(x^\star, -\sqrt{c}\, u^\star)$.
2. The state-regulon pair $(x_h(\cdot), u_h(\cdot))$ starts from $(x^\star, -\sqrt{c}\, u^\star)$ at time $\frac{2u^\star}{\sqrt{c}}$ by taking the velocity of the regulon equal to $+c$. It ranges over the graph of $-\sqrt{c}\, r^\flat$ on the time interval $\left[\frac{2u^\star}{\sqrt{c}}, \frac{3u^\star}{\sqrt{c}}\right]$ until it reaches the equilibrium $(a, 0)$ and next, keeping the velocity of the regulon equal to $+c$, it ranges over the graph of $+\sqrt{c}\, r^\flat$ on the time interval $[\frac{3u^\star}{\sqrt{c}}, \frac{4u^\star}{\sqrt{c}}]$ until it reaches the state-regulon pair $(x^\star, \sqrt{c}\, u^\star)$.

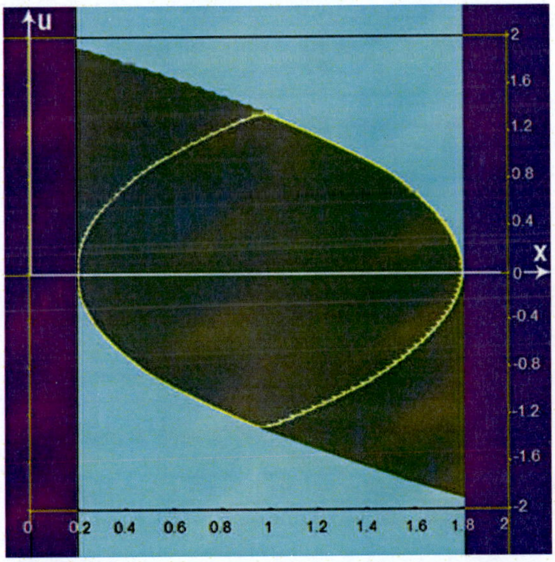

Fig. 6.5 Inert Hysteresis Cycle.
The graph of the smooth evolution is shown in Fig. 6.6. The Algorithm computes the graph of the regulation map R_c (which is a viability kernel) and the inert hysteresis loop.

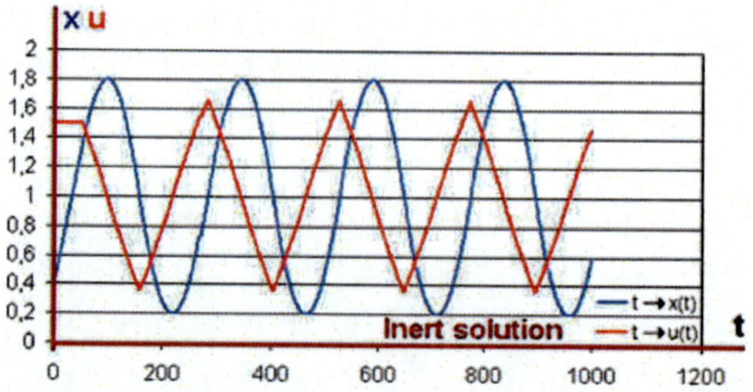

Fig. 6.6 Graph of the Inert Hysteresis Cycle.
Both the graphs of the smooth inert evolution (in blue*) and of its regulon (in* red*) are plotted. They are not computed from the analytical formulas, but extracted from the Viability Kernel Algorithm. The velocity of the regulon oscillates from* $+u^\star$ *to* $-u^\star$*. The evolution is then cyclic, alternatively increasing and decreasing from the* lower bound *of the constrained interval to its* upper bound.

30 *This very simple mathematical metaphor implies that two excitatory/inhibitory simple mechanism of a DNA site with bounds on the quantities and their accelerations are sufficient to explain the production of an isolated protein increasing up to a given viability bound and then, decreasing to disappear and being produced again according to a clock, the cyclicity of which is concealed in this very simple viability oscillator, triggering a biological clock.*

6.6 Controlling Constraints and Targets

Consider parameterized system (6.1), p. 200:

$$x'(t) = f(x(t), u(t)) \text{ where } u(t) \in U(x(t))$$

We study in this section the case when not only the dynamics depends upon the controls, but also when the environment $K(u)$ and the target $C(u)$ depend upon these controls.

Definition 6.6.1 *[Controlled-Dependent Viability Kernels] Let us consider the control system (6.1), p. 200 and controlled-dependent environment $u \rightsquigarrow K(u)$ and target $u \rightsquigarrow C(u)$. The controlled-dependent viability kernel* $\text{Viab}_{(6.1)}(\text{Graph}(K), \text{Graph}(C))$ *is the subset of state-control pairs (x, u) from which start at least an evolution satisfying the control-dependent viability constraints $x(t) \in K(u(t))$ forever or until a finite time $t^\star$ when $x(t^\star)$ reaches the control-dependent target $C(u(t^\star))$. The controlled-dependent capture basin* $\text{Capt}_{(6.1)}(\text{Graph}(K), \text{Graph}(C))$ *is the subset of state-control pairs (x, u) from which start at least an evolution satisfying the control-dependent viability constraints $x(t) \in K(u(t))$ until a finite time $t^\star$ when $x(t^\star)$ reaches the control-dependent target $C(u(t^\star))$.*

Introducing metasystems allows us to characterize this subset regulate such systems in order that they satisfy the control-dependent viability constraints forever or until they reach the control-dependent target in finite time:

Definition 6.6.2 *[Controlled-Dependent Inertia Function] Let us consider control-dependent environment $K(u)$ and targets $C(u)$. We denote by $\mathcal{P}_{(K,C)}(x, u)$ the set of evolutions of state-control $(x(\cdot), u(\cdot))$ governed by the control system (6.1) starting at (x, u) such that either*

$$\forall t \geq 0,\ x(t) \in K(u(t))$$

or such that there exists a finite time $t^\star$ such that

$$\begin{cases} (i)\ \ x(t^\star) \in C(u(t^\star)) \\ (ii)\ \forall t \in [0, t^\star],\ x(t) \in K(u(t)) \end{cases}$$

The associated inertia function $\beta_{(K,C)} : X \times \mathcal{U} \mapsto \mathbb{R} \cup \{+\infty\}$ is defined by

$$\beta_{(K,C)}(x, u) := \inf_{(x(\cdot),u(\cdot)) \in \mathcal{P}_{(K,C)}} \ \inf_{T \in \mathbb{R} \cup \{+\infty\}} \ \sup_{t \in [0,T]} \|u'(t)\| \tag{6.21}$$

As for the case of control-independent constraints, the epigraph of the inertia function is the viability kernel of a meta-environment with a meta-target under the metasystem

$$\begin{cases} (i) & u'(t) = v(t) \\ (ii) & x'(t) = f(x(t), u(t)) \\ (iii) & y'(t) = 0 \\ & \text{where } \|v(t)\| \leq y(t) \end{cases} \tag{6.22}$$

Theorem 6.6.3 ***[Characterization of Controlled-Dependent Viability Kernels by Inertia Functions]*** *The controlled-dependent viability kernel* $\text{Viab}_{(6.1)}(\text{Graph}(K), \text{Graph}(C))$ *is characterized by the viability kernel of the meta-environment* $\text{Graph}(K)$ *with meta-target* $\text{Graph}(C)$ *under auxiliary meta-system (6.22), p. 244 in the following way:*

$$\beta_{(K,C)}(x,u) = \inf_{(u,x,y)\in \text{Viab}_{(6.22)}(\text{Graph}(K)\times\mathbb{R}_+, \text{Graph}(C)\times\mathbb{R}_+)} y \tag{6.23}$$

In other words, the controlled-dependent viability kernel is the domain of the inertia function $\beta_{(K,C)}$.

Proof. Indeed, to say that (u, x, y) belongs to $\text{Viab}_{(6.22)}(\text{Graph}(K) \times \mathbb{R}_+, \text{Graph}(C) \times \mathbb{R}_+)$ amounts to saying that there exists an evolution $t \mapsto (x(t), u(t))$ governed by (6.1) such that $t \mapsto (x(t), u(t), y)$ is governed by control system (6.22), p. 244 because $y(t) \equiv y$. In other words, there exists a solution $(x(\cdot), u(\cdot)) \in \mathcal{P}_{(K,C)}(x, u)$ and $t^\star \in \mathbb{R} \cup \{+\infty\}$ satisfying

$$\forall t \geq 0, \; x(t) \in K(u(t))$$

(when $t^\star + \infty$) or such that there exists a finite time $t^\star$ such that

$$\begin{cases} (i) & x(t^\star) \in C(u(t^\star)) \\ (ii) & \forall t \in [0, t^\star], \; x(t) \in K(u(t)) \end{cases}$$

and

$$\forall t \in [0, t^\star], \; \|u'(t)\| \leq y$$

so that $\beta_{(K,C)}(x,u) \leq \sup_{t\in[0,t^\star]} \|u'(t)\| \leq y$. Hence

$$\beta_{(K,C)}(x,u) \leq V(x,u) := \inf_{(u,x,y)\in \text{Viab}_{(6.22)}(\text{Graph}(K)\times\mathbb{R}_+, \text{Graph}(C)\times\mathbb{R}_+)} y$$

Conversely, if $\beta_{(K,C)}(x,u) < +\infty$, we can associate with any $\varepsilon > 0$ an evolution $(x_\varepsilon(\cdot), u_\varepsilon(\cdot)) \in \mathcal{P}_{(K,C)}(x,u)$ and $t_\varepsilon \in \mathbb{R} \cup \{+\infty\}$ such that

$$\forall t \in [0, t_\varepsilon], \; \|u'_\varepsilon(t)\| \leq \beta_{(K,C)}(x,u) + \varepsilon =: y_\varepsilon$$

Therefore, setting $v_\varepsilon(t) := u'_\varepsilon(t)$ and $y_\varepsilon(t) = y_\varepsilon$, we observe that $t \mapsto (x_\varepsilon(t), u_\varepsilon(t), y_\varepsilon)$ is a solution to the auxiliary system (6.22) viable in $\text{Graph}(K) \times \mathbb{R}_+$ forever (if $t_\varepsilon = +\infty$), or until a finite time $t_\varepsilon < +\infty$ when $x(t_\varepsilon) \in C(u(t_\varepsilon))$, i.e., when $(x_\varepsilon(t_\varepsilon), u_\varepsilon(t_\varepsilon), y_\varepsilon(t_\varepsilon)) \in \text{Graph}(C) \times \mathbb{R}_+$. This implies such that (u, x, y_ε) belongs to $\text{Viab}_{(6.22)}(\text{Graph}(K) \times \mathbb{R}_+, \text{Graph}(C) \times \mathbb{R}_+)$. Hence

$$\begin{cases} V(x,u) := \inf_{(u,x,y) \in \text{Viab}_{(6.22)}(\text{Graph}(K)\times\mathbb{R}_+, \text{Graph}(C)\times\mathbb{R}_+)} y \\ \leq y_\varepsilon := \beta_{(K,C)}(x,u) + \varepsilon \end{cases}$$

and it is enough to let ε converge to 0 to prove the opposite inequality $V(x,u) \leq \beta_{(K,C)}(x,u)$. □

The metasystem (6.22) is Marchaud (see Definition 10.3.2, p. 384) whenever the single-valued map f is continuous, Lipschitz (see Definition 10.3.5, p. 385) whenever the single-valued map f is Lipschitz and the meta-environements and meta-targets are closed whenever the graph of the set-valued maps $u \rightsquigarrow K(u)$ and $u \rightsquigarrow C(u)$ are closed. Hence it inherits of the properties of Marchaud and, Lipschitz systems respectively. □

Chapter 7
Management of Renewable Resources

7.1 Introduction

This chapter is devoted to some problems dealing with births, growth, and survival of populations.

Section 7.2, p. 248 starts with a simple model of the growth of the biomass of a renewable resource, fish population, for instance, following the track of Sect. 6.4, p. 207 of Chap. 6. We first review some dynamical growth systems devised from Malthus to Verhulst and beyond, up to recent growth models. They all share the same *direct approach* (see Box 1, p. 5) where growth models are proposed and the evolutions they govern are studied.

Instead, we suggest to follow the *inverse approach* to the same problem in the framework of this simple example (see Box 2, p. 5): we *assume only the knowledge or viability constraints and inertia thresholds*. This provides, through the inertia function, the "Mother of all viable feedbacks" (since the term "matrix" is already used) governing the evolutions satisfying these two requirements. Among them, we find the Verhulst feedback driving the logistic curves, the trajectories of which are "S-shaped". Furthermore, we are tempted to look for the two "extreme feedbacks". Each of them provides what we call inert evolutions, with velocity of the growth rates equal to the inertial threshold. When, alternatively combined (amalgamated), they govern *cyclic* viable evolutions. This may provide explanations of economic cycles and biological cycles and clocks without using periodic systems.

Combined with the Malthusian feedback, which does not belong the progeny of this Mother, we obtain heavy evolutions, the growth rate of which is kept constant (Malthusian evolution) as long as possible and then, switched to an inert evolution.

Section 7.3, p. 262 introduces fishermen. Their industrial activity depletes the growth rate of their renewable resource. They thus face a double challenge: maintain the economic viability of their enterprize by fishing enough resources, *and* not fishing too much for keeping the resource alive, even

J.-P. Aubin et al., *Viability Theory*, DOI 10.1007/978-3-642-16684-6_7,

though below the level of economic viability. Without excluding the double catastrophe: fishes disappear forever.

Even though the real problem is complex, involving too many variables to be both computable and reasonably relevant, the illustration of the concepts of permanence kernels and crisis functions in the framework of this simple example is quite instructive. It offers a mathematical metaphor showing that these two types of constraints, economical and ecological, produce evolutionary scenarios which can be translated in terms of viability concepts, sharing their properties.

7.2 Evolution of the Biomass of a Renewable Resource

We illustrate some of the results of Sect. 6.4, p. 207 with the study, for example, of the evolution of the biomass of one population (of renewable resources, such as fish in fisheries) in the framework of simple one-dimensional regulated systems. The mention of biomass is just used to provide some intuition to the mathematical concepts and results, but not the other way around, as a "model" of what happens in this mysterious and difficult field of management of renewable resources. Many other interpretations of the variables and controls presented could have been chosen, naturally. In any case, whatever the chosen interpretation, these one-dimensional systems are far too simplistic for their conclusions to be taken seriously.

We assume that there is a constant supply of resources, no predators and limited space: at each instant $t \geq 0$, the biomass $x(t)$ of the population must remain confined in an interval $K := [a, b]$ describing the *environment* (where $0 < a < b$). The maximal size b that the biomass can achieve is called the *carrying capacity* in the specialized literature.

The dynamics governing the evolution of the biomass is unknown, really.

7.2.1 *From Malthus to Verhulst and Beyond*

Several models have been proposed to describe the evolution of these systems. They are all particular cases of a general dynamical system of the form

$$x'(t) = \widetilde{u}(x(t))x(t) \tag{7.1}$$

where $\widetilde{u} : [a, b] \mapsto \mathbb{R}$ is a *feedback*, i.e., a mathematical translation of the *growth rate* of the biomass of the population *feeding back* on the biomass (specialists in these fields prefer to study growth rates than velocities, as in mechanics or physics).

The scarcity of resources sets a limit to population growth: this is typically a viability problem. The question soon arose to know whether the environment $K := [a, b]$ is viable under differential equation (7.1) associated with such or such feedback $\widetilde{u}$ proposed by specialists in population dynamics.

Another question, which we answer in this chapter, is in some sense "inverse problem" (see box 2, p. 5): Given an environment, the viability property and maybe other properties required on the evolutions, *what are all the feedbacks $\widetilde{u}$ under which these properties are satisfied*? Answering the second question automatically answers the first one.

Thomas Robert Malthus (1766–1834)

Pierre François Verhulst (1804–1849)

Raymond Pearl (1879–1940)

Alfred J. Lotka (1880–1949)

1. *Thomas Malthus* was the first one to address this viability problem and came up with a negative answer. He advocated in 1798 to choose a constant positive growth rate $\widetilde{u}_0(x) = r > 0$, leading to an exponential evolution $x(t) = xe^{rt}$ starting at x. It *leaves the interval* $[a, b]$ *in finite time* $t^\star := \frac{1}{r}\log\left(\frac{b}{x}\right)$ (see left column of Fig. 7.1, p. 251). In other words, no bounded interval can be viable under Malthusian dynamics. This is the price to pay for linearity of the dynamics of the population: "*Population, when unchecked, increases in a geometrical ratio*", as he concluded in his celebrated *An essay on the principle of population* (1798). He thus was worried by the great poverty of his time, so that he finally recommended "moral restraint" to stimulate savings, diminish poverty, maintain wages above the minimum necessary, and catalyze happiness and prosperity.
   For overcoming this pessimistic conclusions, other explicit feedbacks have next been offered for providing evolutions growing fast when the population is small and declining when it becomes large to compensate for the never ending expansion of the Malthusian model.
2. The Belgian mathematician *Pierre-François Verhulst* proposed in 1838 the *Verhulst feedback* of the form

$$\widetilde{u}_1(x) := r(b - x) \text{ where } r > 0$$

after he had read Thomas Malthus' Essay. It was rediscovered in 1920 by *Raymond Pearl* and again in 1925 by *Alfred Lotka* who called it the *law of population growth.*

The environment K is viable under the associated *purely logistic* Verhulst *equation* $x'(t) = rx(t)(b - x(t))$. The solution starting from $x \in [a, b]$ is equal to the "sigmoid" $x(t) = \frac{bx}{x + (b - x)e^{-rt}}$. It has the famous S-shape, remains confined in the interval $[a, b]$ and converges to the carrying capacity b when $t \mapsto +\infty$ (see center column of Fig. 7.1, p. 251). The logistic model and the S-shape graph of its solution became very popular since the 1920s and stood as the evolutionary model of a large manifold of growths, from the tail of rats to the size of men.

3. However, other examples of explicit feedbacks have been used in population dynamics. For instance, the environment K is viable under the following feedbacks:

- $\widetilde{u}_2(x) := e^{r(b-x)} - 1$, a continuous counterpart of a discrete time model by model proposed by Ricker and May,
- $\widetilde{u}_3(x) := r(b - x)^{\alpha}$, a continuous counterpart of a discrete-time model proposed by Hassel and May,
- the feedback $\widetilde{u}_4(x) := r\left(\frac{1}{\sqrt{x}} - \frac{1}{\sqrt{b}}\right)$,
- and several other feedbacks can be constructed similarly,

These feedbacks provide *increasing* evolutions which reach the upper bound b of the environment asymptotically. The three next feedbacks provide viable evolutions which reach b in finite time:

1. **Inert feedbacks**

$$\widetilde{u}_5(x) := r\sqrt{2\log\left(\frac{b}{x}\right)}$$

govern evolutions reaching b in finite time with a vanishing velocity so that the state may remain at b forever. It will be studied in more details in Sect. 7.3.1.

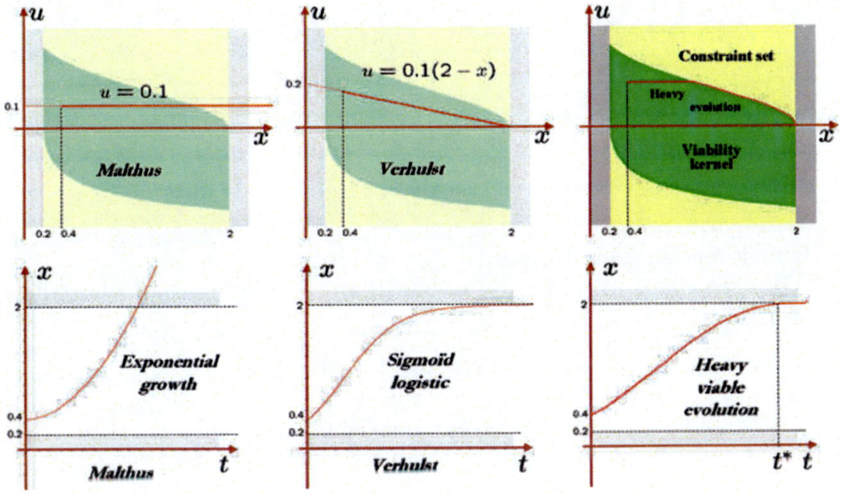

Fig. 7.1 Malthus, Verhulst and Heavy Feedbacks.
The three figures of the top row illustrate the curves representing three examples of feedbacks. The left figures shows the (constant) Malthusian feedback, the center ones the affine Verhulst feedback and the right one the heavy feedback. The Malthus and Verhulst feedbacks are given a priori. The Malthus feedback is not viable in the environment $[0.2, 2]$ *depicted in the vertical axis, but the Verhulst feedback is. The heavy feedback is computed a posteriori from viability, flexibility and inertia requirements. For that purpose, we have to compute the viability kernel obtained with the Viability Kernel Algorithm. The three figures of the bottom row provide the corresponding evolutions. The coordinates represent the time t and the biomass x. The evolution is exponential in the Malthusian case, and thus, leaves the viability interval, the logistic sigmoid in the Verhulst case, which converges asymptotically to the upper bound of the viability interval. It is reached in finite time* $t^\star$ *by a heavy evolution, starting as an exponential, and next slowed down to reach the upper bound with a growth rate equal to zero (an equilibrium).*

2. **Heavy feedbacks** are hybrid feedbacks obtained by "concatenating" the Malthusian and inert feedbacks:

$$\widetilde{u}_6(x) := \begin{cases} r & \text{if } a \le x \le be^{-\frac{r^2}{2c}} \\ r\sqrt{2\log\left(\frac{b}{x}\right)} & \text{if } be^{-\frac{r^2}{2c}} \le x \le b \end{cases}$$

They govern evolutions combining Malthusian and inert growth: a heavy solution evolves (exponentially) with constant regulon r until the instant when the state reaches $be^{-\frac{r^2}{2c}}$. *This is the last time until which the growth rate could remain constant before being changed by taking*

$$\widetilde{u}(x) = c\sqrt{2\log\left(\frac{b}{x}\right)}$$

Then the evolution follows the inert solution starting and reaches b *in finite time*

$$t^\star := \frac{\log\left(\frac{b}{x}\right)}{r} + \frac{r}{2c}$$

It may remain there forever.

3. **Allee inert feedbacks** are obtained by "concatenating" the following feedbacks:

$$\widetilde{u}_7(x) := \begin{cases} r\sqrt{2\log\left(\frac{x}{a}\right)} & \text{if } a \leq x \leq \sqrt{ab} \\ r\sqrt{2\log\left(\frac{b}{x}\right)} & \text{if} \sqrt{ab} \leq x \leq b \end{cases}$$

They govern evolutions combining positive and negative inert growths: An Allee inert evolution increases with positive increasing growth under feedback until the instant when the state reaches $\sqrt{ab}$. *This is the last time until which the growth rate could increase before being switched* to the second feedback $r\sqrt{2\log\left(\frac{b}{x}\right)}$, continuing to increase despite the fact that the positive feedback decreases (the Titanic effect, see Definition 6.5.3, p. 239). Then the evolution reaches b *in finite time.* It may remain there forever, but not forced to do so (see Sect. 7.2.3).

The growth rate feedbacks $\widetilde{u}_i$, $i = 0, \ldots, 7$ are always positive on the interval $[a, b]$, so that the velocity of the population is always positive, even though it slows down. Note that $\widetilde{u}_0(b) = r > 0$ is strictly positive at b whereas the values $\widetilde{u}_i(b) = 0$, $i = 1, \ldots, 7$ for all other feedbacks presented above. The growth rates $\widetilde{u}_i$, $i = 0, \ldots, 6$ are decreasing whereas the Allee inert growth rate $\widetilde{u}_7$ is (strictly) increasing on a sub-interval. □

7.2.2 The Inverse Approach

Instead of finding one feedback $\widetilde{u}$ satisfying the above viability requirements by trial and error, as most scientists historically did, viability theory enables us to proceed systematically for designing feedbacks by leaving the choice of the growth rates open, regarding them as *regulons* (regulation parameters) of the control system

$$x'(t) = u(t)x(t) \tag{7.2}$$

where the control $u(t)$ is chosen at each time t for governing evolutions confined in the interval $[a, b]$.

1. ***viability constraints*** on the biomass by requiring that

$$\forall t \geq 0, \ \ x(t) \ \in \ [a, b]$$

2. ***inertia thresholds*** imposing a *speed limit* on the evolutions of the regulons:

$$\forall t \geq 0, \ \ u'(t) \ \in \ [-c, +c]$$

As we saw in Chap. 6, p. 199, these two requirements are enough to delineate the set of viable feedbacks satisfying inertia threshold thanks to the inertia function.

Lemma 7.2.1 *[Inertia Function] We denote by $\mathcal{P}(x, u)$ the set of solutions $(x(\cdot), u(\cdot))$ to system (7.2)* viable *in the interval $[a, b]$ starting at (x, u).*

The inertia function *is defined by*

$$\alpha(x, u) := \inf_{(x(\cdot),u(\cdot)) \in \mathcal{P}(x,u)} \sup_{t \geq 0} |u'(t)|$$

The domain $\mathrm{Dom}(\alpha)$ of the inertia function of system $x'(t) = u(t)x(t)$ confronted to the environment $K := [a, b]$ is equal to

$$\mathrm{Dom}(\alpha) := (\{a\} \times \mathbb{R}_+) \cup (]a, b[\times \mathbb{R}) \cup (\{b\} \times \mathbb{R}_-)$$

and the inertia function is equal to:

$$\alpha(x, u) \ := \begin{cases} \dfrac{u^2}{2 \log\left(\frac{b}{x}\right)} & \text{if } a \leq x < b \ \& \ u \geq 0 \\ \dfrac{u^2}{2 \log\left(\frac{x}{a}\right)} & \text{if } a < x \leq b \ \& \ u \leq 0 \end{cases}$$

The epigraph $\mathcal{E}p(\alpha)$ of the inertia function is closed. However, its domain is neither closed nor open (and not even locally compact). The restriction of the inertia function to its domain is continuous.

Remark: The associated Hamilton–Jacobi–Bellman Equation. The inertia function is the unique lower semicontinuous solution (in the generalized sense of Barron–Jensen & Frankowska, Definition 17.4.2, p. 701) to the Hamilton-Jacobi partial differential equation

$$\begin{cases} -\dfrac{\partial\alpha(x,u)}{\partial x}ux + \alpha(x,u)\dfrac{\partial\alpha(x,u)}{\partial u} = 0 \text{ if } a \leq x < b \;\&\; u \geq 0 \\ -\dfrac{\partial\alpha(x,u)}{\partial x}ux - \alpha(x,u)\dfrac{\partial\alpha(x,u)}{\partial u} = 0 \text{ if } a < x \leq b \;\&\; u \leq 0 \end{cases}$$

on $\text{Dom}(\alpha)$ with **discontinuous** coefficients. Indeed, one can a posteriori check that the partial derivatives of the inertia function α are equal to

$$\frac{\partial\alpha(x,u)}{\partial x} := \begin{cases} \dfrac{u^2}{2x\left(\log\left(\frac{b}{x}\right)\right)^2} & \text{if } u \geq 0 \\ -\dfrac{u^2}{2x\left(\log\left(\frac{x}{a}\right)\right)^2} & \text{if } u \leq 0 \end{cases} \;\&\; \frac{\partial\alpha(x,u)}{\partial u} := \begin{cases} \dfrac{u}{\log\left(\frac{b}{x}\right)} & \text{if } u \geq 0 \\ \dfrac{u}{\log\left(\frac{x}{a}\right)} & \text{if } u \leq 0 \end{cases}$$

defined on the graph of U.

Observe that $\frac{\partial\alpha(x,u)}{\partial u}$ is positive when $u > 0$ and negative when $u < 0$. □

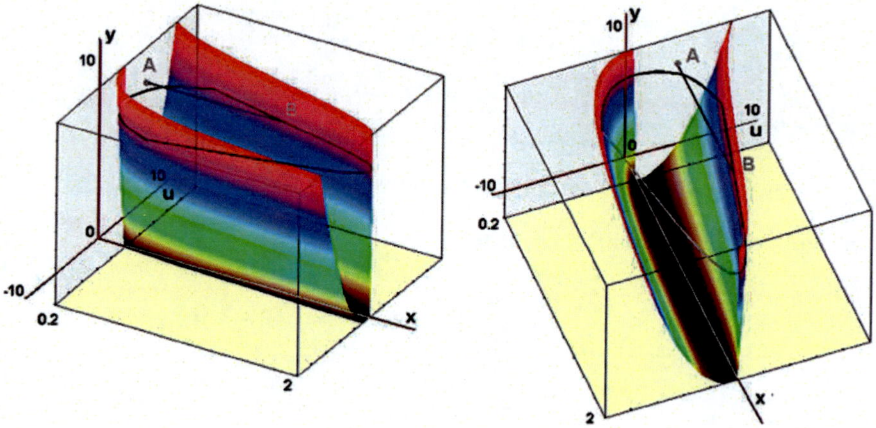

Fig. 7.2 Inertia Function.
Two views of the inertia function, a section of which being displayed in Fig. 7.3, p. 256.

We deduce from Lemma 7.2.1, p. 253 the analytical formulas of the inert regulation and critical maps:

Proposition 7.2.2 ***[Inert Regulation Map]*** *For system* $x'(t) = u(t)x(t)$, *the* inert regulation map *defined by*

$$(c,x) \rightsquigarrow R_c(x) := \{u \in \mathbb{R} \text{ such that } \alpha(x,u) \leq c\}$$

is equal to

$$R_c(x) := \begin{cases} \left[0, \sqrt{2c \log\left(\frac{b}{a}\right)}\right] & \text{if } x = a \\ \left[-\sqrt{2c \log\left(\frac{x}{a}\right)}, \sqrt{2c \log\left(\frac{b}{x}\right)}\right] & \text{if } a < x < b \\ \left[-\sqrt{2c \log\left(\frac{b}{a}\right)}, 0\right] & \text{if } x = b \end{cases}$$

The critical map $(c, u) \rightsquigarrow \Xi_c(u) := \{x \in [a, b]$ *such that* $\alpha(x, u) = c\}$ *is equal to*

$$\Xi_c(u) := \begin{cases} be^{-\frac{u^2}{2c}} & \text{if } u > 0 \\ ae^{\frac{u^2}{2c}} & \text{if } u < 0 \end{cases}$$

if $c > 0$ *and to*

$$\Xi(0, u) := \begin{cases} [a, b] & \text{if } u = 0 \\ \emptyset & \text{if } u \neq 0 \end{cases}$$

if $c = 0$

Since the epigraph of the inertia function is the viability kernel of the "metaenvironment" $\mathcal{K} := [a, b] \times \mathbb{R}_+ \times \mathbb{R}_+$ under the metasystem

$$\begin{cases} (i) \;\; x'(t) = u(t)x(t) \\ (ii) \; u'(t) = v(t) \\ (ii) \; y'(t) = 0 \\ \quad\quad \text{where } |v(t)| \leq y(t) \end{cases} \tag{7.3}$$

the Viability Theorem 11.3.4, p. 455 provides the analytical formula of the adjustment map $(x, u, c) \rightsquigarrow G(x, u, c)$ associating with any auxiliary state (x, u, c) the set $G(x, u, c)$ of control velocities governing the evolution of evolutions with finite inertia:

1. Case when $\alpha(x, u) < c$. Then

$$G(x, u, c) := \begin{cases} [0, \alpha(a, u)] & \text{if } x = a \\ [-\alpha(x, u), +\alpha(x, u)] & \text{if } a < x < b \\ [-\alpha(b, u), 0] & \text{if } x = b \end{cases}$$

2. Case when $\alpha(x, u) = c$. Then

$$G(x, u, c) := \begin{cases} -\alpha(x, u) & \text{if } u \geq 0 \;\&\; a \leq x < b \\ \alpha(x, u) & \text{if } u \leq 0 \;\&\; a < x \leq b \end{cases}$$

The minimal selection $g^\circ(x, u, c) \in G(x, u, c)$ is equal to $g^\circ(x, u, c) = 0$ if $\alpha(x, u) < c$ and to

$$g^{\circ}(x,u,\alpha(x,u)) := \begin{cases} -\alpha(x,u) \text{ if } u \geq 0 \;\&\; a \leq x < b \\ \alpha(x,u) \quad \text{if } u \geq 0 \;\&\; a < x \leq b \end{cases}$$

if $\alpha(x,u) = c$, i.e., if $x \in \Xi_c(u)$ is located in the critical zone of the control u at inertia threshold c.

Although the minimal selection g° is not continuous, for any initial pair $(x,u) \in \mathrm{Dom}(\alpha)$ in the domain of the inertia function, the system of differential equations

$$\begin{cases} (i) \;\; x'(t) = u(t)x(t) \\ (ii) \; u'(t) = g^{\circ}(x(t),u(t),c) \end{cases} \tag{7.4}$$

has solutions which are called *heavy viable evolutions* of initial system (7.2). The trajectory of this heavy evolution is shown on the graph of the inertia function displayed in Fig. 7.2 and Fig. 7.3.

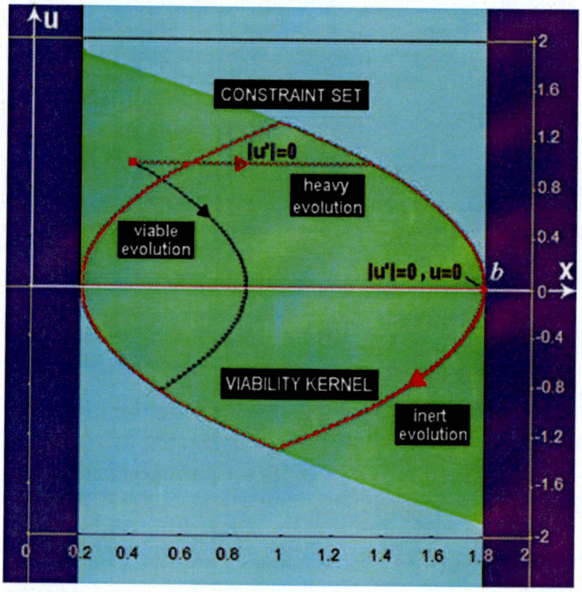

Fig. 7.3 Viability Kernel and Inert Evolution.
The Viability Kernel Algorithm computes the viability kernel (which is the graph of the regulation map) on a sequence of refined grids, provides an arbitrary viable evolution, the heavy evolution minimizing the velocity of the controls and which stops at equilibrium b, and the inert evolutions going back and forth from a to b in an hysteresis cycle.

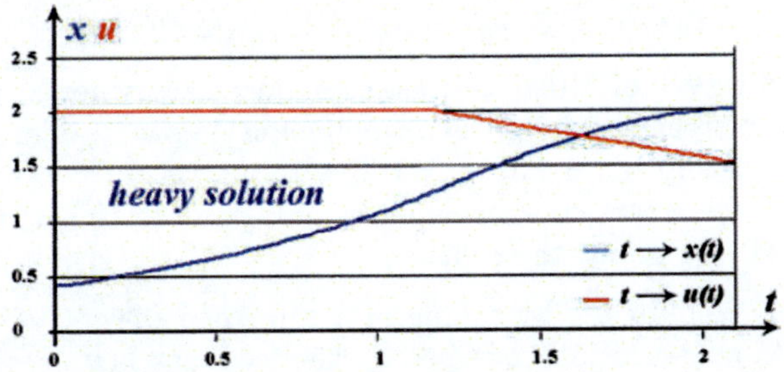

Fig. 7.4 Heavy evolution.
The evolution of the growth rate (in blue) *of the heavy evolution starting at* (x, u) *such that* $\alpha(x, u) < c$ *and* $u > 0$ *is constant until the time* (*kairos*) $\frac{1}{u}\log\left(\frac{b}{x}\right) - \frac{u}{2c}$ *at which the evolution reaches the critical zone* $\Xi_c(u) = [be^{-\frac{u^2}{2c}}, b]$.
During this period, the state (in blue) *follows an exponential (Malthusian) growth* xe^{ut}. *After, the growth rate decreases linearly until the time* $\frac{1}{u}\log\left(\frac{b}{x}\right) + \frac{u}{2c}$ *when it vanishes and when the evolution reaches the upper bound* b. *During this period, the inertia* $\alpha(x(t), u(t)) = c$ *remains equal to the inertia threshold until the evolution reaches the* upper bound b *with a velocity equal to* 0. *This is an equilibrium at which the evolution may remain forever.*

7.2.3 The Inert Hysteresis Cycle

As was shown in Sect. 6.5, p. 233, the inertia feedback provides hysterons and cyclic evolutions.

We observe that the graphs of the feedbacks $\sqrt{c}r^\sharp$ and $\sqrt{c}r^\flat$ intersect at the point $(x^\star, \sqrt{c}u^\star)$ where

$$x^\star := \sqrt{ab} \;\&\; u^\star := \sqrt{\log\left(\frac{b}{a}\right)}$$

Therefore, the warning time is equal to

$$\tau^\star = \tau(x^\star, \sqrt{c}u^\star) = 2\frac{\log\left(\frac{b}{x^\star}\right)}{\sqrt{c}u^\star} = \sqrt{\frac{\log\left(\frac{b}{a}\right)}{c}}$$

The inert evolution $(x(\cdot), u(\cdot))$ starting at $(x^\star, \sqrt{c}u^\star)$ is governed by the regulons

$$\forall t \in [0, \tau^\star],\ u(t) = \widetilde{u}_6(x(t)) := \sqrt{c}\, r^\sharp(x(t))$$

During this interval of time, the regulon decreases whereas the biomass continues to increase: this is a phenomenon known under the *Titanic syndrome* (see Definition 6.5.3, p. 239) due to inertia.

It reaches the state-control equilibrium pair $(b, 0)$ at time $\tau^\star := \tau(x^\star, \sqrt{c}u^\star)$. At this point, the solution:

1. may stop at equilibrium by taking $u(t) \equiv 0$ when $t \geq \tau^\star$,
2. or switch to an evolution governed by the feedback law

$$\forall t \geq \tau^\star,\ u(t) = \widetilde{u}_8(x(t)) := -\sqrt{c}\, r^\sharp(x(t))$$

among (many) other possibilities to find evolutions starting at $(b, 0)$ remaining viable while respecting the velocity limit on the regulons because $(b, 0)$ lies on the boundary of $[a, b] \times \mathbb{R}$.

Using the feedback $\widetilde{u}_8$ for instance, for $t \geq \tau^\star$, we the evolutions $x(t)$ is still defined by

$$\forall t \geq \tau^\star,\ x(t) = xe^{ut - \frac{u^2t^2}{4\log\left(\frac{b}{x}\right)}}$$

and is associated with the regulons

$$\forall t \geq \tau^\star,\ u(t) = u\left(1 - \frac{ut}{2\log\left(\frac{b}{x}\right)}\right)$$

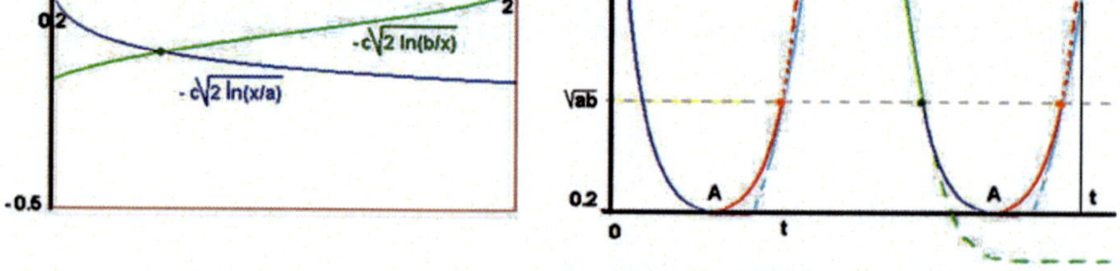

Fig. 7.5 Graph of the Inert Hysteretic Evolution Computed Analytically. *This Figure explains how the inert hysteretic evolution can be obtained by piecing together four feedbacks. Starting from* $(x^\star, \sqrt{c}u^\star)$ *at time* 0 *with the velocity of the regulon equal to* $-c$*, the evolution is governed by the inert feedback* $\widetilde{u}_6(x) := \sqrt{c}\sqrt{2\log\left(\frac{b}{x}\right)}$ *until it reaches* b*, next governed by the*

feedback $\widetilde{u}_8(x) := -\sqrt{c}\sqrt{2\log\left(\frac{b}{x}\right)}$ *until it reaches* $x^\star$, *next governed by the inert feedback* $\widetilde{u}_9(x) := -\sqrt{c}\sqrt{2\log\left(\frac{x}{a}\right)}$ *until it reaches* a *and last governed by the feedback* $\widetilde{u}_{10}(x) :=\sqrt{c}\sqrt{2\log\left(\frac{x}{a}\right)}$ *until it reaches the point* $x^\star$ *again. It can be generated automatically by the Viability Kernel Algorithm, so that inert hysteretic evolutions can be computed for non tractable examples. See Fig. 7.6.*

Let us state

$$\begin{cases} r^\sharp(x) & := \sqrt{2c\log\left(\frac{b}{a}\right)} \\ r^\flat(x) & := -\sqrt{2c\log\left(\frac{b}{a}\right)} \end{cases}$$

The state-control pair $(x(\cdot), u(\cdot))$ is actually governed by the metasystem

$$\begin{cases} (i)\ \ x'(t) = u(t)x(t) \\ (ii)\ u'(t) = -c \end{cases}$$

It ranges over the graph of the map $\sqrt{c}\, r^\sharp(\cdot)$ between 0 and $\tau^\star$ and over the graph of the map $-\sqrt{c}\, r^\sharp(\cdot)$ between $\tau^\star$ and $2\tau^\star$. During this interval of time, both the regulon $u(t)$ and the biomass $x(t)$ starts decreasing. The velocity of the negative regulon is constant and still equal to $-\alpha(x,u)$.

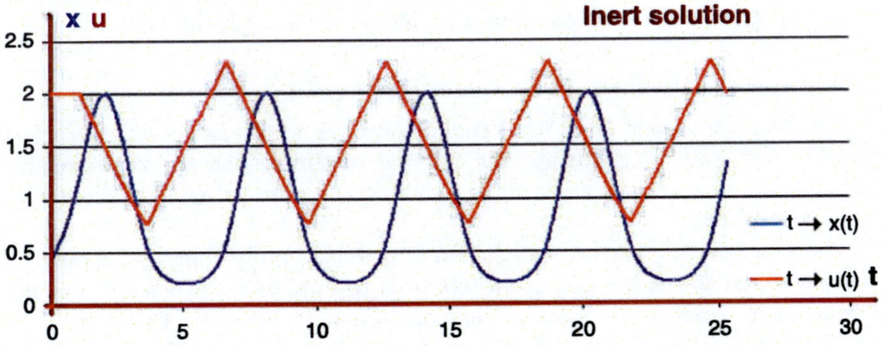

Fig. 7.6 Graph of the Inert Hysteretic Evolution and its Control Computed by the Viability Kernel Algorithm.
Both the graphs of the inert hysteretic evolution (in blue*) and of its control (in* red*) are plotted. They are not computed from the analytical formulas as in Fig. 7.5, but extracted from the Viability Kernel Algorithm. The velocity of the control remains constant until the trajectory of the solution hits the boundary of the viability kernel and then switches to the other extremal*

control with opposite sign and so on. The evolution is then cyclic (contrary to logistic evolutions), alternatively increasing and decreasing from the lower bound *of the constrained interval to its* upper bound.

But it is no longer viable on the interval $[a,b]$, because with such a strictly negative velocity $-\alpha(x,u)$, $x(\cdot)$ leaves $[a,b]$ in finite time. Hence regulons have to be switched before the evolution leaves the graph of U_c by crossing through the graph of $-\sqrt{c}\,r^{\flat}(\cdot)$ when $-\sqrt{c}\,r^{\flat}(x^{\star}) = -\sqrt{c}\,r^{\sharp}(x^{\star})$ at time $2\tau^{\star}$.

Therefore, in order to keep the evolution viable, it is the last instant to switch the velocity of the regulon from $-c$ to $+c$.

Starting at $(x^{\star},\sqrt{c}u^{\star})$ at time $2\tau^{\star}$, we let the state-control pair $(x(\cdot),u(\cdot))$ evolve according the metasystem

$$\begin{cases} (i) \;\; x'(t) = u(t)x(t) \\ (ii) \;\; u'(t) = +c \end{cases}$$

It is governed by the regulons

$$\forall t \in [0,\tau^{\star}],\;\; u(t) = \widetilde{u}_9(x(t)) := -\sqrt{c}\,r^{\flat}(x(t))$$

and ranges over the graph of the map $-\sqrt{c}\,r^{\flat}(\cdot)$ between $2\tau^{\star}$ and $3\tau^{\star}$. During this interval of time, the regulon increases whereas the biomass continues to decrease (the Titanic syndrome again due to inertia) and stops when reaching the state-control equilibrium pair $(a,0)$ at time $3\tau^{\star}$. Since $(a,0)$ lies on the boundary of $[a,b]\times\mathbb{R}$, there are (many) other possibilities to find evolutions starting at $(a,0)$ remaining viable while respecting the velocity limit on the regulons. Therefore, we continue to use the above metasystem with velocity $+c$ starting at $3\tau^{\star}$. The evolutions $x(t)$ obtained through the feedback law

$$\forall t \geq \tau^{\star},\;\; u(t) = \widetilde{u}_{10}(x(t)) := +\sqrt{c}\,r^{\flat}(x(t))$$

The state-control pair $(x(\cdot),u(\cdot))$ ranges over the graph of the map $\sqrt{c}\,r^{\flat}(\cdot)$ between $3\tau^{\star}$ and $4\tau^{\star}$. During this interval of time, both the regulon $u(t)$ and the biomass $x(t)$ increase until reaching the pair $(x^{\star},\sqrt{c}u^{\star})$, the initial state-control pair.

Therefore, letting the heavy solution bypass the equilibrium by keeping its velocity equal to $+c$ instead of switching it to 0, allows us to build a cyclic evolution by taking velocities of regulons equal successively to $-c$ and $+c$ on the intervals $[2n\tau^{\star},(2n+1)\tau^{\star}]$ and $[(2n+1)\tau^{\star},(2n+2)\tau^{\star}]$ respectively. We obtain in this way a cyclic evolution of period $4\tau^{\star}$ showing an *hysteresis property*: *The evolution oscillates between a and b back and forth by ranging alternatively two different trajectories on the metaenvironment* $[a,b]\times\mathbb{R}$. The evolution of the state is governed by concatenating four feedbacks, $\widetilde{u}_6 := +\sqrt{c}r^{\sharp}$ on $[x^{\star},b]$, $\widetilde{u}_8 := -\sqrt{c}r^{\sharp}$ on $[x^{\star},b]$, $\widetilde{u}_9 := -\sqrt{c}r^{\flat}$ on $[a,x^{\star}]$ and $\widetilde{u}_{10} := +\sqrt{c}r^{\flat}$ on $[a,x^{\star}]$.

Remark. Note also that not only this evolution is cyclic, but obeys a *quantized mode* of regulation: We use only two control velocities $-c$ and $+c$ to control the metasystem (instead of an infinite family of open loop controls $v(\cdot) := u'(\cdot)$, as in the control of rockets in space). This is also an other advantage of replacing a control system by its metasystem: use a finite number (quantization) of controls... to the price of increasing the dimension of the system by replacing it by its metasystem. □

We can adapt the inert hysteresis cycle to the heavy case when we start with a given regulon $u < \sqrt{c}\, u^\star = \sqrt{c \log\left(\frac{b}{a}\right)}$. We obtain a cyclic evolution by taking velocities of regulons equal successively to 0, $-c$, 0, $+c$, and so on showing an *hysteresis property*: *The evolution oscillates between a and b back and forth by taking two different routes.*

It is described in the following way. We denote by $a_c(u)$ and $b_c(u)$ the roots

$$a_c(u) = ae^{\frac{u^2}{2c}} \;\&\; b_c(u) = be^{-\frac{u^2}{2c}}$$

of the equations $r^\flat(x) = u$ and $r^\sharp(x) = u$ and we set

$$\tau^\star(u) = 2\frac{\log\left(\frac{b}{a}\right)}{u}$$

1. The state-control pair $(x(\cdot), u(\cdot))$ starts from $(a_c(u), u)$ by taking the velocity of the regulon equal to 0. It remains viable on the time interval $[0, \tau^\star(u) - \frac{u}{2c}]$ until it reaches the state-control pair $(b_c(u), u)$.
2. The state-control pair $(x_h(\cdot), u_h(\cdot))$ starts from $(b_c(u), u)$ at time $\tau^\star(u) - \frac{u}{2c}$ by taking the velocity of the regulon equal to $-c$. It is regulated by the metasystem
$$\begin{cases} (i) & x'(t) = u(t)x(t) \\ (ii) & u'(t) = -c \end{cases}$$
ranging successively over the graphs of $\sqrt{c}\, r^\sharp$ and $-\sqrt{c}\, r^\sharp$ on the time interval $[\tau^\star(u) - \frac{u}{2c}, \tau^\star(u) + \frac{3u}{2c}]$ until it reaches the state-control pair $(b_c(u), -u)$.
3. The state-control pair $(x_h(\cdot), u_h(\cdot))$ starts from $(b_c(u), -u)$ at time $\tau^\star(u) + \frac{3u}{2c}$ by taking the velocity of the regulon equal to 0. It remains viable on the time interval $[\tau^\star(u) + \frac{3u}{2c}, 2\tau^\star(u) + \frac{u}{c}]$ until it reaches the state-control pair $(a_c(u), -u)$.
4. The state-control pair $(x_h(\cdot), u_h(\cdot))$ starts from $(a_c(u), -u)$ at time $2\tau^\star(u) + \frac{u}{c}$ by taking the velocity of the regulon equal to $+c$. It is regulated by the metasystem
$$\begin{cases} (i) & x'(t) = u(t)x(t) \\ (ii) & u'(t) = +c \end{cases}$$

ranging successively over the graphs of $-\sqrt{c}\ r^{\flat}$ and $\sqrt{c}\ r^{\flat}$ on the time interval $[\tau^{\star}(u) - \frac{u}{2c}, \tau^{\star}(u) + \frac{3u}{2c}]$ until it reaches the state-control pair $(a_c(u), u)$.

In summary, the study of inertia functions and metasystems allowed us to discover several families of feedbacks or concatenation of feedbacks providing several cyclic viable evolutions, using two (for the inert hysteresis cycle) or three (for the heavy hysteresis cycle) "control velocities" $+c$, $-c$ and, for the heavy cycle, $+c$, $-c$ and 0.

7.3 Management of Renewable Resources

Let us consider a given positive growth rate feedback $\widetilde{u}$ governing the evolution of the biomass of a renewable resource $x(t) \geq a > 0$, through differential equation: $x'(t) = \widetilde{u}(x(t))x(t)$. We shall take as examples the Malthusian feedback $u_0(x) := u$, the *Verhulst* feedback $u_1(x) := r(x-b)$ and the inert feedbacks $\widetilde{u}_5(x) := r\sqrt{2\log\left(\frac{b}{x}\right)}$ and $\widetilde{u}_9(x) := -\sqrt{c}\sqrt{2\log\left(\frac{x}{a}\right)}$.

The evolution is slowed down by industrial activity which depletes it, such as fisheries.

We denote by $v \in \mathbb{R}_+$ the industrial effort for exploiting the renewable resource, playing now the role of the control. Naturally, the industrial effort is subjected to state-dependent constraints $V(x)$ describing economic constraints.

We integrate the ecological constraint by setting $V(x) = \emptyset$ whenever $x < a$. Hence the evolution of the biomass is regulated by the control system

$$\begin{cases} (i)\ \ x'(t) = x(t)\left(\widetilde{u}(x(t)) - v(t)\right) \\ (ii)\ v(t) \in V(x(t)) \end{cases} \tag{7.5}$$

We denote by $\mathcal{Q}_{\widetilde{u}}(x, v)$ the set of solutions $(x(\cdot), v(\cdot))$ to system (7.5). The *inertia function* is defined by

$$\beta_{\widetilde{u}}(x, v) := \inf_{(x(\cdot),v(\cdot))\in\mathcal{Q}_{\widetilde{u}}(x,v)} \sup_{t\geq 0} |v'(t)|$$

This function is characterized as the viability kernel of a subset under an auxiliary system, known as its "metasystem". It inherits the properties of the viability kernel of an environment and can be computed by the Viability Kernel Algorithm.

The epigraph of the inertia function is characterized as the viability kernel of the "metaenvironment" $\mathcal{K} := \text{Graph}(V) \times \mathbb{R}_+$ under the metasystem

$$\begin{cases} (i) \ x'(t) &= (\widetilde{u}(x(t)) - v(t))x(t) \\ (ii) \ v'(t) &= w(t) \\ (ii) \ y'(t) &= 0 \\ \quad \text{where } |w(t)| \leq y(t) \end{cases} \tag{7.6}$$

The metasytem (7.6) is regulated by the *velocities* of the former regulons. In other words, the metasystem regulates the evolution of the initial system by acting on the velocities of the controls instead of controlling them directly. The component y of the "auxiliary state" is the "inertia threshold" setting an upper bound to the velocities of the regulons. Therefore, metasystem (7.6) governs the evolutions of the state $x(t)$, the control $v(t)$ and the inertia threshold $y(t)$ by imposing constraints on the velocities of the regulons

$$\forall t \geq 0, \ |v'(t)| \leq y(t)$$

called control velocities and used as auxiliary regulons.

Unfortunately, the state-control environment $\text{Graph}(V) \times \mathbb{R}_+$ is obviously not viable under the above metasystem: Every solution starting from $(a(u), v, c)$ with $u < v$ leaves it immediately.

Assume from now on that there exists a decreasing positive map $\mathbf{v}: [a, b] \mapsto [0, \overline{v}]$ defining the set-valued map V

$$\forall x \in [a, \infty[, \ V(x) \ := \ [\mathbf{v}(x), \overline{v}] \tag{7.7}$$

The epigraph $\mathcal{E}p(\beta_{\widetilde{u}})$ of the inertia function $\beta_{\widetilde{u}}$ is equal to the viability kernel $\text{Viab}_{(7.6)}(\text{Graph}(V) \times \mathbb{R}_+)$ of the state-control environment $\text{Graph}(V) \times \mathbb{R}_+$ under metasystem (7.6).

We observe that the inertia function vanishes on the *equilibrium line*:

$$\beta_{\widetilde{u}}(x, \widetilde{u}(x)) = 0$$

It is identically equal to 0 if for any $x \geq a$, $\widetilde{u}(x) \geq \mathbf{v}(x)$ and identically infinite if for any $x \geq a$, $\widetilde{u}(x) < \mathbf{v}(x)$.

The fundamental economic model was originated by *Michael Graham* in 1935 and taken up by *Milner Schaefer* in 1954. They assumed that the exploitation rate is proportional to the biomass and the economic activity: viability constraints are described by *economic constraints*

$$\forall t \geq 0, \ cv(t) + C \ \leq \ \gamma v(t) \, x(t)$$

where $C \geq 0$ is a fixed cost, $c \geq 0$ the unit cost of economic activity and $\gamma \geq 0$ the price of the resource. We also assume that

$$\forall t \geq 0, \ 0 \leq v(t) \leq \overline{v}$$

where $\overline{v} > \frac{C}{\gamma x - c}$ is the maximal exploitation effort. Hence the Graham-Schaeffer constraints are summarized under the set-valued map $V : [a, \infty[\rightsquigarrow \mathbb{R}_+$ defined by

$$\forall x \geq a, \; V(x) := \left[\frac{C}{\gamma x - c}, \overline{v}\right]$$

In any case, the epigraph of the inertia function being a viability kernel, it can be computed by the Viability Kernel Algorithm. Figure 7.7 provides the level sets of the inertia function for the *Verhulst* and inert feedbacks respectively.

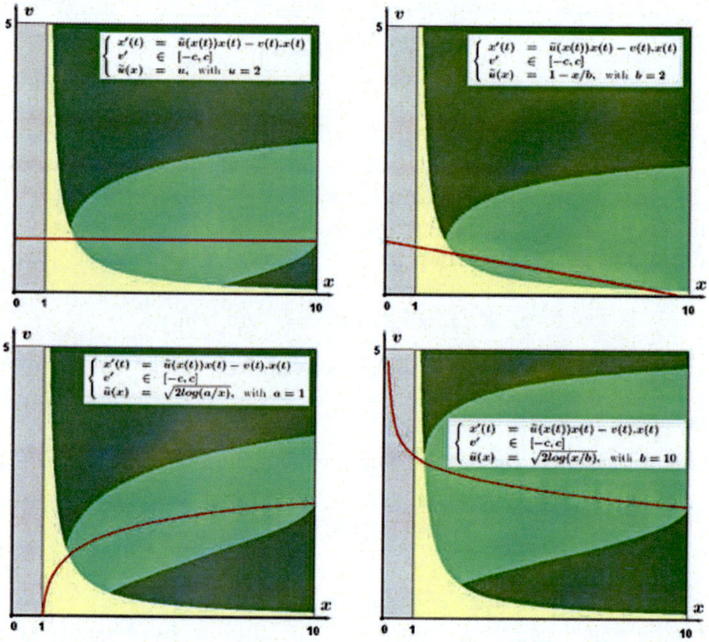

Fig. 7.7 Sections of the Inertia Function for Several Growth Rate Feedbacks. *This figure displays the computation of the section od level* 0 *of the inertia function for Schaeffer models coupled with four growth rate feedbacks* $\widetilde{u}$*: a constant growth rate, the Verhulst one (first row) and, the two growth rates feedbacks* $\sqrt{2\log\left(\frac{x}{a}\right)}$ *and* $\sqrt{2\log\left(\frac{b}{x}\right)}$. *The equilibrium lines are the graphs of the feedbacks. Heavy evolutions stop when their trajectories hit the equilibrium line.*

Using the Malthusian (constant) feedbacks $\widetilde{u}_0(x) \equiv u$ for the growth of the renewable resource allows us to provide analytical formula of the inertia function for any decreasing exploitation function $\mathbf{v}(x)$ such as the Graham-Schaeffer one. Let us define by $\nu(u)$ the root of the equation $\mathbf{v}(x) = u$ and set $a(u) := \max(a, \gamma(u))$.

The inertia function is equal to:

$$\beta_u(x, v) = \begin{cases} \dfrac{(v-u)^2}{2\log\left(\frac{c}{a(u)}\right)} & \text{if } v \geq u \text{ and } x \geq a(u) \\ 0 & \text{if } \mathbf{v}(x) \leq v \leq u \text{ and } x \geq a(u) \end{cases} \tag{7.8}$$

where the function $\mathbf{v}$ was introduced in (7.7), p. 263.

The epigraph $\mathcal{E}p(\beta_u)$ of the inertia function is closed. However, its domain is neither closed nor open (and not even locally compact). The restriction of the inertia function to its domain is continuous.

Remark: The associated Hamilton–Jacobi–Bellman Equation. The inertia function is a solution to the Hamilton–Jacobi partial differential equation

$$\forall\, v \geq u, \quad \frac{\partial \beta_u(x,v)}{\partial x}(u-v)x - \beta_u(x,v)\frac{\partial \beta_u(x,v)}{\partial v} = 0$$

Indeed, the partial derivatives of these two inertia functions are equal to

$$\frac{\partial \beta_u(x,v)}{\partial x} := -\frac{(v-u)^2}{2x\left(\log\left(\frac{x}{a(u)}\right)^2\right)} \quad \& \quad \frac{\partial \beta_u(x,v)}{\partial v} := \frac{v-u}{\log\left(\frac{x}{a(u)}\right)}$$

Observe that $\frac{\partial \beta_u(x,v)}{\partial v}$ is positive when $v > u$ and negative when $v < u$. □

Proposition 7.3.1 *[Regulation and Critical Maps] For system $x'(t) = (u - v(t))x(t)$, the* inert regulation map

$$(c, x) \rightsquigarrow R_c(x) = \{v \in \mathbb{R} \text{ such that } \beta_u(x,v) \leq c\}$$

associated with the inertia function β_u defined in (7.8), p. 265, is equal to

$$R_c(x) := \left[\mathbf{v}(x), u + \sqrt{2c\log\left(\frac{bx}{a(u)}\right)}\right] \text{ if } a(u) \leq x$$

where $\mathbf{v}$ is defined in (7.7), p. 263.

The critical map $(c, v) \rightsquigarrow \Xi_c(v) := \{x \in [a, b] \text{ such that } \beta_u(x, v) = c\}$ *is equal to*

$$\Xi(c, v) = [a(u), \xi(c, v)] \text{ where } \xi(c, v) := a(u)e^{\frac{(v-u)^2}{2c}}$$

if $c > 0$ *and to*

$$\Xi(0, v) := \begin{cases} [a(u), +\infty[& \text{if } \mathbf{v}(x) \le v \le u \\ \emptyset & \text{if } v > u \end{cases}$$

if $c = 0$.

The Viability Theorem 11.3.4, p. 455 provides the analytical formula of the adjustment map $(x, v, c) \rightsquigarrow G(x, v, c)$ associating with any auxiliary state (x, v, c) the set $G(x, v, c)$ of control velocities governing the evolution of evolutions with finite inertia:

1. Case when $\beta_u(x, v) < c$. Then
$$G(x, v, c) := [-\beta_u(x, v), +\beta_u(x, v)]$$
2. Case when $\beta_u(x, v) = c$ and $v > u$. Then
$$G(x, v, c) := -\beta_u(x, v)$$

The minimal selection $g^0(x, v, c)$ is defined by $g^0(x, v, c) = 0$ if $\mathbf{v}(x) \le v < u + \sqrt{2c \log\left(\frac{bx}{a(u)}\right)}$ and by $g^0(x, v, c) = -\beta_u(x, v)$ whenever $v < u + \sqrt{2c \log\left(\frac{bx}{a(u)}\right)}$.

7.3.1 Inert Evolutions

An evolution $(\overline{x}(\cdot), \overline{u}(\cdot))$ *is said to be* inert *on a time interval* $[t_0, t_1]$ *if it is regulated by an affine open-loop controls of the form* $v(t) := v + wt$, *the velocities* $v'(t) = w$ *of which are constant.*

The inertia function remains constant over an *inert evolution* as long as the evolution is viable: On an adequate interval, we obtain

$$\forall t \in [0, \overline{t}], \ \beta_u(\overline{x}(t), \overline{v}(t)) = \beta_u(x, v) = c$$

Let us consider the case when $v > u$.

The velocity governing the inert evolution is constant and equal to $\overline{v}'(t) = -\beta_u(x, v)$, so that

$$\overline{v}(t) = v\frac{(v-u)^2 t}{2\log\left(\frac{x}{a(u)}\right)}$$

and

$$\overline{x}(t) = xe^{-(v-u)t - \frac{(v-u)^2 t^2}{4\log\left(\frac{x}{a(u)}\right)}}$$

The state decreases until it reaches the lower bound $a(u)$ at time

$$\tau(x,v) = 2\frac{\log\left(\frac{x}{a(u)}\right)}{v-u}$$

and decreases until it reaches $a(u)$ in finite time.

This inert evolution is governed by the feedback:

$$\widetilde{v}(x) := u + \sqrt{2\log\left(\frac{x}{a(u)}\right)}.$$

7.3.2 Heavy Evolutions

Let $c > 0$ an *inertia threshold*. Heavy *evolutions $x_c(\cdot)$ are obtained when the absolute value $|w(t)| := |v'(t)|$ of the velocity $w(t) := v'(t)$ of the regulon is minimized at each instant. In particular, whenever the velocity of the regulon is equal to 0, the regulon is kept constant, and if not, it changes as slowly as possible.*

The "heaviest" evolutions are thus obtained by constant regulons. This is not always possible, because, by taking $v > u$ for instance, the solution $x(t) = xe^{-(v-u)t}$ is viable for $t \leq \frac{\log\left(\frac{x}{a(u)}\right)}{v-u}$. At that time, the regulon should be changed immediately (with infinite velocity) to any regulon $v \leq u$. This brutal and drastic measure – which is found in many natural systems – is translated in mathematics by impulse control.

In order to avoid such abrupt changes of regulons, *we add the requirement that the velocity of the regulons is bounded by a velocity bound $c > \beta_u(x,v)$*.

Starting from (x,v), the state $x_c(\cdot)$ of an heavy evolution evolves according

$$x_c(t) = xe^{-(v-u)t}$$

and reaches a at time $\frac{\log\left(\frac{x}{a(u)}\right)}{v-u}$.

The inertia function β_u provides the velocity of the regulons and increases over the heavy evolution according to

$$\forall t \in \left[0, \frac{\log\left(\frac{x}{a(u)}\right)}{v-u}\right], \; \beta_u(x_c(t), v) = \frac{(v-u)^2}{2\left(\log\left(\frac{x}{a(u)}\right) - (v-u)t\right)}$$

The derivatives of the inertia function over the inert evolutions are equal to

$$\frac{d\beta_u(x_c(t), v)}{dt} = \frac{(v-u)^3}{2\left(\log\left(\frac{x}{a(u)}\right) - (v-u)t\right)^2}$$

The inertia function reaches the given velocity limit $c > \beta_u(x, v)$ at

$$\begin{cases} \textbf{warning state } \xi(c, u) = a(u)e^{\frac{(u-v)^2}{2c}} \\ \textbf{warning time } \sigma_c(x, v) := \frac{\log\left(\frac{x}{a(u)}\right)}{v-u} - \frac{v-u}{2c} \end{cases}$$

Hence, once a velocity limit c is fixed, the heavy solution evolves with constant regulon v until the last instant $\sigma_c(x, v)$ when the state reaches $\xi_c(v)$ and the velocity of the regulon $\beta_u(\xi_c(v), v) = c$. *This is the last time when the regulon remains constant* and has to changed by taking

$$v_c(t) = v - c\left(t - \frac{\log\left(\frac{x}{a(u)}\right)}{v-u} + \frac{v-u}{2c}\right)$$

Then the evolution $(x_c(\cdot), v_c(\cdot))$ follows the inert solution starting at $(\xi_c(v),)c$. It reaches equilibrium $(a(u), u)$ at time

$$t^\star := \frac{\log\left(\frac{x}{a(u)}\right)}{v-u} + \frac{v-u}{2c}$$

Taking $x(t) \equiv a(u)$ and $v(t) \equiv u$ when $t \geq t^\star$, the solution may remain at $a(u)$ forever.

For a given inertia bound $c > \beta_u(x, v)$, the *heavy evolution* $(x_c(\cdot), v_c(\cdot))$ is associated with the heavy feedback $\widetilde{v_c}$ defined by

$$\widetilde{v_c}(x) := \begin{cases} v & \text{if } \xi(c, u) \leq x \\ u + \sqrt{2c\log\left(\frac{x}{a(u)}\right)} & \text{if } a(u) \leq x \leq \xi(c, u) \end{cases}$$

7.3.3 The Crisis Function

If we regard $[a,b] \times \mathbb{R}$ as an ecological environment and Graph(V) as an economical sub-environment (instead of a target), the *crisis function* (see Definition 9.2.6, p. 327) assigns to any (prey-predator) state (x,v) the smallest time spent outside the economical sub-environment by an evolution starting from it. It is equal to 0 in the viability kernel of its environment, finite on the *permanence kernel* (see Definition 9.2.3, p. 323) and infinite outside.

The Inert-Schaeffer Metasystem $x'(t) = x(t)\left(\sqrt{\alpha}\sqrt{2\log(\frac{b}{x(t)})} - v(t)\right)$ modelling the evolution of renewable resources depleted by an economic activity $v(t)$. The velocities $|v'(t)| \leq d$ of economic activity are bounded by a constant d. The environment $\left\{(x,v) \in [a,b] \times [0,\overline{v}] \mid v \in \left[\frac{C}{\gamma x - c}, \overline{v}\right]\right\}$ translates economic constraints.

1. The viability kernel of the target (*Zone 1*): the level of economic activity v is not in crisis and can be regulated in order to maintain the viability of both the biological and economical systems, i.e., the survival of fishes and fisheries,
2. The permanence kernel, made of the union of (*Zones 1, 2 and 5*), where economic activities are no longer necessarily viable, but the fishes will survive: (*Zone 2*): The level of economic activity v is not in crisis now, but the enterprize will eventually bankrupt, whereas the fishes survive. (*Zone 5*): The economy is in a recoverable crisis, from which it is possible to go back to the viability kernel of the target (*Zone 1*).
3. The complement of the permanence kernel, made of the union of (*Zone 3 and 4*) is the zone from which both economic activity and renewable resources will eventually die out: (*Zone 3*): The level of economic activity v is not in crisis now, but will end up in a nonrecoverable crisis leading to (*Zone 4*): The economy is bankrupted, and fishes eventually disappear.

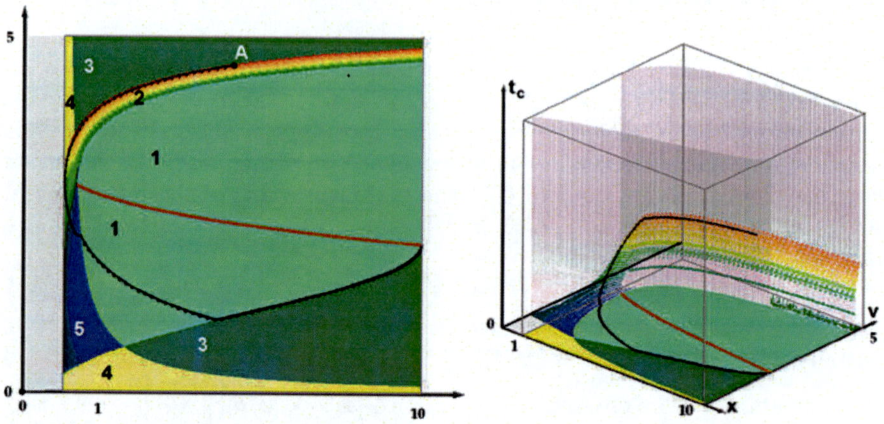

Fig. 7.8 Crisis Function.
Crisis Function under the Inert-Schaeffer Metasystem.

The *left figure* represents the graph of the crisis function, equal to zero on the viability kernel (*Zone 1*), taking finite values at states on the permanence kernel (*Zones 2 and 5*) and infinite values elsewhere (*Zones 3 and 4*), from which it is impossible to reach the environment.

Note that viability algorithm provides the computation of the regulation policy $v(\cdot)$. It can be used to compute particular evolutions. Starting from A in *Zone 2*, the economy ends up in a crisis *Zone 5*, coming back to *Zone 1* (viability kernel if the economic environment).

The case when the feedback is the Verhulst feedback is presented in the Introduction.

Part II
Mathematical Properties of Viability Kernels

Chapter 8
Connection Basins

8.1 Introduction

Until now, we presented and studied evolutions in positive time, or *forward evolutions*, and the associated concepts of (forward) viability kernels and basins. We were looking from the present to the future, without taking into account the past or the history of the evolution. In this chapter, we offer ways to extend this framework by studying evolutions from $-\infty$ to $+\infty$.

In order to achieve this goal, we have to properly define the meaning of an evolution arriving at a final time instead of evolutions starting at AN initial time. In other words, we consider not only the future, as we did until now, but also the past, and *histories*, defined as evolutions in the past.

For that purpose, we shall split in Sect. 8.2, p. 275 the evolution in forward time evolutions and backward time evolutions, or in negative times, going from 0 to $-\infty$. So, in Sect. 8.3, p. 279, we investigate the concept of bilateral viability kernel of an environment, the subset of initial states through which passes at least one evolution viable in this environment.

For instance, up to now, we have only considered *capture basins* of targets C viable in K, which are the subsets of initial states in K from which starts at least one evolution viable in K until it reaches the *target* C in finite time. In Sect. 8.4, p. 284, we shall consider the "backward" case when we consider another subset $B \subset K$, regarded as a *source*, and study the "reachable maps" from the source, subsets of final states in K at which arrives in finite time at least one evolution *viable* in K starting from the *source* B. This leads us to the concepts of reachable maps, detection tubes and Volterra inclusions in Sect. 8.4, p. 284.

Knowing how to arrive from a source and to reach a target, we are led to consider jointly problems when *in the same time* a source B and a target C are given: we shall introduce in Sect. 8.5, p. 291 the *connection basin*, the subset of elements $x \in K$ through which passes at least one viable evolution starting from the source B and arriving in finite time at target C. As a

J.-P. Aubin et al., *Viability Theory*, DOI 10.1007/978-3-642-16684-6_8,

particular case, this leads us to consider evolutions connecting in finite time a state y to another state z by viable evolutions. The issue arises to select one such connecting evolution by optimizing an intertemporal criterion, as we did in Chap. 4, p. 125. For instance, in minimal time, this is the brachistochrone problem, or in minimal length, this is the problem of (viable) geodesics. It turns out that such optimal solutions can be obtained following a strategy going back to Eupalinos 2,500 years ago: start at the same time from both the initial and final states until the two evolutions meet in the middle. This is the reason we attribute the name of this genius to the optimization of evolutions connecting two states.

Actually, this is a particular case of the collision problem studied in Sect. 8.6, p. 298. In this case, two evolutions governed by two different evolutionary systems starting from two different point must collide at some future time. The set of pairs of initial states from which start two colliding evolutions is the collision kernel. Knowing it, we can select among the colliding evolutions the ones which optimize an intertemporal criteria.

We studied connection basins from a source to a target, but, if we regard them as two "cells", one initial, the other final, among a sequence of other cells, we investigate in Sect. 8.8, p. 302 how an evolution can visit a sequence of cells in a given order (see *Analyse qualitative*, [85, Dordan]).

We present here the important results of *Donald Saari* dealing with this issue. Actually, once the "visiting kernels" studied, we adapt Saari's theorems to the case of evolutionary systems. Given a finite sequence of cells, given any arbitrary infinite sequence of orders of visits of the cell, under Saari's assumption, one can always find one initial state from which at least one evolution will visit these cells in prescribed order. This is not a very quite and stable situation, which is another mathematical translation of the polysemous word "chaos", in the sense that "everything can happen". One can translate this vague "everything" in the following way: each cell is interpreted as *qualitative cell.* It describes the set of states sharing a given property characterizing this cell. In comparative economics as well as in qualitative physics, the issue is not so much to know precisely one evolution, but rather, to know what would be the qualitative consequences of at least one evolution starting from a given cell. If this cell is a viability kernel, it can be regarded as "qualitative equilibrium", because at least one evolution remains in the cell. Otherwise, outside of its viability kernel, all evolutions will leave this cell to enter another one. This fact means that the first cell is a qualitative cause of the other one. Saari's Theorem states that under its assumptions, whatever the sequence of properties, there is one possibility that each property "implies" the next one, in some weak but rigourously defined sense. Section 8.9, p. 305, explains how the concepts of invariance kernels and capture basin offer a conceptual basis for "non consistent" logics, involving time delays and some indeterminism. This issue is just presented but not developed in this book.

For simplicity, we assumed until now that the systems were time independent. This is not reasonable, and the concepts which we have met all along this book should hold true for time dependent systems, constraints and targets. This is the case, naturally, thanks a well-known standard "trick" allowing us to treat time-dependent evolutionary systems as time-independent ones. It consists in introducing an auxiliary "time variable" whose velocity is equal to one. So, Sect. 8.10, p. 309 is devoted to the implementation of this procedure, especially regarding capture basins and detection tubes.

Section 8.11, p. 316 grasps the following question: how much information about the current state is contained in past measurements of the states when the initial conditions are not accessible, but replaced by some observations on the past. The question boils down to this one: knowing a control system and a tube (obtained, for instance, as the set of states whose measures are at each instant in a set-valued map), can we recover the evolutions governed by this evolutionary system and *satisfying these past observations?* This is a question which motivated the concept of detector in a time dependent context, and which offers, as the Volterra inclusions (see Sect. 8.4.3, p. 289), quite interesting perspectives for future research on evolutions governed by an evolutionary system when the initial state is unknown.

8.2 Past and Future Evolutions

Until now, evolutions $x(\cdot) \in \mathcal{C}(0,+\infty;X)$ were meant to be "future" evolutions starting from $x(0)$ at "present" time 0, regarded as an initial time.

In order to avoid duplicating proofs of results, the idea is to split the "full evolution" $x(\cdot) \in \mathcal{C}(-\infty,+\infty;X)$ into two (future) evolutions:

1. the *"backward part"* $\overleftarrow{x}(\cdot) \in \mathcal{C}(0,+\infty;X)$ defined by

$$\forall\, t \geq 0, \;\; \overleftarrow{x}(t) := x(-t) \in$$

2. the *"forward part"* $\overrightarrow{x}(\cdot) \in \mathcal{C}(0,+\infty;X)$ defined by

$$\forall\, t \geq 0, \;\; \overrightarrow{x}(t) := x(t)$$

both defined on positive times. Observe that, for negative times,

$$\forall\, t \leq 0, \;\; x(t) \; = \; \overrightarrow{x}(-t) \tag{8.1}$$

Conversely, knowing the forward part $\overrightarrow{x}(\cdot)$ and backward part $\overleftarrow{x}(\cdot)$ of a full evolution, we recover it by formula

$$x(t) \; := \begin{cases} \overleftarrow{x}(-t) \text{ if } t \leq 0 \\ \overrightarrow{x}(+t) \text{ if } t \geq 0 \end{cases} \tag{8.2}$$

The symmetry operation $x(\cdot) \mapsto \varsigma(x(\cdot))(\cdot)$ defined by

$$\varsigma(x(\cdot))(t) = x(-t)$$

is a bijection between the spaces $\mathcal{C}(0,+\infty;X)$ of evolutions and $\mathcal{C}(-\infty,0;X)$ of histories, as well as a bijection $x(\cdot) \in \mathcal{C}(-\infty,+\infty;X) \mapsto \varsigma(x(\cdot))(\cdot) \in \mathcal{C}(-\infty,+\infty;X)$. It is obviously idempotent: $\varsigma(\varsigma(x(\cdot)))(\cdot) = x(\cdot)$. Note that the symmetry operation is also denoted by $\overset{\vee}{x}(\cdot) := \varsigma(x(\cdot))(\cdot)$.

Definition 8.2.1 ***[Histories, Past and Future]*** *Functions* $x(\cdot) \in \mathcal{C}(-\infty,+\infty;X)$ *are called* "full evolutions". *We reserve the term of (future)* "evolutions" *for functions* $x(\cdot) \in \mathcal{C}(0,+\infty;X)$. *The space* $\mathcal{C}(-\infty,0;X)$ *is the space of* "histories". The history of a full evolution is the symmetry of its backward part and the backward part is the symmetry of its history.

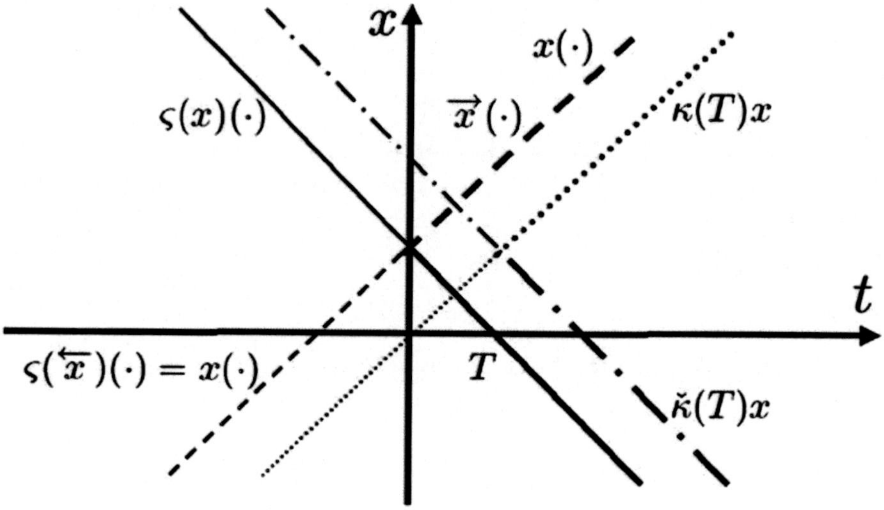

Fig. 8.1 Operations on Evolutions.
Symmetry and decomposition of a full evolution in the its backward and forward parts.

Recall that the translation $\kappa(T)x(\cdot) : \mathcal{C}(-\infty,+\infty;X) \mapsto \mathcal{C}(-\infty,+\infty;X)$ of an evolution $x(\cdot)$ is defined by $(\kappa(T)x(\cdot))(t) := x(t-T)$ (see Definition 2.8.1, p. 69). It is a translation to the right if T is positive and to the left if T is negative, satisfying $\kappa(T+S) = \kappa(T) \circ \kappa(S) = \kappa(S) \circ \kappa(T)$.

Definition 8.2.2 ***[Backward Shift of an Evolution]*** *The* T-backward shift operator $\overset{\vee}{\kappa}(T)$ *associates with any evolution* $x(\cdot)$ *its* T-backward shift evolution $\overset{\vee}{\kappa}(T)(x(\cdot))$ *defined by*

$$\forall t \in \mathbb{R},\ \overset{\vee}{\kappa}(T)(x(\cdot))(t) := x(T-t) \tag{8.3}$$

It is easy to observe that the operator $\overset{\vee}{\kappa}(T)$ the operator is idempotent:

$$\forall x(\cdot) \in \mathcal{C}(-\infty,+\infty;X),\ (\overset{\vee}{\kappa}(T)(\overset{\vee}{\kappa}(T)x(\cdot))) = x(\cdot)$$

and that $\overset{\vee}{\kappa}(T) := \kappa(T) \circ \varsigma = \varsigma \circ \kappa(-T)$.

Let $x(\cdot) : \mathbb{R} \mapsto X$ be a full evolution. Then for all $T \geq 0$, the restriction to $]-\infty,0]$ of the translation $\kappa(-T)(x(\cdot))(\cdot) \in \mathcal{C}(-\infty,0;X)$ can be regarded as *"encoding the history of the full evolution up to time T of the evolution $x(\cdot)$"*. The space $\mathcal{C}(-\infty,0;X)$ allows us to study the evolution of history dependent (or path dependent) systems governing the evolution $T \mapsto \kappa(-T)x(\cdot) \in \mathcal{C}(-\infty,0;X)$ of histories of evolutions. The terminology "path-dependent" is often used, in economics, in particular, but inadequately in the sense that paths are trajectories of evolutions.

A "full" evolutionary system $\mathcal{S} : X \mapsto \mathcal{C}(-\infty,+\infty;X)$ associates with any $x \in X$ an evolution $x(\cdot) \in \mathcal{C}(-\infty,+\infty;X)$ passing through x at time 0. Its backward system : $X \mapsto \mathcal{C}(-\infty,+\infty;X)$ is defined by

$$\overleftarrow{\mathcal{S}}(x) := \{\overleftarrow{x}(\cdot)\}_{x(\cdot)\in\mathcal{S}(x)}$$

We observe that for all $x(\cdot) \in \mathcal{S}(x)$,

$$\forall\, t \leq 0,\ x(t) = \overleftarrow{x}(-t) \text{ where } \overleftarrow{x}(\cdot) \in \overleftarrow{\mathcal{S}}(x)$$

Splitting evolutions allows us to decompose a full evolution passing through a given state at a given time into its backward and forward parts both governed by backward and forward evolutionary systems:

In particular, consider the system

$$\begin{cases} (i)\ x'(t) = f(x(t),u(t)) \\ (ii)\ u(t) \in U(x(t)) \end{cases} \tag{8.4}$$

The *backward system* $\overleftarrow{\mathcal{S}} : X \rightsquigarrow \mathcal{C}(0,+\infty;X)$ associates with any x the set of evolutions $\overleftarrow{x}(\cdot) \in \mathcal{C}(0,+\infty;X)$ governed by system

$$\begin{cases} (i)\ \overleftarrow{x}'(t) = -f(\overleftarrow{x}(t),\overleftarrow{u}(t)) \\ (ii)\ \overleftarrow{u}(t) \in U(\overleftarrow{x}(t)) \end{cases} \tag{8.5}$$

Lemma 8.2.3 *[Splitting Full Evolutions of a Control System] We denote by $\mathcal{S}(T,x)$ the subset of full solutions $x(\cdot)$ to system (8.4) passing through x at time T. We observe that a full evolution $x(\cdot)$ belongs to $\mathcal{S}(T,x)$ if and only if:*

1. *its forward part* $\overrightarrow{x}(\cdot) := (\kappa(T)(x(\cdot)))(\cdot)$ *at time T defined by* $\kappa(T)(x(\cdot))(t) = x(t-T)$ *is a solution to*
$$\overrightarrow{x}'(t) = f(\overrightarrow{x}(t), \overrightarrow{u}(t)) \text{ in which } \overrightarrow{u}(t) \in U(\overrightarrow{x}(t))$$
satisfying $\overrightarrow{x}(0) = x$,
2. *its backward part* $\overleftarrow{x}(\cdot) := (\overset{\vee}{\kappa}\ (T)x(\cdot))(\cdot)$ *at time T defined by* $(\overset{\vee}{\kappa}\ (T)x(\cdot))(t) = x(T-t)$ *is a solution to differential inclusion*
$$\overleftarrow{x}'(t) = -f(\overleftarrow{x}(t), \overleftarrow{u}(t)) \text{ where } \overleftarrow{u}(t) \in U(\overleftarrow{x}(t))$$
satisfying $\overleftarrow{x}(0) = x$.

Therefore, the full evolution $x(\cdot) \in \mathcal{S}(T,x)$ can be recovered from its backward and forward parts by formula

$$x(t) = \begin{cases} \overleftarrow{x}(T-t) \ \ (= (\overset{\vee}{\kappa}\ (T)\overleftarrow{x})(t)) \text{ if } t \leq T \\ \overrightarrow{x}(t-T) \ \ (= (\kappa(T)\overrightarrow{x})(t)) \text{ if } t \geq T \end{cases}$$

As a general rule in this chapter, all concepts introduced in the previous chapters (viable or locally viable evolutions, viability and invariance kernels, capture and absorption basins dealing with the forward part $\overrightarrow{x}(\cdot)$ of an evolution governed by the evolutionary system $\overrightarrow{\mathcal{S}}$ will be qualified of "forward" and those dealing with the backward part $\overleftarrow{x}(\cdot)$ governed by the backward system will be qualified of "backward", taking into account that both forward and backward evolutions are defined on $[0,+\infty[$.

As an example, we single out the concepts backward viability or invariance:

Definition 8.2.4 *[Backward Viability and Invariance] We shall say that a subset K is* backward viable (resp. invariant) *under $\mathcal{S}$ if for every $x \in K$, at least one backward evolution (resp. all backward evolutions) $\overleftarrow{x}(\cdot)$ starting from x is (resp. are) viable in K, or, equivalently, at least one evolution $x(\cdot)$ arriving at $x = x(t)$ at some finite time $t \geq 0$ is (resp. all evolutions are) viable in K.*

8.3 Bilateral Viability Kernels

Definition 8.3.1 *[Bilateral Viability Kernel] Let $B \subset K$ be a subset regarded as a source, $C \subset K$ be a subset regarded as a target and $\mathcal{S}$ be an evolutionary system. The* backward viability kernel $\text{Viab}_{\overleftarrow{\mathcal{S}}}(K,B)$ *(respectively, the* backward capture basin $\text{Capt}_{\overleftarrow{\mathcal{S}}}(K,B)$*) is the viability kernel (resp. capture basin) under the backward system $\overleftarrow{\mathcal{S}}$. The* bilateral viability kernel

$$\overleftrightarrow{\text{Viab}}_{\mathcal{S}}(K,(B,C)) \;=\; \text{Viab}_{\overleftarrow{\mathcal{S}}}(K,B) \cap \text{Viab}_{\mathcal{S}}(K,C)$$

of K between a source B and a target C is the subset of states $x \in K$ such that there exists one evolution $x(\cdot) \in \mathcal{C}(-\infty,+\infty;X)$ passing through $x = x(0)$ at time 0 and two times $\overleftarrow{T} \in [0,+\infty]$ and $\overrightarrow{T} \in [0,+\infty]$ such that $x(\cdot)$ is

1. *viable in K on $]-\infty,+\infty[$,*
2. *or viable in K on $[-\overleftarrow{T},+\infty[$, with $x(-\overleftarrow{T}) \in B$,*
3. *or viable in K on $]-\infty,+\overrightarrow{T}]$ with $x(\overrightarrow{T}) \in C$,*
4. *or viable in K on $[-\overleftarrow{T},+\overrightarrow{T}]$ with $x(-\overleftarrow{T}) \in B$ and $x(\overrightarrow{T}) \in C$.*

When $B = \emptyset$ and $C = \emptyset$ are empty,

$$\overleftrightarrow{\text{Viab}}_{\mathcal{S}}(K) := \overleftrightarrow{\text{Viab}}_{\mathcal{S}}(K,(\emptyset,\emptyset))$$

is the set of elements $x \in K$ through which passes one evolution at time 0 viable on $]-\infty,+\infty[$, called the bilateral viability kernel *of K.*

Observe that a closed subset K connecting a closed source B to a closed target C is both forward locally viable on $K \setminus C$ and backward locally viable on $K \setminus B$ under the evolutionary system (see Definition 2.13.1, p. 94 and Proposition 10.5.2, p. 400).

The *viability partition* of the environment K is made of the four following subsets:

- the bilateral viability kernel

$$\overleftrightarrow{\text{Viab}}_{\mathcal{S}}(K,(B,C)) \;=\; \text{Viab}_{\overleftarrow{\mathcal{S}}}(K,B) \cap \text{Viab}_{\mathcal{S}}(K,C)$$

- the complement $\text{Viab}_{\overleftarrow{\mathcal{S}}}(K,B) \backslash \text{Viab}_{\mathcal{S}}(K,C)$ of the forward viability kernel in the backward viability kernel,
- the complement $\text{Viab}_{\mathcal{S}}(K,C) \backslash \text{Viab}_{\overleftarrow{\mathcal{S}}}(K,B)$ of the backward viability kernel in the forward viability kernel,

- the complement $K \setminus (\text{Viab}_{\overleftarrow{S}}(K,B) \cup \text{Viab}_S(K,C))$.

The following statement describes the viability properties of evolutions starting in each of the subsets of this partition:

Theorem 8.3.2 ***[The Viability Partition of an Environment]*** *Let us consider the viability partition of the environment K under an evolutionary system S:*

- *The bilateral viability kernel $\overleftrightarrow{\text{Viab}}_S(K,(B,C))$ is the set of initial states such that at least one evolution passing through it is bilaterally viable in K outside of B and C.*
- *The subset $\text{Viab}_{\overleftarrow{S}}(K,B)\setminus\text{Viab}_S(K,C)$ is the subset of initial states x from which* all evolutions $x(\cdot) \in S(x)$ *leave K in finite time $\tau_K(x(\cdot)) :=$* $\inf\{t \mid x(t) \notin K\}$ *and are viable in $\text{Viab}_{\overleftarrow{S}}(K,B)\setminus\text{Viab}_S(K,C)$ on the finite interval $[0, \tau_K(x(\cdot))]$.*
- *The subset $\text{Viab}_S(K,C)\setminus\text{Viab}_{\overleftarrow{S}}(K,B)$ is the subset of initial states x from which* all backward evolutions $\overleftarrow{x}(\cdot) \in \overleftarrow{S}(x)$ *passing through x enter K in finite time $\tau_K(\overleftarrow{x}(\cdot)) :=$* $\inf\{t \mid \overleftarrow{x}(t) \notin K\}$ *and are viable in $\text{Viab}_S(K,C)\setminus\text{Viab}_{\overleftarrow{S}}(K,B)$ on the finite interval $[0, \tau_K(\overleftarrow{x}(\cdot))]$ (see property (10.5.5)(iii), p. 410 of Theorem 2.15.4, p. 101).*
- *The set $K \setminus (\text{Viab}_{\overleftarrow{S}}(K,B) \cup \text{Viab}_S(K,C))$ is the subset of initial states x such that all evolutions passing through x are viable in $K \setminus (\text{Viab}_{\overleftarrow{S}}(K,B) \cup \text{Viab}_S(K,C))$ from the finite instant when it enters K to the finite instant when it leaves it.*

If furthermore, the subset $\text{Viab}_{\overleftarrow{S}}(K,B)\setminus\text{Capt}_S(K,C) \subset \text{Int}(K)$, then it is forward invariant.

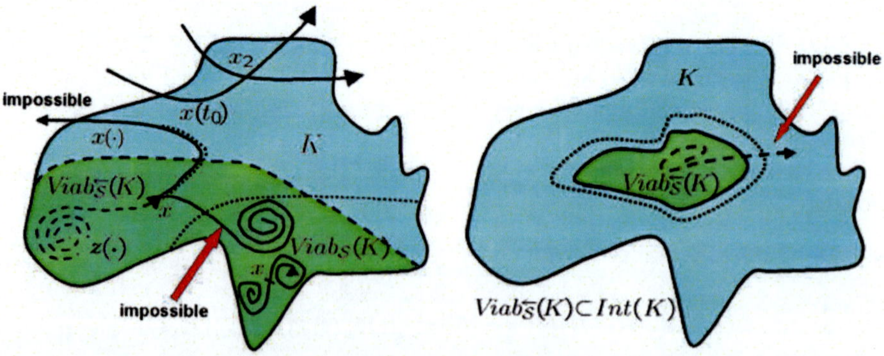

Fig. 8.2 Illustration of the Proof of Theorem 8.3.2.
Left. Proof of parts (2) and (4) of Theorem 8.3.2. *Right.* Proof of the last statement.

Proof. The first statement is obvious and the second and third ones are symmetric. Let us prove the second and fourth ones.

1. Let x belong to $\text{Viab}_{\overleftarrow{\mathcal{S}}}(K,B)\backslash\text{Viab}_{\mathcal{S}}(K,C)$ and $\overrightarrow{x}(\cdot) \in \mathcal{S}(x)$ be any evolution starting at x. It is viable in $\text{Viab}_{\overleftarrow{\mathcal{S}}}(K,B)\backslash\text{Viab}_{\mathcal{S}}(K,C)$ until it *must* leave $\text{Viab}_{\overleftarrow{\mathcal{S}}}(K,B)$ at some time $t^\star \leq \tau_K(\overrightarrow{x}(\cdot))$, where $\tau_K(\overrightarrow{x}(\cdot)) := \inf\{t \mid \overrightarrow{x}(t) \notin K\}$ (which is finite because x does not belong to the forward viability kernel of K with target C). We observe that actually $\tau_K(\overrightarrow{x}(\cdot)) = \tau_K^\sharp(x)$. Otherwise, there would exist t_0 such that $\tau_K(\overrightarrow{x}(\cdot)) < t_0 \leq \tau_K^\sharp(x)$ where $\overrightarrow{x}(t_0) \in K\backslash\text{Viab}_{\overleftarrow{\mathcal{S}}}(K,B)$. Let $\overleftarrow{z}(\cdot) \in \overleftarrow{\mathcal{S}}(x)$ be a backward evolution starting at x and viable in K on $[0,\overleftarrow{T}[$, where $\overleftarrow{T}$ is either infinite or finite, and in this case, $\overleftarrow{z}(\overleftarrow{T}) \in B$. Such an evolution exists since x belongs to the backward viability kernel $\text{Viab}_{\overleftarrow{\mathcal{S}}}(K,B)$. The evolution $\overleftarrow{y}(\cdot)$ defined by
$$\overleftarrow{y}(t) := \begin{cases} \overrightarrow{x}(t_0 - t) \text{ if } t \in [0,t_0] \\ \overleftarrow{z}(t - t_0) \text{ if } t \in [t_0, \overleftarrow{T}] \end{cases}$$
would be a viable evolution of the backward evolutionary system starting at $\overleftarrow{y}(0) = \overrightarrow{x}(t_0) \in K\backslash\text{Viab}_{\overleftarrow{\mathcal{S}}}(K,B)$. This would imply that $\overrightarrow{x}(t_0)$ belongs to the backward viability kernel, a contradiction. Hence $\overrightarrow{x}(\cdot)$ is viable in $\text{Viab}_{\overleftarrow{\mathcal{S}}}(K,B)\backslash\text{Viab}_{\mathcal{S}}(K,C)$ on the finite interval $[0,\tau_K(\overrightarrow{x}(\cdot))]$.
2. For the fourth subset of the viability partition, take any evolution $x(\cdot) \in \mathcal{S}(x)$. Let us set $S := \tau_K(\overleftarrow{x}(\cdot))$ and $T := \tau_K(x(\cdot))$. Then $x(\cdot)$ enters K in finite time $-S$, passes through x at time 0 and leaves K in finite time T. Its translation $y(\cdot) := (\kappa(S)x(\cdot))(\cdot) \in \mathcal{S}(x(-S))$ defined by $y(t) := x(t-S)$ is viable in the complement of $\text{Viab}_{\mathcal{S}}(K,C)$ until it leaves K at time $T+S$. Then it is viable in the complement of $\text{Viab}_{\mathcal{S}}(K,C)$. In the same way, the evolution $\overleftarrow{z}(\cdot) := (\overset{\vee}{\kappa}(T)x(\cdot))(\cdot) \in \overleftarrow{\mathcal{S}}(x(T))$ defined by $\overleftarrow{z}(t) := x(T-t)$ is viable in the complement of $\text{Viab}_{\overleftarrow{\mathcal{S}}}(K,B)$. Then the evolution $x(\cdot) = (\kappa(-S)x)(\cdot) = (\overset{\vee}{\kappa}(T)\overleftarrow{z}(\cdot))(\cdot)$ is viable both in the complement of $\text{Viab}_{\mathcal{S}}(K,C)$ and in the complement of $\text{Viab}_{\overleftarrow{\mathcal{S}}}(K,B)$, and thus, in the complement $K \setminus (\text{Viab}_{\overleftarrow{\mathcal{S}}}(K,B) \cup \text{Viab}_{\mathcal{S}}(K,C))$.
3. If $\text{Viab}_{\overleftarrow{\mathcal{S}}}(K,B)\backslash\text{Capt}_{\mathcal{S}}(K,C) \subset \text{Int}(K)$, then $\text{Viab}_{\overleftarrow{\mathcal{S}}}(K,B)\backslash\text{Capt}_{\mathcal{S}}(K,C)$ is forward invariant. Indeed, any evolution $\overrightarrow{x}(\cdot) \in \mathcal{S}(x)$ starting at $x \in \text{Viab}_{\overleftarrow{\mathcal{S}}}(K,B) \setminus \text{Capt}_{\mathcal{S}}(K,C)$ is viable in $\text{Viab}_{\overleftarrow{\mathcal{S}}}(K,B) \setminus \text{Capt}_{\mathcal{S}}(K,C)$. Otherwise, there would exist $t_0 > 0$ such that $\overrightarrow{x}(t_0) \in \text{Int}(K)\backslash\text{Viab}_{\overleftarrow{\mathcal{S}}}(K,B)$ because $\overrightarrow{x}(t) \notin \text{Capt}_{\mathcal{S}}(K,C)$ since $\text{Capt}_{\mathcal{S}}(K,C)$ is isolated. Associating with it the backward evolution $\overleftarrow{y}(\cdot)$ defined above, we would deduce that $\overrightarrow{x}(t_0) \in \text{Viab}_{\overleftarrow{\mathcal{S}}}(K,B)$, a contradiction. Therefore, $\text{Viab}_{\overleftarrow{\mathcal{S}}}(K,B)\backslash\text{Capt}_{\mathcal{S}}(K,C)$ is forward invariant, and thus, contained in $\text{Inv}_{\mathcal{S}}(K \setminus \text{Capt}_{\mathcal{S}}(K,C))$. □

We shall use the following consequence for localizing attractors, and in particular, Lorenz attractors (see Theorem 9.3.12, p. 352):

Proposition 8.3.3 *[Localization of Backward Viability Kernels] If the backward viability kernel of K is contained in the interior of K, then it is forward invariant and thus, contained in the invariance kernel of K.*

Proof. Take $B = C = \emptyset$, the statement ensues. □

8.3.1 Forward and Backward Viability Kernels under the Lorenz System

Usually, the attractor, defined as the union of limit sets of evolutions, is approximated by taking the union of the "tails of the trajectories" of the solutions that provides an illustration of the shape of the attractor, *although it is not the attractor.* Here, we use the viability kernel algorithm for computing the backward viability kernel, *which contains the attractor.*

Let us consider the Lorenz system (2.6), p. 57

$$\begin{cases} (i) \quad x'(t) = \sigma y(t) - \sigma x(t) \\ (ii) \quad y'(t) = rx(t) - y(t) - x(t)z(t) \\ (iii) \; z'(t) = \; x(t)y(t) - bz(t) \end{cases}$$

(see of Sect. 2.4.2, p. 56).

Figure 8.3, p. 283 displays a numerical computation of the forward and backward viability kernel of the cube $K := [-30, +30] \times [-30, +30] \times [-0, +53]$ and of the cylinder $C := \{(x, y, z) \in [-100, +100] \times [-100, +100] \times [-20, +80] \mid y^2 + (z - r)^2 \leq 35^2\}$.

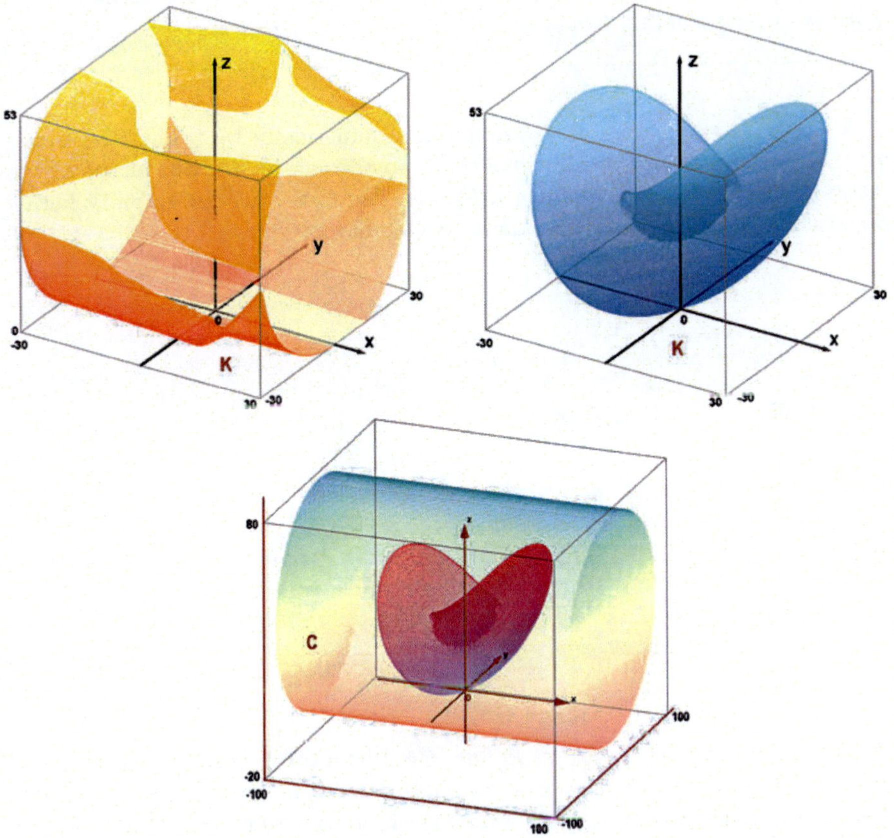

Fig. 8.3 Example of viability kernels for the forward and backward Lorenz systems when $\sigma > b + 1$.
Up: *The figure displays both the forward viability kernel of the cube* $K := [-30, +30] \times [-30, +30] \times [-0, +53]$ *(left) and the backward viability kernel contained in it (right).* ***Down:****. The figure displays both the forward viability kernel of the cylinder* $C := \{(x, y, z) \in [-100, +100] \times [-100, +100] \times [-20, +80] \mid y^2 + (z - r)^2 \leq 35^2\}$ *which coincides with C itself, meaning that C is a viability domain, and the backward viability kernel contained in it which coincides with the backward viability kernel of the upper figure. Indeed, Proposition 8.3.3, p. 282 states that if the backward viability kernel is contained in the interior of K, the backward viability kernel is also contained in the forward viability kernel. The famous attractor is contained in the backward viability kernel.*

Since the backward viability kernel is contained in the interior of K, Proposition 8.3.3, p. 282 implies the following consequence:

Corollary 8.3.4 *[Lorenz Attractor] The Lorenz limit set is contained in this backward viability kernel.*

Usually, the attractor is approximated numerically by taking the union of the trajectories of the solutions that provides an idea of the shape of the attractor, *although it is not the attractor*. Here, we use the viability kernel algorithm for computing the backward viability kernel.

8.4 Detection Tubes

8.4.1 Reachable Maps

When $f : X \mapsto X$ is a Lipschitz single-valued map, it generates a deterministic evolutionary system $\mathcal{S}_f : X \mapsto \mathcal{C}(0, +\infty; X)$ associating with any initial state x the (unique) solution $x(\cdot) = \mathcal{S}_f(x)$ to the differential equation $x'(t) = f(x(t))$ starting at x. The single-valued map $t \mapsto \mathcal{S}_f(x)(t) = \{x(t)\}$ from $\mathbb{R}_+ \mapsto X$ is called the *flow* or *semi-group* associated with f. A flow exhibits the *semi-group property*

$$\forall t \geq s \geq 0, \;\; \mathcal{S}_f(x)(t) \;=\; \mathcal{S}_f(\mathcal{S}_f(x)(s))(t-s)$$

For deterministic systems, studying the dynamical system amounts to studying its associated flow or semi-group, even though when they are not necessarily associated with a dynamical system. Although this will no longer be the case for nondeterministic evolutionary systems, it is worth introducing the semigroup analogues, called "reachable maps" in the control literature:

Definition 8.4.1 *[Reachable Maps and Tubes] Let $\mathcal{S} : X \rightsquigarrow \mathcal{C}(0, +\infty; X)$ be an evolutionary system and $B \subset K \subset X$ be a source contained in the environment. Recall that $\mathcal{S}^K : K \rightsquigarrow \mathcal{C}(0, \infty; K)$ denotes the evolutionary system associating with any initial state $x \in K$ the subset of evolutions governed by $\mathcal{S}$ starting at x viable in K. The* reachable map (or set-valued flow) $\text{Reach}_{\mathcal{S}}^K(\cdot; x)$ *viable in K is defined by*

$$\forall\, x \in X, \; \forall t \geq 0, \; \text{Reach}_{\mathcal{S}}^K(t; x) := \{x(t)\}_{x(\cdot) \in \mathcal{S}^K(x)}$$

When $K := X$ is the whole space, we set $\text{Reach}_{\mathcal{S}}(t; x) := \text{Reach}_{\mathcal{S}}^K(t; x)$. *We associate with the source B the* (viable) reachable tube $t \rightsquigarrow$

$\text{Reach}_{\mathcal{S}}^{K}(t;B)$ *defined by*

$$\text{Reach}_{\mathcal{S}}^{K}(t;B) := \left\{\text{Reach}_{\mathcal{S}}^{K}(t;x)\right\}_{x\in B}$$

For simplifying the notations, we may drop the lower index $\mathcal{S}$ in the notation of reachable tubes, and mention it only when several systems are considered (the system and the backward system, for example).

We obtain the following properties:

Proposition 8.4.2 *[The Semi-Group Property] The reachable map* $t \rightsquigarrow \text{Reach}_{\mathcal{S}}^{K}(t;x)$ *exhibits the* semi-group property*:*

$$\forall t \geq s \geq 0,\ \ \text{Reach}_{\mathcal{S}}^{K}(t;x) = \text{Reach}_{\mathcal{S}}^{K}(t-s;\text{Reach}_{\mathcal{S}}^{K}(s;x))$$

Furthermore,

$$(\text{Reach}_{\mathcal{S}}^{K}(t;\cdot))^{-1} \ := \ \text{Reach}_{\overleftarrow{\mathcal{S}}}^{K}(t;\cdot)$$

Proof. The first statement is obvious. To say that $x \in \text{Reach}_{\mathcal{S}}^{K}(T;y)$ means that there exists a viable evolution $x(\cdot) \in \mathcal{S}^{K}(x)$ such that $x(T) = y$. The evolution $\overleftarrow{y}(\cdot)$ defined by $\overleftarrow{y}(t) := x(T-t)$ belongs to $\overleftarrow{\mathcal{S}}(x)$, is viable in K on $[0,T]$ and $\overleftarrow{y}(T) = y$. This means that $y \in \text{Reach}_{\overleftarrow{\mathcal{S}}}^{K}(T;x)$. □

When a time-independent evolutionary system $\mathcal{S} : X \mapsto \mathcal{C}(0,+\infty;X)$ is deterministic, one can identify the (unique) evolution $x(\cdot) = \mathcal{S}(x)$ starting from x with the reachable (single-valued) map $t \in \mathbb{R} \mapsto \text{Reach}_{\mathcal{S}}(t;x)$. This is why in many cases, the classical study of deterministic systems is reduced to the *flows* $t \in \mathbb{R} \mapsto \text{Reach}_{\mathcal{S}}(t;x)$.

Important Remark: Reachable Maps and Evolutionary Systems. Even though (set-valued) reachable maps $\text{Reach}_{\mathcal{S}}(\cdot;x)$ play an important role, *they no longer characterize a time-independent nondeterministic evolutionary system* $\mathcal{S}$: Knowing a state $y \in \text{Reach}_{\mathcal{S}}(t;x)$, we know that an evolution starting from x passes through y at time t, but this does not guarantee that this evolution passes through any arbitrary $x_s \in \text{Reach}_{\mathcal{S}}(s;x)$ at time s.

This is why, for nondeterministic evolutionary systems, the convenient general setting is to regard it as a set-valued map $\mathcal{S} : X \mapsto \mathcal{C}(0,+\infty;X)$ *instead of a set-valued semi-group or flow.*

The graph of the reachable tube is itself a capture basin under an auxiliary system, and thus, exhibits all the properties of capture basins:

Proposition 8.4.3 ***[Viability Characterization of Reachable Tubes]*** *Let us consider the backward auxiliary system*

$$\begin{cases} (i) \ \tau'(t) = -1 \\ (ii) \ x'(t) = -f(x(t), u(t)) \\ \quad \text{where } u(t) \in U(x(t)) \end{cases} \tag{8.6}$$

and a source B. The graph of the viable reachable tube $\text{Reach}_S^K((\cdot); B) : T \rightsquigarrow \text{Reach}_S^K(T; B)$ *is the capture basin of $\mathbb{R}_+ \times K$ with target $\{0\} \times B$ under the auxiliary system (8.6), p. 286:*

$$\text{Graph}(\text{Reach}_S^K(\cdot; B)) = \text{Capt}_{(8.6)}(\mathbb{R}_+ \times K, \{0\} \times B)$$

Proof. Indeed, to say that (T, x) belongs to the capture basin of target $\{0\} \times B$ viable in $\mathbb{R}_+ \times K$ under the auxiliary system (8.6) means that there exist an evolution $\overleftarrow{x}(\cdot)$ to the backward system (8.6), p. 286:

$$\begin{cases} (i) \ \tau'(t) \ = \ -1 \\ (ii) \ x'(t) \ = \ -f(x(t), u(t)) \\ \quad \text{where } u(t) \ \in \ U(x(t)) \end{cases}$$

starting at $\overleftarrow{x}(0) := x$ and a time $t^\star \geq 0$ such that

$$\begin{cases} (i) \ \ \forall t \in [0, t^\star], \ (T - t, \overleftarrow{x}(t)) \in \mathbb{R}_+ \times K \\ (ii) \qquad\qquad (T - t^\star, \overleftarrow{x}(t^\star)) \in \{0\} \times B \end{cases}$$

The second condition means that $t^\star = T$ and that $\overleftarrow{x}(T)$ belongs to B. The first one means that for every $t \in [0, T]$, $\overleftarrow{x}(t) \in K$. This amounts to saying that the evolution $x(\cdot) := \overleftarrow{x}(T - \cdot)$ is a solution to system (8.4) starting at $\overleftarrow{x}(T) \in B$, satisfying $x(T) = x$ and

$$\forall t \in [0, T], \ x(t) \in K \qquad \square$$

8.4.2 Detection and Cournot Tubes

Definition 8.4.4 ***[Detection Basins and Tubes]*** *Let $B \subset K$ be a subset regarded as a source. The* detection basin $\text{Det}_S(K, B)$ *is the subset of final*

states in K *at which arrives in finite time at least one evolution viable in* K *starting from the* source B. *The subset* K *is said to* detect B *if* $K = \mathrm{Det}_{\mathcal{S}}(K, B)$.
The T-detection basin $\mathrm{Det}_{\mathcal{S}}(K, B)(T)$ *is the subset of final states in* K *at which arrives* before T *at least one evolution viable in* K *starting from the* source B *and the set-valued maps* $T \rightsquigarrow \mathrm{Det}_{\mathcal{S}}(K, B)(T)$ *is called the* detection tube *of* B.

We first point-out the links between capture and detecting basins:

Lemma 8.4.5 *[Capture and Detection Basins] The (forward) detection basin* $\mathrm{Det}_{\mathcal{S}}(K, B)$ *of the source* B *under* $\mathcal{S}$ *is equal to the backward capture basin* $\mathrm{Capt}_{\overleftarrow{\mathcal{S}}}(K, B)$ *under* $\overleftarrow{\mathcal{S}}$ *of the source* B *regarded as a target:*

$$\mathrm{Det}_{\mathcal{S}}(K, B) = \mathrm{Capt}_{\overleftarrow{\mathcal{S}}}(K, B)$$

and thus, exhibits all the properties of the capture basins.

Proof. Indeed, to say that x belongs to $\mathrm{Det}_{\mathcal{S}}(K, B)$ amounts to saying that there exist an initial state $x_0 \in B$, an evolution $x(\cdot) \in \mathcal{S}(x_0)$ and some $T \geq 0$ such that $x(T) = x$ and $x(t) \in K$ for all $t \in [0, T]$. Then the evolution $\overleftarrow{x}(\cdot)$ defined by $\overleftarrow{x}(t) := x(T - t)$ is a solution to the backward system $\overleftarrow{\mathcal{S}}(x)$ starting at x and viable in K until time T when $\overleftarrow{x}(T) = x(0) = x_0 \in B$. This means that x belongs to $\mathrm{Capt}_{\overleftarrow{\mathcal{S}}}(K, B)$. □

The detection tube can be expressed in terms of "reachable maps":

Proposition 8.4.6 *[Detection Tubes and Reachable maps] Let* $\mathcal{S} : X \rightsquigarrow \mathcal{C}(0, +\infty; X)$ *be an evolutionary system and* $B \subset K \subset X$ *be a source contained in the environment. Then the detection tube can be written in the form*

$$\mathrm{Det}_{\mathcal{S}}(K, B)(T) := \bigcup_{t \in [0, T]} \mathrm{Reach}_{\mathcal{S}}^{K}(t; B)$$

We also deduce from Theorem 4.3.2, p. 133 a viability characterization of the detection tube:

Proposition 8.4.7 *[Viability Characterization of Detection Tubes] Let us consider the backward auxiliary system (8.6), p. 286:*

$$\begin{cases} (i)\ \tau'(t) &= -1 \\ (ii)\ x'(t) &= -f(x(t), u(t)) \\ \quad\text{where } u(t) &\in U(x(t)) \end{cases}$$

Then the graph of the viable-capturability tube $\mathrm{Det}_{S}(K,B)(\cdot)$ *is the viable-capture basin of* $\mathbb{R}_{+} \times B$ *viable in* $\mathbb{R}_{+} \times K$ *under the system (8.6):*

$$\mathrm{Graph}(\mathrm{Det}_{S}(K,B)(\cdot)) = \mathrm{Capt}_{(8.6)}(\mathbb{R}_{+} \times K, \mathbb{R}_{+} \times B)$$

Proof. The proof is analogous to the one of Proposition 8.4.3, p. 286. □

Detection tubes provide the set of final states at which arrive at least one evolution emanating from B. The question arises whether we can find the subset of these initial states. This is connected with a concept of uncertainty suggested by *Augustin Cournot* as the meeting of two independent causal series: *"A myriad partial series can coexist in time: they can meet, so that a single event, to the production of which several events took part, come from several distinct series of generating causes."* The search for causes amounts in this case to reversing time in the dynamics and to look for "retrodictions" (so to speak) instead of predictions.

We suggest to combine this Cournot approach uncertainty with the Darwinian view of contingent uncertainty for facing necessity (viability constraints) by introducing the concept of Cournot map.

Definition 8.4.8 *[Cournot Map] The* Cournot map $\mathrm{Cour}_{(\mathbf{K},B)}$: $\mathrm{Graph}(\mathbf{K}) \rightsquigarrow B$ *associates with any* $T \geq 0$ *and* $x \in \mathbf{K}(T)$ *the (possibly empty) subset* $\mathrm{Cour}_{(\mathbf{K},B)}(T,x)$ *of initial causes* $x_0 \in B$ *from which* $x := x(T)$ *can be reached by an* actual evolution $x(\cdot) \in S(x_0)$ *viable in the tube:*

$$\forall t \in [0,T],\ x(t) \in \mathbf{K}(t) \tag{8.7}$$

At time T*, the state* x *is thus the result of past viable evolutions starting from all causal states* $x_0 \in \mathrm{Cour}_{(\mathbf{K},B)}(T,x)$*. The size of the set* $\mathrm{Cour}_{(\mathbf{K},B)}(T,x)$ *could be taken as* a measure of Cournot's concept of uncertainty *for the event* x *at time* T*.*

The set-valued map $T \rightsquigarrow \mathrm{Im}(\mathrm{Cour}_{(\mathbf{K},B)}(T,\cdot))$ *is decreasing*, refining the set of causal states of B from which at least one evolution has been selected through the tube $\mathbf{K}(\cdot)$ as time goes on.

We shall characterize the Cournot map as a viability kernel under an adequate auxiliary system.

Theorem 8.4.9 *[Viability Characterization of Cournot Maps] Let us set* $\boldsymbol{I}_B := \{(x,x)\}_{x\in B}$ *and introduce the auxiliary system*

$$\begin{cases} (i) \ \ \tau'(t) \ = \ -1 \\ (ii) \ \ x'(t) \ = \ -f(x(t),u(t)) \\ \quad\quad \text{where } u(t) \ \in \ U(x(t)) \\ (iii) \ y'(t) \ = \ 0 \end{cases} \tag{8.8}$$

The graph of the Cournot Map $\mathrm{Cour}_{(\mathbf{K},B)}$ *is given by the formula*

$$\mathrm{Graph}(\mathrm{Cour}_{(\mathbf{K},B)}) := \mathrm{Capt}_{(8.9)}(\mathrm{Graph}(\mathbf{K}\times B, \{0\}\times \boldsymbol{I}_B)) \tag{8.9}$$

Proof. To say that (T,x,x_0) belongs to $\mathrm{Capt}_{(8.9)}(\mathrm{Graph}(\mathbf{K}\times B,\{0\}\times \mathbf{I}_B))$ means that there exists an evolution $\overleftarrow{x}(\cdot)$ starting at x and a time $t^\star \geq 0$ such that

$$\begin{cases} (i) \ \ \forall t\in[0,t^\star], \ (T-t,\overleftarrow{x}(t),x_0)\in \mathrm{Graph}(\mathbf{K})\times B \\ (ii) \quad\quad\quad\quad\quad\ (T-t^\star,\overleftarrow{x}(t^\star),x_0)\in\{0\}\times \mathbf{I}_B \end{cases}$$

The second condition means that $t^\star = T$ and that $\overleftarrow{x}(T) = x_0$ belongs to B. The first one means that for every $t\in[0,T]$, $\overleftarrow{x}(t)\subset \mathbf{K}(T-t)$. This amounts to saying that the evolution $x(\cdot) := \overleftarrow{x}(T-\cdot)$ where x_0 belongs to B satisfies $x(T) = x$ and the viability conditions (8.7), i.e., that x_0 belongs to $\mathrm{Cour}_{(\mathbf{K},B)}(T,x)$. □

The issue is pursued in Sect. 13.8, p. 551 at the Hamilton-Jacobi level.

8.4.3 Volterra Inclusions

The standard paradigm of evolutionary system that we adopted is the initial-value (or Cauchy) problem. It assumes that the present is frozen, as well as the initial state from which start evolutions governed by an evolutionary system $\mathcal{S}$.

But the present time evolves, too, and consequences of earlier evolutions accumulate. Therefore, the questions of "gathering" present consequences of all earlier initial states arises.

There are two ways of mathematically translating this idea. The first one, the most familiar, is to take the *sum* of the number of these consequences: This leads to equations bearing the name of *Volterra*, of the form

$$\forall T \geq 0, \;\; x(T) \;=\; \int_0^T \theta(T-s;x(s))ds$$

A particular case is obtained for instance when the "kernel" $\theta(\cdot,\cdot)$ is itself the flow of a deterministic system $y'(t) = f(y(t))$. A solution $x(\cdot)$ to the Volterra equation, if it exists, provides at each ephemeral $T \geq 0$ the sum of the states obtained at time T from the state $x(s)$ at earlier time $T-s \in [0,T]$ of the solution by differential equation $y'(t) = f(y(t))$ starting at time 0 at a given initial state x. Then $\int_0^T \theta(T-s;x(s))ds$ denotes the sum of consequences at time T of a flow of earlier evolving initial conditions, for instance.

This is a typical situation that is met in traffic problems or in biological neuron networks. It is not enough to study the consequences of an initial condition, a vehicle or a neurotransmitter, since they arrive continuously at the entrance of the highway or of the neuron.

In the set-valued case, "gathering" the subsets of consequences at ephemeral time T of earlier initial conditions is mathematically translated by taking their *union*. Hence the map similar to the Volterra equation would be to find a tube $\mathbf{D} : t \rightsquigarrow \mathbf{D}(t)$ and to check whether it satisfies

$$\forall T \geq 0, \;\; \mathbf{D}(T) \;=\; \bigcup_{s\in[0,T]} \theta(T-s;\mathbf{D}(s))$$

where $(t,K) \mapsto \theta(t,K) \subset X$ is a set-valued "kernel".

The particular example of kernel is the reachable map $(t,K) \mapsto \text{Reach}_S^K(t,K)$, a solution to an initial value problem, in the spirit of *Cauchy*. Then, if a tube $\mathbf{D} : t \rightsquigarrow \mathbf{D}(t)$ is given, the set

$$\forall T \geq 0, \;\; \bigcup_{s\in[0,T]} \text{Reach}_S^K(T-s;\mathbf{D}(s))$$

of cumulated consequences gathers the consequences at time T of the evolutions at time T of evolutions starting at time $T-s$ from $\mathbf{D}(s)$. We shall prove that there exist solutions to the set-valued Volterra equation, that we shall call with a slight abuse of language, *Volterra inclusion*

$$\forall T \geq 0, \;\; \mathbf{D}(T) \;=\; \bigcup_{s\in[0,T]} \text{Reach}_S^K(T-s;\mathbf{D}(s)) \qquad (8.10)$$

The reachable tube $\text{Reach}_S^K(\cdot;B)$ is obviously a solution to such a set-valued Volterra equation: This is nothing other than the semi-group property.

We shall see that this is the ***unique*** *viable tube satisfying the semi-group property contained in K and starting at B.*

We shall also prove that the *detection tube* $\mathrm{Det}_{\mathcal{S}}^{K}(\cdot, B)$ *is the* ***unique*** *viable tube solution to the set-valued Volterra equation "Volterra inclusion (8.10)" contained in K and starting at B.*

For that purpose, we have to slightly extend the concept of detection tube of subsets to detection tubes of tubes (see Theorem 8.10.6, p. 314).

8.5 Connection Basins and Eupalinian Kernels

8.5.1 Connection Basins

Let us consider an environment $K \subset X$ and an evolutionary system $\mathcal{S} : X \rightsquigarrow \mathcal{C}(-\infty, +\infty; X)$.

Definition 8.5.1 *[****Connection Basins****] Let $B \subset K$ be a subset regarded as a source, $C \subset K$ be a subset regarded as a target. The* connection basin $\mathrm{Conn}_{\mathcal{S}}(K, (B, C))$ *of K between B and C is the subset of states $x \in K$ through which passes at least one viable evolution starting from the source B and arriving in finite time at target C.*

The subset K is said to connect B to C *if* $K = \mathrm{Conn}_{\mathcal{S}}(K, (B, C))$.

We refer to Sect. 10.6.2, p. 413 for topological and other viability characterization of connection basins.

The set-valued map which associates with any $x \in \mathrm{Conn}_{\mathcal{S}}(K, (B, C))$ of the connection basin the pair $(x(\varpi_{(K,B)}(\overleftarrow{x}(\cdot)), \varpi_{(K,C)}(\overrightarrow{x}(\cdot)))) \in B \times C \subset K \times K$ of end-values of viable evolutions $x(\cdot)$ connecting B to C has, by definition, nonempty values.

The question arises whether we can invert this set-valued map: given a pair $(y, z) \in K \times K$, does there exist an evolution $x(\cdot)$ viable in K linking y to z in finite time in the sense where $x(0) = y$ and $x(T) = z$ for some finite time T? This is an instantiation of the problem of studying the connection basin $\mathrm{Conn}_{\mathcal{S}}(K, (\{y\}, \{z\}))$ when the pairs (y, z) range over the subset $\mathcal{E} \subset K \times K$ of pairs of end-values of viable evolutions $x(\cdot)$ connecting B to C.

In other words, the question boils down whether we can *a priori* know the subset $\mathcal{E} \subset K \times K$ of pairs (y, z) such that

$$\mathrm{Conn}_{\mathcal{S}}(K, (B, C)) = \bigcup_{(y,z)\in\mathcal{E}} \mathrm{Conn}_{\mathcal{S}}(K, (\{y\}, \{z\}))$$

Therefore, the study of connection basins amounts to finding this subset $\mathcal{E}$, that we shall call the *Eupalinian kernel* of K under $\mathcal{S}$, and to characterize it as capture basins of an auxiliary capturability problems.

8.5.1.1 Eupalinian Kernels

Eupalinos, a Greek engineer, excavated around 550 BC a 1,036 m. long tunnel 180 m. below Mount Kastro for building an aqueduct supplying Pythagoreion (then the capital of Samos) with water on orders of tyrant Polycrates. *He started to dig simultaneously the tunnel from both sides by two working teams who met in the center* of the channel and they had only 0.6 m. error. There is still no consensus on how he did it. However[1], this *"Eupalinian strategy"* has been used ever since for building famous tunnels (under the British Channel or the Mont-Blanc) or bridges: it consists in starting the construction at the same time from both end-points x and y and proceed until they meet, by continuously monitoring the progress of the construction.

Such models can also be used as mathematical metaphors in negotiation procedures when both actors start from opposite statements and try to reach a consensus by making mutual concessions step by step, continuously bridging the remaining gap.

This question arose in numerical analysis and control under the name of "shooting" methods, which, whenever the state is known at initial and final time, consists in integrating differential equations at the same time at both initial and final states and matching in the middle.

We suggest a mathematical metaphor for explaining such an Eupalinian strategy.

Definition 8.5.2 ***[Eupalinian Kernels]*** *Let* $\mathcal{S} : X \rightsquigarrow \mathcal{C}(0, +\infty; X)$ *and* $K \subset X$ *be an environment. We denote by* $\mathcal{S}^K(y, z) := \mathrm{Conn}_{\mathcal{S}}(K, (\{y\}, \{z\}))$ *the set of* Eupalinian evolutions $x(\cdot)$ *governed by the evolutionary system* $\mathcal{S}$ *viable in* K *connecting* y *to* z*, i.e., the set of evolutions* $x(\cdot) \in \mathcal{S}(y)$ *such that there exists a finite time* $T \geq 0$ *satisfying* $x(T) = z$ *and, for all* $t \in [0, T]$, $x(t) \in K$.

The Eupalinian kernel $\mathcal{E} := \mathrm{Eup}_{\mathcal{S}}(K) \subset K \times K$ *is the subset of pairs* (y, z) *such that there exists at least one viable evolution* $x(\cdot) \in \mathcal{S}^K(y)$ *connecting* y *to* z *and viable in* K.

[1] The authors thank Hélène Frankowska for communicating them this historical information.

We can characterize the Eupalinian kernel as a capture basin:

Proposition 8.5.3 ***[Viability Characterization of Eupalinian Kernels]*** *Let us denote by* $\text{Diag}(K) := \{(x,x)\}_{x \in K} \subset K \times K$ *the* diagonal of K. *The Eupalinian kernel* $\text{Eup}_{\mathcal{S}}(K)$ *of* K *under the evolutionary system* $\mathcal{S}$ *associated with the system*

$$x'(t) \; = \; f(x(t), u(t)) \text{ where } u(t) \; \in \; U(x(t))$$

is the capture basin

$$\text{Eup}_{\mathcal{S}}(K) \; = \; \text{Capt}_{(8.11)}(K \times K, \text{Diag}(K))$$

of the diagonal of K *viable in* $K \times K$ *under the auxiliary system*

$$\begin{cases} (i) \; y'(t) \; = \; f(y(t), u(t)) \; \text{ where } u(t) \; \in \; U(y(t)) \\ (ii) \; z'(t) = \; -f(z(t), v(t)) \text{ where } v(t) \; \in \; U(z(t)) \end{cases} \quad (8.11)$$

We "quantify" the concept of Eupalinian kernel with the concept of several Eupalinian intertemporal optimization problems. The domains of their value functions are the Eupalinian kernels, so that Proposition 8.5.3, p. 293 follows from the forthcoming Theorem 8.5.6. p. 295.

Since we shall minimize an intertemporal criterion involving controls, as we did in Chap. 4, p. 125, we denote by:

1. $\mathcal{P}^K(x)$ the set of state-control evolutions $(x(\cdot), u(\cdot))$ where $x(0) = x$ and regulated by the system.

$$x'(t) \; = \; f(x(t), u(t)) \text{ where } u(t) \; \in \; U(x(t))$$

   and viable in K
2. $\mathcal{P}^K(y,z)$ the set of state-control evolutions $(x(\cdot), u(\cdot))$ where $x(0) = y$ and $x(t^\star) = z$ for some finite time $t^\star$ governed by this and viable in K.

We introduce a cost function $\mathbf{c} : X \times X \mapsto \mathbb{R} \cup \{+\infty\}$ (regarded as a *connection cost*) and a Lagrangian $\mathbf{l} : (x,u) \rightsquigarrow \mathbf{l}(x,u)$.

We consider the Eupalinian optimization problem

$$\begin{cases} \mathbf{E}_{(\mathbf{c},\mathbf{l})}(y,z) \\ = \inf_{(x(\cdot),u(\cdot)) \in \mathcal{P}^K(y,z), t^\star \geq 0 \; | \; x(2t^\star)=z} \left(\mathbf{c}(x(t^\star), x(t^\star)) \mid \int_0^{2t^\star} \mathbf{l}(x(t), u(t)) dt \right) \end{cases}$$

- By taking $\mathbf{c} \equiv 0$ and $\mathbf{l}(x,u) \equiv 1$, we find the following Eupalinian minimal time problem:

Definition 8.5.4 *[Eupalinian Distance and Brachistochrones]*
The Eupalinian distance $\epsilon_K(y,z)$

$$\epsilon_K(y,z) := 2 \inf_{(x(\cdot),u(\cdot))\in\mathcal{P}^K(y,z),\; t^\star \geq 0 \mid x(2t^\star)=z} t^\star \in [0,+\infty[\quad (8.12)$$

measures the minimal time needed for connecting the two states y and z by a evolution viable in K. *Let* $B \subset K$ *a source and* $C \subset K$ *be a target. The function*

$$\epsilon_K(B,C) := \inf_{y\in B,\, z\in C} \epsilon_K(y,z)$$

is called the Eupalinian distance *between the source* B *and the target* C. *Viable evolutions in* K *connecting* y *to* z *in* minimal time *are called* brachistochrones.

Their existence and computation was posed as a challenge by *Johann Bernoulli* in 1696, challenge met by *Isaac Newton*, *Jacob Bernoulli*, *Gottfried Leibnitz* and *Guillaume de L'Hopital* in a particular case.

- By taking $\mathbf{c} \equiv 0$ and $\mathbf{l}(x,u) = \|f(x,u)\|$, we obtain the *viable geodesic* connecting two states by a viable evolution in *minimal length function* $\gamma_K(x) : X \mapsto \mathbb{R}_+ \cup \{+\infty\}$ defined by

$$\gamma_K(x) := \inf_{(x(\cdot),u(\cdot))\in\mathcal{P}^K(x)} \int_0^\infty \|f(x(t),u(t))\|dt$$

(see Definition 4.4.1, p. 140).

Definition 8.5.5 *[Geodesics] We denote by* geodesic distance

$$\widehat{\gamma}_K(y,z) := \inf_{(x(\cdot),u(\cdot))\in\mathcal{P}^K(y,z),\ t^\star\geq 0 \mid x(2t^\star)=z} \int_0^{2t^\star} \|f(x(t),u(t))\|dt \tag{8.13}$$

measuring the minimal length needed for connecting the two states y and z by a evolution viable in K. *Any viable evolution* $(x(\cdot),u(\cdot)) \in \mathcal{P}^K(y,z)$ *achieving the minimum* $\widehat{\gamma}_K(y,z) := \int_0^{2t^\star} \|f(x(t),u(t))\|dt$ *is called a* viable geodesic. *The function*

$$\widetilde{\gamma}_K(B,C) := \inf_{y\in B,\ z\in C} \gamma_K(y,z)$$

is called the geodesic distance *between the source B and the target C.*

We shall prove that

Theorem 8.5.6 *[Eupalinian Optimization Theorem] Let us consider the auxiliary control system*

$$\begin{cases} (i)\ \ y'(t) = f(y(t),u(t)) \text{ where } u(t) \in U(y(t)) \\ (ii)\ \ z'(t) = -f(z(t),v(t)) \text{ where } v(t) \in U(z(t)) \\ (iii)\ \lambda'(t) = -\mathbf{l}(y(t),u(t)) - \mathbf{l}(z(t),v(t)) \end{cases} \tag{8.14}$$

Then

$$\mathbf{E}_{(\mathbf{c},\mathbf{l})}(y,z) = \inf_{(y,z,\lambda)\in \mathrm{Capt}_{(8.14)}(K\times K\times\mathbb{R}_+,\mathcal{E}p(\mathbf{c})\cap(\mathrm{Diag}(K)\times\mathbb{R}_+))} \lambda$$

where $\mathrm{Diag}(K) := \{(x,x)\}_{x\in K} \subset K\times K$ *is the* diagonal of K.

Proof. Let $(y,z,\lambda) \in \mathrm{Capt}_{(8.14)}(K\times K\times\mathbb{R}_+,\mathcal{E}p(\mathbf{c})\cap(\mathrm{Diag}(K)\times\mathbb{R}_+))$ belong to the capture basin. This means that there exist one forward evolution $\overrightarrow{y}(\cdot) \in \mathcal{S}^K(y)$ viable in K, one backward evolution $\overleftarrow{z}(\cdot) \in \overleftarrow{\mathcal{S}}^K(z)$ viable in K, the evolution $\lambda(t) := \lambda - \int_0^t \mathbf{l}(\overrightarrow{y}(t),\overrightarrow{u}(t))dt - \int_0^t \mathbf{l}(\overleftarrow{z}(t),\overleftarrow{v}(t))dt$ and a time $t^\star$ such that:

- for all $t \in [0,t^\star]$, $\overrightarrow{y}(t) \in K$, $\overleftarrow{z}(t) \in K$,

$$\lambda - \int_0^t \mathbf{l}(\overrightarrow{y}(t), \overrightarrow{u}(t))dt - \int_0^t \mathbf{l}(\overleftarrow{z}(t), \overleftarrow{v}(t))dt \ \geq 0$$

- and $\overrightarrow{y}(t^\star) = \overleftarrow{z}(t^\star)$ and

$$\lambda - \int_0^{t^\star} \mathbf{l}(\overrightarrow{y}(t), \overrightarrow{u}(t))dt - \int_0^{t^\star} \mathbf{l}(\overleftarrow{z}(t), \overleftarrow{v}(t))dt \ \geq \ \mathbf{c}(\overrightarrow{y}(t^\star), \overleftarrow{z}(t^\star))$$

Let us introduce the evolution $x(t)$ defined by $x(t) := \overrightarrow{y}(t)$ for $t \in [0, t^\star]$ and $x(t) := \overleftarrow{z}(2t^\star - t)$ for $t \in [t^\star, 2t^\star]$. This evolution $x(\cdot)$ is continuous at $t^\star$ because $x(t^\star) = \overrightarrow{y}(t^\star) = \overleftarrow{z}(t^\star)$, belongs to $\mathcal{S}^K(y, z)$ since $x(0) = \overrightarrow{y}(0) = y$, $x(2t^\star) = \overleftarrow{z}(0) = z$ and is governed by the differential inclusion starting at y. Furthermore,

$$\begin{cases} \lambda - \left(\int_0^{2t^\star} \mathbf{l}(x(t), u(t))dt\right) \\ = \ \lambda - \left(\int_0^{t^\star} \mathbf{l}(\overrightarrow{y}(t), \overrightarrow{u}(t))dt + \int_0^{t^\star} \mathbf{l}(\overleftarrow{z}(t), \overleftarrow{v}(t))dt\right) \\ \geq \ \mathbf{c}(x(t^\star), x(t^\star)) \end{cases}$$

This means that there exist $x(\cdot) \in \mathcal{S}^K(y, z)$ and $t^\star \geq 0$ such that

$$\mathbf{c}(x(t^\star), x(t^\star)) + \int_0^{2t^\star} \mathbf{l}(x(t), u(t))dt \ \leq \ \lambda$$

This implies in particular that

$$\begin{cases} \mathbf{E}_{(\mathbf{c},\mathbf{l})}(y, z) \ := \ \inf_{(x(\cdot),u(\cdot))\in\mathcal{P}^K(y,z),\ t^\star} \left(\mathbf{c}(x(t^\star), x(t^\star)) + \int_0^{2t^\star} \mathbf{l}(x(t), u(t))dt\right) \\ \leq \ \inf_{(y,z,\lambda)\in\mathrm{Capt}_{(8.14)}(\mathbb{R}_+\times K\times K, \mathcal{E}p(\mathbf{c})\cap(\mathbb{R}_+\times\mathrm{Diag}(K)))} \lambda \end{cases}$$

For proving the opposite inequality, we associate with any $\varepsilon > 0$ an evolution $x_\varepsilon(\cdot) \in \mathcal{S}^K(y, z)$, a control $u_\varepsilon(\cdot)$ and $t_\varepsilon^\star \ \geq \ 0$ such that

$$\left(\mathbf{c}(x_\varepsilon(t_\varepsilon^\star), x_\varepsilon(t_\varepsilon^\star)) + \int_0^{2t_\varepsilon^\star} \mathbf{l}(x_\varepsilon(t), u_\varepsilon(t))dt\right) \ \leq \ \mathbf{E}_{(\mathbf{c},\mathbf{l})}(y, z) + \varepsilon$$

and the function

$$\lambda_\varepsilon(t) \ := \ \mathbf{E}_{(\mathbf{c},\mathbf{l})}(y, z) + \varepsilon - \int_0^{2t} \mathbf{l}(x_\varepsilon(t), u_\varepsilon(t))dt$$

Introducing the forward parts $\overrightarrow{y}_\varepsilon(t) := x_\varepsilon(t)$ and $\overrightarrow{u}_\varepsilon(t) := u_\varepsilon(t)$ for $t \in [0, t_\varepsilon^\star]$ and backward parts $\overleftarrow{z}_\varepsilon(t) := x_\varepsilon(2t_\varepsilon^\star - t)$ and $\overleftarrow{v}_\varepsilon(t) := u_\varepsilon(2t_\varepsilon^\star - t)$, we observe that $(\overrightarrow{y}_{\varepsilon}(t), \overleftarrow{z}_{\varepsilon}(t), \lambda_{\varepsilon}(t))$ is a solution to the auxiliary system (8.14) starting at $(y, z, \mathbf{E}_{(\mathbf{c},\mathbf{l})}(y, z) + \varepsilon)$, viable in $K \times K \times \mathbb{R}_{+}$ and satisfying

$$\begin{cases} \lambda_{\varepsilon}(t) := \mathbf{E}_{(\mathbf{c},\mathbf{l})}(y, z) + \varepsilon - \int_{0}^{2t_{\varepsilon}^{\star}} \mathbf{l}(x_{\varepsilon}(t), u_{\varepsilon}(t))dt \\ \geq \mathbf{c}(\overrightarrow{y}_{\varepsilon}(t_{\varepsilon}^{\star}), \overleftarrow{z}_{\varepsilon}(t_{\varepsilon}^{\star})) \end{cases}$$

This implies that $(y, z, \mathbf{E}_{(\mathbf{c},\mathbf{l})}(y, z) + \varepsilon)$ belongs to the capture basin $\mathrm{Capt}_{(8.14)}(K \times K \times \mathbb{R}_{+}, \mathcal{E}p(\mathbf{c}) \cap (\mathrm{Diag}(K) \times \mathbb{R}_{+}))$. Hence

$$\inf_{(y,z,\lambda) \in \mathrm{Capt}_{(8.14)}(K \times K \times \mathbb{R}_{+}, \mathcal{E}p(\mathbf{c}) \cap (\mathrm{Diag}(K) \times \mathbb{R}_{+}))} \lambda \leq \mathbf{E}_{(\mathbf{c},\mathbf{l})}(y, z) + \varepsilon$$

and it is enough to let ε converge to 0. □

Remark: Eupalinian Graphs. The Eupalinian kernel is a graph in the sense of "graph theory" where the points are regarded as "vertices" or "nodes", a set of pairs (y, z) connected by at least one evolution and, for a given intertemporal optimization problem, the set of "edges" or "arcs" $\mathbf{E}_{(\mathbf{c},\mathbf{l})}(y, z)$ linking y to z. □

Remark: The associated Hamilton-Jacobi-Bellman Equation. The tangential and normal characterizations of capture basins imply that the bilateral value function is the solution to the bilateral Hamilton-Jacobi-Bellman partial differential equation

$$\inf_{u \in U(x)} \left(\left\langle \frac{\partial \mathbf{E}}{\partial x}, f(x, u) \right\rangle + \mathbf{l}(x, u) \right) - \sup_{v \in U(y)} \left(\left\langle \frac{\partial \mathbf{E}}{\partial y}, f(y, v) \right\rangle - \mathbf{l}(y, v) \right) = 0 \tag{8.15}$$

in a generalized sense (see Chap. 17, p. 681) satisfying the *diagonal condition*

$$\forall x \in K, \ \mathbf{E}_{(\mathbf{c},\mathbf{l})}(x, x) = \mathbf{c}(x, x)$$

Even though the solution to this partial differential equation provides the Eupalinian value function, we do not need to approximate this partial differential equation for finding this Eupalinian value function since the Viability Kernel Algorithm provides it and the optimal Eupalinian evolutions. □

Remark: Regulation of Optimal Eupalinian Solutions. We introduce the two following forward and backward maps:

$$\begin{cases} (i) \ \overrightarrow{R}(x,p) := \{\overrightarrow{u} \in U(x) \mid \langle p, f(x,\overrightarrow{u})\rangle + \mathbf{l}(x,\overrightarrow{u}) \\ \quad = \inf_{u\in U(x)} (\langle p, f(x,u)\rangle + \mathbf{l}(x,u))\} \\ (ii) \ \overleftarrow{R}(y,q) := \{\overleftarrow{v} \in U(y) \mid \langle q, f(y,\overleftarrow{v})\rangle - \mathbf{l}(y,\overleftarrow{v}) \\ \quad = \sup_{v\in U(y)} (\langle q, f(y,v)\rangle - \mathbf{l}(y,v))\} \end{cases} \tag{8.16}$$

depending only on the dynamics f and U of the control system and of the transient cost function $\mathbf{l}$.

In order to find and regulate the optimal evolution, we plug into them the partial derivatives $p := \frac{\partial \mathbf{E}(x,y)}{\partial x}$ and $q := \frac{\partial \mathbf{E}(x,y)}{\partial y}$ of the bilateral value function (actually, when constraints K are involved or when the function $\mathbf{c}$ is only lower semicontinuous, the bilateral value function is lower semicontinuous and we have to replace the partial derivatives by subgradients $(p_x, q_y) \in \partial\mathbf{E}(x,y)$ of the bilateral value-function, as indicated in Chap. 17, p. 681).

Knowing the Eupalinian value function and its partial derivatives (or subgradients), one can thus derive from the results of Sect. 17.4.3, p. 704 that the optimal Eupalinian evolution $x(\cdot)$ linking y at time 0 and z at minimal time $2T$ is governed by the control system

$$\begin{cases} (i) \quad x'(t) = f(x(t),u(t)) \\ (ii) \quad \text{where} \\ u(t) \in \begin{cases} \overrightarrow{R}\left(x(t), \dfrac{\partial \mathbf{E}(x(t),x(2T-t))}{\partial x}\right) & \text{if } t \in [0,T] \\ \overleftarrow{R}\left(x(t), \dfrac{\partial \mathbf{E}(x(2T-t),x(t))}{\partial y}\right) & \text{if } t \in [T,2T] \end{cases} \end{cases} \tag{8.17}$$

In other words, the controls regulating an optimal evolution linking y to z *"feed" both at current sate* $x(t)$ *and at state* $x(2T-t)$ *at time* $2T-t$, forward if $t \in [0,T]$ and backward if $t \in [T,2T]$. In other words, optimal evolutions can be governed by "forward and backward retroactions", keeping an eye on the current state and the other one on the state at another instant. In particular, the initial control depends upon both the initial and final states. □

8.6 Collision Kernels

Eupalinian kernels are particular cases of *collision kernels* associated with a pair of evolutionary systems denoted by $\mathcal{S}$ and $\mathcal{T}$ associated with control systems

$$\begin{cases} (i)\ y'(t) &= f(y(t),u(t)) \text{ where } u(t) \in U(y(t)) \\ (ii)\ z'(t) &= g(z(t),v(t)) \text{ where } v(t) \in V(z(t)) \end{cases} \tag{8.18}$$

Definition 8.6.1 *[Collision Kernels] Let $\mathcal{S} : X \rightsquigarrow \mathcal{C}(0,+\infty;X)$ and $\mathcal{T} : X \rightsquigarrow \mathcal{C}(0,+\infty;X)$ be two evolutionary systems, $K \subset X$ and $L \subset X$ be two intersecting environments. We denote by $\mathcal{S}^K(y) \times \mathcal{T}^L(z)$ the set of evolutions $(y(\cdot),z(\cdot)) \in \mathcal{S}^K(y) \times \mathcal{T}^L(z)$ governed by the pair of evolutionary systems $\mathcal{S}$ and $\mathcal{T}$ viable in $K \times L$. We say that they* collide *if there exists a finite* collision time *$t^\star \geq 0$ such that $y(t^\star) = z(t^\star) \in K \cap L$.*
The collision kernel $\mathrm{Coll}_{\mathcal{S},\mathcal{T}}(K,L) \subset K \times L$ *is the subset of pairs $(y,z) \in K \times L$ such that there exist at least two viable colliding evolutions $(y(\cdot),z(\cdot)) \in \mathcal{S}^K(y) \times \mathcal{T}^L(z)$.*

Remark. Eupalinian kernels are obtained when $g = -f$, $U = V$ and $L = K$, or, equivalently, when the evolutionary system $\mathcal{R} = \overleftarrow{\mathcal{S}}$ is the backward evolutionary system. □

We can characterize the collision kernel as the capture basin of an auxiliary problem, so that it inherits the properties of capture basins:

Proposition 8.6.2 *[Viability Characterization of Collision Kernels] Recall that $\mathrm{Diag}(K \cap L) := \{(x,x)\}_{x \in K \cap L}$ denotes the* diagonal *of $K \cap L$. The collision kernel $\mathrm{Coll}_{\mathcal{S},\mathcal{T}}(K,L)$ of $K \cap L$ under the evolutionary systems $\mathcal{S}$ and $\mathcal{T}$ associated with the systems (8.18), p. 299 is the capture basin*

$$\mathrm{Coll}_{\mathcal{S},\mathcal{T}}(K,L) = \mathrm{Capt}_{(8.18)}(K \times L, \mathrm{Diag}(K \cap L))$$

of the diagonal of $K \cap L$ viable in $K \times L$ under the auxiliary system (8.18), p. 299.

We now "quantify" the concept of collision kernel with the concept of several collision intertemporal optimization problems. The domains of their value functions are the collision kernels, so that Proposition 8.6.2, p. 299 follows from Theorem 8.6.3, p. 300 below.

We introduce a cost function $\mathbf{c} : X \times X \mapsto \mathbb{R} \cup \{+\infty\}$ (regarded as a *collision cost*) and a Lagrangian $\mathbf{l} : (y,z,u,v) \rightsquigarrow \mathbf{l}(y,z,u,v)$.

The optimal viable collision problem consists in finding colliding viable evolutions $y(\cdot) \in \mathcal{S}^K(y)$ and $z(\cdot) \in \mathcal{T}^L(z)$ and a time $t^\star \geq 0$ minimizing

$$\begin{cases} \mathbf{W}_{(\mathbf{c},\mathbf{l})}(y,z) = \inf_{(y(\cdot),z(\cdot)) \in \mathcal{S}^K(y) \times \mathcal{T}^L(z),\ t^\star \mid y(t^\star)=z(t^\star)} \\ \left(\mathbf{c}(y(t^\star),z(t^\star)) + \int_0^{t^\star} \mathbf{l}(y(t),z(t),u(t),v(t))dt\right) \end{cases}$$

By taking $\mathbf{c} \equiv 0$ and $\mathbf{l}(y, z, u, v) \equiv 1$, we find the problem of governing two evolutions in minimal time.

We shall prove that

Theorem 8.6.3 *[Collision Optimization Theorem] Let us consider the auxiliary control system*

$$\begin{cases} (i) \ \ y'(t) \ = \ f(y(t), u(t)) \text{ where } u(t) \ \in \ U(y(t)) \\ (ii) \ \ z'(t) \ = \ g(z(t), v(t)) \text{ where } v(t) \ \in \ V(z(t)) \\ (iii) \ \lambda'(t) = -\mathbf{l}(y(t), z(t), u(t), v(t)) \end{cases} \tag{8.19}$$

Then

$$\mathbf{W}_{(\mathbf{c},\mathbf{l})}(y, z) \ = \ \inf_{(y,z,\lambda) \in \mathrm{Capt}_{(8.19)}(K \times K \times \mathbb{R}_+, \mathcal{E}p(\mathbf{c}) \cap (\mathrm{Diag}(K) \times \mathbb{R}_+))} \lambda$$

where $\mathrm{Diag}(K) := \{(x, x)\}_{x \in K} \subset K \times K$ *is the* diagonal of K.

Proof. Let $(y, z, \lambda) \in \mathrm{Capt}_{(8.19)}(K \times L \times \mathbb{R}_+, \mathcal{E}p(\mathbf{c}) \cap (\mathrm{Diag}(K) \times \mathbb{R}_+))$ belong to the capture basin. This means that there exist one evolution $y(\cdot) \in \mathcal{S}^K(y)$ viable in K, one evolution $z(\cdot) \in \mathcal{T}^L(z)$ viable in L, the evolution $\lambda(t) := \lambda - \int_0^t \mathbf{l}(y(s), z(s), u(s), v(s))ds$ and a time $t^\star$ such that:

- for all $t \in [0, t^\star]$, $y(t) \in K$, $z(t) \in L$,

$$\lambda - \int_0^t \mathbf{l}(y(s), z(s), u(s), v(s))ds \ \geq \ 0$$

- $y(t^\star) = z(t^\star)$
- and

$$\lambda - \int_0^{t^\star} \mathbf{l}(y(s), z(s), u(s), v(s))ds \ \geq \ \mathbf{c}(y(t^\star), z(t^\star))$$

This implies that

$$\mathbf{W}_{(\mathbf{c},\mathbf{l})}(y, z) \ \leq \ \mathbf{c}(y(t^\star), z(t^\star)) + \int_0^{t^\star} \mathbf{l}(y(s), z(s), u(s), v(s))ds \ \leq \ \lambda$$

and thus, that

$$\mathbf{W}_{(\mathbf{c},\mathbf{l})}(y, z) \ \leq \ \inf_{(y,z,\lambda) \in \mathrm{Capt}_{(8.19)}(K \times K \times \mathbb{R}_+, \mathcal{E}p(\mathbf{c}) \cap (\mathrm{Diag}(K) \times \mathbb{R}_+))} \lambda$$

For proving the opposite inequality, we associate with any $\varepsilon > 0$ two colliding evolutions $y_\varepsilon(\cdot) \in \mathcal{S}^K(y)$ and $z_\varepsilon(\cdot) \in \mathcal{T}^L(z)$ at some time $t_\varepsilon^\star \geq 0$, controls $u_\varepsilon(\cdot)$ and $v_\varepsilon(\cdot)$ such that

$$\left(\mathbf{c}(y_\varepsilon(t_\varepsilon^\star), z_\varepsilon(t_\varepsilon^\star)) + \int_0^{t_\varepsilon^\star} \mathbf{l}(y_\varepsilon(t), z_\varepsilon(t), u_\varepsilon(t), v_\varepsilon(t))dt\right) \leq \mathbf{W}_{(\mathbf{c},\mathbf{l})}(y,z) + \varepsilon$$

and the function

$$\lambda_\varepsilon(t) := \mathbf{W}_{(\mathbf{c},\mathbf{l})}(y,z) + \varepsilon - \int_0^t \mathbf{l}(y_\varepsilon(s), z_\varepsilon(s), u_\varepsilon(s), v_\varepsilon(s))ds$$

By construction,

$$\lambda_\varepsilon(t_\varepsilon^\star) \geq \mathbf{c}(y(t_\varepsilon^\star), z(t_\varepsilon^\star)) \text{ and } y(t_\varepsilon^\star) = z(t_\varepsilon^\star)$$

This implies that $(y, z, \mathbf{W}_{(\mathbf{c},\mathbf{l})}(y,z) + \varepsilon)$ belongs to the capture basin $\text{Capt}_{(8.19)}(K \times K \times \mathbb{R}_+, \mathcal{E}p(\mathbf{c}) \cap (\text{Diag}(K) \times \mathbb{R}_+))$. Hence

$$\inf_{(y,z,\lambda)\in\text{Capt}_{(8.19)}(K\times K\times\mathbb{R}_+,\mathcal{E}p(\mathbf{c})\cap(\text{Diag}(K)\times\mathbb{R}_+))} \lambda \leq \mathbf{W}_{(\mathbf{c},\mathbf{l})}(y,z) + \varepsilon$$

and it is enough to let ε converge to 0. $\square$

8.7 Particular Solutions to a Differential Inclusion

Consider a pair of evolutionary systems $\mathcal{S}$ and $\mathcal{T}$ associated with control systems (8.18), p. 299:

$$\begin{cases} (i) \ y'(t) = f(y(t), u(t)) \text{ where } u(t) \in U(y(t)) \\ (ii) \ z'(t) = g(z(t), v(t)) \text{ where } v(t) \in V(z(t)) \end{cases}$$

We look for *common solutions* $x(\cdot)$ of these two evolutionary systems (8.18). Whenever the control system (8.18)(i) is simpler to solve than the differential inclusion (8.18)(ii), the solutions of which are interpreted as *"particular" solutions, one can regard such common solutions to (8.18)(i) and (8.18)(ii) as particular solutions to the differential inclusions (8.18)(i)* and (8.18)(ii).

For instance,

- taking $g(z, v) := 0$, the common solutions are equilibria of (8.18)(i),
- taking for $g(z, v) = v$ a constant velocity, then common solutions are affine functions of time t,
- taking for $g(z, v) = -mz$, then common solutions are exponential functions of time ze^{-mt}

and so on. The problem is to detect what are the initial states y which are equilibria, from which starts an affine evolution or from which starts an exponential solution.

In other words, finding particular solutions amounts to finding the set of the initial states from which common solutions do exist.

Lemma 8.7.1 *[**Extraction of Particular Solutions**] Denote by* $\text{Diag}(X) := \{(x,x)\}_{x \in X}$ *the "diagonal" of* $X \times X$*. Then the set of points from which start common solutions to the control systems is the viability kernel* $\text{Viab}_{(8.18)}(\text{Diag}(X))$ *of the diagonal under (8.18).*

Proof. Indeed, to say that $x(\cdot) \in \mathcal{S}(x) \cap \mathcal{T}(x)$ is a common solution to control systems (8.18), p. 299 amounts to saying that the pair $(x(\cdot), x(\cdot))$ is a solution to system (8.18) viable in the diagonal $\text{Diag}(K)$, so that (x,x) belongs to the viability kernel $\text{Viab}_{(8.18)}(\text{Diag}(X))$. Conversely, to say that (x,x) belongs to this viability kernel amounts to saying that there exist evolutions $(y(\cdot), z(\cdot)) \in \mathcal{S}(x) \times \mathcal{T}(y)$ viable in the diagonal $\text{Diag}(X)$, so that, for all $t \geq 0$, $y(t) = z(t)$ is a common solution. □

Being a viability kernel, the subset of initial states from which start particular evolutions inherits the properties of viability kernels.

8.8 Visiting Kernels and Chaos À la Saari

The fundamental problem of *qualitative analysis* (and in particular, of *qualitative physics* in computer sciences and *comparative statics* in economics) is the following: subsets $C_n \subset K$ are assumed to describe "qualitative properties", and thus are regarded as *qualitative cells* or, simply, cells. Examples of such qualitative cells are the *monotonic cells* (see Definition 9.2.14, p. 332). Given such an ordered sequence of overlapping qualitative cells, the question arises whether there exist viable evolutions visiting successively these qualitative cells in prescribed order. These are questions treated by *Donald Saari* that we partially cover here (see [182–184, Saari]).

We answer here the more specific question of the existence of such viable evolutions visiting not only finite sequences of cells, but also infinite sequences. Existence of viable visiting cells require some assumptions.

Definition 8.8.1 *[**Visiting Kernels**] Let us consider a sequence of nonempty closed subsets* $C_n \subset K$ *such that* $C_n \cap C_{n+1} \neq \emptyset$*. An evolution* $x(\cdot) \in \mathcal{S}(x)$ *is* visiting *the subsets* C_n *successively in the following sense: there exists a sequence of finite duration* $\tau_n \geq 0$ *such that, starting with*

t_0, *for all* $n \geq 0$,

$$\begin{cases} (i)\ t_{n+1} = t_n + \tau_n \\ (ii)\ \forall\, t \in [t_n, t_{n+1}],\ \ x(t) \in C_n \text{ and } x(t_{n+1}) \in C_{n+1} \end{cases} \tag{8.20}$$

The set $\text{Vis}_{\mathcal{S}}(K, \overrightarrow{C})$ *of initial states* $x \in K$ *such that there exists an evolution* $x(\cdot) \in \mathcal{S}(x)$ *visiting successfully the cells* C_n *is called the* visiting kernel *of the sequence* $\overrightarrow{C}$ *of cells* C_n *viable in* K *under the evolutionary system* $\mathcal{S}$.
The T*-visiting kernel* $\text{Vis}_{\mathcal{S}}(K, \overrightarrow{C})(T)$ *is the set of initial states from which starts at least one viable evolution visiting the cells with duration* τ_n *bounded by* T.

We begin by considering the case of a finite sequence of cells and a result due to *Donald Saari*:

Lemma 8.8.2 ***[Existence of Evolutions Visiting a Finite Number of Cells]*** *Let us consider a finite sequence of subsets* $C_n \subset K$ $(n = 0, \ldots, N)$ *such that*

$$T := \sup_{n=0,\ldots,N-1} \sup_{y \in C_{n+1}} \inf_{z \in C_n} \epsilon_{C_n}(y, z) < +\infty \tag{8.21}$$

where ϵ_{C_n} *is the* Eupalinian function *viable in* C_n *(see Definition 8.5.4, p. 294).*

Then, the T*-visiting kernel* $\text{Vis}_{\mathcal{S}}(K, C_1, \ldots, C_N)(T)$ *of this finite sequence is not empty.*

Proof. We set

$$M_{N-1}^N := \text{Capt}_{\mathcal{S}}(C_{N-1}, C_{N-1} \cap C_N)(T)$$

which is the subset of $x \in C_{N-1}$ such that there exist $\tau \in [0, T]$ and an evolution $x(\cdot) \in \mathcal{S}(x)$ viable in C_{N-1} on $[0, T]$ such that $x(\tau) \in C_N$. For $j = N-2, \ldots, 0$, we define recursively the cells:

$$M_j^N := \text{Capt}_{\mathcal{S}}(C_j, C_j \cap M_{j+1}^N)(T)$$

which can be written

$$M_j^N = \{x \in C_j \mid \exists\, \tau_j \in [0, T],\ \exists\, x(\cdot) \in \mathcal{S}^{C_j}(x) \text{ such that } x(\tau_j) \in M_{j+1}^N\}$$

Therefore the set $\text{Vis}_{\mathcal{S}}(K, C_0, \ldots, C_N)(T) = M_0^N$ is the set of initial states $x_0 \in C_0$ from which at least one solution will visit successively the cells C_j, $j = 0, \ldots, N$. □

We shall prove here the existence of viable evolutions visiting a infinite number of cells C_n, although they require to take the limit, and therefore, to use theorems of Chap. 10, p. 375 (the proof can thus be omitted in a first reading).

Proposition 8.8.3 ***[Existence of Evolutions Visiting an Infinite Sequence of Cells]*** *Let K be a closed subset viable under an upper semicompact evolutionary system $\mathcal{S}$. We consider a sequence of compact subsets $C_n \subset K$ and we assume that*

$$T := \sup_{n \geq 0} \sup_{y \in C_{n+1}} \inf_{z \in C_n} \epsilon_{C_n}(y, z) < +\infty \tag{8.22}$$

Then, the T-visiting kernel $\text{Vis}_{\mathcal{S}}(K, \overrightarrow{C})(T)$ of this infinite sequence is not empty.

Proof. We shall prove that the intersection

$$K_\infty := \bigcap_{n \geq 0} \text{Vis}_{\mathcal{S}}(K, C_1, \ldots, C_n)(T) \subset \text{Vis}_{\mathcal{S}}(K, \overrightarrow{C})(T)$$

is not empty and contained in the visiting kernel $\text{Vis}_{\mathcal{S}}(K, \overrightarrow{C})(T)$. Lemma 8.8.2, p. 303 implies that the visiting kernels $\text{Vis}_{\mathcal{S}}(K, C_1, \ldots, C_n)(T)$ are not empty, and closed since the evolutionary system is upper semicompact (see Theorem 10.3.14, p. 390). Since the family of subsets $\text{Vis}_{\mathcal{S}}(K, C_1, \ldots, C_n)(T)$ form a decreasing family and since K is compact, the intersection K_∞ is nonempty.

It remains to prove that it is contained in the visiting kernel $\text{Vis}_{\mathcal{S}}(K, \overrightarrow{C})(T)$. Let us take an initial state x in K_∞ and fix n. Hence there exist $x_n(\cdot) \in \mathcal{S}(x)$ and a sequence of $t_n^j \in [0, jT]$ such that

$$\forall j = 1, \ldots, n,\ x_n(t_n^j) \in M_j^n \subset C_j \text{ and } \forall t \in [t_n^{j-1}, t_n^j],\ x_n(t) \in C_j$$

Indeed, there exist $y_1(\cdot) \in \mathcal{S}(x)$ and $\tau_1^n \in [0, T]$ such that $y_1(\tau_1^n)$ belongs to M_1^n. We set $t_1^n := \tau_1^n$, $x_1^n = y_1(t_1^n)$ and $x_n(t) := y_1(t)$ on $[0, t_1^n]$.

Assume that we have built $x_n(\cdot)$ on the interval $[0, t_j^n]$ such that $x_n(t_j^n) \in M_j^n \subset C_j$ for $j = 1, \ldots, k$. Since $x_n(t_k^n)$ belongs to M_k^n, there exist $y_{k+1}(\cdot) \in \mathcal{S}(x_n(t_k^n))$ and $\tau_{k+1}^n \in [0, T]$ such that

$$y_{k+1}(\tau_{k+1}^n) \in M_{k+1}^n$$

We set

$$t_{k+1}^n := t_k^n + \tau_{k+1}^n \ \& \ x_n(t+\tau_k^n) := y_{k+1}(t)$$

on $[t_k^n, t_{k+1}^n]$. When $k = n$, we extend $x_n(\cdot)$ to $[t_n^n, +\infty[$ by any evolution starting at $x_n(t_n^n)$ at time t_n^n.

Since the evolutionary system is assumed to be upper semicompact, the Stability Theorem 10.3.3, p. 385 implies that a subsequence (again denoted $x_n(\cdot)$) of the sequence $x_n(\cdot) \in \mathcal{S}(x)$ converges (uniformly on compact intervals) to some evolution $x(\cdot) \in \mathcal{S}(x)$ starting at x. By extracting successive converging subsequences of $\tau_j^n \in [0,T]$ converging to τ_j when $n \geq j \to +\infty$ and setting $t_{j+1} := t_j + \tau_j$, we infer that $x(t_j) \in C_j$. □

As a consequence, we obtain an extension to evolutionary systems of a theorem on "chaos" due to *Donald Saari*:

Theorem 8.8.4 ***[Chaotic Behavior à la Saari]*** *Let K be a compact subset viable under an upper semicompact evolutionary system $\mathcal{S}$. We assume that K is covered by a family of closed subsets K_a $(a \in \mathcal{A})$ satisfying the following assumption:*

$$T := \sup_{a\in\mathcal{A}} \sup_{y\in K} \inf_{z\in K_a} \epsilon_{K_a}(y,z) < +\infty \tag{8.23}$$

Then, for any sequence $a_0, a_1, \ldots, a_n, \ldots$, there exists at least one evolution $x(\cdot) \in \mathcal{S}(x)$ and an increasing sequence of elements $t_j \geq 0$ such that for all $j \geq 0$, $\forall t \in [t_j, t_{j+1}]$, $x(t) \in K_{a_j}$ and $x(t_{j+1}) \in K_{a_{j+1}}$.

Proof. We associate with the sequence $a_0, a_1, \ldots, a_n, \ldots$ the sequence of subsets $C_n := K_{a_n}$ and we observe that assumption (8.22) of Proposition 8.8.3, p. 304 is satisfied. □

8.9 Contingent Temporal Logic

By identifying a subset $K \subset X$ with the subset of elements $x \in X$ satisfying the property $\mathcal{P}_K(x)$ of belonging to K, i.e., $\mathcal{P}_K(x)$ if and only if $x \in K$, we know that we *can identify implication and negation with inclusion and complementation*:

$$\begin{cases} (i) \ \mathcal{P}_K \Rightarrow \mathcal{P}_L \ (\mathcal{P}_K \text{ implies } \mathcal{P}_L) \text{ if and only if } K \subset L \\ (ii) \ \neg\mathcal{P}_K(x) \ (\text{not } \mathcal{P}_K(x)) \text{ if and only if } x \in \complement K \end{cases}$$

These axioms have been relaxed in many ways to define other logics. We adapt to the "temporal case" under uncertainty the concept of atypic logic introduced by Michel de Glas.

Taking into account time and uncertainty into logical operations requires an evolutionary system $\mathcal{S}$, associating with any x for instance the set $\mathcal{S}(x)$ of solutions $x(\cdot)$ to a differential inclusion $x' \in F(x)$ starting at x.

Definition 8.9.1 ***[Eventual Consequences]*** *Given an evolutionary system, we say that y is an* eventual consequence *of x – and write $y \succeq x$ – if there exist an evolution $x(\cdot) \in \mathcal{S}(x)$ starting from x and a time $T \geq 0$ such that $y = x(T)$ is reached by this evolution.*

This binary relation $y \succeq x$ is the contingent temporal preorder *associated with the evolutionary system $\mathcal{S}$,* temporal *because evolution is involved,* contingent *because this evolution is contingent.*

It is obvious that the binary relation $y \succeq x$ is:

1. **reflexive:** $x \succeq x$ and
2. **transitive:** if $y \succeq x$ and $z \succeq y$, then $z \succeq x$,

so that it is, by definition, a *preorder* on X.

Definition 8.9.2 ***[Contingent Temporal Implications and Falsification]*** *Let us consider an evolutionary system $\mathcal{S}$. We say that x:*

1. satisfies typically *($\overline{\wedge}\mathcal{P}_K(x)$) property $\mathcal{P}_K$ if* ***all eventual consequences*** *of x satisfy property $\mathcal{P}_K$:*
2. satisfies atypically *($\underline{\vee}\mathcal{P}_K(x)$) property $\mathcal{P}_K$ if* ***at least one eventual consequence*** *of x satisfies property $\mathcal{P}_K$,*
3. falsifies *($\dashv\mathcal{P}_K(x)$) property $\mathcal{P}_K$ if* ***at least one eventual consequence*** *of x does not satisfy property $\mathcal{P}_K$,*

Falsification. *The contingent temporal preorder allows us to define a concept of falsification (in French, réfutation), which translates mathematically a weaker concept of negation which Karl Popper (1902–1994) made popular.*

These definitions can readily be formulated in terms of capture basins and invariance kernels:

Lemma 8.9.3 *[Viability Formulation of Contingent Temporal Operations] The formulas below relate logical operations to invariance kernels and capture basins:*

$$
\begin{cases}
(i) & \barwedge \mathcal{P}_K(x) \text{ if and only if } x \in \mathrm{Inv}(K) := \mathrm{Inv}(K, \emptyset) \\
& x \textit{ satisfies typically } \text{property } \mathcal{P}_K \\
(ii) & \veebar \mathcal{P}_K(x) \text{ if and only if } x \in \mathrm{Capt}(K) := \mathrm{Capt}(X, K) \\
& x \textit{ satisfies atypically } \text{property } \mathcal{P}_K \\
(iii) & \dashv \mathcal{P}_K(x) \text{ if and only if } x \in \mathrm{Capt}(\complement(K)) \\
& x \textit{ falsifies } \text{(or } \textit{does not typically satisfies}\text{) property } \mathcal{P}_K
\end{cases}
$$

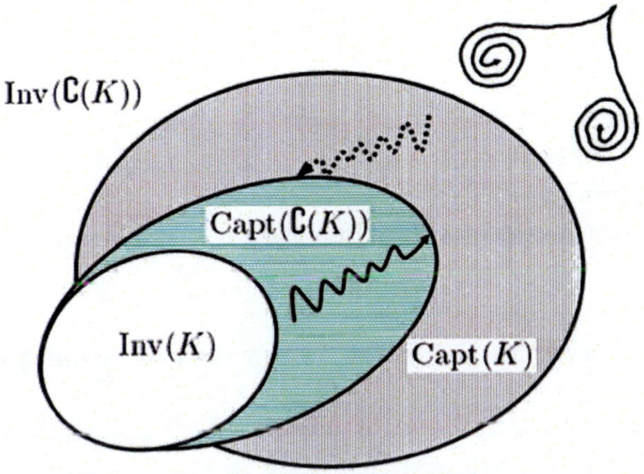

Fig. 8.4 Typical, atypical and falsifying elements.
This figure illustrates the above concepts: the subset K of elements satisfying property $\mathcal{P}_K$ is partitioned in the set $\mathrm{Inv}(K)$ *of* typical elements *satisfying property $\mathcal{P}_K$, and in the set* $\mathrm{Capt}(\complement(K))$ *of elements* falsifying *property $\mathcal{P}_K$. The capture basin* $\mathrm{Capt}(K)$ *of K is the set of* atypical elements *satisfying property $\mathcal{P}_K$.*

We now translate some elementary properties of capture basins and invariance kernels: For instance, contingent temporal logics are nonconsistent in the following sense:

Proposition 8.9.4 *[Non-Consistency of Contingent Temporal Logics] The contingent temporal logic is not consistent in the sense that:*

1. $\barwedge \mathcal{P}_K(x) \vee \dashv \mathcal{P}_K(x)$ is always true,

2. $\veebar \mathcal{P}_K(x) \wedge \dashv \mathcal{P}_K(x)$ *may be true (or is not false):* $\veebar \mathcal{P}_K(x) \wedge \dashv \mathcal{P}_K(x)$ *if and only if* x *both atypically satisfies and falsifies property* $\mathcal{P}_K$
3. *The falsification of the falsification of property* $\mathcal{P}_K$ *is the set of element satisfying extensively and intensively this property:*

$$\dashv\dashv \mathcal{P}_K(x) \Leftrightarrow \veebar \barwedge \mathcal{P}_K(x)$$

The relationships with conjunction and disjunction become

$$\begin{cases} (i) \ \dashv(\mathcal{P}_{K_1} \wedge \mathcal{P}_{K2}) \text{ if and only if } \dashv\mathcal{P}_{K_1} \vee \dashv\mathcal{P}_{K_2} \\ (ii) \ \dashv(\mathcal{P}_{K_1} \vee \mathcal{P}_{K2}) \text{ implies } \dashv\mathcal{P}_{K_1} \wedge \dashv\mathcal{P}_{K_2} \end{cases}$$

Definition 8.9.5 *[**Contingent Temporal Implications**] With an evolutionary system* $\mathcal{S}$*, we associate the following logical operations:*

1. Intensive contingent temporal implication

$$\mathcal{P}_K \rightrightarrows \mathcal{P}_L$$

means that ***all eventual consequences*** *of elements satisfying property* $\mathcal{P}_K$ *satisfy property* $\mathcal{P}_L$

2. Extensive contingent temporal implication

$$\mathcal{P}_K \Rrightarrow \mathcal{P}_L$$

means that whenever ***at least one eventual consequence*** *of an element satisfies property* $\mathcal{P}_K$*, it satisfies property* $\mathcal{P}_L$*.*

We observe that the *intensive and extensive contingent temporal implications imply the usual implication.*

Lemma 8.9.6 *[**Viability Characterization of Implications**] Extensive and intensive implications are respectively formulated in this way:*

$$\begin{cases} (i) \ \mathcal{P}_K \Rrightarrow \mathcal{P}_L \ (\mathcal{P}_K \text{ extensively } \textit{implies } \mathcal{P}_L) \textit{ if and only if } K \subset \mathrm{Inv}(L) \\ (ii) \ \mathcal{P}_K \rightrightarrows \mathcal{P}_L \ (\mathcal{P}_K \text{ intensively } \textit{implies } \mathcal{P}_L) \textit{ if and only if } \mathrm{Capt}(K) \subset L \end{cases}$$

and weak extensive and intensive implications defined respectively by

$$\begin{cases} (i)\ \mathcal{P}_K \rightharpoonup \mathcal{P}_L \text{ } (\mathcal{P}_K \text{ weakly extensively } \textit{implies } \mathcal{P}_L) \textit{ if and only if} \\ \quad \text{Capt}(K) \subset \text{Capt}(L) \\ (ii)\ \mathcal{P}_K \rightharpoondown \mathcal{P}_L \text{ } (\mathcal{P}_K \text{ weakly intensively } \textit{implies } \mathcal{P}_L) \textit{ if and only if} \\ \quad \text{Inv}(K) \subset \text{Inv}(L) \end{cases}$$

We infer the following

Proposition 8.9.7 *[Contraposition Properties] The following statements are equivalent:*

1. property $\mathcal{P}_K$ intensively implies $\mathcal{P}_L$:

$$\mathcal{P}_K \rightrightarrows \mathcal{P}_L$$

2. negation of property $\mathcal{P}_L$ extensively implies the negation of property $\mathcal{P}_K$:

$$\neg\mathcal{P}_L \Rrightarrow \neg\mathcal{P}_K$$

3. falsification of property $\mathcal{P}_L$ implies the negation of property $\mathcal{P}_K$:

$$\dashv\mathcal{P}_L \Rightarrow \neg\mathcal{P}_K$$

8.10 Time Dependent Evolutionary Systems

8.10.1 Links Between Time Dependent and Independent Systems

Consider the time-dependent system

$$x'(t) = f(t, x(t), u(t)) \text{ where } u(t) \in U(t, x(t))$$

Definition 8.10.1 *[Time-Dependent Systems] When the dynamics*

$$\begin{cases} (i)\ \ x'(t) = f(t, x(t), u(t)) \\ (ii)\ u(t) \in U(t, x(t)) \end{cases} \tag{8.24}$$

of a system depend upon the time, we denote by $\mathcal{S} : \mathbb{R} \times X \rightsquigarrow \mathcal{C}(-\infty, +\infty; X)$ the time-dependent evolutionary system *associating with any (T, x) the set* *of evolutions* $x(\cdot) \in \mathcal{C}(-\infty,+\infty;X)$ *governed by this time-dependent system passing through* x *at time* T*:* $x(T) = x$. *Whenever* $\mathbf{K} : t \rightsquigarrow K(t)$ *is a tube, we denote by* $\mathcal{S}^{\mathbf{K}}(x)$ *the set of evolutions* $x(\cdot) \in \mathcal{S}(x)$ *such that*

$$\forall\, t \geq 0,\ \ x(t)\ \in\ K(t)$$

Splitting evolutions allows us to decompose a full evolution passing through a given state at present time 0 into its backward and forward parts both governed by backward and forward evolutionary systems:

The *backward time-dependent system* $\overleftarrow{\mathcal{S}} : \mathbb{R} \times X \rightsquigarrow \mathcal{C}(-\infty,+\infty;X)$ associates with any (T,x) the set of evolutions $x(\cdot) \in \mathcal{C}(-\infty,+\infty;X)$ passing through x at time T: $x(T) = x$ and governed by

$$\begin{cases} (i)\ \ \overleftarrow{x}'(t) = -f(-t, \overleftarrow{x}(t), \overleftarrow{u}(t)) \\ (ii)\ \overleftarrow{u}(t) \in U(-t, \overleftarrow{x}(t)) \end{cases} \tag{8.25}$$

We observe that $x(\cdot) \in \mathcal{S}(T,x)$ if and only if:

1. its forward part $\overrightarrow{x}(\cdot) := \kappa(T)(x(\cdot))(\cdot)$ at time T defined by $\kappa(T)(x(\cdot))(t) = x(t-T)$ is a solution to differential inclusion
$$\overrightarrow{x}'(t)\ =\ f(T+t, \overrightarrow{x}(t), \overrightarrow{u}(t)) \text{ where } \overrightarrow{u}(t) \in U(T+t, \overrightarrow{x}(t))$$
satisfying $\overrightarrow{x}(0)\ =\ x$.
2. its backward part $\overleftarrow{x}(\cdot) := (\check{\kappa}\ (T)x(\cdot))(\cdot)$ at time T defined by $(\check{\kappa}\ (T)x\ (\cdot))(t) = x(T-t)$ is a solution to differential inclusion
$$\overleftarrow{x}'(t)\ =\ f(T-t, \overleftarrow{x}(t), \overleftarrow{x}(t)) \text{ where } \overleftarrow{u}(t) \in U(T-t, \overleftarrow{x}(t))$$
satisfying $\overleftarrow{x}(0)\ =\ x$.

This implies that when the system is time-independent, the *backward time-independent system* $\overleftarrow{\mathcal{S}} : X \rightsquigarrow \mathcal{C}(0,+\infty;X)$ associates with any $x \in X$ the set of evolutions $\overleftarrow{x}(\cdot) \in \mathcal{C}(0,+\infty;X)$ passing through x at time T: $x(T) = x$ and governed by (8.25), p. 310, which boils down to

$$\begin{cases} (i)\ \ \overleftarrow{x}'(t)\ =\ -f(\overleftarrow{x}(t), \overleftarrow{u}(t)) \\ (ii)\ \overleftarrow{u}(t)\ \in\ U(\overleftarrow{x}(t)) \end{cases}$$

This also allows us to introducing an auxiliary "time variable" τ the absolute value of the velocity of which is equal to one: $+1$ for forward time, -1 for backward time (time to horizon, time to maturity in finance, etc.).

Definition 8.10.2 *[Time-Independent Auxiliary System] We associate with the time-dependent evolutionary system (8.24),* *p. 309* *the* time-independent auxiliary system $\mathcal{A}_{\mathcal{S}}$ *associated with:*

$$\begin{cases} (i) \ \tau'(t) = 1 \\ (ii) \ x'(t) = f(\tau(t), x(t), u(t)) \\ \quad\quad \text{where } u(t) \in U(\tau(t), x(t)) \end{cases} \tag{8.26}$$

and the backward time-independent auxiliary system $\mathcal{A}_{\overleftarrow{\mathcal{S}}}$ *associated with*

$$\begin{cases} (i) \ \tau'(t) = -1 \\ (ii) \ x'(t) = -f(\tau(t), x(t), u(t)) \\ \quad\quad \text{where } u(t) \in U(\tau(t), x(t)) \end{cases} \tag{8.27}$$

We can reformulate the evolutions of the time-dependent system in terms of evolutions of the auxiliary time-independent systems $\mathcal{A}_{\mathcal{S}}$:

Lemma 8.10.3 ***[Links Between Evolutionary Systems and their Auxiliary Systems]***

1. *an evolution $x_+(\cdot) \in \mathcal{S}(T, x)$ is a solution to system (8.24),* p. 309 *starting from x at time T and defined on the interval $[T, +\infty[$ if and only if $x_+(\cdot) = \kappa(T)\overrightarrow{x}(\cdot)$ where $(\overrightarrow{\tau}(\cdot), \overrightarrow{x}(\cdot))$ is a solution of the auxiliary time-independent system (8.26) starting at (T, x).*
2. *an evolution $x_-(\cdot) \in \mathcal{S}(T, x)$ is a solution to system (8.24) arriving at x at time T and defined on the interval $]-\infty, T]$ if and only if $x_-(\cdot) = \check{\kappa}(T)\overleftarrow{x(\cdot)}(\cdot)$ wherc $(\overleftarrow{\tau}(\cdot), \overleftarrow{x}(\cdot))$ is a solution to the backward auxiliary time-independent system (8.27) starting at (T, x).*

In other words, an evolution $x(\cdot) \in \mathcal{S}(T, x)$ governed by time-dependent system (8.24), p. 309

$$x'(t) = f(t, x(t), u(t)) \text{ where } u(t) \in U(t, x(t))$$

can be split in the form

$$x(t) := \begin{cases} \overleftarrow{x}(-t) \text{ if } t \leq 0 \\ \overrightarrow{x}(t) \quad \text{ if } t \geq 0 \end{cases}$$

where $\overleftarrow{x}(\cdot) \in \mathcal{A}_{\overleftarrow{\mathcal{S}}}(T, x)$ and $\overrightarrow{x}(\cdot) \in \mathcal{A}_{\mathcal{S}}(T, x)$.

Proof. Indeed, let $x_+(\cdot)$ satisfying $x_+(T) = x$ and $x'_+(t) = f(t, x_+(t), u_+(t))$. Therefore, $\overrightarrow{x}(\cdot) := \kappa(-T)x_+(\cdot)$ defined by $\overrightarrow{x}(t) := x_+(t+T)$ satisfies $\overrightarrow{x}(0) := x_+(T) = x$ and $\overrightarrow{x}'(t) := x'_+(t+T) = f(t+T, x'_+(t+T), u_+(t+T)) =$

$f(\overrightarrow{\tau}(t), \overrightarrow{x}(t), \overrightarrow{u}(t))$ where $\overrightarrow{\tau}(t) := t + T$. This means that $(\overrightarrow{\tau}(\cdot), \overrightarrow{x}(\cdot))$ is a solution of the auxiliary time-independent system (8.26) starting at (T, x).

In the same way, let $x_-(\cdot)$ satisfying $x_-(T) = x$ and $x_-'(t) = f(t, x_-(t), u_-(t))$. Therefore, $\overleftarrow{x}(\cdot) := (\overset{\vee}{\kappa}(T)x_-(\cdot))(\cdot)$ defined by $\overleftarrow{x}(t) := x_-(T-t)$ satisfies $\overleftarrow{x}(0) := x_-(T) = x$ and $\overleftarrow{x}'(t) := -x_-'(T-t) = f(T-t, x_-'(T-t), u_-(T-t)) = f(\overleftarrow{\tau}(t), \overleftarrow{x}(t), \overleftarrow{u}(t))$ where $\overleftarrow{\tau}(t) := T - t$. This means that $(\overleftarrow{\tau}(\cdot), \overleftarrow{x}(\cdot))$ is a solution of the backward auxiliary time-independent system (8.27) starting at (T, x). □

Consequently, we just have to transfer the properties of the forward and backward systems of time-independent systems in forward time for obtaining the properties of time-dependent systems.

8.10.2 Reachable Maps and Detectors

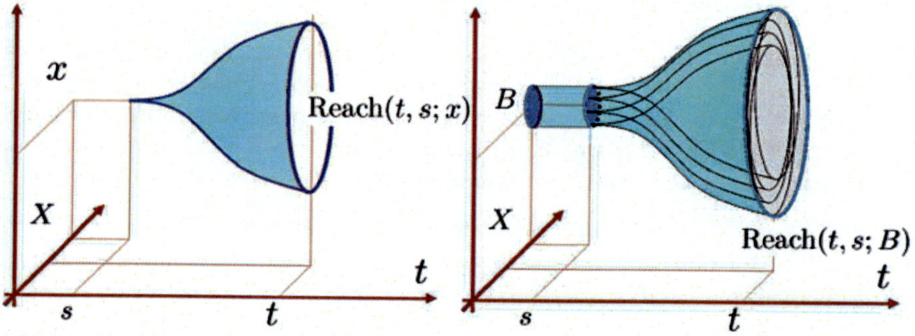

Fig. 8.5 Reachable Maps.
Left: Illustration of the reachable map $\mathrm{Reach}(t, s; x)$ associated with a point x between the times s and t. *Right:* Illustration of the reachable tube $\mathrm{Reach}(t, s; B)$ associated with a set B between the times s and t.

Definition 8.10.4 ***[Viable Reachable Tubes]*** *Let us consider a tube* $\mathbf{K}$ *regarded as the tube of time-dependent environments* $\mathbf{K}(t)$ *and a source* $B \subset X$. *Let* $\mathcal{S}^{\mathbf{K}} : \mathbb{R} \times X \rightsquigarrow \mathcal{C}(0, +\infty; X)$ *be the evolutionary system associated with (8.24), p. 309*

$$x'(t) = f(t, x(t), u(t)) \text{ where } u(t) \in U(t, x(t))$$

the set of evolutions viable in the tube $\mathbf{K}$. *The* viable reachable map $\mathrm{Reach}_{\mathcal{S}}^{\mathbf{K}}((\cdot), s; x) : t \rightsquigarrow \mathrm{Reach}_{\mathcal{S}}^{\mathbf{K}}(t, s; x)$ *associating with any* $x \in \mathbf{K}(s)$ *the* *set of $x(t)$ when $x(\cdot) \in \mathcal{S}^{\mathbf{K}}(s,x)$ ranges over the set of evolutions starting from x at time $s \leq t$ and viable in the tube:*

$$\forall\, x \in X,\ \forall t \geq s \geq 0,\ \text{Reach}_{\mathcal{S}}^{\mathbf{K}}(t,s;B) \ := \ \{x(t)\}_{x(\cdot)\in\mathcal{S}^{\mathbf{K}}(s,B)}$$

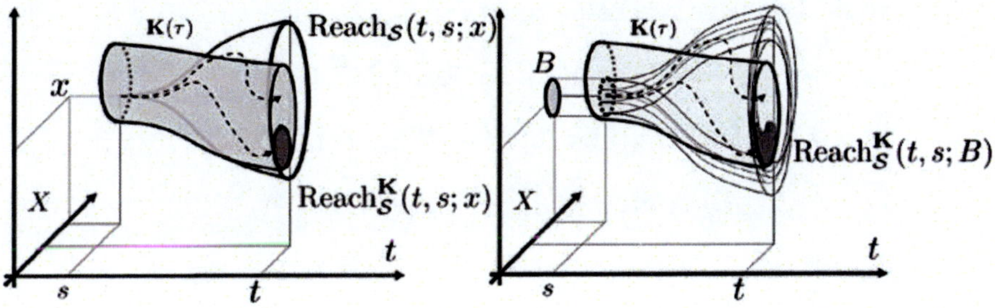

Fig. 8.6 Reachable Tubes.
Left: Illustration of $\text{Reach}_{\mathcal{S}}^{\mathbf{K}}(t,s;x)$ *as defined in Definition 8.10.4. It depicts the reachable tube* $\text{Reach}_{\mathcal{S}}(t,s;x)$ *without constraints and the tube representing the evolving environment* $\mathbf{K}(\cdot)$*. The dark area at time* t *is the viable reachable set* $\text{Reach}_{\mathcal{S}}^{\mathbf{K}}(t,s;B)$*. It is contained in the intersection of the constrained tube and the reachable tube without constraints. Right: Illustration of* $\text{Reach}_{\mathcal{S}}^{\mathbf{K}}(t,s;B)$*, obtained by taking the union of the tubes* $\text{Reach}_{\mathcal{S}}^{\mathbf{K}}(t,s;x)$ *when* x *ranges over* B.

Definition 8.10.5 *[**Detectors**] Consider an evolutionary system* $\mathcal{S} : \mathbb{R} \times X \rightsquigarrow \mathcal{C}(-\infty,+\infty;X)$ *and two tubes* $\mathbf{K}(\cdot) : t \rightsquigarrow \mathbf{K}(t)$ *and* $\mathbf{B}(\cdot) : t \rightsquigarrow \mathbf{B}(t) \subset \mathbf{K}(t)$*. The* detector $\text{Det}_{\mathcal{S}}(\mathbf{K},\mathbf{B}) : \mathbb{R}_+ \rightsquigarrow X$ *associates with any* $T \geq 0$ *the (possibly empty) subset* $\text{Det}_{\mathcal{S}}(\mathbf{K},\mathbf{B})(T)$ *of states* $x \in \mathbf{K}(T)$ *which can be reached by at least one evolution* $x(\cdot)$ *starting at some finite earlier time* $\tau \leq T$ *from* $\mathbf{B}(\tau)$ *and* viable *on the interval* $[\tau,T]$ *in the sense that*

$$\forall t \in [\tau,T],\ x(t) \in \mathbf{K}(t) \tag{8.28}$$

In other words, it is defined by formula:

$$\text{Det}_{\mathcal{S}}(\mathbf{K},\mathbf{B})(T) \ := \ \bigcup_{s\leq T} \text{Reach}_{\mathcal{S}}^{\mathbf{K}}(T,s;\mathbf{B}(s)) \tag{8.29}$$

Observe that when the system is time-independent, the formula for detectors boils down to

$$\mathrm{Det}_{\mathcal{S}}(\mathbf{K},\mathbf{B})(T) = \bigcup_{0\leq s\leq T} \mathrm{Reach}_{\mathcal{S}}^{\mathbf{K}}(T-s,0;\mathbf{B}(s))$$

By taking for tube $\mathbf{B}_{\emptyset}(\cdot)$ the tube defined by

$$\mathbf{B}_{\emptyset}(t) := \begin{cases} B \text{ if } t=0 \\ \emptyset \text{ if } t>0 \end{cases}$$

we recognize the viable reachable tube

$$\mathrm{Det}_{\mathcal{S}}(\mathbf{K},\mathbf{B}_{\emptyset})(T) := \mathrm{Reach}_{\mathcal{S}}^{\mathbf{K}}(T,0;B)$$

and by taking the constant tube $\mathbf{B}_0 : t \rightsquigarrow B$, we obtain

$$\mathrm{Det}_{\mathcal{S}}(\mathbf{K},\mathbf{B}_0)(T) := \mathrm{Det}_{\mathcal{S}}(\mathbf{K},B)(T)$$

An illustration of a detector is shown in Fig. 8.7. This figures relates to Theorem 8.10.6 below.

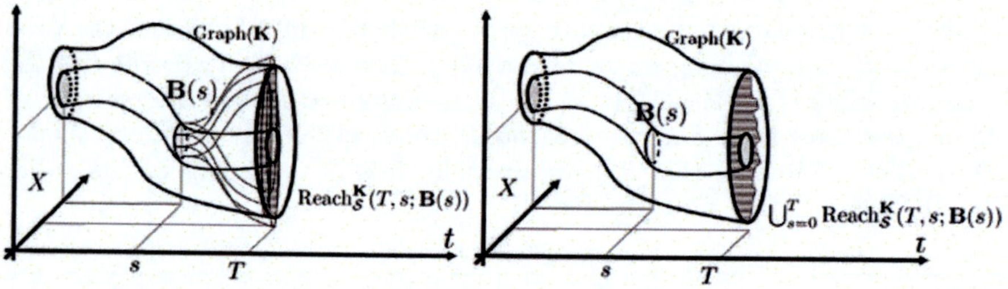

Fig. 8.7 Detectors.
Left: Illustration of $\mathrm{Reach}_{\mathcal{S}}^{\mathbf{K}}(T,s;\mathbf{B}(s))$. ***Right:*** *Illustration of* $\bigcup_{s=0}^{T} \mathrm{Reach}_{\mathcal{S}}^{\mathbf{K}}(T,s;\mathbf{B}(s))$ *which is the detector* $\mathrm{Det}_{\mathcal{S}}(\mathbf{K},\mathbf{B})(T)$.

As for the viable reachable map, the graph of the detector is the viability kernel of the graph of the tube $\mathbf{K}(\cdot)$ with the target chosen to be the graph of the source tube $\mathbf{B}(\cdot)$ under the auxiliary evolutionary system (8.27).

Theorem 8.10.6 ***[Viability Characterization of Detectors]*** *The graph of the detector* $\mathrm{Det}_{\mathcal{S}}(\mathbf{K},\mathbf{B})$ *is the capture basin of the target* $\mathrm{Graph}(\mathbf{B})$ *viable in the graph* $\mathrm{Graph}(\mathbf{K})$ *under the auxiliary system (8.27):*

$$\mathrm{Graph}(\mathrm{Det}_{\mathcal{S}}(\mathbf{K},\mathbf{B})) = \mathrm{Capt}_{(8.27)}(\mathrm{Graph}(\mathbf{K}),\mathrm{Graph}(\mathbf{B}))$$

Furthermore, the detector is the unique tube **D** *between the tubes* **B** *and* **K** *satisfying the "bilateral fixed tube" property:*

$$\mathbf{D}(T) = \bigcup_{s \leq T} \mathrm{Reach}_S^{\mathbf{D}}(T, s; \mathbf{B}(s))$$

and the Volterra property

$$\mathbf{D}(T) = \bigcup_{s \leq T} \mathrm{Reach}_S^{\mathbf{K}}(T, s; \mathbf{D}(s)) \tag{8.30}$$

Proof. Indeed, to say that (T, x) belongs to the capture basin of target $\mathrm{Graph}(\mathbf{B})$ viable in $\mathrm{Graph}(\mathbf{K})$ under the auxiliary system (8.27) means that there exist an evolution $\overleftarrow{x}(\cdot)$ to the backward system

$$\begin{cases} (i) & \overleftarrow{x}'(t) = -f(T-t, \overleftarrow{x}(t), \overleftarrow{u}(t)) \\ (ii) & \overleftarrow{u}(t) \in U(T-t, \overleftarrow{x}(t)) \end{cases}$$

starting at $\overleftarrow{x}(0) := x$ and a time $t^\star \geq 0$ such that

$$\begin{cases} (i) & \forall t \in [0, t^\star], \; (T-t, \overleftarrow{x}(t)) \in \mathrm{Graph}(\mathbf{K}) \\ (ii) & \qquad\qquad (T-t^\star, \overleftarrow{x}(t^\star)) \in \mathrm{Graph}(\mathbf{B}) \end{cases}$$

The second condition means that $\overleftarrow{x}(t^\star)$ belongs to $\mathbf{B}(T-t^\star)$. The first one means that for every $t \in [t^\star, T]$, $\overleftarrow{x}(t) \in \mathbf{K}(T-t)$. This amounts to saying that the evolution $x(\cdot) := \overset{\vee}{\kappa}(T)\overleftarrow{x}(\cdot) = \overleftarrow{x}(T - \cdot)$ is a solution to the parameterized system (8.24), p. 309

$$x'(t) = f(t, x(t), u(t)) \text{ where } u(t) \in U(t, x(t))$$

starting at $\overleftarrow{x}(T-t^\star) \in \mathbf{B}(T-t^\star)$, satisfying $x(T) = x$ and

$$\forall t \in [T-t^\star, T], \; x(t) \in \mathbf{K}(t)$$

Setting $s^\star := T - t^\star$, this means that $x \in \mathrm{Det}_S(\mathbf{K}, \mathbf{B})(T)$. Hence $x \in \mathrm{Reach}_S^{\mathbf{K}}(T, s^\star; x(s^\star))$. This proves formula (8.29).

Theorem 10.2.5 implies that the graph of the detector is the unique graph $\mathrm{Graph}(\mathbf{D})$ of a set-valued map $\mathbf{D}$ between $\mathrm{Graph}(\mathbf{B})$ and $\mathrm{Graph}(\mathbf{K})$ satisfying

$$\begin{cases} \mathrm{Graph}(\mathbf{D}) = \mathrm{Capt}_{(8.27)}(\mathrm{Graph}(\mathbf{D}), \mathrm{Graph}(\mathbf{B})) \\ = \mathrm{Capt}_{(8.27)}(\mathrm{Graph}(\mathbf{K}), \mathrm{Graph}(\mathbf{D})) \end{cases}$$

and thus formula (8.30). □

We shall extend this concept of detection tubes to travel time tubes useful in transportation engineering or in population dynamics.

We do not provide here more illustrations of straightforward adaptations to the time-dependent case of other results gathered in this book to time-independent case.

8.11 Observation, Measurements and Identification Tubes

Detectors are extensively used in control theory, under various names, motivated by different problems dealing with observations, measurements and questions revolving around these issues.

For instance, there are situations when the initial state is not known: We only know the evolutionary system, associated, for instance, with a time-dependent control system

$$\begin{cases} (i) \;\; x'(t) = f(t, x(t), u(t)) \\ (ii) \; u(t) \in U(t, x(t)) \end{cases} \tag{8.31}$$

The question arises to compensate for the ignorance of initial conditions. Among various ways to do it, we investigate here the case when we have access to some observation $y(t) = h(x(t))$ up to a given present time T, where $h : X \mapsto Y$ is regarded as a *measurement map* (or a *sensor map*, an *observation map*). In other words, we do not have direct access to the state $x(t)$ of the system, but to some measurements of observations $y(t) \in Y$ of the state.

The questions arises whether we can find at each present time T an evolution $x(\cdot)$ governed by control system (8.31) satisfying

$$\forall t \in [0, T], \;\; y(t) = h(x(t))$$

More generally, we can take into account "contingent noise" in the measurements, and assume instead that the measurement map is a set-valued map $H : X \rightsquigarrow Y$ associates with the

$$\forall t \in [0, T], \; y(t) \in H(x(t)) \tag{8.32}$$

In summary, we have to detect evolutions governed by an evolutionary system satisfying the "time-dependent viability conditions" (8.32) on each time interval $[0, T]$. This answers questions raised by *David Delchamps* in 1989:

Information Contained in Past Measurements. *David Delchamps* (see *State Space and Input-Output Linear Systems*, [78, Delchamps]) regards measurements *as a deterministic memoryless entity that gives us a limited* *amount of information about the states.* The problem is formulated as follows: *How much information about the current state is contained in a long record of past (quantized) measurements of the system's output? Furthermore, how can the inputs to the system be manipulated so as to make the system's output record more informative about the state evolution than might appear possible based on a cursory appraisal?*

We shall study this problem in a more general setting, since condition (8.32) can be written in the form

$$\forall t \in [0,T],\ x(t) \in \mathbf{K}(t) := H^{-1}(y(t))$$

Since the solutions we will bring to this problem depends upon the "tube" $t \rightsquigarrow \mathbf{K}(t)$ and not on the fact that it is derived from a measurement map, this is in this context that we shall look for the set $\text{Reach}^{\mathbf{K}}_{(8.4)}(T, \mathbf{K}(0))$ of sates $x(T)$ where $x(\cdot)$ is an evolution governed by (8.32) satisfying the "time-dependent viability conditions" (8.32).

Furthermore, we may need also to regulate the evolutions satisfying the above viability property by a regulation law associating with the observations $y(t)$ up to time T the controls $u(t)$ performing such a task. This is a solution to the *parameter identification* problem where controls are regarded as state-dependent parameters to be identified, problems also called "inverse problems" (see Sect. 10.9, p. 427).

8.11.1 Anticipation Tubes

Another example of tube $\mathbf{K}(t)$ is provided not only by past measurements, but also by taking into account expectations made at each instant t for future dates $s := t + a$, $a \geq 0$. For each current time $t \geq 0$, we assume not only that we know (through measurements, for instance) that the state $x(t)$ belongs to a subset $P(t)$, but also that from $x(t)$ starts a prediction $a \mapsto x(t;a)$ made at time t, solution to a differential inclusion $\frac{d}{da}x(t;a) \in G(t;x(a))$ (parameterized by time t) satisfying constraints of the form

$$\forall a \geq 0,\ x(t;a) \in P(t)$$

In other words, we take for tube the viability kernel

$$\mathbf{K}(t) := \text{Viab}_{G(t,\cdot)}(P(t))$$

Taking such a tube $\mathbf{K}(\cdot)$ as an evolving environment means that the decision maker involves at each instant t *predictions* on the future viewed at time t concerning future evolutions $a \mapsto x(t;a)$ governed by an anticipated dynamical system $\frac{d}{da}x(t;a) \in G(t;x(a))$ viable on anticipated constraints $P(t)$. These anticipations are taken into account in the viability kernel $\mathbf{K}(t)$ only, but are not implemented in the differential inclusion $x'(t) \in F(t, x(t))$. When the dynamics depend upon the past, we have to study viability problems under historical differential inclusions.

Chapter 9
Local and Asymptotic Properties of Equilibria

9.1 Introduction

This chapter expands to the case of control and regulated systems some central concepts of "dynamical systems", dealing with local and asymptotic stability, attractors, Lyapunov stability and sensitivity to initial conditions, equilibria, Newton's methods for finding them and the inverse function theorem for studying their stability.

1. We begin by singling out two important and complementary concepts in Sect. 9.2, p. 321. *Permanence*, a concept introduced in biomathematics by *Josef Hofbauer* and *Karl Sigmund* is a stronger concept than the capture of a target: not only we wish to reach the target in finite time, but stay there forever. The permanence kernel is the subset of initial states from which such an evolution exits (see [122, Hofbauer & Sigmund]). In some sense, this is a "double viability" concept, so to speak: to remain in the environment for some finite time, and, next, in the target forever.
   In this framework, the interpretation of the subset of the environment as a target is no longer suitable, it is better to regard it as a *sub-environment, defined by stronger or harder constraints*, as a support for interpretation. Therefore, the evolution either remains in this sub-environment forever. when it reaches its viability kernel, or faces an alternative: either it *fluctuates* in the sense that it leaves the sub-environment for some finite time, it even fluctuates around it, but it finally reaches the viability kernel of the sub-environment to remain there forever. Otherwise, all evolutions leave both environments in finite time.
   The *crisis function* introduced by Luc Doyen and Patrick Saint-Pierre (see Sect. 9.2.2, p. 326) measures the minimal *time spent outside the sub environment* by evolutions starting from an initial state.
   It turns out that these concepts of permanence and fluctuation are not only interesting in themselves, but also quite useful for deriving several interesting consequences: concepts of heteroclines asymptotically linking

J.-P. Aubin et al., *Viability Theory*, DOI 10.1007/978-3-642-16684-6_9,

equilibria, or even, compact viable subsets. For Lorenz systems, for instance, we compute numerically the fluctuation basin of the Lorenz system. We provide in Sect. 9.2.4, p. 331 an important example of qualitative cells, the monotonic cells. Assigning to each component a monotonic behavior (for instance, the first component is decreasing, the second one is increasing, the third one is increasing, etc.), the monotonic cells are the subsets of states for which these monotonic properties are satisfied, of which we consider the fluctuation basin (see *Analyse qualitative*, [85, Dordan]).
Permanence and fluctuation will be used in the rest of this chapter.

2. Section 9.3, p. 344 deals with the concepts of *limit sets, attractors, attraction basins*, their viability properties and some criteria implying the *non emptiness of viability kernels.*
   The main concept introduced is the *limit map* associating with any evolution its limit set, made of cluster points of the evolution when time goes to infinity. The core of a subset, most often, reduced to an equilibrium, under the limit map is the set of evolutions converging to this set, and the inverse image of this core is the *attraction basin* of this set, which enjoys viability properties. These properties allow us to define in a rigorous way the biological concept of *spike evolutions*, which converges to an equilibrium, but *after* leaving a neighborhood of this equilibrium.
   The *attractor map* assigns to each initial state the image by the limit set map of all evolutions starting from this state, i.e., the collection of all cluster points of all evolutions, which is called here the attractor (this word is polysemous).
   It enjoys viability properties that we shall uncover as well as *localization* properties providing "upper set-valued estimates", so to speak. For instance, if the attractor of a set is contained in its interior, it is backward invariant.
3. Next, we focus our attention in Sect. 9.4, p. 354 to the concepts of *(local) Lyapunov stability around an equilibrium* or, more generally, around a compact subset viable under the evolutionary system. We above all measure the sensitivity on initial conditions by a *sensitivity function*, which can be characterized in terms of viability kernels: the sensitivity function assigns to each initial state a measure of its sensitivity to initial conditions around it.
4. Since stability or instability results deal with equilibria, we devote the short Sect. 9.5, p. 360 to their existence. We recall an equivalent statement to the Brouwer Fixed Point theorem stating that *viability implies stationarity*: it states that any *compact convex* subset *viable* under a system contains, under adequate assumptions, an equilibrium.
5. The question arises whether we can adapt the *Newton method* for finding equilibria to control systems. It turns out that this question is closely related to viability techniques, as we shall see in Sect. 9.6, p. 361. They allow us to find alternatives to the Newton method by characterizing maps

different from the derivatives, which enjoy the same goal (convergence to the equilibrium), the graph of which are viability kernels. They can thus be computed by the Viability Kernel Algorithm.

6. We conclude this section with the adaptation to set-valued maps of the *Inverse Function Theorem*, a cornerstone theorem of differential calculus. It is not only an existence theorem, but also a stability result. An equilibrium $\overline{x}$ is the solution to the equation $f(\overline{x}) = 0$ or to the inclusion $0 \in F(\overline{x})$. The Inverse Function Theorem states that for right hand sides $y \approx 0$ small enough, not only there exists a solution x to the equation $f(x) = y$ or to the inclusion $F(x) \ni y$, but also that there exists a constant c such that $\|\overline{x} - x\| \leq c\|y\|$, which is a *stability result*.
The price to pay to obtain this conclusion is to assume that the derivative of the map is continuous and invertible. Thanks to differential calculus of set-valued maps, we adapt this result to set-valued maps by also assuming that adequately defined derivatives of set-valued maps are (lower semi-) continuous and invertible. Section 9.7, p. 365 provides the more recent statement of a long series of results on metric regularity, proved thanks to viability tools.

9.2 Permanence and Fluctuation

Reaching the target $C \subset K$ of an environment in finite time is not the end of the story: What happens next? If the target C is viable, then the evolution may stay in C forever, whereas it has to leave the target in finite time if the target is a repeller. *In some sense, the time spent outside the subset C measures the duration of* crisis *of not staying in C*. This is what the *crisis function* does, instead of measuring the minimal time spent to reach the target C by the minimal time function $\varpi^{\flat}_{(K,C)}$ (see Definition 4.3.4, p. 135).

If an evolution $x(\cdot)$ reaches the target in finite time at a point $x \in \partial K \cap \text{Viab}_{\mathcal{S}}(C)$, it may remain in C forever and, otherwise, if the evolution $x(\cdot)$ reaches C at a point $x \in \partial K \backslash \text{Viab}_{\mathcal{S}}(C)$, the evolution will leave C in finite time and enters a new period of crisis. This crisis may be endless if the evolution enters the complement of the capture basin $\text{Capt}_{\mathcal{S}}(K, C)$ of C. Otherwise, the same scenario is played again, i.e., the evolution captures C.

Hence the complement $C \backslash \text{Viab}_{\mathcal{S}}(C)$ can itself been partitioned in two subsets, one from which the evolutions will never return to the target (before leaving K), the other from which at least one evolution returns and remains in the viability kernel of the target after a crisis lasting for a finite time of crisis. *Luc Doyen* and *Patrick Saint-Pierre* introduced and studied the concept of *crisis function to measure the time spent in K but outside C by evolutions $x(\cdot) \in \mathcal{S}(x)$.*

9.2.1 Permanence Kernels and Fluctuation Basins

Evolutions $x(\cdot) \in \mathcal{K}(K,C)$ viable in K until they reach the target C in finite time (see Definition 2.2.3, p. 49) do not tell us what happens after they hit the target. Here, the idea is not so much to regard $C \subset K$ as a target, but rather as a *cell* (see Sect. 9.2.4, p. 331) or a sub-environment defined by stronger viability constraints and to investigate interesting behaviors of evolutions regarding the sub-environment C.

We adapt the concept of *permanent evolution* introduced in biomathematics by *Josef Hofbauer* and *Karl Sigmund* and the concept of *fluctuating evolution* around C introduced by *Vlastimil Krivan* in the ecological framework of carrying capacity:

Definition 9.2.1 *[Permanent and Fluctuating Evolutions] Let us consider two subsets $C \subset K \subset X$. We denote by:*

1. *$\mathcal{P}(K,C) \subset \mathcal{C}(0,+\infty;X)$ the subset of evolutions viable in K until they reach C in finite time and then, are viable in C forever, called* permanent evolutions *in C viable in K:*

$$\exists\, T \geq 0 \text{ such that } x(s) \in K \text{ if } s \in [0,T] \text{ and } x(t) \in C \text{ if } t \geq T$$

2. *$\mathcal{F}(K,C) \subset \mathcal{C}(0,+\infty;X)$ the subset of evolutions viable in K leaving C in finite time whenever they reach it, called* fluctuating evolutions *around C viable in K:*

$$\forall\ t \geq 0,\ x(t)\ \in\ K \text{ and } \exists\, T \geq t \text{ such that } x(T)\ \in\ K \setminus C$$

For the sake of brevity, we shall not always mention that permanent evolutions in C or fluctuating around C are viable in K: this will be assumed implicitly whenever no risk of confusion happens.

Remark: Bursting and Recurrent Evolutions.
The concept of fluctuating evolution conveys the notion introduced in neuro-sciences and bio-mathematics under the name *bursting evolution*, which have to leave the given cell in finite time (*to burst*, to break open or apart suddenly) again and again (see [101, Françoise] for more details).

By setting $D := K \backslash C$, fluctuating evolutions around C are called *recurrent evolutions* in D in the dynamical system literature. Since viable evolutions which are not permanent in C are fluctuating around C and vice-versa, and since the concept of fluctuating evolution is prevalent in bio-mathematics, we chose to exchange the concept of recurrent evolution in a set with the concept of fluctuating evolution around its complement. □

Lemma 9.2.2 ***[Partition of the Set of Viable Evolutions]*** *Recall that* $\mathcal{V}(K) \subset \mathcal{C}(0,+\infty;X)$ *denotes the set of evolutions viable in* $K \subset X$*. Then, for every subset* $C \subset K$*, the families* $\mathcal{P}(K,C)$ *of permanent evolutions in* C *and* $\mathcal{F}(K,C)$ *of fluctuating evolutions around* C *form a* partition *of the family* $\mathcal{V}(K)$ *of viable evolutions in* K*:*

$$\mathcal{V}(K) \; = \; \mathcal{P}(K,C) \cup_{\emptyset} \mathcal{F}(K,C)$$

Recall that a partition of $\mathcal{V}(K) = \mathcal{P}(K,C) \cup_{\emptyset} \mathcal{F}(K,C)$ means

$$\mathcal{V}(K) \; = \; \mathcal{P}(K,C) \cup \mathcal{F}(K,C) \text{ and } \mathcal{P}(K,C) \cap \mathcal{F}(K,C) \; = \; \emptyset$$

As for viability kernels and absorption basins which are respectively the *inverse image* of the family of viable evolutions in K and the *core* of the set of capturing evolutions under the evolutionary system (see Definition 18.3.3, p. 720), we suggest to use the Partition Lemma 18.3.4, p. 721 for introducing the inverse image of the family of permanent evolutions in C and the core of the family of fluctuating evolutions around C. Indeed, it states that the partition $\mathcal{V}(K) = \mathcal{P}(K,C) \cup_{\emptyset} \mathcal{F}(K,C)$, with $\mathcal{P}(K,C) \cap \mathcal{F}(K,C) = \emptyset$ implies that the inverse image $\mathcal{S}^{-1}(\mathcal{P}(K,C))$ of $\mathcal{P}(K,C)$ and the core $\mathcal{S}^{\ominus 1}(\mathcal{F}(K,C))$ form a partition of the core $\mathcal{S}^{\ominus 1}(\mathcal{V}(K))$ of $\mathcal{V}$:

$$\begin{cases} \mathcal{S}^{\ominus 1}(\mathcal{V}(K)) \; = \; \mathcal{S}^{-1}(\mathcal{P}(K,C)) \cup \mathcal{S}^{\ominus 1}(\mathcal{F}(K,C)) \\ \text{and} \\ \mathcal{S}^{-1}(\mathcal{P}(K,C)) \cap \mathcal{S}^{\ominus 1}(\mathcal{F}(K,C)) \; = \; \emptyset \end{cases}$$

We recognize the invariance kernel $\mathrm{Inv}_{\mathcal{S}}(K) := \mathcal{S}^{\ominus 1}(\mathcal{V}(K))$ in the left-hand side (see Definition 2.11.2, p. 89). It is convenient to attribute names to the subsets appearing in the right hand side:

Definition 9.2.3 ***[Permanence Kernel and Fluctuation Basins of a Set]*** *The* permanence kernel $\mathrm{Perm}_{\mathcal{S}}(K,C)$ *of a nonempty subset* $C \subset K$ *(viable in* K*) under the evolutionary system* $\mathcal{S}$ *is the subset of initial states* $x \in K$ *from which starts at least one permanent evolution in* C *viable in* K*. In other words, it is the set of initial states from which starts* at least one evolution viable in K and eventually viable in C forever.

The fluctuation basin $\text{Fluct}_{\mathcal{S}}(K, C)$ *of a nonempty subset* $C \subset K$ *(viable in* K*) under the evolutionary system* $\mathcal{S}$ *is the subset of initial states* $x \in K$ *from which all evolutions* $x(\cdot) \in \mathcal{S}(x)$ *are viable in* K *and fluctuate around* C *forever (see Definition 9.2.1, p. 322).*

We observe that

$$\text{Viab}_{\mathcal{S}}(C) \subset \text{Perm}_{\mathcal{S}}(K, C) = \text{Capt}_{\mathcal{S}}(K, \text{Viab}_{\mathcal{S}}(C)) \subset \text{Viab}_{\mathcal{S}}(K) \quad (9.1)$$

These definitions readily imply that

Lemma 9.2.4 ***[Viability Characterization of Permanence Kernels and Fluctuation Basins]*** *The permanence and fluctuation basins form a partition of the invariance kernel of* K*:*

$$\text{Inv}_{\mathcal{S}}(K) = \text{Perm}_{\mathcal{S}}(K, C) \cup_{\emptyset} \text{Fluct}_{\mathcal{S}}(K, C) \quad (9.2)$$

If K *is assumed to be invariant, it is the unique partition* $K = P \cup_{\emptyset} Q$ *such that*

$$\text{Viab}_{\mathcal{S}}(C) \subset P, \ \ P \text{ is viable and } Q \text{ is invariant}$$

In this case,

$$\text{Fluct}_{\mathcal{S}}(K, C) = \text{Inv}_{\mathcal{S}}(K \setminus \text{Viab}_{\mathcal{S}}(C))$$

Recall that (9.2) means that

$$\begin{cases} \text{Inv}_{\mathcal{S}}(K) = \text{Perm}_{\mathcal{S}}(K, C) \cup \text{Fluct}_{\mathcal{S}}(K, C) \\ \text{and} \\ \text{Perm}_{\mathcal{S}}(K, C) \cap \text{Fluct}_{\mathcal{S}}(K, C) = \emptyset \end{cases}$$

Proof. The first statement is a consequence of Partition Lemma 18.3.4, p. 721 and the second is the translation in plain language of $\text{Capt}_{\mathcal{S}}(K, \text{Viab}_{\mathcal{S}}(C))$.

By Theorem 10.2.7, p. 382, the capture basin $\text{Capt}_{\mathcal{S}}(K, \text{Viab}_{\mathcal{S}}(C))$ is the unique subset P between $\text{Viab}_{\mathcal{S}}(C)$ and K such that $P = \text{Capt}_{\mathcal{S}}(P, \text{Viab}_{\mathcal{S}}(C))$ and $\text{Capt}_{\mathcal{S}}(K, P) = P$. By taking complements, this means that $Q := K \setminus P = \text{Inv}_{\mathcal{S}}(\complement P, \complement K)$. Since K is assumed to be invariant, this means that $K = \text{Inv}_{\mathcal{S}}(K)$. Hence

$$\text{Viab}_{\mathcal{S}}(C) \subset P, \ \ P \text{ is viable and } K \text{ is invariant}$$

When K is invariant, then

$$\begin{cases} \text{Fluct}_{\mathcal{S}}(K,C) = K \cap (\complement\text{Capt}_{\mathcal{S}}(K, \text{Viab}_{\mathcal{S}}(C))) \\ = K \cap \text{Inv}_{\mathcal{S}}(\complement\text{Viab}_{\mathcal{S}}(C), \complement K) = \text{Inv}_{\mathcal{S}}(K \setminus \text{Viab}_{\mathcal{S}}(C)) \end{cases}$$

because, to say that $x \in K \cap \text{Inv}_{\mathcal{S}}(\complement\text{Viab}_{\mathcal{S}}(C), \complement K)$ means that all evolutions starting from x are either viable in $\complement\text{Viab}_{\mathcal{S}}(C)$ forever or until they reach K in finite time. Since K is assumed to be invariant, this possibility never happens. □

Proposition 9.2.5 ***[Case when the viability kernel and permanence kernels of the target coincide]*** *Let us assume that the subsets C and K are closed, that C has a nonempty interior and that K is invariant. The following statements are equivalent:*

$$\begin{cases} (i) \;\; \text{Perm}_{\mathcal{S}}(K,C) = \text{Viab}_{\mathcal{S}}(C) \\ (ii) \;\; \text{Viab}_{\mathcal{S}}(C) \subset \text{Int}(C) \\ (iii) \;\; \partial C \subset \text{Fluct}_{\mathcal{S}}(K,C) \\ (iv) \;\; K \setminus \text{Viab}_{\mathcal{S}}(C) = \text{Inv}_{\mathcal{S}}(K \setminus \text{Viab}_{\mathcal{S}}(C)) \text{ is invariant} \\ (v) \;\; \text{Viab}_{\mathcal{S}}(C) \text{ is backward invariant with respect to } K \\ \quad\quad \text{(see Definition 10.5.3, p. 401.)} \end{cases} \tag{9.3}$$

Proof. 1. Assume that $\text{Viab}_{\mathcal{S}}(C) \subset \text{Int}(C)$. We shall prove that

$$\partial C \cup (K \setminus \text{Viab}_{\mathcal{S}}(C)) \subset \text{Fluct}_{\mathcal{S}}(K,C)$$

Indeed, let us take $x \in K \setminus \text{Viab}_{\mathcal{S}}(C)$ and any evolution $x(\cdot) \in \mathcal{S}(x)$. It is viable in K since K is assumed to be invariant. Let $\varpi_{(K,C)}(x(\cdot))$ the first time when $x(\cdot)$ reaches C at its boundary ∂C. Since $\text{Viab}_{\mathcal{S}}(C) \subset \text{Int}(C)$, we deduce that $x^\star := x(\varpi_{(K,C)}(x(\cdot)))$ does not belong to the viability kernel of C, so that every evolution starting from $x^\star$ leaves C in finite time, including the evolution $x(\cdot) \in \mathcal{S}(x)$. This implies that $x \in \text{Fluct}_{\mathcal{S}}(K,C)$.
Furthermore, since $\partial C \subset K \setminus \text{Int}(C) \subset K \setminus \text{Viab}_{\mathcal{S}}(C)$, we infer that

$$\partial C \subset \text{Fluct}_{\mathcal{S}}(K,C)$$

2. Condition $\partial C \subset \text{Fluct}_{\mathcal{S}}(K,C)$ implies that the viability kernel of C is contained in its interior. If not, there would exist some $\overline{x} \in \text{Viab}_{\mathcal{S}}(C) \cap \partial C$. Since $\overline{x} \in \text{Viab}_{\mathcal{S}}(C)$, there exists $x(\cdot) \in \mathcal{S}(\overline{x})$ such that, for all $t \geq 0$, $x(t) \in K$. Since $\overline{x} \in \partial C \subset \text{Fluct}_{\mathcal{S}}(K,C)$, all evolutions starting from $\overline{x}$, and in particular $x(\cdot) \in \mathcal{S}(\overline{x})$, leave C in finite time. We obtained a contradiction.

3. Consequently, condition $\partial C \subset \text{Fluct}_{\mathcal{S}}(K,C)$ implying that the viability kernel of C is contained in its interior, it implies that $K \setminus \text{Viab}_{\mathcal{S}}(C) \subset \text{Fluct}_{\mathcal{S}}(K,C)$.
4. Condition $K \setminus \text{Viab}_{\mathcal{S}}(C) \subset \text{Fluct}_{\mathcal{S}}(K,C)$ is equivalent, by complementarity, to $\text{Perm}_{\mathcal{S}}(K,C) \subset \text{Viab}_{\mathcal{S}}(C)$. Actually, equality $\text{Perm}_{\mathcal{S}}(K,C) = \text{Viab}_{\mathcal{S}}(C)$ holds true because $\text{Viab}_{\mathcal{S}}(C) \subset \text{Perm}_{\mathcal{S}}(K,C)$ by (9.1), p. 324.
5. This equality implies that $K \setminus \text{Viab}_{\mathcal{S}}(C) = \text{Fluct}_{\mathcal{S}}(K,C)$. Lemma 9.2.4, p. 324 implies that since K in invariant, $\text{Fluct}_{\mathcal{S}}(K,C) = \text{Inv}_{\mathcal{S}}(K \setminus \text{Viab}_{\mathcal{S}}(C))$.
6. Equality $\text{Viab}_{\mathcal{S}}(C) = \text{Perm}_{\mathcal{S}}(K,C) = \text{Capt}_{\mathcal{S}}(K, \text{Viab}_{\mathcal{S}}(C))$ is equivalent to say that the viability kernel of C is backward invariant relatively to K, thanks to Theorem 10.5.6, p. 402. □

Remark: Barriers. Barrier Theorem 10.5.19, p. 409 implies that whenever the evolutionary system is upper semicompact and lower semicontinuous:

1. $\text{Int}(C) \cap \partial\text{Viab}_{\mathcal{S}}(C)$ exhibits the barrier property: For every $x \in \text{Int}(C) \cap \partial\text{Viab}_{\mathcal{S}}(C)$, all evolutions viable in the viability kernel $\text{Viab}_{\mathcal{S}}(C)$ are actually viable in its boundary as long as they remain in the interior of C,
2. $\text{Int}(K \setminus \text{Viab}_{\mathcal{S}}(C)) \cap (\partial\text{Perm}_{\mathcal{S}}(K,C) \setminus \text{Viab}_{\mathcal{S}}(C))$ exhibits the barrier property: For every $x \in \text{Int}(K \setminus \text{Viab}_{\mathcal{S}}(C)) \cap \partial\text{Perm}_{\mathcal{S}}(K,C)$, all evolutions viable in the permanence kernel $\text{Perm}_{\mathcal{S}}(K,C)$ are actually viable in its boundary as long as they remain in the interior of $K \setminus \text{Viab}_{\mathcal{S}}(C)$. □

9.2.2 The Crisis Function

In the same way that the *exit functions* $\tau_K^\sharp$ and the *minimal time functions* $\varpi_{(K,C)}^\flat$ "quantify" the concepts of viability kernels and capture basins, the *crisis function* $\upsilon_{(K,C)}$ "quantifies" the concept of *permanence kernel*, which is its domain (see Definition 9.2.1, p. 322).

> **31** ***[Wei Ji = Danger-Opportunity]*** *In Chinese, the two ideograms viability* 危机 *translating the word crisis have an interesting meaning in terms of viability: The first ideogram, wei-xian, means "danger", the second one, "ji hui", means "opportunity". The Greek etymology "Krisis" means decision.*

Consider a target $C \subset K$ contained in the environment K assumed to be invariant. Starting from the viability kernel, the minimal time spent outside C is equal to zero, starting from the permanence kernel outside the viability kernel, it will be strictly positive, but finite, and infinite outside the permanence kernel. The crisis function measures this minimal finite time of an evolution in the permanence kernel.

Definition 9.2.6 *[**Crisis Function**] The* crisis function $\upsilon_{(K,C)}(x) : X \mapsto \mathbb{R}_+ \cup \{+\infty\}$ *associates with* $x(\cdot)$ *its* crisis time *defined by*

$$\upsilon_{(K,C)}(x) := \inf_{x(\cdot)\in\mathcal{S}(x)} \text{meas}\{t \geq 0 \mid x(t) \in K\backslash C\}$$

By introducing the characteristic function $\chi_{K\backslash C}$ of the complement of $K\backslash C$, we observe that

$$\upsilon_{(K,C)}(x) = \inf_{x(\cdot)\in\mathcal{S}(x)} \int_0^{+\infty} \chi_{K\backslash C}(x(\tau))d\tau$$

This is an intertemporal minimization problem in infinite horizon.

As a consequence, Theorem 4.7.2, p. 156 implies that the crisis function can be characterized in terms of the viability kernel of the auxiliary system:

$$\begin{cases} (i) \;\; x'(t) = f(x(t), u(t)) \\ (ii) \; y'(t) = -\chi_{K\backslash C}(x(t)) \\ \qquad \text{where } u(t) \in U(x(t)) \end{cases} \tag{9.4}$$

subject to the constraint

$$\forall t \geq 0, \;\; (x(t), y(t)) \in K \times \mathbb{R}_+$$

Proposition 9.2.7 *[**Viability Characterization of the Crisis Function**] The crisis function is related to the viability kernel of* $K \times \mathbb{R}_+$ *under auxiliary system (9.4) by the following formula*

$$\upsilon_{(K,C)}(x) = \inf_{(x,y)\in\text{Viab}_{(9.4)}(K\times\mathbb{R}_+)} y$$

Remark: The associated Hamilton–Jacobi–Bellman Equation. The **crisis function** $\upsilon_{(K,C)}$ is the smallest positive solution to the Hamilton–Jacobi–Bellman partial differential equation

$$\begin{cases} (i)\ \forall x \in K\backslash C, & \inf_{u\in U(x)} \sum_{i=1}^{n} \frac{\partial \mathbf{v}(x)}{\partial x_i} f_i(x,u) + 1 = 0 \\ (ii)\ \forall x \in C, & \inf_{u\in U(x)} \sum_{i=1}^{n} \frac{\partial \mathbf{v}(x)}{\partial x_i} f_i(x,u) = 0 \ \square \end{cases}$$

Example. The epigraph of the crisis function under the Verhulst-Schaeffer metasystem $x'(t) = x(t)\left(\sqrt{\alpha}\sqrt{2\log\left(\frac{b}{x(t)}\right)} - v(t)\right)$ and some evolutions are described in Fig. 7.8 in Chap. 7, p. 247.

9.2.3 Cascades of Environments

From now on, we shall assume once and for all in this section that the environment K is invariant, so that we will be able to use its partition $K = \text{Perm}_{\mathcal{S}}(K,C) \cup_{\emptyset} \text{Fluct}_{\mathcal{S}}(K,C)$ in two disjoint subsets $\text{Perm}_{\mathcal{S}}(K,C)$ and $\text{Fluct}_{\mathcal{S}}(K,C)$. The above partition can be interpreted as follows:

Lemma 9.2.8 ***[Paradise, Purgatory and Hell]*** *Assume that K is invariant under an evolutionary system $\mathcal{S}$. Given a subset $C \subset K$ regarded as a cell, it can be divided into:*

1. *the* **Hell:** $C \cap \text{Fluct}_{\mathcal{S}}(K,C)$ *from which all evolutions $x(\cdot)$ fluctuate around C forever,*
2. *the* **Purgatory:** $(C \cap \text{Perm}_{\mathcal{S}}(K,C))\backslash\text{Viab}_{\mathcal{S}}(C)$ *from which at least one evolution leaves C in finite time before returning to C in finite time and then remaining in C forever, but some evolutions may leave it to enter hell* $\text{Fluct}_{\mathcal{S}}(K,C)$ *without being able to return (semi-permeability property),*
3. *the* **Paradise:** $\text{Viab}_{\mathcal{S}}(C)$, *from which starts at least one evolution viable in C forever, but some evolution may leave it (semi-permeability property).*

The Paradise is eternal, but does not guarantee that all evolutions remain in the paradise eternally and without interruption. Paradise $\text{Viab}_{\mathcal{S}}(C)$ is lost whenever C is a repeller. The Purgatory is lost whenever $\text{Viab}_{\mathcal{S}}(C) = \text{Perm}_{\mathcal{S}}(K,C)$, so that, in this case, outside Paradise is Hell, never allowing an evolution to enter C and remaining in it forever.

We can extend this result to the case in which the environment K is covered by an *(increasing) cascade of $N+1$ sub-environments* $K_0 \subset K_1 \subset$

$\cdots \subset K_i \subset \cdots \subset K_N := K$. Each cell K_i can describe more severe constraints increasingly. One assumes in this context that violating the constraints defining a sub-environment is dangerous, but not necessarily lethal. The candidate to be the "paradise" of this cascade is naturally the viability kernel $C_0 := \text{Viab}_S(K_0)$ of K_0, since from each state $x \in C_0$ starts at least one evolution viable in K_0, and thus, in all other cells K_i for $i \geq 1$.

However, there are several ways to reach this paradise from K_0 according to the position of the initial state outside K_0:

Proposition 9.2.9 ***[Viability Kernel of a Cascade of Cells]*** *Let us associate with an* (increasing) cascade *of "cells"* $K_0 \subset K_1 \subset \cdots \subset K_i \subset \cdots \subset K_N := K$ *of* $N+1$ *environments:*

- *the increasing sequence of subsets* $C_i := \text{Capt}_S(K_i, C_{i-1})$ *between* C_{i-1} *and* K_i
- *the sequence of disjoint subsets* $P_i := (C_{i+1} \setminus C_i) \cap K_i$ *of* K_i.

Then the ultimate environment K_0 *can be partitioned as the union of* C_0, *the subsets* $P_j \cap K_0$ *and* $K_0 \setminus C_N$. *From* C_0, *at least one evolution remains viable in* C_0 *forever. Starting from* $P_j \cap K_0$, $j = 0, \ldots, N-1$, *at least one evolution will leave successively the cells* $P_j \cap K_l$ *in increasing order for* $l = 0, \ldots, j$, *reach the cell* C_{j+1} *and then, all cells* $C_k \cap P_{k-1}$ *in decreasing order* $k = j, \ldots, 1$ *and reach the paradise in which it can remain viable forever. Starting from* $K_0 \setminus C_N$, *all evolutions fluctuate around* K_0.

Proof. Indeed, starting from C_k, at least one evolution will cross successively for periods of finite duration the m-th order "purgatorics" in decreasing order to reach finally the paradise C_0 in which he can stay forever.

Starting from $P_j \cap K_i$ for $0 \leq j \leq i$, at least one evolution starting from $P_j \cap K_i$ will cross successively for periods of finite duration the lth-order purgatories in increasing order $i \leq l \leq j$ to reach the $(j+1)$-th purgatory from which the evolution will go down through better and better purgatories before reaching in finite time C_0 in which he can remain viable forever.

Therefore, the ultimate environment K_0 can be partitioned in the paradise C_0, a sequence of k-th order purgatories $P_j \cap K_0$ and the ultimate hell $K_0 \setminus C_N$. From C_0, at least one evolution remains viable in C_0 forever. Starting from $P_j \cap K_0$, $j = 0, \ldots, N-1$, at least one evolution will leave successively the cells $P_j \cap K_l$ in increasing order for $l = 0, \ldots, j$, reach the cell C_{j+1} and then, all cells $C_k \cap P_{k-1}$ in decreasing order $k = j, \ldots, 1$ and reach the paradise in which it can remain viable forever.

The environment K being assumed invariant, the complement of C_N in K is the set of initial states from which all evolutions are fluctuating around C_{N-1}, and thus, in K_0. □

Remark. Barrier Theorem 10.5.19, p. 409 implies that whenever the evolutionary system is upper semicompact and lower semicontinuous, then, for all $i = 1, \ldots, N$, $\mathrm{Int}(K_i \backslash C_{i-1}) \cap (\partial C_i \backslash C_{i-1})$ exhibits the barrier property: For every $x \in \mathrm{Int}(K_i \setminus C_{i-1}) \cap \partial C_i$, all evolutions viable in C_i are actually viable in its boundary as long as they remain in the interior of $K_i \setminus C_{i-1}$. □

The concepts of fluctuation basins around one subset can easily be extended to the concept of fluctuation basins between two, or even, a finite number of cells K_i covering K (in the sense that $K = K_1 \cup K_2$ or $K = \bigcup_{i=1}^{n} K_i$), regarded as "cells" (as *qualitative cells* of Sect. 9.2.4, p. 331 for example), because:

Proposition 9.2.10 ***[Fluctuation Between Two Subsets]*** *Let $K_1 \subset K$ and $K_2 \subset K$ be two closed subsets covering an invariant subset K: $K = K_1 \cup K_2$. Then the intersection*

$$\begin{cases} \mathrm{Fluct}_{\mathcal{S}}(K; K_1, K_2) := \mathrm{Fluct}_{\mathcal{S}}(K, K_1) \cap \mathrm{Fluct}_{\mathcal{S}}(K, K_2) \\ = \mathrm{Inv}_{\mathcal{S}}(K \setminus (\mathrm{Viab}_{\mathcal{S}}(K_1) \cup \mathrm{Viab}_{\mathcal{S}}(K_2))) \end{cases}$$

of the fluctuation basins of each or the cells K_i $(i = 1, 2)$ is the set of initial states $x \in K$ from which all evolutions $x(\cdot) \in \mathcal{S}(x)$ viable in K fluctuate back and forth between K_1 to K_2 in the sense that the evolution leaves successively K_1 and K_2 in finite time.

If we assume furthermore that $\mathrm{Viab}_{\mathcal{S}}(K_i) \subset \mathrm{Int}(K_i)$, $i = 1, 2$, *then*

$$\mathrm{Fluct}_{\mathcal{S}}(K; K_1, K_2) := K \setminus (\mathrm{Viab}_{\mathcal{S}}(K_1) \cup \mathrm{Viab}_{\mathcal{S}}(K_2))$$

We can extend the situation where the subset K is covered by two cells to the case when $K = \bigcup_{i=1}^{n} K_i$ is covered by n "cells" $K_i \subset K$:

Proposition 9.2.11 ***[Fluctuation between Several Cells]*** *Assume that an invariant environment $K = \bigcup_{i=1}^{n} K_i$ is covered by n "cells" $K_i \subset K$. Then the intersection*

$$\begin{cases} \mathrm{Fluct}_{\mathcal{S}}(K; K_1, \ldots K_i \ldots, K_n) := \bigcap_{i=1}^{n} \mathrm{Fluct}_{\mathcal{S}}(K, K_i) \\ = \mathrm{Inv}_{\mathcal{S}}(K \setminus \bigcup_{i=1}^{n} (\mathrm{Viab}_{\mathcal{S}}(K_i))) \end{cases}$$

of the fluctuation basins of the cells K_i is the set of initial states $x \in K$ such that, for any evolution $x(\cdot) \in \mathcal{S}(x)$, for any cell K_i, $i = 1, \ldots, n$ and for any $x(t) \in K_i$, there exist a different cell K_j and a finite time $t_i^j > t$ such that $x(t_i^j) \in K_j$. If we assume furthermore that $\mathrm{Viab}_{\mathcal{S}}(K_i) \subset \mathrm{Int}(K_i)$, $i =$

1, ..., n, then

$$\mathrm{Fluct}_{\mathcal{S}}(K; K_1, \ldots, K_n) := K \setminus \bigcup_{i=1}^{n} (\mathrm{Viab}_{\mathcal{S}}(K_i))$$

Proof. To say that x belongs to the intersection of the fluctuation basins of the cells K_i amounts to saying that for every cell K_i, any evolution $x(\cdot) \in \mathcal{S}(x)$ fluctuates around K_i, i.e., that whenever $x(t) \in K_i$ at some t, there exists some $t^\star \geq t$ such that $x(t^\star) \in K \setminus K_i$. Since $K = \bigcup_{j=1}^{n} K_j$, we infer that $K \setminus K_i = \bigcup_{j \neq i}^{n} (K_j \setminus K_i)$. This means that there exists $j \neq i$ such that $x(t^\star) \in K_j \setminus K_i$.

The second statement follows from Proposition 9.2.5, p. 325 stating that whenever the viability kernel $\mathrm{Viab}_{\mathcal{S}}(K_i) \subset \mathrm{Int}(K_i)$ is contained in the interior of K_i, then $\mathrm{Fluct}_{\mathcal{S}}(K, K_i) = \mathrm{Inv}_{\mathcal{S}}(K \setminus \mathrm{Viab}_{\mathcal{S}}(K_i))$. □

This justifies the introduction of the definition

Definition 9.2.12 ***[Fluctuation Basins]*** *Assume that an invariant environment $K = \bigcup_{i=1}^{n} K_i$ is covered by n "cells" $K_i \subset K$. The intersection*

$$\mathrm{Fluct}_{\mathcal{S}}(K; K_1, \ldots K_i \ldots, K_n) := \bigcap_{i=1}^{n} \mathrm{Fluct}_{\mathcal{S}}(K, K_i)$$

of the fluctuation basins of the cells K_i is called the fluctuation basin *of the covering $K = \bigcup_{i=1}^{n} K_i$ by n "cells" $K_i \subset K$.*

9.2.4 Fluctuation Between Monotonic Cells

9.2.4.1 Monotonic Cells

Consider the control system (f, U) of the form (2.10):

$$x'(t) = f(x(t), u(t)) \text{ where } u(t) \in U(x(t))$$

Definition 9.2.13 *[The Equilibrium Map]* A state e is an equilibrium *if there exists at least one regulon $u \in U(e)$ such that the associated velocity $f(e,u) = 0$ vanishes to 0, so that the stationary evolution $t \mapsto x(t) := e$ is governed by this system. Given a subset K, we say that an equilibrium e is viable in K if it belongs to K. The set-valued map $U_{\propto} : X \rightsquigarrow \mathcal{U}$ defined by*

$$U_{\propto}(x) := \{u \in U(x) \text{ such that } f(x,u) = 0\}$$

associates with any state x the (possibly empty) subset $U_{\propto}(x)$ of regulons for which x is an equilibrium, called the equilibrium map.

Observe that a state e is an equilibrium if and only if the associated *singleton $\{e\}$ is viable under the system (f, U)* (so that they are minimal viable sets, see Definition 10.7.10, p. 421). The study of the inverse $U_{\propto}^{-1}$ of the equilibrium map, associating with any regulon the set of associated equilibria, is the topic of bifurcation and of catastrophe theories.

Monotonic cells are subsets on which the monotonicity behavior of the evolutions is the same, i.e., on which the sign of the velocities are the same.

For that purpose, we introduce the family of n-signs $a \in \mathcal{A} := \{-,+\}^n$ with which we associate the cones

$$\mathbb{R}^n_a := \{u \in \mathbb{R}^n \mid \text{sign}(u_i) = a_i \text{ or } 0, \ i = 1, \ldots, n\}$$

Definition 9.2.14 *[Monotonic Cell Maps] For studying the functions $t \mapsto \text{sign}(x'(t))$ associated with solutions $x(\cdot)$ of (2.10), we associate with any n-sign $a \in \mathcal{A} := \{-,+\}^n$:*

1. *the* monotonic map
$$U_a(x) := \{u \in U(x) \text{ such that } f(x,u) \in \mathbb{R}^n_a\}$$
2. *its domain, called the* monotonic cell
$$K(a) := \{x \in K \text{ such that } U_a(x) \neq \emptyset\}$$

Indeed, the "quantitative" states $x(\cdot)$ evolving in a given monotonic cell $K(a)$ share the same monotonicity properties because, as long as $x(t)$ remains in $K(a)$,

$$\forall i = 1, \ldots, n, \ \text{sign}\left(\frac{dx_i(t)}{dt}\right) = a_i$$

Hence,

$$\begin{cases} (i)\ K = \bigcup_{a\in\mathcal{A}} K(a), \text{ and } \forall x \in K,\ U(x) = \bigcup_{a\in\mathcal{A}} U_a(x) \\ (ii)\ K_0 := \bigcap_{a\in\mathcal{A}} K(a) \text{ is the set of (viable) equilibria} \\ \quad U_{\propto}(x) = \bigcap_{a\in\mathcal{A}} U_a(x) \text{ is the equilibrium map} \end{cases}$$

Indeed, to say that x belongs to K_0 means that, for all sign vectors $a \in \mathcal{A} := \{-,+\}^n$, $f(x,u) \in \mathbb{R}^n_a$, so that, $f(x,u) = 0$, i.e., that x is an equilibrium.

Lemma 9.2.15 *[Viable Evolutions in Monotonic Cells]* *Assume that K is compact. Then whenever $x \in \text{Viab}_{\mathcal{S}}(K(a))$, any evolution $x(\cdot) \in \mathcal{S}(x)$ viable in the monotonic cell $K(a)$ converges to an equilibrium.*

Proof. Indeed, since the evolution $x(\cdot)$ is viable in a monotonic cell $K(a)$, the signs of the derivatives $x_i'(t)$ of the components $x_i(t)$ of the evolutions are constant, so that the numerical functions $t \mapsto x_i(t)$ are monotone and bounded, and thus, convergent. □

The same proof than the one of Proposition 9.2.11 implies the following

Proposition 9.2.16 *[Monotonic Fluctuations]* *Let us consider the covering of $K = \bigcup_{a\in\{-,+\}^n} K(a)$ by the "monotonic cells" $K(a) \subset K$. Then*

$$\text{Fluct}_{\mathcal{S}}\left(K; \{K_a\}_{a\in\mathcal{A}}\right) = \text{Inv}_{\mathcal{S}}\left(K \setminus \bigcup_{a\in\{-,+\}^n} \text{Viab}_{\mathcal{S}}(K(a))\right)$$

is the set of initial states $x \in K$ such that, for all evolutions $x(\cdot) \in \mathcal{S}(x)$ viable in K, for all $t \geq 0$, there is a finite time $s > t$ such that the sign of the derivative of one component of the evolution changes.
When K is invariant, so is the subset

$$\text{Fluct}_{\mathcal{S}}\left(K; \{K_a\}_{a\in\mathcal{A}}\right) = K \setminus \bigcup_{a\in\{-,+\}^n} \text{Viab}_{\mathcal{S}}(K(a))$$

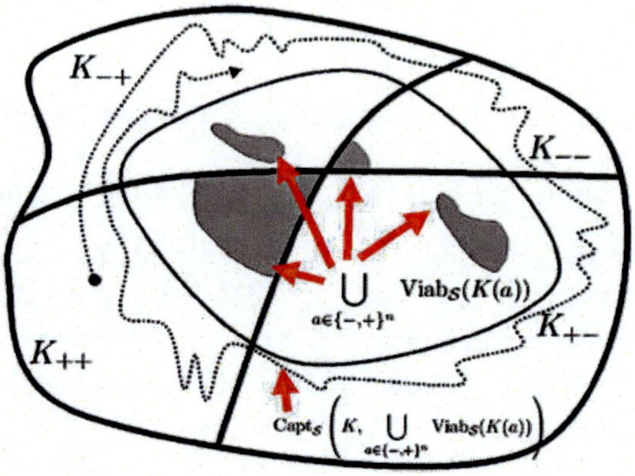

Fig. 9.1 Monotonic Fluctuations, Proposition 9.2.16.

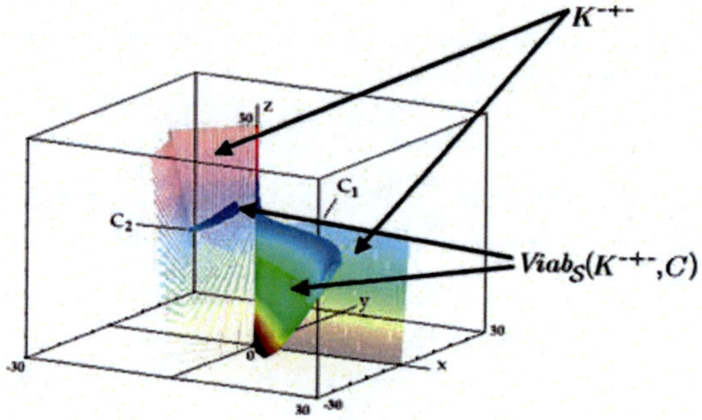

Fig. 9.2 The Viability Kernel of the Monotonic Cell K^{-+-} under the Lorenz System.
This Figure shows K^{+-+} (in light*), and the corresponding* $\mathrm{Viab}_S(K^{-+-}, C)$ *for the cell. Points in* $\mathrm{Viab}_S(K^{-+-}, C)$ *(in* dark*) are viable in the set K^{+-+} (in* light*).*

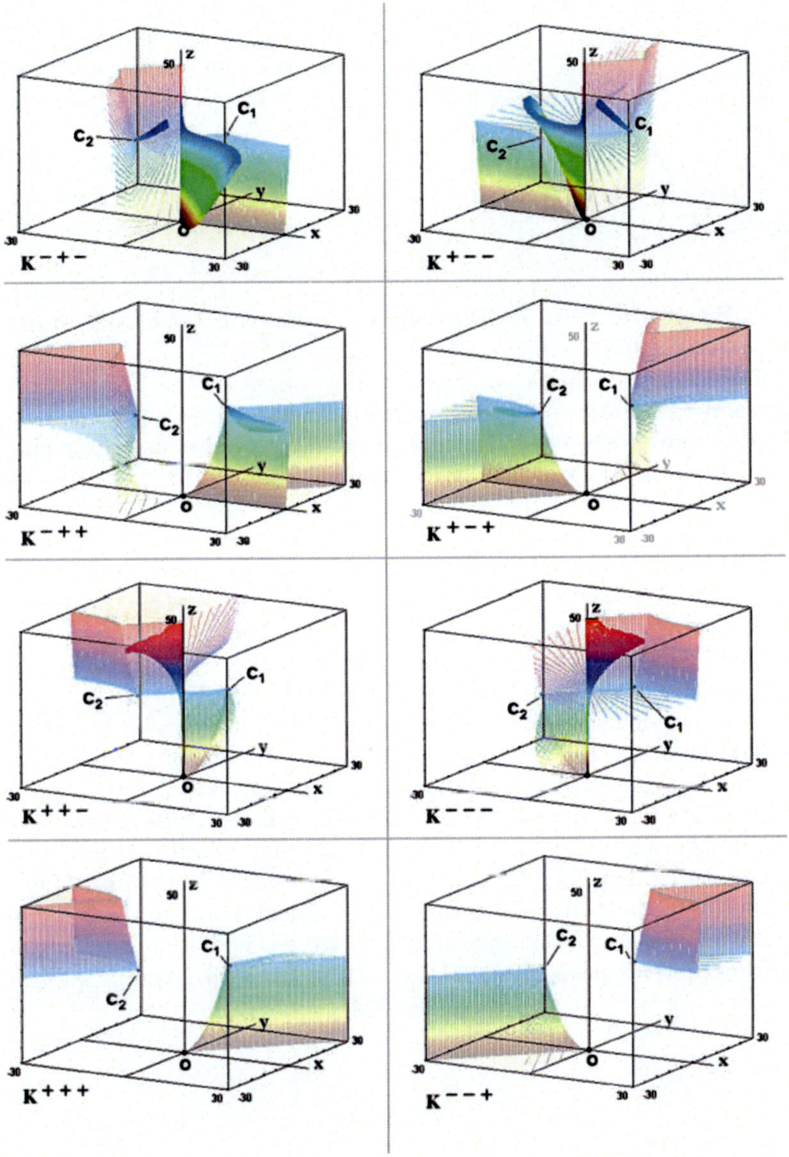

Fig. 9.3 Viability Kernels of Monotonic Cells under the Lorenz System. *Each of the subfigures displays the 8 viability kernels* $\text{Viab}_S(K(a))$ *of the monotonic cells of the cube under the Lorenz system. The viability kernels of the monotonic cells* K_{+++} *and* K_{--+} *are reduced to the equilibrium set made of the three equilibria. The solutions starting from the viability kernel of a monotonic cell converge to an equilibrium and the solutions starting outside leave the cell in finite time.*

Figure 9.3 shows the computation of $\text{Viab}_S(K(a),C)$ for $K(a) := K^{+++}$, K^{+-+}, K^{--+}, etc. Each subfigure shows the set $K(a)$ for the corresponding a (light set). For example for the top right subfigure K^{-+-} this set represents the set of points x such $U_{-+-}(x) \neq \emptyset$. The dark set corresponds to $\text{Viab}_S(K^{-+-},C)$.

Thus, a point in the complement of the capture basin of the union of all dark sets is the starting point of an evolution which fluctuates between all orthants K^{+++}, K^{+-+}, K^{--+}, etc.

9.2.4.2 Example: Fluctuation Basin Under the Lorenz System

One of the striking features of evolutions governed by the Lorenz system is the property that, *apparently*, all of them "fluctuate" between the half-cube containing one of the nontrivial equilibria to the half-cube containing the other equilibrium. Actually, it is the main flavor or the most spectacular feature encapsulated in the concept of "chaotic property" of the Lorenz system, a concept not rigourously defined in the literature. This fluctuation property is not true for all initial states: none of the three equilibria fluctuate! Hence the desire to know what is the fluctuation basin between those two half-cubes arises. The Viability Kernel Algorithms allowing to compute viability kernels and capture basins, hence permanence kernels, and thus, fluctuation basins, one can compute indeed this fluctuation basin.

The reason why *apparently* all evolutions fluctuate is due to the strong sensitivity on initial conditions (see Definition 9.4.3, p. 355). Using standard numerical methods, even if an approximate evolution belongs to the complement of the fluctuation basin, the next steps have a high chance to enter the fluctuation basin. However, as displayed in the following Fig. 9.4, p. 337, the Viability Kernel Algorithm allows us to precisely locate an initial state in the union of the permanent basins and to illustrate that these evolutions do not fluctuate at all, but, rather, that they are viable in the complement of the fluctuation basin. Otherwise, starting from the complement of the permanence kernel, the evolution does fluctuate.

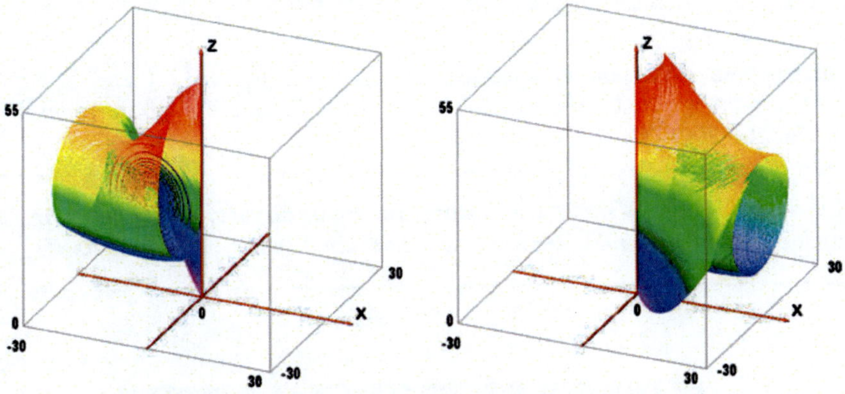

Fig. 9.4 Fluctuation Basin under the Lorenz System.
Forward Viability Kernel of K_1^- and K_1^+. For clarity, the viability kernels of K_1^- and K_1^+ are separated in this figure. The evolutions starting from each of these viability kernels remain in them, and do not fluctuate by construction. Observe the structure of these viability kernels. Starting outside the viability kernel of C_1^-, the evolutions reach C_1^+ in finite time. Either they reach it in the viability kernel of C_1^+, and they do not fluctuate anymore, or they reach the capture basin of C_1^- viable in K, and they fluctuate once more. They fluctuate back and forth from C_1^- to C_1^+ if they do not belong to the capture basin of the union of the two viability kernels, thanks to Proposition 9.2.10.

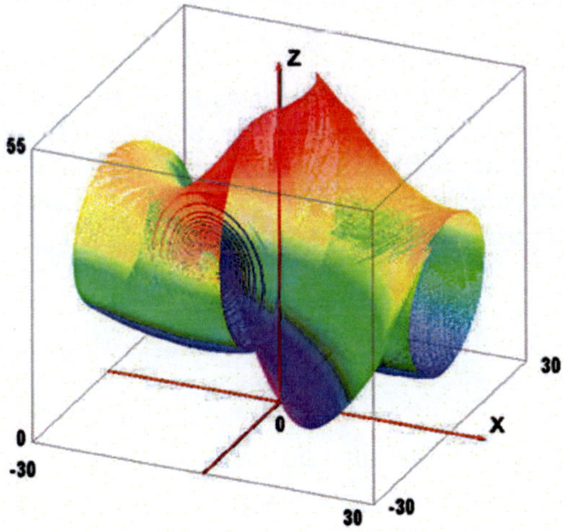

Fig. 9.5 Fluctuation Basin under the Lorenz System.
Union of the Forward Viability Kernel of K_1^- and K_1^+. The fluctuation is the complement of the invariant envelope $\mathrm{Inv}_S(K \setminus (\mathrm{Viab}_S(K_1) \cup \mathrm{Viab}_S(K_2)))$ of the union of viability kernels of the two half-cubes.

9.2.5 Hofbauer-Sigmund Permanence

Consider an evolutionary system $\mathcal{S} : X \rightsquigarrow \mathcal{C}(0, +\infty; X)$ and a closed environment $K \subset X$. We first observe that the viability of the boundary implies the viability of the environment:

> **Lemma 9.2.17** *[Viable or Invariant Boundaries] If the boundary of a closed subset is viable (respectively invariant), so is the subset itself.*

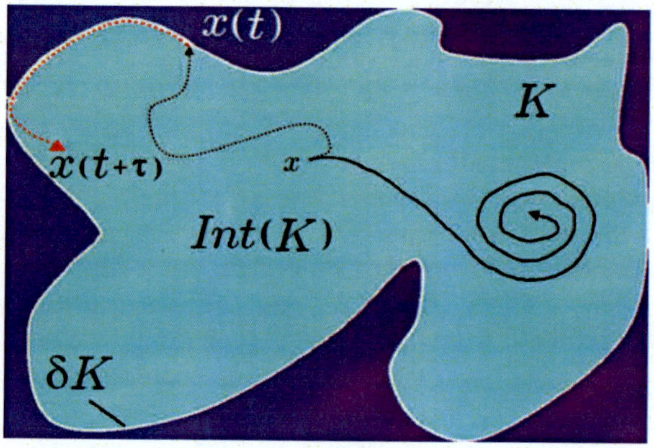

Fig. 9.6 Proof of Lemma 9.2.17.

Proof. *Indeed, if $x \in \text{Int}(K)$, any evolution $x(\cdot) \in \mathcal{S}(x)$ starting from x is either viable in the interior of K, and thus in K, or otherwise, leaves K at a finite time T at a point $x(T)$ of its boundary. Then we can concatenate the evolution with an evolution viable in ∂K, so that the concatenated evolution is viable in K. Then K is viable.*

The converse is not true, naturally. However, Theorem 9.2.18, p. 339 expresses the boundary as viability kernel of some closed subset C such that $\partial K \subset C \subset K$ under adequate assumptions. Hence we study in more details the viability properties of the boundary $\partial K := K \backslash \overset{\circ}{K} = K \cap \complement \overset{\circ}{K} = K \cap \overline{\complement(K)}$ of K where $\overset{\circ}{K} := \text{Int}(K)$ denotes the interior of K (see Definition 10.5.14,

p. 407). In order to avoid the trivial case $\partial K = K$, we assume once and for all in this section that the *interior of* K *is not empty.*

The problem to solve it to characterize such subsets C satisfying *the property* $\partial K = \text{Viab}_{\mathcal{S}}(C)$.

This problem is important for several purposes:

- First, if this property is true, then the boundary ∂K of K inherits the properties of viability kernels presented in this book, and in particular, can be computed (see the computation of Julia sets, which are the boundaries of fill-in Julia sets, in Fig. 2.5, p. 79).
- Second, the characterization is interesting in itself, since it is (up to a minor detail) equivalent to saying that the interior $\overset{\circ}{K}$ of K *fluctuates* around $\overset{\circ}{K} \cap C$: $\overset{\circ}{K} = \text{Fluct}_{\mathcal{S}}(K, \overset{\circ}{K} \cap C)$. This means that the interior $\overset{\circ}{K}$ is invariant and that every evolution $x(\cdot) \in \mathcal{S}(x)$ starting from $x \in \overset{\circ}{K}$ remains in the interior of K and for any $t \geq 0$, there exists a *finite* time $t^\star \geq t$ such that $x(t^\star) \notin \overset{\circ}{K} \cap C$.

Theorem 9.2.18 *[Characterization of Viable Boundaries] Let us consider an evolutionary system* $\mathcal{S}$ *and two closed subsets* $C \subset K$ *such that the boundary* $\partial K \subset C$. *Then the two following statements are equivalent:*

- *the boundary* ∂K *is equal to the viability kernel of* C:

$$\partial K = \text{Viab}_{\mathcal{S}}(C) \tag{9.5}$$

- *the boundary of* K *is contained in the viability kernel of* C *and the interior* $\overset{\circ}{K} := \text{Int}(K)$ *of* K *is invariant and fluctuates around* $\overset{\circ}{K} \cap C$ *of* C *in the interior of* K:

$$\begin{cases} (i) \ \partial K \subset \text{Viab}_{\mathcal{S}}(C) \\ (ii) \ \overset{\circ}{K} = \text{Fluct}_{\mathcal{S}}(\overset{\circ}{K}, \overset{\circ}{K} \cap C) \end{cases} \tag{9.6}$$

Observe that whenever the viability of the boundary is satisfied, then, property (9.6)(i) is satisfied for any subset C such that $\partial K \subset C \subset K$, because whenever the boundary is viable, it is contained in the viability kernel of any larger set C:

$$\partial K = \text{Viab}_{\mathcal{S}}(\partial K) \text{ implies that } \partial K \subset \text{Viab}_{\mathcal{S}}(C)$$

Proof.

- Assume that $\partial K = \text{Viab}_{\mathcal{S}}(C)$ and derive that $\overset{\circ}{K} \subset \text{Abs}_{\mathcal{S}}(\overset{\circ}{K}, \overset{\circ}{K} \backslash C)$. Il not, there would exist some $x \in \overset{\circ}{K} \backslash \text{Abs}_{\mathcal{S}}(\overset{\circ}{K}, \overset{\circ}{K} \backslash C) \subset \overset{\circ}{K}$. By Lemma 2.12.2, p. 92, we use equality

$$\complement(\text{Abs}_{\mathcal{S}}(\overset{\circ}{K}, \overset{\circ}{K} \backslash C)) = \text{Viab}_{\mathcal{S}}(C \cup \complement \overset{\circ}{K}, \complement \overset{\circ}{K})$$

to derive that x belongs to $\overset{\circ}{K} \cap \text{Viab}_{\mathcal{S}}(C \cup \complement \overset{\circ}{K}, \complement \overset{\circ}{K}) \subset C$. Therefore, there exists an evolution $x(\cdot) \in \mathcal{S}(x)$ viable in C forever or up to some time T when $x(T)$ belongs to the boundary ∂K. In this case, since $\partial K \subset \text{Viab}_{\mathcal{S}}(C)$ by assumption, the evolution can be extended as the concatenation of $x(\cdot)$ viable in C up to time T and an evolution $y(\cdot) \in \mathcal{S}(x(T))$ viable in C forever, so that this concatenated evolution starting from x is viable in C. Hence x belongs to $\text{Viab}_{\mathcal{S}}(C)$, which, by assumption, is contained in ∂K. This is a contradiction to the fact that x was chosen in $\overset{\circ}{K}$. We thus have proved that property $\partial K = \text{Viab}_{\mathcal{S}}(C)$ implies that $\overset{\circ}{K} \subset \text{Abs}_{\mathcal{S}}(\overset{\circ}{K}, \overset{\circ}{K} \backslash C)$, and actually, $\overset{\circ}{K} = \text{Abs}_{\mathcal{S}}(\overset{\circ}{K}, \overset{\circ}{K} \backslash C)$ on one hand, and that, on the other hand, the obvious inclusion $\partial K \subset \text{Viab}_{\mathcal{S}}(C)$ holds true.
- To prove the converse statement, assume that both inclusions $\overset{\circ}{K} \subset \text{Abs}_{\mathcal{S}}(\overset{\circ}{K}, \overset{\circ}{K} \backslash C)$ and inclusion $\partial K \subset \text{Viab}_{\mathcal{S}}(C)$ hold true and derive that $\text{Viab}_{\mathcal{S}}(C) \subset \partial K$, and thus, that $\partial K = \text{Viab}_{\mathcal{S}}(C)$. Indeed, since

$$\text{Viab}_{\mathcal{S}}(C \cup \complement \overset{\circ}{K}, \complement \overset{\circ}{K}) = \complement(\text{Abs}_{\mathcal{S}}(\overset{\circ}{K}, \overset{\circ}{K} \backslash C)) \subset \complement \overset{\circ}{K}$$

then $K \cap \text{Viab}_{\mathcal{S}}(C \cup \complement \overset{\circ}{K}, \complement \overset{\circ}{K}) \subset K \cap \complement \overset{\circ}{K} =: \partial K$ whenever K is closed. Obviously, since $\text{Viab}_{\mathcal{S}}(C) \subset K$ and $\text{Viab}_{\mathcal{S}}(C) \subset \text{Viab}_{\mathcal{S}}(C \cup \complement \overset{\circ}{K}, \complement \overset{\circ}{K})$, we infer that $\text{Viab}_{\mathcal{S}}(C) \subset K \cap \text{Viab}_{\mathcal{S}}(C \cup \complement \overset{\circ}{K}, \complement \overset{\circ}{K})$, and thus, that $\text{Viab}_{\mathcal{S}}(C) \subset \partial K$. Since we assumed the opposite inclusion, we deduce that $\text{Viab}_{\mathcal{S}}(C) = \partial K$.
- In this case, $\overset{\circ}{K} = \text{Fluct}_{\mathcal{S}}(\overset{\circ}{K}, \overset{\circ}{K} \cap C)$. Indeed, whenever the evolution $x(\cdot)$ leaves $\overset{\circ}{K} \cap C$ at $x(t^\star)$ a given time $t^\star$ (which always happens in finite time), either $x(t^\star) \in \text{Viab}_{\mathcal{S}}(\overset{\circ}{K} \backslash C)$, and thus, $x(\cdot)$ remains in $\overset{\circ}{K} \backslash C$ forever, or it enters $\overset{\circ}{K} \cap C \subset \overset{\circ}{K}$ in finite time t_1. Hence there exists another finite instant $t_1^\star \geq t_1$ such that $x(t_1^\star) \in \overset{\circ}{K} \backslash C$, from which either the evolution remains in $\overset{\circ}{K} \backslash C$ forever or until a finite time $t_2 \geq t_1^\star$ where $x(t)_2^\star \in \overset{\circ}{K} \cap C$ and so on. □

Remark. Proposition 10.5.17, p. 409 also implies that K exhibits the backward barrier property: every backward evolution starting from the boundary ∂K viable in K is actually viable in the boundary ∂K. □

We illustrate Theorem 9.2.18, p. 339 by computing Julia sets in Fig. 2.5, p. 79, which are by Definition 2.9.6, p. 76 boundaries of viability kernels, and thus, can be computed as the viability kernel of the complement of a smaller absorbing ball in K (in the sense that $K = \text{Viab}(K \setminus C)$).

As a consequence, we obtain:

Corollary 9.2.19 ***[Permanence and Fluctuation Properties]*** *Let $C \subset K$ be a closed subset of a closed environment K such that $\partial K = \text{Viab}_{\mathcal{S}}(C)$. Then the interior of K is invariant and:*

1. *If $\overset{\circ}{K} \setminus C$ is invariant, all evolutions starting from the interior of K reach $\overset{\circ}{K} \setminus C$ in finite time and remain in $\overset{\circ}{K} \setminus C$, (*Permanence Property*)*
2. *If $\overset{\circ}{K} \setminus C$ is a repeller, all evolutions starting from the interior of K "fluctuate" between $\overset{\circ}{K} \cap C$ and $\overset{\circ}{K} \setminus C$ in the sense that they alternatively reach and leave $\overset{\circ}{K} \setminus C$ in finite time (*Fluctuation Property*).*

9.2.6 Heteroclines and Homoclines

Definition 9.2.20 ***[Heteroclinic and Homoclinic Basins]*** *Let $B \subset K$ be a subset regarded as a source, $C \subset K$ be a subset regarded as a target. The connection basin*

$$\text{Cline}_{\mathcal{S}}(K, (B, C)) := \text{Conn}_{\mathcal{S}}(K, (\text{Viab}_{\overleftarrow{\mathcal{S}}}(B), \text{Viab}_{\mathcal{S}}(C)))$$

of K between the backward viability kernel of B and the forward viability kernel of C (see Definition 8.5.1, p. 291) is called the heteroclinic basin *viable in K under the evolutionary system $\mathcal{S}$. The viable evolutions passing through initial states in the hetoroclinic basin are called* heteroclines. *If $B = C$, the heteroclinic basin and heteroclines are called* homoclinic basin *and* homoclines *respectively.*

They exhibit the cumulated properties of connection basins (see Sect. 10.6.2, p. 413) and permanence kernels since

$$\text{Cline}_{\mathcal{S}}(K, (B, C)) = \text{Perm}_{\overleftarrow{\mathcal{S}}}(K, B) \cap \text{Perm}_{\mathcal{S}}(K, C) \tag{9.7}$$

We begin by studying an example:

Proposition 9.2.21 *[Example of Clines] Assume that* $B \subset K$ *and* $C \subset K$ *are disjoint and that* $\text{Viab}_{\overleftarrow{S}}(K,B) \subset \text{Int}(B)$ *(see Proposition 9.2.5, p. 325). Then the cline* $\text{Cline}_{S}(K,(B,C)) = \emptyset$ *is empty and the permanence kernel*

$$\text{Perm}_{S}(K,C) \subset \text{Fluct}_{\overleftarrow{S}}(K,B) \tag{9.8}$$

is contained in the backward fluctuation basin of B*. In other words, through any* $x \in \text{Perm}_{S}(K,C)$ *passes at time 0 one evolution fluctuating around* B *before 0 and permanent in* C *after, which is, so to speak, a cline connecting evolutions fluctuating around* B *to the viability kernel of* C*.*

Consequently the cline is equal to

$$\text{Cline}_{S}(K,(B,C)) = \text{Fluct}_{\overleftarrow{S}}(K,B) \cap \text{Perm}_{S}(K,C)$$

"connects evolutions fluctuating around B *in the past and arriving in finite time to the viability kernel of* B*".*

Proof. Since $\text{Perm}_{\overleftarrow{S}}(K,B) := \text{Viab}_{\overleftarrow{S}}(B)$ by Proposition 9.2.5, p. 325, then $\text{Cline}_{S}(K,(B,C)) = \text{Viab}_{\overleftarrow{S}}(B) \cap \text{Perm}_{S}(K,C)$. Let us assume that there exists some x in this intersection. There exists one backward evolution $\overleftarrow{x}(\cdot) \in \overleftarrow{S}(x)$ viable in $\text{Viab}_{\overleftarrow{S}}(B)$ and one forward evolution $\overrightarrow{x}(\cdot) \in S(x)$ viable in K and reaching $\text{Viab}_{S}(C)$ in finite time T at some point y. Since $\text{Viab}_{S}(C) \cap \text{Viab}_{\overleftarrow{S}}(B) = \emptyset$ because we have assumed that $B \cap C = \emptyset$, then y belongs to $K \setminus \text{Viab}_{\overleftarrow{S}}(B) = \text{Fluct}_{\overleftarrow{S}}(K,B)$. Therefore, all backward evolutions starting from y fluctuate around C and cannot reach $\text{Viab}_{\overleftarrow{S}}(B)$ in finite time. On the other hand, the backward evolution $\overleftarrow{y}(\cdot) \in \overleftarrow{S(y)}$ defined by

$$\overleftarrow{y}(t) = \begin{cases} \overrightarrow{x}(T-t) \text{ if } t \in [0,T] \\ \overleftarrow{x}(t-T) \text{ if } t \geq T \end{cases}$$

starts from y and arrives at finite time T at $x \in \text{Viab}_{\overleftarrow{S}}(B)$. Hence, we have derived a contradiction. Therefore, the permanent kernel $\text{Perm}_{S}(K,C)$ of C is contained in the backward fluctuation $\text{Fluct}_{\overleftarrow{S}}(K,B)$ basin of B. □

Next, we consider the case of homoclines of an union of disjoint cells:

Proposition 9.2.22 *[Decomposing Clines] Let us consider a finite family of closed disjoint subsets* C_i *and set* $C := \bigcup_{i=1,\ldots,n} C_i$*. Then the homoclinic basin of* C *is the union of the heteroclinic basins between the pairs of disjoint* *subsets* C_i *and* C_j*:*

$$\text{Cline}_{\mathcal{S}}\left(K, \bigcup_{i=1,\ldots,n} C_i, \bigcup_{i=1,\ldots,n} C_i\right) = \bigcup_{i,\, j=1,\ldots,n} \text{Cline}_{\mathcal{S}}(K, (C_i, C_j)) \quad (9.9)$$

Proof. Indeed, since the subsets are closed and disjoints, there exist disjoint open subsets U_i containing the subsets C_i. Therefore, no continuous function connecting C_i to C_j can be viable in $C_i \cup C_j$ since it has to cross the nonempty open subset $(U_i \setminus C_i) \cup (U_j \setminus C_j)$. Hence

$$\text{Viab}_{\mathcal{S}}\left(\bigcup_{i=1,\ldots,n} C_i\right) = \bigcup_{i=1,\ldots,n} \text{Viab}_{\mathcal{S}}(C_i)$$

is the union of the disjoint viability kernels, so that

$$\text{Perm}_{\mathcal{S}}\left(K, \bigcup_{i=1,\ldots,n} C_i\right) = \bigcup_{i=1,\ldots,n} \text{Perm}_{\mathcal{S}}(K, C_i) \quad (9.10)$$

Consider now the homoclinic basin of the union of the subsets C_i. We obtain:

$$\begin{cases} \text{Cline}_{\mathcal{S}}\left(K, \bigcup_{i=1,\ldots,n} C_i, \bigcup_{i=1,\ldots,n} C_i\right) \\ = \text{Perm}_{\overleftarrow{\mathcal{S}}}\left(K, \bigcup_{i=1,\ldots,n} C_i\right) \cap \text{Perm}_{\mathcal{S}}\left(K, \bigcup_{j=1,\ldots,n} C_j\right) \\ = \left(\bigcup_{i=1,\ldots,n} \text{Perm}_{\overleftarrow{\mathcal{S}}}(K, C_i)\right) \cap \left(\bigcup_{j=1,\ldots,n} \text{Perm}_{\mathcal{S}}(K, C_j)\right) \\ = \bigcup_{i,j=1,\ldots,n} \left(\text{Perm}_{\overleftarrow{\mathcal{S}}}(K, C_i) \cap \text{Perm}_{\mathcal{S}}(K, C_j)\right) \\ = \bigcup_{i,j=1,\ldots,n} \left(\text{Cline}_{\mathcal{S}}(K, (C_i, C_j))\right) \ \square \end{cases}$$

Hence, we have to study the heteroclinic and homoclinic basins between the subsets C_i.

The viability kernels and backward viability kernels of compact neighborhoods of equilibria describe their local stability properties. For instance, from any initial state x of the viability kernel of a compact equilibrium C of an equilibrium e, for any evolution $x(\cdot) \in \mathcal{S}(x)$ viable in C, some subsequence

$x(t_n)$ converges to the equilibrium e. Since e is the unique cluster point in a compact set, this implies that $x(t)$ converges to e. Its trajectory is therefore viable. The viability kernel of this compact neighborhood of an equilibrium is regarded as a "local stable manifold" around the equilibrium. Taking the backward viability kernel, we obtain a "local unstable manifold" around the equilibrium.

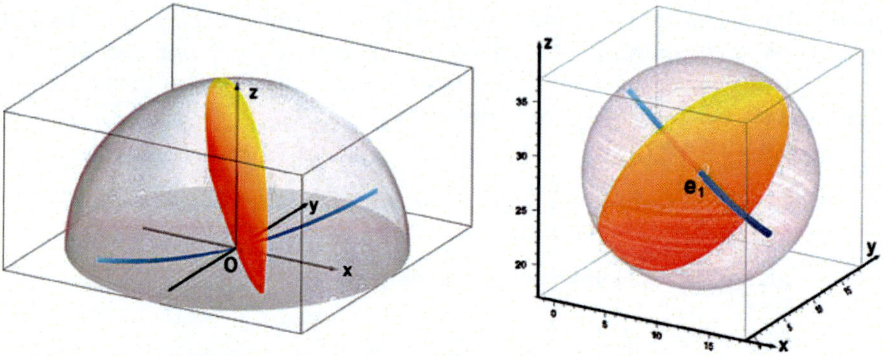

Fig. 9.7 Viability Kernels around Equilibria.
Left*: Forward and backward viability kernels of a neighborhood of the trivial equilibrium* O. Right*: Forward and backward viability kernels of a neighborhood of the nontrivial equilibrium* e_1*. They are given by the viability kernel algorithm, not computed analytically.*

9.3 Asymptotic Behavior: Limit Sets, Attractors and Viability Kernels

9.3.1 Limit Sets and Attractor

The asymptotic behavior of evolutions is rooted in the concept of limit sets of evolutions:

Definition 9.3.1 *[The Limit Set of an Evolution] Let* $x(\cdot)$ *be a function from* $\mathbb{R}$ *to* X*. We say that the (closed) subsets*

$$\begin{cases} \omega(x(\cdot)) := \bigcap_{T>0} \overline{x([T,\infty[)} = \mathrm{Limsup}_{t\to+\infty}\{x(t)\} \\ \alpha(x(\cdot)) := \bigcap_{T>0} \overline{x(]-\infty,-T])} = \mathrm{Limsup}_{t\to-\infty}\{x(t)\} \end{cases}$$

*(where "*Limsup*" denotes the* Painlevé-Kuratowski upper limit, *Definition 18.4.1, p. 728) of the cluster points when* $t \to +\infty$ *and* $t \to -\infty$

are respectively the ω-limit set *of* $x(\cdot)$ *and the* α-limit set *of the evolution* $x(\cdot)$. *We regard* $\omega(\cdot) : \mathcal{C}(0,+\infty;X) \rightsquigarrow X$ *as the* limit map *associating with any evolution* $x(\cdot)$ *its* ω*-limit set* $\omega(x(\cdot))$.

Remark. Observe that limit sets of evolutions viable in a compact subset are not empty. Note also that $\alpha(x(\cdot)) = \omega(\overleftarrow{x}(\cdot))$. □

We shall study successively the *inverse images and cores* of environments under the limit map $\omega : \mathcal{C}(0,+\infty;X) \rightsquigarrow X$, and next, after introducing an evolutionary system $\mathcal{S} : X \rightsquigarrow \mathcal{C}(0,+\infty;X)$, the composition product $\mathcal{A} := \omega \circ \mathcal{S}$ of ω and $\mathcal{S}$, the *attractor map*, as well as the inverse images and cores under $\mathcal{S}$ of the cores and the inverse images of environments under the limit map. We recover in this way numerous concepts of dynamical systems, and show that they are closely related to the concepts of permanence kernels and fluctuation basins, inheriting their properties.

We recall (see Definition 9.2.1, p. 322) that $\mathcal{P}(K,C)$ and $\mathcal{F}(K,C)$ denotes the families of evolutions *permanent in and fluctuating around* a subset C respectively. We can reformulate the definitions of cluster points and limits in the following form

Lemma 9.3.2 ***[Inverse and Core of the Limit Set Map]*** *For any compact subset* $E \subset X$*:*

- *the inverse image* $\omega^{-1}(E) \subset \mathcal{C}(0,+\infty;X)$ *is the set of evolutions* $x(\cdot)$ *such that one of its cluster point belongs to* E *or, equivalently, such that* $\liminf_{t\to+\infty} d(x(t),E) = 0$*, or, again,*

$$\omega^{-1}(E) = \bigcap_{n\geq 0} \mathcal{F}(X, \complement B_K(E,1/n))$$

  where $B_K(E,1/n)$ *is the ball of radius* $\frac{1}{n}$ *around* E.
- *the core* $\omega^{\ominus 1}(E) \subset \mathcal{C}(0,+\infty;X)$ *is the set of evolutions* $x(\cdot)$ *such that all its cluster points belong to* E*, or, equivalently, such that* $\lim_{t\to+\infty} d(x(t),E) = 0$ *or, again,*

$$\omega^{\ominus 1}(E) = \bigcap_{n\geq 0} \mathcal{P}(X, B_K(E,1/n))$$

If $E := \{e\}$ *is a singleton, the core* $\omega^{\ominus 1}(e)$ *is the set of evolutions converging to* e.

Proof. Indeed, to say that $\liminf_{t\mapsto+\infty} d(x(t),E) = 0$ means that for any $n \geq 0$ and for any $t \geq 0$, there exists $t_n \geq t$ such that $d(x(t_n),E) \leq 1/n$, i.e., such that $x(t_n) \in B_K(E,1/n) = \complement(\complement(B_K(E,1/n)))$. By the very definition of fluctuating evolutions, this means that for any $n \geq 0$, the evolution $x(\cdot)$ is fluctuating around the complement of the ball $B_K(E,1/n)$. This also means that there exists a subsequence $t_n \to +\infty$ and $y_n \in E$ such that $\|x(t_n) - y_n\| = d(x(t_n),E)$ converges to 0. Since E is compact, this implies that a subsequence (again denoted by) y_n converges to some e, which is also the limit of some subsequence of $x(t_n)$, and thus, belongs to $\omega(E)$.

On the other hand, to say that $\lim_{t\mapsto+\infty} d(x(t),E) = 0$ means that for any $n \geq 0$, there exists $T \geq 0$ such that for any $t \geq T$, $d(x(t),E) \leq 1/n$, i.e., such that $x(t) \in B_K(E,1/n)$. By the very definition of permanent evolutions, this means that for any $n \geq 0$, the evolution $x(\cdot)$ is permanent in the ball $B_K(E,1/n)$. Property $\lim_{t\mapsto+\infty} d(x(t),E) = 0$ implies that for any $e \in \omega(\cdot)$, there exists a subsequence $t_n \to +\infty$ such that $x(t_n)$ converges to e. We have to prove that e belongs to E. Let $y_n \in E$ such that $\|x(t_n)-y_n\| = d(x(t_n),E)$. Since it converges to 0, the sequence $y_n \in E$ converges also to e, which thus belongs to E. This means that $\omega(x(\cdot)) \subset E$. Conversely, if the limit set of $x(\cdot)$ is contained in E, $d(x(t_n),E) \to 0$ for all subsequences t_n. This implies that $d(x(t),E)$ converges to 0. □

The other important trivial property that we have to single out is that

$$\forall T \in \mathbb{R},\ \omega(\kappa(-T)x(\cdot)) = \omega(x(\cdot)) \text{ and } \omega(\overset{\vee}{\kappa}(T)x(\cdot)) = \alpha(x(\cdot)) \tag{9.11}$$

because the cluster points of the evolution $t \mapsto \kappa(-T)x(t) := x(T+t)$ are the same than the ones of $x(\cdot)$ and the cluster points of the map $t \mapsto (\overset{\vee}{\kappa}(T)x(t)) := x(T-t)$ when $t \mapsto +\infty$ are the same as the ones of $x(\cdot)$ when $t \mapsto -\infty$.

This implies that

$$\forall T \geq 0,\ \kappa(-T)\omega^{-1}(E) = \omega^{-1}(E) \text{ and } \kappa(-T)\omega^{\ominus 1}(E) = \omega^{\ominus 1}(E) \tag{9.12}$$

The sets $\omega^{-1}(E)$ and $\omega^{\ominus 1}(E)$ are invariant under translation and thus, by Proposition 2.13.3, p. 95, $\omega^{-1}(E)$ is viable and $\omega^{\ominus 1}(E)$ is invariant under $\mathcal{S}$.

9.3.2 *Attraction and Cluster Basins*

Let us consider an evolutionary system $\mathcal{S} : X \rightsquigarrow \mathcal{C}(0,+\infty;X)$ and a nonempty compact subset E viable under $\mathcal{S}$. We observe that, *for any evolution* $x(\cdot)$

viable in E, its ω-limit set $\omega(x(\cdot)) \subset E$ is contained in E. This is in particular the (trivial) case when $E := \{e\}$ is an equilibrium.

Definition 9.3.3 *[**Attraction and Cluster Basins**] Let $\mathcal{S} : X \rightsquigarrow \mathcal{C}(0,+\infty;X)$ be an evolutionary system and $\mathcal{S}^K(x)$ denote the set of evolutions starting at $x \in K$ viable in K. We consider a nonempty compact subset E viable under $\mathcal{S}$ and we denote by:*

- $\mathrm{Lim}_{\mathcal{S}}(K,E) := \mathcal{S}^{-1}((\mathcal{V}(K) \cap \omega^{\ominus 1}(E)))$ *the* attraction basin *of E viable in K under $\mathcal{S}$, the subset of initial states $x \in K$ such that there exists an evolution $x(\cdot) \in \mathcal{S}(x)$ viable in K such that* $\lim_{t \to +\infty} d(x(t), E) = 0$,
- $\mathrm{Clust}_{\mathcal{S}}(K,E) := \mathcal{S}^{\ominus 1}((\mathcal{V}(K) \cap \omega^{-1}(F)))$ *the* cluster basin *of E viable in K under $\mathcal{S}$, the subset of initial states $x \in K$ from which all evolutions $x(\cdot) \in \mathcal{S}(x)$ are viable in K and satisfy* $\liminf_{t \to +\infty} d(x(t), E) = 0$.

We observe that whenever E is viable and compact, then

$$E \subset \mathrm{Lim}_{\mathcal{S}}(K,E) \subset \mathrm{Clust}_{\mathcal{S}}(K,E) \tag{9.13}$$

We deduce at once the following consequence of Partition Lemma 18.3.4, p. 721:

Lemma 9.3.4 *[**Partition of the Invariance Kernels in terms of Attraction and Cluster Basins**] Assume that the environment $K := K_1 \cup_{\emptyset} K_2$ is partitioned into two subsets K_i, $i = 1,2$. Then its invariance kernel is partitioned*

$$\begin{cases} \mathrm{Inv}_{\mathcal{S}}(K) = \mathrm{Lim}_{\mathcal{S}}(K,K_1) \cup_{\emptyset} \mathrm{Clust}_{\mathcal{S}}(K,K_2) \\ = \mathrm{Lim}_{\mathcal{S}}(K,K_2) \cup_{\emptyset} \mathrm{Clust}_{\mathcal{S}}(K,K_1) \end{cases}$$

by the attraction basin of K_1 and the cluster basin of K_2. In particular, if K is invariant under $\mathcal{S}$, then, for any subset $E \subset K$,

$$K = \mathrm{Lim}_{\mathcal{S}}(K,E) \cap \mathrm{Clust}_{\mathcal{S}}(K, K \setminus E) \tag{9.14}$$

Proof. The Partition Lemma 18.3.4, p. 721 implies that

$$\omega^{\ominus 1}(K) = \omega^{\ominus 1}(K_1) \cup_{\emptyset} \omega^{-1}(K_2)$$

and therefore, that

$$\mathcal{V}(K) = \mathcal{V}(K) \cap \omega^{\ominus 1}(K) = (\mathcal{V}(K) \cap \omega^{\ominus 1}(K_1)) \cup (\mathcal{V}(K) \cap_{\emptyset} \omega^{-1}(K_2))$$

By taking the core of this partition, we deduce that

$$\mathrm{Inv}_{\mathcal{S}}(K) = \mathcal{S}^{-1}(\mathcal{V}(K) \cap \omega^{\ominus 1}(K_1)) \cup_{\emptyset} \mathcal{S}^{\ominus 1}(\mathcal{V}(K) \cap \omega^{-1}(K_2)) \quad \square$$

Property (9.11), p. 346 implies the viability properties of the limit and cluster basins:

Proposition 9.3.5 ***[Viability of Limit Sets]*** *The attraction basin* $\mathrm{Lim}_{\mathcal{S}}(X, E)$ *is viable and the cluster basin is invariant under* $\mathcal{S}$.

Proof. Indeed, if x belongs to the attraction basin $\mathrm{Lim}_{\mathcal{S}}(X, E)$, there exists an evolution $x(\cdot) \in \mathcal{S}(x)$ such that $\omega(x(\cdot)) \subset E$, and thus, by property (9.11), p. 346, such that, for any $T \geq 0$, $\omega(\kappa(-T)x(\cdot)) \subset E$. Hence, $x(T) \in \mathrm{Lim}_{\mathcal{S}}(X, E)$ for all $T \geq 0$, and thus, the attraction basin is viable. The proof of the invariance of the cluster basin is analogous. $\square$

If C is closed, evolutions $x(\cdot)$ viable in C satisfy $x(0) \in C$ and $\omega(x(\cdot)) \subset C$. However, neuroscientists have singled out evolutions starting from C, such that their limit set is contained in C but which *are not viable* in C. The shape of their graphs suggests to call such evolutions "spike" evolutions (see [101, Françoise] for more details):

Definition 9.3.6 ***[Spike Evolutions]*** *An evolution* $x(\cdot)$ *such that* $x(0) \in C$, $\omega(x(\cdot)) \subset C$ *and such that* $x(t) \notin C$ *for some finite time* $t > 0$ *is called a* spike evolution.

Lemma 9.3.7 ***[Spike Evolutions Governed by an Evolutionary System]*** *Assume that the subset* C *is compact and viable under* $\mathcal{S}$. *Then:*

- *from any initial state* $x \in \mathrm{Lim}_{\mathcal{S}}(K, C)$ *starts at least one evolution viable in* C *and converging to* C,
- *from any initial state* $x \in C \cap (\mathrm{Lim}_{\mathcal{S}}(K, C) \backslash \mathrm{Viab}_{\mathcal{S}}(C))$ *starts at least one spike evolution in* C, *whereas no evolution can be viable in* C. *However, some evolutions starting from the viability kernel* $\mathrm{Viab}_{\mathcal{S}}(C)$ *may be spike evolutions in* C *(semi-permeability property).*

Proof. Starting from $x \in \mathrm{Viab}_{\mathcal{S}}(C)$, one evolution $x(\cdot) \in \mathcal{S}(x)$ is viable in C, so that, C being compact, its limit set $\omega(x(\cdot)) \subset C$ is not empty.

Starting from $x \in C \cap (\mathrm{Lim}_{\mathcal{S}}(K, E) \setminus \mathrm{Viab}_{\mathcal{S}}(C))$, all evolutions leave C in finite time, and at least one evolution $x(\cdot) \in \mathcal{S}(x)$ is viable in K, satisfies $\lim_{t \to +\infty} d(x(t), C) = 0$. □

We can also compose the evolutionary system $\mathcal{S}^K : K \rightsquigarrow \mathcal{C}(0, +\infty; K)$ (see Notation 40, p. 716) with the limit map $\omega : \mathcal{C}(0, +\infty; X) \rightsquigarrow X$ to obtain the map $\mathcal{A}_{\mathcal{S}}^K : K \rightsquigarrow K$:

Definition 9.3.8 *[Attractor Map and The Attractor] We denote by*

$$\mathcal{A}_{\mathcal{S}}^K(x) := \bigcup_{x(\cdot) \in \mathcal{S}^K(x)} \omega(x(\cdot)) \ \& \ \overleftarrow{\mathcal{A}}_{\mathcal{S}}^K(x) := \bigcup_{x(\cdot) \in \mathcal{S}^K(x)} \alpha(x(\cdot))$$

the ω-attractor map or simply attractor map *and the α-attractor map or* backward attractor *of the subset K under $\mathcal{S}$ respectively.*

Their images

$$\mathrm{Attr}_{\mathcal{S}}(K) \ := \ \bigcup_{x \in K} \mathcal{A}_{\mathcal{S}}^K(x) \ \& \ \overleftarrow{\mathrm{Attr}}_{\mathcal{S}}(K) \ := \ \bigcup_{x \in K} \overleftarrow{\mathcal{A}}_{\mathcal{S}}^K(x)$$

are called the ω-attractor or simply attractor *and the α-attractor or* backward attractor *of the subset K under $\mathcal{S}$ respectively.*

Remarks on terminology. The *attractor of K under $\mathcal{S}$* is sometimes defined as the closure of the attractor as we defined it.

Usually, a closed viable subset $E \subset \mathrm{Int}(K)$ is said to be *attracting* if there exists a viable neighborhood $C \subset K$ of E such that $E \subset \mathrm{Lim}_{\mathcal{S}}(X, C)$, i.e., such that all evolutions satisfy $\lim_{t \to +\infty} d(x(t), E) = 0$. A subset is *repelling* if it is attracting under the backward system.

We defined here THE attractor as the union of limit sets of evolutions. In many books and papers, subsets that are both attracting and *topologically transitive* (see Definition 9.4.7, p. 358) are called attractors (there may exist several of them). □

Remark. We deduce at once from property (9.11), p. 346 that

$$\forall \, x(\cdot) \in \mathcal{S}(x), \ \forall \, T \ \geq \ 0, \ \ \mathcal{A}_{\mathcal{S}}^K(x(T)) \ = \ \mathcal{A}_{\mathcal{S}}^K(x) \tag{9.15}$$

i.e., that *the attractor map is constant over the trajectories of any evolution* $x(\cdot) \in \mathcal{S}(x)$.

Indeed, by Definition 2.8.2, p. 70 of evolutionary systems, for any $x(\cdot) \in \mathcal{S}(x)$, $\kappa(-T)x(\cdot) \in \mathcal{S}(x(T))$, so that, by applying the limit map ω, we obtain the above property. □

The next question we ask is whether one can express the attraction basin in terms of viability kernels and capture basins, so that we could benefits of their properties and compute them by the Viability Kernel Algorithms. The answer is positive:

Proposition 9.3.9 ***[Viability Characterization of Limit Sets]*** *Let us consider two subsets $K \subset X$ and $C \subset K$. Inclusion $\mathrm{Perm}_{\mathcal{S}}(K, C) \subset \mathrm{Lim}_{\mathcal{S}}(K, \mathrm{Attr}_{\mathcal{S}}(C))$ is always true. The converse inclusion holds true if we assume that $\mathrm{Attr}_{\mathcal{S}}(C) \subset \mathrm{Int}(C)$ is contained in the interior of C:*

$$\mathrm{Lim}_{\mathcal{S}}(K, \mathrm{Attr}_{\mathcal{S}}(C)) \;=\; \mathrm{Perm}_{\mathcal{S}}(K, C)$$

Proof. Let x belong to $\mathrm{Perm}_{\mathcal{S}}(K, C)$. There exist an evolution $x(\cdot)$ and a time T such that $x(T) \in \mathrm{Viab}_{\mathcal{S}}(C)$. Hence its limit set $\omega(\kappa(-T)x(\cdot)) = \omega(x(\cdot))$ (by property (9.11), p. 346) is contained in $\mathrm{Attr}_{\mathcal{S}}(C)$. Therefore x belongs to the attraction basin of the attractor $\mathrm{Attr}_{\mathcal{S}}(C)$ of C. Conversely, assume that $E := \mathrm{Attr}_{\mathcal{S}}(C) \subset \overset{\circ}{C}$. Therefore, there exists $\varepsilon > 0$ such that $B_K(E, \varepsilon) \subset C$. By Lemma 9.3.2, p. 345, we know that $\omega^{\ominus 1}(E) \subset \mathcal{P}(K, B_K(E, \varepsilon)) \subset \mathcal{P}(K, C)$. Therefore, taking the intersection with $\mathcal{V}(K)$ and their inverse images under $\mathcal{S}$, this inclusion implies that $\mathrm{Lim}_{\mathcal{S}}(K, E)$ is contained in $\mathrm{Perm}_{\mathcal{S}}(K, C)$. Hence the equality. □

Lemma 9.2.4, p. 324 provides a useful "localization property" of the attractor of K:

Proposition 9.3.10 ***[Localization Property of the Attractor]*** *Let us assume that K is invariant under $\mathcal{S}$. For any subset $C \subset K$,*

$$\begin{cases} \mathrm{Attr}_{\mathcal{S}}(K) \;=\; \mathrm{Attr}_{\mathcal{S}}(\mathrm{Perm}_{\mathcal{S}}(K, C)) \cup_{\emptyset} \mathrm{Attr}_{\mathcal{S}}(\mathrm{Fluct}_{\mathcal{S}}(K, C)) \\ \subset \mathrm{Viab}_{\mathcal{S}}(C) \cup \overline{K \setminus C} \end{cases}$$

Proof. By Lemma 9.2.4, p. 324:

- either $x \in \mathrm{Fluct}_{\mathcal{S}}(K,C)$: Since it is invariant, the limit sets of all evolutions starting from x are contained in $\mathrm{Fluct}_{\mathcal{S}}(K,C)$, so that $\mathcal{A}_{\mathcal{S}}^{K}(x) \subset \mathrm{Fluct}_{\mathcal{S}}(K,C)$,
- or $x \in \mathrm{Perm}_{\mathcal{S}}(K,C)$: Since it is viable, at least one evolution is viable in the permanence kernel, and the limit sets of all evolutions starting from x viable in $\mathrm{Perm}_{\mathcal{S}}(K,C)$ are contained in it, or some evolution may leave the permanence kernel to enter the fluctuation basin, and its limit set is contained in $\mathrm{Fluct}_{\mathcal{S}}(K,C)$, so that $\mathcal{A}_{\mathcal{S}}^{K}(x) \subset \mathcal{A}_{\mathcal{S}}^{K}(\mathrm{Perm}_{\mathcal{S}}(K,C)) \cup \mathcal{A}_{\mathcal{S}}^{K}(\mathrm{Fluct}_{\mathcal{S}}(K,C))$. The permanence kernel and fluctuation basin being disjoint, the union of their attractors is actually a partition of the attractor of K.
- The inclusion follows from the fact that

$$\mathrm{Attr}_{\mathcal{S}}(\mathrm{Perm}_{\mathcal{S}}(K,C)) \subset \mathrm{Viab}_{\mathcal{S}}(K)$$

and that

$$\mathrm{Attr}_{\mathcal{S}}(\mathrm{Fluct}_{\mathcal{S}}(K,C)) \subset \overline{K \setminus C}$$

This concludes the proof. □

9.3.3 Viability Properties of Limit Sets and Attractors

Limit sets provide examples of forward and backward viable subsets:

Theorem 9.3.11 ***[Viability of Limit Sets]*** *If $\mathcal{S}$ is upper semicompact (see Definition 18.4.3, p. 729), the ω-limit set $\omega(x(\cdot))$ of an evolution $x(\cdot) \in \mathcal{S}(x)$ is always forward and backward viable under $\mathcal{S}$:*

$$\omega(x(\cdot)) = \mathrm{Viab}_{\mathcal{S}}(\omega(x(\cdot))) = \mathrm{Viab}_{\overleftarrow{\mathcal{S}}}(\omega(x(\cdot)))$$

The same statement holds for α-limit sets:

$$\alpha(x(\cdot)) = \mathrm{Viab}_{\mathcal{S}}(\alpha(x(\cdot))) = \mathrm{Viab}_{\overleftarrow{\mathcal{S}}}(\alpha(x(\cdot)))$$

When an evolution has a limit $\overline{x}$ when $t \to +\infty$, the subset $\{\overline{x}\}$ is viable, and thus, $\overline{x}$ is an equilibrium.

Proof. Let $\overline{x} \in \omega(x(\cdot))$ belong to the limit set of an evolution $x(\cdot) \in \mathcal{S}(x_0)$ starting at some element x_0. It is the limit of a sequence of elements $x(t_n)$ when $t_n \to +\infty$.

1. We first introduce the functions $y_n(\cdot) := \kappa(-t_n)(x(\cdot))$ defined by $y_n(t) := x(t + t_n)$ belonging to $\mathcal{S}(x(t_n))$. Since $\mathcal{S}$ is upper semicompact, a

subsequence (again denoted) $y_n(\cdot)$ converges uniformly on compact intervals to an evolution $y(\cdot) \in \mathcal{S}(\overline{x})$. On the other hand, for all $t > 0$,

$$y(t) = \lim_{n \to +\infty} y_n(t) = \lim_{n \to +\infty} x(t + t_n) \in \omega(x(\cdot))$$

i.e., $y(\cdot)$ is *viable* in the limit set $\omega(x(\cdot))$.

2. Next, we associate the functions $\overleftarrow{z}_n(\cdot) := \check{\kappa}(t_n)x(\cdot)$ defined by $\overleftarrow{z}_n(t) := x(t_n - t)$ belonging to $\overleftarrow{\mathcal{S}}(x(t_n))$. Since $\overleftarrow{\mathcal{S}}$ is upper semicompact, a subsequence (again denoted) $\overleftarrow{z}_n(\cdot)$ converges uniformly on compact intervals to an evolution $\overleftarrow{z}(\cdot) \in \overleftarrow{\mathcal{S}}(\overline{x})$. On the other hand, for all $t > 0$,

$$\overleftarrow{z}(t) = \lim_{n \to +\infty} \overleftarrow{z}_n(t) = \lim_{n \to +\infty} x(t_n - t) \in \omega(x(\cdot))$$

i.e., $\overleftarrow{z}(\cdot)$ is *viable* in the limit set $\omega(x(\cdot))$. □

Theorem 9.3.12 ***[Properties of the Attractor]*** *Assume that the evolutionary system $\mathcal{S}$ is upper semicompact and that K is closed. The forward and backward attractors of K under $\mathcal{S}$, as well as their closures are respectively subsets viable and backward viable under the evolutionary system:*

$$\mathrm{Attr}_{\mathcal{S}}(K) = \mathrm{Viab}_{\mathcal{S}}(\mathrm{Attr}_{\mathcal{S}}(K)) = \mathrm{Viab}_{\overleftarrow{\mathcal{S}}}(\mathrm{Attr}_{\mathcal{S}}(K))$$

and

$$\mathrm{Attr}_{\overleftarrow{\mathcal{S}}}(K) = \mathrm{Viab}_{\mathcal{S}}(\mathrm{Attr}_{\overleftarrow{\mathcal{S}}}(K)) = \mathrm{Viab}_{\overleftarrow{\mathcal{S}}}(\mathrm{Attr}_{\overleftarrow{\mathcal{S}}}(K))$$

They are consequently contained in the bilateral viability kernel $\overleftrightarrow{\mathrm{Viab}}_{\mathcal{S}}(K) = \mathrm{Viab}_{\overleftarrow{\mathcal{S}}}(K, B) \cap \mathrm{Viab}_{\mathcal{S}}(K, C)$ *(see Definition 8.3.1, p. 279):*

$$\mathrm{Attr}_{\mathcal{S}}(K) \cup \mathrm{Attr}_{\overleftarrow{\mathcal{S}}}(K) \subset \overleftrightarrow{\mathrm{Viab}}_{\mathcal{S}}(K)$$

Furthermore

$$\mathrm{Attr}_{\mathcal{S}}(K \backslash \mathrm{Viab}_{\overleftarrow{\mathcal{S}}}(K)) \subset \mathrm{Viab}_{\mathcal{S}}(K) \cap \partial \mathrm{Viab}_{\overleftarrow{\mathcal{S}}}(K)$$

Consequently, if $\mathrm{Viab}_{\overleftarrow{\mathcal{S}}}(K) \subset \mathrm{Int}(K)$, *then*

$$\mathrm{Attr}_{\mathcal{S}}(K) \subset \mathrm{Viab}_{\overleftarrow{\mathcal{S}}}(K) \subset \mathrm{Inv}_{\mathcal{S}}(K)$$

Proof. Proposition 9.3.11 stating that the limit subsets of evolutions viable in K are forward and backward viable and contained in K implies that the attractor is contained in the intersection of the forward and backward viability kernels of K. As an union of viable subsets, it is also viable, and since $\mathcal{S}$ is upper semicompact, its closure are forward and backward viable.

Consider now the case when $x \in \text{Viab}_{\mathcal{S}}(K)\backslash\text{Viab}_{\overleftarrow{\mathcal{S}}}(K)$. Theorem 8.3.2, p. 280 states that any viable evolution in K is viable in $\overline{\text{Viab}_{\mathcal{S}}(K)\backslash\text{Viab}_{\overleftarrow{\mathcal{S}}}(K)}$. Consequently, the attractor is also contained in the closure of the complement of the backward viability kernel, and thus, in its boundary.

The last statement follows from Proposition 8.3.3. □

9.3.4 Nonemptyness of Viability Kernels

We infer from Theorem 9.3.12 that a compact backward invariant subset has a nonempty viability kernel:

Proposition 9.3.13 ***[Forward and Backward Viability Kernels of Compact Subsets]*** *Assume that $\mathcal{S}$ is upper semicompact and that K is compact. Then the viability kernel $\text{Viab}_{\mathcal{S}}(K)$ is nonempty if and only if the backward viability kernel $\text{Viab}_{\overleftarrow{\mathcal{S}}}(K)$ is nonempty.*

Consequently, if K is a compact subset backward invariant under $\mathcal{S}$, then the viability kernel $\text{Viab}_{\mathcal{S}}(K)$ is nonempty and is furthermore both viable and backward invariant.

Proof. Indeed, if the backward viability kernel $\text{Viab}_{\overleftarrow{\mathcal{S}}}(K)$ is not empty, then there exists a backward viable evolution viable in $\text{Viab}_{\overleftarrow{\mathcal{S}}}(K)$, the α-limit set of which is not empty, because K is assumed to be compact, and viable, thanks to Theorem 9.3.12. Since the α-limit set is also forward viable, this implies that the viability kernel $\text{Viab}_{\mathcal{S}}(K)$ is not empty.

In the case when K is both compact and backward invariant, the forward viability kernel $\text{Viab}_{\mathcal{S}}(K)$ is not empty. It remains to prove that it is backward invariant: indeed, let x belong to $\text{Viab}_{\mathcal{S}}(K)$, $x(\cdot) \in \mathcal{S}(x)$ an evolution viable in K, which exists by assumption, and $\overleftarrow{y}(\cdot) \in \overleftarrow{\mathcal{S}}(x)$ any backward evolution starting at x. It is viable in $\text{Viab}_{\mathcal{S}}(K)$ because, otherwise, there would exist some finite time T such that $\overleftarrow{y}(T) \in K\backslash\text{Viab}_{\mathcal{S}}(K)$. Since the evolution $z_T(\cdot)$ defined by $z_T(t) := \overleftarrow{y}(T-t)$ belongs to $\mathcal{S}(\overleftarrow{y}(T))$, we can concatenate it with $x(\cdot)$ to obtain an evolution starting at $\overleftarrow{y}(T)$ viable in K, so that $\overleftarrow{y}(T)$ would belong to $\text{Viab}_{\mathcal{S}}(K)$, a contradiction. □

Remark. When the evolutionary system $\mathcal{S} := \mathcal{S}_F$ is associated with a differential inclusion $x'(t) \in F(x)$, subsets which are both viable and backward invariant under $\mathcal{S}_F$ are "morphological equilibria" of the *morphological equation* associated with F, the solution of which is the *reachable map* $t \mapsto \text{Reach}_{\mathcal{S}}(t, K)$ (Definition 8.4.1, p. 284). They thus play an important role in the context of morphological equations (see *Mathematical Methods of Game and Economic Theory*, [23, Aubin]). □

9.4 Lyapunov Stability and Sensitivity to Initial Conditions

9.4.1 Lyapunov Stability

If C is a neighborhood of E such that

$$E \subset D := \text{Int}(\text{Viab}_{\mathcal{S}}(C))$$

(where D is another smaller neighborhood of E) then from any initial state x in D, there exists at least one evolution $x(\cdot) \in \mathcal{S}(x)$ remaining in the neighborhood C of E. We recognize a property involved in the concept of Lyapunov stability:

Definition 9.4.1 *[Lyapunov Stability] A subset E is* stable in the sense of Lyapunov *if for every neighborhood C of E, there exists a neighborhood D of E such that for every initial state $x \in D$, there exists at least one evolution $x(\cdot) \in \mathcal{S}(x)$ viable in the neighborhood C.*

Therefore, this Lyapunov stability property can be reformulated in terms of viability kernels:

Proposition 9.4.2 *[Viability Characterization of Lyapunov Stability] Let E be viable under an evolutionary system. It is* stable *in the sense of Lyapunov if and only if*

$$E \subset \bigcap_{C \text{ neighborhood of } E} \text{Int}(\text{Viab}_{\mathcal{S}}(C))$$

9.4.2 The Sensitivity Functions

The sensitive dependence on initial conditions is one prerequisite of chaotic behavior of evolutions, whatever the meaning given to this polysemous nowadays fashionable word. This means roughly speaking that solutions starting from initial conditions close to each other move farther form each other away with time passing. When the dynamics is singled-valued and smooth, the basic method goes back to *Alexandr Lyapunov* again, with the concept of "Lyapunov exponents".

However, one can also capture *globally* and no longer locally this phenomenon through inequalities between evolutions $x(\cdot)$ and $y(\cdot)$ of the form

$$\forall t \geq 0, \ \|x(t) - y(t)\| \ \geq \ \lambda e^{mt}$$

where $m \geq 0$ and $\lambda > 0$. This suggests to introduce largest constant λ providing the function defined by

$$\lambda(x, y) \ := \sup_{x(\cdot)\in\mathcal{S}(x),\ y(\cdot)\in\mathcal{S}(y)} \inf_{t\geq 0} e^{-mt}\|x(t) - y(t)\|$$

regarded as a "sensitivity function".

Hence the knowledge of the sensitivity function allows us to detect the initial conditions x such that the set $\{y \mid \lambda(x, y) > 0\}$ is not empty.

Actually, any positive extended function $\mathbf{u} : X \times X \mapsto \mathbb{R} \cup \{+\infty\}$ could be used to measure sensitivity, not only the function $(x, y) \mapsto \|x - y\|$.

Definition 9.4.3 ***[Sensitivity Function]*** *The* strong sensitivity function $\lambda_{\mathbf{u}}(x, y) : X \times X \mapsto \mathbb{R}_+ \cup \{+\infty\}$ *associated with a function* $\mathbf{u}$ *is defined by*

$$\lambda_{\mathbf{u}}(x, y) := \sup_{x(\cdot)\in\mathcal{S}(x),\ y(\cdot)\in\mathcal{S}(y)} \inf_{t\geq 0} e^{-mt}\mathbf{u}(x(t), y(t))$$

and the (weak) sensitivity function $\varsigma_{\mathbf{u}}(x, y) \ : \ X \times X \ \mapsto \ \mathbb{R}_+ \cup \{+\infty\}$ *associated with a function* $\mathbf{u}$ *is defined by*

$$\varsigma_{\mathbf{u}}(x, y) := \sup_{x(\cdot)\in\mathcal{S}(x),\ y(\cdot)\in\mathcal{S}(y)} \sup_{t\geq 0} e^{-mt}\mathbf{u}(x(t), y(t))$$

Therefore, whenever $\lambda_{\mathbf{u}}(x, y) > 0$*, and for any* ε *such that* $0 < \varepsilon < \lambda_{\mathbf{u}}(x, y)$*, there exist evolutions* $x(\cdot) \in \mathcal{S}(x)$ *and* $y(\cdot) \in \mathcal{S}(y)$ *such that*

$$\forall t \geq 0, \ \mathbf{u}(x(t), y(t)) \ \geq \ e^{mt}(\lambda_{\mathbf{u}}(x, y) - \varepsilon)$$

The sensitivity function $\lambda_{\mathbf{u}}$ can be characterized in terms of the viability kernel of the hypograph of the function $\mathbf{u}$ under the auxiliary system:

$$\begin{cases} (i) & x'(t) = f(x(t),u(t)) \text{ where } u(t) \in U(x(t)) \\ (i) & y'(t) = f(y(t),v(t)) \text{ where } v(t) \in U(y(t)) \\ (iii) & z'(t) = mz(t) \end{cases} \tag{9.16}$$

subject to the constraint

$$\forall t \geq 0, \quad (x(t),y(t),z(t)) \in \mathcal{H}yp(\mathbf{u})$$

Proposition 9.4.4 *[Viability Characterization of the Sensitivity Functions] The strong sensitivity function is related to the viability kernel of the hypograph $\mathcal{H}yp(\mathbf{u})$ of the function $\mathbf{u}$ under the auxiliary system (9.16) by the following formula*

$$\lambda_{\mathbf{u}}(x,y) = \sup_{(x,y,z)\in \mathrm{Viab}_{(9.16)}(\mathcal{H}yp(\mathbf{u}))} z$$

and the weak sensitivity function is related to the capture basin of the hypograph $\mathcal{H}yp(\mathbf{u})$ of the function $\mathbf{u}$ under the auxiliary system (9.16), p. 356 by the following formula

$$\varsigma_{\mathbf{u}}(x,y) = \sup_{(x,y,z)\in \mathrm{Capt}_{(9.16)}(X\times X\times \mathbb{R}_+,\mathcal{H}yp(\mathbf{u}))} z$$

Proof. Indeed, to say that (x,y,z) belongs to the viability kernel of $\mathcal{H}yp(\mathbf{u})$ under the auxiliary system (9.16) amounts to saying that there exists an evolution $t \mapsto (x(t),y(t),z(t))$ governed by the auxiliary system such that, for all $t \geq 0$, $u(t) \in U(x(t))$, $v(t) \in U(y(t))$ and $z(\cdot) = ze^{m\cdot}$. By definition of (9.16), we know that for all $t \geq 0$, this evolution satisfies also for all $t \geq 0$,

$$\mathbf{u}(x(t),y(t)) \geq z(t) = e^{mt}z$$

Therefore

$$\inf_{t\geq 0} e^{-mt}\mathbf{u}(x(t),y(t)) \geq z$$

and thus, $\lambda_{\mathbf{u}}(x,y) \geq \overline{\lambda} := \sup_{(x,y)\in \mathrm{Viab}_{(9.16)}(\mathcal{H}yp(\mathbf{u}))} z$. For proving the converse inequality, we associate with any $\varepsilon > 0$ evolutions $x_\varepsilon(\cdot) \in \mathcal{S}(x)$ and $y_\varepsilon(\cdot) \in \mathcal{S}(y)$ such that

$$\inf_{t\geq 0} e^{-mt}\mathbf{u}(x(t),y(t)) \geq \lambda_{\mathbf{u}}(x,y) - \varepsilon$$

Setting $z_\varepsilon(t) := (\lambda_{\mathbf{u}}(x,y) - \varepsilon)e^{mt}$, we infer that $(x_\varepsilon(\cdot), y_\varepsilon(\cdot), z_\varepsilon(\cdot))$ is a solution to the auxiliary system starting at $(z,y,\lambda_{\mathbf{u}}(x,y) - \varepsilon)$ viable in the hypograph of $\mathbf{u}$, and thus, that $(z,y,\lambda_{\mathbf{u}}(x,y) - \varepsilon)$ belongs to the viability kernel of its hypograph. Hence $\lambda_{\mathbf{u}}(x,y) - \varepsilon \leq \overline{\lambda}$ and letting ε converge to 0, we have proved the equality.

The proof for the weak sensitivity function is similar: to say that (x,y,z) belongs to the capture basin of $\mathcal{H}yp(\mathbf{u})$ under the auxiliary system (9.16) amounts to saying that there exist an evolution $t \mapsto (x(t), y(t), z(t))$ governed by the auxiliary system, and a finite time $t^\star$ such that, $u(t^\star) \in U(x(t^\star))$, $v(t^\star) \in U(y(t^\star))$ and $z(\cdot) = ze^{m\cdot}$ belonging to the hypograph of $\mathbf{u}$:

$$\mathbf{u}(x(t^\star), y(t^\star)) \geq z(t^\star) = e^{mt}z$$

Therefore

$$\sup_{t\geq 0} e^{-mt}\mathbf{u}(x(t), y(t)) \geq z$$

and thus, $\varsigma_{\mathbf{u}}(x,y) \geq \overline{\lambda} := \sup_{(x,y)\in \text{Capt}_{(9.16)}(\mathcal{H}yp(\mathbf{u}))} z$. The proof of the converse inequality is standard. □

In order to recover some of the concepts of "sensitivity" found in the literature, we associate with the two-variable sensitivity functions λ and ς the following one-variable functions:

Definition 9.4.5 *[**Sensitivity Basins**] We introduce the following functions*

$$\lambda_{\mathbf{u}}^{\flat}(x) := \liminf_{y\to x} \lambda_{\mathbf{u}}(x,y) \ \& \ \varsigma_{\mathbf{u}}^{\flat}(x) := \liminf_{y\to x} \varsigma_{\mathbf{u}}(x,y)$$

The subsets

$$\Lambda_{\mathbf{u}} := \{x \text{ such that } \lambda_{\mathbf{u}}^{\flat}(x) > 0\} \ \& \ \Sigma_{\mathbf{u}} := \{x \text{ such that } \varsigma_{\mathbf{u}}^{\flat}(x) > 0\}$$

are called the strong *and* weak *sensitivity basin respectively of the parameterized system $x'(t) = f(x(t), u(t))$ where $u(t) \in U(x(t))$.*

The definition of "liminf" implies the following

Lemma 9.4.6 *[**Characterization of the Sensitivity Basins**] The strong sensitivity basin $\Lambda_{\mathbf{u}}$ (respectively the weak sensitivity basin $\Sigma_{\mathbf{u}}$) is the set of initial points such that, for any neighborhood $\mathcal{U}$ of x, there exists an initial state $y \in X$ from which start evolutions $x(\cdot) \in \mathcal{S}(x)$ and $y(\cdot) \in \mathcal{S}(y)$ such that*

$$\forall t \geq 0, \ \mathbf{u}(x(t), y(t)) \geq e^{mt}\frac{1}{2}\lambda_{\mathbf{u}}^{\flat}(x)$$

and

$$\exists\, T \geq 0 \text{ such that } \mathbf{u}(x(T), y(T)) \;\geq\; e^{mT}\frac{1}{2}\lambda_{\mathbf{u}}^{\flat}(x)$$

respectively.

Sensitivity to initial conditions has been claimed to be common behavior rather than exception. But one should be careful to distinguish *between sensitivity to initial conditions of deterministic systems and* redundancy – *availability of a set of evolutions* – met in many systems arising in biology and social sciences. The situation is quite orthogonal to the chaotic deterministic pattern, indeed: There are many evolutions, but only those "stable" – actually, viable – persist in time, are permanent.

That physical systems in fluid dynamics, meteorology, astrophysics, electrical circuits, etc., exhibit a deterministic chaotic behavior of some of its evolutions (unpredictable in the sense of sensitive dependence on initial conditions) is thus an issue different (however, complementary to) the study of viability properties, although viability techniques may help us to bring supplementary results to sensitivity analysis on initial conditions.

Remark: The associated Hamilton–Jacobi–Bellman Equation. The **sensitivity function** $\lambda_{\mathbf{u}}(x, y)$ associated with a function **u** is the largest solution to the Hamilton–Jacobi–Bellman partial differential equation

$$\inf_{u \in U(x)} \left(\sum_{i=1}^{n} \frac{\partial \mathbf{v}(x,y)}{\partial x_i} f_i(x,u) \right) + \inf_{v \in U(y)} \left(\sum_{i=1}^{n} \frac{\partial \mathbf{v}(x,y)}{\partial y_i} f_i(y,v) \right) + m\mathbf{v}(x,y) = 0$$

smallest than or equal to **u**. □

9.4.3 Topologically Transitive Sets and Dense Trajectories

Definition 9.4.7 *[**Topologically Transitive Sets**] A closed subset K invariant under an evolutionary system $\mathcal{S}$ is said to be* topologically transitive *if for any pair of nonempty subsets $\overset{\circ}{C} \subset K$ and $\overset{\circ}{D} \subset K$ open in K (for its relative topology), there exist $x \in \overset{\circ}{C}$ such that all evolutions $x(\cdot) \in \mathcal{S}(x)$ are viable in K and reach $\overset{\circ}{D}$ in finite time.*

In other words, to say that a subset K is topologically transitive amounts to saying that for any open subset $\overset{\circ}{D}\subset K$, $\text{Abs}_{\mathcal{S}}(\overset{\circ}{K},\overset{\circ}{D})$ is dense in K, because, for any nonempty open subset $\overset{\circ}{C}$, the intersection $\overset{\circ}{C}\cap\text{Abs}_{\mathcal{S}}(\overset{\circ}{K},\overset{\circ}{D})$ is not empty. We deduce the following property of topologically transitive subsets:

Proposition 9.4.8 *[Existence of Evolutions with Dense Trajectories] Assume that the evolutionary system is upper semicompact and lower semicontinuous and that K is closed and topologically transitive. Then:*

- *there exists an initial state from which the trajectories of all evolutions are dense in K*
- *the set of such initial states is itself dense in K.*

Proof. Since the evolutionary system $\mathcal{S}$ is upper semicompact, Theorem 10.3.12, p. 389 implies that the absorption basins $\text{Abs}_{\mathcal{S}}(\overset{\circ}{K},\overset{\circ}{D})$ are open. Therefore, if K is topologically transitive, they are also dense in K.

On the other hand, since K is a closed subset of a finite dimensional vector space, there exists a denumerable subset $D := \{a_1,\ldots,a_n,\ldots\}$ dense in K. Let us introduce the denumerable sequence of dense open subsets $\overset{\circ}{D^p_n}:=\overset{\circ}{B}_K\left(a_n,\frac{1}{p}\right)$. The Baire Category Theorem implies that the intersection $\bigcap_{n\geq 0}\text{Abs}_{\mathcal{S}}(\overset{\circ}{K},\overset{\circ}{D^p_n})$ is dense in K:

$$K=\overline{\bigcap_{n\geq 0}\text{Abs}_{\mathcal{S}}(\overset{\circ}{K},\overset{\circ}{D^p_n})}$$

Therefore, for any x in K and any open subset $\overset{\circ}{C}$ of K, there exists $x^\star\in C\cap\bigcap_{n\geq 0}\text{Abs}_{\mathcal{S}}(\overset{\circ}{K},\overset{\circ}{D^p_n})$.

It remains to prove that for any evolution $x^\star(\cdot)\in\mathcal{S}(x^\star)$, its trajectory (or orbit) $\{x^\star(t)\}_{t\geq 0}$ is dense in K. Indeed, for any $y\in K$ and any $\varepsilon>0$, there exist $p>\frac{2}{\varepsilon}$ and $a_{n_p}\in D:=\{a_1,\ldots,a_n,\ldots\}$ such that $y\in D^p_n:=\overset{\circ}{B}\left(a_{n_p},\frac{1}{p}\right)$. Therefore, since $x^\star\in\text{Abs}_{\mathcal{S}}(\overset{\circ}{K},\overset{\circ}{D^p_n})$, there exist $t^\star<+\infty$ such that $x^\star(t^\star)\in C^p_n$, is such a way that $d(y,x^\star(t^\star))\leq d(y,a_{n_p})+d(a_{n_p},x^\star(t^\star))\leq\frac{2}{p}\leq\varepsilon$. Hence, the trajectory of the evolution $x^\star(\cdot)\in\mathcal{S}(x^\star)$ is dense in K, and the set of initial states such that the trajectories of the evolutions starting from it are dense in K, is itself dense in K. □

9.5 Existence of Equilibria

We emphasize that there exists a basic and curious link between viability theory and general equilibrium theory: the Fundamental Equilibrium Theorem – an equivalent version of the 1910 *Brouwer Fixed Point Theorem* – states that *viability implies stationarity*:

Theorem 9.5.1 *[The **Fundamental Equilibrium Theorem**] Assume that X is a finite dimensional vector-space and that $F : X \rightsquigarrow X$ is Marchaud. If $K \subset X$ is **viable** and **convex compact** under F, then there exists an equilibrium of F belonging to K.*

Proof. We refer to *Optima and Equilibria*, [19, Aubin], or *Mathematical Methods of Game and Economic Theory*, [17, Aubin] for instance, where one can find other theorems on the existence of equilibria of set-valued maps. □

One can derive from the nonemptyness of the viability kernel of K a comparable statement, where the assumption of the convexity of K is replaced by the convexity of the image $F(K)$ of K by F:

Theorem 9.5.2 *[**Equilibrium Theorem # 2**] Assume that X is a finite dimensional vector-space and that $F : X \rightsquigarrow X$ is Marchaud. If $F(K)$ is compact convex and if the viability kernel* $\mathrm{Viab}_F(K)$ *of K under F is not empty, it contains an equilibrium of F.*

Proof. Assume that there is no equilibrium. This means that 0 does not belong to the closed convex subset $F(K)$, so that the Separation Theorem 18.2.4, p. 715 implies the existence of some $p \in X^\star$ and $\varepsilon > 0$ such that

$$\sup_{x\in K, v\in F(x)} \langle v, -p\rangle = \sigma(F(K), -p) \; < \; -\varepsilon$$

Take any $x_0 \in \mathrm{Viab}_F(K)$, which is not empty by assumption, from which starts one evolution $x(\cdot) \in \mathcal{S}(x)$ viable in K. We deduce that

$$\forall t \geq 0, \; \langle -p, x'(t)\rangle \leq -\varepsilon$$

so that, integrating from 0 to t, we infer that

$$\varepsilon\, t \; \leq \; \langle p, x(t) - x(0)\rangle \; = \; \langle p, x(t) - x_0\rangle$$

But K being bounded, we thus derive a contradiction. □

9.6 Newton's Methods for Finding Equilibria

We revisit the famous Newton algorithm with viability concepts and results. Viewed from this angle, we are able to devise several ways to conceive other types of Newton methods for approximating solutions $\overline{x} \in K$ to equations

$$0 = f(\overline{x}) \text{ (or } \overline{x} \in K \cap f^{-1}(0))$$

or inclusions

$$0 \in F(\overline{x}) \text{ (or } \overline{x} \in K \cap F^{-1}(0))$$

When $X := \mathbb{R}$ and $f : \mathbb{R} \mapsto \mathbb{R}$ is differentiable, recall that the Newton method is the algorithm described by

$$x_{n+1} = x_n - \frac{f(x_n)}{f'(x_n)}$$

which can be written in the form

$$f'(x_n)(x_{n+1} - x_n) = -f(x_n) \text{ where } x_0 \text{ is given}$$

Written in this form, it can also be defined for differentiable maps $f : X \mapsto X$ where X is a finite dimensional vector space. Newton's method is known to converge to a solution $\bar{x}$ to equation $f(\overline{x}) = 0$ when the derivatives $f'(x)$ exist, satisfy $f'(x) \neq 0$ (and thus, are invertible) and when the initial point x_0 "is not too far" from this solution.

9.6.1 Behind the Newton Method

This discrete Newton algorithm, the prototype of many further sophisticated extensions that came later, is the discretization of the continuous version of Newton's algorithm given by the differential equation

$$f'(x(t))x'(t) = -f(x(t)), \text{ where } x(0) = x_0 \text{ is given} \tag{9.17}$$

which makes sense when $f'(y)$ is invertible, at least in a neighborhood of the (unknown) equilibrium. We observe that $y(t) := f(x(t))$ is a solution to the differential equation $y'(t) = f'(x(t))x'(t) = -y(t)$ and thus, that it is equal to y_0e^{-t}, so that the cluster points $x^\star := \lim_{t_n \to +\infty} f(x(t_n))$ of $x(t)$ when $t \to +\infty$ are equilibria of f because $f(x(t_n)) = y_0e^{-t_n}$ converges to 0.

This remark leads us to throw a "graphical glance" at this problem. Assuming that the derivatives $f'(x)$ are invertible and setting

$$\mathbf{N}_f(x,y) := -f'(x)^{-1}y \tag{9.18}$$

we rewrite the continuous version of the Newton method under the form

$$\begin{cases} (i) \;\; x'(t) = \mathbf{N}_f(x(t),y(t)) \\ (ii) \; y'(t) = -y(t) \end{cases} \tag{9.19}$$

starting at initial states $(x_0, f(x_0)) \in \text{Graph}(f)$. By construction, we observe that

$$\forall (x,y),\; (\mathbf{N}_f(x,y), -y) \in T_{\text{Graph}(f)}(x(t),y(t))$$

where $T_{\text{Graph}(f)}(x,y)$ denotes the tangent cone to the graph of f at a pair (x,y) of its graph, since the graph of the differential $f'(x)$ of f at x is the tangent space to the graph of f at the point (x,y) where $y = f(x)$ (see Theorem 18.5.4, p. 739). This tangential condition characterizes the fact that the solutions $(x(\cdot), y(\cdot))$ governed by the above system of differential equations are viable in the graph of f thanks to the Nagumo Theorem 11.2.3, p. 444.

Therefore, looking behind and beyond the Newton Method, and letting aside for the time being the benefits of approximating locally the nonlinear map f by its derivatives $f'(x)$, we may note and underline the following points:

1. to say that $x^\star$ is an equilibrium of f amounts to saying that $(x^\star, 0) \in \text{Graph}(f)$,
2. to say that $x(t)$ converges to the equilibrium $x^\star$ amounts to saying that, setting, $y(t) = f(x(t))$,

$$(x(t), y(t)) \in \text{Graph}(f) \text{ converges to } (x^\star, 0)$$

3. $y(t) = f(x_0)e^{-t}$ is the solution to the differential equation $y'(t) = -y(t)$ starting at $f(x_0)$ which converges to 0 when $t \to +\infty$.

We see at once how we can generalize this idea for approximating an equilibrium $x^\star$ of any set-valued map $F : X \rightsquigarrow Y$, i.e., a solution to the inclusion $0 \in F(x^\star)$ or an element $x^\star \in F^{-1}(0)$: It is sufficient to build pairs $(x(\cdot), y(\cdot))$ of functions such that:

1. the function $t \mapsto (x(t), y(t))$ is viable in the graph of F (i.e., $y(t) \in F(x(t))$),
2. the limit when $t \to +\infty$ of $y(t)$ is equal to 0

We obtain in this way a continuous version of Newton's algorithm, in the sense that the ω-limit set of $x(\cdot)$ is contained in the set $F^{-1}(0)$ of equilibria of F whenever the graph of F is closed.

9.6.2 Building Newton Algorithms

The problem of approximating equilibria of a given set-valued map $F : X \rightsquigarrow X$ thus boils down to building differential equations or differential inclusions governing the evolution of $(x(\cdot), y(\cdot))$ satisfying the above requirements.

- We can choose at will any differential equation governing the evolution of $y(\cdot)$ in such a way that $y(t)$ converges to 0 asymptotically (or reaches it in finite time). Among many other choices that we shall not develop in this book, the linear equation $y' = -y$ provides evolutions converging exponentially to 0,
- Next, build a map $\mathbf{N} : X \times Y \rightsquigarrow X$ (hopefully a single-valued map $\mathbf{n}$) governing the evolution of $(x(\cdot), y(\cdot))$ through the *Newton system* of differential inclusions

$$\begin{cases} (i)\ x'(t) \in \mathbf{N}(x(t), y(t)) \\ (ii)\ y'(t) = -y(t) \end{cases} \tag{9.20}$$

describing the associated algorithm, generalizing the concept of Newton algorithm. Solutions $(x(\cdot), y(\cdot))$ governed by this system have $y(t)$ converging to 0, so the issue that remains to be settled is to find whether or not it provides evolutions *viable in the graph of F*.

We are now facing an alternative *for building such a Newton map* $\mathbf{N}$:

1. **Either**, take the set-valued version of the Newton algorithm by introducing the set-valued map $\mathbf{N}_F$ defined by

$$\mathbf{N}_F(x, y) := D^{\star\star}F(x, y)^{-1}(x, -y) \tag{9.21}$$

where $D^{\star\star}F(x, y) : X \rightsquigarrow Y$ is the *convexified derivative* of the set-valued map F at (x, y) defined by

$$T^{\star\star}_{\mathrm{Graph}(F)}(x, y) =: \mathrm{Graph}(D^{\star\star}F(x, y))$$

where $T^{\star\star}_K(x)$ is the closed convex hull to the tangent cone $T_K(x)$ to K at x (see Definition 18.5.5, p. 740). This is the natural extension of the map $(x, y) \mapsto \mathbf{N}_f(x, y) := -f'(x)^{-1}y$ defined by (9.18), p. 362. The continuous set-valued version of the Newton algorithm (9.20), p. 363 can be written in the form

$$\begin{cases} (i)\ x'(t) \in \mathbf{N}_F(x(t), y(t)) \\ (ii)\ y'(t) = -y(t),\ \ (x_0, y_0) \in \mathrm{Graph}(F) \end{cases} \tag{9.22}$$

To justify this claim, we have check that we can apply the Fundamental Viability Theorem 11.3.4, p. 455, which provides an evolution $(x(\cdot), y(\cdot))$

viable in the graph of F, so that the cluster points of its component $x(\cdot)$ are equilibria of F.
The assumptions of the Fundamental Viability Theorem 11.3.4, p. 455 assumptions require that the set-valued map $\mathbf{N}_F$ is Marchaud (see Definition 10.3.2, p. 384) and that the tangential condition:

$$\forall (x,y) \in \text{Graph}(F),\ (\mathbf{N}_F(x,y) \times \{-y\}) \cap T_{\text{Graph}(F)}(x,y) \neq \emptyset \quad (9.23)$$

where $T^{\star\star}_{\text{Graph}(F)}(x,y)$ denotes the closed convex hull of tangent cone to the graph of F at a pair (x,y) of its graph.
By the very definition of the derivative $D^{\star\star}F(x,y)$ of F at (x,y), condition (9.23), p. 364 can be rewritten in the form

$$\forall (x,y) \in \text{Graph}(F),\ \mathbf{N}_F(x,y) \cap D^{\star\star}F(x,y)^{-1}(x,-y) \neq \emptyset \quad (9.24)$$

2. **Or**, and more importantly, in order to avoid the difficult problem of computing the derivative of the set-valued map for practical purposes, we start with an arbitrary map $\mathbf{N} : X \times Y \rightsquigarrow Y$ given independently of F, but chosen for solving easily system (9.20), p. 363.
The simplest example of set-valued map is the map defined by $\mathbf{N}(x,y) \equiv B$ where B is the unit ball of Y: this example allows us to leave open the choice of the directions $x'(t)$ with which the state will evolve. This is the set-valued version of the *Montagnes Russes Algorithm* (see Sect. 4.5.3, p. 149).
The drawback of adopting this approach is that there is no reason why, starting from any initial pair $(x_0, y_0) \in \text{Graph}(F)$, there would exist an evolution governed by system (9.20), p. 363 viable in the graph of F. Hence the idea is to replace the set-valued map $\mathbf{N}$ by another one, depending both of $\mathbf{N}$ and F, which can be computed by the Viability Kernel Algorithm, and provide evolutions $(x(\cdot), y(\cdot))$ viable in the graph of F and converging to some state $(x^\star, 0) \in \text{Graph}(F)$, i.e., to solution of $0 \in F(x^\star)$. The solution is obvious: take the map the graph of which is the viability kernel of the graph of F under system (9.20), p. 363.

Definition 9.6.1 ***[The Newton Kernel of a Map]*** *Let $F : X \rightsquigarrow Y$ be a set-valued map and $\mathbf{N} : X \times Y \rightsquigarrow X$ a set-valued map defining a Newton type algorithm. The* Newton kernel $\text{New}_{\mathbf{N}}(F) : X \rightsquigarrow Y$ *of F by $\mathbf{N}$ is the set-valued map defined by*

$$\text{Graph}(\text{New}_{\mathbf{N}}(F)) := \text{Viab}_{(9.20)}(\text{Graph}(F))$$

Its domain is called the Newton Basin *of the set-valued map F.*

Therefore, we have proven the following:

Proposition 9.6.2 *[General Newton Type Algorithms] Let $F : X \rightsquigarrow Y$ be a set-valued map and $\mathbf{N} : X \times Y \rightsquigarrow X$ a set-valued map defining a Newton algorithm. Then from any initial state $x_0 \in \mathrm{Dom}(\mathrm{New}_{\mathbf{N}}(F))$ and $y_0 \in \mathrm{New}_F(\mathbf{N})(x_0)$ starts at least one evolution of the Newton algorithm*

$$x'(t) \in \mathrm{New}_{\mathbf{N}}(F)(x(t), e^{-t}y_0)$$

the cluster points of which are equilibria of F.

The Viability Kernel Algorithm provides the Newton kernel of $\mathbf{N}$ associated with a Newton map F, as well as the feedback governing the evolutions $(x(\cdot), y(\cdot))$ viable in the graph of F. Therefore, the second component $t \mapsto y_0 e^{-t}$ converges to 0 so that the cluster points of the first component $x(\cdot)$ are equilibria of F.

Remark. We never used here the fact that $F : X \rightsquigarrow X$ maps X to itself (although this is the case of interest in this book since we deal with equilibria of a differential inclusion). Hence the above results dealing with the Newton type methods can be extended to the search of solutions to inclusions of the type $y \in F(x)$ when $F : X \rightsquigarrow Y$. □

The next problem is to approximate solutions to system (9.20) of differential inclusions *viable in the graph of F* (standard discretizations of even differential equations generally do not provide viable evolutions, they have to be modified to provide viable evolutions). We shall not address this issue in this book, referring to the many monographs and papers dealing with these issues.

9.7 Stability: The Inverse Set-Valued Map Theorem

9.7.1 The Inverse Set-Valued Map Theorem

The Inverse Function Theorem for single-valued maps in Banach spaces goes back to *Graves* and to *Lazar Ljusternik* in the 1930s. It states that if f is continuously differentiable at x_0 and if $f'(x_0)$ is surjective, then:

1. the equation $f(x) = y$ has a solution for any right-hand side y close enough to $f(x_0)$,

2. The set-valued map f^{-1} behaves in a Lipschitz way.

The first statement is an *existence* theorem: If the equation $f(x) = y_0$ has at least one solution x_0, the equation $f(x) = y$ has still a solution when the right hand side y ranges over a neighborhood of y_0. The second statement is as important, but different: This is a *stability* result, expressing that if x_0 is a solution to $f(x) = y_0$, one can find solutions x_y to the equation $f(x) = y$ depending in a continuous way (actually, a Lipschitz way) on the right-hand side.

This is an abstract version of the celebrated *Lax Principle* of *Peter Lax* stating that convergence of the right hand side and stability implies the convergence of the solution (actually, in numerical analysis, the map f is also approximated), proved in the linear case, a consequence of the Banach Inverse Mapping Theorem.

It had been extended to set-valued maps. This is possible by using the concept of *convexified derivative* $D^{\star\star}F(x,y)$ of the set-valued map F at (x,y) is defined by (see Definition 18.5.5, p. 740):

$$T^{\star\star}_{\mathrm{Graph}(F)}(x,y) \;=:\; \mathrm{Graph}(D^{\star\star}F(x,y))$$

Theorem 9.7.1 ***[Inverse Set-Valued Map Theorem]*** *Let $F : X \rightsquigarrow Y$ be a set-valued map with closed graph, $y_0 \in Y$ a right hand side and $x_0 \in F^{-1}(y_0)$ a solution to inclusion $y_0 \in F(x_0)$. Assume that a set-valued map F satisfies*

$$\begin{cases} (i) \ \mathrm{Im}(D^{\star\star}F(x_0,y_0)) = Y \ \text{(i.e., } D^{\star\star}F(x_0,y_0) \text{ is surjective)} \\ (ii) \ (x,y) \rightsquigarrow \mathrm{Graph}(D^{\star\star}F(x,y)) \text{ is lower semicontinuous at } (x_0,y_0) \end{cases} \tag{9.25}$$

There exist $\nu > 0$ and $\eta > 0$ such that, for any $(x,y) \in \overset{\circ}{B}((x_0,y_0),\eta)$, and for any $y_1 \in \overset{\circ}{B}(y, \eta - \|y-y_0\|)$:

1. *there exists a solution $x_1 \in F^{-1}(y_1)$ solution to inclusion $y_1 \in F(x_1)$*
2. *satisfying the stability condition*

$$\|x_1 - x\| \leq \nu\|y_1 - y\|$$

We shall deduce this theorem from the Local Viability Theorem 19.4.3, p. 783.

This stability property appearing in the second statement, weaker than the Lipschitz property (see Definition 10.3.5, p. 385), has been introduced in the early 1980s under the name of *pseudo-Lipschitzianity* of F^{-1} around (x_0, y_0), and is now known and studied under the name of *Aubin's property*.

9.7.2 Metric Regularity

The Banach Inverse Mapping (on the stability of the inverse of the linear operator) is equivalent to the Banach Closed Graph Theorem. Theorem 9.7.1, p. 366 extends it to the case of set-valued maps: the sufficient conditions has been proved in 1982 from the Gérard Lebourg version of Ekeland's Variational Principle (see Theorem 18.6.15, p. 751), the necessary condition in 2004 with the proof given in this book, and, later, with another proof by Asen Dontchev and Marc Quincampoix (see for instance *Applied nonlinear analysis*, [26, Aubin & Ekeland] and *Implicit Functions and Solution Mappings*, Springer, [81, Dontchev & Rockafellar] specifically dedicated to inverse theorems and *metric regularity*). The link between inverse function theorems and viability tools goes back to Luc Doyen.

For the sake of simplicity, we shall prove the statement of Theorem 9.7.1, p. 366 when F^{-1} is replaced by F:

Theorem 9.7.2 *[Viability Characterization of Aubin's Property]* *Let $F : X \rightsquigarrow Y$ be a set-valued map with closed graph and $x_0 \in \mathrm{Dom}(F)$. Let $y_0 \in F(x_0)$ belong to $F(x_0)$. If a set-valued map F satisfies*

$$\begin{cases} (i) & \mathrm{Dom}(DF(x_0, y_0)) = X \\ (ii) & (x, y) \rightsquigarrow \mathrm{Graph}(DF(x, y)) \text{ is lower semicontinuous at } (x_0, y_0) \end{cases} \tag{9.26}$$

then there exist $\alpha > 0$ and $\eta > 0$ such that, for any $(x, y) \in \overset{\circ}{B}((x_0, y_0), \eta)$, and for any $x_1 \in \overset{\circ}{B}(x, \eta - \|x - x_0\|)$:

1. *there exists $y_1 \in F(x_1)$*
2. *satisfying*

$$\|y_1 - y\| \leq \alpha \|x_1 - x\|$$

From the Banach Closed Graph Theorem for continuous linear operators to the above statement, we have first to prove the extension of Banach theorem to "linear set-valued maps", called *closed convex processes* by *Terry Rockafellar, the graph of which are closed convex cones* (see Definition 18.3.7, p. 722). They enjoy most of the properties of continuous linear operators, including the concept of transposition.

We shall transfer the *global continuity* properties of a closed convex process to *local continuity* properties of any set-valued map F whenever its "derivative" is a closed convex process, in the same spirit than passing from the original Banach Closed Graph theorem to the Inverse Function Theorem, the cornerstone of classical differential calculus and analysis, or, actually, to its metric regularity formulation. This is possible by using the concept of *convexified derivative* $D^{\star\star}F(x, y)$ of the set-valued map F at (x, y) is defined by (see Definition 18.5.5, p. 740):

$$T^{\star\star}_{\mathrm{Graph}(F)}(x,y) \;=:\; \mathrm{Graph}(D^{\star\star}F(x,y))$$

It always exists, and, by construction, is a closed convex process.

The set-valued adaptation of the Closed Graph Theorem was due to *Corneliu Ursescu* and *Stephen Robinson*:

Theorem 9.7.3 *[The Robinson–Ursescu Theorem] If the domain of the closed convex process $D^{\star\star}F(x,y)$ is the whole space X, then it is Lipschitz: $\lambda(D^{\star\star}F(x,y)) < +\infty$.*

Proof. We provide the proof when the spaces X and Y are finite dimensional vector spaces. Since the domain of $DF(x,y) = X$ is the whole space X, the function $u \mapsto d(0, D^{\star\star}F(x,y)(u))$ is finite. Hence X is the union of the sections

$$S_n \;:=\; \{u \mid d(0, D^{\star\star}F(x,y)(u)) \leq n\}$$

which are closed because $u \mapsto d(0, D^{\star\star}F(x,y)(u))$ is lower semicontinuous. Baire's Theorem implies that the interior of one of these sections is not empty, and actually, that there exists a ball ηB of radius $\eta > 0$ contained in S_n because the function $u \mapsto d(0, D^{\star\star}F(x,y)(u))$ is convex and positively homogeneous. Therefore,

$$\forall\, u \in B,\;\; d(0, D^{\star\star}F(x,y)(u)) \;\leq\; \frac{n}{\eta}$$

and thus

$$\lambda(D^{\star\star}F(x,y)) \;:=\; \sup_{\|u\|=1} d(0, D^{\star\star}F(x,y)(u)) \;\leq\; \frac{n}{\eta}\|u\|$$

This means that the inf-norm $\lambda(D^{\star\star}F(x,y))$ is finite. □

The classical concept of Lipschitz (set-valued) map is global, in the sense that the Lipschitz λ constant does not depend upon the pairs of elements chosen in the inequality

$$\forall\,(x_i,y_i) \in \mathrm{Graph}(F),\; i=1,2,\;\; \|x_1 - x_2\| \;\leq\; \lambda\|y_1 - y_2\|$$

where $\lambda := \sup_{x_1 \neq x} \dfrac{d(y, F(x_1))}{\|x_1 - x\|}$ is the *Lipschitz constant.* This may be too demanding to require that the Lipschitz constant is independent of all pairs points $(x,y) \in \mathrm{Graph}(F)$, so that we have to suggest a weaker concept strong enough for the inverse mapping theorem for set-valued maps to hold true and to characterize it in terms of "inf-norm" of the derivatives of F.

We set $\lim_{x \to_K y} f(x) := \lim_{x \in K, x \to y} f(x)$.

Definition 9.7.4 *[Aubin Constant] Let us associate with any pair* $(x, y) \in \text{Graph}(F)$

$$\delta_F(T; x, y) = \sup_{\|x_1 - x\| \leq T \ \& \ x_1 \neq x} \frac{d(y, F(x_1))}{\|x_1 - x\|} \tag{9.27}$$

and the function α_F *defined on the graph of* F *by*

$$\begin{cases} \alpha_F(x_0, y_0) := \limsup_{(x,y) \mapsto_{\text{Graph}(F)} (x_0, y_0),\ T \to 0+} \delta_F(T; x, y) \\ := \inf_{\eta > 0} \sup_{T \leq \eta \& \max(\|x - x_0\|, \|y - y_0\|) \leq \eta} \delta_F(T; x, y) \end{cases}$$

called the pseudo-Lipschitz modulus *or* Aubin constant *of* F *at* (x_0, y_0).

This definition is motivated by the following property:

Proposition 9.7.5 *[Aubin's Property] If* $\alpha_F(x_0, y_0) < +\infty$*, then the* Aubin's property*of* F *at* $(x_0, y_0) \in \text{Graph}(F)$ *holds true: For any* $\varepsilon > 0$*, there exists* $\eta > 0$ *such that*

$$\begin{cases} \forall\, x, x_1 \in B\left(x_0, \frac{\eta}{2}\right),\ \forall\, y \in F(x) \cap B(y_0, \eta), \\ d(y, F(x_1)) \leq (\alpha_F(x_0, y_0) + \varepsilon)\|x_1 - x\| \end{cases}$$

Proof. Indeed, to say that $\alpha_F(x_0, y_0) < +\infty$ amounts to saying that for any ε, there exist η such that for any $(x, y) \in \text{Graph}(F) \cap B((x_0, y_0), \eta)$ and $T \leq \eta$,

$$\delta_F(T; x, y) := \sup_{\|x_1 - x\| \leq T \ \& \ x_1 \neq x} \frac{d(y, F(x_1))}{\|x_1 - x\|} \leq \alpha_F((x_0, y_0)) + \varepsilon$$

Therefore, if both x and x_1 belong to $B\left(x_0, \frac{\eta}{2}\right)$ and if $y \in B(y_0, \eta)$, then $x_1 \in B(x, \eta)$ and we infer that

$$d(y, F(x_1)) \leq (\alpha_F((x_0, y_0)) + \varepsilon)\|x_1 - x\| \ \square$$

9.7.3 Local and Pointwise Norms of Graphical Derivatives

The first step of our study is the proof of the following characterization of Aubin's property in terms of *convexified derivatives* $D^{\star\star}F(x,y)$ of F at points $(x,y) \in \text{Graph}(F)$ in a neighborhood of (x_0, y_0) (see Definition 18.5.5, p. 740).

Definition 9.7.6 *[inf-Norm of Derivatives] Let* $S := \{u \in X, \ \|u\| = 1\}$ *denote the unit sphere. We associate with any pair* $(x,y) \in \text{Graph}(F)$ *the* pointwise inf-norm

$$\lambda(D^{\star\star}F(x,y)) \ := \ \sup_{u\in S} d(0, D^{\star\star}F(x,y)(u)) \tag{9.28}$$

and the local inf-norm λ_F *of* $DF^{\star\star}(x_0,y_0)$ *at* (x_0,y_0) *defined by*

$$\begin{cases} \lambda_F(x_0,y_0) \ := \ \limsup_{(x,y)\to_{\text{Graph}(F)}(x_0,y_0)} \lambda(D^{\star\star}F(x,y)) \\ := \ \inf_{\eta>0} \sup_{\max(\|x-x_0\|,\|y-y_0\|)\leq\eta} \lambda(D^{\star\star}F(x,y)) \end{cases}$$

By definition,

$$\lambda(D^{\star\star}F(x_0,y_0)) \ \leq \ \limsup_{(x,y)\to_{\text{Graph}(F)}(x_0,y_0)} \lambda(D^{\star\star}F(x,y)) \ =: \ \lambda_F(x_0,y_0)$$

They are equal under adequate continuity assumptions:

Lemma 9.7.7 *[Local and Pointwise inf-Norms] If* $(x,y) \rightsquigarrow \text{Graph}(DF(x,y))$ is lower semicontinuous at (x_0,y_0), *then*

$$\lambda(D^{\star\star}F(x_0,y_0)) \ = \ \limsup_{(x,y)\to_{\text{Graph}(F)}(x_0,y_0)} \lambda(D^{\star\star}F(x,y)) \ \leq \ \lambda_F(x_0,y_0)$$

Proof. Since $(x,y) \rightsquigarrow \text{Graph}(DF(x,y))$ is lower semicontinuous at (x_0,y_0), Theorem 18.4.10, p. 733 implies that $DF(x_0,y_0) = D^{\star\star}F(x_0,y_0)$. We infer that

$$\begin{cases} \lambda_F(x_0,y_0) \ := \\ \limsup_{(x,y)\to_{\text{Graph}(F)}(x_0,y_0)} \lambda(D^{\star\star}F(x,y)) \ \leq \ \lambda(D^{\star\star}F(x_0,y_0)) \end{cases}$$

Indeed, we associate with any $u \in S$ an element $v_0 \in DF(x_0, y_0)(u) = D^{\star\star}F(x_0, y_0(u))$ achieving $\|v_0\| = d(0, D^{\star\star}F(x_0, y_0)(u))$. Since the graph of DF is lower semicontinuous at (x_0, y_0), we can associate with any sequence (x_n, y_n) converging to (x_0, y_0) elements (u_n, v_n) converging to (u, v_0). Hence $d(0, D^{\star\star}F(x_n, y_n)(u_n)) \leq \|v_n\| \leq \|v_0\| + \varepsilon$ for n large enough. Consequently, $d(0, D^{\star\star}F(x, y)(u))$ is upper semicontinuous at (x_0, y_0), and so achieves its supremum over the compact sphere $u \in S$. □

9.7.4 Norms of Derivatives and Metric Regularity

The first question is to compare the *Aubin constant* $\alpha_F(x_0, y_0)$ and the *inf-norm* $\lambda_F(x_0, y_0)$ of $DF^{\star\star}(x_0, y_0)$ at some point (x_0, y_0).

Theorem 9.7.8 ***[Differential Characterization of Aubin's Constant]*** *Assume that the graph of F is closed. Then*

$$\alpha_F(x_0, y_0) = \lambda_F(x_0, y_0)$$

Consequently, F is Aubin around $(x_0, y_0) \in$ Graph(F) if and only if $\lambda_F(x_0, y_0) < +\infty$.

Proof. 1. We assume that if $\alpha_F(x_0, y_0) < +\infty$, then $\lambda_F(x_0, y_0) \leq \alpha_F(x_0, y_0)$. By definition of $\alpha_F(x_0, y_0)$, for any $\varepsilon > 0$, there exist η and $T \leq \eta$ such that for any $(x, y) \in \text{Graph}(F) \cap B((x_0, y_0), \eta)$,

$$\delta_F(T; x, y) := \sup_{\|x_1 - x\| \leq T \ \& \ x_1 \neq x} \frac{d(y, F(x_1))}{\|x_1 - x\|} \leq \alpha_F((x_0, y_0)) + \varepsilon$$

Then, for any $u \in S$ and any $0 < t \leq T$, associating elements $x_1 := x + tu$ and $y_1 := y_t \in F(x_1)$, or by associating with x_1 and $y_1 \in F(x_1)$ elements $u := \dfrac{x_1 - x}{\|x_1 - x\|}$, $t := \|x_1 - x\|$, $y_1 := y\|x_1 - x\| \in F(x(\|x_1 - x\|)) = F(x_1)$, we observe that

$$\frac{d(y, F(x + tu))}{t} = \frac{d(y, F(x_1))}{\|x_1 - x\|}$$

Therefore, for all $h \leq T$, $d(y, F(x + hu)) \leq h(\delta + \varepsilon)$. This means that there exists $v_h \in B$ such that $y + (\alpha_F(x_0, y_0) + \varepsilon)v_h \in F(x + hu)$, a subsequence $v_{h_n} \in B$ of which converges to some $v \subset B$ since the unit sphere is compact. This implies that $(\alpha_F(x_0, y_0) + \varepsilon)v \in DF(x, y)(u) \subset D^{\star\star}F(x, y)(u)$. Therefore,

$$d(0, DF^{\star\star}(x, y)(u)) \leq \|(\alpha_F(x_0, y_0) + \varepsilon)v\| \leq \alpha_F(x_0, y_0) + \varepsilon$$

from which the desired inequality ensues by letting ε converge to 0.

2. We assume that if $\lambda_F(x_0, y_0) < +\infty$, then $\alpha_F(x_0, y_0) \leq \lambda_F(x_0, y_0)$. Indeed, by definition, for any $\varepsilon > 0$, there exist η and $T_\eta \leq \eta$ such that for any $(x, y) \in \mathrm{Graph}(F) \cap B((x_0, y_0), \eta)$ such that, for every $u \in S$ and $t \leq T_\eta$ such that $d(0, D^{\star\star}F(x, y)(u)) \leq \lambda_\eta := \lambda_F(x_0, y_0) + \varepsilon$. By definition of the *convexified derivatives* $D^{\star\star}F(x, y)$ of F at points $(x, y) \in \mathrm{Graph}(F)$

$$\begin{cases} (x, y) \in \mathrm{Graph}(F) \cap B((x_0, y_0), \eta), \forall\, u \in S, \;\; \exists\, v \in \lambda_\eta B \text{ such that} \\ \qquad (u, v) \;\in\; T^{\star\star}_{\mathrm{Graph}(F)}(x, y) \end{cases}$$

Let us set

$$\begin{cases} \mathcal{F}_\eta := \mathrm{Graph}(F) \cap B((x_0, y_0), \eta) \\ \Phi_{(u,\lambda_\eta)} := \{u\} \times \lambda_\eta B \\ T(\eta) \;:=\; \frac{\eta}{2\max(1,\lambda_\eta)} \end{cases}$$

Therefore, the tangential condition

$$\forall\, (x, y) \in \mathcal{F}_\eta, \;\; \Phi_{(u,\lambda_\eta)} \cap T^{\star\star}_{\mathcal{F}_\eta}(x, y) \;\neq\; \emptyset$$

holds true. Local Viability Theorem 19.4.3, p. 783 implies that there exists $\tau \mapsto v(\tau) \in B$ such that the evolution

$$(x(\cdot), y(\cdot)) \,:\; t \mapsto \left(x + tu, y + \lambda_\eta \int_0^t v(\tau)d\tau \right)$$

governed by differential inclusion $(x'(t), y'(t)) \in \Phi_{(u,\lambda_\eta)}$ is viable in $\mathcal{F}_\eta$ on the interval $[0, T(\eta)]$ because $B\left((x, y), \frac{\eta}{2}\right) \subset B((x_0, y_0), \eta)$ whenever $(x, y) \in B\left((x_0, y_0), \frac{\eta}{2}\right)$ and $\|\Phi_{(u,\lambda_\eta)}\| \leq \max(1, \lambda_\eta)$. This implies that

$$\begin{cases} \forall\, (x, y) \in B\left((x_0, y_0), \frac{\eta}{2}\right), \; \forall\, t \leq T(\eta), \; \forall\, u \in S, \\ \dfrac{d(y, F(x + tu))}{t} \;\leq\; \lambda_\eta \;:=\; \lambda_F(x_0, y_0) + \varepsilon \end{cases}$$

In other words,

$$\begin{cases} \displaystyle\sup_{(x,y)\in B\left((x_0,y_0),\frac{\eta}{2}\right)} \delta_F(T(\eta); (x, y)) \\ \displaystyle := \sup_{(x,y)\in B\left((x_0,y_0),\frac{\eta}{2}\right)} \; \sup_{t\leq T(\eta)} \; \sup_{u\in S} \frac{d(0, DF(x + tu))}{t} \\ \leq \sup_{(x,y)\in \mathrm{Graph}(F)\cap B((x_0,y_0),\eta)} \lambda(D^{\star\star}F(x, y)) \leq \lambda_\eta := \lambda_F(x_0, y_0) + \varepsilon \end{cases}$$

By letting ε converges to 0, we infer that

$$\begin{cases} \alpha_F(x_0,y_0) := \limsup_{(x,y)\to_{\mathrm{Graph}(F)}(x_0,y_0),\, T\mapsto 0} \delta_F(T;(x,y)) \\ \leq \limsup_{(x,y)\to_{\mathrm{Graph}(F)}(x_0,y_0)} \lambda(D^{\star\star}F(x,y)) =: \lambda_F(x_0,y_0) \end{cases}$$

This completes the proof. □

This theorem allows us to connect stability properties described by Aubin's property in terms of properties of the (convexified) derivatives of the set-valued map which appear in the assumptions of Theorems 9.7.2, p. 367 and 9.7.1, p. 366.

Proof. (Theorem 9.7.2, p. 367) For F to enjoy the Aubin's property around $(x_0,y_0) \in \mathrm{Graph}(F)$, we have to assume that $\alpha_F(x_0,y_0)$ is finite. Actually, Theorem 9.7.8, p. 371 implies that $\alpha_F(x_0,y_0) = \lambda_F(x_0,y_0)$ and, since $(x,y) \rightsquigarrow \mathrm{Graph}(DF(x,y))$ is lower semicontinuous, Lemma 9.7.7, p. 370 implies that $\alpha_F(x_0,y_0) = \lambda(D^{\star\star}F(x_0,y_0)) < +\infty$. This follows from the assumption that the derivative $DF(x,y)(u) \neq \emptyset$ for all $u \in X$.

This is a consequence of an extension to the celebrated Banach Closed Graph Theorem stating that the norm of a linear operator with closed graph is finite, i.e., that it is continuous, and even, Lipschitz.

Indeed, since $(x,y) \rightsquigarrow \mathrm{Graph}(DF(x,y)) := T_{\mathrm{Graph}(F)}(x,y)$ is lower semicontinuous at (x_0,y_0), Theorem 18.4.10, p. 733 states that the graph of

$$DF(x_0,y_0) = T_{\mathrm{Graph}(F)}(x_0,y_0) = T^{\star\star}_{\mathrm{Graph}(F)}(x_0,y_0) = DF^{\star\star}(x_0,y_0)$$

is a closed convex cone closed, whereas the graph of a continuous linear operator is a closed vector subspace.

The Robinson–Ursescu Theorem 9.7.3, p. 368 states that the Lipschitz constant of $D^{\star\star}F(x_0,y_0)$ is the point-wise inf-norm $\lambda(D^{\star\star}F(x_0,y_0)) := \sup_{u\in S} d(0,DF(x_0,y_0)(u))$.

Therefore, $\alpha_F(x_0,y_0) \leq \lambda(D^{\star\star}F(x_0,y_0))$ is finite, and thus, F is Aubin around (x_0,y_0). □

Replacing F by its inverse F^{-1}, Theorem 9.7.2 is nothing else than the 1982 Inverse Theorem 9.7.1 for set-valued maps stated in the introduction.

Chapter 10
Viability and Capturability Properties of Evolutionary Systems

10.1 Introduction

This chapter presents properties proved at the level of evolutionary systems, whereas Chap. 11, p. 437 focuses on specific results on evolutionary systems generated by control systems based on the Viability Theorem 11.3.4, p. 455 and Invariance Theorem 11.3.7, p. 457 involving tangential conditions. Specific results of the same nature are presented in Sect. 12.3, p. 503 for *impulse* systems, in Chap. 11 of the first edition of [18, Aubin] for evolutionary systems generated by *history dependent* (or path dependent) systems and in [23, Aubin] for *mutational* and *morphological* systems, which will inherit properties uncovered in this chapter.

This chapter is mainly devoted to the first and second fundamental viability characterizations of kernels and basins. The first one, in Sect. 10.2, p. 377, characterizes them as bilateral fixed points. The second one, in Sect. 10.5, p. 399, translates these fixed point theorems in terms of viability properties which will be exploited in Chap. 11, p. 437. The first one states that *the viability kernel is the largest subset viable outside the target* and the second one that *it is the smallest isolated subset*, and thus, the *unique* one satisfying both. This uniqueness theorem plays an important role, in particular for deriving the uniqueness property of viability episolutions of Hamilton–Jacobi–Bellman partial differential equations in Chap. 17, p. 681.

For that purpose, we uncover the topological properties of evolutionary systems in Sect. 10.3, p. 382 for the purpose of proving that under these topological properties, kernels and basins are closed. We need to define in Sect. 10.3.1, p. 382 two concepts of semicontinuity of evolutionary systems: "upper semicompact" evolutionary systems, under which viability properties hold true, and "lower semicontinuous" evolutionary systems, under which invariance (or tychastic) properties are satisfied. These assumptions are realistic, because they are respectively satisfied for systems generated by the

J.-P. Aubin et al., *Viability Theory*, DOI 10.1007/978-3-642-16684-6_10,

"Marchaud control systems" for the class of "upper semicompact" evolutionary systems and "Lipschitz" ones for "lower semicontinuous" evolutionary systems.

This allows us to prove that exit and minimal time functions are semicontinuous in Sect. 10.4, p. 392 and that the optimization problems defining them are achieved by "persistent and minimal time evolutions." Some of these properties are needed for proving the viability characterization in Sect. 10.5, p. 399, the most useful one of this chapter. It characterizes:

1. subsets viable outside a target by showing that the complement of the target in the environment is *locally viable*, a viability concept which can be characterized further for specific evolutionary systems, or, equivalently, that its exit set is contained in the target,
2. isolated subsets as subsets *backward invariant* relatively to the environment, again a viability concept which can be exploited further. Therefore, viability kernels being the unique isolated subset viable outside a target, they are the unique ones satisfying such local viability and backward invariance properties.

Section 10.6, p. 411 presents such characterizations for invariance kernels and connection basins.

We pursue by studying in Sect. 10.7, p. 416 under which conditions the capture basin of a (Painlevé–Kuratowski) limit of targets is the limit of the capture basins of those targets. This is quite an important property which is studied in Sect. 10.7.1, p. 416.

The concepts of viability and invariance kernels of environments are defined as the *largest subsets* of the environments satisfying either one of these properties. The question arises whether it is possible to define the concepts of viability and invariance envelopes of given subsets, which are the *minimal subsets* containing an environment which are viable and invariant respectively. This issue is dealt with in Sect. 10.7.2, p. 420. In the case of invariance kernels, this envelope is unique: it is the intersection of invariant subsets containing it. In the case of viability envelopes, we obtain, under adequate assumptions, the existence of nonempty viability envelopes. Equilibria are viable singletons which are necessarily minimal. They do not necessarily exist, except in the case of compact convex environments. For plain compact environments, minimal viability envelopes are not empty, and they enjoy a singular property, a weaker property than equilibria, which are the asymptotic limits of evolutions. *Minimal viability envelopes are the subsets coinciding with the limit sets of their elements*, i.e., are made of limit sets of evolutions instead of limit points (equilibria) which may not exist.

We briefly uncover without proofs the links between invariance of an environment under a tychastic system and stochastic viability in Sect. 10.10, p. 433. They share the same underlying philosophy: the viability property is satisfied *by all* evolutions of a tychastic system (tychastic viability), *by almost all* evolutions under a stochastic system. This is made much more precise by using the Strook–Varadhan Theorem, implying, so to speak, that stochastic viability is a very particular case in comparison to tychastic viability (or invariance). Hence the results dealing with invariant subsets can bring another point of view on the mathematical translation of this type uncertainty: *either by stochastic systems, or by tychastic systems.*

Exit sets also play a crucial role for regulating viable evolutions with a finite number of non viable feedbacks (instead of a viable feedback), but which are, in some sense made precise, "collectively viable": this is developed in Sect. 10.8, p. 422 for regulating viable punctuated evolutions satisfying the "hard" version of the inertia principle.

Section 10.9, p. 427 is devoted to *inverse problems* of the following type: assuming that both the dynamics $F(\lambda, (\cdot))$, the environment $K(\lambda)$ and the target $C(\lambda)$ depend upon a parameter λ, and given any state x, what is the set of parameters λ for which x lies in the viability kernel $\mathrm{Viab}_{F(\lambda,(\cdot))}(K(\lambda), C(\lambda))$?

This is a prototype of a *parameter identification problem.* It amounts to inverting the viability kernel map $\lambda \rightsquigarrow \mathrm{Viab}_{F(\lambda,(\cdot))}(K(\lambda), C(\lambda))$. For that purpose, we need to know the graph of this map, since the set-valued map and its inverse share the same "graphical properties." It turns out that the graph of the viability kernel map is itself the viability kernel of an auxiliary map, implying that both the viability kernel map and its inverse inherits the properties of viability kernels. When the parameters $\lambda \in \mathbb{R}$ are scalar, under some monotonicity condition, the inverse of this viability kernel map is strongly related to an extended function associating with any state x the best parameter λ, as we saw in many examples of Chaps. 4, p. 125 and 6, p. 199.

10.2 Bilateral Fixed Point Characterization of Kernels and Basins

We begin our investigation of viability kernels and capture basins by emphasizing simple algebraic properties of utmost importance, due to the collaboration with Francine Catté, which will be implicitly used all along the book. We begin by reviewing simple algebraic properties of viability invariance kernels as maps depending on the evolutionary system, environment and the target.

Lemma 10.2.1 *[Monotonicity Properties of Viability and Invariance Kernels] Let us consider the maps* $(\mathcal{S}, K, C) \mapsto \mathrm{Viab}_{\mathcal{S}}(K, C)$ *and* $(\mathcal{S}, K, C) \mapsto \mathrm{Inv}_{\mathcal{S}}(K, C)$. *Assume that* $\mathcal{S}_1 \subset \mathcal{S}_2$, $K_1 \subset K_2$, $C_1 \subset C_2$. *Then*

$$\begin{cases} (i) \ \mathrm{Viab}_{\mathcal{S}_1}(K_1, C_1) \subset \mathrm{Viab}_{\mathcal{S}_2}(K_2, C_2) \\ (ii) \ \mathrm{Inv}_{\mathcal{S}_2}(K_1, C_1) \subset \mathrm{Inv}_{\mathcal{S}_1}(K_2, C_2). \end{cases} \tag{10.1}$$

The consequences of these simple observations are important:

Lemma 10.2.2 ***[Union of Targets and Intersection of Environments]***

$$\begin{cases} (i) \ \text{Viab}_{\mathcal{S}}\left(K, \bigcup_{i\in I} C_i\right) = \bigcup_{i\in I} \text{Viab}_{\mathcal{S}}(K, C_i) \\ (ii) \ \text{Inv}_{\mathcal{S}}\left(\bigcap_{i\in I} K_i, C\right) = \bigcap_{i\in I} \text{Inv}_{\mathcal{S}}(K_i, C) \end{cases} \quad (10.2)$$

Evolutionary systems $\mathcal{R} \subset \mathcal{S}$ satisfying equality $\text{Inv}_{\mathcal{R}}(K,C) = \text{Viab}_{\mathcal{S}}(K,C)$ enjoy the following monotonicity property:

Lemma 10.2.3 ***[Comparison between Invariance Kernels under Smaller Evolutionary Systems and Viability Kernels of a Larger System]*** *Let us assume that there exists an evolutionary system* $\mathcal{R}$ *contained in* $\mathcal{S}$ *such that* $\text{Inv}_{\mathcal{R}}(K,C) = \text{Viab}_{\mathcal{S}}(K,C)$. *Then, for all evolutionary systems* $\mathcal{Q} \subset \mathcal{R}$, $\text{Inv}_{\mathcal{Q}}(K,C) = \text{Viab}_{\mathcal{S}}(K,C)$.

Proof. Indeed, by the monotonicity property with respect to the evolutionary system, we infer that

$$\text{Inv}_{\mathcal{R}}(K,C) \subset \text{Inv}_{\mathcal{Q}}(K,C) \subset \text{Viab}_{\mathcal{Q}}(K,C) \subset \text{Viab}_{\mathcal{S}}(K,C) = \text{Inv}_{\mathcal{R}}(K,C)$$

Hence equality $\text{Inv}_{\mathcal{Q}}(K,C) = \text{Viab}_{\mathcal{S}}(K,C)$ ensues. □

Next, we need the following properties.

Lemma 10.2.4 ***[Fundamental Properties of Viable and Capturing Evolutions]*** *Let* $\mathcal{S} : X \rightsquigarrow \mathcal{C}(0,+\infty; X)$ *be an evolutionary system,* $K \subset X$ *be an environment and* $C \subset K$ *be a target.*

1. *Every evolution* $x(\cdot) \in \mathcal{S}(x)$ *viable in* K*, forever or until it reaches* C *in finite time, is actually viable in the viability kernel* $\text{Viab}_{\mathcal{S}}(K,C)$,
2. *Every evolution* $x(\cdot) \in \mathcal{S}(x)$ *viable in* K *forever or which captures the viability kernel* $\text{Viab}_{\mathcal{S}}(K,C)$ *in finite time, remains viable in* $\text{Viab}_{\mathcal{S}}(K,C)$ *until it captures also the target* C *in finite time.*

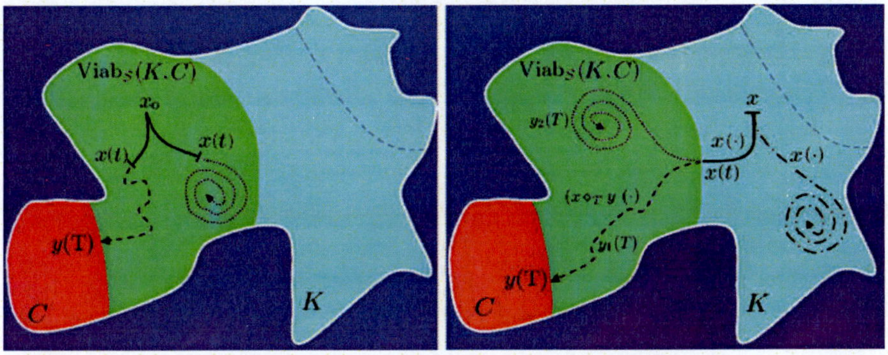

Fig. 10.1 Illustration of the proof of Lemma 10.2.4, p. 378.
Left: *an evolution* $x(\cdot)$ *viable in* K *is viable in* Viab$_{\mathcal{S}}(K,C)$ *forever* (spiral) *or reaches* C *at time* T. Right: *an evolution* $x(\cdot)$ *viable in* K *forever or which captures* Viab$_{\mathcal{S}}(K,C)$ *in finite time remains viable in* Viab$_{\mathcal{S}}(K,C)$ *until it captures the target at* T (dotted trajectory).

Proof. The first statement follows from the translation property. Let us consider an evolution $x(\cdot) \in \mathcal{S}(x)$ viable in K, forever or until it reaches C in finite time T. Therefore, for all $t \in [0,T[$, the translation $y(\cdot) := \kappa(-t)x(\cdot)$ of $x(\cdot)$ defined by $y(\tau) := x(t+\tau)$ is an evolution $y(\cdot) \in \mathcal{S}(x(t))$ starting at $x(t)$ and viable in K until it reaches C at time $T-t$. Hence $x(t)$ does belong to Viab$_{\mathcal{S}}(K,C)$ for every $t \in [0,T[$.

The second statement follows from the concatenation property because it can be concatenated with an evolution either remaining in Viab$_{\mathcal{S}}(K,C) \subset K$ or reaching the target C in finite time. □

10.2.1 Bilateral Fixed Point Characterization of Viability Kernels

We shall start our presentation of kernels and basins properties by a simple and important algebraic property:

Theorem 10.2.5 ***[The Fundamental Characterization of Viability Kernels]*** *Let* $\mathcal{S} : X \rightsquigarrow \mathcal{C}(0,+\infty;X)$ *be an evolutionary system,* $K \subset X$ *be an environment and* $C \subset K$ *be a target. The viability kernel* Viab$_{\mathcal{S}}(K,C)$ *of* K *with target* C *(see Definition 2.10.2, p. 86) is the* ***unique*** *subset between* C *and* K *that is both:*

1. ***viable outside*** *C (and is the largest subset $D \subset K$ viable outside C),*
2. ***isolated in*** *K (and is the smallest subset $D \supset C$ isolated in K):*

$$\mathrm{Viab}_{\mathcal{S}}(K, \mathrm{Viab}_{\mathcal{S}}(K,C)) = \mathrm{Viab}_{\mathcal{S}}(K,C) = \mathrm{Viab}_{\mathcal{S}}(\mathrm{Viab}_{\mathcal{S}}(K,C),C) \tag{10.3}$$

The viability kernel satisfies the properties of both the subsets viable outside a target and of isolated subsets in a environment, and is the unique one to do so.

This statement is at the root of uniqueness properties of solutions to some Hamilton–Jacobi–Bellman partial differential equations whenever the epigraph of a solution is a viability kernel of the epigraph of a function outside the epigraph of another function.

Proof. We begin by proving the two following statements:

1. The translation property implies that the viability kernel $\mathrm{Viab}_{\mathcal{S}}(K,C)$ is viable outside C:
$$\mathrm{Viab}_{\mathcal{S}}(K,C) \subset \mathrm{Viab}_{\mathcal{S}}(K, \mathrm{Viab}_{\mathcal{S}}(K,C)) \subset \mathrm{Viab}_{\mathcal{S}}(K,C)$$
Take $x_0 \in \mathrm{Viab}_{\mathcal{S}}(K,C)$ and prove that there exists an evolution $x(\cdot) \in \mathcal{S}(x_0)$ starting at x_0 viable in $\mathrm{Viab}_{\mathcal{S}}(K,C)$ until it possibly reaches C. Indeed, there exists an evolution $x(\cdot) \in \mathcal{S}(x_0)$ viable in K until some time $T \geq 0$ either finite when it reaches C or infinite. Then the first statement of Lemma 10.2.4, p. 378 implies that x_0 belongs to the viability kernel $\mathrm{Viab}_{\mathcal{S}}(\mathrm{Viab}_{\mathcal{S}}(K,C),C)$ of the viability kernel $\mathrm{Viab}_{\mathcal{S}}(K,C)$ of K with target C.
2. The concatenation property implies that the viability kernel $\mathrm{Viab}_{\mathcal{S}}(K,C)$ is isolated in K:
$$\mathrm{Viab}_{\mathcal{S}}(K, \mathrm{Viab}_{\mathcal{S}}(K,C)) \subset \mathrm{Viab}_{\mathcal{S}}(K,C)$$
Let x_0 belong to $\mathrm{Viab}_{\mathcal{S}}(K, \mathrm{Viab}_{\mathcal{S}}(K,C))$. There exists at least one evolution $x(\cdot) \in \mathcal{S}(x_0)$ that would either remain in K or reach the viability kernel $\mathrm{Viab}_{\mathcal{S}}(K,C)$ in finite time. Lemma 10.2.4, p. 378 implies that $x_0 \in \mathrm{Viab}_{\mathcal{S}}(K,C)$. □

We now observe that the map $(K,C) \mapsto \mathrm{Viab}_{\mathcal{S}}(K,C)$ satisfies

$$\begin{cases} (i)\ C \subset \mathrm{Viab}_{\mathcal{S}}(K,C) \subset K \\ (ii)\ (K,C) \mapsto \mathrm{Viab}_{\mathcal{S}}(K,C) \text{ is increasing} \end{cases} \tag{10.4}$$

in the sense that if $K_1 \subset K_2$ and $C_1 \subset C_2$, then $\text{Viab}_{\mathcal{S}}(K_1, C_1) \subset \text{Viab}_{\mathcal{S}}(K_2, C_2)$.

Setting $\mathcal{A}(K,C) := \text{Viab}_{\mathcal{S}}(K,C)$, the statements below follow from general algebraic Lemma 10.2.6 below.

Lemma 10.2.6 *[Uniqueness of Bilateral Fixed Points] Let us consider a map $\mathcal{A} : (K,C) \mapsto \mathcal{A}(K,C)$ satisfying*

$$\begin{cases} (i)\ C \subset \mathcal{A}(K,C) \subset K \\ (ii)\ (K,C) \mapsto \mathcal{A}(K,C) \text{ is increasing} \end{cases} \quad (10.5)$$

1. *If $\mathcal{A}(K,C) = \mathcal{A}(\mathcal{A}(K,C),C)$, it is the largest fixed point of the map $D \mapsto \mathcal{A}(D,C)$ between C and K,*
2. *If $\mathcal{A}(K,C) = \mathcal{A}(K,\mathcal{A}(K,C))$, it is the smallest fixed point of the map $E \mapsto \mathcal{A}(K,E)$ between C and K.*

Then, any subset D between C and K satisfying

$$D = \mathcal{A}(D,C) \text{ and } \mathcal{A}(K,D) = D$$

is the unique bilateral fixed point D *between C and K of the map $\mathcal{A}$ in the sense that:*

$$\mathcal{A}(K,D) = D = \mathcal{A}(D,C)$$

and is equal to $\mathcal{A}(K,C)$.

Proof. If $D = \mathcal{A}(D,C)$ is a fixed point of $D \mapsto \mathcal{A}(D,C)$, we then deduce that $D = \mathcal{A}(D,C) \subset \mathcal{A}(K,C)$, so that *whenever* $\mathcal{A}(K,C) = \mathcal{A}(\mathcal{A}(K,C),C)$, *we deduce that $\mathcal{A}(K,C)$ is the largest fixed point of $D \mapsto \mathcal{A}(D,C)$* contained in K. In the same way, *if $\mathcal{A}(K,\mathcal{A}(K,C)) = \mathcal{A}(K,C)$, then $\mathcal{A}(K,C)$ is the smallest fixed points of $E \mapsto \mathcal{A}(K,E)$* containing C. Furthermore, equalities

$$\mathcal{A}(K,D) \ = \ D \ = \ \mathcal{A}(D,C)$$

imply that $D = \mathcal{A}(K,C)$ because the monotonicity property implies that

$$\mathcal{A}(K,C) \subset \mathcal{A}(K,D) \subset D \subset \mathcal{A}(D,C) \subset \mathcal{A}(K,C) \ \square$$

10.2.2 Bilateral Fixed Point Characterization of Invariance Kernels

This existence and uniqueness of a "bilateral fixed point" is shared by the invariance kernel with target, the capture basin and the absorption basin of a target that satisfy property (10.5), and thus, the conclusions of Lemma 10.2.6:

Theorem 10.2.7 ***[Characterization of Kernels and Basins as Unique Bilateral Fixed Point]*** *Let $\mathcal{S} : X \rightsquigarrow \mathcal{C}(0,+\infty;X)$ be an evolutionary system, $K \subset X$ be a environment and $C \subset K$ be a target.*

1. *The viability kernel* $\mathrm{Viab}_{\mathcal{S}}(K,C)$ *of a subset K with target $C \subset K$ is the* unique *bilateral fixed point D between C and K of the map $(K,C) \mapsto$* $\mathrm{Viab}_{\mathcal{S}}(K,C)$ *in the sense that*
$$D = \mathrm{Viab}_{\mathcal{S}}(K,D) = \mathrm{Viab}_{\mathcal{S}}(D,C)$$
2. *The invariance kernel* $\mathrm{Inv}_{\mathcal{S}}(K,C)$ *of a subset K with target $C \subset K$ is the* unique *bilateral fixed point D between C and K of the map $(K,C) \mapsto$* $\mathrm{Inv}_{\mathcal{S}}(K,C)$ *in the sense that*
$$D = \mathrm{Inv}_{\mathcal{S}}(K,D) = \mathrm{Inv}_{\mathcal{S}}(D,C)$$

The same properties are shared by the maps $(K,C) \mapsto$ $\mathrm{Capt}_{\mathcal{S}}(K,C)$ *and $(K,C) \mapsto$* $\mathrm{Abs}_{\mathcal{S}}(K,C)$.

10.3 Topological Properties

We begin this section by introducing adequate semicontinuity concepts for evolutionary systems in Sect. 10.3.1, p. 382 for uncovering topological properties of kernels and basins in Sect. 10.3.2, p. 387.

10.3.1 Continuity Properties of Evolutionary Systems

In order to go further in the characterization of viability and invariance kernels with targets in terms of properties easier to check, we need to bring in the forefront some continuity requirements on the evolutionary system $\mathcal{S} : X \rightsquigarrow \mathcal{C}(0,+\infty;X)$. First, both the state space X and the evolutionary $\mathcal{C}(0,+\infty;X)$ have to be complete topological spaces.

32 ***[The Evolutionary Space]*** *Assume that the state space X is a complete metric space. We supply the space $\mathcal{C}(0,+\infty;X)$ of continuous evolutions with the "compact topology": A sequence of continuous evolutions $x_n(\cdot) \in \mathcal{C}(0,+\infty;X)$ converges to the continuous evolution $x(\cdot)$ as $n \to +\infty$ if for every $T > 0$, the sequence $\sup_{t\in[0,T]} d(x_n(t), x(t))$ converges to 0. It is a complete metrizable space. The Ascoli Theorem states that a subset $\mathcal{H}$ is compact if and only if it is closed, equicontinuous and for any $t \in \mathbb{R}_+$, the subset $\mathcal{H}(t) := \{x(t)\}_{x(\cdot)\in\mathcal{H}}$ is compact in X.*

Stability, a polysemous word, means formally that the solution of a problem depends "continuously" upon its data. Here, for evolutionary systems, the data are principally the initial states: In this case, stability means that the set of solutions depends "continuously" on the initial state. We recall that a deterministic system $\mathcal{S} := \{\mathbf{s}\} : X \mapsto \mathcal{C}(0,+\infty;X)$ is continuous at some $x \in X$ if it maps any sequence $x_n \in X$ converging to x to a sequence $\mathbf{s}(x_n)$ converging to $\mathbf{s}(x)$.

However, when the evolutionary system $\mathcal{S} : X \rightsquigarrow \mathcal{C}(0,+\infty;X)$ is no longer single-valued, there are several ways of describing the convergence of the set $\mathcal{S}(x_n)$ to the set $\mathcal{S}(x)$. We shall use in this book only two of them, that we present in the context of evolutionary systems (see Definition 18.4.3, p. 729 and other comments in the Appendix 18, p. 713). We begin with the notion of upper semicompactness:

Definition 10.3.1 ***[Upper Semicompactness]*** *Let $\mathcal{S} : X \rightsquigarrow \mathcal{C}(0,+\infty;X)$ be an evolutionary system, where both the state space X and the evolutionary space $\mathcal{C}(0,+\infty;X)$ are topological spaces. The evolutionary system is said to be* upper semicompact *at x if for every sequence $x_n \in X$ converging to x and for every sequence $x_n(\cdot) \in \mathcal{S}(x_n)$, there exists a subsequence $x_{n_p}(\cdot)$ converging to some $x(\cdot) \in \mathcal{S}(x)$. It is said to be* upper semicompact *if it is upper semicompact at every point $x \in X$ where $\mathcal{S}(x)$ is not empty.*

Before using this property, we need to provide examples of evolutionary system exhibiting it: this is the case for Marchaud differential inclusions:

Definition 10.3.2 ***[Marchaud Set-Valued Maps]*** *We say that F is a* Marchaud map *if*

$$\left\{\begin{array}{ll} (i) & \textit{the graph and the domain of } F \textit{ are nonempty and closed} \\ (ii) & \textit{the values } F(x) \textit{ of } F \textit{ are convex} \\ (iii) & \exists\, c > 0 \text{ such that } \forall x \in X,\ \|F(x)\| := \sup_{v \in F(x)} \|v\| \leq c(\|x\|+1) \end{array}\right. \tag{10.6}$$

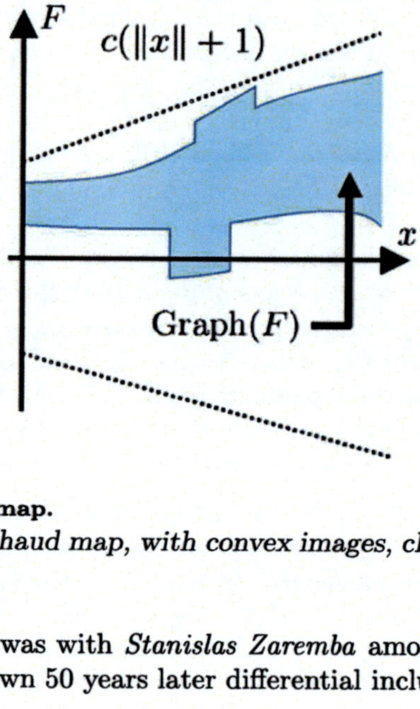

Fig. 10.2 Marchaud map.
Illustration of a Marchaud map, with convex images, closed graph and linear growth.

André Marchaud was with *Stanislas Zaremba* among the firsts to study what did become known 50 years later differential inclusions:

André Marchaud [1887–1973]. *After entering École Normale Supérieure in 1909, he fought First World War, worked in ministry of armement and industrial reconstruction and became professor and Dean at Faculté des Sciences de Marseille from 1927 to 1938 before being Recteur of several French Universities. He was a student of Paul Montel, was in close relations with Georges Bouligand and Stanislas Zaremba, and was a mentor of André Lichnerowicz [1915–1998]. His papers dealt with analysis and differentiability.*

The (difficult) Stability Theorem states that the set of solutions depends continuously upon the initial states in the upper semicompact sense:

> **Theorem 10.3.3** *[Upper Semicompactness of Marchaud Evolutionary Systems] If $F : X \rightsquigarrow X$ is Marchaud, the solution map S is an upper semicompact evolutionary system.*

Proof. This a consequence of the Convergence Theorem 19.2.3, p. 771. □

The other way to take into account the idea of continuity in the case of evolutionary systems is by introducing the following concept:

> **Definition 10.3.4** *[Lower Semicontinuity of Evolutionary Systems] Let $S : X \rightsquigarrow \mathcal{C}(0, +\infty; X)$ be an evolutionary system, where both the state space X and the evolutionary space $\mathcal{C}(0, +\infty; X)$ are topological spaces. The evolutionary system is said to be* lower semicontinuous *at x if for every sequence $x_n \in X$ converging to x and for every sequence $x(\cdot) \in S(x)$ (thus assumed to be nonempty), there exists a sequence $x_n(\cdot) \in S(x_n)$ converging to $x(\cdot) \in S(x)$. It is said to be* lower semicontinuous *if it is lower semicontinuous at every point $x \in X$ where $S(x)$ is not empty.*

Warning: An evolutionary system can be upper semicompact at x without being lower semicontinuous and lower semicontinuous at x without being upper semicompact. If the evolutionary system is deterministic, lower semicontinuity coincides with continuity and upper semicompactness coincides with "properness" of single-valued maps (in the sense of Bourbaki). Note also the unfortunate confusions between the semicontinuity of numerical and extended functions (Definition 18.6.3, p. 744) and the semicontinuity of set-valued maps (Definition 18.4.3, p. 729).

Recall that a single-valued map $f : X \mapsto Y$ is said to be λ-*Lipschitz* if for any $x_1,\ x_2 \in X$, $d(f(x_1), f(x_2)) \le \lambda d(x_1, x_2)$. In the case of normed vector spaces, denoting by B the unit ball of the vector space, this inequality can be translated in the form $f(x_1) \in f(x_2) + \lambda\|x_1 - x_2\|B$. This is this formulation which is the easiest to adapt to set-valued maps in the case of (finite) dimensional vector spaces:

> **Definition 10.3.5** *[Lipschitz Maps] A set-valued map $F : X \rightsquigarrow Y$ is said to be* λ-Lipschitz *(or Lipschitz for the constant $\lambda > 0$) if*

$$\forall x_1,\ x_2,\ F(x_1) \subset F(x_2) + \lambda \|x_1 - x_2\| B$$

The Lipschitz norm $\|F\|_\Lambda$ *of a map* $F : x \rightsquigarrow Y$ *is the smallest Lipschitz constants of* F. *The evolutionary system* $\mathcal{S} : X \rightsquigarrow \mathcal{C}(0, +\infty; X)$ *associated with a Lipschitz set-valued map is called a* Lipschitz evolutionary system.

The Filippov Theorem 11.3.9, p. 459 implies that Lipschitz systems are lower semicontinuous:

Theorem 10.3.6 *[Lower Semicontinuity of Lipschitz Evolutionary Systems] If* $F : X \rightsquigarrow X$ *is Lipschitz, the associated evolutionary system* $\mathcal{S}$ *is lower semicontinuous.*

Under appropriate topological assumptions, we can prove that inverse images and cores of closed subsets of evolutions are closed.

Definition 10.3.7 *[Closedness of Inverse Images] Let* $\mathcal{S} : X \rightsquigarrow \mathcal{C}(0, +\infty; X)$ *be an upper semicompact evolutionary system. Then for any subset* $\mathcal{H} \subset \mathcal{C}(0, +\infty; X)$,

$$\overline{\mathcal{S}^{-1}(\mathcal{H})} \subset \mathcal{S}^{-1}(\overline{\mathcal{H}})$$

Consequently, the inverse images $\mathcal{S}^{-1}(\mathcal{H})$ *under* $\mathcal{S}$ *of any closed subset* $\mathcal{H} \subset \mathcal{C}(0, +\infty; X)$ *are closed.*

Furthermore, the evolutionary system $\mathcal{S}$ *maps compact sets* $K \subset X$ *to compact sets* $\mathcal{H} \subset \mathcal{C}(0, +\infty; X)$.

Proof. Let us consider a subset $\mathcal{H} \subset \mathcal{C}(0, +\infty; X)$, a sequence of elements $x_n \in \mathcal{S}^{-1}(\mathcal{H})$ converging to some x and prove that x belongs to $\mathcal{S}^{-1}(\mathcal{H})$. Hence there exist elements $x_n(\cdot) \in \mathcal{S}(x_n) \cap \mathcal{H}$. Since $\mathcal{S}$ is upper semicompact, there exists a subsequence $x_{n_p}(\cdot) \in \mathcal{S}(x_{n_p})$ converging to some $x(\cdot) \in \mathcal{S}(x)$. It belongs also to the closure of $\mathcal{H}$, so that $x \in \mathcal{S}^{-1}(\overline{\mathcal{H}})$.

Take now any compact subset $K \subset X$. For proving that $\mathcal{S}(K)$ is compact, take any sequence $x_n(\cdot) \in \mathcal{S}(x_n)$ where $x_n \in K$. Since K is compact, a subsequence $x_{n'}$ converges to some $x \in K$ and since $\mathcal{S}$ is upper semicompact, a subsequence $x_{n''}(\cdot) \in \mathcal{S}(x_{n''})$ converges to some $x(\cdot) \in \mathcal{S}(x)$. □

Example: Let us consider an upper semicompact evolutionary system $\mathcal{S} : X \rightsquigarrow \mathcal{C}(0,+\infty;X)$.
If $K \subset X$ is a closed subset, then the set of equilibria of the evolutionary system that belong to K is closed: Indeed, it is the inverse images $\mathcal{S}^{-1}(\mathcal{K})$ of the set $\mathcal{K}$ of stationary evolutions in K, which is closed whenever K is closed.
In the same way, the set of points through which passes at least one T-periodic evolution of an upper semicompact evolutionary system is closed, since it is the inverse images $\mathcal{S}^{-1}(\mathcal{P}_T(X))$ of the set $\mathcal{P}_T(X)$ of T-periodic evolutions, which is closed.
If a function $\mathbf{v} : X \mapsto \mathbb{R}$ is continuous, the set of initial states from which starts at least one evolution of the evolutionary system monotone along the function $\mathbf{v}$ is closed, since it is the inverse images of the set $\mathcal{M}_{\mathbf{v}}$ of monotone evolutions, which is closed.

For cores, we obtain

Theorem 10.3.8 *[Closedness of Cores] Let $\mathcal{S} : X \rightsquigarrow \mathcal{C}(0,+\infty;X)$ be a lower semicontinous evolutionary system. Then for any subset $\mathcal{H} \subset \mathcal{C}(0,+\infty;X)$,*

$$\overline{\mathcal{S}^{\ominus 1}(\mathcal{H})} \subset \mathcal{S}^{\ominus 1}(\overline{\mathcal{H}})$$

Consequently, the core $\mathcal{S}^{\ominus 1}(\mathcal{H})$ under $\mathcal{S}$ of any closed subset $\mathcal{H} \subset \mathcal{C}(0,+\infty;X)$ is closed.

Proof. Let us consider a closed subset $\mathcal{H} \subset \mathcal{C}(0,+\infty;X)$, a sequence of elements $x_n \in \mathcal{S}^{\ominus 1}(\mathcal{H})$ converging to some x and prove that x belongs to $\mathcal{S}^{\ominus 1}(\mathcal{H})$. We have to prove that any $x(\cdot) \in \mathcal{S}(x)$ belongs to $\mathcal{H}$. But since $\mathcal{S}$ is lower semicontinuous, there exists a sequence of elements $x_n(\cdot) \in \mathcal{S}(x_n) \subset \mathcal{H}$ converging to $x(\cdot) \in \overline{\mathcal{H}}$. Therefore $\mathcal{S}(x) \subset \overline{\mathcal{H}}$, i.e., $x \in \mathcal{S}^{\ominus 1}(\overline{\mathcal{H}})$. □

10.3.2 Topological Properties of Viability Kernels and Capture Basins

Recall that the set $\mathcal{V}(K,C)$ of evolutions *viable in K outside C* is defined by (2.5), p. 49:

$$\begin{cases} \mathcal{V}(K,C) := \{x(\cdot) \text{ such that } \forall t \geq 0,\ x(t) \in K \\ \text{or } \exists\, T \geq 0 \text{ such that } x(T) \in C \;\&\; \forall t \in [0,T],\ x(t) \in K\} \end{cases}$$

Lemma 10.3.9 ***[Closedness of the Subset of Viable Evolutions]*** *Let us consider a environment $K \subset X$ and a (possibly empty) target $C \subset K$. Then*

$$\overline{\mathcal{V}(K,C)} \subset \mathcal{V}(\overline{K},\overline{C})$$

and consequently, if C and K are closed, the set $\mathcal{V}(K,C)$ of evolutions that are viable in K forever or until they reach the target C in finite time is closed.

Proof. Let us consider a sequence of evolutions $x_n(\cdot) \in \mathcal{V}(K,C)$ converging to some evolution $x(\cdot)$. We have to prove that $x(\cdot)$ belongs to $\mathcal{V}(\overline{K},\overline{C})$, i.e., that it is viable in $\overline{K}$ forever or until it reaches the target $\overline{C}$ in finite time. Indeed:

1. either for any $T > 0$ and any $N > 0$, there exist $n \geq N$, $t_n \geq T$ and an evolution $x_n(\cdot)$ for which $x_n(t) \in K$ for every $t \in [0, t_n]$,
2. or there exist $T > 0$ and $N > 0$ such that for any $t \geq T$ and $n \geq N$ and any evolution $x_n(\cdot)$, there exists $t_n \leq t$ such that $x_n(t_n) \notin K$.

In the first case, we deduce that for any $T > 0$, $x(T) \in \overline{K}$, so that the limit $x(\cdot)$ is viable in $\overline{K}$ forever.

In the second case, all the solutions $x_n(\cdot)$ leave K before T. This is impossible if evolutions $x_n(\cdot)$ are viable in K forever. Therefore, since $x_n(\cdot) \in \mathcal{V}(K,C)$, they have to reach C before leaving K: There exist $s_n \leq T$ such that

$$x_n(s_n) \in C \;\&\; \forall t \in [0, s_n],\; x_n(t) \in K$$

Then some subsequence $s_{n'}$ converges to some $S \in [0,T]$. Therefore, for any $s < S$, then $s < s_{n'}$ for n' large enough, so that $x_{n'}(s) \in K$. By taking the limit, we infer that for every $s < S$, $x(s) \in \overline{K}$. Furthermore, since $x_n(\cdot)$ converges to $x(\cdot)$ uniformly on the compact interval $[0,T]$, then $x_n(s_n)$ converges to $x(S)$, that belongs to $\overline{C}$.

This shows that the limit $x(\cdot)$ belongs to $\mathcal{V}(\overline{K},\overline{C})$. □

Consequently, the viability kernel of a closed subset with a closed target under an upper semicompact evolutionary subset is closed:

Theorem 10.3.10 ***[Closedness of the Viability Kernel]*** *Let $\mathcal{S} : X \rightsquigarrow \mathcal{C}(0,+\infty;X)$ be an upper semicompact evolutionary system. Then for any environment $K \subset X$ and any target $C \subset K$,*

$$\overline{\mathrm{Viab}_{\mathcal{S}}(K,C)} \subset \mathrm{Viab}_{\mathcal{S}}(\overline{K},\overline{C})$$

Consequently, if $C \subset K$ *and* K *are closed, so is the viability kernel* $\mathrm{Viab}_{\mathcal{S}}(K,C)$ *of* K *with target* C*. Furthermore, if* $K \backslash C$ *is a repeller, the capture basin* $\mathrm{Capt}_{\mathcal{S}}(K,C)$ *of* C *viable in* K *under* $\mathcal{S}$ *is closed.*

Proof. Since the viability kernel $\mathrm{Viab}_{\mathcal{S}}(K,C) := \mathcal{S}^{-1}(\mathcal{V}(K,C))$ is the inverse image of the subset $\mathcal{V}(K,C)$ by Definition 2.10.2, the closedness of the viability kernel follows from Theorem 10.3.7 and Lemma 10.3.9. □

Theorem 10.3.8 implies the closedness of the invariance kernels:

Theorem 10.3.11 *[**Closedness of Invariance Kernels**] Let* $\mathcal{S} : X \rightsquigarrow \mathcal{C}(0,+\infty;X)$ *be a lower semicontinuous evolutionary system. Then for any environment* $K \subset X$ *and any target* $C \subset K$*,*

$$\overline{\mathrm{Inv}_{\mathcal{S}}(K,C)} \subset \mathrm{Inv}_{\mathcal{S}}(\overline{K},\overline{C})$$

Consequently, if $C \subset K$ *and* K *are closed, so is the invariance kernel* $\mathrm{Inv}_{\mathcal{S}}(K,C)$ *of* K *with target* C*.*

Therefore, if $K \backslash C$ *is a repeller, the absorption basin* $\mathrm{Abs}_{\mathcal{S}}(K,C)$ *of* C *invariant in* K *under* $\mathcal{S}$ *is closed.*

As for interiors of capture and absorption basins, we obtain the following statements:

Theorem 10.3.12 *[**Interiors of Capture and Absorption Basins**] For any environment* $K \subset X$ *and any target* $C \subset K$*:*

- *if* $\mathcal{S} : X \rightsquigarrow \mathcal{C}(0,+\infty;X)$ *is lower semicontinuous, then*

$$\mathrm{Capt}_{\mathcal{S}}(\mathrm{Int}(K),\mathrm{Int}(C)) \subset \mathrm{Int}(\mathrm{Capt}_{\mathcal{S}}(K,C))$$

- *if* $\mathcal{S} : X \rightsquigarrow \mathcal{C}(0,+\infty;X)$ *is upper semicontinuous, then*

$$\mathrm{Abs}_{\mathcal{S}}(\mathrm{Int}(K),\mathrm{Int}(C)) \subset \mathrm{Int}(\mathrm{Abs}_{\mathcal{S}}(K,C))$$

Consequently, if $C \subset K$ *and* K *are open, so are the capture basin* $\mathrm{Capt}_{\mathcal{S}}(K,C)$ *and the absorption basin* $\mathrm{Capt}_{\mathcal{S}}(K,C)$ *whenever the evolutionary system is respectively lower semicontinuous and upper semicompact.*

Proof. Observe that, taking the complements, Lemma 2.12.2 implies that if $\mathcal{S} : X \rightsquigarrow \mathcal{C}(0, +\infty; X)$ is lower semicontinuous, then Theorem 10.7.8, p. 420 implies that

$$\mathrm{Capt}_{\mathcal{S}}(\mathrm{Int}(K), \mathrm{Int}(C)) \subset \mathrm{Int}(\mathrm{Capt}_{\mathcal{S}}(K, C))$$

since the complement of an invariance kernel is the capture basin of the complements and since the complement of a closure is the interior of the complement, and Theorem 10.3.10, p. 388 imply the similar statement for absorption basins. □

For capture basins, we obtain another closedness property based on backward invariance (see Definition 8.2.4, p. 278):

Proposition 10.3.13 *[Closedness of Capture Basins] If the set-valued map $\overleftarrow{\mathcal{S}}$ is lower semicontinuous and if K is backward invariant, then for any closed subset $C \subset K$,*

$$\mathrm{Capt}_{\mathcal{S}}(\overline{K}, \overline{C}) \subset \overline{\mathrm{Capt}_{\mathcal{S}}(K, C)} \tag{10.7}$$

Proof. Let us take $x \in \mathrm{Capt}_{\mathcal{S}}(\overline{K}, \overline{C})$ and an evolution $x(\cdot) \in \mathcal{S}(x)$ viable in $\overline{K}$ until it reaches the target $\overline{C}$ at time $T < +\infty$ at $c := x(T) \in \overline{C}$. Hence the function $t \mapsto y(t) := x(T - t)$ is an evolution $y(\cdot) \in \overleftarrow{\mathcal{S}}(c)$.

Let us consider a sequence of elements $c_n \in C$ converging to c. Since $\overleftarrow{\mathcal{S}}$ is lower semicontinuous, there exist evolutions $y_n(\cdot) \in \overleftarrow{\mathcal{S}}(c_n)$ converging uniformly over compact intervals to $y(\cdot)$. These evolutions $y_n(\cdot)$ are viable in K, since K is assumed to be backward invariant. The evolutions $x_n(\cdot)$ defined by $x_n(t) := y_n(T - t)$ satisfy $x_n(0) = y_n(T) \in K$, $x_n(T) = c_n$ and, for all $t \in [0, T]$, $x_n(t) \in K$. Therefore $x_n(0) := y_n(T)$ belongs to $\mathrm{Capt}_{\mathcal{S}}(K, C)$ and converges to $x := x(0)$, so that $x \in \overline{\mathrm{Capt}_{\mathcal{S}}(K, C)}$. □

As a consequence, we obtain the following topological regularity property (see Definition 18.2.2, p. 714) of capture basins:

Proposition 10.3.14 *[Topological Regularity of Capture Basins] If the set-valued map $\mathcal{S}$ is upper semicompact and the set-valued map $\overleftarrow{\mathcal{S}}$ is lower semicontinuous, if $K = \overline{\mathrm{Int}(K)}$ and $C = \overline{\mathrm{Int}(C)}$, if $K \backslash C$ is a repeller and if $\mathrm{Int}(K)$ is backward invariant, then*

$$\mathrm{Capt}_{\mathcal{S}}(K, C) = \overline{\mathrm{Capt}_{\mathcal{S}}(\mathrm{Int}(K), \mathrm{Int}(C))} = \overline{\mathrm{Int}(\mathrm{Capt}_{\mathcal{S}}(K, C))} \tag{10.8}$$

Proof. Since $K = \overline{\mathrm{Int}(K)}$ and $C = \overline{\mathrm{Int}(C)}$, since $\overleftarrow{\mathcal{S}}$ is lower semicontinuous and since $\mathrm{Int}(K)$ is backward invariant, Proposition 10.3.14, p. 390 implies that

$$\mathrm{Capt}_{\mathcal{S}}(K,C) \;=\; \mathrm{Capt}_{\mathcal{S}}(\overline{\mathrm{Int}(K)},\overline{\mathrm{Int}(C)}) \;\subset\; \overline{\mathrm{Capt}_{\mathcal{S}}(\mathrm{Int}(K),\mathrm{Int}(C))}$$

Inclusion

$$\mathrm{Capt}_{\mathcal{S}}(\mathrm{Int}(K),\mathrm{Int}(C)) \;\subset\; \mathrm{Int}(\mathrm{Capt}_{\mathcal{S}}(K,C))$$

follows from Theorem 10.3.12, p. 389. On the other hand, since $\mathcal{S}$ is upper semicompact and $K \setminus C$ is a repeller, Theorem 10.3.10, p. 388 implies that

$$\overline{\mathrm{Capt}_{\mathcal{S}}(\mathrm{Int}(K),\mathrm{Int}(C))} \;\subset\; \overline{\mathrm{Int}(\mathrm{Capt}_{\mathcal{S}}(K,C))} \;\subset\; \mathrm{Capt}_{\mathcal{S}}(K,C)$$

so that $\mathrm{Capt}_{\mathcal{S}}(K,C) = \overline{\mathrm{Int}(\mathrm{Capt}_{\mathcal{S}}(K,C))}$. □

We turn now our attention to connectedness properties of viability kernels:

Lemma 10.3.15 *[The connectedness Lemma] Assume that the evolutionary system $\mathcal{S}$ is upper semicompact. Let K be a closed environment and $C_1 \subset K$ and $C_2 \subset K$ be nonempty closed disjoint targets. If the viability kernel* $\mathrm{Viab}_{\mathcal{S}}(K, C_1 \cup C_2)$ *is connected, then the intersection*

$$\mathrm{Viab}_{\mathcal{S}}(K,C_1) \cap \mathrm{Viab}_{\mathcal{S}}(K,C_2)$$

is closed and not empty. Consequently, if we assume further that $K \setminus C_1$ and $K \setminus C_2$ are repellers, we infer that

$$\mathrm{Capt}_{\mathcal{S}}(K,C_1) \cap \mathrm{Capt}_{\mathcal{S}}(K,C_2) \;\neq\; \emptyset$$

Proof. This follows from the definition of connectedness since $\mathcal{S}$ being upper semicompact, the viability kernels $\mathrm{Viab}_{\mathcal{S}}(K,C_1)$ and $\mathrm{Viab}_{\mathcal{S}}(K,C_2)$ are closed, nonempty (they contain their nonempty targets) and cover the viability kernel with the union of targets :

$$\mathrm{Viab}_{\mathcal{S}}(K,C_1 \cup C_2) \;=\; \mathrm{Viab}_{\mathcal{S}}(K,C_1) \cup \mathrm{Viab}_{\mathcal{S}}(K,C_2)$$

Since this union is assumed connected, the intersection $\mathrm{Viab}_{\mathcal{S}}(K,C_1) \cap \mathrm{Viab}_{\mathcal{S}}(K,C_2)$ must be empty (and is closed). □

Motivating Remark. The intersection

$$\mathrm{Capt}_{\mathcal{S}}(K,C_1) \cap \mathrm{Capt}_{\mathcal{S}}(K,C_2) \;\neq\; \emptyset$$

being the subset of states from which at least one volution reaches one target in finite time and another one reaches the other target also in finite time, could be use as a proto-concept of a *"watershed"*. It is closed when the evolutionary system is upper semicompact and when the environment and the targets are closed (see *Morphologie Mathématique*, [187, Schmitt M. & Mattioli], and *Mathematical Morphology*, [166, Najman]).

10.4 Persistent Evolutions and Exit Sets

This section applies the above topological results to the study of exit and minimal time functions initiated in Sect. 4.3, p. 132. We shall begin by proving that these functions are respectively upper and lower semicontinuous and that persistent and minimal time evolutions exist under upper semicompact evolutionary systems. We next study the exit sets, the subset of states at the boundary of the environment from which all evolutions leave the environment immediately. They play an important role in the characterization of local viability, of *transversality*.

10.4.1 Persistent and Minimal Time Evolutions

Let us recall the definition of the *exit function* of K defined by

$$\tau_K(x(\cdot)) := \inf\{t \in [0,\infty[\mid x(t) \notin K\} \text{ and } \tau_K^\sharp(x) := \sup_{x(\cdot)\in\mathcal{S}(x)} \tau_K(x(\cdot))$$

and of *minimal time function* $\varpi_{(K,C)}$ defined by

$$\varpi_{(K,C)}(x(\cdot)) := \inf\{t \geq 0 \mid x(t) \in C \ \&\ \forall s \in [0,t],\ x(s) \in K\ \}$$

and

$$\varpi^\flat_{(K,C)}(x) := \inf_{x(\cdot)\in\mathcal{S}(x)} \varpi_{(K,C)}(x(\cdot))$$

We summarize the semi-continuity properties of the exit and minimal time functions in the following statement:

Theorem 10.4.1 ***[Semi-Continuity Properties of Exit and Minimal Time Functions]*** *Let us assume that the evolutionary system is upper semicompact and that the subsets K and $C \subset K$ are closed. Then:*

1. the hypograph of the exit function $\tau_K^\sharp(\cdot)$ is closed,

2. *the epigraph of the minimal time function $\varpi^{\flat}_{(K,C)}(\cdot)$ is closed*
This can be translated by saying that the exit function is upper semicontinuous and the minimal time function is lower semicontinuous.

Proof. The first statements follow from Theorems 4.3.6 and 10.3.10. □

Actually, in several applications, we would like to maximize the exit functional and minimize the minimal time or minimal time functional. Indeed, when an initial state $x \in K$ does not belong to the viability kernel, all evolutions $x(\cdot) \in \mathcal{S}(x)$ leave K in finite time. The question arises to select the "persistent evolutions" in K which persist to remain in K as long as possible:

Definition 10.4.2 *[Persistent Evolutions] Let us consider an evolutionary system $\mathcal{S} : X \rightsquigarrow \mathcal{C}(0,+\infty;X)$ and a subset $K \subset X$.*
The solutions $x^{\sharp}(\cdot) \in \mathcal{S}(x)$ which maximize the exit time function

$$\forall x \in K,\ \tau_K(x^{\sharp}(\cdot)) = \tau^{\sharp}_K(x) := \max_{x(\cdot)\in\mathcal{S}(x)} \tau_K(x(\cdot)) \tag{10.9}$$

are called persistent evolutions *in K (Naturally, when $x \in \mathrm{Viab}_{\mathcal{S}}(K)$, persistent evolutions starting at x are the viable ones).*
We denote by $\mathcal{S}^{K^{\sharp}} : K \rightsquigarrow \mathcal{C}(0,+\infty;X)$ the evolutionary system $\mathcal{S}^{K^{\sharp}} \subset \mathcal{S}$ associating with any $x \in K$ the set of persistent evolutions in K.

In a symmetric way, we single out the evolutions which minimize the minimal time to a target:

Definition 10.4.3 *[Minimal Time Evolutions] Let us consider an evolutionary system $\mathcal{S} : X \rightsquigarrow \mathcal{C}(0,+\infty;X)$ and subsets $K \subset X$ and $C \subset K$.*
The evolutions $x^{\flat}(\cdot) \in \mathcal{S}(x)$ which minimize the minimal time function

$$\forall x \in K,\ \varpi_{(K,C)}(x^{\flat}(\cdot)) = \varpi^{\flat}_{(K,C)}(x) := \min_{x(\cdot)\in\mathcal{S}(x)} \varpi_{(K,C)}(x(\cdot)) \tag{10.10}$$

are called minimal time evolutions *in K.*

Persistent evolutions and minimal time evolutions exist when the evolutionary system is upper semicompact:

Theorem 10.4.4 *[Existence of Persistent and Minimal Time Evolutions] Let $K \subset X$ be a closed subset and $\mathcal{S} : X \rightsquigarrow \mathcal{C}(0, +\infty; X)$ be an upper semicompact evolutionary system. Then:*

1. *For any $x \notin \mathrm{Viab}_{\mathcal{S}}(K)$, there exists at least one persistent evolution $x^\sharp(\cdot) \in \mathcal{S}^{K^\sharp}(x) \subset \mathcal{S}(x)$ viable in K on the interval $[0, \tau_K^\sharp(x)]$.*
2. *For any $x \in \mathrm{Capt}_{\mathcal{S}}(K, C)$, there exists at least one evolution $x^\flat(\cdot) \in \mathcal{S}(x)$ reaching C in minimal time while being viable in K.*

Proof. Let $t < \tau_K^\sharp(x)$ and $n > 0$ such that $t < \tau_K^\sharp(x) - \frac{1}{n}$. Hence, by definition of the supremum, there exists an evolution $x_n(\cdot) \in \mathcal{S}(x)$ such that $\tau_K(x_n(\cdot)) \geq \tau_K^\sharp(x) - \frac{1}{n}$, and thus, such that $x_n(t) \in K$. Since the evolutionary system $\mathcal{S}$ is upper semicompact, we can extract a subsequence of evolutions $x_{n'}(\cdot) \in \mathcal{S}(x)$ converging to some evolution $x_\star(\cdot) \in \mathcal{S}(x)$. Therefore, we infer that $x_\star(t)$ belongs to K because K is closed. Since this is true for any $t < \tau_K^\sharp(x)$ and since the evolution $x_\star(\cdot)$ is continuous, we infer that $\tau_K^\sharp(x) \leq \tau_K(x_\star(\cdot))$. We deduce that such an evolution $x_\star(\cdot) \in \mathcal{S}(x)$ is persistent in K because $\tau_K(x_\star(\cdot)) \leq \tau_K^\sharp(x)$ by definition.

By definition of $T := \varpi^\flat_{(K,C)}(x)$, for every $\varepsilon > 0$, there exists N such that for $n \geq N$, there exists an evolution $x_n(\cdot) \in \mathcal{S}(x_n)$ and $t_n \leq T + \varepsilon$ such that $x_n(t_n) \in C$ and for every $s < t_n$, $x_n(s) \in K$. Since $\mathcal{S}$ is upper semicompact, a subsequence (again denoted by) $x_n(\cdot)$ converges uniformly on compact intervals to some evolution $x(\cdot) \in \mathcal{S}(x)$. Let us also consider a subsequence (again denoted by) t_n converging to some $T^\star \leq T + \varepsilon$. By taking the limit, we infer that $x(T^\star)$ belongs to C and that, for any $s < T^\star$, $x(s)$ belongs to K. This implies that

$$\varpi^\flat_{(K,C)}(x) \leq \varpi_{(K,C)}(x(\cdot)) \leq T^\star \leq T + \varepsilon$$

We conclude by letting ε converge to 0: The evolution $x(\cdot)$ obtained above achieves the infimum. □

We deduce the following characterization of viability kernels and viable-capture basins:

Proposition 10.4.5 *[Sections of Exit and Minimal Time Functions] Let $\mathcal{S} : X \rightsquigarrow \mathcal{C}(0, +\infty; X)$ be a strict upper semicompact evolutionary system and C and K be two closed subsets such that $C \subset K$. Then the* viability kernel *is characterized by*

$$\mathrm{Viab}_{\mathcal{S}}(K) = \{x \in K \mid \tau_K^\sharp(x) = +\infty\}$$

and the viable-capture basin

$$\text{Capt}_{\mathcal{S}}(K,C) = \{x \in K \mid \varpi^{\flat}_{(K,C)}(x) \ < \ +\infty\}$$

is the domain of the (constrained) minimal time function $\varpi^{\flat}_{(K,C)}$.

Furthermore, for any $T \geq 0$*, the viability kernel and capture basin tubes defined in Definition 4.3.1, p. 133 can be characterized by exit and minimal time functions:*

$$\begin{cases} \text{Viab}_{\mathcal{S}}(K)(T) & := \left\{x \in K \mid \tau^{\sharp}_{K}(x) \ \geq \ T\right\} \\ \text{Capt}_{\mathcal{S}}(K,C)(T) & := \left\{x \in X \mid \varpi^{\flat}_{(K,C)}(x) \leq T\right\} \end{cases} \tag{10.11}$$

Proof. Inclusions

$$\begin{cases} \text{Viab}_{\mathcal{S}}(K) \ \subset \ \{x \in K \mid \tau^{\sharp}_{K}(x) = +\infty\} \\ \text{Capt}_{\mathcal{S}}(K,C) \ \subset \ \{x \in K \mid \varpi^{\flat}_{(K,C)}(x) \ < \ +\infty\} \end{cases}$$

as well as

$$\begin{cases} \text{Viab}_{\mathcal{S}}(K)(T) & \subset \left\{x \in K \mid \tau^{\sharp}_{K}(x) \ \geq \ T\right\} \\ \text{Capt}_{\mathcal{S}}(K,C)(T) & \subset \left\{x \in X \mid \varpi^{\flat}_{(K,C)}(x) \leq T\right\} \end{cases}$$

are obviously always true.

Equalities follow from Theorem 10.4.4 by taking one persistent evolution $x^{\sharp}(\cdot) \in \mathcal{S}(x)$ when $T \leq \tau^{\sharp}_{K}(x) \leq +\infty$, since we deduce that $T \leq \tau^{\sharp}_{K}(x) = \tau_{K}(x^{\sharp}(\cdot))$, so that $x(\cdot)$ is viable in K on the interval $[0,T]$. In the same way, taking one minimal time evolution $x^{\flat}(\cdot) \in \mathcal{S}(x)$ when $\varpi^{\flat}_{(K,C)}(x) \leq T < +\infty$, we deduce that $\varpi_{(K,C)}(x^{\flat}(\cdot)) = \varpi^{\flat}_{(K,C)}(x) \leq T$, so that $x(\cdot)$ is viable in K before it reaches C at T. □

The viability kernel is in some sense the paradise for viability, which is lost whenever an environment K is a repeller. Even though there is no viable evolutions in K, one can however look for an ersatz of viability kernel, which is the subset of evolutions which survive with the highest life expectancy:

Proposition 10.4.6 ***[Persistent Kernel]*** *If* K *is a compact repeller under a upper semicompact evolutionary system* $\mathcal{S}$*, then there exists a nonempty compact subset of initial states which maximize their exit time. This set can be regarded as a* persistent kernel.

Proof. By Theorem 10.4.1, p. 392, the exit functional $\tau_K^\sharp$ is upper semicontinuous. Therefore, it achieves its maximum whenever the environment K is compact. □

10.4.2 Temporal Window

Let $x \in K$ and $x(\cdot) \in \mathcal{S}(x)$ be a (full) evolution passing through x (see Sect. 8.2, p. 275). The sum of the exit times $\tau_K(\overrightarrow{x}(\cdot))$ of the forward part of the evolution and of the exit time $\tau_K(\overleftarrow{x}(\cdot))$ of its backward part can be regarded as the "*temporal window*" of the evolution in K. One can observe that the *maximal temporal window* is the sum of the exit time function of its backward time and of its forward part since

$$\sup_{x(\cdot)\in\mathcal{S}(x)} (\tau_K(\overleftarrow{x}(\cdot))+\tau_K(\overrightarrow{x}(\cdot))) = \sup_{\overleftarrow{x}(\cdot)\in\overleftarrow{\mathcal{S}}(x)} (\tau_K(\overleftarrow{x}(\cdot)))+ \sup_{\overrightarrow{x}(\cdot)\in\mathcal{S}(x)} (\tau_K(\overrightarrow{x}(\cdot)))$$

Any (full) evolution $x^\sharp(\cdot) \in \mathcal{S}(x)$ passing through $x \in K$ maximizing the temporal window is still called *persistent*. It is the concatenation of the persistent forward part $\tau_K(\overrightarrow{x}^\sharp(\cdot))$ and of its backward part $\tau_K(\overleftarrow{x}^\sharp(\cdot))$. The maximal temporal window of a (full) evolution viable in K is infinite, and the converse is true whenever the evolutionary system is upper semicompact (see Proposition 10.4.6, p. 395). If the subset $K \setminus B$ is a backward repeller and the subset $K \setminus C$ is a forward repeller, the bilateral viability kernel is empty, but the subset of states $x \in K$ maximizing their temporal window function is not empty and can be called the persistent kernel of K.

10.4.3 Exit Sets and Local Viability

We continue the study of local viability initiated in Sect. 2.13, p. 94 by characterizing it in terms of exit sets:

Definition 10.4.7 *[**Exit Subsets**] Let us consider an evolutionary system $\mathcal{S} : X \rightsquigarrow \mathcal{C}(0,+\infty;X)$ and a subset $K \subset X$. The* exit subset $\mathrm{Exit}_{\mathcal{S}}(K)$ *is the (possibly empty) subset of elements $x \in \partial K$ which leave K immediately:*

$$\mathrm{Exit}_{\mathcal{S}}(K) := \left\{x \in K \textit{ such that } \tau_K^\sharp(x) = 0\right\}$$

Exit sets characterize viability and local viability of environments. Recall that Definition 2.13.1, p. 94 states that a subset D is said *locally viable* under $\mathcal{S}$ if from any initial state $x \in D$, *there exists at least one evolution* $x(\cdot) \in \mathcal{S}(x)$ and a strictly positive $T_{x(\cdot)} > 0$ such that $x(\cdot)$ is viable in D on the nonempty interval $[0, T_{x(\cdot)}[$.

Proposition 10.4.8 *[Local Viability Kernel] The subset* $K \setminus \mathrm{Exit}_{\mathcal{S}}(K)$ *is the largest locally viable subset of* K *(and thus, can be regarded as the* "local viability kernel *of* K"*).*

Proof. Let $D \subset K$ be locally viable. If an evolution $x(\cdot) \in \mathcal{S}(x)$ starting from $x \in D$ is locally viable in D, it is clear that $\tau_K^\sharp(x) \geq \tau_D^\sharp(x) \geq \tau_D(x(\cdot)) > 0$, so that $x \in K \setminus \mathrm{Exit}_{\mathcal{S}}(K)$. Furthermore, the subset $K \setminus \mathrm{Exit}_{\mathcal{S}}(K)$ itself is locally viable because to say that $x \in K \setminus \mathrm{Exit}_{\mathcal{S}}(K)$ means that $\tau_K^\sharp(x) > 0$. Hence for any $0 < \lambda < \tau_K^\sharp(x)$, there exists $x(\cdot) \in \mathcal{S}(x)$ such that $0 < \lambda \leq \tau_K(x(\cdot))$, i.e., such that $x(\cdot)$ is viable in K on the nonempty interval $[0, \tau_K(x(\cdot))]$. □

If an environment K is not viable, the subset K can be covered in the following way:

$$K = \mathrm{Viab}_{\mathcal{S}}(K) \cup \mathrm{Abs}_{\mathcal{S}}(K, \mathrm{Exit}_{\mathcal{S}}(K))$$

because, starting outside the viability kernel of K, all solutions leave K in finite time through the exit set.

We also observe that $K \setminus (\mathrm{Viab}_{\mathcal{S}}(K) \cup \mathrm{Exit}_{\mathcal{S}}(K))$ is the set of initial states from which starts at least one evolution locally viable in K, but not viable in K.

Proposition 10.4.8, p. 397 implies

Proposition 10.4.9 *[Locally Viable Subsets] The following statements are equivalent:*

1. *the complement* $K \setminus C$ *of a target* $C \subset K$ *in the environment* K *is locally viable*
2. $\mathrm{Exit}_{\mathcal{S}}(K) \subset C$,
3. $C \cap \mathrm{Exit}_{\mathcal{S}}(K) \subset \mathrm{Exit}_{\mathcal{S}}(C)$.

In particular, K *is locally viable if and only if its exit set* $\mathrm{Exit}_{\mathcal{S}}(K) = \emptyset$ *is empty.*

Proof. Indeed, $K \setminus C$ is locally viable if and only if $K \setminus C \subset K \setminus \mathrm{Exit}_{\mathcal{S}}(K)$ is contained in the local viability kernel $K \setminus \mathrm{Exit}_{\mathcal{S}}(K)$, i.e., if and only if

$\text{Exit}_{\mathcal{S}}(K) \subset C$. On the other hand, since $C \cap \text{Exit}_{\mathcal{S}}(K) \subset \text{Exit}_{\mathcal{S}}(C)$. Hence the three statements are equivalent. □

There is a close link between the closedness of exit sets and the continuity of the exit function:

Theorem 10.4.10 *[Closedness of Exit Sets and Continuity of Exit Functions] Let us assume that the evolutionary system is upper semicompact and that the subset K is closed. Then the epigraph of the exit function $\tau_K^\sharp(\cdot)$ is closed if and only if the exit subset $\text{Exit}_{\mathcal{S}}(K)$ is closed.*

Proof. Since the exit function is upper semicontinuous, its continuity is equivalent to its lower semicontinuity, i.e., to the closedness of its epigraph. The lower semicontinuity of the exit function implies the closedness of the exit subset

$$\text{Exit}_{\mathcal{S}}(K) := \left\{x \in K \text{ such that } \tau_K^\sharp(x) = 0\right\}$$

because the lower sections of a lower semicontinuous function are closed. Let us prove the converse statement. Consider a sequence (x_n, y_n) of the epigraph of the exit function converging to some (x, y) and prove that the limit belongs to its epigraph, i.e., that $\tau_K^\sharp(x) \leq y$.

Indeed, since $t_n := \tau_K^\sharp(x_n) \leq y_n \leq y + 1$ when n is large enough, there exists a subsequence (again denoted by) t_n converging to $t_\star \leq y + 1$. Since the evolutionary system is assumed to be upper semicompact, there exists a persistent evolution $x_n^\sharp(\cdot) \in \mathcal{S}(x_n)$ such that $t_n := \tau_K(x_n^\sharp(\cdot))$. Furthermore, a subsequence (again denoted by) $x_n^\sharp(\cdot)$ converges to some evolution $x_\star(\cdot) \in \mathcal{S}(x)$ uniformly on the interval $[0, y+1]$. By definition of the persistent evolution, for all $t \in [0, t_n]$, $x_n^\sharp(t) \in K$ and $x_n(t_n) \in \text{Exit}_{\mathcal{S}}(K)$, which is closed by assumption. We thus infer that for all $t \in [0, t_\star]$, $x_\star(t) \in K$ and $x_\star(t_\star) \in \text{Exit}_{\mathcal{S}}(K)$. This means that $t_\star = \tau_K(x_\star(\cdot))$ and consequently, that $\tau_K(x_\star(\cdot)) \leq y$. This completes the proof. □

We single out the important case in which the evolutions leaving K cross the boundary at a single point:

Definition 10.4.11 *[Transverse Sets] Let $\mathcal{S}$ be an evolutionary system and K be a closed subset. We shall say that K is transverse to $\mathcal{S}$ if for every $x \in K$ and for every evolution $x(\cdot) \in \mathcal{S}(x)$ leaving K in finite time, $\tau_K(x(\cdot)) = \varpi_{\partial K}(x(\cdot))$.*

Transversality of an environment means that all evolutions governed by an evolutionary system cross the boundary as soon as they reach it to leave the environment immediately.

We deduce the following consequence:

Proposition 10.4.12 *[**Continuity of the Exit Function of a Transverse Set**] Assume that the evolutionary system $\mathcal{S}$ is upper semicompact and that the subset K is closed and transverse to $\mathcal{S}$. Then the exit function $\tau_K^\sharp$ is continuous and the exit set $\text{Exit}_{\mathcal{S}}(K)$ of K is closed.*

10.5 Viability Characterizations of Kernels and Basins

We shall review successively the viability characterizations of viable subsets outside a target, introduce the concept of relative backward invariance for characterizing isolated systems before proving the second viability characterizations of viability kernels and capture basins. They enjoy semi-permeable barrier properties investigated at the end of this section.

10.5.1 Subsets Viable Outside a Target

We now provide a characterization of a subset D viable outside a target C in terms of local viability of $D \backslash C$:

Proposition 10.5.1 *[**Characterization of Viable Subsets Outside a Target**] Assume that $\mathcal{S}$ is upper semicompact. Let $C \subset D$ and D be closed subsets. The following conditions are equivalent:*

1. *D is viable outside C under $\mathcal{S}$ ($\text{Viab}_{\mathcal{S}}(D, C) = D$ by Definition 2.2.3, p. 49),*
2. *$D \setminus C$ is locally viable under $\mathcal{S}$,*
3. *The exit set of D is contained in the exit set of C: $C \cap \text{Exit}_{\mathcal{S}}(D) \subset \text{Exit}_{\mathcal{S}}(C)$*

In particular, a closed subset D is viable under $\mathcal{S}$ if and only if its exit set is empty:

$$\text{Viab}_{\mathcal{S}}(D) = D \text{ if and only if } \text{Exit}_{\mathcal{S}}(D) = \emptyset$$

Proof.

1. First, assume that $\text{Viab}_{\mathcal{S}}(D, C) = D$ and derive that $D \backslash C$ is locally viable. Take $x_0 \in D \backslash C$ and prove that there exists an evolution $x(\cdot) \in \mathcal{S}(x_0)$ starting at x_0 viable in $D \backslash C$ on a nonempty interval. Indeed, since C is closed, there exists $\eta > 0$ such that $B(x_0, \eta) \cap C = \emptyset$, so that $x(t) \in B(x_0, \eta) \cap D \subset D \backslash C$ on some nonempty interval. This means that $\text{Viab}_{\mathcal{S}}(D, C) \backslash C$ is locally viable.
2. Assume that $D \setminus C$ is locally viable and derive that $\text{Viab}_{\mathcal{S}}(D, C) = D$. Take any $x \in D \setminus C$ and, since the evolutionary system is assumed to be semicompact, at least one persistent evolution $x^{\sharp}(\cdot) \in \mathcal{S}(x)$, thanks to Theorem 10.4.4. Either this persistent evolution is viable forever, and thus $x \in \text{Viab}_{\mathcal{S}}(D) \subset \text{Viab}_{\mathcal{S}}(D, C)$, or else, it leaves D in finite time $\tau_D^{\sharp}(x)$ at $x^{\Rightarrow} := x^{\sharp}(\tau_D^{\sharp}(x)) \in \partial D$.
   Such an element $x^{\Rightarrow}$ belongs to C because, otherwise, since $D \backslash C$ is locally viable and C is closed, one could associate with $x^{\Rightarrow} \in D \backslash C$ another evolution $y(\cdot) \in \mathcal{S}(x^{\Rightarrow})$ and $T > 0$ such that $y(\tau) \in D \backslash C$ for all $\tau \in [0, T]$, so that $\tau_D^{\sharp}(x^{\Rightarrow}) = T > 0$, contradicting the fact that $x^{\sharp}(\cdot)$ is a persistent evolution.
3. The equivalence between the second and third statement follows from Propositions 10.4.9, p. 397 on exit sets. □

As a consequence, Proposition 10.5.2, p. 400 and Theorem 10.3.10, p. 388 (guaranteeing that the viability kernels $\text{Viab}_{\mathcal{S}}(D, C)$ are closed) Theorem 2.15.2 imply the following:

Theorem 10.5.2 ***[Characterization of Viable Subsets Outside a Target]*** *Assume that $\mathcal{S}$ is upper semicompact. Let $C \subset K$ and K be closed subsets.*

Then the viability kernel $\text{Viab}_{\mathcal{S}}(K, C)$ *of K with target C under $\mathcal{S}$ is:*

- *either the* largest closed subset $D \subset K$ *containing C such that $D \backslash C$ is locally viable,*
- *or, equivalently, the* largest closed subset *satisfying*

$$C \cap \text{Exit}_{\mathcal{S}}(D) \subset \text{Exit}_{\mathcal{S}}(C) \subset C \subset D \subset K \tag{10.12}$$

Therefore, under these equivalent assumptions (10.12), p. 400, inclusion

$$D \cap \text{Exit}_{\mathcal{S}}(K) \subset C \subset D \tag{10.13}$$

holds true. In particular, the viability kernel $\text{Viab}_{\mathcal{S}}(K)$ *of K is the largest closed viable subset contained in K.*

Remark. We shall see that inclusion (10.13), p. 400 is the "mother of boundary conditions" when the subsets C, K and D are graphs of set-valued maps or epigraphs of extended functions. □

10.5.2 Relative Invariance

We characterize further isolated subsets in terms of backward invariance properties – discovered by Hélène Frankowska in her investigations of Hamilton-Jacobi equations associated with value functions of optimal control problems under state constraints. They play a crucial role for enriching the Characterization Theorem 10.2.5 stating that the viability kernel of an environment with a target is the smallest subset containing the target and isolated in this environment. We already introduced the concept of backward relative invariance (see Definition 2.15.3, p. 100):

Definition 10.5.3 *[Relative Invariance] We shall say that a subset $C \subset K$ is* (backward) invariant relatively to K *under $\mathcal{S}$ if for every $x \in C$, all (backward) evolutions starting from x and viable in K on an interval $[0,T[$ are viable in C on the same interval $[0,T[$.*

If K is itself (backward) invariant, any subset (backward) invariant relatively to K is (backward) invariant.

If $C \subset K$ is (backward) invariant relatively to K, then $C \cap \mathrm{Int}(K)$ is (backward) invariant.

Proposition 10.5.4 *[Capture Basins of Relatively Invariant Targets] Let $C \subset D \subset K$ three subsets of X.*

1. *If D is backward invariant relatively to K, then $\mathrm{Capt}_{\mathcal{S}}(K,C) = \mathrm{Capt}_{\mathcal{S}}(D,C)$,*
2. *If C is backward invariant relatively to K, then $\mathrm{Capt}_{\mathcal{S}}(K,C) = C$.*

Proof. Since $\mathrm{Capt}_{\mathcal{S}}(D,C) \subset \mathrm{Capt}_{\mathcal{S}}(K,C)$, let us consider an element $x \in \mathrm{Capt}_{\mathcal{S}}(K,C)$, an evolution $x(\cdot)$ viable in K until it reaches C in finite time $T \geq 0$ at $z := x(T) \in C$. Setting $\overleftarrow{y}(t) := x(T-t)$, we observe that $\overleftarrow{y}(\cdot) \in \overleftarrow{\mathcal{S}}(x(T))$, satisfies $\overleftarrow{y}(T) = x \in K$ and is viable in K on the interval $[0,T]$. Since D is backward invariant relatively to K, we infer that this evolution $\overleftarrow{y}(\cdot)$ is viable in D on the interval $[0,T]$, so that $x(t) = \overleftarrow{y}(T-t)$ belongs to D for all $t \in [0,T]$. This implies that x belongs to $\mathrm{Capt}_{\mathcal{S}}(D,C)$.

Taking $C := D$, then $\text{Capt}_{\mathcal{S}}(D,C) = C$, so that $\text{Capt}_{\mathcal{S}}(K,C) = \text{Capt}_{\mathcal{S}}(D,C) = C$. □

Capture basins of targets viable in environments are backward invariants relatively to this environment:

Proposition 10.5.5 ***[Relative Backward Invariance of Capture Basins]*** *The capture basin* $\text{Capt}_{\mathcal{S}}(K,C)$ *of a target* C *viable in the environment* K *is backward invariant relatively to* K.

Proof. We have to prove that for every $x \in \text{Capt}_{\mathcal{S}}(K,C)$, every backward evolution $\overleftarrow{y}(\cdot) \in \overleftarrow{\mathcal{S}}(x)$ viable in K on some interval $[0,T]$ is actually viable in $\text{Capt}_{\mathcal{S}}(K,C)$ on the same interval.

Since x belongs to $\text{Capt}_{\mathcal{S}}(K,C)$, there exists an evolution $z(\cdot) \in \mathcal{S}(x)$ and $S \geq 0$ such that $z(S) \in C$ and, for all $t \in [0,S]$, $z(t) \in K$. We associate with it the evolution $\overrightarrow{x}_T(\cdot) \in \mathcal{S}(\overleftarrow{y}(T))$ defined by

$$\overrightarrow{x}_T(t) := \begin{cases} \overleftarrow{y}(T-t) \text{ if } t \in [0,T] \\ \overrightarrow{z}(t-T) \text{ if } t \in [T, T+S] \end{cases}$$

starting at $y(T) \in K$. It is viable in K until it reaches C at time $T+S$. This means that $y(T)$ belongs to $\text{Capt}_{\mathcal{S}}(K,C)$ and this implies that for every $t \in [0, T+S]$, $\overrightarrow{x}_T(t)$ belongs to the capture basin $\text{Capt}_{\mathcal{S}}(K,C)$. This is in particular the case when $t \in [0,T]$: then $\overleftarrow{y}(t) = \overrightarrow{x}_T(T-t)$ belongs to the capture basin. Therefore, the backward evolution $\overleftarrow{y}(\cdot) \in \overleftarrow{\mathcal{S}}(x)$ is viable in $\text{Capt}_{\mathcal{S}}(K,C)$ on the interval $[0,T]$. □

We deduce that a subset $C \subset K$ is backward invariant relatively to K if and only if K is the capture basin of C:

Theorem 10.5.6 ***[Characterization of Relative Invariance]*** *A subset* $C \subset K$ *is backward invariant relatively to* K *if and only if* $\text{Capt}_{\mathcal{S}}(K,C) = C$.

Proof. First, Proposition 10.5.5, p. 402 implies that whenever $\text{Capt}_{\mathcal{S}}(K,C) = C$, C is backward invariant relatively to K. Conversely, assume that C is backward invariant relatively to K and we shall derive a contradiction by assuming that there exists $x \in \text{Capt}_{\mathcal{S}}(K,C) \backslash C$: in this case, there would exist a forward evolution denoted $\overrightarrow{x}(\cdot) \in \mathcal{S}(x)$ starting at x and viable in

K until it reaches C at time $T > 0$ at $c = x(T)$. Let $\overleftarrow{z}(\cdot) \in \overleftarrow{\mathcal{S}}(x)$ be any backward evolution starting at x and viable in K on some interval $[0, T]$. We associate with it the function $\overleftarrow{y}(\cdot)$ defined by

$$\overleftarrow{y}(t) := \begin{cases} \overrightarrow{x}(T - t) \text{ if } t \in [0, T] \\ \overleftarrow{z}(t - T) \text{ if } t \geq T \end{cases}$$

Then $\overleftarrow{y}(\cdot) \in \overleftarrow{\mathcal{S}}(c)$ and is viable in K on the interval $[0, T]$. Since C is assumed to be backward invariant relatively to K, then $\overleftarrow{y}(t) \in C$ for all $t \in [0, T]$, and in particular $\overleftarrow{y}(T) = x$ belongs to C. We have obtained a contradiction since we assumed that $x \notin C$. Therefore $\text{Capt}_{\mathcal{S}}(K, C) \backslash C = \emptyset$, i.e., $\text{Capt}_{\mathcal{S}}(K, C) = C$. □

As a consequence of Proposition 10.5.6, we obtain:

Proposition 10.5.7 *[Backward Invariance of the Complement of an Invariant Set] A subset C is backward invariant under an evolutionary system $\mathcal{S}$ if and only if its complement $\complement C$ is invariant under $\mathcal{S}$.*

Proof. Applying Proposition 10.5.6 with $K := X$, we infer that C is backward invariant if and only if $C = \text{Capt}_{\mathcal{S}}(X, C)$, which is equivalent, by Lemma 2.12.2, to the statement that $\complement C = \text{Inv}_{\mathcal{S}}(\complement C, \emptyset) =: \text{Inv}_{\mathcal{S}}(\complement C)$ is invariant. □

10.5.3 Isolated Subsets

The following Lemma is useful because it allows isolated subsets to be also characterized by viability properties:

Lemma 10.5.8 *[Isolated Subsets] Let D and K be two subsets such that $D \subset K$. Then the following properties are equivalent:*

1. *D is isolated in K under $\mathcal{S}$:* $\text{Viab}_{\mathcal{S}}(K, D) = D$,
2. $\text{Viab}_{\mathcal{S}}(K) = \text{Viab}_{\mathcal{S}}(D)$ *and* $\text{Capt}_{\mathcal{S}}(K, D) = D$,
3. *$K \backslash D$ is a repeller and* $\text{Capt}_{\mathcal{S}}(K, D) = D$.

Proof. Assume that D is isolated in K. This amounts to writing that,

1. by definition,

$$D = \mathrm{Viab}_{\mathcal{S}}(K, D) = \mathrm{Viab}_{\mathcal{S}}(K) \cup \mathrm{Capt}_{\mathcal{S}}(K, D)$$

and thus, equivalently, that $\mathrm{Capt}_{\mathcal{S}}(K, D) = D$ and $\mathrm{Viab}_{\mathcal{S}}(K) \subset D$. Since $D \subset K$, inclusion $\mathrm{Viab}_{\mathcal{S}}(K) \subset D$ is equivalent to $\mathrm{Viab}_{\mathcal{S}}(K) = \mathrm{Viab}_{\mathcal{S}}(D)$.

2. by formula (2.26),

$$D = \mathrm{Viab}_{\mathcal{S}}(K, D) = \mathrm{Viab}_{\mathcal{S}}(K \backslash D) \cup \mathrm{Capt}_{\mathcal{S}}(K, D)$$

and thus, equivalently, that $\mathrm{Capt}_{\mathcal{S}}(K, D) = D$ and that $\mathrm{Viab}_{\mathcal{S}}(K \backslash D) \subset D$. Since $D \cap \mathrm{Viab}_{\mathcal{S}}(K \backslash D) = \emptyset$, this implies that $\mathrm{Viab}_{\mathcal{S}}(K \backslash D) = \emptyset$. □

We derive the following characterization:

Theorem 10.5.9 *[Characterization of Isolated Subsets] Let us consider a closed subset $D \subset K$. Then D is isolated in K by $\mathcal{S}$ if and only if:*

1. *D is backward invariant relatively to K,*
2. *either $K \backslash D$ is a repeller or* $\mathrm{Viab}_{\mathcal{S}}(K) = \mathrm{Viab}_{\mathcal{S}}(D)$.

We provide now another characterization of isolated subsets involving complements:

Proposition 10.5.10 *[Complement of an Isolated Subset] Let us assume that K and $D \subset K$ are closed.*

1. *If the evolutionary system $\mathcal{S}$ is lower semicontinuous and if $D = \mathrm{Capt}_{\mathcal{S}}(K, D)$, then either one of the following equivalent properties:*

$$\begin{cases} (i)\ \ \overline{\complement D} = \mathrm{Inv}_{\mathcal{S}}(\overline{\complement D}, \overline{\complement K})\ (\overline{\complement D} \text{ is invariant outside } \overline{\complement K}) \\ (ii)\ \mathrm{Int}(D) = \mathrm{Capt}_{\mathcal{S}}(\mathrm{Int}(K), \mathrm{Int}(D)) \\ (iii)\ \mathrm{Int}(D) \text{ is backward invariant relatively to } \mathrm{Int}(K) \end{cases} \quad (10.14)$$

hold true.

2. *Conversely, if $\mathrm{Int}(K)$ is backward invariant and if the set-valued map $\overleftarrow{\mathcal{S}}$ is lower semicontinuous, then any of the equivalent properties (10.14), p. 404 implies that* $\overline{\mathrm{Int}(D)} = \mathrm{Capt}_{\mathcal{S}}(\overline{\mathrm{Int}(K)}, \overline{\mathrm{Int}(D)})$.

Proof. Lemma 2.12.2 implies that $\text{Capt}_{\mathcal{S}}(K, D) = D$ if and only if

$$\complement D = \text{Inv}_{\mathcal{S}}(\complement D, \complement K)$$

Since $\mathcal{S}$ is assumed to be lower semicontinuous, we deduce from Theorem 10.3.7 that

$$\begin{cases} \complement(\text{Int}(D)) = \overline{\complement D} = \overline{\text{Inv}_{\mathcal{S}}(\complement D, \complement K)} \\ \subset \text{Inv}_{\mathcal{S}}(\overline{\complement D}, \overline{\complement K}) = \text{Inv}_{\mathcal{S}}(\complement(\text{Int}(D)), \complement(\text{Int}(K))) \subset \complement(\text{Int}(D)) \end{cases}$$

so that the closure of the complement of D is invariant outside the closure of the complement of K. Observe that, taking the complements, Lemma 2.12.2 states that this is equivalent to property $\text{Int}(D) = \text{Capt}_{\mathcal{S}}(\text{Int}(K), \text{Int}(D))$, which, by Theorem 10.5.6, p. 402, amounts to saying that the interior of D is relatively backward invariant relatively to the interior of K.

For proving the converse statement, Proposition 10.3.14, p. 390 states that under the assumptions of the theorem, condition $\text{Int}(D) = \text{Capt}_{\mathcal{S}}(\text{Int}(K), \text{Int}(D))$ implies that

$$\overline{\text{Int}(D)} \subset \text{Capt}_{\mathcal{S}}(\overline{\text{Int}(K)}, \overline{\text{Int}(D)}) \subset \overline{\text{Capt}_{\mathcal{S}}(\text{Int}(K), \text{Int}(D))} = \overline{\text{Int}(D)} \quad \square$$

10.5.4 The Second Fundamental Characterization Theorem

Putting together the characterizations of viable subsets and isolated subsets, we reformulate Theorem 10.2.5 characterizing viability kernels with targets in the following way:

Theorem 10.5.11 ***[Viability Characterization of Viability Kernels]*** *Let us assume that $\mathcal{S}$ is upper semicompact and that the subsets $C \subset K$ and K are closed. The viability kernel* $\text{Viab}_{\mathcal{S}}(K, C)$ *of a subset K with target C under $\mathcal{S}$ is the* ***unique*** *closed subset satisfying $C \subset D \subset K$ and*

$$\begin{cases} (i) \ D \backslash C \text{ is } \textit{locally viable} \text{ under } \mathcal{S}, \text{ i.e., } D = \text{Viab}_{\mathcal{S}}(D, C) \\ (ii) \ D \textit{ is backward invariant relatively to } K \text{ under } \mathcal{S}, \\ (iii) \ K \backslash D \text{ is a } \textit{repeller} \text{ under } \mathcal{S}, \text{ i.e., } \text{Viab}_{\mathcal{S}}(K \setminus D) = \emptyset. \end{cases} \quad (10.15)$$

We mentioned in Sect. 2.15, p. 98 the specific versions for viability kernels (Theorem 2.15.4, p. 101) and capture basin (Theorem 2.15.5, p. 101). However, Theorem 10.5.11 implies that when $K \backslash C$ is a repeller, the above theorem implies a characterization of the viable-capture basins in a more general context:

Theorem 10.5.12 ***[Characterization of Capture Basins]*** *Let us assume that $\mathcal{S}$ is upper semicompact and that a closed subset $C \subset K$ satisfies property*

$$\text{Viab}_{\mathcal{S}}(K \backslash C) = \emptyset \tag{10.16}$$

Then the viable-capture basin $\text{Capt}_{\mathcal{S}}(K, C)$ *is the* ***unique*** *closed subset D satisfying $C \subset D \subset K$ and*

$$\begin{cases} (i) & D \backslash C \text{ is } \textit{locally viable} \text{ under } \mathcal{S} \\ (ii) & D \textit{ is backward invariant relatively to } K \text{ under } \mathcal{S} \end{cases} \tag{10.17}$$

We deduce from Proposition 10.5.10, p. 404 another characterization of capture basins that provide existence and uniqueness of viscosity solutions to some Hamilton–Jacobi–Bellman equations:

Theorem 10.5.13 ***["Viscosity" Characterization of Capture Basins]*** *Assume that the evolutionary system $\mathcal{S}$ is both upper semicompact and lower semicontinuous, that K is closed, that* $\text{Int}(K) \neq \emptyset$ *is backward invariant, that* $\text{Viab}_{\mathcal{S}}(K \backslash C) = \emptyset$*, that* $\overline{\text{Int}(K)} = K$ *and that* $\overline{\text{Int}(C)} = C$.

Then the capture basin $\text{Capt}_{\mathcal{S}}(K, C)$ *is the* ***unique*** *subset topologically regular subset D between C and K satisfying*

$$\begin{cases} (i) & D \backslash C \text{ is } \textit{locally viable} \text{ under } \mathcal{S}, \\ (ii) & \overline{\complement D} \text{ is } \textit{invariant} \text{ outside } \overline{\complement K} \text{ under } \mathcal{S}. \end{cases} \tag{10.18}$$

Proof. Since $\text{Viab}_{\mathcal{S}}(K \backslash C) = \emptyset$, the viability kernel and the capture basin are equal. By Theorem 10.2.5, p. 379, the capture basin is the unique subset D between C and K such that:

1. the largest subset $D \subset K$ such that $\text{Capt}_{\mathcal{S}}(D, C) = D$,
2. the smallest subset $D \supset C$ such that $\text{Capt}_{\mathcal{S}}(K, D) = D$.

The evolutionary system being upper semicompact, the first condition amounts to saying that $D \backslash C$ is locally viable.

By Proposition 10.5.10, p. 404, property $\text{Capt}_{\mathcal{S}}(K, D) = D$ implies that $\overline{\complement D}$ is invariant outside $\overline{\complement K}$, as well as the other properties (10.14), p. 404.

Conversely, let D satisfy those properties (10.14). Proposition 10.7.6, p. 419 implies that, under the assumptions of the theorem, the capture basin $\text{Capt}_{\mathcal{S}}(K, C)$ is topologically regular whenever K and C are topologically regular. Let D satisfy properties (10.18), p. 406. By Proposition 10.5.10, p. 404, property $\text{Capt}_{\mathcal{S}}(K, D) = D$ implies that $\text{Capt}_{\mathcal{S}}(K, D) = D$. Therefore, Theorem 10.2.5, p. 379 implies that $D = \text{Capt}_{\mathcal{S}}(K, C)$. □

Remark. We shall see that whenever the environment $K := \mathcal{E}p(\mathbf{k})$ and the target $C := \mathcal{E}p(\mathbf{c})$ are epigraphs of functions $\mathbf{k} \leq \mathbf{c}$, the capture basin under adequate dynamical system is itself the epigraph of a function $\mathbf{v}$. Theorem 10.5.13, p. 406 implies that $\mathbf{v}$ is a viscosity solution to a Hamilton–Jacobi–Bellman equation. □

10.5.5 The Barrier Property

Roughly speaking, an environment exhibits the barrier property if all viable evolutions starting from its boundary are viable on its boundary, so that no evolution can enter the interior of this environment: this is a *semi-permeability property* of the boundary.

For that purpose, we need to define the concept of boundary:

Definition 10.5.14 *[**Boundaries**] Let $C \subset K \subset X$ be two subsets of X. The subsets*

$$\partial_K C := \overline{C} \cap \overline{K \backslash C} \ \& \ \overset{\circ}{\partial}_K C := C \cap \overline{K \backslash C}$$

are called respectively the boundary and the pre-boundary of the subset C relatively to K. When $K := X$, we set

$$\partial C := \overline{C} \cap \overline{\complement C} \ \& \ \overset{\circ}{\partial} C := C \cap \overline{\complement C}$$

In other words, the interior of a set D and its pre-boundary form a partition of $D = \text{Int}(D) \cup \overset{\circ}{\partial} D$. Pre-boundaries are useful because of the following property:

Lemma 10.5.15 *[**Pre-boundary of an intersection with an open set**] Let $\Omega \subset X$ be an open subset and $D \subset X$ be a subset. Then*

$$\overset{\circ}{\partial} (\Omega \cap D) = \Omega \cap \overset{\circ}{\partial} D$$

In particular, if $C \subset D$ is closed, then

$$\mathrm{Int}(D) \setminus C = \mathrm{Int}(D \setminus C) \text{ and } \overset{\circ}{\partial} (D \setminus C) = \overset{\circ}{\partial} (D) \setminus C$$

Proof. Indeed, $D = \mathrm{Int}(D) \cup \overset{\circ}{\partial} D$ being a partition of D, we infer that $D \cap \Omega = \mathrm{Int}(D \cap \Omega) \cup \overset{\circ}{\partial} D \cap \Omega$ is still a partition. By definition, $D \cap \Omega = \mathrm{Int}(D \cap \Omega) \cup \overset{\circ}{\partial} (D \cap \Omega)$ is another partition of $D \cap \Omega$. Since Ω is open, $\mathrm{Int}(D \cap \Omega) = \mathrm{Int}(D) \cap \mathrm{Int}(\Omega) = \mathrm{Int}(D) \cap \Omega$, so that $\overset{\circ}{\partial} (D \cap \Omega) = \Omega \cap \overset{\circ}{\partial} D$. □

Definition 10.5.16 ***[Barrier Property]*** *Let $D \subset X$ be a subset and $\mathcal{S}$ be an evolutionary system. We shall say that D exhibits the* barrier property *if its pre-boundary $\overset{\circ}{\partial} D$ is relatively invariant with respect to D itself. In other words, starting from any $x \in \overset{\circ}{\partial} D$, all evolutions viable in D on some time interval $[0, T[$ are actually viable in $\overset{\circ}{\partial} D$ on $[0, T[$.*

Remark. The barrier property of an environment is a *semi-permeability property* of D, since no evolution can enter the interior of D from the boundary (whereas evolutions may leave D). This is very important in terms of interpretation. Viability of a subset D having the barrier property is indeed a *very fragile property*, which cannot be restored from the outside, or equivalently, no solution starting from outside the viability kernel can cross its boundary from outside. In other words, starting from the pre-boundary of the environment, *love it or leave it...* The "barrier property" played an important role in control theory and the theory of differential games, because their boundaries could be characterized as solutions of first-order partial differential equations under (severe) regularity assumptions. *Marc Quincampoix* made the link at the end of the 1980s between this property and the boundary of the viability kernel: every solution starting from the boundary of the viability kernel can either remain in the boundary or leave the viability kernel, or equivalently, no solution starting from outside the viability kernel can cross its boundary. □

We deduce from Theorem 10.5.6, p. 402 that a subset D exhibits the barrier property if and only if its interior is backward invariant:

Proposition 10.5.17 ***[Backward Invariance of the interior and Barrier Property]*** *A subset D exhibits the barrier property if and only if its interior* $\mathrm{Int}(D)$ *is backward invariant.*

Proof. Theorem 10.5.6, p. 402 states that the pre-boundary $\overset{\circ}{\partial} D \subset D$ is invariant relatively to D if and only if $\mathrm{Capt}_{\overleftarrow{S}}(D, \overset{\circ}{\partial} D) = \overset{\circ}{\partial} D$. Therefore, from every $x \in \mathrm{Int}(D) = D \backslash \overset{\circ}{\partial} D = D \setminus \mathrm{Capt}_{\overleftarrow{S}}(D, \overset{\circ}{\partial} D)$, all backward evolutions are viable in $\mathrm{Int}(D) = D \backslash \overset{\circ}{\partial} D$ as long as they are viable in D. Such evolutions always remain in $\mathrm{Int}(D)$ because they can never reach $x(t) \in \overset{\circ}{\partial} D$ in some finite time t. □

Viability kernels exhibit the barrier property whenever the evolutionary system is both upper and lower semicontinuous:

Theorem 10.5.18 ***[Barrier Property of Boundaries of Viability Kernels]*** *Assume that K is closed and that the evolutionary system $\mathcal{S}$ is lower semicontinuous. Then the intersection* $\mathrm{Viab}_{\mathcal{S}}(K, C) \cap \mathrm{Int}(K)$ *of the viability kernel of K with the interior of K exhibits the* barrier property *and the interior* $\mathrm{Int}(\mathrm{Viab}_{\mathcal{S}}(K))$ *of the viability kernel of K is backward invariant.*

If $\mathrm{Viab}_{\mathcal{S}}(K) \subset \mathrm{Int}(K)$*, then* $\mathrm{Viab}_{\mathcal{S}}(K)$ *exhibits the barrier property, and thus, its interior is backward invariant.*

In some occasions, the boundary of the viability kernel can be characterized as the viability kernel of the complement of a target, and in this case, exhibits the properties of viability kernels, in particular, can be computed by the Viability Kernel Algorithm: see Theorem 9.2.18, p. 339.

Actually, Theorem 10.5.18, p. 409 is a consequence of the Barrier Theorem 10.5.19, p. 409 of viability kernels with nonempty targets:

Theorem 10.5.19 ***[Barrier Property of Viability Kernels with Targets]*** *Assume that K and $C \subset K$ are closed and that the evolutionary system $\mathcal{S}$ is lower semicontinuous. Then the intersection* $\mathrm{Viab}_{\mathcal{S}}(K, C) \cap \mathrm{Int}(K \setminus C)$ *of the viability kernel of K with target $C \subset K$ under $\mathcal{S}$ with the interior of $K \setminus C$ exhibits the* barrier property.

Furthermore, $\mathrm{Int}(\mathrm{Viab}_{\mathcal{S}}(K,C)) \setminus C$ *is backward invariant.*
In particular, if $\mathrm{Viab}_{\mathcal{S}}(K,C) \subset \mathrm{Int}(K \setminus C)$*, then* $\mathrm{Int}(\mathrm{Viab}_{\mathcal{S}}(K,C)) \setminus C$ *exhibits the barrier property.*

Proof. Let us set $D := \mathrm{Viab}_{\mathcal{S}}(K,C)$. Theorem 10.5.11, p. 405 implies that D satisfies (10.5.5), p. 410:

$$\begin{cases} (i) & D\backslash C \text{ is } \textit{locally viable} \text{ under } \mathcal{S} \\ (ii) & D \textit{ is backward invariant relatively to } K \text{ under } \mathcal{S} \\ (iii) & K\backslash D \text{ is a } \textit{repeller} \text{ under } \mathcal{S} \text{ or } \mathrm{Viab}_{\mathcal{S}}(K) = \mathrm{Viab}_{\mathcal{S}}(D). \end{cases}$$

and Proposition 10.5.10, p. 404 states that if the evolutionary system $\mathcal{S}$ is lower semicontinuous, then condition $D = \mathrm{Capt}_{\mathcal{S}}(K,D)$ implies that

$$\overline{\complement D} = \mathrm{Inv}_{\mathcal{S}}(\overline{\complement D}, \overline{\complement K}) \; (\overline{\complement D} \text{ is invariant outside } \overline{\complement K})$$

Lemma 10.5.15, p. 407 states that, since the target C is assumed to be closed,

$$\overset{\circ}{\partial} (\mathrm{Int}(K \setminus C) \cap D) = \mathrm{Int}(K \setminus C) \cap \overset{\circ}{\partial} D = (\overset{\circ}{\partial} D \cap \mathrm{Int}(K)) \setminus C$$

because the interior of a finite intersection of subsets is the intersection of their interiors.

Let x belong to $\mathrm{Int}(K \setminus C) \cap \overset{\circ}{\partial} (\mathrm{Viab}_{\mathcal{S}}(K,C))$. Since $x \in D := \mathrm{Viab}_{\mathcal{S}}(K,C)$, there exists at least one evolution belonging to $\mathcal{S}(x)$ viable in K forever or until it reaches C in finite time. Take any such evolution $x(\cdot) \in \mathcal{S}(x)$. Since $x \in \overline{\complement D} := \mathrm{Inv}_{\mathcal{S}}(\overline{\complement D}, \overline{\complement K})$, this evolution $x(\cdot)$, as well as every evolution starting from x, remains viable in $\overline{\complement D}$ as long as $x(t) \in \mathrm{Int}(K)$. Therefore, it remains viable in $\mathrm{Int}(K \setminus C) \cap \overset{\circ}{\partial} (D)$ as long as $x(t) \in \mathrm{Int}(K) \setminus C = \mathrm{Int}(K \setminus C)$ (since C is assumed to be closed, thanks to the second statement of Lemma 10.5.15, p. 407).

Proposition 10.5.17, p. 409 implies that the interior $\mathrm{Int}(D \cap (K \setminus C)) = \mathrm{Int}(D) \setminus C$ is backward invariant. □

Remark. If we assume furthermore that $\mathcal{S}$ is upper semicompact, then the viability kernel with target is closed, so that its pre-boundary coincides with its boundary. □

10.6 Other Viability Characterizations

10.6.1 Characterization of Invariance Kernels

We now investigate the viability property of invariance kernels.

Proposition 10.6.1 ***[Characterization of Invariant Subsets Outside a Target]*** *Assume that $\mathcal{S}$ is upper lower semicontinuous. Let $C \subset K$ be closed subsets.*

Then the invariance kernel $\text{Inv}_{\mathcal{S}}(K, C)$ *of K with target C under $\mathcal{S}$ is the* largest closed subset *$D \subset K$ containing C such that $D \backslash C$ is locally invariant.*

In particular, K is invariant outside C if and only if $K \backslash C$ is locally invariant.

Proof. First, we have to check that if $D \supset C$ is invariant outside C, then $D \backslash C$ is locally invariant: take $x_0 \in D \backslash C$ and prove that all evolutions $x(\cdot) \in \mathcal{S}$ starting at x_0 are viable in $D \backslash C$ on a nonempty interval. Indeed, since C is closed, there exists $\eta > 0$ such that $B(x_0, \eta) \cap C = \emptyset$, so that $x(t) \in B(x_0, \eta) \cap D \subset D \backslash C$ on some nonempty interval.

In particular, $\text{Inv}_{\mathcal{S}}(K, C) \backslash C$ is locally invariant and the invariance kernel $\text{Inv}_{\mathcal{S}}(K, C)$ of K with target C under $\mathcal{S}$ is closed by Theorem 10.7.8.

Let us prove now that any subset D between C and K such that $D \backslash C$ is locally invariant is contained in the invariance kernel $\text{Inv}_{\mathcal{S}}(K, C)$ of K with target C under $\mathcal{S}$.

Since $C \subset \text{Inv}_{\mathcal{S}}(K, C)$, let us pick any x in $D \backslash C$ and show that it belongs to $\text{Inv}_{\mathcal{S}}(K, C)$. Let us take any evolution $x(\cdot) \in \mathcal{S}(x)$. Either it is viable in D forever or, if not, leaves D in finite time $\tau_D(x(\cdot))$ at $\overline{x} := x(\tau_D(x(\cdot)))$: there exists a sequence $t_n \geq \tau_D(x(\cdot))$ converging to $\tau_D(x(\cdot))$ such that $x(t_n) \notin D$. Actually, this element $\overline{x}$ belongs to C. Otherwise, since $D \backslash C$ is locally invariant, this evolution remains in D in some nonempty interval $[\tau_D(x(\cdot)), T]$, a contradiction. □

Further characterizations require properties of the invariance kernels in terms of closed viable or invariant subsets. For instance:

Proposition 10.6.2 ***[Invariance Kernels]*** *Let us assume that $C \subset K$ and K are closed, that $K \backslash C$ is a repeller and that the evolutionary system $\mathcal{S}$ is both upper semicompact and lower semicontinuous. Then the invariance kernel* $\text{Inv}_{\mathcal{S}}(K, C)$ *is a closed subset D between C and K satisfying*

$$\begin{cases} (i) \ D = \text{Inv}_{\mathcal{S}}(D, C) \\ (ii) \ \overline{\complement D} = \text{Capt}_{\mathcal{S}}(\overline{\complement D}, \overline{\complement K}) \end{cases} \tag{10.19}$$

Furthermore, condition (10.19)(ii), p. 412 is equivalent to

$$\text{Int}(D) = \text{Inv}_{\mathcal{S}}(\text{Int}(K), \text{Int}(D)) \text{ is invariant in Int}(K) \text{ outside Int}(D).$$

Proof. Let us consider the invariance kernel $D := \text{Inv}_{\mathcal{S}}(K, C)$. By Theorem 10.2.7, p. 382, it is the unique subset between C and K such that $D = \text{Inv}_{\mathcal{S}}(D, C)$ and $D = \text{Inv}_{\mathcal{S}}(K, D)$. Thanks to Lemma 2.12.2, the latter condition is equivalent to

$$\complement\text{Inv}_{\mathcal{S}}(K, D) = \text{Capt}_{\mathcal{S}}(\complement D, \complement K)$$

Since $\mathcal{S}$ is upper semicompact and since $\complement C \backslash \complement K = K \backslash C$ is a repeller, we deduce from Theorem 10.3.10 that

$$\overline{\complement D} = \overline{\text{Capt}_{\mathcal{S}}(\complement D, \complement K)} \subset \text{Capt}_{\mathcal{S}}(\overline{\complement D}, \overline{\complement K}) \subset \overline{\complement D}$$

and thus, that $\complement \overset{\circ}{D} = \text{Capt}_{\mathcal{S}}(\complement \overset{\circ}{D}, \complement \overset{\circ}{K})$. By Lemma 2.12.2, this amounts to saying that $\text{Int}(D) = \text{Inv}_{\mathcal{S}}(\text{Int}(K), \text{Int}(D))$. □

Lemma 10.6.3 ***[Complement of a Separated Subset]*** *Let us assume that the evolutionary system $\mathcal{S}$ is upper semicompact and that a closed subset $D \subset K$ is separated from K. Then $\text{Int}(K \setminus D) \backslash \text{Int}(D)$ is locally viable under $\mathcal{S}$. In particular, if $C \subset K$ is closed, $\text{Int}(K) \backslash \text{Int}(\text{Inv}_{\mathcal{S}}(K, C))$ is locally viable.*

Proof. Let $x \in \text{Int}(K) \backslash \text{Int}(D)$ be given and $x_n \in \text{Int}(K) \backslash D$ converge to x. Since $D = \text{Inv}_{\mathcal{S}}(K, D)$ is separated by assumption, for any n, there exists $x_n(\cdot) \in \mathcal{S}(x_n)$ such that

$$T_n := \varpi_{\partial K}(x_n(\cdot)) \leq \tau_K(x_n(\cdot)) \leq \varpi_D(x_n(\cdot))$$

because $x_n \in K \backslash D$ and $\varpi_{\partial K}(x_n(\cdot)) \leq \tau_K(x_n(\cdot)) < +\infty$. Therefore, for any $t < \varpi_{\partial K}(x_n(\cdot))$, $x_n(t) \in \text{Int}(K) \setminus D$.

Since $\mathcal{S}$ is upper semicompact, a subsequence (again denoted by) $x_n(\cdot)$ converges to some $x(\cdot) \in \mathcal{S}(x)$. Since the functional $\varpi_{\partial K}$ is lower semicontinuous, we know that for any $t < \varpi_{\partial K}(x(\cdot))$, we have $t < \varpi_{\partial K}(x_n(\cdot))$ for n large enough. Consequently, $x_n(t) \in \complement D$, and, passing to the limit, we infer that for any $t < \varpi_{\partial K}(x(\cdot))$, $x(t) \in \overline{\complement D}$. This solution is thus locally viable in $\text{Int}(K) \backslash \text{Int}(D)$. □

The boundary of the invariance kernel is locally viable:

Theorem 10.6.4 ***[Local Viability of the Boundary of an Invariance Kernel]*** *If $C \subset K$ and K are closed and if S is upper semicompact, then, for every $x \in (\overset{\circ}{\partial}(\mathrm{Inv}_{\mathcal{S}}(K,C)) \cap \mathrm{Int}(K)) \setminus C$, there exists at least one solution $x(\cdot) \in \mathcal{S}(x)$ locally viable in*

$$(\overset{\circ}{\partial}(\mathrm{Inv}_{\mathcal{S}}(K,C)) \cap \mathrm{Int}(K)) \setminus C$$

Proof. Let x belong to $\overset{\circ}{\partial}\,\mathrm{Inv}_{\mathcal{S}}(K,C) \cap \mathrm{Int}(K \setminus C)$. Lemma 10.6.3, p. 412 states there exists an evolution $x(\cdot)$ viable in $\mathrm{Int}(K) \setminus (\mathrm{Inv}_{\mathcal{S}}(K,C))$ because the invariance kernel is separated from K. Since x belongs to the invariance kernel, it is viable in $\mathrm{Inv}_{\mathcal{S}}(K,C)$ until it reaches the target C, and thus viable in $\overset{\circ}{\partial}\,\mathrm{Inv}_{\mathcal{S}}(K,C)$ as long as it is viable in the interior of $K \setminus C$. □

10.6.2 Characterization of Connection Basins

The *connection basin* $\mathrm{Conn}_{\mathcal{S}}(K,(B,C))$ of K between B and C (see Definition 8.5.1, p. 291) can be written

$$\mathrm{Conn}_{\mathcal{S}}(K,(B,C)) = \mathrm{Det}_{\mathcal{S}}(K,B) \cap \mathrm{Capt}_{\mathcal{S}}(K,C) = \mathrm{Capt}_{\overleftarrow{\mathcal{S}}}(K,B) \cap \mathrm{Capt}_{\mathcal{S}}(K,C)$$

because $\mathrm{Det}_{\mathcal{S}}(K,B) := \mathrm{Capt}_{\overleftarrow{\mathcal{S}}}(K,B)$ thanks to Lemma 8.4.5, p. 287.

We begin by proving a statement analogous to Theorem 10.2.5, p. 379 for viability kernels:

Theorem 10.6.5 ***[Characterization of Connection Basins]*** *Let $\mathcal{S} : X \rightsquigarrow \mathcal{C}(-\infty,\infty;X)$ be an evolutionary system, $K \subset X$ be a environment, and $B \subset K$ be a source and $C \subset K$ be a target. The connection basin $\mathrm{Conn}_{\mathcal{S}}(K,(B,C))$ is the intersection of the detection and capture basin*

$$\mathrm{Conn}_{\mathcal{S}}(K,(B,C)) = \mathrm{Det}_{\mathcal{S}}(K,B) \cap \mathrm{Capt}_{\mathcal{S}}(K,C)$$

The connection basin is the ***largest*** *subset $D \subset K$ of K that is connecting B to C* viable *in D, i.e., the* ***largest*** *fixed point of the map $D \mapsto \mathrm{Conn}_{\mathcal{S}}(D,(B \cap D, C \cap D))$ contained in K.*

Furthermore, all evolutions connecting B to C viable in K are actually viable in $\mathrm{Conn}_{\mathcal{S}}(D,(B \cap D, C \cap D))$.

Proof. Let us set $\overline{D} := \text{Conn}_{\mathcal{S}}(K, (B, C))$.

If $\overline{D} = \emptyset$, and since $\emptyset = \text{Conn}_{\mathcal{S}}(\emptyset, (B \cap \emptyset, C \cap \emptyset))$, the empty set is a fixed point of $D \mapsto \text{Conn}_{\mathcal{S}}(D, (B \cap D, C \cap D))$.

Otherwise, we shall prove that

$$\overline{D} \subset \text{Det}_{\mathcal{S}}(\overline{D}, \overline{D} \cap B) \cap \text{Capt}_{\mathcal{S}}(\overline{D}, \overline{D} \cap C)$$

and thus, since $\overline{D} \subset \text{Det}_{\mathcal{S}}(\overline{D}, \overline{D} \cap B) \cap \text{Capt}_{\mathcal{S}}(\overline{D}, \overline{D} \cap C) \subset \text{Conn}_{\mathcal{S}}(K, (B, C)) =: \overline{D}$, that $\overline{D}$ is a fixed point of the map $D \mapsto \text{Conn}_{\mathcal{S}}(D, (B \cap D, C \cap D))$.

Indeed, let x belong to the connection basin $\overline{D}$. By Definition 8.5.1, p. 291, there exist an evolution $x(\cdot) \in \mathcal{S}(x)$ passing through x and times $\overleftarrow{T} \geq 0$ and $\overrightarrow{T} \geq 0$ such that

$$\forall t \in [-\overleftarrow{T}, +\overrightarrow{T}], \;\; x(t) \in K, \;\; x(\overleftarrow{T}) \in B \text{ and } x(\overrightarrow{T}) \in C$$

Now, let us consider any such evolution $x(\cdot) \in \mathcal{S}(x)$ connecting B to C and viable in K and prove that it is viable in $\text{Conn}_{\mathcal{S}}(\overline{D}, (B \cap \overline{D}, C \cap \overline{D}))$.

Let us consider the evolution $y(\cdot) := (\kappa(\overleftarrow{T})x(\cdot))(\cdot) \in \mathcal{S}(x(\overleftarrow{T}))$ defined by $y(t) := x(t - \overleftarrow{T})$, viable in K until it reaches the target C in finite time $\overleftarrow{T} + \overrightarrow{T}$ at $y(\overleftarrow{T} + \overrightarrow{T}) = x(\overrightarrow{T}) \in C$. This implies that $x(\overleftarrow{T}) \in \text{Capt}_{\mathcal{S}}(K, C)$. Since all evolutions capturing C viable in K are actually viable in $\text{Capt}_{\mathcal{S}}(K, C)$ by Lemma 10.2.4, p. 378, this implies that $y(\cdot)$ is viable in $\text{Capt}_{\mathcal{S}}(K, C)$ on the interval $[0, \overleftarrow{T} + \overrightarrow{T}]$. Hence the evolution $x(\cdot) = (\kappa(-\overleftarrow{T})y(\cdot))(\cdot) \in \mathcal{S}(x)$ is viable in $\text{Capt}_{\mathcal{S}}(K, C)$ on the interval $[-\overleftarrow{T}, +\overrightarrow{T}]$. We prove in the same way that the evolution $x(\cdot)$ is viable in $\text{Det}_{\mathcal{S}}(K, B)$ on the interval $[-\overleftarrow{T}, +\overrightarrow{T}]$.

Therefore, this evolution $x(\cdot)$ is connecting B to C in the connection basin $\text{Conn}_{\mathcal{S}}(\overline{D}, (B \cap \overline{D}, C \cap \overline{D}))$ itself. Therefore, we deduce that $\overline{D}$ is a fixed point $\overline{D} = \text{Conn}_{\mathcal{S}}(\overline{D}, (B \cap \overline{D}, C \cap \overline{D}))$ and the largest one, obviously. □

Proposition 10.6.6 ***[Relative Bilateral Invariance of Connection Basins]*** *The connection basin* $\text{Conn}_{\mathcal{S}}(K, (B, C))$ *between a source* B *and a target* C *viable in the environment* K *is both forward and backward invariant relatively to* K.

Proof. We have to prove that for every $x \in \text{Conn}_{\mathcal{S}}(K, (B, C))$, every evolution $x(\cdot) \in \mathcal{S}(x)$ connecting B to C, viable in K on some time interval $[S, T]$, is actually viable in $\text{Conn}_{\mathcal{S}}(K, (B, C))$ on the same interval.

Since $x(S)$ belongs to $\text{Det}_{\mathcal{S}}(K, B)$, there exist an element $b \in B$, a time $T_b \geq 0$ and an evolution $z_b(\cdot) \in \mathcal{S}(b)$ viable in K until it reaches $x(S) = z_b(T_b)$ at time T_b. Since $x(T)$ belongs to $\text{Capt}_{\mathcal{S}}(K, C)$, there exist $T_c \geq 0$, an element

$c \in C$ and an evolution $z_c(\cdot) \in \mathcal{S}(x(T))$ such that $z_c(T_c) = c \in C$ and, for all $t \in [0, S]$, $z_c(t) \in K$. We associate with these evolutions their concatenation $y(\cdot) \in \mathcal{S}(b)$ defined by

$$y(t) := \begin{cases} z_b(t) & \text{if } t \in [0, T_b] \\ x(t + S - T_b) & \text{if } t \in [T_b, T_b + T - S] \\ z_c(t - T_b + S - T) & \text{if } t \in [T_b + T - S, T_b + T - S + T_c] \end{cases}$$

starting at $b \in B$ is viable in K until it reaches C at time $T_b + S + T + T_c$. This means that b belongs to $\text{Conn}_{\mathcal{S}}(K, (B, C))$ and this implies that for every $t \in [0, T_b+S+T+T_c]$, $y(t)$ belongs to the connection basin $\text{Conn}_{\mathcal{S}}(K, (B, C))$. This is in particular the case when $t \in [S, T]$: then $x(t) = y(t + T_b - S)$ belongs to the capture basin. Therefore, the evolution $x(\cdot)$ is viable in $\text{Conn}_{\mathcal{S}}(K, (B, C))$ on the same interval $[S, T]$. □

Theorem 10.6.7 ***[Characterization of Bilateral Relative Invariance]*** *A subset $D \subset K$ is bilaterally invariant relatively to K if and only if* $\text{Conn}_{\mathcal{S}}(K, (D, D)) = D$.

Proof. First, Proposition 10.6.6, p. 414 implies that whenever $\text{Conn}_{\mathcal{S}}(K, (D, D)) = D$, D is bilaterally invariant relatively to K.

Conversely, assume that D is bilaterally invariant relatively to K and we shall derive a contradiction by assuming that there exists $x \in \text{Conn}_{\mathcal{S}}(K, (D, D)) \backslash D$. Indeed, there would exist an evolution $x(\cdot) \in \mathcal{S}(x)$ through x, times $T_b \geq 0$ and $T_c \geq 0$ and elements $b \in D$ and $c \in D$ such that $x(-T_b) = b$, $x(T_c) = c$ and viable in K on the interval $[-T_b, +T_c]$. Since D is bilaterally viable and since $x(\cdot)$ is bilaterally viable in K, it is bilaterally viable in D by assumption. Therefore, for all $t \in [-T_b, +T_c]$, $x(t)$ belongs to D, and in particular, for $t = 0$: then $x(0) = x$ belongs to D, the contradiction we were looking for. □

Theorem 10.6.8 ***[Characterization of Connection Basins as Unique Bilateral Fixed Point]*** *Let $\mathcal{S} : X \rightsquigarrow \mathcal{C}(0, +\infty; X)$ be an evolutionary system, $K \subset X$ be a environment and $C \subset K$ be a target. The connection basin* $\text{Conn}_{\mathcal{S}}(K, (D, D))$ *between subset C and itself is the* unique *bilateral fixed point between C and K of the map $(L, D) \mapsto$* $\text{Conn}_{\mathcal{S}}(L, (D, D))$ *in the sense that*

$$D = \text{Conn}_{\mathcal{S}}(D, (C, C)) = \text{Conn}_{\mathcal{S}}(K, (D, D))$$

Proof. Let us consider the map $(K,C) \mapsto \mathcal{A}(K,C) := \mathrm{Conn}_{\mathcal{S}}(K,(C,C))$. It satisfies properties (10.5), p. 381:

$$\begin{cases} (i) \;\; C \subset \mathcal{A}(K,C) \subset K \\ (ii) \;\; (K,C) \text{ is increasing} \end{cases}$$

Theorem 10.6.5, p. 413 states that $\mathcal{A}(K,C) := \mathrm{Conn}_{\mathcal{S}}(K,(C,C))$ is a fixed point of $L \mapsto \mathcal{A}(L,C)$ and Theorem 10.6.7, p. 415 that $\mathcal{A}(K,C)$ is a fixed point of $D \mapsto \mathcal{A}(K,D)$. Then $\mathcal{A}(K,C)$ is the unique bilateral fixed point of the map D between C and K of the map $\mathcal{A}$: $D = \mathcal{A}(D,C) = \mathcal{A}(K,D)$ thanks to the Uniqueness Lemma 10.2.6, p. 381. □

10.7 Stability of Viability and Invariance Kernels

In this section we study conditions under which kernels and basins of limit of a sequence of environments and/or of targets is the limit of these kernels and basins, and apply these results to the existence of viability envelopes in Sect. 10.7.2, p. 420.

10.7.1 Kernels and Basins of Limits of Environments

Let us consider a sequence of environments $K_n \subset X$, of targets $C_n \subset K_n$, and of viability kernels $\mathrm{Viab}_{\mathcal{S}}(K_n, C_n)$ of K_n with targets C_n under a given evolutionary system $\mathcal{S}$.

A natural and important question arises whether we can "take the limit" and *compare the limit of the viability kernels and the viability kernels of the limits.*

Answers to such questions require first an adequate concept of limit. Here, dealing with subsets, the natural concept of limit is the one of the Painlevé–Kuratowski upper limit of subsets. We recall the Definition 18.4.1, p. 728:

Definition 10.7.1 *[Upper Limit of Sets] Let $(K_n)_{n\in\mathbf{N}}$ be a sequence of subsets of a metric space E. We say that the subset*

$$\mathrm{Limsup}_{n\to\infty} K_n := \left\{ x \in E \mid \liminf_{n\to\infty} d(x, K_n) = 0 \right\}$$

is the upper limit *of the sequence K_n.*

We would like to derive formulas of the type

$$\text{Limsup}_{n\to+\infty}\text{Viab}_{\mathcal{S}}(K_n, C_n) \subset \text{Viab}_{\mathcal{S}}(\text{Limsup}_{n\to+\infty}K_n, \text{Limsup}_{n\to+\infty}C_n)$$

and analogous formulas for invariance kernels.

10.7.1.1 Upper Limit of Subsets of Viable Evolutions

It is worth recalling that the viability kernel

$$\text{Viab}_{\mathcal{S}}(K, C) = \mathcal{S}^{-1}(\mathcal{V}(K, C))$$

is the inverse image of the subset $\mathcal{V}(K, C) \subset \mathcal{C}(0, +\infty; X)$ of evolutions *viable in* K *outside* C defined by (2.5):

$$\begin{cases} \mathcal{V}(K, C) := \{x(\cdot) \text{ such that } \forall t \geq 0, \;\; x(t) \in K \\ \text{or } \exists\, T \geq 0 \text{ such that } x(T) \in C \;\&\; \forall t \in [0, T], \; x(t) \; \in \; K\} \end{cases}$$

and that the invariance kernel

$$\text{Inv}_{\mathcal{S}}(K, C) = \mathcal{S}^{\ominus 1}(\mathcal{V}(K, C))$$

is the core of the subset $\mathcal{V}(K, C) \subset \mathcal{C}(0, +\infty; X)$.

Hence, we begin by studying the upper limits of subsets $\mathcal{V}(K_n, C_n)$:

Lemma 10.7.2 *[Upper Limit of Subsets of Viable Evolutions] For any sequence of environments $K_n \subset X$ and any target $C_n \subset K_n$,*

$$\text{Limsup}_{n\to+\infty}\mathcal{V}(K_n, C_n) \subset \mathcal{V}(\text{Limsup}_{n\to+\infty}K_n, \text{Limsup}_{n\to+\infty}C_n)$$

Proof. The proof is a slight generalization of the proof of Lemma 10.3.9, p. 388. Let us consider a sequence of evolutions $x_n(\cdot) \in \mathcal{V}(K_n, C_n)$ converging to some evolution $x(\cdot)$. We have to prove that $x(\cdot)$ belongs to $\mathcal{V}(\text{Limsup}_{n\to+\infty}K_n, \text{Limsup}_{n\to+\infty}C_n)$, i.e., that is viable in $\text{Limsup}_{n\to+\infty}K_n$ forever or until it reaches the target $\text{Limsup}_{n\to+\infty}C_n$ in finite time.

Indeed:

1. either for any $T > 0$ and any $N > 0$, there exist $n \geq N$, $t_n \geq T$ and an evolution $x_n(\cdot)$ for which $x_n(t) \in K_n$ for every $t \in [0, t_n]$,
2. Or there exist $T > 0$ and $N > 0$ such that for any $n \geq N$ and any evolution $x_n(\cdot)$, there exists $t_n \leq T$ such that $x_n(t_n) \notin K_n$.

In the first case, we deduce that for any $T > 0$, $x(T) \in \text{Limsup}_{n\to+\infty}K_n$, so that the limit $x(\cdot)$ is viable in $\text{Limsup}_{n\to+\infty}K_n$ forever. In the second case, all the solutions $x_n(\cdot)$ leave K_n before T. This is impossible if evolutions $x_n(\cdot)$ are viable in K_n forever. Therefore, since $x_n(\cdot) \in \mathcal{V}(K_n, C_n)$, they have to reach C_n before leaving K_n: there exists $s_n \leq T$ such that

$$x_n(s_n) \in C_n \;\&\; \forall t \in [0, s_n],\; x_n(t) \in K_n$$

Then a subsequence $s_{n'}$ converges to some $S \in [0,T]$. Therefore, for any $s < S$, then $s < s_{n'}$ for n' large enough, so that $x_{n'}(s) \in K_n$. By taking the limit, we infer that for every $s < S$, $x(s) \in \text{Limsup}_{n\to+\infty} K_n$. Furthermore, since $x_n(\cdot)$ converges to $x(\cdot)$ uniformly on the compact interval $[0,T]$, then $x_n(s_n) \in C_n$ converges to $x(S)$, which therefore belongs to $\text{Limsup}_{n\to+\infty} C_n$.

This shows that the limit $x(\cdot)$ belongs to $\mathcal{V}(\text{Limsup}_{n\to+\infty} K_n, \text{Limsup}_{n\to+\infty} C_n)$. □

10.7.1.2 Upper Limits of Inverse Images and Cores

Stability problems amount to study the upper limits of inverse images and cores of subsets $\mathcal{H}_n \subset \mathcal{C}(0,+\infty;X)$ of evolutions, such as the subsets $\mathcal{V}(K_n, C_n)$ defined by (2.5), p. 49.

Theorem 10.7.3 *[Upper Limit of Inverse Images] Let $\mathcal{S} : X \rightsquigarrow \mathcal{C}(0,+\infty;X)$ be an upper semicompact evolutionary system. Then for any sequence of subsets $\mathcal{H}_n \subset \mathcal{C}(0,+\infty;X)$,*

$$\text{Limsup}_{n\to+\infty} \mathcal{S}^{-1}(\mathcal{H}_n) \subset \mathcal{S}^{-1}(\text{Limsup}_{n\to+\infty} \mathcal{H}_n)$$

Proof. Let $x \in \text{Limsup}_{n\to+\infty} \mathcal{S}^{-1}(\mathcal{H}_n)$ be the limit of a sequence of elements $x_n \in \mathcal{S}^{-1}(\mathcal{H}_n)$. Hence there exist evolutions $x_n(\cdot) \in \mathcal{S}(x_n) \in \mathcal{H}_n$. Since $\mathcal{S}$ is upper semicompact, there exists a subsequence of evolutions $x_{n'}(\cdot) \in \mathcal{S}(x_{n'})$ starting at $x_{n'}$ and converging to some $x(\cdot) \in \mathcal{S}(x)$. It also belongs to the upper limit $\text{Limsup}_{n\to+\infty} \mathcal{H}_n$ of the subsets $\mathcal{H}_n$, so that $x \in \mathcal{S}^{-1}(\text{Limsup}_{n\to+\infty} \mathcal{H}_n)$. □

For cores, we obtain

Theorem 10.7.4 *[Upper Limit of Cores] Let $\mathcal{S} : X \rightsquigarrow \mathcal{C}(0,+\infty;X)$ be a lower semicontinous evolutionary system. Then for any sequence of subsets $\mathcal{H}_n \subset \mathcal{C}(0,+\infty;X)$,*

$$\text{Limsup}_{n\to+\infty} \mathcal{S}^{\ominus 1}(\mathcal{H}_n) \subset \mathcal{S}^{\ominus 1}(\text{Limsup}_{n\to+\infty} \mathcal{H}_n)$$

Proof. Let us consider a sequence of subsets $\mathcal{H}_n \subset \mathcal{C}(0,+\infty;X)$ and a sequence of elements $x_n \in \mathcal{S}^{\ominus 1}(\mathcal{H}_n)$ converging to some $x \in \text{Limsup}_{n\to+\infty}\mathcal{S}^{\ominus 1}(\mathcal{H}_n)$. We have to prove that any $x(\cdot) \in \mathcal{S}(x)$ belongs to $\text{Limsup}_{n\to+\infty}\mathcal{H}_n$. Indeed, since $\mathcal{S}$ is lower semicontinuous, there exists a sequence of elements $x_n(\cdot) \in \mathcal{S}(x_n) \subset \mathcal{H}_n$ converging to $x(\cdot)$. Therefore the evolution $x(\cdot)$ belongs to the upper limit $\text{Limsup}_{n\to+\infty}\mathcal{H}_n$ of the subsets $\mathcal{H}_n$. Since the evolution $x(\cdot)$ was chosen arbitrarily in $\mathcal{S}(x)$, we infer that $x \in \mathcal{S}^{\ominus 1}(\text{Limsup}_{n\to+\infty}\mathcal{H}_n)$. □

10.7.1.3 Upper Limits of Viability and Invariance Kernels

Theorem 10.7.3 and Lemma 10.7.2 imply

Theorem 10.7.5 ***[Upper Limit of Viability Kernels]*** *Let $\mathcal{S} : X \rightsquigarrow \mathcal{C}(0,+\infty;X)$ be an upper semicompact evolutionary system. Then for any sequence of environments $K_n \subset X$ and of targets $C_n \subset K_n$,*

$$\text{Limsup}_{n\to+\infty}\text{Viab}_{\mathcal{S}}(K_n, C_n) \subset \text{Viab}_{\mathcal{S}}(\text{Limsup}_{n\to+\infty}K_n, \text{Limsup}_{n\to+\infty}C_n)$$

For capture basins, we obtain another property:

Lemma 10.7.6 ***[Upper Limit of Capture Basins]*** *If the set-valued map $\overleftarrow{\mathcal{S}}$ is lower semicontinuous and if K is backward invariant, then for any closed subset $C \subset K$,*

$$\text{Capt}_{\mathcal{S}}(\text{Limsup}_{n\to+\infty}K_n, \text{Limsup}_{n\to+\infty}C_n) \subset \text{Limsup}_{n\to+\infty}\text{Capt}_{\mathcal{S}}(K_n, C_n) \tag{10.20}$$

Proof. Let us take $x \in \text{Capt}_{\mathcal{S}}(\text{Limsup}_{n\to+\infty}K_n, \text{Limsup}_{n\to+\infty}C_n)$ and an evolution $x(\cdot) \in \mathcal{S}(x)$ viable in $\text{Limsup}_{n\to+\infty}K_n$ until it reaches the target $\text{Limsup}_{n\to+\infty}C_n$ at time $T < +\infty$ at $c := x(T) \in \text{Limsup}_{n\to+\infty}C_n$. Hence the function $t \mapsto y(t) := x(T-t)$ is an evolution $y(\cdot) \in \overleftarrow{\mathcal{S}}(c)$. Let us consider a sequence of elements $c_n \in C_n$ converging to c. Since $\overleftarrow{\mathcal{S}}$ is lower semicontinuous, there exist evolutions $y_n(\cdot) \in \overleftarrow{\mathcal{S}}(c_n)$ converging uniformly over compact intervals to $y(\cdot)$. These evolutions $y_n(\cdot)$ are viable in K_n, since K_n is assumed to be backward invariant, so that $x_n(0)$ belongs to $\text{Capt}_{\mathcal{S}}(K_n, C_n)$. Therefore $x_n(0) := y_n(T)$ converges to $x := x(0)$. □

Putting together these results, we obtain the following useful theorem on the stability of capture basins:

Theorem 10.7.7 ***[Stability Properties of Capture Basins]*** *Let us consider a sequence of closed subsets* C_n *satisfying* $\text{Viab}_{\mathcal{S}}(K) \subset C_n \subset K$ *and*

$$\text{Lim}_{n \to +\infty} C_n := \text{Limsup}_{n \to +\infty} C_n = \text{Liminf}_{n \to +\infty} C_n$$

If the evolutionary system $\mathcal{S}$ *is upper semicompact and lower semicontinuous and if* K *is closed and backward invariant under* $\mathcal{S}$, *then*

$$\text{Lim}_{n \to +\infty} \text{Capt}_{\mathcal{S}}(K, C_n) = \text{Capt}_{\mathcal{S}}(K, \text{Lim}_{n \to +\infty} C_n) \tag{10.21}$$

For invariance kernels, we deduce from Theorem 10.7.4 and Lemma 10.7.2 the stability theorem:

Lemma 10.7.8 ***[Upper Limit of Invariance Kernels]*** *Let* $\mathcal{S} : X \rightsquigarrow \mathcal{C}(0, +\infty; X)$ *be a lower semicontinuous evolutionary system. Then for any sequence of environments* $K_n \subset X$ *and any target* $C_n \subset K_n$,

$$\text{Limsup}_{n \to +\infty} \text{Inv}_{\mathcal{S}}(K_n, C_n) \subset \text{Inv}_{\mathcal{S}}(\text{Limsup}_{n \to +\infty} K_n, \text{Limsup}_{n \to +\infty} C_n)$$

10.7.2 Invariance and Viability Envelopes

Since the intersection of sets that are invariant under an evolutionary system is still invariant, it is natural to introduce the smallest invariant subset containing a given set:

Definition 10.7.9 ***[Invariance Envelope]*** *We shall say that the smallest invariant subset containing C is the* invariance envelope of C *and that the smallest subset of K containing C invariant outside C is the* invariance envelope of K outside C.

However, an intersection of subsets viable under an evolutionary system is not necessarily viable. Nevertheless, we may introduce the concept of minimal subsets viable outside a target:

Definition 10.7.10 ***[Viability Envelope]*** *Let L be any subset satisfying $C \subset L \subset \text{Viab}_{\mathcal{S}}(K, C)$. A (resp. closed)* viability envelope *of K with target C is any (resp. closed) set $L^\star \supset L$ viable outside C such that there is no strictly smaller subset $M \supset L$ viable outside C.*

We prove the existence of viability envelopes:

Proposition 10.7.11 ***[Existence of Viability Envelopes]*** *Let K be a closed subset viable under an upper semicompact evolutionary system $\mathcal{S}$. Then any closed subset $L \subset K$ is contained into a* viability envelopes of L *under $\mathcal{S}$.*

Proof. We apply Zorn's lemma for the inclusion order on the family of nonempty closed subsets viable under $\mathcal{S}$ between L and K. For that purpose, consider any decreasing family of closed subsets M_i, $i \in I$, viable under $\mathcal{S}$ and their intersection $M_\star := \bigcap_{i \in I} M_i$. It is a closed subset viable under $\mathcal{S}$ thanks to the Stability Theorem 10.7.7. Therefore every subset $L \subset K$ is contained in a minimal element for this preorder. □

When $L = \emptyset$, we have to assume that K is compact to guarantee that the intersection of any decreasing family of nonempty closed subset viable under $\mathcal{S}$ is not empty. In this case, we obtain the following

Proposition 10.7.12 ***[Non emptiness of Viability Envelopes]*** *Let K be a nonempty compact subset viable under an upper semicompact evolutionary system $\mathcal{S}$. Then nonempty minimal closed subsets M viable under $\mathcal{S}$ exist and are made of limit sets of viable evolutions. Actually, they exhibit the following property:*

$$\forall x \in M,\ \exists\, x(\cdot) \in \mathcal{S}(x) \mid x \in M = \omega(x(\cdot))$$

where, by Definition 9.3.1, p. 344, $\omega(x(\cdot)) := \bigcap_{T>0} \overline{(x([T,\infty[))}$ *is the* ω-limit set *of* $x(\cdot)$.

Proof. Let $M \subset K$ be a minimal closed subset viable under $\mathcal{S}$. We can associate with any $x \in M$ a viable evolution $x(\cdot) \in \mathcal{S}(x)$ starting at x. Hence its limit set $\omega(x(\cdot))$ is contained in M. But limit sets being closed subsets viable under $\mathcal{S}$ by Theorem 9.3.11 and M being minimal, it is equal to $\omega(x(\cdot))$, so that $x \in \omega(x(\cdot))$. □

10.8 The Hard Version of the Inertia Principle

Exit sets also play a crucial role for regulating viable evolutions with a finite number of feedbacks instead of the unique feedback, which, whenever it exists, regulates viable evolutions. However, even when its existence is guaranteed and when the Viability Kernel Algorithm allows us to compute it, it is often preferable to use available and well known feedbacks derived from a long history than computing a new one. Hence, arises the question of "quantized retroactions" using a subset of the set of all available retroactions (see Sect. 6.4, p. 207 for fixed degree open loop controls). In this section, we are investigating under which conditions a given finite subset of available feedbacks suffices to govern viable evolutions. We have to give a precise definition of the concept of "amalgams" of feedbacks for producing other ones, in the same way that a finite number of monomials generates the class of fixed degree polynomials. Once this operation which governs *concatenations* of evolutions defined, we can easily characterize a condition involving the exit set of the system under each of the finite class of systems. They govern specific evolutions satisfying the ***hard version*** of the ***inertia principle***.

33 ***[Quantized Controls.]*** *Recent important issues in control theory are known under the name of* "quantized controls"*, where, instead of finding adequate retroactions for governing evolutions satisfying such and such properties (viability, capturability, optimality, etc.), we are restricting the regulation of these evolutions by a smaller class of retroactions generated in some way by a* finite number *of feedbacks. Indeed, the* regulation map *(see Definition 2.14.3, p. 98) using the entire family of controls* $u \in U(x)$ *may be too difficult to construct.* Quantized control *combines only a finite number of retroactions to regulate viable, capturing or optimal evolutions.*

Chapter 6, p. 199 provided examples of such quantized systems where the retroactions are open loops controls made of polynomials of fixed degree m. The regulation by the *amalgam* of a finite number of given feedbacks provides another answer to the issue of quantization. Indeed, let us consider control system

$$x'(t) = f(x(t), u(t)) \text{ where } u(t) \in U(x(t))$$

We introduce a *finite family* of closed loop feedbacks $\widetilde{u}_i : x \rightsquigarrow \widetilde{u}_i(x) \in U(x)$ and $i \in I$ where I is a finite number of indices. They define a finite number of evolutionary systems $\mathcal{S}_i$ associated with differential equations

$$x'(t) = f(x(t), \widetilde{u}_i(x(t)))$$

Each of these a priori feedbacks is not especially designed to regulate viable evolutions in an arbitrary set, for instance. A compromise is obtained by "amalgamating" those closed loop feedbacks for obtaining the following class of retroactions (see Definition 2.7.2, p. 65):

Recall (Definition 18.3.12, p. 724) that the *mark* $\Xi_{[s,t]\times A} := \Xi^{\mathcal{U}}_{[s,t]\times A} : \mathbb{R} \times X \rightsquigarrow \mathcal{U}$ of a subset $[s,t] \times A$ is defined by

$$\Xi_{[s,t]\times A}(\tau, x) := \begin{cases} \mathcal{U} \text{ if } (\tau, x) \in [s,t] \times A \\ \emptyset \text{ if } (\tau, x) \notin [s,t] \times A \end{cases} \quad (10.22)$$

and plays the role of a "characteristic set-valued map of a subset". Therefore, for any $u \in \mathcal{U}$

$$u \cap \Xi_{[s,t]\times A}(\tau, x) := \begin{cases} \{u\} \text{ if } (\tau, x) \in [s,t] \times A \\ \emptyset \quad \text{ if } (\tau, x) \notin [s,t] \times A \end{cases}$$

Definition 10.8.1 *[Amalgam of Feedbacks] Let us consider a family of feedbacks $\widetilde{u}_i : X \rightsquigarrow \mathcal{U}$, a covering $X = \bigcup_{i \in I} A_i$ of X and an increasing sequence of instants t_i, $i = 0, \ldots, n$. The associated* amalgam *of these feedbacks is the retroaction*

$$\widetilde{u} := \bigcup_{i \geq 0} \widetilde{u}_i \cap \Xi_{[t_i, t_{i+1}[\times A_i}$$

defined by

$$\widetilde{u}(t, x) := \widetilde{u}_i(x) \text{ if } t \in [t_i, t_{i+1}[\text{ and } x \in A_i$$

Amalgams of feedbacks play an important role in the regulation of control and regulated systems.

Proposition 10.5.1, p. 399 characterizes environments viable outside a target C under an evolutionary system $\mathcal{S}$ if and only if $\mathrm{Exit}_{\mathcal{S}}(K) \subset C$.

What happens if a given environment is not viable under any of the evolutionary systems $\mathcal{S}_i$ of a finite family $i \in I$? Is it possible to restore the viability by letting these evolutionary system cooperate?

To say that K is not viable outside C under $\mathcal{S}_i$ means that $\mathrm{Exit}_{\mathcal{S}_i}(K) \cap \complement C \neq \emptyset$. However, even though K is not viable under each of the system $\mathcal{S}_i$, it may be possible that amalgated together, the collective condition

$$\bigcap_{i \in I} \mathrm{Exit}_{\mathcal{S}_i}(K) \subset C$$

weaker than the individual condition $\mathrm{Exit}_{\mathcal{S}}(K) \subset C$ may be enough to regulate a control system. This happens to be the case, if, for that purpose, we define the "cooperation" between evolutionary systems $\mathcal{S}_i$ by "amalgamating" them. For control systems, amalgamating feedbacks amounts to amalgamating the associated evolutionary system.

The examination of the exit sets of each of the evolutionary systems allows us to answer this important practical question by using the notion of the *amalgam* $\mathcal{S}^{\ddagger}$ of the evolutionary systems $\mathcal{S}_i$:

Definition 10.8.2 ***[Amalgam of a Family of Evolutionary Systems]*** *The amalgamated system $\mathcal{S}^{\ddagger}$ of the evolutionary systems $\mathcal{S}_i$ associates with any $x \in K$ concatenated evolutions $x(\cdot)$ associated with sequences of indices i_p, $p \in \mathbb{N}$, of times $\tau_{i_p} > 0$ and of evolutions $x_{i_p}(\cdot) \in \mathcal{S}_{i_p}(x_p)$ such that, defining*

$$t_0 = 0, \ \ t_{p+1} := t_p + \tau_{i_p}$$

the evolution $x(\cdot)$ is defined by

$$\forall \, p \geq 0, \ \forall \, t \in [t_p, t_{p+1}], \ x(t) := x_{i_p}(t - t_p) \text{ and } x_{i_p}(t_{p+1}) = x_{p+1}$$

where $x_0 := x$ and $x_p := x_{i_{p-1}}(t_p)$, $p \geq 1$.

We derive a viability criterion allowing us to check whether a target $C \subset K$ can be captured under the amalgated system:

Theorem 10.8.3 ***[Viability Under Amalgams of Evolutionary Systems]*** *Let us consider a finite set I of indices and a family of upper semicompact evolutionary systems $\mathcal{S}_i$. Assume that K is a repeller under each evolutionary system $\mathcal{S}_i$ and that*

$$\bigcap_{i \in I} \text{Exit}_{\mathcal{S}_i}(K) \subset C \qquad (10.23)$$

then K is viable outside C under the amalgam $\mathcal{S}^{\ddagger}$ of the evolutionary systems $\mathcal{S}_i$

$$\text{Viab}_{\mathcal{S}^{\ddagger}}(K, C) = K$$

Proof. For simplicity, we set $\mathcal{S}^{\sharp} := \mathcal{S}^{K^{\sharp}} \subset \mathcal{S}$ the sub-evolutionary system generating persistent evolutions (see Definition 10.4.2, p. 393).

Let us set $E_i := \text{Exit}_{\mathcal{S}_i}(K)$. Assumption $\bigcap_{i \in I} E_i \subset C$ amounts to saying that

$$K \setminus C = \bigcup_{i \in I} (K \setminus E_i)$$

Therefore, we associate with any $x \in K \setminus C$ the set $I(x) \subset I$ of indices i such that

$$\tau_K^{\mathcal{S}_i^{\sharp}}(x) := \max_{k \in I} \tau_K^{\mathcal{S}_k^{\sharp}}(x)$$

achieving the maximum of the exit times for each evolutionary system $\mathcal{S}_k^{\sharp}$. For each index $i \in I$, and $e_i \in E_i$, one can observe that $\max_{j \in I} \tau_K^{\mathcal{S}_j^{\sharp}}(e_i) = \max_{j \in I(e_i)} \tau_K^{\mathcal{S}_j^{\sharp}}(e_i)$. We next define the smallest of the largest exit times of states ranging the exit sets of each evolutionary system $\mathcal{S}_j^{\sharp}$:

$$\overline{\tau} := \min_{i \in I} \sup_{e_i \in \text{Exit}_{\mathcal{S}_i}(K)} \sup_{j \in I(e_i)} \tau_K^{\mathcal{S}_j^{\sharp}}(e_i)$$

Since the set I of indices is finite, assumption $\bigcap_{i \in I} E_i \subset C$ implies that $0 < \overline{\tau} < +\infty$.

This being said, we can build a concatenated evolution $x(\cdot)$ made of an increasing sequence of times t_p and of "pieces" of *persistent evolutions* $x_{i_p}^{\sharp}(\cdot) \in \mathcal{S}_{i_p}(x_{i_p})$ defined by

$$\forall p \geq 0, \ \forall t \in [t_p, t_{p+1}], \ x(t) := x_{i_p}^{\sharp}(t - t_p) \text{ and } x(t_{p+1}) = x_{p+1}$$

which is viable in K forever or until a finite time when it reaches C.

Indeed, we associate with any evolutionary system $\mathcal{S}_i$ and any $x_i \in K$ a *persistent evolution* $x_i^{\sharp}(\cdot) \in \mathcal{S}_i(x_i)$, its exit time $\tau_i^{\sharp} > 0$ (since we assumed that K is a repeller under each evolutionary system $\mathcal{S}_i$) and an exit state $e_i^{\sharp} \in E_i := \text{Exit}_{\mathcal{S}_i}(K)$.

To say that $\bigcap_{i \in I} E_i \subset C$ amounts to saying that

$$K \setminus C = \bigcup_{i \in I} (K \setminus E_i)$$

Therefore, starting with any initial state $x_0 \in K \setminus C$, we infer from the assumption $\bigcap_{i \in I} E_i \subset C$ that there exists $i_0 \in I$ such that $x \in K \setminus E_{i_0}$. Hence, we associate $x^\sharp_{i_0}(\cdot) \in \mathcal{S}_{i_0}(x_{i_0})$, its exit time $\tau^\sharp_{i_0} > 0$ and an exit state $e^\sharp_{i_0} \in E_{i_0}$. Setting $x_1 := e^\sharp_{i_0}$, either $x_1 \in C$, and the evolution $x(\cdot) := x^\sharp_{i_0}(\cdot)$ reaches C in finite time 0, or $x_1 \in K \setminus C$, and our assumption implies the existence of $i_1 \in I$ such that $x \in K \setminus E_{i_1}$, so that we can find a $x^\sharp_{i_1}(\cdot) \in \mathcal{S}_{i_1}(x_1)$, its exit time $\tau^\sharp_{i_1} > 0$ and an exit state $e^\sharp_{i_1} \in E_{i_1}$. And so on, knowing that $e^\sharp_{i_{p-1}} \in E_{i_{p-1}} \in K \setminus C$, we choose an index $i_p \in I$ such that $x_{i_p} \in K \setminus E_{i_p}$ and built recursively evolutions $x^\sharp_{i_p}(\cdot) \in \mathcal{S}_{i_p}(x_{i_p})$, its exit time $\tau^\sharp_{i_p} > 0$ and an exit state $e^\sharp_{i_p} \in E_{i_p}$.

We associate with this sequence of evolutions the sequence of times defined by

$$t_0 = 0, \;\; t_{p+1} \; := \; t_p + \tau^\sharp_{i_p}$$

and evolutions

$$x_p(t) := (\kappa(\tau^\sharp_{i_p}) x^\sharp_{i_p}(\cdot))(t) = x^\sharp_{i_p}(t - \tau^\sharp_{i_p}) \text{ where } t \in [t_p, t_{p+1}] \text{ and } x_p(t_{p+1}) = x_{p+1}$$

and their concatenation $x(\cdot)$ defined by

$$\forall\, p \geq 0, \;\; \forall\, t \in [t_p, t_{p+1}], \;\; x(t) \; := \; x_p(t)$$

Since the set I of indices is assumed to be finite, then $\overline{\tau} > 0$, so that the concatenated evolution is defined on $\mathbb{R}_+$ because $\sum_{p=0}^{+\infty} \tau^\sharp_{i_p} = +\infty$. Hence the concatenated evolution of persistent evolutions is viable forever or until it reaches the target C in finite time. □

We mentioned in Sect. 6.4, p. 207 the concept of the "soft" version of the inertia principle. Persistent evolutions and Theorem 10.8.3, p. 424 provide the "hard version" of this principle:

> **34** *[**The Hard Inertia Principle**] Theorem 10.8.3, p. 424 provides another answer to the* inertia principle *(see Sect. 6.4.4, p. 217) without inertia threshold:* When, where and how *change the available feedbacks (among them, constant controls) to maintain the viability of a system. Starting with a finite set of regulons, the system uses them successively as long as possible (*persistent evolutions*), up to the exit time (warning signal) and its exit set, which is its* critical zone *(see Definition 6.4.9, p. 216).*

In summary, when the viability is at stakes:

1. The *hard* version of the inertia principle requires that whenever the evolution reaches the boundary, then, and not before, the state has to switch instantaneously to a *new initial state* and a *new feedback* has to be chosen,
2. The *soft* version of the inertia principle involves an inertia threshold determining when, at the right time, the kairos, where, in the critical zone, the regulon only has to *evolve and how*.

10.9 Parameter Identification: Inverse Viability and Invariance Maps

When the differential inclusion (or parameterized system) $F(\lambda,\cdot)$, the environment $K(\lambda)$ and the target $C(\lambda)$ depend upon a parameter $\lambda \in Y$ ranging over a finite dimensional vector space Y, a typical example of *inverse problem* (see Comment 2, p. 5) is to associate with any state x the subset of parameters λ such that we know, for instance, that x belongs to the viability kernel $\mathbb{V}(\lambda) := \mathrm{Viab}_{F(\lambda,\cdot)}(K(\lambda), C(\lambda))$.

The set of such parameters λ is equal to $\mathbb{V}^{-1}(x)$, where $\mathbb{V}^{-1} : X \rightsquigarrow Y$ is the inverse of the set-valued map $\mathbb{V} : Y \mapsto X$ associating with λ the viability kernel $\mathbb{V}(\lambda) := \mathrm{Viab}_{F(\lambda,\cdot)}(K(\lambda), C(\lambda))$.

In control terminology, the search of those parameters λ such that a given x belongs to $\mathbb{V}(\lambda) := \mathrm{Viab}_{F(\lambda,\cdot)}(K(\lambda), C(\lambda))$ is called a *parameter identification problem* formulated for viability problems. This covers as many examples as problems which can be formulated in terms of kernels and basins, as the ones covered in this book. As we shall see, most of the examples covered in Chaps. 4, p. 125 and 6, p. 199 are examples of inverse viability problems.

10.9.1 Inverse Viability and Invariance

It turns out that for these types of problems, the solution can be obtained by viability techniques. Whenever we know the graph of a set-valued map, we know both this map and its inverse (see Definition 18.3.1, p. 719). The graphs of such maps associating kernels and basins with those parameters are also kernels and basins of auxiliary environments and targets under auxiliary systems. Therefore, *they inherit their properties, which are then shared by both the set-valued map and its inverse.* This simple remark is quite useful.

Let us consider the parameterized differential inclusion

$$x'(t) \in F(\lambda, x(t)) \tag{10.24}$$

when environments $K(\lambda)$ and targets $C(\lambda)$ depend upon a parameter $\lambda \in Y$ ranging over a finite dimensional vector space Y. We set $F(\lambda,\cdot) : x \rightsquigarrow F(\lambda, x)$.

The problem is to *invert* the set-valued maps

$$\mathbb{V} : \lambda \rightsquigarrow \mathrm{Viab}_{F(\lambda,\cdot)}(K(\lambda), C(\lambda)) \text{ and } \mathbb{I} : \lambda \rightsquigarrow \mathrm{Inv}_{F(\lambda,\cdot)}(K(\lambda), C(\lambda))$$

For that purpose, we shall characterize the graphs of these maps:

Proposition 10.9.1 ***[Graph of the Viability Map]*** *The graph of the map* $\mathbb{V} : \lambda \rightsquigarrow \mathrm{Viab}_{F(\lambda,\cdot)}(K(\lambda), C(\lambda))$ *is equal to the viability kernel*

$$\mathrm{Graph}(\mathbb{V}) \;=\; \mathrm{Viab}_{(10.25)}(\mathcal{K}, \mathcal{C})$$

of the graph $\mathcal{K} := \mathrm{Graph}(\lambda \rightsquigarrow K(\lambda))$ *with target* $\mathcal{C} := \mathrm{Graph}(\lambda \rightsquigarrow C(\lambda))$ *under the auxiliary system*

$$\begin{cases} (i) \;\; \lambda'(t) \;=\; 0 \\ (ii) \;\; x'(t) \;\in\; F(\lambda(t), x(t)) \end{cases} \tag{10.25}$$

Proof. The proof is easy: to say that (λ, x) belongs to the viability kernel $\mathrm{Viab}_{(10.25)}(\mathcal{K}, \mathcal{C})$ amounts to saying that there exists a solution $t \mapsto (\lambda(t), x(t))$ viable in $\mathcal{K} := \mathrm{Graph}(K(\cdot))$ of (10.25) until it possibly reaches $\mathcal{C} := \mathrm{Graph}(C(\cdot))$, i.e., since $\lambda(t) = \lambda$ and $(\lambda(\cdot), x(\cdot)) \in \mathcal{S}_{\{0\}\times F}(\lambda, x)$ such that $x(t) \in K(\lambda)$ forever or until it reaches $C(\lambda)$. This means that (λ, x) belongs to the graph of the viability map $\mathbb{V}$. □

In the same way, one can prove the analogous statement for the invariance map:

Proposition 10.9.2 ***[Graph of the Invariance Map]*** *The graph of the map* $\mathbb{I} : \lambda \rightsquigarrow \mathrm{Inv}_{F(\lambda,\cdot)}(K(\lambda), C(\lambda))$ *is equal to the invariance kernel*

$$\mathrm{Graph}(\mathbb{I}) \;=\; \mathrm{Inv}_{(10.25)}(\mathcal{K}, \mathcal{C})$$

of the graph $\mathcal{K} := \mathrm{Graph}(\lambda \rightsquigarrow K(\lambda))$ *with target* $\mathcal{C} := \mathrm{Graph}(\lambda \rightsquigarrow C(\lambda))$ *under the auxiliary system (10.25), p. 428.*

Consequently, the inverses $\mathbb{V}^{-1}$ and $\mathbb{I}^{-1}$ of the set-valued maps $\mathbb{V}$ and $\mathbb{I}$ associate with any $x \in X$ the subsets of parameters $\lambda \in Y$ such that the pairs (λ, x) belong to the viability and invariance kernels of the graph

$\mathcal{K} := \text{Graph}(\lambda \rightsquigarrow K(\lambda))$ with target $\mathcal{C} := \text{Graph}(\lambda \rightsquigarrow C(\lambda))$ under the auxiliary system (10.25) respectively.

10.9.2 Level Tubes of Extended Functions

When the parameters $\lambda \in \mathbb{R}$ are scalars, the set-valued maps $\lambda \rightsquigarrow \text{Graph}(F(\lambda,\cdot))$, $\lambda \rightsquigarrow \text{Graph}(K(\lambda))$ and $\lambda \rightsquigarrow \text{Graph}(C(\lambda))$, $\mathbb{V} : \lambda \rightsquigarrow \mathbb{V}(\lambda)$ and the viability and invariance maps $\mathbb{I} : \lambda \rightsquigarrow \mathbb{I}(\lambda)$ are tubes (see Fig. 4.3, p. 132).

We shall study the monotonicity properties of tubes:

Definition 10.9.3 *[Monotone Tubes] A tube is* increasing *(resp.* decreasing*) if whenever $\mu \leq \nu$, then $K(\mu) \subset K(\nu)$ (resp. $K(\nu) \subset K(\mu)$). A* monotone tube *is a tube which is either increasing or decreasing.*

The monotonicity properties of the tubes $\lambda \rightsquigarrow \mathbb{V}(\lambda)$ and $\lambda \rightsquigarrow \mathbb{I}(\lambda)$ depend upon the monotonicity properties of the tubes $\lambda \rightsquigarrow \text{Graph}(F(\lambda,\cdot))$, $\lambda \rightsquigarrow \text{Graph}(K(\lambda))$ and $\lambda \rightsquigarrow \text{Graph}(C(\lambda))$:

Lemma 10.9.4 *[Monotonicity of the Viability and Invariance Maps] The map $(F, K, C) \rightsquigarrow \text{Viab}_F(K, C)$ is increasing, the map $(K, C) \rightsquigarrow \text{Inv}_F(K, C)$ is increasing and the map $F \rightsquigarrow \text{Inv}_F(K, C)$ is decreasing.*

Recall, a tube is characterized by its *graph* (see Definition 18.3.1, p. 719): The graph of the tube $K : \mathbb{R} \rightsquigarrow X$ is the set of pairs (λ, x) such that x belongs to $K(\lambda)$:

$$\mathcal{K} := \text{Graph}(K) = \{(\lambda, x) \in \mathbb{R} \times X \text{ such that } x \in K(\lambda)\}$$

Monotonic tubes can be characterized by their epilevel and hypolevel functions whenever the tubes $\lambda \rightsquigarrow \mathbb{V}(\lambda)$ and $\lambda \rightsquigarrow \mathbb{I}(\lambda)$ are monotone:

We then introduce the concepts of lower and upper level sets or sections of an extended function:

Definition 10.9.5 *[Levels Sets or Sections of a Function] Let $\mathbf{v} : X \mapsto \overline{\mathbb{R}}$ be an extended function. The* lower level map $\mathbf{L}^{\leq}_{\mathbf{v}}$ *associates with any $\lambda \in \mathbb{R}$ the* λ-lower section *or* λ-lower level set

$$\mathbf{L}_{\mathbf{v}}^{\leq}(\lambda) := \{x \in K \textit{ such that } \mathbf{v}(x) \leq \lambda\}$$

We define in the same way the strictly lower, exact, upper and strictly upper level maps $\mathbf{L}_{\mathbf{v}}^{\star}$ *which associate with any* λ *the* λ*-level sets*

$$\mathbf{L}_{\mathbf{v}}^{\star}(\lambda) := \{x \in K \textit{ such that } \mathbf{v}(x) \star \lambda\}$$

where $\star$ *denotes respectively the signs* $<$, $=$, $\geq$ *and* $>$.

We next introduce the concept of level function of a tube:

Definition 10.9.6 *[Level Function of a Tube] Let us consider a tube* $K : \mathbb{R} \rightsquigarrow X$. *The* epilevel function $\Lambda_K^{\uparrow}$ *of the tube* K *is the extended function defined by*

$$\Lambda_K^{\uparrow}(x) := \inf\{\lambda \textit{ such that } x \in K(\lambda)\} = \inf_{(\lambda,x)\in \mathrm{Graph}(K)} \lambda \quad (10.26)$$

and its hypolevel function $\Lambda_K^{\downarrow}$ *is the extended function defined by*

$$\Lambda_K^{\downarrow}(x) := \sup\{\lambda \textit{ such that } x \in K(\lambda)\} = \sup_{(\lambda,x)\in \mathrm{Graph}(K)} \lambda \quad (10.27)$$

We observe that level set map $\lambda \rightsquigarrow \mathbf{L}_{\mathbf{v}}^{\star}(\lambda)$ is a tube from $\mathbb{R}$ to the vector space X. For instance, *the lower level map* $\lambda \rightsquigarrow \mathbf{L}_{\mathbf{v}}^{\leq}(\lambda)$ *is an increasing tube*:

$$\text{If } \lambda_1 \leq \lambda_2, \quad \text{then } \mathbf{L}_{\mathbf{v}}^{\leq}(\lambda_1) \subset \mathbf{L}_{\mathbf{v}}^{\leq}(\lambda_2)$$

and that the upper level map $\lambda \rightsquigarrow \mathbf{L}_{\mathbf{v}}^{\geq}(\lambda)$ is a decreasing tube. Lemma 18.6.3, p. 744 implies that *the level set map* $\mathbf{L}_{\mathbf{v}}^{\leq}$ *of a lower semicontinuous function is a closed tube.*

We observe at once that the images $K(\lambda)$ are contained in the λ-lower level sets:

$$\forall \lambda \in \mathbb{R}, \; K(\lambda) \subset \mathbf{L}_{\Lambda_K^{\uparrow}}^{\leq}(\lambda)$$

The question arises whether the converse is true: is an increasing tube the lower level map of an extended function $\Lambda_K^{\uparrow}$, called the epilevel function of the tube? This means that we can represent the images $K(\lambda)$ of the tube in the form

$$\forall \lambda, \; K(\lambda) = \left\{x \text{ such that } \Lambda_K^{\uparrow}(x) \leq \lambda\right\}$$

This property can be reformulated as $K(\lambda) = \mathbf{L}^{\leq}_{\Lambda^{\uparrow}_K}(\lambda)$, stating that the inverse of the set-valued map $x \rightsquigarrow K^{-1}(x)$ of the tube is the map $x \rightsquigarrow \Lambda^{\uparrow}_K(x) + \mathbb{R}_+$.

The answer is positive for closed monotonic tubes.

The equality between these two subsets is (almost) true for increasing tubes (a necessary condition) and really true when, furthermore, the tube is closed:

Proposition 10.9.7 ***[Inverses of Monotone Tubes and their Level Functions]*** *Let us assume that the tube K is increasing. Then it is related to its epilevel function by the relation*

$$\forall\, \lambda \in \mathbb{R},\ \ \mathbf{L}^{<}_{\Lambda^{\uparrow}_K}(\lambda) \subset K(\lambda) \subset \mathbf{L}^{\leq}_{\Lambda^{\uparrow}_K}(\lambda) \tag{10.28}$$

Furthermore, if the graph of the tube is closed, then $x \in K(\lambda)$ if and only if $\Lambda^{\uparrow}_K(x) \leq \lambda$, i.e.,

$$\forall\, \lambda \in \mathbb{R},\ \ K(\lambda) = \mathbf{L}^{\leq}_{\Lambda^{\uparrow}_K}(\lambda) =: \{x \mid \Lambda^{\uparrow}_K(x) \leq \lambda\} \tag{10.29}$$

Proof. By the very definition of the infimum, to say that $\Lambda^{\uparrow}_K(x) = \inf_{(\lambda,x)\in \mathrm{Graph}(K)} \lambda$ amounts to saying that for any $\lambda > \Lambda^{\uparrow}_K(x)$, there exists $(\mu, x) \in \mathrm{Graph}(K)$ such that $\mu \leq \lambda$. To say that $x \in \mathbf{L}^{<}_{\Lambda^{\uparrow}_K}(\lambda)$ means $\lambda > \Lambda^{\uparrow}_K(x)$. Hence there exists $(\mu, x) \in \mathrm{Graph}(K)$, and there exist $\mu \leq \lambda$ and $x \in K(\mu)$. Since the tube K is decreasing, we deduce that $x \in K(\lambda)$. The first inclusion is thus proved, the other one being always obviously true.

If the graph of K is closed, then letting $\lambda > \Lambda^{\uparrow}_K(x)$ converge to $\Lambda^{\uparrow}_K(x)$ and knowing that (λ, x) belongs to $\mathrm{Graph}(K)$, we deduce $(\Lambda^{\uparrow}_K(x), x)$ belongs to the graph of K, and thus, that $x \in K(\Lambda^{\uparrow}_K(x))$. □

The counterpart statement holds true for decreasing tubes and their hypolevel functions: If a tube is decreasing, then

$$\forall\, \lambda \in \mathbb{R},\ \ \mathbf{L}^{>}_{\lambda^{\downarrow}_K}(\lambda) \subset K(\lambda) \subset \mathbf{L}^{\geq}_{\Lambda^{\downarrow}_K}(\lambda) \tag{10.30}$$

and if it is closed

$$\forall\, \lambda \in \mathbb{R},\ \ K(\lambda) = \mathbf{L}^{\geq}_{\Lambda^{\downarrow}_K}(\lambda) \tag{10.31}$$

In this case, when the graph of the tube is closed, (10.29) and (10.31) can be written in terms of epigraphs and hypographs (see Definition 4.2.2, p. 131) in the form

$$\text{Graph}(K^{-1}) = \mathcal{E}p(\Lambda_K^{\uparrow}) \tag{10.32}$$

and

$$\text{Graph}(K^{-1}) = \mathcal{H}yp(\Lambda_K^{\downarrow}) \tag{10.33}$$

respectively.

In the scalar case, Theorem 10.9.7, p. 431 implies that these tubes are characterized by their epilevel or hypolevel functions (see Definition 10.9.6, p. 430). For instance, the epilevel function of the viability tube is defined by $\Lambda_{\mathbb{V}}^{\uparrow}(x) := \inf_{(\lambda,x)\in \text{Graph}(\mathbb{V})} \lambda$ whenever this map is increasing. In this case, if the graph of the viability tube is closed,

$$\mathbb{V}(\lambda) = \left\{ x \text{ such that } \Lambda_{\mathbb{V}}^{\uparrow}(x) \leq \lambda \right\}$$

If the tube is decreasing, the hypolevel function defined by $\Lambda_{\mathbb{V}}^{\downarrow}(x) := \sup_{(\lambda,x)\in \text{Graph}(\mathbb{V})} \lambda$ characterizes the tube in the sense that

$$\mathbb{V}(\lambda) = \left\{ x \text{ such that } \Lambda_{\mathbb{V}}^{\downarrow}(x) \geq \lambda \right\}$$

whenever the tube is closed.

For instance, for the viability map, we derive the following statement from Proposition 10.9.1, p. 428:

Proposition 10.9.8 ***[Level Functions of the Viability Tube]*** *Let us assume that the tubes $\lambda \mapsto \text{Graph}(F(\lambda,\cdot))$, $\lambda \mapsto K(\lambda)$ and $\lambda \mapsto C(\lambda)$ are increasing. Then the tube $\mathbb{V}$ is characterized by its epilevel function*

$$\Lambda_{\mathbb{V}}^{\uparrow}(x) := \inf_{(\lambda,x)\in \text{Graph}(\mathbb{V})} \lambda := \inf_{(\lambda,x)\in \text{Viab}_{(10.25)}(\mathcal{K},\mathcal{C})} \lambda \tag{10.34}$$

If $\lambda \mapsto \text{Graph}(F(\lambda,\cdot))$, $\lambda \mapsto K(\lambda)$ and $\lambda \mapsto C(\lambda)$ are increasing, the tube $\mathbb{V}$ is characterized by its hypolevel function

$$\Lambda_{\mathbb{V}}^{\downarrow}(x) := \sup_{(\lambda,x)\in \text{Graph}(\mathbb{V})} \lambda = \sup_{(\lambda,x)\in \text{Viab}_{(10.25)}(\mathcal{K},\mathcal{C})} \lambda \tag{10.35}$$

The counterpart statement holds true for the invariance tubes.

10.10 Stochastic and Tychastic Viability

The invariance kernel is an example of the core $\mathcal{S}^{\ominus 1}(\mathcal{H})$ of a subset $\mathcal{H} \subset \mathcal{C}(0,\infty;\mathbb{R}^d)$ for $\mathcal{H} = \mathcal{K}_{(K,C)}$ being the set of evolutions viable in K reaching the target C in finite time.

Let us consider *random events* $\omega \in \Omega$, where $(\Omega, \mathcal{F}, \mathbb{P})$ is a probability space, instead of tyches $v(\cdot)$ ranging over the values $V(x(\cdot))$ of a tychastic map V (see (2.23), p. 89).

A stochastic system is a specific parameterized evolutionary system described by a map $\mathbb{X} : (x,\omega) \in \mathbb{R}^d \times \Omega \mapsto \mathbb{X}(x,\omega) \in \mathcal{C}(0,\infty;\mathbb{R}^d)$ (usually denoted by $t \mapsto \mathbb{X}_\omega^x$ in the stochastic literature) where $\mathcal{C}(0,\infty;\mathbb{R}^d)$ is the space of continuous evolutions. In other words, it defines evolutions $t \mapsto \mathbb{X}(x,\omega)(t) := \mathbb{X}_\omega^x(t) \in \mathbb{R}^d$ starting at x at time 0 and parameterized by random events $\omega \in \Omega$ satisfying technical requirements (measurability, filtration, etc.) that are not relevant to involve at this stage of the exposition. The initial state x being fixed, the random variable $\omega \mapsto \mathbb{X}(x,\omega) := \mathbb{X}_\omega^x(\cdot) \in \mathcal{C}(0,+\infty;\mathbb{R}^d)$ is called a stochastic process. A subset $\mathcal{H} \subset \mathcal{C}(0,\infty;\mathbb{R}^d)$ of evolutions sharing a given property being chosen, it is natural, as we did for tychastic systems, to introduce the *stochastic core* of $\mathcal{H}$ under the stochastic system: it is the subset of initial states x from which starts a stochastic process $\omega \mapsto \mathbb{X}(x,\omega)$ such that for almost all $\omega \in \Omega$, $\mathbb{X}(x,\omega) \in \mathcal{H}$:

$$\mathrm{Stoc}_{\mathbb{X}}(\mathcal{H}) := \{x \in \mathbb{R}^d \mid \text{for almost all } \omega \in \Omega,\ \mathbb{X}(x,\omega) := \mathbb{X}_\omega^x(\cdot) \in \mathcal{H}\} \tag{10.36}$$

Regarding a stochastic process as a set-valued map $\mathbb{X}$ associating with any state x the family $\mathbb{X}(x) := \{\mathbb{X}(x,\omega)\}_{\omega\in\Omega}$, the definitions of stochastic cores (10.36) of subsets of evolution properties are similar in spirit to definition:

$$\mathcal{S}^{\ominus 1}(\mathcal{H}) := \{x \in \mathbb{R}^d \mid \text{ for all } v(\cdot) \in Q(x(\cdot)),\ x_{v(\cdot)}(\cdot) \in \mathcal{H}\}$$

under a tychastic system

$$x'(t) = f(x(t), v(t)) \text{ where } v(t) \in Q(x(t))$$

Furthermore, *the parameters ω are constant in the stochastic case, whereas the tychastic uncertainty $v(\cdot)$ is dynamic* in nature and involves a state dependence, two more realistic assumptions in the domain of life sciences.

There is however a deeper similarity that we mention briefly. When the stochastic system $(x,\omega) \mapsto \mathbb{X}(x,\omega)$ is derived from a stochastic differential equation, the Strook-Varadhan Support Theorem (see [201, Stroock &Varadhan]) states that there exists a tychastic system $(x,v) \mapsto \mathcal{S}(x,v)$ such that, whenever $\mathcal{H}$ is closed, the stochastic core of $\mathcal{H}$ under the stochastic system $\mathbb{X}$ and its tychastic core under the associated tychastic system $\mathcal{S}$ coincide:

$$\mathrm{Stoc}_{\mathbb{X}}(\mathcal{H}) = \mathcal{S}^{\ominus 1}(\mathcal{H})$$

To be more specific, let $\mathbb{X}(x,\omega)$ denote the solution to the stochastic differential equation

$$dx = \gamma(x)dt + \sigma(x)dW(t)$$

starting at x, where $W(t)$ ranges over $\mathbb{R}^c$, the drift $\gamma : \mathbb{R}^d \mapsto \mathbb{R}^d$ and the diffusion $\sigma : \mathbb{R}^d \mapsto \mathcal{L}(\mathbb{R}^c, \mathbb{R}^d)$ are smooth and bounded maps. Let us associate with them the Stratonovitch drift $\widehat{\gamma}$ defined by $\widehat{\gamma}(x) := \gamma(x) - \frac{1}{2}\sigma'(x)\sigma(x)$. The Stratonovitch stochastic integral is an alternative to the Ito integral, and easier to manipulate. Unlike the Ito calculus, Stratonovich integrals are defined such that the chain rule of ordinary calculus holds. It is possible to convert Ito stochastic differential equations to Stratonovich ones.

Then, the associated tychastic system is

$$x'(t) = \widehat{\gamma}(x(t)) + \sigma(x(t))v(t) \text{ where } v(t) \in \mathbb{R}^c \tag{10.37}$$

where *the tychastic map is constant and equal to* $\mathbb{R}^c$.

Consequently, *the tychastic system associated with a stochastic one by the Strook–Varadhan Support Theorem is very restricted: there are no bounds at all on the tyches, whereas general tychastic systems allow the tyches to range over subsets $Q(x)$ depending upon the state x, describing so to speak a state-dependent uncertainty*:

$$x'(t) = \widehat{\gamma}(x(t)) + \sigma(x(t))v(t) \text{ where } v(t) \in Q(x(t))$$

This state-dependent uncertainty, unfortunately absent in the mathematical representation of uncertainty in the framework of stochastic processes, is of utmost importance for describing uncertainty in problems dealing with living beings.

When $\mathcal{H}$ is a Borelian of $\mathcal{C}(0,\infty;\mathbb{R}^d)$, we denote by $\mathbb{P}_{\mathbb{X}(x,\cdot)}$ the *law* of the random variable $\mathbb{X}(x,\cdot)$ defined by

$$\mathbb{P}_{\mathbb{X}(x,\cdot)}(\mathcal{H}) \ := \ \mathbb{P}(\{\omega \mid \mathbb{X}(x,\omega) \in \mathcal{H}\}) \tag{10.38}$$

Therefore, we can reformulate the definition of the stochastic core of a set $\mathcal{H}$ of evolutions in the form

$$\mathrm{Stoc}_{\mathbb{X}}(\mathcal{H}) = \{x \in \mathbb{R}^d \mid \mathbb{P}_{\mathbb{X}(x,\cdot)}(\mathcal{H}) = 1\} \tag{10.39}$$

In other words, the stochastic core of $\mathcal{H}$ is the set of initial states x such that the subset $\mathcal{H}$ has probability one under the law of the stochastic process $\omega \mapsto \mathbb{X}(x,\omega) \in \mathcal{C}(0,+\infty;\mathbb{R}^d)$ (if $\mathcal{H}$ is closed, $\mathcal{H}$ is called the *support* of the law $\mathbb{P}_{\mathbb{X}(x,\cdot)}$). The Strook–Varadhan Support Theorem states that under regularity assumptions, this support is the core of $\mathcal{H}$ under the tychastic system (10.37). It furthermore provides a characterization of stochastic viability in terms of tangent cones and general curvatures of the environments (see the contributions of Giuseppe da Prato, Halim Doss, Hélène Frankowska and Jerzy Zabczyk among many other).

These remarks further justify our choice of privileging tychastic systems because, as far as the properties of initial states of evolution are concerned, stochastic systems are just (very) particular cases of tychastic systems.

Chapter 11
Regulation of Control Systems

11.1 Introduction

This chapter is devoted to viability properties specific to evolutionary systems generated by control systems of the form

$$x'(t) = f(x(t), u(t)) \text{ where } u(t) \in U(x(t))$$

These systems share all the properties gathered in Chap. 10, p. 375 regarding the viability characterization of viability kernels of environments and capture basins of a target viable in an environment.

It is specifically dedicated to the tangential characterizations of viability and invariance properties.

1. The tangential characterizations are presented in Sect. 11.2, p. 440.
2. The main Viability and Invariance Theorems 11.3.4, p. 455 and 11.3.7, p. 457 are quoted in Sect. 11.3, p. 453 (we do not reproduce the proof of Filippov Theorem 11.3.9, p.459, which is by now available in many monographs).
3. The rest of this chapter is devoted to the study or the *regulation map*. Indeed, we need to go one step further: not only study the properties of these viability kernels and capture basin, but ***how** to regulate the evolutions which are viable in the environment forever or until they reach the target.*
   Therefore, in the framework of control systems (f, U), we investigate how it is possible to carve in the set-valued map U a sub-map $R_K \subset U$, the *regulation map*. We built from the knowledge of
   - the set-valued map U and the right hand side $(x, u) \in \text{Graph}(U) \mapsto f(x, u) \in X$ of the control system (f, U)
   - the environment K and the target $C \subset K$

J.-P. Aubin et al., *Viability Theory*, DOI 10.1007/978-3-642-16684-6_11,

in such a way that viable evolutions are governed by the new *regulated system*

$$x'(t) = f(x(t), u(t)) \text{ where } u(t) \in R_K(x(t))$$

on the viability kernel with target.
Observe that this regulation map is the "mother of all viable feedbacks" $r_K : x \mapsto r_K(x) \in R_K(x)$ governing the evolution of viable evolutions through the dynamical system

$$x'(t) = f(x(t), r_K(x(t)))$$

Hence, instead of "guessing" what kind of feedback $\widetilde{u} : x \mapsto \widetilde{u}(x)$ should be used without being sure it would drive viable evolutions (for instance, affine feedbacks), the knowledge of the regulation map allows us to find all the viable ones.
However, given an arbitrary feedback $\widetilde{u}$ the viability kernel under system

$$x'(t) = f(x(t), \widetilde{u}(x(t)))$$

provides a measure of the "safety" of $\widetilde{u}$ for governing evolutions viable in this environment, by comparing it with the viability kernel under the initial control system

$$x'(t) = f(x(t), u(t)) \text{ where } u(t) \in U(x(t))$$

Now that the ultimate objective of this chapter is fixed, we provide a third type of characterization of viability kernels and capture basins in terms of the regulation map.
In the discrete case, Lemma 2.14.3, p. 98 provides the clue. Indeed, to be of value, the task of finding the viability kernel *should be done without checking the existence of viable solutions* for each initial state ("shooting methods"). Even for deterministic discrete systems, as we saw in Sect. 2.14, p. 96, these methods provide at best an approximation of the viability kernel instead of the viability kernel itself. This has already been mentioned in Sect. 2.9.3, p. 75.
As was mentioned in Box 14, p. 98, the idea is to check whether some conditions relating the geometry of the environment K and the right hand side of the differential inclusion are satisfied. This is a much easier task.
This idea turns out to be quite intuitive: *at each point on the boundary of the environment, where the viability of the system is at stake, there should exist a velocity which in some sense is* tangent *to the environment for allowing the solution to bounce back in the environment and remain inside it.* This is, in essence, what the Viability Theorem 11.3.4, p. 455 states.

- For that purpose, Sect. 11.2, p. 440 provides a rigorous definition (see Definition 11.2.1, p. 442) of *tangency* to any closed subset, even non convex, even nonsmooth.

- This allows us to compute in Sect. 11.3.2, p. 454 the *regulation map*

$$\forall x \in K, \ \ R_K(x) \ := \ \{u \in U(x) \ \text{such that} \ f(x,u) \ \in \ T_K(x)\}$$

in terms of subsets $T_K(x)$ of tangent velocities to K at x.
- The main Viability and Invariance Theorems 11.3.4, p. 455 and 11.3.7, p. 457 are couched in terms of the regulation map: in essence, and under adequate conditions,
  - an environment is viable if and only if

$$\forall x \in K, \ \ R_K(x) \neq \emptyset$$

  - an environment is invariant if and only if

$$\forall x \in K, \ \ R_K(x) := U(x)$$

- Since we know how to characterize viability and invariance properties in terms of the regulation map, we use them conjointly with the characterizations of viability kernels and capture basin obtained at the level of evolutionary systems. This leads us to characterize viability kernels as the ***unique*** subsets D between C and K satisfying the Frankowska property defined by (11.19), p. 462.
- When the right hand side of a control system (f,U) does not depend upon the state, but uniquely on the controls, and under convexity assumption, it is possible to give a very simple formula of the capture basin (see Definition 4.3.1, p. 133). It turns out that for a certain class of Hamilton–Jacobi partial differential equations approached by viability techniques as we shall do in Chap. 17, p. 681, we obtain the famous Lax–Hopf "explicit" analytical formula for computing solutions to these partial differential equations. This is presented in Sect. 11.5, p. 465.
- The tangential characterizations are enriched in Sect. 11.6, p. 475 by providing an equivalent Hamiltonian formulation of the Frankowska property. It involves the concept of *Hamiltonian* associated with a control system (f,U). We shall introduce a property of a Hamiltonian, which, in dynamic economic theory, is related to the *Walras law*. This is the reason we use this term to denote it. The Walras law allows us to provide an elegant characterization of the regulation map as well as a criterion to guarantee that it has nonempty values, in other words, a useful viability criterion. The dual formulation of the Frankowska property, when applied to epigraphs of functions, provides right away the concept of Barron–Jensen/Frankowska solutions to Hamilton–Jacobi–Bellman partial differential equations.

4. The regulation map being computed and characterized, and knowing that all feedbacks governing viable evolutions are necessarily contained in the regulation map (such single-valued maps are called *selections*),

the question arises to built such feedbacks, both static and dynamic. Section 11.7, p. 480 summarizes some of the results presented and proved in Chap. 6 of the first edition of *Viability Theory* [18, Aubin].

11.2 Tangential Conditions for Viability and Invariance

As mentioned in the introduction of this chapter, we have to choose a mathematical implementation of the concept of tangency inherited from *Pierre de Fermat.*

Pierre de Fermat [1601–1665]. *The author of* Maxima et minima *(1638) and the founder of differential and integral calculus, called tangents* "touchantes", *by opposition to* "sécantes". *He defined and computed tangents to several curves (parabola, ellipses, cycloid, etc,) by using his "Fermat Rule" for determining minima, maxima and inflexion points. This was the starting point of the differential calculus developed by* Gottfried Leibniz *and* Isaac Newton. René Descartes, *who called the Councillor of the King in the Parliament of Toulouse the "conseiller des Minimis", and who discovered analytical geometry independently of Fermat, called them* contingentes, *coming from the Latin* contingere, *to touch on all sides. Descartes, who thought that his rival was inadequate as a mathematician and a thinker, reluctantly admitted his misjudgment and eventually wrote Fermat* "... seeing the last method that you use for finding tangents to curved lines, I can reply to it in no other way than to say that it is very good and that, if you had explained it in this manner at the outset, I would have not contradicted it at all". *From a more friendly side,* Blaise Pascal, *who shared with him the creation of the mathematical theory of probability, wrote him that* "he had a particular veneration to those who bear the name of the first man of the world". *His achievements in number theory overshadowed his other contributions. He was on top of that a poet, a linguist,... and made his living as a lawyer.*

It is impossible to restrict ourselves to environments that are smooth manifolds without boundaries because viability theory deals with questions raised when the evolutions reach their boundaries in order to remain viable. The simplest example of a constrained environment, $K := \mathbb{R}_+ \subset X := \mathbb{R}$, would thereby be ruled out, as well as environments defined by inequality constraints instead of equality constraints (as for balls, that possess distinct boundaries, the spheres, which do not posses boundaries). Furthermore, we are no longer free of choosing environments and targets, some of them being provided as results of other problems (such as viability kernels and capture basins of other sets).

So, we need to "implement" the concept of a direction v "tangent" to any subset K at $x \in K$, which should mean that starting from x in the direction v, "we do not go too far" from K.

To convert this intuition into a rigorous mathematical definition, we shall choose from among the many ways there are to translate what it means to be "not too far" the one independently suggested in the beginning of the 1930s by *Georges Bouligand* and *Francesco Severi.*

Georges Bouligand [1889–1979]. *Orphan of his father when he was six years old, he succeeded through the French Education System of this time to have the highest rank when graduating at École Polytechnique in 1909, but chose to enter École Normale Supérieure, where he started to work with* Jacques Hadamard *in hydrodynamics. He was "half of the mathematical faculty" at the university of Poitiers from 1920–1938, when he was appointed to the Faculté des Sciences de Paris. He authored around ten books, including his* Introduction à la Géométrie infinitésimale directe *and books on mathematical philosophy and epistemology.* Norbert Wiener *wrote in* Ex-Prodigy "... because I had recently seen a series of articles by Lebesgue and a young mathematician named Georges Bouligand, which were getting uncomfortably near to the complete solution of the problem in which I was interested, and which would eliminate the problem [Generalized Dirichlet problem] from the literature... This incident was the start of a friendship between Bouligand and myself which has lasted to the present day". Gustave Choquet, *whose thesis' title was* Application des propriétés descriptives de la fonction contingent, *concluded 20 years after his death a Bouligand bibliographical notice with these words:* "But above all, Bouligand's name will remain attached to his contingent and paratingent (1932). Surely, these had ancestors: Severi in 1927–1928 had defined analogous notions for studying algebraic manifolds, but did not pass this particular case... The success of these tools had been and still is considerable; for instance... they are indispensable to set-valued analysis. This is the most beautiful application of the interior idea that Bouligand has attempted to describe in his philosophical essays. *His eighth child, Yves Bouligand, is a biologist, working in a field where the tools forged by his father could be used.*

Actually, *Francesco Severi* discovered few months earlier the concepts of *semitangenti* and of *corde improprie* to a set at a point of its closure which are equivalent to the concepts of *contingentes* and *paratingentes.*

35 *Francesco Severi's Semitangenti.* *Severi explains for the second time that he had discovered these concepts developed by Bouligand in* "suo interessante libro recente" *and comments:* (It obviously escaped to the eminent geometer that his own research started later than mine. But I will not reprimand him, because I too cannot follow with care the literature and I rather read an article or a book after having thought for myself about the argument.)

This important statement deserves to be quoted.

Francesco Severi [1879–1961]. *Orphan of his father when he was nine years old, he studied at Turin University to become a geometer under the influence of Corrado Segre, under the supervision of whom he obtained his doctoral degree in 1900. He next pursued his investigations in algebraic geometry, with Castelnuovo and Enriques. He carried wonderful mathematical achievements despite numerous other activities, including political ones. He founded in 1939 in Roma the Istituto Nazionale di Alta Matematica which now bears his name.*

Definition 11.2.1 ***[Tangent (Contingent) Cone to a Subset]*** *Let $K \subset X$ be a subset and $x \in K$ an element of K. A direction v is* contingent *(or, more simply, "tangent") to K at $x \in K$ if it is a limit of a sequence of directions v_n such that $x + h_n v_n$ belongs to K for some sequence $h_n \to 0+$. The collection of such contingent directions constitutes a closed cone $T_K(x)$, called the* contingent cone *to K at x, or more simply,* tangent cone.

We refer to Definition 18.4.8, p. 732 for an equivalent formulation in terms of Painlevé–Kuratowski limits of "set difference quotients" $\frac{K-x}{h}$.

Since, by definition of the interior of a subset, any element $x \in \text{Int}(K)$ can be surrounded by a ball $B(x, \eta) \subset K$ with $\eta > 0$, then

$$\forall\, x \in \text{Int}(K),\;\; T_K(x) \;=\; X \tag{11.1}$$

so that the non trivial properties of the tangent cone happen on the boundary $\partial K := \overline{K} \setminus \text{Int}(K)$ of the subset K.

Except if K is a smooth manifold, the set of tangent vectors is no longer a vector-space, but this discomfort is compensated by advances in set-valued analysis providing a calculus of tangent cones allowing us to compute them. Closed convex set are *sleek*: this means that the set-valued map $x \rightsquigarrow T_K(x)$ is lower semicontinuous: in this case, $T_K(x)$ is a closed *convex* cone.

Tangent cones allow us to "differentiate" viable evolutions: Indeed, we observe readily the following property.

Lemma 11.2.2 *[**Derivatives of Viable Evolutions**] Let $x(\cdot) : t \in \mathbb{R}_+ \mapsto x(t) \in X$ be a differentiable evolution viable in K on an open interval $\mathbb{I}$: $\forall t \in \mathbb{I},\ x(t) \in K$. Then*

$$\forall t \in \mathbb{I},\ x'(t) \in T_K(x(t))$$

Proof. Indeed, since $x(s) \in K$ for s in a neighborhood of t, then $x(t+h) = x(t)+hx'(t)+h\varepsilon(h) \in K$ where $\varepsilon(h) \to 0$ when $h \to 0$. Hence $x'(t) \in T_K(x(t))$ by Definition 11.2.1, p. 442. □

These simple observations suggest us that the concept of tangent cone plays a crucial role for characterizing viability and invariance properties when the state space $X := \mathbb{R}^d$ is a finite-dimensional space.

11.2.1 The Nagumo Theorem

The Viability Theorem 11.3.4, p. 455 has a long history. It began in the case of differential equations in 1942 with the Japanese mathematician Nagumo in a paper written in German (who however did not relate its tangential condition to the Bouligand–Severi tangent cone). The Nagumo Theorem has been rediscovered many times since.

Mitio Nagumo [1905-]. *Mitio Nagumo provided a criterion for uniqueness of a solution to differential equations, known as the Nagumo condition, in his very first paper written when he was 21. He then visited Götingen from 1932 to 1934, a visit which influenced his mathematical work since. His first papers were written in German, and after, in Esperanto, a language he was trying to promote. He then spent 32 years of his research life at Osaka, where he founded the "Nagumo School" then to Sophia University in Tokyo until he retired at 70. He founded with Hukuara a private journal in Japanese in which all contributors introduced new ideas without referees. Besides his important contributions in the fields of ordinary and partial differential equations, Nagumo contributed to clarify the Leray–Schauder theory, as well as many related fields. as early as 1935, for example, he proposed the basic ideas of Banach algebras, in a paper written... in Japanese. (See Collected papers, [165, Nagumo]).*

It is convenient to place on a same footing open and closed subsets: in finite dimensional vector spaces, they are both *locally compact subsets*: Recall that K is *locally compact* if for any $x \in K$, there exists $r > 0$ such that the ball $B_K(x,r) := K \cap (x + rB)$ is *compact.*

Theorem 11.2.3 *[The Nagumo Theorem] Let us assume that*

$$\begin{cases} (i) \ K \ \textit{is locally compact} \\ (ii) \ f \ \textit{is continuous from } K \textit{ to } X \end{cases} \tag{11.2}$$

Then K is locally viable under f if and only if the tangential condition

$$\forall x \in K, \ \ f(x) \ \in \ T_K(x)$$

holds true.

Since the tangent cone to an open subset is equal to the whole space (see (11.1)), an open subset satisfies the tangential condition whatever the map. So, it enjoys the viability property because any open subset of a finite dimensional vector-space is locally compact. The Peano Existence Theorem is then a consequence of Theorem 11.2.3, p. 444:

Theorem 11.2.4 *[The Peano Theorem] Let Ω be an open subset of a finite dimensional vector-space X and $f : \Omega \mapsto X$ be a continuous map.*

Then, for every $x_0 \in \Omega$, there exists $T > 0$ such that the differential equation $x'(t) = f(x(t))$ has a solution viable in Ω on the interval $[0,T]$ starting at x_0.

We shall extend the Nagumo Theorem to the case of differential inclusions $x'(t) \in F(x(t))$. In this case, the tangential condition

$$\forall x \in K, \ \ f(x) \ \in \ T_K(x)$$

splits in two:

$$\forall x \ \in \ K, \begin{cases} (i) \ \ F(x) \cap T_K(x) \ \neq \ \emptyset \ \ \text{(Viability Tangential Condition)} \\ (ii) \ \ \ \ F(x) \subset T_K(x) \ \ \ \ \text{(Invariance Tangential Condition)} \end{cases}$$

Actually, we shall improve these conditions, but for that purpose, we need a technical tool, amounting to integrate differential inequalities involving lower semicontinuous functions. This technical lemma can actually be derived from the Nagumo Theorem.

11.2.2 Integrating Differential Inequalities

The Nagumo Theorem allows us to extend to the class of lower semicontinuous (and thus, non necessarily differentiable) functions the classical integration of the differential inequality

$$\forall\, t \in [0,T[,\ \ \varphi'(t)\ \le\ \alpha'(t)\varphi(t)+\beta(t)$$

implying

$$\forall\, t \in [0,T[,\ \ \varphi(t)\ \le\ e^{\alpha(t)}\left(\varphi(0)+\int_0^t e^{-\alpha(s)}\beta(s)ds\right)$$

where α is continuously differentiable and β is continuous.

This is an easy and important technical tool, that we shall need not only for differentiable functions φ, but also for lower semicontinuous functions.

If $\varphi : \mathbb{R} \mapsto \mathbb{R} \cup \{+\infty\}$, we define its epiderivative by its epigraph (see Definition 4.2.2, p. 131): The epigraph of the epiderivative $D_{\uparrow}\varphi(t)(\cdot)$ is equal to the tangent cone to the epigraph of φ at $(t, \varphi(t))$:

$$\mathcal{E}p(D_{\uparrow}\varphi(t)) = T_{\mathcal{E}p(\varphi)}(t,\varphi(t))$$

(see Definition 18.6.9, p. 747 and Proposition 18.6.10, p. 748).

Proposition 11.2.5 ***[Integration of Linear Differential Inequalities]*** *Let α be continuously differentiable and vanishing at 0, β be continuous. Let us consider a lower semicontinuous function $\varphi : \mathbb{R} \mapsto \mathbb{R} \cup \{+\infty\}$ satisfying,*

$$\forall\, t \in [0,T[,\ \ D_{\uparrow}\varphi(t)(1)\ \le\ \alpha'(t)\varphi(t)+\beta(t)$$

Therefore, an upper bound on the growth of φ is estimated in the following way:

$$\exists\, S \in]0,T[,\ \textit{such that}\ \forall\, t \in [0,S],\ \ \varphi(t)\ \le\ e^{\alpha(t)}\left(\varphi(0)+\int_0^t e^{-\alpha(s)}\beta(s)ds\right)$$

This proposition follows from a more general one, when we replace the function $\alpha'(t)y + \beta(t)$ by a general continuous function $f(t,y)$. We assume for simplicity that for any $y \in \mathbb{R}_+$, there exists a unique solution $\mathcal{S}_f(y)(\cdot)$ to the differential equation $y'(t) = f(t,y(t))$ satisfying $y(0) = y$. This is the case when $f(t,y) := \alpha'(t)y + \beta(t)$ because, in this case,

$$\mathcal{S}_f(y)(t) \;=\; e^{\alpha(t)}\left(y + \int_0^t e^{-\alpha(s)}\beta(s)ds\right)$$

Theorem 11.2.6 *[Integration of Differential Inequalities] Let us assume that for any $y \geq 0$, there exists a unique solution $\mathcal{S}_f(y)(\cdot)$ to the differential equation $y'(t) = f(t,y(t))$ satisfying $y(0) = y$. Let us consider a lower semicontinuous function $\varphi : \mathbb{R} \mapsto \mathbb{R} \cup \{+\infty\}$ satisfying,*

$$\forall\, t \in [0,T[,\;\; D_{\uparrow}\varphi(t)(1) \;\leq\; f(t,\varphi(t))$$

Therefore, φ is estimated in the following way:

$$\exists\, S \in]0,T[,\; \textit{such that}\, \forall\, t \in [0,S],\;\; \varphi(t) \;\leq\; \mathcal{S}_f(\varphi(0))(t) \tag{11.3}$$

Proof. This statement follows from the Nagumo Theorem 11.2.3, p. 444 in the case when $X := \mathbb{R}^2$ and when the differential equation is (11.4)

$$\begin{cases} (i) \;\; \tau'(t) \;=\; 1 \\ (ii) \; y'(t) \;=\; f(\tau(t),y(t)) \end{cases} \tag{11.4}$$

The solution of this differential equation starting at $(0,y)$ is $t \mapsto (t,\mathcal{S}_f(y)(t))$.

By Lemma 18.6.3, p. 744, to say that φ is lower semicontinuous amounts to saying that its epigraph $\mathcal{E}p(\varphi) \subset \mathbb{R}^2$ is closed. The product $[0,T[\times\mathbb{R}$ is locally compact, so is the intersection $K := \mathcal{E}p(\varphi) \cap ([0,T[\times\mathbb{R})$. We observe that, thanks to the statement

$$\mathcal{E}p(D_{\uparrow}\varphi(t)) = T_{\mathcal{E}p(\varphi)}(t,\varphi(t))$$

characterizing epiderivatives, condition

$$\forall\, t \in [0,T[,\;\; D_{\uparrow}\varphi(t)(1) \;\leq\; f(t,\varphi(t))$$

is equivalent to

$$\forall\, t \in [0,T[,\;\; (1,f(t,\varphi(t))) \;\in\; T_{\mathcal{E}p(\varphi)}(t,\varphi(t))$$

Conversely, this inequality implies that $\mathcal{E}p(\varphi)$ is viable under (11.4), p. 446, i.e., that

$$\forall\, t \in [0,T[,\ \ \forall\, y \geq \varphi(t),\ \ (1, f(t,\varphi(t))) \in T_{\mathcal{E}p(\varphi)}(t,y)$$

This is true for $y = \varphi(t)$ by definition, and follows from Lemma 18.6.18, p. 753 when $y > \varphi(t)$ because 1 belongs to the domain of $D_{\uparrow}\varphi(t)$.

In other words, the assumption

$$\forall\, t \in [0,T[,\ \ D_{\uparrow}\varphi(t)(1) \leq f(t,\varphi(t))$$

is equivalent to

$$\forall\, (t,y) \in K,\ \ (1, f(t,y)) \in T_K(t,y)$$

Hence the Nagumo Theorem 11.2.3, p. 444 implies that starting from $(0,\varphi(0))$, the solution $t \mapsto (t, \mathcal{S}_f(y)(t))$ is viable in $K := \mathcal{E}p(\varphi) \cap ([0,T[\times \mathbb{R})$ on some nonempty interval $[0,S]$, i.e., that for any $t \in [0,S]$, $\varphi(t) \leq \mathcal{S}_f(\varphi(0))(t)$ and that $S < T$. □

11.2.3 Characterization of the Viability Tangential Condition

Tangent cones are not necessarily convex. However, we point out that $T_K(x)$ is convex when K is convex (or, more generally, when K is sleek i.e., the tangent cone map $T_K(\cdot)$ is lower semicontinuous. See Definition 18.4.8). As it turns out for characterizing viability and invariance, we shall be able to replace the tangent cones by their *convexified tangent cone* $T_K^{\star\star}(\cdot)$ (see Definition 18.4.8, p. 732).

Then the lack of convexity of the tangent cone can be, under adequate assumption, compensated using the convexified tangent cone $T_K^{\star\star}(\cdot)$ in the viability tangential condition:

Theorem 11.2.7 *[Viability Tangential Conditions] Let $K \subset X$ be a nonempty closed subset of a finite dimensional vector space. Let x_0 belong to K. Assume that the set-valued map $F : K \rightsquigarrow X$ is upper semicontinuous (see Definition 18.4.3, p. 729) with convex compact values. Then the two following properties are equivalent:*

$$\begin{cases} (i)\ \forall x \in K \cap \overset{\circ}{B}(x_0,\alpha),\ F(x) \cap T_K(x) \neq \emptyset \\ (ii)\ \forall x \in K \cap \overset{\circ}{B}(x_0,\alpha),\ F(x) \cap T_K^{\star\star}(x) \neq \emptyset \end{cases} \tag{11.5}$$

(where $\overset{\circ}{B}(x_0, \alpha)$ *denotes the open ball).*
Furthermore, assume that K *is compact and that*

$$\forall x \in K, \; F(x) \cap T_K^{\star\star}(x) \neq \emptyset$$

Then, for all $\varepsilon > 0$*, and for any "graphical approximation"* F_ε *(in the sense that* $\mathrm{Graph}(F_\varepsilon) \subset \mathrm{Graph}(F) + \varepsilon(B \times B)$*) of* F*, then*

$$\exists\, \eta(\varepsilon) > 0 \text{ such that } \forall x \in K, \; \forall h \in [0, \eta(\varepsilon)], \; (x + hF_\varepsilon(x)) \cap K \neq \emptyset \quad (11.6)$$

Remark. Property (19.18) means that the discrete explicit schemes $x \rightsquigarrow x + hF_\varepsilon(x)$ associated with the graphical approximation F_ε of F are viable in K uniformly with respect to the discretization step h on compact sets. It is very useful for approximating solutions of differential inclusion $x'(t) \in F(x)$ viable in K (see Chap. 19, p.769). □

Remark. This theorem has been announced in 1985 by Guseinov, Subbotin and Ushakov. Hélène Frankowska provided a proof in 1990 which was reproduced in the first edition of *Viability Theory*, [18, Aubin]. We use here a simpler proof designed by Hélène Frankowska which appeared in 2000. □

Proof. Since (11.5)(i) implies (11.5)(ii), assume that (11.5)(ii) holds true. Fix any pair of elements $x, \; y \in K$. Since the vector space X is finite dimensional and since the subsets K and $F(y)$ are closed, there exist $v_t \in F(y)$ and $x_t \in K$ achieving the minimum in

$$\varphi_{(x,y)}(t) \;:=\; d(x + tF(y), K) \;=\; \|x + tv_t - x_t\|$$

Since $x \in K$, we infer that

$$\|x + tv_t - x_t\| \;=\; d(x + tF(y), K) \;\leq\; \|x + tv_t - x\| \;=\; t\|v_t\|$$

from which we obtain inequality

$$\begin{cases} \|x - x_t\| \;\leq\; \|x + tv_t - x_t\| + t\|v_t\| \\ \leq\; \|x + tv_t - x\| + t\|v_t\| \;=\; 2t\|v_t\| \;\leq\; 2t\|F(y)\| \end{cases} \quad (11.7)$$

Choose now any $u \in F(y)$ and any $w \in T_K(x_t)$.

Observe that for any $u \in F(y)$, $tv_t + hu \in tF(y) + hF(y) \subset (t + h)F(y)$ *because* $F(y)$ *is convex.* Recall that for all $w \in T_K(x_t)$, there exists a sequence $e(h_n)$ converging to 0 such that $x_t + h_n w + h_n e(h_n) \in K$. Therefore

$$\frac{\varphi_{(x,y)}(t+h_n)-\varphi_{(x,y)}(t)}{h_n} \leq \frac{\|x+tv_t-x_t+h_n(u-w-e(h_n))\|-\|x+tv_t-x_t\|}{h_n}$$

Recalling that $D_{\uparrow}\|x\|(u) \leq \left\langle \frac{x}{\|x\|}, u \right\rangle$, dividing by h_n and passing to the limit, we infer that for all $u \in F(y)$ and $w \in T_K(x_t)$,

$$D_{\uparrow}\varphi_{(x,y)}(t)(1) \leq \left\langle u-w, \frac{x+tv_t-x_t}{\|x+tv_t-x_t\|} \right\rangle \leq \|u-w\|$$

and, consequently, that, this inequality holds true for all $u \in F(y)$ and $w \in T_K^{\star\star}(x_t)$. Therefore, we have proved that

$$D_{\uparrow}\varphi_{(x,y)}(t)(1) \ \leq \ d(F(y), T_K(x_t)) \tag{11.8}$$

Furthermore, F being upper semicontinuous, we can associate with any $y \in K$ and $\varepsilon > 0$ an $\eta(\varepsilon, y) \leq \varepsilon$ such that

$$\forall z \in B(y, \eta(\varepsilon, y)), \ F(z) \ \subset \ F(y) + \varepsilon B$$

Let x_0 belong to K mentioned in Theorem 11.2.7, p.447. By taking $y = x \in \overset{\circ}{B}(x_0, \alpha)$ and $z := x_t$, we deduce from (11.7) that whenever $t \leq \frac{\min(\eta(\varepsilon, x), \alpha - \|x_0 - x\|)}{2\|F(x)\|}$, then

$$\|x - x_t\| < \min(\eta(\varepsilon, x), \alpha - \|x_0 - x\|) \ \& \ \|x_0 - x_t\| < \alpha$$

and thus, that $F(x_t) \subset F(x) + \varepsilon B$.

Assumption (11.5)(ii) implies that there exists $w_t \in F(x_t) \cap T_K^{\star\star}(x_t)$, so that there exists $u_t \in F(y)$ such that $\|u_t - w_t\| \leq \varepsilon$. Therefore, inequality (11.8) implies that $\forall t \in \left[0, \frac{\min(\eta(\varepsilon, x), \alpha - \|x_0 - x\|)}{2\|F(x)\|}\right]$,

$$\begin{aligned} D_{\uparrow}\varphi_{(x,x)}(t)(1) &\leq d(F(x), T_K(x_t)) \\ &\leq \|u_t - w_t\| \leq \varepsilon \end{aligned}$$

The function $\varphi_{(x,y)}(\cdot)$ being lower semicontinuous, Proposition 11.2.5, p. 445 implies that $\forall t \in \left[0, \frac{\min(\eta(\varepsilon, x), \alpha - \|x_0 - x\|)}{2\|F(x)\|}\right]$,

$$\varphi_{(x,x)}(t) \ = \ d\,(x + tF(x), K) \ = \ \|x + tv_t - x_t\| \ \leq \ t\varepsilon$$

which can be written in the form $\frac{d\,(x + tF(x), K)}{t} = \left\| v_t - \frac{x_t - x}{t} \right\| \leq \varepsilon$. Since $\|v_t\| \leq \|F(x)\|$, the sequence $\frac{x_t - x}{t}$ is bounded, so that a subsequence

$v_{t_n} := \frac{x_{t_n} - x}{t_n}$ converges to some $v \in F(x)$. But since the subsequence v_{t_n} belongs to $\frac{K - x}{t_n}$ and converges to v, we infer that v also belongs to the tangent cone $T_K(x)$.

For proving the second statement, let us fix $\varepsilon > 0$, take any $y \in K$ and the associated $\eta(\varepsilon, y)$ and derive that

$$\begin{cases} \forall t \in \left[0, \frac{\eta(\varepsilon,y)}{4\|F(y)\|}\right], \ \forall x \in K \cap B\left(y, \frac{\eta(\varepsilon,y)}{2}\right), \\ D_\uparrow \varphi_{(x,y)}(t)(1) \le d(F(y), T_K^{\star\star}(x_t)) \le \|u_t - w_t\| \le \varepsilon \end{cases} \tag{11.9}$$

Indeed, if $\|x - y\| \le \frac{\eta(\varepsilon, y)}{2}$, we associate with it $v_t \in F(y)$ and $x_t \in K$ achieving the minimum in

$$\varphi_{(x,y)}(t) \ := \ d(x + tF(y), K) \ = \ \|x + tv_t - x_t\|$$

as in the first part of the proof. Then there exists $w_t \in F(x_t) \cap T_K^{\star\star}(x_t)$ by assumption (11.5)(ii).

Property (11.7) implies that $\|x - x_t\| \le 2t\|F(y)\| \le \frac{\eta(\varepsilon, y)}{2}$ whenever $t \le \frac{\eta(\varepsilon, y)}{4\|F(y)\|}$, so that $\|y - x_t\| \le \|x - y\| + \|x - x_t\| \le \eta(\varepsilon, y)$ because $\|x - y\| \le \frac{\eta(\varepsilon, y)}{2}$. Since F is upper semicontinuous, we deduce that there exists $u_t \in F(y)$ such that $\|u_t - w_t\| \le \varepsilon$, so that (11.9) holds true.

Since $\varphi_{(x,y)}(\cdot)$ is lower semicontinuous, Proposition 11.2.5, p. 445 implies that

$$\forall t \in \left[0, \frac{\eta(\varepsilon, y)}{4\|F(y)\|}\right], \forall x \in K \cap B\left(y, \frac{\eta(\varepsilon, y)}{2}\right), d\left(x + tF(y), K\right) = \varphi_{(x,y)}(t) \le \varepsilon t$$

The subset K being compact, it can be covered by a finite number of balls $B\left(y_j, \frac{\eta(\varepsilon,y_j)}{2}\right)$. Setting $\eta(\varepsilon) := \min_j \frac{\eta(\varepsilon,y_j)}{4\|F(y_j)\|} > 0$, we infer that

$$\begin{cases} \forall \varepsilon > 0, \ \exists \, \eta(\varepsilon) > 0 \text{ such that } \forall x \in K, \ \exists \, y_j \in B\left(x, \frac{\eta(\varepsilon,y_j)}{2}\right) \text{ such that} \\ \forall t \le \eta(\varepsilon), \ d\left(x + tF(y_j), K\right) = \varphi_{(x,y)}(t) \le \varepsilon t \end{cases}$$

This means that there exist some $v_j \in F(y_j)$ and $z_j \in K$ such that $\|z_j - x - tv_j\| \le \varepsilon t$. On the other hand,

$$\left(x, \frac{z_j - x}{t}\right) = (y_j, v_j) + \left(x - y_j, \frac{z_j - x}{t} - v_j\right) \in \text{Graph}(F) + \varepsilon(B \times B)$$

Consequently, taking any set-valued map F_ε such that $\text{Graph}(F_\varepsilon) \subset \text{Graph}(F) + \varepsilon(B \times B)$, we have proved (19.18):

$$\exists\, \eta(\varepsilon) > 0 \text{ such that } \forall x \in K,\ \forall h \in [0, \eta(\varepsilon)],\ (x + hF_\varepsilon(x)) \cap K \neq \emptyset$$

This completes the proof. □

11.2.4 Characterization of the Invariance Tangential Condition

We characterize in the same way the invariance tangential condition by replacing the tangent cone by its convex hull:

Theorem 11.2.8 *[Invariance Tangential Conditions] Let $K \subset X$ be a nonempty closed subset of a finite dimensional vector space. Assume that the set-valued map $F : K \rightsquigarrow X$ is Lipschitz on K with bounded values $F(x)$ (see Definition 10.3.5, p. 385).*

Let x_0 belong to K and $\alpha > 0$. Then the two following properties are equivalent:

$$\begin{cases} (i)\ \forall x \in K \cap \mathring{B}(x_0, \alpha),\ F(x) \subset T_K(x) \\ (ii)\ \forall x \in K \cap \mathring{B}(x_0, \alpha),\ F(x) \subset T_K^{\star\star}(x) \end{cases} \tag{11.10}$$

Furthermore, if F is locally bounded *(i.e., bounded on a neighborhood of each element), if K is compact and if*

$$\forall x \in K,\ F(x) \subset T_K^{\star\star}(x) \neq \emptyset$$

then, for all $\varepsilon > 0$,

$$\exists\, \eta(\varepsilon) > 0 \text{ such that } \forall x \in K,\ \forall h \in [0, \eta(\varepsilon)],\ x + hF(x) \subset K + \varepsilon h B \neq \emptyset \tag{11.11}$$

Remark. Property (11.11) means that the discrete explicit schemes $x \rightsquigarrow x + hF(x)$ associated with F are invariant in approximations $K + \varepsilon hB$ uniformly with respect to the discretization step h on compact sets. □

Proof. Assume that (11.10)(ii) holds true. We associate with any $v \in F(x)$ the function

$$\varphi_{(x,v)}(t) := d(x + tv, K) = \|x + tv - x_t\|$$

where $x_t \in K$ achieves the minimum. Recall that for all $w \in T_K(x_t)$, there exists a sequence $e(h_n)$ converging to 0 such that such that $x + h_n w + h_n e(h_n) \in K$. Therefore

$$\left\{\begin{array}{l}\frac{\varphi_{(x,v)}(t+h_n)-\varphi_{(x,v)}(t)}{h_n}\\ \leq \frac{\|x+tv+h_n(v-w)-x_t-h_ne(h_n)\|-\|x+tv-x_t\|}{h_n}\end{array}\right.$$

Taking the limit, we infer that

$$D_{\uparrow}\varphi_{(x,v)}(t)(1)\leq\left\langle v-w,\frac{x+tv-x_t}{\|x+tv-x_t\|}\right\rangle$$

and thus, that this inequality holds true for all $w \in T_K^{\star\star}(x_t)$.

Furthermore, we can associate with any $v \in F(x)$ an element $u_t \in F(x+tv)$ such that $\|v-u_t\| = d(v, F(x+tv))$ and, F being Lipschitz, we associate with $u_t \in F(x+tv)$ an element $w_t \in F(x_t) \subset T_K^{\star\star}(x_t)$ such that $\|u_t - w_t\| \leq \lambda\|x+tv-x_t\|$, where λ denotes the Lipschitz constant of F, so that

$$\left\langle u_t-w_t,\frac{x+tv-x_t}{\|x+tv-x_t\|}\right\rangle \leq \lambda\|x+tv-x_t\| = \lambda\varphi_{(x,v)}(t)$$

Therefore,

$$D_{\uparrow}\varphi_{(x,v)}(t)(1)\leq\left\langle v-w_t,\frac{x+tv-x_t}{\|x+tv-x_t\|}\right\rangle\leq\lambda\varphi_{(x,v)}(t)+d(v,F(x+tv))$$

Since $\varphi_{(x,v)}$ is lower semicontinuous, Proposition 11.2.5, p. 445 implies that

$$\forall t\geq 0,\ \varphi_{(x,v)}(t)\leq e^{\lambda t}\left(\varphi_{(x,v)}(0)+\int_0^t e^{-\lambda s}\sup_{v\in F(x)}d(v,F(x+sv))ds\right)$$

Therefore, for every $x \in K$ and any $v \in F(x)$, $\varphi_{(x,v)}(0) = 0$, so that, for any $v \in F(x)$,

$$\left\{\begin{array}{l}\forall t\geq 0,\ \left\|v-\frac{x_t-x}{t}\right\| \leq \sup_{s\in[0,t]}\sup_{v\in F(x)}d(v,F(x+sv))\int_0^t e^{\lambda(t-\tau)}d\tau\\ =\ \sup_{s\in[0,t]}\sup_{v\in F(x)}d(v,F(x+sv))\frac{e^{\lambda t}-1}{\lambda t}\end{array}\right.$$

Actually, since F is Lipschitz with bounded images,

$$\sup_{s\in[0,t]}\sup_{v\in F(x)}d(v,F(x+sv))\ \leq\ t\lambda\|F(x)\|$$

so that, setting $\eta_{(\varepsilon,x)} := \frac{1}{\lambda}\log\left(\frac{\varepsilon+\|F(x)\|}{\|F(x)\|}\right)$, for any $v \in F(x)$, for any $t \in \left[0,\eta_{(\varepsilon,x)}\right]$,

$$\left\| v - \frac{x_t - x}{t} \right\| \leq (e^{\lambda t} - 1)\|F(x)\| \leq (e^{\lambda \eta_{(\varepsilon,x)}} - 1)\|F(x)\| = \varepsilon$$

We infer that for any $\varepsilon > 0$, there exists $\eta_{(\varepsilon,x)}$ such that, for all $t \in \left[0, \eta_{(\varepsilon,x)}\right]$, for all $v \in F(x)$, $x + tv \in x_t + t\varepsilon B \subset K + t\varepsilon B$. In other words,

$$\forall t \in \left[0, \eta_{(\varepsilon,x)}\right],\ x + tF(x) \subset K + t\varepsilon B$$

Furthermore, $\|v - \frac{x_t - x}{t}\|$ converges to 0 with t. Since $x_t \in K$, we infer that v belongs also to the tangent cone $T_K(x)$.

Since F is locally bounded means that for any $x \in K$, there exists $\alpha_{(\varepsilon,x)}$ such that

$$\sup_{y \in B(x, \alpha_{(\varepsilon,x)})} \|F(y)\| \leq M(x) < +\infty$$

we infer that, setting $T_{(\varepsilon,x)} := \frac{1}{\lambda} \log \left(\frac{\varepsilon + M(x)}{M(x)} \right)$,

$$\forall t \in \left[0, T_{(\varepsilon,x)}\right], \forall y \in B(x, \alpha_{(\varepsilon,x)}) \cap K,\ y + tF(y) \subset K + t\varepsilon B$$

Consequently, if K is compact, it can be covered by a finite number n of balls $B(x_i, \alpha_{(\varepsilon,x_i)}) \cap K$. Therefore, setting $\eta(\varepsilon) := \min_{i=1,\ldots,n} T_{(\varepsilon,x)_i}$, we infer that for any $\varepsilon > 0$, there exists $\eta(\varepsilon)$ such that

$$\forall h \in [0, \eta_\varepsilon],\ \forall x \in K,\ x + hF(x) \subset K + \varepsilon h B$$

This completes the proof. □

11.3 Fundamental Viability and Invariance Theorems for Control Systems

11.3.1 The Regulation Map for Control Systems

The Regulation Map translates tangential conditions given in Theorems 11.2.7, p. 447 and 11.2.8, p. 451 to the case of control systems (f, U) defined by

$$x'(t) = f(x(t), u(t)) \text{ where } u(t) \in U(x(t)) \tag{11.12}$$

We introduce:

Definition 11.3.1 *[The Regulation Map] We associate with the dynamical system described by (f, U) and with the constraints described by K the* (set-valued) regulation map R_K*: it maps any state $x \in K$ to the (possibly empty) subset $R_K(x)$ consisting of controls $u \in U(x)$ which are* viable *in* *the sense that $f(x,u)$ is tangent to K at x*

$$R_K(x) := \{u \in U(x) \mid f(x,u) \in T_K^{\star\star}(x)\}$$

We translate the tangential conditions given in Theorems 11.2.7, p. 447 and 11.2.8, p. 451 for differential inclusions in terms of control systems:

Lemma 11.3.2 ***[Regulation Map and Tangential Conditions]*** *The viability tangential condition (11.5), p. 447 amounts to saying that*

$$\forall x \in K \backslash C, \;\; R_K(x) \neq \emptyset$$

and the invariance tangential condition (11.10), p. 451 amounts to saying that

$$\forall x \in K \backslash C, \;\; R_K(x) = U(x)$$

11.3.2 The Fundamental Viability Theorem

Results on properties satisfied by ***at least*** one evolution governed by control system (f,U)

$$x'(t) \;=\; f(x(t),u(t)) \text{ where } u(t) \;\in\; U(x(t))$$

(such as local viability, capturability, intertemporal optimality, etc.), hold true for the class of Marchaud set-valued maps and control systems:

Definition 11.3.3 ***[Marchaud Control Systems]*** *We say that the* control system (f,U) *is* Marchaud *if*

- *the set-valued map $U : X \rightsquigarrow \mathcal{U}$ is Marchaud,*
- *$f : X \times \mathcal{U} \mapsto X$ is continuous and affine with respect to the control,*
- *f and U satisfy the growth condition*

$$\forall (x,u) \in \text{Graph}(U), \;\; \|f(x,u)\| \leq c(\|x\| + \|u\| + 1)$$

Therefore, a Marchaud control system (f,U) provides a Marchaud evolutionary system.

We state the Viability Theorem:

Theorem 11.3.4 *[The Fundamental Viability Theorem for Control Systems] Let $K \subset X$ and $C \subset K$ be two closed subsets. Assume that the system (f, U) is Marchaud. Then the two following statements are equivalent*

1. *$K = \mathrm{Viab}_{\mathcal{S}}(K, C)$ is viable outside C under control system (f, U),*
2. *The regulation map*

$$\forall x \in K \backslash C, \quad R_K(x) \neq \emptyset \tag{11.13}$$

has nonempty values on the complement of the target in the environment.

In this case, every evolution $x(\cdot) \in \mathcal{S}(x)$ viable in K outside C is regulated by the system

$$x'(t) = f(x(t), u(t)) \text{ where } u(t) \in R_K(x(t)) \tag{11.14}$$

In other words, the set-valued map U involved in the original system (f, U) (11.12), p. 453 is replaced by the regulation map R_K.
Furthermore, the regulation map R_K has convex compact values.

Proof. The proof is given in Sect. 19.4.4, p.785. □

Note that when the control system (f, U) where $f(x, u) = u$ is actually differential inclusion $x'(t) \in F(x(t))$, this theorem boils down to Theorem 2.15.7, p. 103 of the viability survival kit 2.15, p. 98.

The first independent investigations of differential inclusions began in the early 1930s with papers by *Stanislas Zaremba* and *André Marchaud*, collaborators of George Bouligand.

Stanislaw Zaremba [1863–1942]. *His doctoral thesis* "Sur un problème concernant l'état calorifique d'un corp homogène indéfini" *was presented in 1889 Zaremba made many contacts with mathematicians of the French school, in particular with* Paul Painlevé, Henri Poincaré *and Hadamard. He returned to Kraków in 1900. In 1930, Hadamard wrote:* "The profound generalization due to him has recently transformed the foundations of potential theory and immediately became the starting point of research by young mathematicians of the French school." *In particular, he invited Georges Bouligand in Krakow in 1925 to deliver lectures which became known as the famous* "Leçons de Cracovie".

Theorem 11.3.4, p.455 is derived from Theorem 19.4.3, p.783 couched in terms of differential inclusions stated and proved in Sect. 19, p.769.

The viability Theorem 11.3.4, p. 455 allows us to derive the existence of solutions to implicit differential inclusions:

Theorem 11.3.5 ***[Implicit Differential inclusions]*** *Assume that a set-valued map $\Phi : X \times X \rightsquigarrow Y$ has a closed graph, that there exists a constant $c > 0$ that for every $x \in X$, $d(0, \Phi(x, v)) \leq c(\|x\| + 1)$, that the graph of $v \mapsto \Phi(x, v)$ is convex and that there exists $v \in T_K(x) \cap c(\|x\|+1)B$ such that $0 \in \Phi(x, v)$. Then, for any $x_0 \in K$, there exists a solution $x(\cdot)$ to implicit differential inclusion $0 \in \Phi(x(t), x'(t))$ satisfying $x(0) = x_0$ and viable in K.*

Proof. We associate with Φ the set-valued map $F : X \rightsquigarrow X$ defined by

$$F(x) := \{v \text{ such that } d(0, \Phi(x, v)) = 0\}$$

By assumption, $\|F(x)\| \leq c(\|x\| + 1)$ and $F(x) \cap T_K(x) \neq \emptyset$. The images are convex because the function $v \mapsto d(0, \Phi(x, v))$ is convex whenever the graph of $v \rightsquigarrow \Phi(x, v)$ is convex process, i.e., satisfy inclusions $\sum_i \alpha_i \Phi(x, v_i) \subset \Phi(x, \sum_i \alpha_i v_i)$. They imply that

$$d\left(0, \Phi\left(x, \sum_i \alpha_i v_i\right)\right) \leq d\left(0, \sum_i \alpha_i \Phi(x, v_i)\right) \leq \sum_i \alpha_i d(0, \Phi(x, v_i))$$

Hence, if $v_i \in F(x)$, then $\Phi(x, v_i) = 0$, so that $d\left(0, \Phi\left(x, \sum_i \alpha_i v_i\right)\right) = 0$ and thus, $\sum_i \alpha_i v_i \in F(x)$.

The graph of F is closed: Let $(x_n, u_n) \in \text{Graph}(F)$ converge to (x, u). Let $z_n \in \Phi(x_n, u_n)$ such that $\|z_n\| = d(0, \Phi(x_n, u_n)) \leq c(\|x_n\| + 1) \leq c(\|x\| + 2)$ for n large enough. A subsequence (again denoted by) z_n converges to some $z \in \Phi(x, u)$ (since the graph of Φ is assumed to be closed) satisfying $d(0, \Phi(x, u)) \leq \|z\|$. Consequently, inequalities $-c(\|x_n\| + 1) \leq -\|z_n\| = -d(0, \Phi(x_n, u_n))$ imply, by taking the limit, inequalities $-c(\|x\| + 1) \leq -\|z\| \leq -d(0, \Phi(x, u))$. Hence $(x, u) \in \text{Graph}(F)$.

Therefore, the Viability Theorem 11.3.4, p. 455 implies that from $x \in K$ starts a solution $t \mapsto x(t)$ to the differential inclusion $x'(t) \in F(x(t))$ which can be written $d(0, \Phi(x(t), x'(t))) = 0$, i.e., $0 \in \Phi(x(t), x'(t))$. □

11.3.3 The Fundamental Invariance Theorem

The context when a given property on one evolutionary system hold true for all evolutions under

$$x'(t) = f(x(t), u(t)) \text{ where } u(t) \in U(x(t))$$

we regarded the above system as a *tychastic system* and no longer as a *control system*, where we have to choose at least one control governing an evolution satisfying the required property.

This is the case for properties such as local invariance and absorption, for instance. We saw in Chap. 10, p. 375 that, at the level of evolutionary systems, these properties hold true whenever the evolutionary system is lower semicontinuous. This is the case for Lipschitz tychastic systems, which generate Lipschitz differential inclusions (see Definition 10.3.5, p. 385). By Theorem 10.3.6, p. 386 the lower semicontinuity of Lipschitz semicontinuous evolutionary systems ensues.

Definition 11.3.6 ***[Lipschitz Control Systems]*** *We say that the* control system (f, U) *is* Lipschitz *if*

- *the set-valued map $U : X \rightsquigarrow \mathcal{U}$ is Lipschitz,*
- *$f : X \times \mathcal{U} \mapsto X$ is Lipschitz with respect to the control.*

The Invariance Theorem states that K is invariant under a tychastic system if and only if all velocities are tangent to K, i.e., by Lemma 11.3.2, p.454, if and only if the regulation map coincides with the set-valued map U:

Theorem 11.3.7 ***[The Fundamental Invariance Theorem]*** *Let $K \subset X$ and $C \subset K$ be two closed subsets. Assume that the system (f, U) is Lipschitz with bounded values.*

Then the two following statements are equivalent

1. *K is invariant outside C under (f, U)*
2. *The regulation map coincides with the set-valued map U on $K \setminus C$:*

$$\forall x \in K \backslash C, \; R_K(x) = U(x) \tag{11.15}$$

In particular, when the target C is empty, K is invariant under (f, U) if and only if the above condition (11.10) holds true for any $x \in K$.

The Invariance Theorem has been proved by Frank Clarke in *Optimization and nonsmooth analysis*, [63, Clarke]. We shall prove separately the sufficient and necessary conditions in the framework of the differential inclusion $x'(t) \in F(x(t))$ generated by the control system, defined by $F(x) := f(x, U(x))$.

Proposition 11.3.8 ***[Sufficient Conditions for Invariance]*** *Assume that F is Lipschitz. Then condition*

$$\forall x \in K \backslash C, \; F(x) \subset T_K^{\star\star}(x)$$

implies that K is invariant outside C under F.

Proof. Let us assume that $F(y) \subset T_K^{\star\star}(y)$ in a neighborhood of $x \in K \setminus C$ and let $x(\cdot) \in \mathcal{S}(x_0)$ be any solution to differential inclusion $x'(t) \in F(x(t))$ starting at x_0 in a neighborhood of x and defined on some interval $[0, T]$. Let t be a point such that both $x'(t)$ exists and $x'(t)$ belongs to $F(x(t))$. Then there exists $\varepsilon(h)$ converging to 0 with h such that

$$x(t+h) = x(t) + hx'(t) + h\varepsilon(h)$$

Introduce

$$\varphi(t) := d(x(t), K) = \|x(t) - x_t\|$$

where $x_t \in K$ achieves the minimum. Recall that for all $w \in T_K(x_t)$, there exists a sequence $e(h_n)$ converging to 0 such that such that $x_t + h_n w + h_n e(h_n) \in K$. Therefore

$$\frac{\varphi(t+h_n) - \varphi(t)}{h_n} \leq \frac{\|x(t) + h_n(x'(t) - w) - x_t - h_n e(h_n)\| - \|x(t) - x_t\|}{h_n}$$

Taking the limit, we infer that

$$\forall\, w \in T_K(x_t), \;\; D_\uparrow \varphi(t)(1) \leq \left\langle x'(t) - w, \frac{x(t) - x_t}{\|x(t) - x_t\|} \right\rangle$$

and thus, that this inequality also holds true for all $w \in T_K^{\star\star}(x_t)$.

Since F is Lipschitz, there exists a constant $\lambda \in \mathbb{R}$ and $w_t \in F(x_t) \subset T_K^{\star\star}(x_t)$ such that $\|x'(t) - w_t\| \leq \lambda \|x(t) - x_t\|$. Therefore

$$D_\uparrow \varphi(t)(1) \leq \left\langle x'(t) - w_t, \frac{x(t) - x_t}{\|x(t) - x_t\|} \right\rangle \leq \lambda \|x(t) - x_t\| = \lambda \varphi(t)$$

Then φ is a lower semicontinuous solution to $D_\uparrow \varphi(t)(1) \leq \lambda \varphi(t)$ and thus, by Proposition 11.2.5, p. 445, satisfies inequality $\varphi(t) \leq \varphi(0)e^{\lambda t}$. Since $\varphi(0) = 0$, we deduce that $\varphi(t) = 0$ for all $t \in [0, T]$, and therefore that $x(t)$ is viable in K on $[0, T]$. □

A necessary condition requires the existence of a solution to a differential inclusion starting from both an initial state and initial velocity. For that purpose, we need the Filippov Theorem stating that a Lipschitz set-valued map with closed values satisfies the Filippov property:

Theorem 11.3.9 *[Filippov Theorem] Lipschitz maps $F : X \rightsquigarrow X$ with constant $\lambda \in \mathbb{R}_+$ and closed values satisfy the λ-Filippov property : If for any evolution $\xi(\cdot)$ such that $t \to d(\xi'(t), F(\xi(t)))$ is integrable for the measure $e^{-\lambda s}ds$, there exists a solution $x(\cdot) \in \mathcal{S}_F(x)$ to differential inclusion $x'(t) \in F(x(t))$ such that, for all $t \geq 0$, the Filippov inequality: $\forall t \geq 0$,*

$$\|x(t) - \xi(t)\| \leq e^{\lambda t}\left(\|x - \xi(0)\| + \int_0^t d(\xi'(s), F(\xi(s)))e^{-\lambda s}ds\right) \quad (11.16)$$

holds true.

Proof. We do not provide the proof of this important, but classical result, which can be found in *Differential Inclusions*, [25, Aubin & Cellina]. □

36 ***Alexei Fyodorovich Filippov [1923-].*** *Professor at the department of differential equations in the faculty of mechanics and mathematics. He fought in the Great Patriotic War and was awarded several medals. He won the M.V.Lomonosov Prize for being a brilliant lecturer and for having written important monographs and texts. Together with the Krakow school, he proved the main results dealing with differential inclusion.*

The Filippov Theorem implies readily

Proposition 11.3.10 *Let F be a Lipschitz map. Then, for any velocity $v_0 \in F(x_0)$, there exists a solution $x(\cdot) \in \mathcal{S}_F(x_0)$ satisfying $x'(0) = v_0$.*

Proof. The Filippov property applied to the evolution $\xi(t) := x_0 + tv_0$ implies that there exists an evolution $x(\cdot) \in \mathcal{S}(x_0)$ to differential inclusion $x'(t) \in F(x(t))$ starting at x_0 and satisfying the Filippov inequality

$$\forall t \geq 0,\ \|x(t) - \xi(t)\| \leq e^{\lambda t}\left(\|x - \xi(0)\| + \int_0^t d(\xi'(s), F(\xi(s)))e^{-\lambda s}ds\right) \tag{11.17}$$

which boils down to

$$\|x(t) - x_0 - tv_0\| \leq \int_0^t d(v_0, F(x_0 + sv_0))e^{\lambda(t-s)}ds \ \leq\ t\|v_0\|(e^{\lambda t} - 1)$$

Dividing by $t > 0$, we obtain

$$\left\|\frac{x(t) - x_0}{t} - v_0\right\| \ \leq\ \|v_0\|(e^{\lambda t} - 1)$$

and letting t converge to $0+$, we infer $x'(0) = v_0$. □

This proposition implies the following necessary condition:

Proposition 11.3.11 *[Necessary Conditions for Invariance]* *Assume that for any* $(x_0, v_0) \in \text{Graph}(F)$, *there exists a solution* $x(\cdot) \in \mathcal{S}_F(x_0)$ *satisfying* $x'(0) = v_0$. *If* K *is invariant outside* C *under* F, *then*

$$\forall x \in K\backslash C,\ F(x)\ \subset\ T_K(x)$$

Proof. Let $x_0 \in K\backslash C$. We have to prove that any $v_0 \in F(x_0)$ is tangent to K at x_0. By assumption, for all x_0 and $v_0 \in F(x_0)$, there exists a solution $x(\cdot)$ to the differential inclusion $x'(t) \in F(x(t))$ satisfying $x(0) = x_0$ and $x'(0) = v_0$ viable in $K \setminus C$. Hence v_0, being the limit of $\frac{x(t_n) - x_0}{t_n} \in \frac{K - x_0}{t_n}$, it belongs to $T_K(x_0)$. It follows that $F(x_0)$ is contained in $T_K(x_0)$. □

11.4 Regulation Maps of Kernels and Basins

We shall use those fundamental Viability and Invariance theorems, together with the viability characterizations of viability and invariance, for characterizing viability kernel and capture basins in terms of the *regulation map*.

Proposition 11.4.1 *[Characterization of Viability Kernels with Target]* *Let us assume that the control system* (f, U) *is Marchaud, that the environment* $K \subset X$ *and the target* $C \subset K$ *are closed subsets. Then the viability kernel* $\text{Viab}_S(K, C)$ *is the* ***largest*** *closed subset* D *satisfying*

- $C \subset D \subset K$,

- $K \setminus D$ *is a repeller*
- *and*

$$\forall x \in D \backslash C, \;\; R_D(x) \; := \; \{u \in U(x) \mid f(x,u) \; \in \; T_D^{\star\star}(x)\} \; \neq \; \emptyset$$

In this case,

$$C \cap \mathrm{Exit}_{\mathcal{S}}(K) \; \subset \; C \cap \mathrm{Exit}_{\mathcal{S}}(D) \; \subset \; \mathrm{Exit}_{\mathcal{S}}(C) \; \subset \; C \; \subset \; D \; \subset \; K \quad (11.18)$$

Furthermore, for every $x \in D$,

1. *there exists at least one evolution $x(\cdot) \in \mathcal{S}(x)$ viable in D forever or until it reaches the target C in finite time*
2. *and all evolutions $x(\cdot) \in \mathcal{S}(x)$ viable in D forever or until they reach the target C in finite time are governed by the differential inclusion*

$$x'(t) \; = \; f(x(t), u(t)) \;\text{ where }\; u(t) \; \in \; R_D(x(t))$$

Invariance kernels are also characterized in terms of tangential conditions:

Proposition 11.4.2 *[Characterization of Invariance Kernels] Assume that $(f,U)$is Lipschitz and that K and $C \subset K$ are closed. Then the invariance kernel* $\mathrm{Inv}_F(K,C)$ *of K with target C under control system (f,U) is the* largest closed subset D *between C and K such that*

$$\forall x \in D \backslash C, \; U(x) \; = \; R_D(x)$$

These fundamental theorems characterizing viability kernels and capture basins justify a further study of the regulation map and equivalent ways to characterize it. Actually, using Proposition 11.4.2, p. 461, we can go one step further and characterize viability kernels and capture basins in terms of *Frankowska Property*, stated in two equivalent forms: the *tangential formulation*, expressed in terms of tangent cones, and its *dual version*, expressed in terms of normal cones in Sect. 11.6, p. 475. This property was discovered by Hélène Frankowska in a series of papers at the end of the 1980s and the beginning of the 1990s in the framework of her epigraphical approach of optimal control theory. It happens that they are true at the level of differential inclusions, and thus, useful in other contexts than intertemporal optimization.

Applied to the case of viability kernels and capture basins of epigraphs of functions (see Chaps. 13, p.523, and 17, p. 681), the tangential version is at the origin of theorems stating that value functions in optimal control theory are generalized "episolutions" to Hamilton–Jacobi partial differential equation introduced in the early 1990. Using the dual version, we recover an extension to lower semicontinuous solutions to Hamilton–Jacobi–Bellman of the concept of viscosity solutions (proposed in 1983 by *Michael Crandall* and *Pierre-Louis Lions* for continuous solutions), discovered independently by *Emmanuel Barron* and *Robert Jensen* with partial differential techniques and *Hélène Frankowska* with viability techniques.

We begin with the case of tangential characterization. For that purpose, we introduce the backward regulation map:

Definition 11.4.3 ***[Backward Regulation Map]*** *The* backward regulation map $\overleftarrow{R}_K$ *is defined by*

$$\overleftarrow{R}_K(x) \;:=\{u\in U(x) \text{ such that } -f(x,u)\in T_K(x)\}$$

Definition 11.4.4 ***[Frankowska Property]*** *Let us consider three subsets $C\subset D\subset K$ (where the target C may be empty) and a set-valued map (f,U).*

The set-valued map $F : K \rightsquigarrow X$ satisfies the Frankowska property *on D with respect to (K,C) if*

$$\begin{cases} (i) \;\; \forall\, x\in D\setminus C, \;\; R_D(x) \;\neq\; \emptyset \\ (ii) \;\; \forall\, x\in D\cap \mathring{K}, \;\; U(x) \;=\; \overleftarrow{R}_D(x) \\ (iii) \; \forall\, x\in D\cap\partial K, \;\; \overleftarrow{R}_K(x) \;=\; \overleftarrow{R}_D(x)\end{cases} \tag{11.19}$$

Observe that if K is backward invariant, condition (11.19) boils down to condition

$$\begin{cases} (i) \;\forall\, x\in D\setminus C, \;\; R_D(x) \;\neq\; \emptyset \\ (ii) \;\forall\, x\in D, \;\; U(x) \;=\; \overleftarrow{R}_D(x)\end{cases} \tag{11.20}$$

Note that whenever $K\setminus C$ is a repeller, so are the subsets $K\setminus D$ whenever $D\supset C$, because $\text{Viab}_F(K\setminus D)\subset \text{Viab}_F(K\setminus C)=\emptyset$.

Frankowska property (11.19), p. 462 implies that

$$\forall\, x\in D\cap(\mathring{K}\setminus C), \;\; R_D(x) \;\neq\; \emptyset \text{ and } U(x) \;=\; \overleftarrow{R}_D(x)$$

The Viability and Invariance Theorems imply that

Theorem 11.4.5 *[Frankowska Characterization of Viability Kernels] Let us assume that (f, U) is both Marchaud and Lipschitz (see Definition 10.3.5, p. 385), and that the subset K is closed. The viability kernel* $\text{Viab}_F(K)$ *of a subset K under S is the* **unique** *closed subset satisfying*

- $C \subset D \subset K$,
- $K \setminus D$ *is a repeller*
- *and the* Frankowska property *(11.19), p. 462.*

Furthermore, inclusions (11.18), p. 461

$$C \cap \text{Exit}_{\mathcal{S}}(K) \subset C \cap \text{Exit}_{\mathcal{S}}(D) \subset \text{Exit}_{\mathcal{S}}(C) \subset C \subset D \subset K$$

hold true.

For capture basins, we obtain

Theorem 11.4.6 *[Characterization of Capture Basins] Let us assume that (f, U) is Marchaud and Lipschitz and that the environment $K \subset X$ and the target $C \subset K$ are closed subsets such that $K \setminus C$ is a repeller* ($\text{Viab}_F(K \backslash C) = \emptyset$)*. Then the viable-capture basin* $\text{Capt}_F(K, C)$ *is the* **unique** *closed subset D satisfying*

- $C \subset D \subset K$,
- *and the* Frankowska property *(11.19), p. 462.*

In this case, inclusions (11.18), p. 461 hold true.

It may be useful to translate Theorem 10.5.13, p. 406 in terms of tangential conditions to the complement of the subset D, which happen to imply the existence and uniqueness of continuous value *viscosity solutions* to Hamilton–Jacobi partial differential equation.

Definition 11.4.7 *[Viscosity Property] Let us consider control system (f, U) and two subsets $C \subset K$ and K. We shall say that a subset D between C and K satisfies the* viscosity property *with respect to (f, U) if*

$$\begin{cases} (i) \ \forall x \in D \backslash C, \ R_D(x) \neq \emptyset \\ (ii) \ \forall x \in \text{Int}(K) \backslash \text{Int}(D), \ U(x) = R_{\overline{\complement D}}(x) \end{cases} \quad (11.21)$$

The Fundamental Viability Theorem 11.3.4, p. 455, Invariance Theorem 11.3.7, p. 457 and Theorem 10.5.13, p. 406 imply the following "viscosity characterization" of capture basins:

Proposition 11.4.8 *[The Viscosity Property of Viable-Capture Basins] Let us assume that the control system (f, U) is Marchaud and Lipschitz, that $C \subset K$ and K are closed, that $\mathrm{Int}(K) \neq \emptyset$ is backward invariant, that $\mathrm{Viab}_F(K \backslash C) = \emptyset$, that $\overline{\mathrm{Int}(K)} = K$ and that $\overline{\mathrm{Int}(C)} = C$.*

Then the capture basin $\mathrm{Capt}_F(K, C)$ of the target C viable in K is the ***unique*** *topologically regular subset D between C and K satisfying the viscosity property (11.21).*

In the absence of constraints, we obtain the following

Corollary 11.4.9 *[The Viscosity Property of Capture Basins] Let us assume that the control system (f, U) is Marchaud and Lipschitz, that $\mathrm{Abs}_F(C) = X$ and that $\overline{\mathrm{Int}(C)} = C$.*

Then the capture basin $\mathrm{Capt}_F(K, C)$ of the target C is the ***unique*** *topologically regular subset D containing C and satisfying the viscosity property (11.21).*

Invariance kernels are also characterized in terms of viscosity condition:

Proposition 11.4.10 *[Viscosity Characterization of Invariance Kernels] Assume that the system (f, U) is both Marchaud and Lipschitz. Then the invariance kernel satisfies the two following tangential conditions*

$$\begin{cases} (i) \ \forall x \in D \backslash C, \ \ U(x) = R_D(x) \\ (ii) \ \forall\, x \in \mathrm{Int}(K) \setminus \mathrm{Int}(D), \ \ R_{\overline{\complement D}}(x) \neq \emptyset \end{cases} \tag{11.22}$$

Proof. The first property follows from Invariance Theorem 11.3.7, p. 457 and Proposition 10.6.1, p. 411, stating that the invariance kernel is the largest closed subset such that $D \setminus C$ is locally invariant. The second property is derived from Viability Theorem 11.3.4, p. 455 and Proposition 10.6.2, p. 411, stating that $\overline{\complement D} = \mathrm{Capt}_F(\overline{\complement D}, \overline{\complement K})$, and thus, that $\overline{\complement D} \setminus \mathrm{Int}(K)$ is locally viable. □

11.5 Lax–Hopf Formula for Capture Basins

11.5.1 *P-Convex and Exhaustive Sets*

Viability Theorems require the differential inclusion to be Marchaud, and thus, in particular, that the differential inclusion has convex values. Many examples, among the most important ones, do not satisfy these convexity properties. However, one can "correct" the lack of convexity by, up to the Minkowski addition of a closed convex cone P to the images of the set-valued map, obtain a new set-valued map with convex values, for which Viability Theorem 11.3.4, p. 455 applies. Therefore, the main task remains to check the relations between viability kernels with targets under the first non-Marchaud differential inclusion and the corrected one.

Proceeding further, for differential inclusions with constant right hand side, one can prove a Lax–Hopf formula expressing capture and absorption basins in terms of simple Minkowski sums and differences. In this case, we obtain more interesting consequences.

Why the Lax–Hopf formula? Because, when we shall apply viability techniques in the epigraphical and graphical approaches to first-order partial differential equations, the Lax–Hopf formula for capture basins provides the Lax–Hopf formula for the solutions to these partial differential equations which enjoy this property.

In this section, we shall introduce a closed convex cone $P \subset X$ of the state space. When X is a vector space, a closed convex cone induces a preordering $\succeq_P$ defined by

$$x \succeq_P y \text{ if and only if } x - y \in P$$

Definition 11.5.1 ***[P-exhaustive Sets]*** *Let P be a closed convex cone. For any subset $C \subset X$, the subset $C + P$ is regarded as the P*-exhaustive envelope *of C. A subset C is said to be P*-exhaustive *if $C = C + P$, P-convex if its P- exhaustive envelope is convex, P-closed if its P- exhaustive envelope is closed, etc. Observe that $C = C + \{0\}$, so that we recover the usual concepts by using the trivial cone $P := \{0\}$. We associate with a cone P the preorder relation $C_1 \subset_P C_2$ if $C_1 + P \subset C_2 + P$ and the associated equivalence relation $C_1 \equiv_P C_2$ if $C_1 + P = C_2 + P$.*

In optimization theory, the boundary $\partial(C + P)$ of the *P-exhaustive envelope* of C is called the Pareto subset of C of Pareto minima with respect to the preorder relation $\succeq_P$ induced by the closed convex cone.

Recall that the *Minkowski difference* $B \ominus A$ is the subset of elements x such that $x + A \subset B$ (see Definition 4.6.3, p. 154). Therefore

$$C \text{ is } P\text{-exhaustive if and only if } P \subset C \ominus C$$

We single out some obvious property

- If P is a convex cone and C is a convex subset, then its P-exhaustive envelope is convex;
- If P is a closed cone and C is a compact subset, then its P-exhaustive envelope is closed.

If H is a closed subset, we denote by

$$\mathbb{R}_+ H = \bigcup_{\lambda \geq 0} \lambda H \text{ the cone spanned by the subset } H$$

Lemma 11.5.2 *Let H be a compact convex subset and P a closed convex cone. If $\mathrm{co}(H) \cap P = \emptyset$, then*

$$\mathbb{R}_+ \mathrm{co}(H) - P = \mathbb{R}_+ (\mathrm{co}(H - P)) \text{ is closed} \tag{11.23}$$

Consequently, if $G := H - P$ is closed and convex, then

$$\mathbb{R}_+ \mathrm{co}(H) - P = \mathbb{R}_+ G \tag{11.24}$$

Proof. First, we observe that $\mathbb{R}_+ \mathrm{co}(H) - P = \mathbb{R}_+ \mathrm{co}(H - P)$ because P is a closed convex cone. The subset H being assumed to be compact, its convex hull $\mathrm{co}(H) = \overline{\mathrm{co}(H)}$ is also compact. We have to prove that if $\mathrm{co}(H) \cap P = \emptyset$, then $\mathbb{R}_+ (\mathrm{co}(H) - P)$ is closed. For that purpose, let us take a sequence $\lambda_n > 0$, $x_n \in \mathrm{co}(H)$ and $p_n \in P$ and such that the sequence $y_n := \lambda_n x_n - p_n \in \mathrm{co}(H) - P$ converges to some y. Since x_n ranges over a compact subset, a subsequence (again denoted by) x_n converges to some $x \in \mathrm{co}(H)$. Next, let us prove that the sequence λ_n is bounded. If not, there would exist a subsequence (again denoted by) λ_n going to $+\infty$. Dividing by λ_n, we infer that $x_n = \frac{y_n}{\lambda_n} + q_n$ where $q_n := \frac{p_n}{\lambda_n} \in P$ because P is a cone. Since $\frac{y_n}{\lambda_n}$ converges to 0, we infer that q_n converges to $q = x \in P$, because P is closed. Hence $x \in \mathrm{co}(H) \cap P$, which is impossible. Therefore, the sequence λ_n being bounded, there exists a subsequence (again denoted by) λ_n converging to some $\lambda \geq 0$. Consequently, $p_n = \lambda_n x_n - y_n$ converges to some $p = \lambda x - y \in P$, so that y belongs to $\mathrm{co}(H) - P$, which is then closed.
□

Remark. By taking $P := \{0\}$ and assuming that H is a compact convex subset which does not contain $\{0\}$, the cone $\mathbb{R}_+ H$ spanned by H is a closed convex cone. Such a subset H is called a *sole* of the closed convex cone $\mathbb{R}_+ H$.
□

There are many examples of P exhaustive subsets, among which we mention the following ones:

Examples:

1. The main example of a P-convex map G is obtained when $a := (a_i)_{i=1,\ldots,n} \in \{-1,0,+1\}^n$ is a sign vector and where $P := \mathbb{R}^n_a$ is the closed convex cone of vectors $x := (x_i)_{i=1,\ldots,n}$ such that $x_i \geq 0$ if $a_i = 1$, $x_i \leq 0$ if $a_i = -1$ and $x_i = 0$ if $a_i = 0$ and when $G(u) := \{g_i(u)\}$ is the single-valued map where the components g_i are convex if $a_i = 1$, concave if $a_i = -1$ and affine if $a_i = 0$. It is easy to check in this case, the single-valued map $u \rightsquigarrow G(u)$ is $\mathbb{R}^n_a$-convex. Furthermore, it is $\mathbb{R}^n_a$-closed whenever the components g_i are lower semicontinuous if $a_i = 1$, upper semicontinuous if $a_i = -1$ and continuous if $a_i = 0$.
2. This is for instance the case for epigraphs $\mathcal{E}p(\mathbf{c}) \subset X \times \mathbb{R}$ of extended functions $\mathbf{c} : X \mapsto \mathbb{R} \cup \{+\infty\}$ satisfying
$$\mathcal{E}p(\mathbf{c}) = \mathcal{E}p(\mathbf{c}) + \{0\} \times \mathbb{R}_+$$
for the cone $P := \{0\} \times \mathbb{R}_+$. This property is at the root of the epigraphical approach to a large class of Hamilton–Jacobi–Bellman equations and conservation laws (see Chaps. 14, p. 563 and 17, p. 681), and in particular, of Lax–Hopf formulas of solutions to some Hamilton–Jacobi–Bellman equations;
3. We shall see that an extended function $\mathbf{c} : X \mapsto \mathbb{R} \cup \{+\infty\}$ is decreasing with respect to a closed convex cone P if and only if
$$\mathcal{E}p(\mathbf{c}) = \mathcal{E}p(\mathbf{c}) + P \times \mathbb{R}_+$$
(see Proposition 14.6.1, p.592).
4. To say that an extended function $\mathbf{c}$ is λ-Lipschitz amounts to saying that, denoting by $\mathbf{v}_\lambda$ the function defined by $\mathbf{v}_\lambda(x) := \lambda\|x\|$,
$$\mathcal{E}p(\mathbf{c}) = \mathcal{E}p(\mathbf{c}) + \mathcal{E}p(\mathbf{v}_\lambda)$$
for the cone $P := \mathcal{E}p(\mathbf{v}_\lambda)$.
5. If $\mathbf{v} : X \mapsto \mathbb{R} \cup \{+\infty\}$ is a positively homogenous lower semicontinuous convex function, then its epigraph $\mathcal{E}p(\mathbf{v})$ is a closed convex cone. By definition, the inf-convolution $\mathbf{c} \star \mathbf{v}$ is defined by
$$\mathcal{E}p(\mathbf{c} \star \mathbf{v}) \ := \ \mathcal{E}p(\mathbf{c}) + \mathcal{E}p(\mathbf{v})$$
Under some adequate assumptions, the function $\mathbf{c} \star \mathbf{v}$ is equal to
$$\mathbf{c} \star \mathbf{v}(x) \ := \inf_{x=y+z,\ y \in X,\ z \in X} \mathbf{c}(y) + \mathbf{v}(z)$$
Hence a function $\mathbf{c}$ is $\mathcal{E}p(\mathbf{v})$-exhaustive if and only if $\mathbf{c} = \mathbf{c} \star \mathbf{v}$.

Definition 11.5.3 *[$Q \times P$-Exhaustive and P-Valued Maps] Let us consider a set-valued map $F : X \rightsquigarrow Y$, a closed convex cone $Q \subset X$ and a closed convex cone $P \subset Y$.*

The set-valued map F is said to be $Q \times P$-exhaustive if $\mathrm{Graph}(F) = \mathrm{Graph}(F) + Q \times P$ *and said to be P-valued if for any $x \in \mathrm{Dom}(F)$, $F(x) = F(x) + P$.*

Observe that a P-valued map is $\{0\} \times P$-set-valued map and that the set-valued map G defined by $\mathrm{Graph}(H) := \mathrm{Graph}(F) + Q \times P$ satisfies

$$H(x) = \bigcup_{p \in P,\, q \in Q} (F(x-q)+p)$$

More generally, if $F : X \rightsquigarrow Y$ and $G : X \rightsquigarrow Y$ are set-valued maps from X to Y, we denote by $F \star G$ the set-valued map defined by $\mathrm{Graph}(F \star G) := \mathrm{Graph}(F) + \mathrm{Graph}(G)$ satisfying

$$F \star G(x) = \bigcup_{x=y+z,\, y \in X,\, z \in X} (F(y)+G(z)) = \bigcup_{y \in X} (F(x-y)+G(y))$$

Observe that the union of two P-exhaustive subsets is P-exhaustive and that the sum $C + A$ of an arbitrary subset A with a P-exhaustive subset C is P-exhaustive. This means that the family $\mathbf{E}(P)$ of P-exhaustive subsets $C \subset K$ forms a *Max-Plus ring* where the "scalars" are an arbitrary subset, where the union plays the role of the sum in vector spaces and the sum of a P-exhaustive subset and of an arbitrary A plays the role of the multiplication by the scalars.

A map $\Phi : \mathbf{E}(P) \mapsto \mathbf{E}(P)$ is called a *Max-Plus morphism* if

$$\begin{cases} (i)\ \Phi(C_1 \cup C_2) = \Phi(C_1) \cup \Phi(C_2) \\ (ii)\ \Phi(C + A) = \Phi(C) + A \end{cases}$$

It turns out that the Lax–Hopf formula implies that, for constant set-valued maps H such that $H - P$ is closed and convex, the map $C \mapsto \mathrm{Capt}_H(K, C)$ is a Max-Plus morphism on the ring of P-exhaustive subsets (see Theorem 11.5.6, p.471 below). This algebraic result will play an important role in the analysis of the Moskowitz partial differential equation studied in Chap. 14, p. 563.

11.5.2 Lax–Hopf Formulas

For constant differential inclusions $x'(t) \in G$, we can obtain simple formulas of the capture basins $\mathrm{Capt}_G(K, C)$ when G is a closed convex subset, and even for non convex right hand sides G such that their $-P$-exhaustive envelopes are closed and convex. This allows us to deduce properties of capture basins under $-P$-convex and $-P$-closed constant maps: their capture basins of P exhaustive targets are also P-exhaustive.

Theorem 11.5.4 *[Lax–Hopf Formula for Capture Basins] Assume that the target C is contained in the environment K. The capture basin satisfies the inclusion*

$$\text{Capt}_F(K,C) \subset K \cap (C - \mathbb{R}_+\overline{\text{co}}(\text{Im}(F))) \quad (11.25)$$

If the set-valued map $F(x) := G$ is constant, then

$$K \cap (C - \mathbb{R}_+ G) \subset \text{Capt}_G(\overline{\text{co}}(K), C) \quad (11.26)$$

Consequently, if K is a closed convex subset, $C \subset K$ is closed and G is a constant set-valued map with a closed convex image G, then the capture basin enjoys the Lax–Hopf formula

$$\text{Capt}_G(K,C) = K \cap (C - \mathbb{R}_+ G) \quad (11.27)$$

Proof. First, let us consider an element $x \in \text{Capt}_F(K,C)$. Then x belongs to K and there exist a solution $x(\cdot)$ to the differential inclusion $x'(t) \in F(x(t))$ and $T \geq 0$ such that $x(T) \in C$ and $\forall\, t \in [0,T],\ x(t) \in K$.

Hence

$$\frac{x(T)-x}{T} \in \frac{1}{T}\int_0^T F(x(t))dt \subset \frac{1}{T}\int_0^T \text{Im}(F)dt = \overline{\text{co}}(\text{Im}(F)) \quad (11.28)$$

This implies that

$$x = x(T) - T\frac{x(T)-x}{T} \in C - T\overline{\text{co}}(\text{Im}(F)) \subset C - \mathbb{R}_+\overline{\text{co}}(\text{Im}(F))$$

On the other hand, if the set-valued map $F(x) \equiv G$ is constant, let us take $x \in \overline{\text{co}}(K) \cap (C - \mathbb{R}_+ G)$. Hence, there exist $g \in G$, $T \geq 0$ and $\xi \in C$ such that $x = \xi - Tg$. The evolution $x(\cdot) : t \mapsto x(t) := x + tg$ is a solution to differential equation $x'(t) = g \in G$ starting at $x \in K$ and satisfying $x(T) := x + Tg = \xi \in C$. It is viable in $\overline{\text{co}}(K)$. Indeed, since $x \in K$ and since $\xi = x + Tg \in C \subset K$, then $x(t) := x + tg = \left(1 - \frac{t}{T}\right)x + \frac{t}{T}(x + Tg) \in \overline{\text{co}}(K)$. This means that x belongs to the capture basin $\text{Capt}_F(\overline{\text{co}}(K), C)$.

The last statement follows from inclusions (11.25), p. 469 and (11.26), p. 469 when K is assumed convex and the constant set-valued map G is closed and convex. □

We infer from these definitions the following consequence of Lax–Hopf formula (11.27), p. 469 on exhaustivity.

Theorem 11.5.5 ***[Lax–Hopf Formula for the Sum of Two Targets]*** *Let P be a closed convex cone contained in K. Assume that H is a compact subset which does not contain 0 and satisfies $\text{co}(H) \cap P = \emptyset$. Then, for any target $C \subset K$,*

$$\text{Capt}_H(K,C)+P \subset \text{Capt}_{\text{co}}(H-P)(K+P,C+P) \subset \text{Capt}_H(K+P,C+P)$$

Consequently, if $G := H - P$ is closed and convex, if the closed convex environment K and the closed target C are P-exhaustive, then

$$\text{Capt}_H(K,C) = \text{Capt}_{\text{co}(H-P)}(K,C) = \text{Capt}_H(K,C) + P \tag{11.29}$$

Proof. Formula (11.25), p. 469 of Theorem 11.5.4, p.469 implies that

$$\begin{cases} \text{Capt}_H(K,C) + P \subset K \cap (C - \mathbb{R}_+\overline{\text{co}}(H)) + P \\ \subset (K+P) \cap (C - (\mathbb{R}_+\overline{\text{co}}(H) - P)) \end{cases}$$

By formula (11.23), p. 466 of Lemma 11.5.2, p.466, we infer that $\mathbb{R}_+\overline{\text{co}}(H) - P = \mathbb{R}_+\text{co}(H-P)$, and thus, by formula (11.27), p. 469, that

$$\text{Capt}_H(K,C)+P \subset (K+P)\cap(C-\mathbb{R}_+\text{co}(H-P)) = \text{Capt}_{\text{co}(H-P)}(K+P,C)$$

On the other hand, by formula (11.26), p. 469, we obtain

$$\begin{cases} \text{Capt}_{\text{co}(H-P)}(K+P,C) = (K+P) \cap (C - \mathbb{R}_+(H-P)) \\ = (K+P) \cap (C + P - \mathbb{R}_+(H)) \subset \text{Capt}_H(K+P,C+P) \end{cases}$$

We thus derived inclusions

$$\text{Capt}_H(K,C) + P \subset \text{Capt}_{\text{co}(H-P)}(K+P,C) \subset \text{Capt}_H(K+P,C+P)$$

Consequently, if $K = K + P$ and $C = C + P$ are convex, then

$$\text{Capt}_H(K,C) \subset \text{Capt}_H(K,C)+P \subset \text{Capt}_{\overline{\text{co}}(H-P)}(K,C) \subset \text{Capt}_H(K,C)$$

which implies equation (11.29), p. 470 holds true. □

Theorem 11.5.6 *[Max-Plus Morphism] Let P be a closed convex cone contained in K. Assume that the $-P$-exhaustive envelope $H - P$ of a constant subset H is closed and convex and that the environment K and target $C \subset K$ are P-exhaustive. Then the map $A \mapsto \text{Capt}_H(K, A)$ satisfies*

$$\begin{cases} (i) \ \text{Capt}_H(K, A \cup B) = \text{Capt}_H(K, A) \cup \text{Capt}_H(K, B) \\ (ii) \ \text{Capt}_H(K, C + A) = \text{Capt}_H(K, C) + A \end{cases} \tag{11.30}$$

Proof. The first statement is provided by the first statement of Lemma 10.2, p.378. The second one follows from the Lax–Hopf formula. Observe first that if C is P-exhaustive, then, for any subset A, $C + A$ is P-exhaustive because $(C + A) + P = (C + P) + A = C + A$. Therefore, formula (11.29), p. 470 of Theorem 11.5.5, p.470 implies that

$$\text{Capt}_H(K, C + A) = \text{Capt}_{H-P}(K, C + A)$$

Since $H - P$ is closed and convex by assumption, formula (11.27), p. 469 implies that

$$\begin{cases} \text{Capt}_{H-P}(K, C + A) = (C + A) - \mathbb{R}_+(H - P) \\ = (C - \mathbb{R}_+(H - P)) + A = \text{Capt}_{H-P}(K, C) + A \end{cases}$$

and thus, by (11.29), p. 470 again, that

$$\text{Capt}_H(K, C + A) = \text{Capt}_H(K, C) + A \quad \square$$

As a first example, we derive the following result:

Corollary 11.5.7 *[Lax–Hopf Formula for Capture Basins] Assume that U is a closed convex subset of a vector space and the components g_i of a single-valued map $g : U \mapsto X$ are*

- *convex and lower semicontinuous if $a_i = 1$,*
- *concave and upper semicontinuous if $a_i = -1$*
- *affine and continuous if $a_i = 0$,*

that U is convex and that the target C satisfies $C = C - \mathbb{R}^n_a$. Then the capture basin under control system $x'(t) = g(u(t))$ is equal to

$$\text{Capt}_g(C) = C - \mathbb{R}_+ g(U)$$

11.5.3 Viability Kernels Under P-Valued Micro–Macro Systems

Let $\mathbf{M}(x,u) \in \mathcal{L}(Y,Y)$ is a linear operator from Y to itself depending on (x,u) and a lagrangian $\mathbf{l} : (x,u) \mapsto \mathbf{l}(x,u) \in \mathbb{R} \cup \{+\infty\}$. We restrict our attention to micro–macro systems affine with respect to the macroscopic variable:

$$\begin{cases} (i) \; x'(t) \; = \; f(x(t),u(t)) \text{ where } u(t) \; \in \; U(x(t)) \\ (ii) \; y'(t) \; = \; \mathbf{M}(x(t),u(t))y(t) - \mathbf{l}(x(t),u(t)) \end{cases} \tag{11.31}$$

Even if we assume that the Lagrangian $\mathbf{l}$ is convex with respect to u, the right hand-side of system (11.31), p. 472 is not a Marchaud set-valued map, for which Viability Theorem 11.3.4, p. 455 is proven to be true. We can correct the lack of convexity of the right hand side of the macrosystem by subtracting a closed convex cone P making the right hand side of the corrected system a Marchaud one. We thus propose to consider corrected macrosystems of the following form:

$$\begin{cases} (i) \; x'(t) \; = \; f(x(t),u(t)) \text{ where } u(t) \; \in \; U(x(t)) \\ (ii) \; y'(t) \; \in \; \mathbf{M}(x(t),u(t))y(t) - \mathbf{l}(x(t),u(t)) - P \end{cases} \tag{11.32}$$

Let us introduce two set-valued maps $\mathbf{K} : X \rightsquigarrow Y$ and $\mathbf{C} : X \rightsquigarrow Y$, where $\mathbf{C} \subset \mathbf{K}$.

Theorem 11.5.8 ***[Viability Kernels of Exhaustive Maps]*** *Let us consider a Marchaud control microsystem*

$$x'(t) \; = \; f(x(t),u(t)) \text{ where } u(t) \; \in \; U(x(t)) \tag{11.33}$$

a closed convex cone P assumed to be viable under linear operators $\mathbf{M}(x,u) \in \mathcal{L}(Y,Y)$ bounded and depending continuously on $(x,u) \in \mathrm{Graph}(U)$ and two closed set-valued map $\mathbf{K}$ and $\mathbf{C}$ where $\mathbf{C} \subset \mathbf{K}$. If the microsystem (11.33), p. 472 is Marchaud, if the Lagrangian $\mathbf{l}$ is lower semicontinuous and if the images of (11.32), p. 472 are closed and convex, then the P-envelope of the viability kernel of $\mathbf{K}$ with target $\mathbf{C}$ is closed. Let us define the set-valued map $\mathbf{D}$ by

$$\mathrm{Graph}(\mathbf{D}) := \mathrm{Viab}_{(11.31)}(\mathrm{Graph}(\mathbf{K}), \mathrm{Graph}(\mathbf{C}))$$

If the set-valued maps $\mathbf{K}$ and $\mathbf{C}$ are closed and P-valued ($\mathbf{K}(x) = \mathbf{K}(x) + P$ and $\mathbf{C}(x) = \mathbf{C}(x) + P$), so is the set-valued map $\mathbf{D}$: it is closed and $\mathbf{D}(x) = \mathbf{D}(x) + P$.

Proof. Actually, we shall prove that the P-envelope of the viability kernel of the graph of $\mathbf{K}$ with target equal to the graph of $\mathbf{C}$ is equal to the viability kernel of the P-envelope of the graph of $\mathbf{K}$ with target equal to the P-envelope of the target map $\mathbf{C}$:

$$\left\{\begin{array}{l}\text{Viab}_{(11.31)}(\text{Graph}(\mathbf{K})+\{0\}\times P,\text{Graph}(\mathbf{C})+\{0\}\times P)\\ := \ \text{Viab}_{(11.32)}(\text{Graph}(\mathbf{K}),\text{Graph}(\mathbf{C}))+\{0\}\times P\\ = \ \text{Viab}_{(11.31)}(\text{Graph}(\mathbf{K})+\{0\}\times P,\text{Graph}(\mathbf{C})+\{0\}\times P)\end{array}\right. \tag{11.34}$$

It follows from Lemma 11.5.9, p.473 below because

$$\left\{\begin{array}{l}\text{Viab}_{(11.31)}(\text{Graph}(\mathbf{K})+\{0\}\times P,\text{Graph}(\mathbf{C})+\{0\}\times P)\\ \subset \ \text{Viab}_{(11.32)}(\text{Graph}(\mathbf{K}),\text{Graph}(\mathbf{C}))+\{0\}\times P\end{array}\right.$$

and because the micro–macro system (11.32), p. 472 is Marchaud. This implies that the viability kernel under this map is closed. Furthermore, when $\mathbf{K}$ and $\mathbf{C}$ are P-valued, we infer that

$$\left\{\begin{array}{l}\text{Viab}_{(11.31)}(\text{Graph}(\mathbf{K}),\text{Graph}(\mathbf{C}))+\{0\}\times P\\ = \ \text{Viab}_{(11.32)}(\text{Graph}(\mathbf{K}),\text{Graph}(\mathbf{C}))\\ = \ \text{Viab}_{(11.31)}(\text{Graph}(\mathbf{K}),\text{Graph}(\mathbf{C}))\end{array}\right. \tag{11.35}$$

It remains to prove Lemma 11.5.9, p.473 □

Lemma 11.5.9 ***[Viability Kernels of Exhaustive Maps]*** *Let us consider the Marchaud microsystem (11.33), p. 472:*

$$x'(t) \ = \ f(x(t),u(t)) \text{ where } u(t) \ \in \ U(x(t))$$

a closed convex cone P assumed to be viable under linear operators $\mathbf{M}(x,u) \in \mathcal{L}(Y,Y)$ for all $(x,u) \in \text{Graph}(U)$. Then

$$\left\{\begin{array}{l}\text{Viab}_{(11.32)}(\text{Graph}(\mathbf{K}),\text{Graph}(\mathbf{C}))+\{0\}\times P\\ \subset \ \text{Viab}_{(11.31)}(\text{Graph}(\mathbf{K})+\{0\}\times P,\text{Graph}(\mathbf{C})+\{0\}\times P)\end{array}\right. \tag{11.36}$$

Proof. (Lemma 11.5.9) Let us pick any x and a solution $(x(\cdot),u(\cdot))$ to the microcontrol system

$$x'(t) \ = \ f(x(t),u(t)) \text{ where } u(t) \ \in \ U(x(t))$$

starting at x.

For simplifying notations, we denote by $S(t,s) := e^{\int_s^t \mathbf{M}(x(\tau),u(\tau))d\tau}$ the linear semi-group associated with the time dependent linear operator $t \mapsto \mathbf{M}(x(t),u(t))$. Any solution $z(\cdot)$ to the affine differential equation

$$z'(t) = \mathbf{M}(x(t), u(t))z(t) - \mathbf{l}(x(t), u(t))$$

starting at some ζ can be written

$$z(t) = S(t,0)\zeta - \int_0^t S(t,\tau)\mathbf{l}(x(\tau), u(\tau))d\tau$$

Let us consider a pair $(x, z) \in \text{Viab}_{(11.32)}(\text{Graph}(\mathbf{K}), \text{Graph}(\mathbf{C}))$ and an element $\pi \in P$. Setting $y := z + \pi$, we have to check that (x, y) belongs to $\text{Viab}_{(11.31)}(\text{Graph}(\mathbf{K}) + \{0\} \times P, \text{Graph}(\mathbf{C}) + \{0\} \times P)$. There exist a function $t \mapsto p(t) \in P$, some time $t^\star \leq +\infty$ and the macro evolution

$$z(t) = S(t,0)z - \int_0^t S(t,\tau)\mathbf{l}(x(\tau), u(\tau))d\tau - \int_0^t S(t,\tau)p(\tau)d\tau$$

such that $(x(t^\star), y(t^\star))$ belongs to the target $\text{Graph}(\mathbf{C})$ if $t^\star < +\infty$ is finite and such that, for all $t < t^\star$, $(x(t), y(t)) \in \text{Graph}(\mathbf{K})$.

Since P is viable under the time dependent linear operator $t \mapsto \mathbf{M}(x(t), u(t))$ by assumption, we associate with π the solution $S(t,0)\pi$ which is viable in P.

Therefore, the evolution $y(\cdot)$ defined by

$$\begin{cases} y(t) := z(t) + S(t,0)\pi \\ = S(t,0)(z+\pi) - \int_0^t S(t,\tau)\mathbf{l}(x(\tau), u(\tau))d\tau - \int_0^t S(t,\tau)p(\tau)d\tau \end{cases}$$

is a solution to the affine system (11.32)(ii), p. 472 starting at $z + \pi$. Since $(x(t), z(t))$ is viable in $\text{Graph}(\mathbf{K})$ forever or until it reaches the target $\text{Graph}(\mathbf{C})$ at finite time $t^\star$ and since $S(t,0)\pi$ is viable in P, we infer that $(x(t), y(t))$ is viable in $\text{Graph}(\mathbf{K}) + \{0\} \times P$ forever or until it reaches the target $\text{Graph}(\mathbf{C}) + \{0\} \times P$ at finite time $t^\star$. Therefore, the function

$$\begin{cases} \left(x(t), S(t,0)(z+\pi) - \int_0^t S(t,\tau)\mathbf{l}(x(\tau), u(\tau))d\tau\right) \\ = \left(x(t), y(t) + \int_0^t S(t,\tau)p(\tau)d\tau\right) \end{cases}$$

is a solution of the micro–macro system (11.31), p. 472 starting at (x, y), viable in $\text{Graph}(\mathbf{K}) + \{0\} \times P + \{0\} \times P = \text{Graph}(\mathbf{K}) + \{0\} \times P$ forever or until it reaches the target $\text{Graph}(\mathbf{C}) + \{0\} \times P + \{0\} \times P = \text{Graph}(\mathbf{C}) + \{0\} \times P$ at finite time $t^\star$. □

11.6 Hamiltonian Characterization of the Regulation Map

The dual formulation of the Frankowska property involves duality between

- the finite dimensional vector space X and its dual $X^\star := \mathcal{L}(X, \mathbb{R})$ trough duality pairing $\langle p, x\rangle := p(x)$ on $X^\star \times X$,
- tangent cones $T_K(x)$ and $T_K^{\star\star}(x)$ and "normal cones" defined by

$$N_K(x) \;:=\; T_K(x)^\star \;:=\; \{p \in X^\star \text{ such that } \forall\, v \in T_K(x), \langle p, v\rangle \;\leq\; 0\}$$

  to K at x,
- the dynamics (f, U) of a control system and its Hamiltonian we are about to define:

Definition 11.6.1 *[**Hamiltonian of a Control System**] We associate with the control system (f, U) the* Hamiltonian $\mathbf{h} : X \times X^\star \mapsto \mathbb{R} \cup \{+\infty\}$ *and the* boundary Hamiltonian $\mathbf{h}_K : \partial K \times X^\star \mapsto \mathbb{R} \cup \{+\infty\}$ *defined by*

$$\begin{cases} \forall\, x \in \overset{\circ}{K},\; \forall\, p \in X^\star, & \mathbf{h}(x,p) \;:=\; \inf_{u \in U(x)} \langle p, f(x,u)\rangle \\ \forall\, x \in K,\; \forall\, p \in X^\star, & \mathbf{h}_K(x,p) \;:=\; \inf_{v \in \overleftarrow{R}_K(x)} \langle p, f(x,v)\rangle \end{cases} \quad (11.37)$$

where $\overleftarrow{R}_K$ is the backward regulation map (see Definition 11.4.3, p.462).

We say that the Hamiltonian $\mathbf{h}$ *satisfies the* Walras law *on a subset $D \subset K$ at $x \in D$ if*

$$\forall\, p \in N_D(x),\;\; \mathbf{h}(x,p) \;\leq\; 0$$

and that $p^\star \in N_D(x)$ is a Walras normal *to D at x if*

$$\mathbf{h}(x,p^\star) \;=\; \sup_{p \in N_D(x)} \mathbf{h}(x,p) \;=\; 0$$

The Hamiltonian satisfies the Walras law on D if it satisfies it at every state $x \in D$.

Note that $\mathbf{h}(x,p) = \mathbf{h}_K(x,p)$ whenever $x \in \overset{\circ}{K} := \text{Int}(K)$.

The function $p \mapsto \mathbf{h}(x,p)$ is concave, positively homogeneous and upper semicontinuous, as the infimum of continuous affine functions. The Fenchel Theorem 18.7.3, p. 756 states that, conversely, *any concave, positively homogeneous and upper semicontinuous Hamiltonian $p \mapsto \mathbf{h}(x,p)$ is associated with the set-valued map $F : x \mapsto F(x)$* defined by

$$F(x) \;:=\; \{v \in X \text{ such that } \forall\, p \in X^\star,\;\; \mathbf{h}(x,p) \;\leq\; \langle p, v\rangle\}$$

When $\mathbf{h}(x,p) = \inf_{u\in U(x)} \langle p, f(x,u)\rangle$, we obtain $F(x) = f(x,U(x))$, as expected.

Formulas on support functions of convex analysis imply that whenever $0 \in \text{Int}(F(x) + T_K(x))$, the boundary Hamiltonian is equal to

$$\mathbf{h}_K(x,p) = \sup_{q\in N_K(x)} \inf_{u\in U(x)} \langle p-q, u\rangle = \sup_{q\in N_K(x)} \mathbf{h}(x,p-q)$$

Theorem 11.6.2 ***[Walrasian Characterization of the Regulation Map]*** *Let us assume that the control system (f,U) is Marchaud. Then the regulation map $R_D(x)$ at x is not empty if and only if the Hamiltonian satisfies the Walras law on D at x. In this case, setting*

$$R^-(x,p) := \{u \in U(x) \text{ such that } \langle p, f(x,u)\rangle \le 0\} \tag{11.38}$$

the regulation map can be written in the form

$$\forall\, x \in D, \quad R_D(x) = \bigcap_{p\in N_D(x)} R^-(x,p) \tag{11.39}$$

Proof. • Assume that some $u^\star \in R_D(x)$. Then, $f(x,u^\star) \in T_D^{\star\star}(x)$, so that, for all $p \in N_D(x)$, $\langle p, f(x,u^\star)\rangle \le 0$. In other words,

$$\sup_{p\in N_D(x)} \mathbf{h}_K(x,p) = \inf_{u\in U(x)} \sup_{p\in N_D(x)} \langle p, f(x,u^\star)\rangle \le 0$$

so that the Hamiltonian satisfies the Walras law on D at x and

$$R_D(x) \subset \bigcap_{p\in N_D(x)} R^-(x,p)$$

• Assume that the Hamiltonian satisfies the Walras law on D at x.
Let us consider the function $(u,p) \mapsto \langle p, f(x,u)\rangle$ defined on the product $U(x) \times N_D(x)$ of the compact convex subset $U(x)$ and the convex (closed cone) $N_D(x)$. This function is continuous and affine (or, possibly, convex) with respect to u and concave with respect to p. Therefore, the Lopsided Minimax Theorem 18.7.1, p. 755 implies the existence of some $u^\star \in U(x)$ such that

$$\begin{cases} \inf_{u\in U(x)} \sup_{p\in N_D(x)} \langle p, f(x,u)\rangle = \sup_{p\in N_D(x)} \langle p, f(x,u^\star)\rangle \\ = \sup_{p\in N_D(x)} \inf_{u\in U(x)} \langle p, f(x,u)\rangle = \sup_{p\in N_D(x)} \mathbf{h}_K(x,p) \end{cases} \tag{11.40}$$

This implies that there exists $u^\star \in U(x)$ such that

$$\sup_{p \in N_D(x)} \langle p, f(x,u^\star)\rangle = \sup_{p \in N_D(x)} \mathbf{h}_K(x,p) \le 0$$

since the Hamiltonian satisfies the Walras law. Consequently, we derive the opposite inclusion $u^\star \in \bigcap_{p \in N_D(x)} R^-(x,p)$.

It remains to observe that every $u^\star \in \bigcap_{p \in N_D(x)} R^-(x,p)$ implies that $f(x,u^\star) \in T_D^{\star\star}(x)$. We have proved that $u^\star \in R_D(x)$.

This concludes the proof of the dual characterization of the regulation map. □

The Minimax Theorem 18.7.1, p. 755 provides sufficient conditions for the regulation map to be nonempty:

Theorem 11.6.3 *[Walras Law and Viability] Let us assume that the control system (f,U) is Marchaud and that the normal cone $N_D(x)$ is spanned by a compact convex subset $S_D(x)$ which does not contain 0 (such a cone is called a* sole *of the cone).*
If the Hamiltonian satisfies the Walras law on D at $x \in D$, there exist a control $u^\star \in R_D(x)$ and a Walras normal $p^\star \in N_D(x)$ such that

$$\mathbf{h}(x,p^\star) = \langle p^\star, f(x,u^\star)\rangle = 0 \quad (11.41)$$

Proof. Since $S_D(x)$ and $U(x)$ are convex compact subsets and since the function $(u,p) \mapsto \langle p, f(x,u)\rangle$ is linear with respect to p and affine with respect to u (because, (f,U) is Marchaud), the Von Neuman Minimax Theorem 18.7.1, p. 755 implies that there exists a minimax $(u^\star,p^\star) \in U(x) \times S_D(x)$, i.e., which satisfies

$$\inf_{u \in U(x)} \sup_{p \in S_D(x)} \langle p, f(x,u)\rangle = \langle p^\star, f(x,u^\star)\rangle = \sup_{p \in S_D(x)} \mathbf{h}(x,p)$$

Since the Hamiltonian $\mathbf{h}$ satisfies the Walras law, we deduce that $\sup_{p \in S_D(x)} \langle p, f(x,u^\star)\rangle \le \mathbf{h}(x,p^\star) = \langle p^\star, f(x,u^\star)\rangle \le 0$, and thus, that

$$\langle p^\star, f(x,u^\star)\rangle = \sup_{p \in S_D(x)} \langle p^\star, f(x,u^\star)\rangle \le 0$$

Since $S_D(x)$ spans the normal cone $N_D(x)$, we infer that all $p \in N_D(x)$, $\langle p, f(x,u^\star)\rangle \le 0$. This implies that $f(x,u^\star) \in T_D^{\star\star}(x)$ and that $\langle p^\star, f(x,u^\star)\rangle = 0$. We have proved that $u^\star \in R_D(x)$ and that $p^\star \in N_D(x)$ is a Walras normal. □

Remark. The name "Walras law" we proposed is motivated by economic theory after *Léon Walras*. This is the fundamental assumption under which a Walras equilibrium does exist in the static general equilibrium theory and, in its dynamic formulation, the fundamental assumption under which viable economic evolutions do exist (see *Dynamic Economic Theory* [22, Aubin]).

In this framework, the state x is, roughly speaking, a "commodity" and elements $p \in X^\star$ are regarded as "prices". In this case, prices are taken as controls $u := p \in X^\star$ ranging the dual of the state space and the dynamics are of the form

$$x'(t) \ := \ f(x(t), p(t)) \text{ where } p(t) \in P(x(t))$$

where the right hand side $f(x,p)$ denotes the *(instantaneous) algebraic transaction* of commodity x at price p.

The *Walras law* requires

$$\forall\, p \in X^\star, \ \ \langle p, f(x,p)\rangle \ \le \ 0$$

It means that for any price, the value of the transaction is always negative, i.e., than the value of the new acquired commodity is smaller than or equal to the value of the old one.

Since

$$\mathbf{h}(x,p) \ := \ \inf_{q \in \overleftarrow{R}_K(x)} \ \langle p, f(x,q)\rangle \ \le \ \langle p, f(x,p)\rangle \ \le \ 0$$

the Hamiltonian satisfies the Walras law in the sense of Definition 11.6.1, p. 475 for any subset D such that, for any $x \in D$, for any $p \in N_D(x)$, $\mathbf{h}(x,p) \le 0$.

It is not surprising that sufficient conditions implying the existence of a Walras equilibrium, such as the celebrated *Walras law*, also implies that the regulation map is nonempty. □

We also can formulate the Frankowska property (11.19), p. 462 in terms of normal cones:

Definition 11.6.4 *[**Dual Frankowska Property**] A Hamiltonian satisfies the* dual Frankowska property *on D with respect to (K,C) if*

$$\begin{cases} (i) \ \forall x \in D \cap (\overset{\circ}{K} \setminus C), \ \forall\, p \in N_D(x), \ \ \mathbf{h}(x,p) \ = \ 0 \\ (ii) \ \forall x \in D \cap \partial K, \ \forall\, p \in N_D(x), \ \ \ \mathbf{h}_K(x,p) \ \ge \ 0 \end{cases} \tag{11.42}$$

Therefore, in all forthcoming statements, one can replace "Frankowska property" (11.19), p. 462 by "dual Frankowska property" (11.6.4), p. 478. This will be used only for defining Barron–Jensen/Frankowska viscosity solutions of Hamilton–Jacobi–Bellman partial differential equations in Chap. 17, p. 681.

Lemma 11.6.5 ***[Equivalence between Frankowska Property and its Dual Form]*** *Frankowska property (11.19), p. 462 is equivalent to the dual Frankowska property (11.42), p. 478.*

Proof. Indeed, Frankowska property (11.19), p. 462 is equivalent to property

$$\begin{cases} (i) \ \forall\, x \in D \setminus C, \ \ R_D(x) \neq \emptyset \\ (ii) \ \forall\, x \in D, \ \ \overleftarrow{R}_D(x) \ = \ \overleftarrow{R}_D(x) \end{cases} \tag{11.43}$$

It remains to prove that it is equivalent to

$$\begin{cases} (i) \ \forall\, x \in D \setminus C, \forall\, p \in N_D(x), \ \ \mathbf{h}(x,p) \leq 0 \\ (ii) \ \forall\, x \in D, \forall\, p \in N_D(x), \ \ \mathbf{h}_K(x,p) \geq 0 \end{cases}$$

The equivalence between (i) and (11.43)(i), p. 479 follows from Theorem 11.6.2, p. 476. To say that $\overleftarrow{R}_K(x) = \overleftarrow{R}_D(x)$ means that for all $u \in \overleftarrow{R}_K(x)$, $-f(x,u) \in T_D^{\star\star}(x)$. The later condition means that for any $p \in N_D(x)$ and for any $u \in \overleftarrow{R}_K(x)$, $\langle p, f(x,u)\rangle \geq 0$, or, equivalently, that $\mathbf{h}_K(x,p) := \inf_{u\in \overleftarrow{R}_K(x)} \langle p, f(x,u)\rangle \geq 0$. Since condition (ii) splits into two conditions

$$\begin{cases} (i) \ \forall\, x \in D \in \overset{\circ}{K}, \forall\, p \in N_D(x), \ \ \mathbf{h}(x,p) \geq 0 \\ (ii) \ \forall\, x \in D \cap \partial K, \forall\, p \in N_D(x), \ \ \mathbf{h}_K(x,p) \geq 0 \end{cases}$$

we infer that the Frankowska and the dual Frankowska properties are equivalent. □

The "dual" version of the tangential characterization of viability kernels provided by Theorem 11.4.5, p. 463 is stated in this the following terms:

Theorem 11.6.6 ***[Dual Characterization of Viability Kernels]*** *Let us assume that the control system (f, U) is Marchaud and Lipschitz and that the subset K is closed. The viability kernel* $\mathrm{Viab}_F(K, C)$ *of a subset K under $\mathcal{S}$ is the* ***unique*** *closed subset satisfying*

- $C \subset D \subset K$,
- $K \setminus D$ *is a repeller*
- *and the* dual Frankowska property *(11.42), p. 478.*

The dual characterization of capture basins reads

Theorem 11.6.7 *[Dual Characterization of Capture Basins] Let us assume that* (f, U) *is Marchaud and Lipschitz, that the environment* $K \subset X$ *and the target* $C \subset K$ *are closed subsets such that* $K \setminus C$ *is a repeller* ($\text{Viab}_F(K \setminus C) = \emptyset$)*.*

Then the viable-capture basin $\text{Capt}_F(K, C)$ *is the* ***unique*** *closed subset satisfying*

- $C \subset D \subset K$,
- *and the* dual Frankowska property *(11.42), p. 478.*

11.7 Deriving Viable Feedbacks

11.7.1 Static Viable Feedbacks

Unfortunately, the graph of the regulation map is generally not closed, so that system (11.14) is not Marchaud. However, reasonable assumptions imply that the regulation map is lower semicontinuous.

Viability Theorem 11.3.4, p. 455 states that K is viable outside C if and only if the regulation map satisfies

$$\forall x \in K \backslash C, \ R_K(x) \neq \emptyset$$

Hence, the tangential condition for invariance is satisfied by the system (f, R_K) defined by (11.14). *However, this does not imply that* K *is invariant outside* C, despite the fact that by construction, all velocities of this new system are tangent to K. For K to be invariant outside C under the system (f, R_K), a sufficient condition provided by the Invariance Theorem 11.3.7, p. 457 is that both f and the regulation map R_K are Lipschitz.

Counter-Example: Even in the case of differential equations $x' = f(x)$ where f is continuous and satisfies the tangential condition $f(x) \in T_K(x)$ for all $x \in K$ and where K is a closed subset, for any initial state x from which start several solutions, at leat one is viable in K, but not all, despite the fact that the tangential condition is satisfied. □

Regulating or controlling the system means looking for a subset of controls which provide solutions satisfying viability/capturability properties (which are the issues tackled here).

A *(static) feedback* $r : x \in K\backslash C \mapsto r(x) \in U(x)$ is said to be *viable* in K outside C if K is viable outside C under the differential equation $x' = f(x, r(x))$.

The most celebrated examples of linear feedbacks in linear control theory designed to control a system have no reason to be viable for an arbitrary environment K, and, according to the environment K, the viable feedbacks are not necessarily linear.

However, Viability Theorem 11.3.4, p. 455 implies that a feedback r is viable in $K\backslash C$ if and only if r is a *selection* of the regulation map R_K in the sense that

$$\forall x \in K\backslash C, \; r(x) \in R_K(x) \tag{11.44}$$

Hence, the method for designing viable feedbacks governing evolutions viable in K outside C amounts to finding *selections* $r(x)$ of the regulation map $R_K(x)$. One can design "factories" for designing selections (see Chap. 6 of the first edition of [18, Aubin], for instance) of set-valued maps.

Ideally, a feedback should be continuous to guarantee the existence of a solution to the differential equation $x' = f(x, r(x))$. In this case, the Viability Theorem 11.3.4, p. 455 states that K is viable outside C under the differential equation $x' = f(x, r(x))$.

The problem is to find continuous selections of the regulation map R_K:

Definition 11.7.1 *[Selections] A* selection *of a set-valued map* $U : X \rightsquigarrow \mathcal{U}$ *is a single-valued map* $\widetilde{u} : x \mapsto \widetilde{u}(x)$ *such that*

$$\forall x, \;\; \widetilde{u}(x) \in U(x)$$

But this is not always possible. However, the Michael Selection Theorem states that if R_K is lower semicontinuous with closed convex images, then there exists a continuous selection of R_K, and thus, of a continuous viable feedback. However, the Michael Selection Theorem is not constructive in the sense that it does not tell how to find such a continuous selection.

The question arises whether one can find constructive selections of the regulation map. The simplest example is the *minimal selection* r° of R_K associating with any $x \in K\backslash C$ the control $r^\circ(x) \in R_K(x)$ with minimal norm (which exists whenever the image $R_K(x)$ is closed and convex thanks to the Projection Theorem). This holds true for Marchaud systems, but the problem is that this constructive selection is not necessarily continuous. However, the Falcone-Saint-Pierre Theorem states that despite this lack of continuity, this feedback still provides viable evolutions, under the very same assumptions than the ones of the Michael Selection Theorem:

Definition 11.7.2 *[Slow Feedback] We posit the assumptions of Theorem 11.3.4. The* slow feedback $r_K^\circ \in R_K$ *is the selection with minimal norm of the regulation map* R_K. *The system*

$$x'(t) = f(x(t), r_K^\circ(x(t)))$$

governs the evolution of slow solutions.

If the regulation map R_K *is lower semicontinuous on* $K \backslash C$, *for every initial state* $x \in K \backslash C$, *there exists a slow evolution governed by the differential equation* $x' = f(x, r^\circ(x))$ *viable in* K *outside* C.

Slow feedbacks are a particular case of selection procedures of the regulation map:

Definition 11.7.3 *[Selection Procedure] A* selection procedure of a set-valued map $F : X \rightsquigarrow Y$ *is a set-valued map* $S_F : X \rightsquigarrow Y$ *satisfying*

$$\begin{cases} (i) \ \forall x \in \mathrm{Dom}(F), \ \ S(F(x)) := S_F(x) \cap F(x) \neq \emptyset \\ (ii) \ \textit{the graph of } S_F \ \textit{is closed} \end{cases}$$

The set-valued map $S(F) : x \rightsquigarrow S(F(x))$ *is called the* selection *of* F.

We provide sufficient conditions implying that a selection procedure of a regulation map governing viable evolutions:

Theorem 11.7.4 *[Selection Procedures of a Regulation Map] Assume that the control system is Marchaud and* $K \setminus C$ *is locally viable. Let* S_{R_K} *be a selection of the regulation map* R_K. *Suppose that the values of* S_{R_K} *are convex. Then, for any initial state* $x_0 \in K$, *there exists an evolution starting at* x_0 *viable in* K *until it reaches* C *which is regulated by the selection* $S(R_K)$ *of the regulation map* R_K, *in the sense that*

$$u(t) \ \in \ S(R_K)(x(t)) \ := \ R_K(x(t)) \cap S_{R_K}(x(t))$$

Proof. Since the convex selection procedure S_{R_K} has a closed graph and convex values, we can replace the original control system by the control system

$$\begin{cases} (i) \ x'(t) \ = \ f(x(t), u(t)) \\ (ii) \ u(t) \ \in \ U(x(t)) \cap S_{R_K}(x(t)) \end{cases} \tag{11.45}$$

which satisfies also the assumptions of the Viability Theorem. It remains to check that K is still viable under this new system. But by construction, we know that for all $x \in K$, there exists $u \in S(R_K)(x)$, which belongs to the intersection $U(x) \cap S_{R_K}(x)$ and which is such that $f(x,u)$ belongs to $T_K(x)$.

Hence the new control system enjoys the viability property, so that, for all initial states $x_0 \in K$, there exist a viable solution and a viable control to control system (11.45) which, for almost all $t \geq 0$, are related by

$$\begin{cases} (i) \;\; u(t) \;\; \in \;\; U(x(t)) \cap S_{R_K}(x(t)) \\ (ii) \; f(x(t),u(t)) \in T_K(x(t)) \end{cases}$$

Therefore, $u(t)$ belongs to the intersection of $R_K(x(t))$ and $S_{R_K}(x(t))$, i.e., to the selection $S(R_K)(x(t))$ of the regulation map R_K. □

The selection procedure S_F° of a closed convex valued set-valued map F defined by

$$S_F^\circ(x) \; := \; \{y \in Y \mid \|y\| \; \leq \; d(0, F(x))\}$$

provides slow evolutions, so that Theorem 11.7.2, p.482 ensues.

We can easily provide other examples of selection procedures through optimization thanks to the Maximum Theorem.

11.7.2 Dynamic Viable Feedbacks

Consider parameterized system (2.10):

$$x'(t) \;=\; f(x(t), u(t)) \text{ where } u(t) \;\in\; U(x(t))$$

We slightly generalize the concept of metasystem (see Definition 6.4.6, p. 211) associated with the simple differential inclusion $u'(t) \in B(0,c)$ by replacing it by any differential inclusion $u'(t) \in G(x(t), u(t))$. We join it to the initial differential equation (2.10)(i) to form a more balanced auxiliary system

$$\begin{cases} (i) \;\; x'(t) = f(x(t), u(t)) \\ (ii) \; u'(t) \in G(x(t), u(t)) \end{cases}$$

of differential inclusions. The output-input regulation (2.10)(ii) becomes a viability constraint of the new system.

Definition 11.7.5 ***[General Metasystem]*** *The* metasystem *associated with the control system* (f, U) *and a set-valued map* $G : X \times \mathcal{U} \rightsquigarrow \mathcal{U}$ *is the system*

$$\begin{cases} (i) \;\; x'(t) = f(x(t), u(t)) \\ (ii) \; u'(t) \in G(x(t), u(t)) \end{cases} \tag{11.46}$$

subjected to the viability constraint

$$\forall t \geq 0, \quad (x(t), u(t)) \in \text{Graph}(U)$$

The set-valued map $U_G : X \rightsquigarrow \mathcal{U}$ *defined by*

$$\text{Graph}(U_G) := \text{Viab}_{(f,G)}(\text{Graph}(U))$$

is called the G-regulation map *and the set-valued map* $R_G : X \times \mathcal{U} \rightsquigarrow \mathcal{U}$ *defined by*

$$R_G(x, u) := \{v \in G(x, u) \mid (f(x, u), v) \in T_{\text{Graph}(U_G)}(x, u)\}$$

is called the metaregulation map.

Therefore, we know that this "metaregulation map" R_G regulates the evolutions $(x(\cdot), u(\cdot))$ viable in the graph of U_G.

Theorem 11.7.6 *[The Metaregulation Map] Assume that f is continuous, that G is Marchaud and that the graph of U is closed. Then the evolutions of the state-control pairs are governed by the system of differential inclusions*

$$\begin{cases} (i) \ x'(t) = f(x(t), u(t)) \\ (ii) \ u'(t) \in R_G(x(t), u(t)) \end{cases}$$

and the metaregulation map R_G has nonempty closed convex images $R_G(x, u)$.

Among the dynamic feedbacks, the heavy ones are noteworthy:

Theorem 11.7.7 *[Heavy Feedback] Under the assumptions of Theorem 11.7.6, the* heavy feedback $g_U^\circ \in R_G$ *is the selection with minimal norm of the metaregulation map R_G. The system of differential equations*

$$\begin{cases} (i) \ x'(t) = f(x(t), u(t)) \\ (ii) \ u'(t) = g_U^\circ(x(t), u(t)) \end{cases}$$

governs the evolution of heavy solutions, *i.e., evolutions with* minimal velocity.

If the regulation map R_G is lower semicontinuous on $\text{Graph}(U_G)$, *for every initial state-control pair* $(x, u) \in \text{Graph}(U_G)$, *there exists a heavy evolution.*

Chapter 12
Restoring Viability

12.1 Introduction

There is no reason why an arbitrary subset K should be viable under a given control system. The introduction of the concept of viability kernel does not exhaust the problem of restoring viability by keeping the same dynamics of the control system and "shrinking" the environment to its viability kernel. We devote this chapter to two other methods for restoring viability *without changing the environment*. We want to:

1. Change initial dynamics by introducing regulons that are "viability multipliers";
2. Change the initial conditions by introducing a *reset map* Φ mapping any state of K to a (possibly empty) set $\Phi(x) \subset X$ of new "initialized states" whenever the evolution reaches the domain of the reset map (*impulse control*).

12.2 Viability Multipliers

12.2.1 Definition of Viability Multipliers

Consider the case in which a subset K is not viable under a differential equation $x'(t) \in F(x(t))$. Denoting by $T_K(x)$ the tangent (contingent) cone to K at $x \in K$ and by $T_K^{\star\star}(x)$ its closed convex hull (see Definition 18.4.8, p.732), the Viability Theorem 11.3.4, p.455 tells us that when F is a Marchaud map, the necessary and sufficient tangential condition $F(x) \cap T_K^{\star\star}(x) \neq \emptyset$ which can be written in the form

$$0 \in F(x) - T_K^{\star\star}(x)$$

J.-P. Aubin et al., *Viability Theory*, DOI 10.1007/978-3-642-16684-6_12,

is not satisfied for at least one state $x \in K$.

The question arises whether we can correct the dynamics F in such a way that K becomes viable under the new one.

The first idea that jumps to the mind is to introduce *viability multipliers* $p \in F(x) - T_K^{\star\star}(x)$ measuring the "viability discrepancy" between the set $F(x)$ of velocities and the (closed convex hull) of the cone of tangent directions to K at x.

Definition 12.2.1 ***[Viability Discrepancy and Viability Multipliers]*** *The* viability discrepancy *is the function* $c^\circ : X \mapsto \mathbb{R}_+$ *defined by*

$$c^\circ(x) := d(F(x), T_K^{\star\star}(x)) := \inf_{u \in F(x), v \in T_K^{\star\star}(x)} \|u - v\|$$

Elements

$$p(t) \in F(x(t)) - T_K^{\star\star}(x(t)) \tag{12.1}$$

are called viability multipliers *measuring the viability discrepancy.*

They are used to correct the following differential inclusion $x'(t) \in F(x(t))$ under which K is not viable by replacing it by the control system

$$x'(t) \in F(x(t)) - p(t) \text{ where } p(t) \in F(x(t)) - T_K^{\star\star}(x(t))$$

Actually, we can restrict the choice of viability multipliers by requiring that the evolution is regulated by "viability multipliers" $p(t) \in P(x(t))$ where $P : X \rightsquigarrow X$ is a set-valued map intersecting the viability gap:

$$\forall x \in K, \;\; P(x) \cap (F(x) - T_K^{\star\star}(x)) \neq \emptyset$$

Therefore, we regulate viable evolutions by correcting the initial dynamics through "viability multipliers" measuring the "viability discrepancy" in the following way:

$$\begin{cases} (i) \;\; x'(t) \in F(x(t)) - p(t) \\ (ii) \; p(t) \in P(x(t)) \end{cases} \tag{12.2}$$

The Viability Theorem implies that when F and P are Marchaud, K is viable under this corrected control system (12.2).

The question arises how to find examples of such set-valued maps P.

12.2.2 Viability Multipliers with Minimal Norm

The first class of examples is to take balls $P(x) := c(x)B$ centered at 0 of radius $c(x)$. The function $c(\cdot)$ must be upper semicontinuous for the set-valued map $c(\cdot)B$ to be continuous, larger than or equal to $d(F(x), T_K^{\star\star}(x))$ so that

$$\forall x \in K,\ c(x)B \cap (F(x) - T_K^{\star\star}(x)) \neq \emptyset$$

and smaller than $d(0, F(x))$ for avoiding to find only equilibria.

In particular, the question arises whether we can "minimize the viability discrepancy" by singling out the viability multiplier $p^\circ(x)$ of minimal norm:

$$\|p^\circ(x)\| = \inf_{v \in F(x),\ u \in T_K^{\star\star}(x)} \|v - u\|$$

which exists whenever $F(x)$ is convex and compact thanks to the Projection Theorem 18.4.17: there exist $v^\circ(x) \in F(x)$ and $u^\circ(x) \in T_K^{\star\star}(x)$ such that $p^\circ(x) = v^\circ(x) - u^\circ(x)$, where

$$u^\circ(x) := \Pi_{T_K^{\star\star}(x)}(v^\circ(x))$$

is *the projection*[1] *of the velocity $v^\circ(x)$ onto the (closed convex hull of) the tangent cone to K at x.*

The solutions to the control system

$$x'(t) \in F(x(t)) - p^\circ(x(t)) \tag{12.3}$$

are called *slow evolutions* of the corrected system (12.2).

Even though control system (12.3) was built to meet the tangential conditions of the Viability Theorem 11.3.4, p.455, we observe that it looses the minimal continuity requirements sufficient (but not necessary) for guaranteeing the existence of a solution to (12.3). However, when K is closed and convex[2] and when F is Marchaud and lower semicontinuous, one can prove that *slow solutions do exist.*

This is the case in particular when $F(x) := \{f(x)\}$ is a continuous single-valued map. We observe that when K is closed and convex, control system (12.3) can be written in the form of "variational inequalities":

$$\forall y \in K,\ \langle x'(t) - f(x(t)), x(t) - y\rangle \geq 0 \tag{12.4}$$

because, in this case, $N_K(x) = \{p \in X^\star | \forall y \in K,\ \langle p, y - x\rangle \leq 0\}$ (see Theorem 12.2.3 below).

Remark. This is under the form (12.4) that variational inequalities were introduced in *unilateral mechanics* at the beginning of the 1960s by Jean-Jacques Moreau, Guido Stampacchia and Jacques-Louis Lions for taking into account viability constraints (in the case of partial differential equations).

---

[1] Projecting the velocities of $F(x)$ onto $T_K^{\star\star}(x)$ is not the only possibility. We could use any selector $(x, K) \mapsto r(x, K) \in K$ and replace $v \in F(x)$ by $r(v, T_K^{\star\star}(x))$. We just focus our attention on the projector of best approximation onto a closed convex subset because it is a universal choice.

[2] And more generally, "sleek" (see Definition 18.4.8, p.732).

This was the starting point of a new field in mathematics and mechanics (see [144–146, Lions J.-L.], [36, Bensoussan & Lions J.L.], [3, Alart, Maisonneuve & Rockafellar]). The links with *planning methods* in economics started with Claude Henry in the middle of the 1970s, and later, the issue was taken up by Bernard Cornet. □

12.2.3 Viability Multipliers and Variational Inequalities

Let $K \subset X$ be a closed subset and $F : X \rightsquigarrow X$ be a Marchaud map defining the dynamics of the differential inclusion $x'(t) \in F(x(t))$. Let $c := \sup_{x\in K} \frac{\|F(x)\|}{\|x\|+1}$, which is finite since we assumed F to be Marchaud and $B_\star$ denote the unit ball of the dual space. We observe that

$$\forall x \in K,\ d(F(x), T_K^{\star\star}(x)) \leq d(F(x), T_K(x)) \leq d(0, F(x)) \leq c(\|x\|+1)$$

Therefore, the set-valued map

$$R_K(x) \ := \ (F(x) - T_K^{\star\star}(x)) \cap c(x)B_\star$$

is empty if $c(x) < d(F(x), T_K^{\star\star}(x))$ and does not contain 0 if $c(x) < d(0, F(x))$. The set-valued map R_K is then a good candidate to be a regulation map:

Theorem 12.2.2 *[**Correction by Viability Multipliers**] Let $K \subset X$ be a closed subset and $F : X \rightsquigarrow X$ be a Marchaud map defining the dynamics of the differential inclusion $x'(t) \in F(x(t))$.*

For any upper semicontinuous function $c : K \mapsto \mathbb{R}_+$ satisfying

$$\forall x \in K,\ d(F(x), T_K(x)) \leq c(x) \leq c(\|x\|+1)$$

K is viable under the differential inclusion

$$\begin{cases} (i) \ \ x'(t) \in F(x(t)) - p(t) \\ (ii) \ \|p(t)\|_\star \leq c(x(t)) \end{cases} \tag{12.5}$$

corrected by viability multipliers $p(t) \in R_K(x(t))$ *where* $R_K(x) := (F(x) - T_K^{\star\star}(x)) \cap c(x)B_\star$ *defines the* regulation map.

Proof. The set-valued map $G : X \rightsquigarrow X$ defined by $G(x) := F(x) - c(x)B_\star$ is also Marchaud.

Solutions to the differential inclusion $x'(t) \in G(x(t))$ are obviously solutions to corrected differential inclusion (12.5) where the viability multipliers are required to be bounded by $c(x)$.

Viability Theorem 11.3.4, p.455 states that K is viable under G if and only if $G(x) \cap T_K^{\star\star}(x) \neq \emptyset$. We shall prove this last point by showing that the velocity associated with the viability multiplier $p^\circ(x)$ with the smallest viability discrepancy belongs to the closed convex hull of the tangent cone $T_K(x)$.

Indeed, since $F(x)$ is convex and compact, the subset $F(x) - T_K^{\star\star}(x)$ is closed and convex. The Moreau Projection Theorem 18.4.17 implies the existence of unique elements $v^\circ(x) \in F(x)$ and $u^\circ(x) \in T_K^{\star\star}(x)$ such that $p^\circ(x) := v^\circ(x) - u^\circ(x)$ satisfies

$$\|p^\circ(x)\|_\star = \|v^\circ(x) - u^\circ(x)\| = \inf_{v \in F(x),\, u \in T_K^{\star\star}(x)} \|v - u\| = d(F(x), T_K^{\star\star}(x)) \tag{12.6}$$

This implies in particular that $u^\circ(x) := \Pi_{T_K^{\star\star}(x)} v^\circ(x)$ because

$$\|v^\circ(x) - u^\circ(x)\| = \inf_{u \in T_K^{\star\star}(x)} \|v^\circ(x) - u\| \tag{12.7}$$

Therefore the Moreau Theorem 18.4.17 implies that the viability multiplier $p^\circ(x)$ minimizing the viability discrepancy is the projection $p^\circ(x) = \Pi_{N_K(x)}(v^\circ(x))$ onto the normal cone $N_K(x)$ of $v^\circ(x)$.

Hence

$$\|p^\circ(x)\|_\star \leq d(F(x), (T_K^{\star\star}(x))) \leq c(x)$$

so that the viability multiplier minimizing the viability discrepancy satisfies

$$p^\circ(x) \in N_K(x) \cap c(x) B_\star \tag{12.8}$$

Therefore $u^\circ(x) = v^\circ(x) - p^\circ(x)$ belongs to $G(x) \cap T_K^{\star\star}(x)$. The assumptions of the Viability 11.3.4, p.455 (stated in the survival kit 2.15, p. 98) are satisfied, and we have proved that K is viable under the corrected differential inclusion (12.5). □

The follow up question is to know whether K remains viable under the corrected differential inclusion (12.5) when the viability multiplier is required to obey further constraint.

A natural (at least, classical) one is to require that the viability multipliers $q(t)$ belong to $N_K(x(t))$ at each instant: in this case, the corrected differential inclusion is called a variational inequalities:

Theorem 12.2.3 ***[Variational Inequalities]*** *Let $F : X \rightsquigarrow X$ be a Marchaud map. We assume that K is closed and sleek. Then K is viable* *under* variational inequalities

$$\begin{cases} (i) \;\; x'(t) \in F(x(t)) - p(t) \\ (ii) \; p(t) \in N_K(x(t)) \end{cases} \tag{12.9}$$

When K is assumed to be convex, control system (12.9) can be written in the form

$$\sup_{v \in F(x)} \inf_{y \in K} \langle x'(t) - v, x(t) - y \rangle \geq 0$$

and when $F(x) := \{f(x)\}$ is single-valued, it boils down to variational inequalities (12.4).

Proof. Recall that K is sleek if and only if the graph of the normal cone map $x \rightsquigarrow N_K(x)$ is closed, and that in this case, the tangent cones $T_K(x)$ are convex. The set-valued map G° defined by $G^\circ(x) := F(x) - (N_K(x) \cap c(\|x\| + 1)B_\star)$ is also Marchaud.

We have proved in (12.8) that the viability multiplier $p^\circ(x)$ belongs to $N_K(x) \cap c(\|x\| + 1)B_\star$ in such a way that the associated velocity $u^\circ(x) = v^\circ(x) - p^\circ(x)$ belongs to $G^\circ(x) \cap T_K(x)$. Therefore K is viable under G° thanks to Viability Theorem 11.3.4, p.455.

It remains to observe that the solutions to differential inclusion $x'(t) \in G^\circ(x(t))$ are solutions to the variational inequalities (12.9). □

Remark. We could have minimized $(v, u) \in F(x) \times T_K(x) \mapsto \mathbf{d}(v - u)$ where $\mathbf{d} : X \mapsto \mathbb{R}_+$ is any inf-compact lower semicontinuous convex function replacing the Hilbertian norm we used above and obtain velocities $v^\circ(c) \in F(x)$ and directions $u^\circ(x) \in T_K(x)$ such that:

$$\mathbf{d}(v^\circ(x) - u^\circ(x)) := \inf_{v \in F(x),\; u \in T_K(x)} \mathbf{d}(v - u) \tag{12.10}$$

Since $\mathbf{d}(v^\circ(x) - u^\circ(x)) := \inf_{u \in T_K(x)} \mathbf{d}(v^\circ(x) - u)$, we deduce that there exists a Lagrange multiplier $p^\circ(x) \in \partial \mathbf{d}(v^\circ(x) - u^\circ(x)) \in N_{T_K}(x)$, thus satisfying $v^\circ(x) - u^\circ(x) \in \partial \mathbf{d}^\star(p^\circ)$. The same proof implies that whenever K is sleek, K is viable under the differential inclusion

$$\begin{cases} (i) \;\; x'(t) \in F(x(t)) - \partial \mathbf{d}^\star(p(t)) \\ (ii) \; p'(t) \in N_K(x(t)) \end{cases}$$

This generalization does not change much the fundamental nature of the correction. □

12.2.4 Slow Evolutions of Variational Inequalities

We obtain slow evolutions of both the variational inequalities and the projected differential inclusion when K is sleek and F is lower semicontinuous:

Theorem 12.2.4 ***[Slow Solutions of Variational inequalities]*** *Let $K \subset X$ be a closed sleek subset and $F : X \rightsquigarrow X$ be a Marchaud and lower semicontinuous. Then K is viable under the evolutions of variational inequalities with smallest viability discrepancies*

$$\begin{cases} (i) & x'(t) \in F(x(t)) - p(t) \\ (ii) & p(t) \in N_K(x(t)) \cap d(F(x(t)), T_K(x(t)))B_\star \end{cases} \tag{12.11}$$

(the solutions of which are called slow evolutions).

Proof. Since the set-valued maps $x \rightsquigarrow F(x)$ and $x \rightsquigarrow T_K(x)$ are both lower semicontinuous, we infer that the set-valued map $x \rightsquigarrow d(F(x), T_K(x))B_\star$ is upper semicontinuous. Since F is Marchaud, so is the set-valued map $G^{\circ\circ} : X \rightsquigarrow X$ defined by $G^{\circ\circ}(x) := F(x) - (N_K(x) \cap d(F(x), T_K(x))B_\star)$.

We have proved in (12.8) that the viability multiplier $p^\circ(x)$ belongs to $N_K(x) \cap d(F(x), T_K(x))B_\star$ in such a way that the associated velocity $u^\circ(x) = v^\circ(x) - p^\circ(x)$ belongs to $G^\circ(x) \cap T_K(x)$.

Then Viability Theorem 11.3.4, p.455 implies that K is viable under $G^{\circ\circ}$: From any initial state $x \in K$ starts a solution to $x'(t) \in G^{\circ\circ}(x(t))$ which is a slow solution to (12.11). □

12.2.5 Case of Explicit Constraints

We now investigate the case when the environment $K := h^{-1}(M)$ is more explicitly defined through a continuously differentiable map $h : X \mapsto Y$ and a subset $M \subset Y$ of another finite dimensional vector space Y:

$$K := \{x \in X \text{ such that } h(x) \in M\}$$

In this case, when the tangent and normal cones to $M \subset Y$ are simpler to compute, the tangent and normal cones to K can be couched in terms of tangent and normal cones to M and of the map h.

Theorem 12.2.5 ***[Tangent and Normal Cones to Environments]*** *Let $M \subset Y$ be a closed convex subset and $h : X \mapsto Y$ a continuously differentiable map satisfying either one of the following equivalent formulations* *of the* transversality condition*:*

$$\forall x \in K,\ \mathrm{Im}(h'(x)) - T_M(h(x)) := Y \text{ or } \ker(h'(x)^\star) \cap N_M(h(x)) = \{0\}$$

Then the tangent and normal cones to K *at* $x \in K$ *are given by the formulas*

$$\begin{cases} (i)\ T_K(x) = h'(x)^{-1}(T_M(h(x))) \\ (ii)\ N_K(x) = h'(x)^\star(N_M(h(x))) \end{cases} \tag{12.12}$$

For this purpose, we need other results outside the scope of this book that we are about to describe.

When maps $h'(x)$ are surjective, or, equivalently, when their transposes $h'(x)^\star : Y^\star \mapsto Y^\star$ are injective, the linear operators $h'(x)h'(x)^\star : Y^\star \mapsto Y$ are symmetric and invertible. We denote by $h'(x)^+$ its orthogonal right inverse (see Proposition 18.4.18, p.736). We supply the dual $Y^\star$ on the space Y with the norm $\|q\|^{h'(x)^\star} := \|h'(x)^\star q\|$ depending upon each $x \in K$. If $N \subset Y^\star$ is a closed convex cone, we denote by $\Pi_N^{h'(x)^\star}$ the projector of best approximation onto N when $Y^\star$ is supplied with the norm $\|q\|^{h'(x)^\star}$.

Theorem 12.2.2, p.488 becomes:

Theorem 12.2.6 ***[Corrected Differential Equations under Explicit Constraints]*** *Assume that* f *is continuous with linear growth and that* $K = h^{-1}(M)$ *where* $M \subset Y$ *is a closed sleek subset where* $h : X \mapsto Y$ *is a continuously differentiable map satisfying*

$$\forall x \in K,\ h'(x) \text{ is surjective } \&\ \sup_{x \in K} \|h'(x)^+\| < +\infty \tag{12.13}$$

Then K *is viable under the corrected system*

$$x'(t) = f(x(t)) - h'(x(t))^\star q(t)$$

where the viability multipliers $q(t)$ *obey the* regulation law

$$\text{for almost all } t \geq 0,\ q(t) \in R_M(x(t))$$

where the regulation map R_M *is defined by*

$$R_M(x) := (h'(x)h'(x)^\star)^{-1}(h'(x)f(x) - T_M(h(x)))$$

Actually, K *is viable under the variational inequalities*

$$\begin{cases} (i)\ x'(t) = f(x(t)) - h'(x(t))^\star q(t) \\ (ii)\ q(t) \in N_M(h(x(t))) \end{cases}$$

In particular, these variational inequalities have slow evolutions

$$\begin{cases} (i) \quad x'(t) = f(x(t)) - h'(x(t))^\star q(t) \\ (ii) \; q(t) \in r^\circ_M(x(t)) \end{cases} \tag{12.14}$$

where the feedback map $r^\circ_M : X \mapsto Y^\star$ *is defined by*

$$\forall x \in K, \; r^\circ_M(x) := \Pi^{h'(x)^\star}_{N_M(h(x))}(h'(x)h'(x)^\star)^{-1}h'(x)f(x)$$

When $K = h^{-1}(M)$, it may be wiser to try other regulons $q(t) \in Y^\star$ than the specific regulon $q(t) := r^\circ_M(x(t))$:

Definition 12.2.7 ***[Explicit Viability Multipliers]*** *When* $K = h^{-1}(M)$, *the specific regulon* $q(t) := r^\circ_M(x(t))$ *are called* explicit viability multipliers.

In summary, we were able to associate with a differential equation under which a subset $K := h^{-1}(M)$ *is not viable a specific type of control systems, where the regulons are explicit viability multipliers, under which K is viable.*

Remark: why viability multipliers. *Explicit viability multipliers* $q(t) \in Y^\star$ play a analogous role to *Lagrange multipliers* in optimization under constraints.

Indeed, when $X := \mathbb{R}^n$, $Y := \mathbb{R}^m$, $f(x) := (f_1(x), \ldots, f_n(x))$ and $h(x) := (h_1(x), \ldots, h_m(x))$, this control system can be written more explicitly under the form

$$x'_j(t) = f_j(x(t)) + \sum_{k=1}^{m} \frac{\partial h_k(x(t))}{\partial x_j} q_k(t), \; j = 1, \ldots, m$$

We recall that the minimization of a function $x \mapsto J(x)$ over K is equivalent to the minimization **without constraints** of the function

$$x \mapsto J(x) + \sum_{k=1}^{m} \frac{\partial h_k(x)}{\partial x_j} q_k$$

for an appropriate "Lagrange multiplier" $q \in Y^\star$.

These Lagrange and explicit viability multipliers associated to constraints of the form $h(x) \in M$ share the same interpretations. In economics, for instance, Lagrange multipliers are often regarded as (shadow) prices in optimization or equilibrium models. Therefore, explicit viability multipliers enjoy the same status as prices, slowing down consumption to meet scarcity constraints, boosting production to guarantee given outputs, etc. □

12.2.6 Meta Variational Inequalities

The viability multipliers governed by variational inequalities $x'(t) \in F(x(t)) - N_K(x(t))$ vanish on the interior of the environment K and may be discontinuous on its boundary. For obtaining (absolutely) continuous viability multipliers, we shall use the same idea than the one motivating the introduction of metasystems and heavy evolutions (see Chap. 6, p.199).

Starting with a differential inclusion $x'(t) \in F(x(t))$ and an environment K, we introduce a set-valued map $P : K \rightsquigarrow X^\star$ (or a set-valued map $P : X \rightsquigarrow X^\star$ and take $K := \mathrm{Dom}(P)$). We consider the heavy system

$$\begin{cases} (i) \ \ x'(t) \in F(x(t)) \\ (ii) \ p'(t) = 0 \end{cases}$$

constrained by the metaconstraint

$$\forall t \geq 0, \ p(t) \in P(x(t))$$

We recall that the derivative $DP(x,p) : X \rightsquigarrow X^\star$ is defined by

$$\mathrm{Graph}(DP(x,p)) := T_{\mathrm{Graph}(P)}(x,p)$$

and that its co-derivative $DP(x,p)^\star : X \rightsquigarrow X^\star$ by

$$r \in DP(x,p)^\star(q) \text{ if and only if } (r,-q) \in N_{\mathrm{Graph}(P)}(x,p)$$

We also recall that the *set-valued map* P *is said to be sleek* if and only if its graph is sleek (see Definition 18.4.8, p.732).

Theorem 12.2.3 implies the following statement

Theorem 12.2.8 *[Meta Variational Inequalities] Let* $F : X \rightsquigarrow X$ *be a Marchaud map. We assume that* $P : X \rightsquigarrow X^\star$ *is closed and sleek. Then the graph* Graph(P) *of the set-valued map* P *is viable under meta variational inequalities*

$$\begin{cases} (i) \ \ x'(t) \in F(x(t)) - DP(x(t),p(t))^\star(q(t)) \\ (ii) \ p'(t) = q(t) \text{ where } q(t) \text{ ranges over } X^\star \end{cases} \quad (12.15)$$

Therefore, heavy evolutions, *minimizing the norm of the derivatives* $p'(t) = q(t)$ *of the regulons* $p(t)$, *amounts here to minimize the norm of the elements* $q(t) \in DP(x(t), p(t))^{\star^{-1}}(F(x(t)) - T_K(x(t)))$, *instead minimizing the norm of the elements* $p(t) \in F(x(t)) - T_K(x(t))$ *as for the slow evolutions.*

Proof. Theorem 12.2.3 implies that the variational inequalities associated to

$$\begin{cases} (i) \ x'(t) \in F(x(t)) \\ (ii) \ p'(t) = 0 \end{cases}$$

on the graph of P can be written

$$(x'(t), p'(t)) \in F(x) \times \{0\} - N_{\text{Graph}(P)}(x(t), p(t))$$

In other words, there exist $(r(t), -q(t)) \in N_{\text{Graph}(P)}(x(t), p(t))$ such that $(x'(t), p'(t)) \in F(x(t)) \times \{0\} + (-r(t), q(t))$, i.e., such that $x'(t) \in F(x(t)) - r(t)$ and $p'(t) = q(t)$. It is enough to remark that $r(t) \in DP(x(t), p(t))^{\star}(q(t))$. □

12.2.7 Viability Multipliers for Bilateral Constraints

Let us consider a system of n differential equations (instead of inclusions for simplifying the exposition)

$$\forall i = 1, \ldots, n, \ \forall t \geq 0, \ x_i'(t) \ = \ f_i(x(t)) \tag{12.16}$$

subjected to bilateral constraints

$$\forall t \geq 0, \ \Sigma(x(t)) \ := \ (\sigma_{i,j}(x(t)))_{i,j} \ \in \ \mathcal{M} \ \subset \ \mathcal{L}(\mathbb{R}^n, \mathbb{R}^n) \tag{12.17}$$

where $\mathcal{M}$ is a subset of matrices $(\sigma_{i,j}(x(t)))_{i,j}$ of linear operators.

Naturally, these bilateral constraints are not necessarily viable, so that we correct this system by explicit viability multipliers:

Theorem 12.2.9 ***[Viability Matrices]*** *Bilateral constraints (12.17), p. 495, are viable under a corrected system regulated by* viability matrices $W(t) := \left(w^{i,j}(t)\right)_{i,j}$:

$$\forall i = 1, \ldots, n, \ \forall t \geq 0, \ x_i'(t) \ = \ f_i(x(t)) - \sum_{k,l=1}^{n} \frac{\partial \sigma_{k,l}(x(t))}{\partial x_i(t)} w^{k,l}(t) \tag{12.18}$$

The regulation map *maps state variables x to* non empty *subsets $R_{\mathcal{M}}(x)$ of viability matrices satisfying*

$$\left(\sum_{i,p,q=1}^{n}\frac{\partial\sigma_{k,l}(x)}{\partial x_i}\frac{\partial\sigma_{p,q}(x)}{\partial x_i}w^{p,q}\right)_{k,l} \in T_{\mathcal{M}}(\Sigma(x)) - \Sigma'(x)(f(x)) \quad (12.19)$$

where $T_{\mathcal{M}}(W) \subset \mathcal{L}(\mathbb{R}^n,\mathbb{R}^n)$ is the tangent cone to $\mathcal{M}$ at $W \in \mathcal{L}(\mathbb{R}^n,\mathbb{R}^n)$ and $N_{\mathcal{M}}(W) := T^{\star}_{\mathcal{M}}(W)$ its normal cone.

For example, we consider n vehicles i. Safety regulations (as in aeronautics) require that the positions $x_i \in X := \mathbb{R}^d$ $(d = 2,4)$ must mutually be kept apart at a minimal distance: the entries of the matrix of bilateral constraints are defined by $\sigma_{i,j}(x) := \frac{1}{2}\|x_i - x_j\|^2$. We observe that $\frac{\partial}{\partial x_i}\sigma_{i,j}(x)(u) = \langle x_i - x_j, u_i\rangle$, or, equivalently, that

$$\begin{cases} \frac{\partial\sigma_{k,l}(x)}{\partial x_i}(u_k) &= 0 \text{ if } i \neq k,l \\ \frac{\partial\sigma_{k,l}(x)}{\partial x_k}(u_k) &= \langle x_k - x_l, u_k\rangle, \quad \frac{\partial\sigma_{k,l}(x)}{\partial x_l}(u_l) = -\langle x_k - x_l, u_l\rangle \end{cases}$$

Taking a given matrix $U := (u_{i,j}(x(t)))_{i,j}$ and the set of matrices with larger or equal entries $\mathcal{M} := U + \mathbb{R}^{n^2}$, the environment $\mathcal{K} := \Sigma^{-1}(\mathcal{M})$ is the subset of mutually safe positions defined by

$$\forall\, i,j = 1,\ldots,n, \quad \|x_i - x_j\| \geq u_{i,j}$$

Consequently, the corrected system (12.18), p. 495 can be written in the form

$$\forall\, i = 1,\ldots,n, \ \forall\, t \geq 0, \ x_i'(t) = f_i(x(t)) - \sum_{j=1}^{n}(w^{i,j}(t) - w^{j,i}(t))(x_i(t) - x_j(t)) \quad (12.20)$$

Instead of taking for viability multipliers the coefficients $w^{i,j} \in \mathbb{R}$ of an arbitrary matrix W, it is enough to take the coefficients $q^{i,j} := w^{i,j} - w^{j,i}$ of the Q satisfying $Q^{\star} = -Q$ i.e., $q^{i,j} = -q^{j,i}$. In this case, we obtain

$$\forall\, i = 1,\ldots,n, \ \forall\, t \geq 0, \ x_i'(t) = f_i(x(t)) - \sum_{j=1}^{n} q^{i,j}(t)(x_i(t) - x_j(t)) \quad (12.21)$$

12.2.8 Connectionnist Complexity

As the simplest example of connectionnist complexity, we assume that *disconnected* or autonomous systems are given in the form of a system of differential equations

$$\forall i = 1, \ldots, n, \;\; x_i'(t) \;=\; f_i(x_i(t))$$

We also postulate that the state of the system must comply to *collective viability constraints* of the form

$$\forall t \geq 0, \;\; h(x_1(t), \ldots, x_n(t)) \;\in\; M$$

where M is a subset of $Y := \mathbb{R}^m$.

Since the nature of the constraints is collective and the dynamics are autonomous, the *viability property* requiring that from any initial state satisfying the constraints starts at least one solution obeying them forever is generally not met.

We have seen in Sect. 12.2.5, p.491 how to correct this system by viability multipliers $p_i(t)$

$$\forall i = 1, \ldots, n, \;\; x_i'(t) \;=\; f_i(x_i(t)) - p_i(t)$$

Another method consists in connecting the components of the original dynamics through *connecting matrices* $W := (w_i^j)_{i,j} \in \mathcal{L}(X, X)$:

$$\forall i = 1, \ldots, n, \;\; x_i'(t) \;=\; \sum_{j=1}^{n} w_i^j(t) f_j(x_j(t))$$

Definition 12.2.10 *[**Complexity Index**] If $W \in \mathcal{L}(X, X)$ is a connection matrix, we regard the norm $\|W - \mathbf{I}\|_{\mathcal{L}(X,X)}$ as the complexity index of the connection matrix W. For a given class R of connection matrices, we denote by $W^\circ \in R$ the simplest – least complex – matrix W° defined by*

$$\|W^\circ - \mathbf{I}\| \;=\; \min_{W \in R} \|W - \mathbf{I}\| \tag{12.22}$$

We shall prove that in this context, the connection matrix which minimizes this connectionnist complexity index over the set of *viability matrices* is associated with the regulation mechanism driving *slow* viable evolutions, i.e., that the least complex connection matrix governs slow evolutions.

This further justifies the decentralizing role of prices in economic decentralization mechanisms when the viability multipliers are regarded as (shadow prices): *the decentralization by viable prices with minimum norm coincides with the regulation by simplest viable connection matrices.*

The natural framework to study these connection matrices is *tensor algebra* of matrices because the connection matrix which minimizes the connectionnist complexity index happens to involve the tensor product of the dynamics and the regulons. In this presentation, we refer to *Neural networks and qualitative physics: a viability approach*, [21, Aubin], and mention only the following notations.

Let $X := \mathbf{R}^n$, $Y := \mathbf{R}^p$ be finite dimensional vector spaces. If $p \in X^\star$ and $y \in X$, we denote by $p \otimes y \in \mathcal{L}(X,Y)$ the linear operator defined by

$$p \otimes y : x \mapsto (p \otimes y)(x) \;:=\; \langle p, x \rangle \, y \tag{12.23}$$

It is called the *tensor product* of $p \in X^\star$ and $y \in X$. The transpose $(p \otimes y)^\star$ of $p \otimes y$ is the operator $y \otimes p \in \mathcal{L}(Y^\star, X^\star)$ is equal to

$$(y \otimes p)(q) \;:=\; \langle q, y \rangle \, p$$

More generally, if $x \in X$ and $B \in \mathcal{L}(Y,Z)$, we denote by $x \otimes B \in \mathcal{L}(\mathcal{L}(X,Y),Z)$ the operator defined

$$W \in \mathcal{L}(X,Y) \mapsto (x \otimes B)(W) \;:=\; BWx \;\in\; Z$$

By taking $Z := \mathbb{R}$ and $B := q \in Y^\star := \mathcal{L}(Y,\mathbb{R})$, we deduce that $x \otimes q \in \mathcal{L}(X^\star, Y^\star)$ is the *tensor product*

$$x \otimes q \;:\; p \in X^\star := X \mapsto (x \otimes q)(p) \;:=\; \langle p, x \rangle q$$

the matrix of which is made of entries $(x \otimes q)_i^j = x_i q^j$. Since $x \otimes q$ belongs to $\mathcal{L}(X^\star, Y^\star)$, we observe that for any $W \in \mathcal{L}(X,Y)$,

$$\langle x \otimes q, W \rangle \;:=\; \langle q, Wx \rangle \;=\; := \; \langle W^\star q, x \rangle$$

In particular, if $W := p \otimes y$

$$\langle x \otimes q, p \otimes y \rangle \;:=\; \langle p, x \rangle \, \langle q, y \rangle$$

Let $f : X \mapsto X$ describe the continuous dynamics of the system, $h : X \mapsto Y$ represent the constraint map and $M \subset Y$ the closed environment.

We shall compare:

1. the regulation obtained by subtracting prices to the original dynamical behavior
$$x'(t) \;=\; f(x(t)) - p(t)$$
2. the regulation by connecting the agents of the original dynamics through connecting matrices $W \in \mathcal{L}(X,X)$:
$$x'(t) \;=\; W(t) f(x(t))$$

We first observe the following link between regulons and connection matrices:

Proposition 12.2.11 ***[Connection Matrix Associated with a Viability Multiplier]*** *Let us associate with $p(\cdot)$ the connection matrix $W_{p(\cdot)}(\cdot)$ defined by the formula*

$$W_{p(\cdot)}(t) := \mathbf{I} - \frac{f(x(t))}{\|f(x(t))\|^2} \otimes p(t) \qquad (12.24)$$

the entries of which are equal to

$$w_j^i(t) \;=\; \delta_j^i - \frac{f_i(x(t))}{\|f(x(t))\|^2} p_j(t)$$

The solutions to the differential equations $x'(t) = f(x(t)) - p(t)$ and $x'(t) = W_{p(\cdot)}(t) f(x_{p(\cdot)}(t))$ coincide.

Proof. Indeed, we observe that

$$W_{p(\cdot)}(t) f(x(t)) \;=\; f(x(t)) - \frac{\langle f(x(t)), f(x(t))\rangle}{\|f(x(t))\|^2} p(t) \;=\; f(x(t)) - p(t)$$

so that the two differential equations $x'(t) = W(t)f(x(t))$ and $x'(t) = f(x(t)) - p(t)$ are the same. □

We introduce the new regulation map

$$O_M(x) \;:=\; \{W \in \mathcal{L}(X,X) \mid h'(x)Wf(x) \;\in\; T_M(h(x))\}$$

the connection matrices W regulating viable solutions to $x' = Wf(x)$ are given by the *regulation law*

$$W(t) \;\in\; O_M(x(t))$$

We observe that whenever the regulons $p(\cdot) \in R_M(x(\cdot))$ obey the regulation law, the associated connection matrix $W_{p(\cdot)} \in O_M(x(\cdot))$ because the equation

$$h'(x(t))W(x(t))f(x(t)) \;=\; h'(x(t))(f(x(t)) - p(t)) \;\in\; T_M(h(x))$$

holds true.

The problem of minimal complexity is then to find connection matrices $W^\circ(x) \in O_M(x)$ as close as possible to the identity matrix $\mathbf{I}$:

Theorem 12.2.12 ***[Viable Simplest Connection Matrices]*** *Let us assume that the map f is continuous and bounded and that the constraint* *map h is continuously differentiable and satisfies the uniform surjectivity condition:*

$$\forall x \in K,\ \ h'(x) \text{ is surjective \& } \sup_{x\in K} \frac{\|h'(x)^{+}\|}{\|f(x)\|} < +\infty \qquad (12.25)$$

Let $M \subset Y$ be a closed convex (or more generally, sleek) subset. Then the environment $K := h^{-1}(M)$ is viable under the connected system $x' = Wf(x)$ if and only if for every $x \in K$, the image $O_M(x)$ of the regulation map is not empty.

The solution with minimal complexity subjected to the viability constraints

$$\forall t \geq 0,\ \ h(x(t)) \ \in \ M$$

is governed by the differential equation

$$x'(t) \ = \ W^{\circ}(x(t))f(x(t))$$

where

$$\begin{cases} W^{\circ}(x) = \mathbf{1} - \dfrac{f(x)}{\|f(x)\|^2} \otimes h'(x)^{+}\left(\mathbf{1} - \Pi^{h'(x)}_{T_M(h(x))}\right) h'(x)f(x) \\ \qquad\quad = \mathbf{1} - \dfrac{f(x)}{\|f(x)\|^2} \otimes h'(x)^{\star}\pi^{h'(x)^{\star}}_{N_M(h(x))}(h'(x)h'(x)^{\star})^{-1}h'(x)f(x) \end{cases} \qquad (12.26)$$

Furthermore, the slow viable solutions regulated by $x' = f(x) - p$ and the viable solutions to the connected system $x' = Wf(x)$ under minimal complexity coincide.

Proof. We can apply the same proof as the one of Theorem 12.2.6 where the connected system is a control system $x' = Wf(x)$ and where the role of the regulon p is played by the connection matrix W.

Using the definition of tensor products $(f(x)\otimes h'(x))W := h'(x)Wf(x)$ of linear operators, the regulation map can be written in the form:

$$O_M(x) \ := \ \{W \in \mathcal{L}(X,X) \mid (f(x) \otimes h'(x))W \ \in \ T_M(h(x))\}$$

so that viable solutions are regulated by the regulation law

$$(f(x(t)) \otimes h'(x(t)))W(t) \ \in \ T_M(h(x))$$

Since the map $f(x) \otimes h'(Wx)$ is surjective from the space $\mathcal{L}(X,X)$ of connection matrices to the space Y, we observe that

$$\|(f(x)\otimes h'(x))^+\| = \frac{\|h'(x)^+\|}{\|f(x)\|}$$

so that the assumptions of Theorem 12.2.6, p.492 are satisfied. Hence the regulation map O_M is lower semicontinuous with closed continuous images.

Therefore, the viable connection matrices closest to the identity matrix $\mathbf{I}$ are the solutions W° minimizing the distance $\|W-\mathbf{I}\|$ among the connection matrices satisfying the above constraints.

We infer from Theorem 2.5.6 of *Neural networks and qualitative physics: a viability approach*, [21, Aubin], that since the map $f(x)\otimes h'(Wx)$ is surjective from the space $\mathcal{L}(X,X)$ of connection matrices to the space Y, the solution is given by formula (12.26).

Therefore, we derive from Theorem 6.6.3 of [18, Aubin] the existence of a viable solutions to the differential inclusion $x' = W^\circ(x)f(x)$. □

12.2.9 Hierarchical Viability

Hierarchical systems involve a finite number of n states $x_i \in X_i$ and $n-1$ linear operators W_i^{i-1} cooperating for transforming the state x_1 into a final state $x_n := W_n^{n-1}\cdots W_i^{i-1}\ldots W_2^1 x_1$ subjected to viability constraints.

They describe for instance a production process associating with the input x_1 the intermediate outputs $x_2 := W_2^1 x_1$, which itself produces an output $x_3 := W_2^1 x_2$, and so on, until the final output $x_n := W_n^{n-1}\cdots W_i^{i-1}\ldots W_2^1 x_1$ which must belong to the production set M_n.

Hierarchical systems are particular cases of *Lamarckian systems* where the hierarchical constraint are replaced by a network of constraints, more difficult and fastidious to describe but not to prove. Such systems appear in many economic, biological and engineering systems. This study only opens the gates for future investigations.

Let us consider n vector spaces $X_i := \mathbb{R}^{n_i}$, elements $x_i \in X_i$ and $n-1$ connection matrices $W_{i+1}^i \in \mathcal{L}(X_i, X_{i+1})$.

The constraints are of the form

$$\begin{cases} K := \left\{ (x_1, (W_{i+1}^i)_{i=1,\ldots,n-1}) \in X_1 \times \prod_{i=1}^{n-1} \mathcal{L}(X_i, X_{i+1}) \right. \\ \text{such that } W_n^{n-1}\cdots W_i^{i-1}\ldots W_2^1 x_1 \in M_n \subset X_n \} \end{cases} \quad (12.27)$$

The evolution without constraints of the commodities and the operators is governed by dynamical systems of the form

$$\begin{cases} (i) \;\; x_i'(t) = f_i(x_i(t)) \\ (ii) \;\; \frac{d}{dt}W_{i+1}^i(t) = \alpha_{i+1}^i(W_{i+1}^i(t)) \end{cases}$$

where the functions α^i_{i+1} govern the evolutions of the matrices $t \mapsto W^i_{i+1}(t)$.

Theorem 12.2.13 *[Hierarchical Viability] The constraints*

$$\forall\, t \geq 0,\;\; W^{n-1}_n(t)\cdots W^{i-1}_i(t)\ldots W^1_2(t)x_1(t) \;\in\; M_n$$

are viable under the regulated system

$$\begin{cases} x_1'(t) \;=\; f_1(x_1(t)) + W^1_2(t)^\star(t)p^1(t) & (i=1) \\ x_i'(t) \;=\; f_i(x_i(t)) - p^{i-1}(t) + W^i_{i+1}(t)^\star p^i(t) & (i=1,\ldots,n-1) \\ x_n'(t) \;=\; f_n(x_n(t)) - p^{n-1}(t) & (i=n) \\ \dfrac{d}{dt}W^i_{i+1}(t) \;=\; \alpha^i_{i+1}(W^i_{i+1}(t)) + x_i(t)\otimes p^i(t) & (i=1,\ldots,n-1) \end{cases}$$

involving viability multipliers $p^i(t) \in X_i^\star$ (intermediate "shadow price"). The dynamics of the matrices $W^i_{i+1}(t)$ are corrected by the tensor product of $x_i(t)$ and $p^i(t)$.

In other words, at each level i of the hierarchical organization, the dynamics governing the evolution $x_i(\cdot)$ of the state involves both a viability multiplier $p^i(\cdot)$ at the same level and viability multiplier $p^{i-1}(\cdot)$ at the level $i-1$, a message which makes sense.

The specialists of artificial neural networks will recognize that the evolution of each connection matrix W^i_{i+1} is corrected by the tensor product $x_i(t)\otimes p^i(t)$ of the state and of the viability multiplier at level i, which is an example of a Hebbian law (see for instance *The Organization of Behavior*, [114, Hebb] and *Neural networks and qualitative physics: a viability approach* [21, Aubin] for the links between tensor products and Hebbian laws).

Proof. We introduce

1. the product spaces

$$\mathbb{X} := \prod_{i=1}^{n} X_i \times \prod_{i=1}^{n-1} \mathcal{L}(X_i, X_{i+1}) \text{ and } \mathbb{Y} := \prod_{i=1}^{n} X_i$$

   of sequences $(x_1,\ldots,x_n,W^1_2,\ldots,W^{n-1}_n)$ and $(y_1,\ldots,y_n)$
2. the (bilinear) operator $h : \mathbb{X} \mapsto \mathbb{Y}$ defined by

$$h(x,W){:=}h(x_1,\ldots,x_n,W^1_2,\ldots,W^{n-1}_n){:=}((W^i_{i+1}x_i - x_{i+1})_{i=1,\ldots,n-1}, x_n)$$

3. the set of constraints

$$\mathbf{M} := \{0\}^{(n-1)} \times M_n \subset \mathbb{Y}$$

where $M_n \subset X_n$ is defined in (12.2.13), p.502.

We observe that the environment K defined by (12.27), p.501, is equal to

$$K = \{(x, W) \in \mathbb{X} \text{ such that } h(x, W) \in \mathbf{M} \subset \mathbb{Y}\}$$

The directional derivative $h'(x, W)$ is the linear operator mapping every pair (u, U) to $h'(x, W)(u, U)$ defined by

$$h'(x, W)(u, U) = ((U^i_{i+1} x_i + W^i_{i+1} u_i - u_{i+1})_{i=1,\ldots,n-1}, u_n)$$

Its transpose $h'(x, W)^\star$ map every pair $p := (p^1, \ldots, p^n)$ to $h'(x, W)^\star p$ equal to

$$h'(x, W)^\star p = [(W_2^{1^\star} p^1, (W_{i+1}^{i^\star} p^i - p^{i-1})_{i=2,\ldots,n-1}, p^n), (x_i \otimes p^i)_{i=1,\leq,n-1}]$$

Hence Theorem 12.2.13, p.502 follows from Theorem 12.2.6, p.492. □

Remark. This result has been extended to more complex situations, in the context of cooperative games, for instance, where coalitions $S \subset N := \{1, \ldots, n\}$ are involved and where linear connectionnist operators W^i_{i+1} are replace by multilinear operators W_S associated with coalitions S. □

12.3 Impulse and Hybrid Systems

12.3.1 Runs of Impulse Systems

As mentioned in the introduction of this chapter, there is no reason why an arbitrary subset K should be viable under a control system

$$\begin{cases} (i) \;\; x'(t) = f(x(t), u(t)) \\ (ii) \; u(t) \in U(x(t)) \end{cases}$$

or a differential inclusion $x' \in F(x)$ or actually its evolutionary system $\mathcal{S}(x)$.

One way to restore viability is to change instantaneously the initial conditions when viability is at stakes or when other requirement is asked by using a *reset map* Φ mapping any state of K to a (possibly empty) set $\Phi(x) \subset X$ of new (reinitialized) "reset states". Hence an *impulse differential inclusion* (and in particular, an *impulse control system*) is described by a pair (F, Φ), where the set-valued map $F : X \rightsquigarrow X$ mapping the state space $X := \mathbb{R}^n$ to itself governs the *continuous evolution* components $x(\cdot)$ of the system in

K and where Φ, the *reset map*, governs the *discrete* impulses to new "initial conditions" when the continuous evolution is doomed to leave K.

Such a hybrid evolution, mixing continuous evolution "punctuated" by discontinuous impulses at impulse times is called in the "hybrid system" literature a *"run"* or an *"execution"*, and could be called a *punctuated evolution.*

Many examples coming from different fields of knowledge fit this framework:

1. multiple-phase economic dynamics in economics, as it was proposed by the economist Richard Day back to 1995,
2. stock management in production theory,
3. viability theory, for implementing the extreme version of the "inertia principle",
4. and in particular, evolutions with *punctuated equilibria*, as proposed by Niels Eldredge and Stephen J. Gould in 1972 for describing biological evolution,
5. propagation of the nervous influx along axones of neurons triggering spikes in neurosciences and biological neuron networks proposed in 1907 by Lapicque[3] ("Integrate-and-Fire" models),
6. "threshold" impulse control, when "controls jump" when the threshold is about to be trespassed,
7. issues dealing with "qualitative physics" in artificial intelligence (AI) and/or "comparative statics" in economics,
8. in logics, where connections are made between impulse differential inclusions and the μ-calculus,
9. in manufacturing and economic production systems, when the jumps are governed by Markov processes instead of set-valued maps,
10. and, above all, in automatic control theory where a fast growing literature deals with hybrid "systems".

Hybrid systems are described by a family of control systems and by a family of viability (or state) constraints indexed by parameters e called "locations". Starting with an initial condition in a set associated with an initial location, the control system associated with the initial location governs the evolution of the state in this set for some time until some impulse time resets the system by imposing a new location, and thus, a new control system, a new environment and a new initial condition. One can show that they fit in the above class of impulse systems in a simple and natural way. The use of viability techniques for dealing with impulse systems was trigged by Shankar Sastry, John Lygeros, Claire Tomlin, Marc Quimcampoix, Nicolas Seube, Éva Cruck, Georges Pappas and the authors. This section presents some of these results.

We identify $\mathcal{C}(0,0;X)$ with X and we define "runs" in the following way:

[3] Instead of the continuous time Hodgkin–Huxley type of systems of differential equations inspired by the propagation of electrical current which are the subject of an abundant literature.

Definition 12.3.1 ***[Runs]*** *We say that a (finite or infinite) sequence*

$$\overrightarrow{x}(\cdot) := (\tau_n, x_n(\cdot))_{n\geq 0} \in \prod_{n\geq 0} \mathbb{R}_+ \times \mathcal{C}(0,\tau_n;X)$$

is a run *(or an* execution *or a* punctuated evolution*)* $\overrightarrow{x}(\cdot)$ *made of two finite or infinite sequences of*

1. cadences $\tau(\overrightarrow{x}(\cdot)) := \{\tau_n\}_n$*;*
2. motives $x_n(\cdot) \in \mathcal{C}(0,\tau_n;X)$.

We associate with a run $\overrightarrow{x}(\cdot)$

1. *its* impulse set $\mathcal{T}(\overrightarrow{x}(\cdot)) := \{t_n\}_{n\geq 0}$ *of sequences of* impulse times $t_{n+1} := t_n + \tau_n$, $t_0 = 0$
2. *its* development *defined by*

$$\forall n \geq 0,\ \overrightarrow{x}(t_n) := x_n(0)\ \&\ \forall t \in [t_{n-1}, t_n[,\ \overrightarrow{x}(t) := x_n(t-t_n)$$

We say that the sequence of $x_n := x_n(0) \in X$ *is the sequence of* reinitialized reset states *of the run* $\overrightarrow{x}(\cdot)$.

A run $\overrightarrow{x}(\cdot)$ *is said to be* viable *in* K *on an interval* $I \subset \mathbb{R}_+$ *if for any* $t \in I$, $\overrightarrow{x}(t) \in K$.

Naturally, if $\tau_n = 0$, i.e., if $t_{n+1} = t_n$, we identify the motive $x_n(\cdot)$ with the reset state $x_n(\cdot) \equiv x_n \in \mathcal{C}(0,0;X) \equiv X$, so that runs can be continuous-time evolutions, sequences (discrete time evolutions), or hybrids of them.

Definition 12.3.2 ***[Particular Ends of a Run]*** *A run* $\overrightarrow{x}(\cdot) := (\tau_n, x_n(\cdot))_{n\geq 0}$ *is said to be*

1. discrete *if there exists* $N < +\infty$ *such that for all* $n \geq N$, $\tau_n = 0$ *(the run ends with a sequence),*
2. finite *if the sequence of cadences is finite, i.e., there exists* $N < +\infty$ *such that* the Nth motive $x_N(\cdot)$ is taken on $[0,+\infty[$ *and we set* $\tau_{N+1} = +\infty$ *(the run ends with a continuous-time evolution). The run is said to be* infinite *if its sequence of cadences is infinite.*

A run is said to be proper *if it is neither discrete nor finite.*

If $t \notin \mathcal{T}(\overrightarrow{x}(\cdot))$ is not an impulse time of the run $\overrightarrow{x}(\cdot)$, then we can set $\overrightarrow{x}(t) = x(t)$ without ambiguity. At impulse time $t_n \in \mathcal{T}(\overrightarrow{x}(\cdot))$, we defined

$\overrightarrow{x}(t_n) := x_n(0)$. But we also need to define the value of the run just before it "jumps" at impulse time:

Definition 12.3.3 *[Jumps of a Run] If $t_n \in \mathcal{T}(\overrightarrow{x}(\cdot))$ is an impulse time of the run $\overrightarrow{x}(\cdot)$, we define the state of the evolution just before impulse time t_n by*

$$\overrightarrow{x}(^-t_n) := \begin{cases} \lim_{\tau \mapsto t_n -} x(\tau) & \text{if } t_n > t_{n-1} \\ x_{n-1}(0) := \overrightarrow{x}(^-t_{n-1}) & \text{if } t_n = t_{n-1} \end{cases}$$

We associate with a run $\overrightarrow{x}(\cdot)$ its sequence of jumps $s(\overrightarrow{x}(\cdot)) := (s_n(\overrightarrow{x}(\cdot)))_{n \geq 1}$ *defined by*

$$s_n(\overrightarrow{x}(\cdot)) := x_n(0) - x_{n-1}(\tau_{n-1}) = \overrightarrow{x}(t_n) - \overrightarrow{x}(^-t_n)$$

We now adapt to runs the integral representation $x(t) = x_0 + \int_0^t x'(\tau)d\tau$ of an evolution:

Proposition 12.3.4 *[Integral Representation of a Run] Assume that the motives $x_n(\cdot)$ of a run $\overrightarrow{x}(\cdot)$ are differentiable on the intervals $]0, \tau_n[$ for all n such that the cadence $\tau_n > 0$ is positive. Therefore,*

$$\begin{cases} (i)\ \forall\, n \geq 0,\ \ \forall t \in [t_{n-1}, t_n[, \\ \qquad \overrightarrow{x}(t) = x_0 + \sum_{k=1}^{n-1} s_k(\overrightarrow{x}(\cdot)) + \int_0^t x'(\tau)d\tau \text{ when } t_{n-1} < t_n \\ (ii)\ x_n := \overrightarrow{x}(t_n) = x_0 + \sum_{k=1}^{n} s_k(\overrightarrow{x}(\cdot)) + \int_0^{t_n} x'(\tau)d\tau \end{cases}$$

Proof. We prove this statement recursively. It is obvious for $j = 0$. We assume that we have constructed the viable run $\overrightarrow{x}(\cdot)$ on the interval $[0, t_n[$ through the jumping times $0 \leq t_1 \leq \cdots t_{n-1} \leq t_n \leq \cdots$ satisfying for all $j = 1, \ldots n-1$

$$\forall t \in [t_{j-1}, t_j],\ \overrightarrow{x}(t) \ = \ x_0 + \sum_{k=1}^{j-1} s_k(\overrightarrow{x}(\cdot)) + \int_0^t x'(\tau)d\tau$$

At time t_n, we take a motive $x_n(\cdot)$ that can be written

$$\forall t \in [0, \tau_n],\ x_n(t) = x_n(0) + \int_0^t x_n'(\tau)d\tau = \overrightarrow{x}(^-t_n) + s_n(\overrightarrow{x}(\cdot)) + \int_{t_n}^t x'(\tau)d\tau$$

Taking into account the induction assumption

$$\overrightarrow{x}(^-t_n) = x_0 + \sum_{k=1}^{n-1} \mathrm{s}_k(\overrightarrow{x}(\cdot)) + \int_0^{t_n} x'(\tau)d\tau$$

we infer that for any $t \in [t_n, t_{n+1}]$,

$$\overrightarrow{x}(t) = x_0 + \sum_{k=1}^{n} \mathrm{s}_k(\overrightarrow{x}(\cdot)) + \int_0^{t} x'(\tau)d\tau$$

This completes the proof. □

12.3.2 Impulse Evolutionary Systems

Consider the control system (2.10):

$$\begin{cases} (i) \;\; x'(t) = f(x(t), u(t)) \\ (ii) \; u(t) \in U(x(t)) \end{cases}$$

with state-dependent constraints on the controls, generating the *evolutionary system* denoted by $\mathcal{S}$ associating with any initial state x the subset $\mathcal{S}(x)$ of solutions $x(\cdot)$ to (2.10) starting at x.

Definition 12.3.5 *[Reset Map] We introduce a set-valued map $\Phi : X \rightsquigarrow X$, regarded as a* reset map *governing the resetting of the initial conditions.*

The subset

$$\text{Ingress}_{\Phi}(K) := K \cap \Phi^{-1}(K) \tag{12.28}$$

is called the ingress set *of K under Φ and the subset*

$$\text{Egress}_{\Phi}(K) := K \backslash \Phi^{-1}(K) = K \cap \Phi^{\ominus 1}(\complement K) = K \backslash \text{Ingress}_{\Phi}(K) \tag{12.29}$$

is called the egress set *of K under Φ.*

The graphical restriction *$\Phi|_K^K := \Phi \cap \Xi_K^K : K \rightsquigarrow K$ of Φ to $K \times K$ defined by*

$$\Phi|_K^K(x) := \begin{cases} \Phi(x) \cap K \text{ if } x \in \text{Ingress}_{\Phi}(K) := K \cap \Phi^{-1}(K) \\ \emptyset \qquad\qquad\;\; \text{if } x \in \text{Egress}_{\Phi}(K) := K \backslash \Phi^{-1}(K) \end{cases}$$

(see Definition 18.3.13, p.725). Its graph is equal

$$\text{Graph}(\Phi|_K^K) = \text{Graph}(\Phi) \cap (K \times K)$$

The $\text{Ingress}_{\Phi}(K) := K \cap \Phi^{-1}(K)$ *is the domain of the graphical restriction to K of the reset map Φ. It is known under various names, such as* *the* transition set, stopping set *or* guard *in the control literature. The* discrete egress set $\text{Egress}_\Phi(K) := K\backslash\Phi^{-1}(K)$ *of* K *under* Φ *is called the* continuation set *in optimal impulse control theory. The image of the graphical restriction is equal to* $K \cap \Phi(K)$.

In other words, the *ingress set* is the set of elements subjected to be mapped "impulsively" in the environment. The "egress set" is the set on states from which starts continuous time evolution.

Remark: Extension of Reset Maps. In some problems, the reset map $\Phi : G \rightsquigarrow K$ is given as a set-valued map from a subset $G \subset K$ of K to K. This particular situation fits the general framework since Φ coincides with the graphical restriction $\Phi = \Phi|_K^K$ defined by

$$\Phi|_G^G(x) := \begin{cases} \Phi(x) \text{ if } x \in G \\ \emptyset \quad\ \text{ if } x \notin G \end{cases}$$

and thus, $\text{Ingress}_\Phi(K) = G$ and $\text{Egress}_\Phi(K) = K\backslash G$. □

An impulse evolutionary system is associated with an evolutionary system $\mathcal{S}$ governing the continuous-time components of a run and a reset map Φ governing its discrete components in the following way:

Definition 12.3.6 *[Impulse Evolutionary Systems] Let* $\Phi : X \rightsquigarrow X$ *be the reset map and* $\mathcal{S} : X \rightsquigarrow \mathcal{C}(0,\infty;X)$ *the evolutionary system associated with a control system. We associate with any initial state* $x \in X$ *the subset* $\mathcal{R}(x) := \mathcal{R}_{(\mathcal{S},\Phi)}(x)$ *of runs* $\overrightarrow{x}(\cdot)$ *satisfying*

$$\forall n \geq 0, \begin{cases} (i)\ x_n(\cdot) \in \mathcal{S}(x_n(0)) \qquad \text{or} \quad (x_n(0), x_n(\cdot)) \in \text{Graph}(\mathcal{S}) \\ (ii)\ x_{n+1}(0) \in \Phi(x_n(\tau_n)) \text{ or } (x_n(\tau_n), x_{n+1}(0)) \in \text{Graph}(\Phi) \end{cases} \tag{12.30}$$

Such a run is called a solution to the impulse evolutionary system starting at x.

The set-valued map $\mathcal{R} : X \rightsquigarrow \prod_{n\geq 0} \mathbb{R}_+ \times \mathcal{C}(0,\tau_n;X)$ *is called the* impulse evolutionary system *associated with the pair* $(\mathcal{S},\Phi)$.

A subset K *is said to be viable outside a target* $C \subset K$ *under the impulse evolutionary system* $\mathcal{R}$ *is from every* $x \in K$ *starts at least one run* $\overrightarrow{x}(\cdot) \in \mathcal{R}(\cdot)$ *viable in* K *forever or until it reaches the target* C *in finite time.*

This definition, convenient for the mathematical treatment of impulse systems, is equivalent to its explicit formulation:

Lemma 12.3.7 ***[Equivalent Formulation of Solutions to Impulse Differential Inclusions]*** *Let $\mathcal{R} := \mathcal{R}_{(\mathcal{S},\Phi)}$ be an impulse evolutionary system and consider a subset $K \subset X$. A run $\overrightarrow{x}(\cdot) \in \mathcal{R}_{(\mathcal{S},\Phi)}(x)$ is a solution to the impulse evolutionary system if and only if for every $n \geq 0$, the impulse times are given by $t_{n+1} := t_n + \tau_n$, $t_0 = 0$,*

$$\begin{cases} \overrightarrow{x}(t_n) \in \Phi(\overrightarrow{x}(^-t_n)) \\ \forall t \in [t_n, t_{n+1}[,\ \overrightarrow{x}(t) := x(\cdot) \text{ starts at impulse time } t_n \text{ at } \overrightarrow{x}(t_n) \end{cases}$$

Such a solution can also be written in the form

$$\begin{cases} (i)\ \forall\, n \geq 0,\ \ \forall t \in [t_{n-1}, t_n[, \\ \quad \overrightarrow{x}(t) \ \in\ x_0 + \sum_{k=0}^{n-1} (\Phi - 1)(\overrightarrow{x}(^-t_k)) + \int_0^t F(x(\tau))d\tau \\ (ii)\ \overrightarrow{x}(t_n) \in x_0 + \sum_{k=0}^{n} (\Phi - 1)(\overrightarrow{x}(^-t_k)) + \int_0^{t_n} F(x(\tau))d\tau \end{cases}$$

We begin our study by characterizing subsets viable under an impulse evolutionary system in terms of viability kernels under its continuous-time component $\mathcal{S}$ and its discrete-time component Φ:

Theorem 12.3.8 ***[Characterization of Impulse Viability]*** *Let $\mathcal{R} := \mathcal{R}_{(\mathcal{S},\Phi)}$ be an impulse evolutionary system and two subsets $K \subset X$ and $C \subset K$. We set*

$$\text{Ingress}_{\Phi}(K, C) := C \cup (K \cap \Phi^{-1}(K))$$

A subset $K \subset X$ is viable outside the target C under the impulse evolutionary system $\mathcal{R}$ if and only if K is viable outside $\text{Ingress}_{\Phi}(K, C)$ *under the evolutionary system $\mathcal{S}$:*

$$K = \text{Viab}_{\mathcal{S}}(K, \text{Ingress}_{\Phi}(K, C))$$

If

$$K = \text{Capt}_{\mathcal{S}}(K, \text{Ingress}_{\Phi}(K, C))$$

then from any initial state x starts at least one run $\overrightarrow{x}(\cdot) \in \mathcal{R}(x)$ with finite cadences viable in K until it reaches the target C in finite time. This happens whenever $K \backslash \text{Ingress}_{\Phi}(K, C)$ is a repeller.

Proof. Indeed:

1. If K is viable outside C under the impulse evolutionary system, then, from any initial state $x \in K$ starts a run $\overrightarrow{x}(\cdot) \in \mathcal{R}(x)$ viable in K forever or until some finite time $t^\star$ when $\overrightarrow{x}(t^\star) \in C$. Either $\overrightarrow{x}(\cdot) := x(\cdot) \in \mathcal{S}(x)$ is an evolution, and thus x belongs to the viability kernel $\text{Viab}_{\mathcal{S}}(K, C)$ under the evolutionary system $\mathcal{S}$, or there exist a motive $x_0(\cdot) \in \mathcal{S}(x)$, a cadence $\tau_0 \geq 0$ such that either $x(\tau_0) \in C$ or there exists a reset state $x_1 \in \Phi(x_0(^-\tau_0))$ such that both $x_0(t) \in K$ for $t \in [0, \tau_0]$ and $x_1 \in K$. But this means that x belongs to the capture basin $\text{Capt}_{\mathcal{S}}(K, \text{Ingress}_{\Phi}(K, C))$ of $\text{Ingress}_{\Phi}(K, C)$. Hence x belongs to $\text{Viab}_{\mathcal{S}}(K, \text{Ingress}_{\Phi}(K, C)) = \text{Viab}_{\mathcal{S}}(K) \cup \text{Capt}_{\mathcal{S}}(K, \text{Ingress}_{\Phi}(K, C))$.
2. Conversely, assume that $K = \text{Viab}_{\mathcal{S}}(K, \text{Ingress}_{\Phi}(K, C))$ and fix any initial state $x \in K$. Then there exists an evolution $x_0(\cdot) \in \mathcal{S}(x)$ either viable in K forever or until it reaches the target C in finite time, which is the unique motive of a run viable in K outside C, or viable in K until it reaches the target $\text{Ingress}_{\Phi}(K, C)$ at some state $x_0(\tau_0) \in C \cup (K \cap \Phi^{-1}(K))$ at time $\tau_0 \geq 0$. Hence either $x_0(\tau_0) \in C$, and the run reaches C in finite time, or there exists an element $x_1 \in \Phi(x_0(\tau_0)) \cap K$ which is a reset state. We then proceed by induction to construct a run $\overrightarrow{x}(\cdot) := (\tau_n, x_n(\cdot))_n \in \mathcal{R}(x)$ viable in K outside C. □

12.3.3 Hybrid Control Systems and Differential Inclusions

"Hybrid control systems", as they are called by engineers, or "multiple-phase dynamical economies", as they are called by economists, or "integrate and fire models" of neurobiology may be regarded as auxiliary impulse differential inclusions.

Definition 12.3.9 *[Hybrid Differential Inclusions] A* hybrid differential inclusion (K, F, Φ) *is defined by:*

1. *a finite dimensional vector space* E *of states* e *called* locations *or* events *in the control literature,*
2. *a set-valued map* $K : E \rightsquigarrow X$ *associating with any location* e *a (possibly empty) subset* $K(e) \subset X$*,*
3. *a set-valued map* $F : \text{Graph}(K) \rightsquigarrow X$ *with which we associate the differential inclusion* $x'(t) \in F(e, x(t))$*,*
4. *a reset map* $\Phi : \text{Graph}(K) \rightsquigarrow E \times X$*.*

A hybrid differential inclusion is called a qualitative differential inclusion *if the reset map is defined by* $\Phi(e, x) := \Phi_E(e, x) \times \{x\}$ *where* $\Phi_E(e, x) : E \times X \rightsquigarrow E$*.*

A run $\overrightarrow{x}(\cdot) := (\tau_n, e_n, x_n(\cdot))_n \in \prod_{n\geq 0} \mathbb{R}_+ \times E \times \mathcal{C}(0, \tau_n, X)$ *is a solution to such a hybrid differential inclusion if for every* n,

1. *the motives* $x_n(\cdot)$ *are solutions to differential inclusion* $x_n'(t) \in F(e_n, x_n(t))$ *viable in* $K(e_n)$ *on the interval* $[0, \tau_n]$,
2. $(e_n, x_n(0)) \in \Phi(e_{n-1}, x_{n-1}(\tau_{n-1}))$

Hybrid systems are particular cases of impulse differential inclusions in the following sense:

Lemma 12.3.10 ***[Links between Hybrid and Impulse Systems]*** *A run* $\overrightarrow{x}(\cdot) := (\tau_n, e_n, x_n(\cdot))_n$ *is a solution of the hybrid differential inclusions* (K, F, Φ) *if and only if* $(\overrightarrow{e}(\cdot), \overrightarrow{x}(\cdot)) := (\tau_n, (e_n(\cdot), x_n(\cdot)))_n$ *where* $e_n := e(\tau_n)$ *is a run of the auxiliary system of impulse differential inclusions* $\mathcal{R}_{(G,\Phi)}$ *where* $G : E \times X \rightsquigarrow E \times X$ *defined by* $G(e, x) := \{0\} \times F(e, x)$ *governs the differential inclusion*

$$\begin{cases} (i) \;\; e'(t) = 0 \\ (ii) \; x'(t) \in F(e(t), x(t)) \end{cases}$$

viable in $\text{Graph}(K)$.

Proof. Indeed the motives $e_n(\cdot) = e_n$ of the locations remaining constant in the intervals $[t_{n-1}, t_n[$ since their velocities are equal to 0. □

Theorem 12.3.8 implies a characterization of existence of runs of hybrid differential inclusions:

Theorem 12.3.11 ***[Existence Theorem of Solutions to Impulse Differential Inclusions]*** *Let* (K, F, Φ) *be a hybrid differential inclusion. Consider the set-valued map* $K_1 : E \rightsquigarrow X$ *defined by*

$$\text{Graph}(K_1) := \text{Ingress}_{\Phi}(\text{Graph}(K))$$

Then the hybrid differential inclusion has a solution for every initial state if and only if

$$\forall e \in E,\ K(e) \textit{ is viable under the differential inclusion } \\ x'(t) \in F(e, x(t)) \text{ outside } K_1(e)$$

Remark. If $\Phi(e, x) := \Phi_E(e, x) \times \Phi_X(x)$, the set-valued map K_1 can be written in the form

$$\forall e \in E,\ K_1(e) = K(e) \cap \Phi_X^{-1}(K(\Phi_E(e,x)))$$

In particular,

1. $\Phi(e,x) := \Phi_E(e) \times \Phi_X(x)$, then
$$\forall e \in E,\ K_1(e) = K(e) \cap \Phi_X^{-1}(K(\Phi_E(e)))$$
2. if $\Phi(e,x) := \Phi_E(e,x) \times \{x\}$ defines the reset map of a qualitative differential inclusion, the set-valued map K_1 is defined by
$$\forall e \in E,\ K_1(e) = K(e) \cap K(\Phi_E(e,x))$$
3. if $\Phi(e,x) := \Phi_E(e) \times \{x\}$ defines the reset map of a qualitative differential inclusion, the set-valued map K_1 is defined by
$$\forall e \in E,\ K_1(e) = K(e) \cap K(\Phi_E(e))$$

We also observe that

$$\text{Viab}_{\{0\}\times F}(\text{Graph}(K)) = \text{Graph}(e \rightsquigarrow \text{Viab}_{F(e,\cdot)}(K(e)))$$

and that the graph of K is a repeller if and only if for every e, the set $K(e)$ is a repeller under $F(e,\cdot)$. □

12.3.4 Substratum of an Impulse System

The sequence of reset states of viable runs of an impulse evolutionary system is governed by a backward dynamical systems:

Lemma 12.3.12 *[Underlying Backward Discrete System] The sequence $(x_n)_{n\in\{0,\ldots,N\}}$ is the sequence of reset states $x_n := x_n(0)$ of a run $\overrightarrow{x}(\cdot) := (\tau_n, x_n(\cdot))_{n\geq 0}$ of an impulse evolutionary system $\mathcal{R} := \mathcal{R}_{(\mathcal{S},\Phi)}$ viable in K outside C if and only if it satisfies the backward discrete dynamical system*
$$x_n \in \text{Capt}_{\mathcal{S}}(K, K \cap \Phi^{-1}(x_{n+1}))$$
1. *for every $n \geq 0$ when the run is infinite and viable in K,*
2. *or until some $N \geq 0$ when $x_N \in \text{Viab}_{\mathcal{S}}(K,C)$.*

Proof. Consider a run $\overrightarrow{x}(\cdot) := (\tau_n, x_n(\cdot))_n$ viable in K outside a target C. For every $n \geq 0$, the initial state of the motive is $x_n(0) := x_n$ and the motive $x_n(\cdot)$ belongs to $\mathcal{S}(x_n(0))$. Either $x_n(0) \in \text{Viab}_{\mathcal{S}}(K,C)$ and it is viable in K

forever or until it reaches C in finite time, and thus, belongs to the viability kernel $\text{Viab}_{\mathcal{S}}(K,C)$ or it satisfies the end-point condition $\overrightarrow{x}(\tau_n) \in \Phi^{-1}(x_{n+1})$ thanks to (12.30), p.508. This means that the motive $x_n(\cdot)$ is viable in $K\backslash C$ on the interval $[0,\tau_n]$, i.e., that $x_n \in \text{Capt}_{\mathcal{S}}(K, K \cap \Phi^{-1}(x_{n+1}))$.

Conversely, this condition implies the existence of an impulse times τ_n and of motives $x_n(\cdot)$ satisfying the requirement of a run governed by $\mathcal{R}$ and viable in K. □

This suggests to introduce the inverse of the set-valued map $y \rightsquigarrow \text{Capt}_{\mathcal{S}}(K, K \cap \Phi^{-1}(y))$:

Definition 12.3.13 ***[The Reinitialization Map]*** *The inverse* $\mathbf{I} := \mathbf{I}_{(\mathcal{S},\Phi)} : K \rightsquigarrow K$ *of the map* $y \rightsquigarrow \text{Capt}_{\mathcal{S}}(K, K \cap \Phi^{-1}(y))$ *is called the* reinitialization map *of the impulse evolutionary system restricted to* K.

Thanks to Lemma 12.3.12, we propose another characterization of the viability kernel of a subset K outside a target C under an impulse evolutionary system in terms of the reinitialization map:

Proposition 12.3.14 ***[Role of the Reinitialization Map]*** *Let* $\mathcal{R} := \mathcal{R}_{(\mathcal{S},\Phi)}$ *be an impulse evolutionary system. A subset* $K \subset X$ *is viable outside* C *under the impulse evolutionary system* $\mathcal{R}$ *if and only if* K *is viable outside* $\text{Viab}_{\mathcal{S}}(K,C)$ *under the reinitialization map* $\mathbf{I}_{(\mathcal{S},\Phi)}$*:*

$$K = \text{Viab}_{\mathbf{I}_{(\mathcal{S},\Phi)}}(K, \text{Viab}_{\mathcal{S}}(K,C))$$

Therefore the behavior of a run is "summarized" by the reinitialization map $\mathbf{I}_{(\mathcal{S},\Phi)}$. It is a discrete dynamical system governing the sequence of reset states of infinite runs of the impulse evolutionary system viable in K forever or until it reaches the viability kernel $\text{Viab}_{\mathcal{S}}(K,C)$ of K with target C under the continuous-time evolutionary system, when the last motive of the run is viable forever or reaches the target C in finite time. In other words,

1. either $\mathbf{I}_{(\mathcal{S},\Phi)}(x_n) \cap \text{Viab}_{\mathcal{S}}(K,C) = \emptyset$, and then $\mathbf{I}_{(\mathcal{S},\Phi)}(x_n)$ is the set of new reset states $x_{n+1} \in \Phi(x_n(\tau_n)) \cap K$ when $x_n(\cdot) \in \mathcal{S}(x_n)$ ranges over the set of evolutions starting at $x_n \in K$ viable in K until they reach $\Phi^{-1}(K)$ at time $\tau_n \geq 0$ at $x_n(\tau_n) \in \Phi^{-1}(K)$,
2. or there exists $x_{N+1} \in \mathbf{I}_{(\mathcal{S},\Phi)}(x_N) \cap \text{Viab}_{\mathcal{S}}(K,C)$, and thus a motive $x_N(\cdot) \in \mathcal{S}(x_N)$ such that $x_{N+1} \in \Phi(x_N)(\tau_N) \cap K$. From x_{N+1} starts a motive $x_{N+1}(\cdot) \in \mathcal{S}(x_N)$ which is viable in K forever or until it reaches the target C.

It is then important to characterize the reinitialization map: For that purpose, we introduce the auxiliary system

$$\begin{cases} (i) \;\; x'(t) = f(x(t), u(t)) \text{ where } u(t) \in U(x(t)) \\ (ii) \; y'(t) = 0 \end{cases} \tag{12.31}$$

Proposition 12.3.15 *[Viability Characterization of the Reinitialization Map] The graph of the initialization map $\boldsymbol{I} := \boldsymbol{I}_{(\mathcal{S},\Phi)}$ is the capture basin of* $\mathrm{Graph}(\Phi|_K^K)$ *viable in $K \times K$ under the auxiliary evolutionary system (12.31):*

$$\mathrm{Graph}(\boldsymbol{I}_{(\mathcal{S},\Phi)}) = \mathrm{Capt}_{(12.31)}\left(K \times K, \mathrm{Graph}(\Phi|_K^K)\right)$$

Hence the graph of the reinitialization map inherits the properties of capture basins.

Proof. Indeed, to say (x, y) belongs to the viability kernel $\mathrm{Capt}_{(12.31)}(K \times K, \mathrm{Graph}(\Phi|_K^K))$ means that there exist an evolution $x(\cdot) \in \mathcal{S}(x)$ and $t^\star \geq 0$ such that

$$\begin{cases} (i) \;\; \forall t \in [0, t^\star], \; (x(t), y) \in K \times K \\ (ii) \; (x(t^\star), y) \in \mathrm{Graph}(\Phi|_K^K) \end{cases}$$

i.e., if and only if

$$\begin{cases} (i) \;\; \forall t \in [0, t^\star[, \; x(t) \; \in \; K \\ (ii) \; y \in \Phi(x(t^\star)) \cap K \end{cases}$$

This is equivalent to say that $y \in \mathbf{I}_{(\mathcal{S},\Phi)}(x)$. □

The sequence of successive reset conditions x_n of a viable run $x(\cdot)$ of the impulse evolutionary system $(\mathcal{S}, \Phi)$ – constituting the "discrete component of the run" – is governed by the discrete system $x_n \in \mathbf{I}_{(\mathcal{S},\Phi)}(x_{n-1}) \cap K$ starting at x_0. The knowledge of the sequence of initialized states x_n allows us to reconstitute the "continuous component" of the run governed by the evolutionary system starting at each reset state x_n. This reconstruction needs more information, such as the sequence of cadences, which often is the main information on the run that is needed.

We denote by

$$(t, x) \mapsto \vartheta_{\mathcal{S}}^K(t, x) := \bigcup_{x(\cdot) \in \mathcal{S}^K(x)} \{x(t)\} \text{ and } \vartheta_{\mathcal{S}}^K(t, C) := \bigcup_{x \in C} \vartheta_{\mathcal{S}}^K(t, x)$$

the *reachable maps* of $x \in K$ and $C \subset K$ respectively.

Definition 12.3.16 *[The Substratum of an Impulse Evolutionary System] Let $\mathcal{R}_{(\mathcal{S},\Phi)}$ be an impulsive evolutionary system. The* substratum *of the impulse evolutionary system on K is the set-valued map $\Gamma := \Gamma_{(\mathcal{S},\Phi)} : \mathbb{R}_+ \times K \rightsquigarrow K$ associating with any (t,x) the subset*

$$\Gamma_{(\mathcal{S},\Phi)}(t,x) = \Phi(\vartheta_{\mathcal{S}}^K(t,x) \cap \Phi^{-1}(K)) \cap K$$

of the elements $y \in \Phi(c) \cap K$ where $c \in K \cap \Phi^{-1}(K)$ ranges over the set of elements at which arrives at time t at least one evolution $x(\cdot) \in \mathcal{S}(x)$ starting at x and viable in K until it reaches $c \in \Phi^{-1}(K)$. The set-valued map $\boldsymbol{T}_{(\mathcal{S},\Phi)}$ defined by

$$\boldsymbol{T}_{(\mathcal{S},\Phi)}(x) := \{\tau \geq 0 \ \text{ such that } \ \Gamma_{(\mathcal{S},\Phi)}(\tau,x) \neq \emptyset\}$$

is called the cadence map *of the impulse evolutionary system on K.*

The *reinitialization map* is linked to the substratum and the cadence map by the formula:

$$\mathbf{I}_{(\mathcal{S},\Phi)}(x) = \bigcup_{t \in \mathbf{T}_{(\mathcal{S},\Phi)}(x)} \Gamma_{(\mathcal{S},\Phi)}(t,x)$$

Proposition 12.3.17 *[Construction of Runs with the Substratum] Let $\mathcal{R}_{(\mathcal{S},\Phi)}$ be an impulse evolutionary system defined on a subset K. Knowing the substratum $\Gamma_{(\mathcal{S},\Phi)}$ of $\mathcal{R}_{(\mathcal{S},\Phi)}$, and thus the cadence map $\boldsymbol{T}_{(\mathcal{S},\Phi)}$ and the reinitialization map $\boldsymbol{I}_{(\mathcal{S},\Phi)}$, we can* reconstruct *a viable run $\overrightarrow{x}(\cdot) \in \mathcal{R}_{(\mathcal{S},\Phi)}(x)$ of the impulse evolutionary system through the following algorithm: Given the cadence τ_n and the initial state x_n, we take*

1. *the next cadence $\tau_{n+1} \in \boldsymbol{T}_{(\mathcal{S},\Phi)}(x_n)$,*
2. *the next reset state $x_{n+1} \in \Gamma_{(\mathcal{S},\Phi)}(\tau_{n+1},x_n) \subset \boldsymbol{I}_{(\mathcal{S},\Phi)}(x_n)$,*
3. *the next motive $x_n(\cdot) := x(\cdot + t_n) \in \mathcal{S}(x_n)$ satisfying $x_n(0) = x_n$ and $x_n(\tau_{n+1}) \in \Phi^{-1}(x_{n+1})$.*

Proof. Take any run $\overrightarrow{x}(\cdot)$ associated with a sequence $\mathcal{T}(\overrightarrow{x}(\cdot)) := \{t_n\}$ of impulse times starting at $x_0 \in K$ and viable in K. Then the sequence $\overrightarrow{x} : n \to \overrightarrow{x}(t_n)$ is a solution of the discrete dynamical system $\Gamma_{(\mathcal{S},\Phi)}(t_{n+1}-t_n, x_n)$, obviously viable in K.

Conversely, assume that the substratum $\Gamma_{(\mathcal{S},\Phi)}$ is known. The above method of constructing a run starting at time 0 and state $x_0 \in K$ provides a run $\overrightarrow{x}(\cdot)$ associated with the sequence $\mathcal{T}(\overrightarrow{x}(\cdot)) := \{t_n\}$ of impulse times of the impulse differential inclusion $(\mathcal{S}, \Phi)$ viable in K. □

For characterizing the graph of the substratum $\Gamma_{(\mathcal{S},\Phi)}$, we need the auxiliary system

$$\begin{cases} (i) & \tau'(t) = -1 \\ (ii) & x'(t) = f(x(t), u(t)) \text{ where } u(t) \in U(x(t)) \\ (iii) & y'(t) = 0 \end{cases} \tag{12.32}$$

the evolutionary system of which being the set-valued map associating with (T, x, y) the evolution $t \mapsto (T - t, x(T), y)$ where $x(\cdot) \in \mathcal{S}(x)$.

Proposition 12.3.18 ***[Viability Characterization of the Substratum]*** *The graph of the substratum $\Gamma_{(\mathcal{S},\Phi)}$ of $\mathcal{R}_{(\mathcal{S},\Phi)}$ is the capture basin of $\mathbb{R}_+ \times K \times K$ with target $\{0\} \times \mathrm{Graph}(\Phi|_K^K)$:*

$$\mathrm{Graph}(\Gamma_{(\mathcal{S},\Phi)}) = \mathrm{Capt}_{(12.32)}\left(\mathbb{R}_+ \times K \times K, \{0\} \times \mathrm{Graph}(\Phi|_K^K)\right)$$

Proof. Indeed, to say (τ, x, y) belongs to the capture basin $\mathrm{Capt}_{(12.32)}(\mathbb{R}_+ \times K \times K, \{0\} \times \mathrm{Graph}(\Phi|_K^K))$, means that there exists an solution $x(\cdot) \in \mathcal{S}(x)$ and $t^\star \geq 0$ such that

$$\begin{cases} (i) & \forall t \in [0, t^\star],\ (\tau - t, x(t), y) \in \mathbb{R}_+ \times K \\ (ii) & (\tau - t^\star, x(t^\star), y) \in \{0\} \times \mathrm{Graph}(\Phi|_K^K) \end{cases}$$

i.e., if and only if $t^\star = \tau$ and

$$\begin{cases} (i) & \forall t \in [0, \tau[,\ x(t) \in K \\ (ii) & y \in \Phi(x(\tau)) \cap K \end{cases}$$

This is equivalent to say that $y \in \Gamma_{(\mathcal{S},\Phi)}(\tau, x)$. □

12.3.5 Impulse Viability Kernels

We now associate with an impulse evolutionary system viability kernels of targets in the following way:

Definition 12.3.19 ***[Impulse Viability Kernel]*** *Let K and $C \subset K$ be two subsets. The* impulse viability kernel $\mathrm{ImpViab}_{(\mathcal{S},\Phi)}(K, C)$ *of K under* *the impulse evolutionary system* $(\mathcal{S}, \Phi)$ *is the subset of initial states* $x \in K$ *from which starts* at least *one run viable in* K *forever or until a finite time until it reaches the target* C.

Theorem 12.3.20 ***[Fundamental Characterization of Impulse Viability Kernels]*** *The impulse viability kernel* $\text{ImpViab}_{(\mathcal{S},\Phi)}(K, C)$ *of* K *with a target* $C \subset K$ *is*

1. *the* largest *subset* D *satisfying* $C \subset D \subset K$ *and*
$$D \subset \text{ImpViab}_{(\mathcal{S},\Phi)}(D, C)$$
2. *the* smallest *subset* D *satisfying* $C \subset D \subset K$ *and*
$$\text{ImpViab}_{(\mathcal{S},\Phi)}(K, D) \subset D$$
3. *the* unique *subset* D *satisfying* $C \subset D \subset K$ *and*
$$D = \text{ImpViab}_{(\mathcal{S},\Phi)}(K, D) = \text{ImpViab}_{(\mathcal{S},\Phi)}(D, C)$$

The same holds true for the impulse capture basins.

Proof. We prove that

$$\begin{cases} (i) \ \text{ImpViab}_{(\mathcal{S},\Phi)}(K, C) \ \subset \ \text{ImpViab}_{(\mathcal{S},\Phi)}(\text{ImpViab}_{(\mathcal{S},\Phi)}(K, C), C) \\ (ii) \ \text{ImpViab}_{(\mathcal{S},\Phi)}(K, \text{ImpViab}_{(\mathcal{S},\Phi)}(K, C)) \ \subset \ \text{ImpViab}_{(\mathcal{S},\Phi)}(K, C) \end{cases}$$

and then, derive that the impulse viability kernel is the unique "bilateral fixed point" D described in the Theorem.

1. For proving the first inclusion, take $x_0 \in \text{ImpViab}_{(\mathcal{S},\Phi)}(K, C)$ and prove that there exists a run $\overrightarrow{x}(\cdot) \in \mathcal{R}_{(\mathcal{S},\Phi)}(x_0)$ starting at x_0 viable in $\text{ImpViab}_{(\mathcal{S},\Phi)}(K, C)$ forever or until it reaches C in finite time. Indeed, there exists a run $\overrightarrow{x}(\cdot) \in \mathcal{R}_{(\mathcal{S},\Phi)}(x_0)$ viable in K forever or until some finite time $T \geq 0$ when it reaches C. Then for all $t \in [0, T[$, the run $\overrightarrow{y}(\cdot)$ defined by $\overrightarrow{y}(\tau) := \overrightarrow{x}(t + \tau)$ is a run $\overrightarrow{y}(\cdot) \in \mathcal{R}_{(\mathcal{S},\Phi)}(\overrightarrow{x}(t))$ starting at $\overrightarrow{x}(t)$ and viable in K forever or until it reaches C at time $T - t$. Hence $\overrightarrow{x}(t)$ does belong to $\text{ImpViab}_{(\mathcal{S},\Phi)}(K, C)$ for every $t \in [0, T]$.
2. For proving the second inclusion, let x belong to $\text{ImpViab}_{(\mathcal{S},\Phi)}(K, \text{ImpViab}_{(\mathcal{S},\Phi)}(K, C))$. There exists at least one run $\overrightarrow{x}(\cdot) \in \mathcal{R}_{(\mathcal{S},\Phi)}(x)$ viable in K forever or until it reaches the impulse viability kernel $\text{ImpViab}_{(\mathcal{S},\Phi)}(K, C)$ in finite time. In the latter case, it thus can be

concatenated with a run remaining in $\text{ImpViab}_{(\mathcal{S},\Phi)}(K,C) \subset K$ forever or until it reaches the target C in finite time. This implies that $x \in \text{ImpViab}_{(\mathcal{S},\Phi)}(K,C)$.

The map $(K,C) \mapsto \text{ImpViab}_{(\mathcal{S},\Phi)}(K,C)$ satisfies the property

$$C \subset \text{ImpViab}_{(\mathcal{S},\Phi)}(K,C) \subset K$$

and is increasing in the sense that

$$\text{If } C_1 \subset C_2 \ \& \ K_1 \subset K_2, \text{then } \text{ImpViab}_{(\mathcal{S},\Phi)}(K_1,C_1) \subset \text{ImpViab}_{(\mathcal{S},\Phi)}(K_2,C_2)$$

Therefore Lemma 10.2.6 implies that the impulse viability kernel $\text{ImpViab}_{(\mathcal{S},\Phi)}(K,C)$ is the unique bilateral fixed point between C and K of the map $(K,C) \mapsto \text{ImpViab}_{(\mathcal{S},\Phi)}(K,C)$. □

12.3.6 Stability Properties

We now prove that the set of solutions to an impulse evolutionary system depends continuously upon the initial states "in the upper semicompact sense":

Theorem 12.3.21 *[Upper Semicompactness of Impulse Evolutionary Systems] Let us assume that the subset K is closed, that F is Marchaud, that the graph of Φ is closed, that $K \cap \Phi(K)$ is compact and that*

$$\text{Viab}_{\mathcal{S}}(K) \cap \Phi(K) = \emptyset$$

holds true. Then the solution map $\mathcal{R}^K_{(F,\Phi)}$ is upper semicompact on $K \backslash \text{Viab}_{\mathcal{S}}(K)$: If x_0^ε converges to x_0 and if $\overrightarrow{x}^\varepsilon(\cdot) \in \mathcal{R}_{(\mathcal{S},\Phi)}(x_0^\varepsilon)$ is a solution to the impulse evolutionary system starting at x_0^ε, a subsequence (again denoted by) $\overrightarrow{x}^\varepsilon(\cdot)$ converges to a run $\overrightarrow{x}(\cdot) \in \mathcal{R}_{(\mathcal{S},\Phi)}(x_0)$.

Proof. Let us consider a sequence of initial states $x_0^\varepsilon \in K \backslash \text{Viab}_{\mathcal{S}}(K)$ converging to $x_0 \in K \backslash \text{Viab}_{\mathcal{S}}(K)$ and a sequence of runs

$$\overrightarrow{x}^\varepsilon(\cdot) = \{(\tau_n^\varepsilon, x_n^\varepsilon)\}_{n \geq 0} \in \mathcal{R}_{(\mathcal{S},\Phi)}(x_0^\varepsilon)$$

viable in K.

We can identify the graph $\mathcal{G}$ of the solution map $\mathcal{R}_{(F,\Phi)}$ to a subset

$$\mathcal{G} \subset K \times (\mathbb{R}_+ \times \text{Graph}(\mathcal{S}))^{\mathbf{N}}$$

We supply it with the product topology. Under our assumptions, for any compact subset L of K, Theorem 10.3.3 implies that the graph of $\mathcal{S}|_L$ is compact. Let $\overline{T} := \sup_{x\in K\cap\Phi(K)} \tau^{\sharp}_{(K,\mathrm{Ingress}_{\Phi}(K))}(x)$ which is finite since the exit time function is upper semicompact and since $K \cap \Phi(K)$ is compact.

Hence the graph $\mathcal{G}_L$ of the restriction $\mathcal{R}_{(F,\Phi)}|_L$ to L of the solution map $\mathcal{R}_{(F,\Phi)}$ satisfies

$$\mathcal{G}_L \subset L \times ([0,\overline{T}] \times \mathrm{Graph}(\mathcal{S}|_L))^{\mathbf{N}}$$

which is a product of compacts. By the Tychonoff Theorem, it is itself a compact subset.

Therefore, a subsequence (again denoted by) $\{(\tau_n^{\varepsilon}, x_n^{\varepsilon}(\cdot))\}_{n\geq 0}$ converges to some sequence $\{(\tau_n, x_n(\cdot))\}_{n\geq 0}$. This means that for every $n \geq 0$, τ_n^{ε} converges to $\tau_n \in [0,\overline{T}]$ and $(x_n^{\varepsilon}(0), x_n^{\varepsilon}(\cdot)) \in \mathrm{Graph}(\mathcal{S})$ to some $(x_n, x_n(\cdot)) \in \mathrm{Graph}(\mathcal{S})$ Consequently, the sequences $x_n^{\varepsilon}(\tau_n^{\varepsilon})$ converge $x_n(\tau_n)$ and thus, since the graph of the reset map Φ is closed, inclusions

$$\forall n \geq 0,\ (x_{n+1}^{\varepsilon}(0), x_n^{\varepsilon}(\tau_n^{\varepsilon})) \in \mathrm{Graph}(\Phi)$$

imply that

$$\forall n \geq 0,\ (x_{n+1}(0), x_n(\tau_n)) \in \mathrm{Graph}(\Phi)$$

The sequence $(\tau_n, x_n(\cdot))_{n\geq 0}$ defines a run $\overrightarrow{x}(\cdot) \in \mathcal{R}_{(\mathcal{S},\Phi)}(x)$ of the impulse evolutionary system viable in K. □

12.3.7 Cadenced Runs

The analogue of equilibria for usual evolutionary systems are "cadenced runs", that are discontinuous periodic evolutions:

Definition 12.3.22 *[Cadenced Runs] A run*

$$\overrightarrow{x}(\cdot) := (\tau_n, x_n(\cdot))_{n\geq 0} \in \prod_{n\geq 0} \mathbb{R}_+ \times \mathcal{C}(0,\tau_n;X)$$

is said to be cadenced if the cadences and the motives are constant: for every $n \geq 0$, $\tau_n = \overline{\tau}$ *and* $x_n(\cdot) = \overline{x}(\cdot)$

If the cadence $\overline{\tau}$ of a cadenced run is equal to 0, then $\overline{x}$ is an equilibrium of the reset map Φ.

The following asymptotic property of a run implies the existence of a cadenced run:

Theorem 12.3.23 *[**Existence of a Cadenced Runs**] Assume that F is Marchaud, that the graph of the reset map Φ is closed, that* $\mathrm{Viab}_{\mathcal{S}}(K) = \emptyset$ *and that K is compact. Let $\overrightarrow{x}(\cdot) := (\tau_n, x_n(\cdot))_{n \geq 0} \in \mathcal{R}(x)$ be a run viable in K associated with a sequence $\mathcal{T}(x(\cdot))$ of impulse times t_n viable in K.*

If the sequence of reset states $\overrightarrow{x}(t_n)$ of the run $\overrightarrow{x}(\cdot)$ converges to some $\overline{x}$, then a subsequence of the motives $x_n(\cdot)$ converges to the motive $\overline{x}(\cdot)$ of a cadenced run starting at $\overline{x}$ and viable in K.

Proof. Assume that the sequence $x_n := \overrightarrow{x}(t_n)$ converges to some $\overline{x} \in \Phi(K)$. The motive $x_n(\cdot)$ belongs to $\mathcal{S}(x_n)$ and satisfies $x_n(\tau_{n+1}) \in \Phi^{-1}(\overrightarrow{x}(t_{n+1}))$.

Since K is a repeller by assumption and since the exit time function is upper semicontinuous, all the cadences $\tau_n := t_n - t_{n-1}$ are bounded by finite time $\overline{T} := \max_{x \in K} \tau_K^\sharp(x) < +\infty$.

By Theorem 10.3.3, a subsequence $x_{n_p}(\cdot)$ of motives converges uniformly on the compact interval $[0, \overline{T}]$ to some evolution $\overline{x}(\cdot) \in \mathcal{S}(\overline{x})$ starting at $\overline{x}$. Another subsequence of cadences $\tau_{n_{p_q}+1}$ converges to some $\overline{\tau} \in [0, \overline{T}]$. Hence $x_{n_{p_q}}(\tau_{n_{p_q}+1})$ converges to $\overline{x}(\overline{\tau})$. Since $x_{n_{p_q}}(\tau_{n_{p_q}+1})$ belongs to $\Phi^{-1}(x(t_{n_{p_q}+1}))$, since $x(t_{n_{p_q}+1})$ converges also to $\overline{x}$ by assumption and since the graph of the reset map Φ is closed, we infer that $\overline{x}(\overline{\tau})$ belongs to $\Phi^{-1}(\overline{x})$. Hence a subsequence of the motives $x_n(\cdot) := \overrightarrow{x}(\cdot + t_n)$ of the run $\overrightarrow{x}(\cdot)$ converges to the motive $\overline{x}(\cdot)$ of a cadenced run starting at $\overline{x}$ of cadence $\overline{\tau}$. □

Part III
First-Order Partial Differential Equations

Chapter 13
Viability Solutions to Hamilton–Jacobi Equations

13.1 Introduction

This chapter presents the viability approach to a class of Hamilton–Jacobi equations. We assume not only that the solution depends on time, but on "structured" or "causal" variables. They include age-structured Hamilton–Jacobi–McKendrick equations, useful in population dynamics as well as in transport management (the age variable being replaced by the travel time), as well as Hamilton–Jacobi–Cournot equation, where the "structured" or "causal" variable is the initial state of the underlying control system. Chapters 14, p. 563 and 15, p. 603 apply the results of this chapter to transportation management, finance and economics.

They also are a particular case of Hamilton–Jacobi–Bellman equations of optimal control problems and more generally, intertemporal optimization problems, studied in Chap. 14, p. 563. We already presented in Chap. 4, p. 125 examples of such control problems: minimal time and exit functions, minimal length functions, Lyapunov functions, safety and transgression functions, etc., are value functions. We illustrated the "viability approach" showing that the epigraphs of these functions are the viability kernels or capture basins of epigraphical environments and targets under an adequately defined "characteristic system" (see the epigraphical miracle mentioned in Sect. 4.12.2, p. 172). We only alluded to the fact that they were solutions to Hamilton–Jacobi–Bellman equations.

In this chapter, we start instead with a class of Hamilton–Jacobi equations *the Hamiltonian of which is convex with respect to the gradient of the solution.* We shall prove that their "solution" is the value function of an intertemporal optimization of evolutions governed by a *hidden* underlying control system, where:

1. the hidden controls, called "celerities", range over the state space,

2. the optimality criterion involves the Lagrangian associated with the Hamiltonian by the Fenchel Transform,
3. the evolutions are governed by a specific class of "epigraphical control system" involving the epigraph of the Lagrangian,
4. the map regulating optimal evolutions is associated with the gradient of the solution.

Which solution? We define the concept of "viability solution" *solving at once all the above related problems.* They are "constructive solutions", in the sense that their epigraph is *defined* as the viability kernel or capture basin of an epigraphical environment and target. They inherit their properties which are enough to "translate" in the language of partial differential equations and control theory.

13.2 From Duality to Trinity

Let $t \geq 0$ denotes the time, $x \in X := \mathbb{R}^n$ the *state variable* and $d \in \mathcal{D} \subset \mathbb{R}^m$ the *causal variable* or *structuring variable* of the system.

We introduce:

1. a *causal map* $\varphi : d \in \mathbb{R}^m \mapsto \varphi(d) := (\varphi_i(d))_{i=1,\dots,d} \in \mathbb{R}^d$ depending only on the causal variable d (assumed single-valued for simplicity of the presentation),
2. a *Hamiltonian* function

$$(d, x, p) \in \mathbb{R}^m \times \mathbb{R}^n \times \mathbb{R}^n \mapsto \mathbf{l}^\star(d, x; p) \in \mathbb{R} \cup \{+\infty\}$$

convex with respect to the *"costate" variable* $p \in \mathbb{R}^n$,
3. a *viability constraint* function $(d, x) \mapsto \mathbf{k}(d, x) \in \mathbb{R} \cup \{+\infty\}$ such that $\mathbf{k}(d, x) < +\infty$ implies that $d \in \mathcal{D}$,
4. an *internal condition* function $(d, x) \mapsto \mathbf{c}(d, x) \in \mathbb{R} \cup \{+\infty\}$ such that

$$\mathbf{k}(d, x) \leq \mathbf{c}(d, x)$$

The terminology is motivated by the *asymmetric* role played by the two variables d and x, since φ *only* depends on the causal variable d whereas the Hamiltonian $\mathbf{l}^\star$ may depend on *both* causal and state variables. Hijacking the terminology used in population dynamics, we say that the causal variables *structure* the system.

1. The Macroscopic Approach

The *macroscopic* description of the system requires us to look for the *viable solution* V to the *structured Hamilton–Jacobi equation*

$$\sum_{i=1}^{m}\left\langle \frac{\partial V(d,x)}{\partial d_i}, \varphi_i(d)\right\rangle + \mathbf{l}^\star\left(d,x;\frac{\partial V(d,x)}{\partial x}\right) = 0 \qquad (13.1)$$

satisfying inequalities

$$\mathbf{k}(d,x) \le V(d,x) \le \mathbf{c}(d,x) \qquad (13.2)$$

Examples of Structured Hamilton–Jacobi Equations

(a) *Hamilton–Jacobi equations.* By taking $d := t \in \mathcal{D} := \mathbb{R}_+$ describing time and $\varphi(t) = 1$, we obtain the usual Hamilton–Jacobi partial differential equation

$$\frac{\partial V(t,x)}{\partial t} + \mathbf{l}^\star\left(t,x;\frac{\partial V(t,x)}{\partial x}\right) = 0$$

(b) *Hamilton–Jacobi–McKendrick equations.* By taking $d := (t,a) \in \mathcal{D} := \mathbb{R}^2_+$ describing time and age (in population dynamics) or travel time (in traffic problems for instance) and taking $\varphi(t,a) = (1,1)$, we obtain Hamilton–Jacobi–McKendrick partial differential equations

$$\frac{\partial V(t,a,x)}{\partial t} + \frac{\partial V(t,a,x)}{\partial a} + \mathbf{l}^\star\left(t,a,x;\frac{\partial V(t,a,x)}{\partial x}\right) = 0$$

(c) Taking $d := (t,b)$ and $\varphi(t,b) := (1,\psi(t,b))$, we obtain the following partial differential equation

$$\frac{\partial V(t,b,x)}{\partial t} + \left\langle \frac{\partial V(t,b,x)}{\partial b}, \psi(t,b)\right\rangle + \mathbf{l}^\star\left(t,b,x;\frac{\partial V(t,b,x)}{\partial x}\right) = 0$$

(d) *Hamilton–Jacobi–Cournot equations.* We introduce an other causal variable χ and set $\psi(d,\chi) := (\varphi(d),0)$, independent of the first causal variable χ. we obtain structured Hamilton–Jacobi partial differential equations

$$\sum_{i=1}^{m}\left\langle \frac{\partial V(d,x)}{\partial d_i}, \varphi_i(d)\right\rangle + \mathbf{l}^\star\left(d,x;\frac{\partial V(d,\chi,x)}{\partial x}\right) = 0$$

structured *also* by *constant parameters* χ. They are solutions to the same partial differential equation (13.1), p. 525, but are subjected to conditions

$$\mathbf{k}(d,\chi,x) \le V(d,\chi,x) \le \mathbf{c}(d,\chi,x)$$

depending on χ.
An important example is provided by Hamilton–Jacobi–Cournot partial differential equations parameterized by specific parameters $\chi \in \mathbb{R}^n$ regarded as *initial conditions* and requiring that $\mathbf{c}(d,\chi,x) = +\infty$ whenever $x \neq \chi$ (see Definition 8.4.8, p. 288) and Sect. 13.8, p. 551. □

Examples of Internal and Viability Functions

(a) An internal condition function $\mathbf{c} : (d, x) \mapsto \mathbf{c}(d, x) \in \mathbb{R} \cup \{+\infty\}$ becomes a *boundary condition* function if $\mathbf{c}(d, x) = +\infty$ whenever $d \in \text{Int}(\mathcal{D})$.
(b) Consider the case when the environment in which ranges the state variable x is described by *viability environments* $K(d) \subset X$ of the state variable depending on the structural variable d. This constraint is taken into consideration by *requiring that the constraint function* $\mathbf{k}$ *satisfies* $\mathbf{k}(d, x) = +\infty$ *for all* $x \notin K(d)$ (since this implies automatically that $V(t, x) = +\infty$ whenever $d \notin K(d)$).

Here, we define the concept of "viability solution" *which always exists and can be computed by viability algorithms.* This is a *constructive* approach allowing us to derive some known and new properties of the viability solution from the tools of viability theory (dealing with sets instead of functions). Regarding functions as their epigraphs, we bypass the regularity issues to arrive directly to the concept of Barron–Jensen/Frankowska viscosity solutions. We above all take into account viability constraints and extend classical boundary conditions to other "internal" conditions.

2. **The Variational Approach**

The link between Hamilton–Jacobi equations and the associated variational problem relies on the *Legendre–Fenchel transform* $(d, x, u) \mapsto \mathbf{l}(d, x; u)$ of the Hamiltonian defined by

$$\mathbf{l}(d, x; u) \;:=\; \sup_p [\langle p, u\rangle - \mathbf{l}^\star(d, x; p)]$$

the *Lagrangian*. We denote by

$$F(d, x) \;:=\; \{u \text{ such that } \ell(d, x; u) \;+\infty\} \tag{13.3}$$

the domain of the Lagrangian ℓ. It defines a set-valued map $F : (d, x) \rightsquigarrow F(d, x)$ which will be the right hand side of the differential inclusion governing the evolutions of the microsystem.
Optimal evolutions achieve the minimum in the variational principle

$$V(d, x) \;:=\; \inf_{x(\cdot)} \left(\mathbf{c}(d(0), x(0)) + \int_0^{t^\sharp} \mathbf{l}(d(t), x(t), x'(t))dt \right) \tag{13.4}$$

among all *viable* evolutions $x(\cdot)$ starting at initial time 0 and arriving at x at terminal time $t^\sharp \leq t$ when $d(t^\sharp) = d$. The function V is called the "valuation function" (and not the classical value function which depends upon current time t whereas the valuation function depends upon the terminal time).

They satisfy the dynamic programming equation: $\forall\, t \in [0, t^\sharp]$,

$$V(d,x) \;=\; V(d(t),x(t)) + \int_t^{t^\sharp} \mathbf{l}(d(\tau),x(\tau),x'(\tau))d\tau \qquad (13.5)$$

3. **The Microscopic Approach**

The main purpose of this study is not only to prove that the *viability solution* is the *unique* solution to this partial differential equation in an adequate generalized (weak) sense, but also to *uncover a hidden dual microscopic* equivalent problem allowing us to characterize and *compute* from the solution V the *retroaction map* $(d,x) \rightsquigarrow R(d,x)$ governing *optimal evolutions* of the state through the *microscopic* regulation of structured-viable evolutions $(d(\cdot),x(\cdot))$ satisfying, for any given $d \in \mathcal{D}$ and x and for some $t^\sharp \geq 0$,

$$\begin{cases} (i)\;\; d(t^\sharp) \;=\; d \text{ and } x(t^\sharp) \;=\; x \;\; \text{(terminal condition)} \\ (ii)\; \forall\, t \in [0,t^\sharp],\;\; x'(t) \;\in\; R(d(t),x(t)) \;\; \text{(retroaction law)} \end{cases} \qquad (13.6)$$

In other words, the macroscopic problem (looking for a macroscopic function V) and microscopic problem (looking for the regulation of evolutions $x(\cdot)$) are "dual".

4. **The Viability Solution**

The links between the microscopic, macroscopic and variational approaches are due to a "matchmaker", the "viability solution", which

(a) coincides with the viable solution to the macroscopic structured Hamilton–Jacobi partial differential equation (13.1), p. 525, satisfying the boundary condition (13.2), p. 525,
(b) provides the *retroaction law* microsystem (13.6), p. 527 governing optimal viable evolutions of the state variable to a any given terminal state in optimal time,
(c) Is equal to the valuation function (13.4), p. 526, of the associated intertemporal optimization problem.

13.3 Statement of the Problem

13.3.1 Lagrangian and Hamiltonian

Recall that the *epigraph* of an *extended function* $\mathbf{v} : X \mapsto \mathbb{R} \cup \{+\infty\}$ is the set of pairs $(x,y) \in X \times \mathbb{R}$ such that $\mathbf{v}(x) \leq y$. An extended function is called *lower semicontinuous* if *its epigraph is closed.*

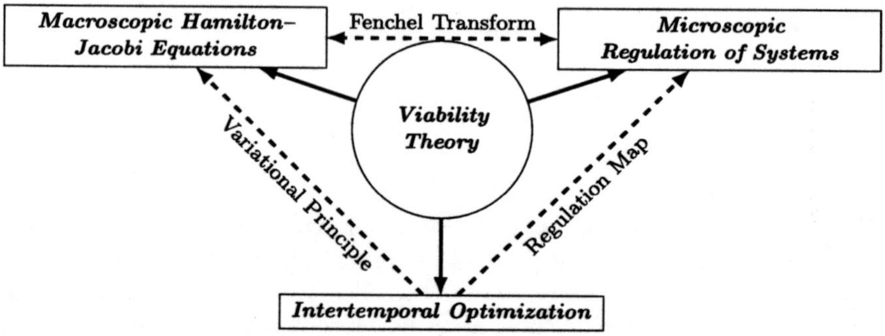

Fig. 13.1 From duality to trinity. This diagram describes the three problems under investigation: the macroscopic approach through first-order partial differential equations, the microscopic version dealing with the regulation of an underlying control system and the intertemporal optimization problem. The links relating optimization problems to Hamilton–Jacobi–Bellman equations and the regulation of control systems has been extensively studied. The tools of viability theory allow us to show that the viability solution solves these three problems at once.

The Lagrangian $\mathbf{l} : (d, x; u) \in \mathbb{R}^m \times \mathbb{R}^n \times \mathbb{R}^n \mapsto \mathbf{l}(d, x; u) \in \mathbb{R} \cup \{+\infty\}$ is assumed once and for all to be *a nontrivial lower semicontinuous function convex with respect to* u: $\mathbf{l}(d, x; \cdot) : u \in x \mapsto \mathbf{l}(d, x; u) \in \mathbb{R} \cup \{+\infty\}$ is convex. Here, x is regarded as the state and u as a *celerity*. For instance, u is a velocity (in mechanics), a transaction (in economics and finance) or a *celerity* in general systems.

We also introduce the *costate* or *dual variable* $p \in \mathbb{R}^{n^\star}$ the dual of $\mathbb{R}^n$ and the duality product $\langle p, u\rangle := p(u)$. For instance, p is regarded as a force, a price or a density, and the duality product $\langle p, u\rangle$ is a power, a value or a flux respectively.

The *conjugate function* $\mathbf{l}^\star(d, x; \cdot)$ is defined on costate variables by

$$\forall\, p \in \mathbb{R}^n,\ \ \mathbf{l}^\star(d, x; p) \ := \ \sup_u[\langle p, u\rangle - \mathbf{l}(d, x; u)]$$

The *conjugate function* $\mathbf{l}^\star(d, x; \cdot)$ *is always a non trivial lower semicontinuous convex function* satisfying the *Fenchel inequality*

$$\langle p, u\rangle \ \leq \ \mathbf{l}(d, x; u) + \mathbf{l}^\star(d, x; p)$$

The biconjugate satisfies $\mathbf{l}^{\star\star}(d, x; u) \leq \mathbf{l}(d, x; u)$. The main result of convex analysis states that $\mathbf{l}^{\star\star}(d, x; \cdot) = \mathbf{l}(d, x; \cdot)$ *if and only if* $\mathbf{l}(d, x; \cdot)$ *is convex, lower semicontinuous and nontrivial.*

The *Legendre property* of the *Fenchel transform* $\mathbf{l}(d, x; \cdot) \mapsto \mathbf{l}^\star(d, x; \cdot)$ implies that the *subdifferentials*

$$\partial_u \mathbf{l}(d, x; u) \ := \ \partial_u \mathbf{l}(d, x; (\cdot)(u))$$

and

$$\partial_p \mathbf{l}^\star(d,x;p) \; := \; \partial_p \mathbf{l}^\star(d,x;(\cdot)(p))$$

of the lower semicontinuous convex functions $\mathbf{l}$ and $\mathbf{l}^\star$ are *defined* by the following equivalent conditions:

$$\begin{cases} (i) \;\; \langle p,u\rangle \; = \; \mathbf{l}(d,x;u) + \mathbf{l}^\star(d,x;p) \\ (ii) \; p \; \in \; \partial_u \mathbf{l}(d,x;u) \\ (iii) \; u \; \in \; \partial_p \mathbf{l}^\star(d,x;p) \end{cases} \tag{13.7}$$

The two equalities (13.7)(ii) and (iii) describe the Legendre property: the inverse of the subdifferential map $u \rightsquigarrow \partial_u \mathbf{l}(d,x;u)$ is the subdifferential map $p \rightsquigarrow \partial_p \mathbf{l}^\star(d,x;p)$.

If the functions $\mathbf{l}$ or $\mathbf{l}^\star$ are differentiable in the classical sense, then

$$\begin{cases} \partial_u \mathbf{l}(d,x;u) \; = \; \left\{ \dfrac{\partial}{\partial u}\mathbf{l}(d,x;u) \right\} \\ \partial_p \mathbf{l}^\star(d,x;p) \; = \; \left\{ \dfrac{\partial}{\partial p}\mathbf{l}^\star(d,x;p) \right\} \end{cases}$$

At this point, we have to make assumptions under which viability properties hold true. In our specific settings, we need to make assumptions either on the Lagrangian $\mathbf{l}$ or on the Hamiltonian $\mathbf{l}^\star$ to fit the Marchaud requirement.

Definition 13.3.1 *[Marchaud Functions] We shall say that*

1. *a Lagrangian* $(d,x,u) \mapsto \mathbf{l}(d,x;u) \; \in \; \mathbb{R} \cup \{+\infty\}$ *is* Marchaud *if it is a lower semicontinuous function convex with respect to* u *and if there exists a finite positive constant* $c > 0$ *such that*

$$\begin{cases} \mathrm{Dom}(\mathbf{l}(d,x;\cdot)) \; \subset \; c(\|x\| + \|d\| + 1)B \text{ and is closed} \\ \forall\, u \; \in \; \mathrm{Dom}(\mathbf{l}(d,x;\cdot)), \quad 0 \leq \; \mathbf{l}(d,x;u) \; \leq \; c(\|x\| + \|d\| + 1) \end{cases} \tag{13.8}$$

2. *a Hamiltonian* $(d,x;p) \mapsto \mathbf{l}^\star(d,x;p) \; \in \; \mathbb{R} \cup \{+\infty\}$ *is* Marchaud *if it is convex and lower semicontinuous with respect to* u*, upper semicontinuous with respect to* (d,x) *and if there exist finite positive constants* c *and* c_0 *such that, for all* $p \in \mathbb{R}^n$*,*

$$\begin{cases} \sigma_{\mathrm{Dom}(\mathbf{l})}(d,x;p) - c(\|x\| + \|d\| + 1) \\ \leq \; \mathbf{l}^\star(d,x;p) \; \leq \; c(\|x\| + \|d\| + 1)\|p\|_\star \end{cases}$$

where $\sigma_K(\cdot)$ *denotes the* support function *of* K *(see Definition 18.2.3, p. 715).*

Lemma 18.7.4, p. 757 states that the Lagrangian $\mathbf{l}$ is Marchaud if the Hamiltonian $\mathbf{l}^\star$ is Marchaud. If the Lagrangian is Marchaud and continuous with respect to (d,x), then the Hamiltonian is Marchaud.

13.3.2 The Viability Solution

Knowing the Hamiltonian $\mathbf{l}^\star$, the viability constraint function $\mathbf{k}$ and the internal condition function $\mathbf{c}$, we define the structured Hamilton–Jacobi problem:

Definition 13.3.2 ***[Structured Hamilton–Jacobi Problem]*** *A function $(d,x) \mapsto V(d,x)$ is said to be a solution to the structured Hamilton–Jacobi equation if*

$$\left\langle \frac{\partial V(d,x)}{\partial d}, \varphi(d) \right\rangle + \mathbf{l}^\star \left(d, x; \frac{\partial V(d,x)}{\partial x} \right) = 0 \qquad (13.9)$$

and a solution to the structured Hamilton–Jacobi problem if, furthermore, it satisfies the conditions

$$\mathbf{k}(d,x) \le V(d,x) \le \mathbf{c}(d,x) \qquad (13.10)$$

We introduce the *structured characteristic system* defined by

$$\begin{cases} (i) \ \delta'(t) = -\varphi(\delta(t)) \\ (ii) \ (\xi'(t), \eta'(t)) \in -\mathcal{E}p(\mathbf{l}(\delta(t), \xi(t); \cdot)) \end{cases} \qquad (13.11)$$

The characterization of the solution to the structured Hamilton–Jacobi problem states that its epigraph is the *viable-capture basin* of the epigraph of $\mathbf{c}$, viable in the epigraph of $\mathbf{k}$, under the structured characteristic system defined by (13.11), p. 530.

Definition 13.3.3 ***[Viability Solution]*** *The viability solution V to structured Hamilton–Jacobi problem 13.3.2, p. 530 is defined by*

$$V(d,x) := \inf_{(d,x,y) \in \mathrm{Capt}_{(13.10)}(\mathcal{E}p(\mathbf{k}), \mathcal{E}p(\mathbf{c}))} y \qquad (13.12)$$

It may seem strange at first glance to solve a well known partial differential equation by a solution of an auxiliary and seemingly artificial viability problem.

Defining (lower semicontinuous) functions through their (closed) epigraphs allows us to treat the functions **l**, **k**, **c** and the viability solution V as subsets, bypassing and avoiding the pointwise characterization of partial differential equations familiar in classical analysis.

But this also allows us to just apply results surveyed and summarized in the "viability survival kit" (Sect. 2.15, p. 98) and proved in Chaps. 10, p. 375 and 11, p. 437 based on the fundamental viability and invariance theorems at the simpler level of set-valued analysis (with much less notations).

Above all, viable-capture basins and their regulation maps can be computed numerically by software using viability algorithms.

The translation of the properties of viable-capture basins in terms of structured problems provides without technical difficulties the properties we shall uncover.

The definition of the viability solution does not involve the concept of derivatives, a strange way for defining solutions to partial differential equations. Actually, it is known that the solution to the structured Hamilton–Jacobi is not differentiable. The lack of regularity happens whenever viability constraints are involved: in this case, the most we can expect is that *the solution is only lower semicontinuous.* However, it is possible to give a meaning to lower semicontinuous solutions to structured Hamilton–Jacobi equation (13.9), p. 530, by weakening the concepts of gradients in the sense of nonsmooth analysis. The viability solutions then becomes a "solution" to this partial differential equation in the sense of Barron–Jensen/Frankowska viscosity solution (Theorem 13.10.3, p. 560).

13.4 Variational Principle

13.4.1 Lagrangian Microsystems

We shall assume that the causal map φ is Lipschitz (or more generally, monotone) for guaranteeing the *uniqueness* of the solution $\delta(\cdot)$ to differential equation

$$\delta'(t) \; = \; -\varphi(\delta(t))$$

starting from any given initial state $d \in \mathcal{D}$.

We further assume that the subset $\mathcal{D}$ is closed and is a *repeller* under $-\varphi$ in the sense that for any $d \in \mathcal{D}$, the evolution $\delta(\cdot)$ leaves $\mathcal{D}$ in finite time

$$\tau^{\sharp}(d) \; := \; \inf_{\delta(t) \in \complement \mathcal{D}} t$$

at $\delta(\tau^{\sharp}(d)) \in \partial \mathcal{D}$.

This is the case for usual Hamilton–Jacobi equations when $d := t \in \mathcal{D} := \mathbb{R}_+$ and $\varphi(t) = 1$: in this case $\tau^\sharp_{\mathcal{D}}(t) = t$, as well as for the Hamilton–Jacobi–McKendrick equations when $d := (t,a) \in \mathcal{D} := \mathbb{R}^2_+$ and $\varphi(t,a) = (1,1)$: in this case $\tau^\sharp_{\mathcal{D}}(t,a) = a$ if $t \geq a \geq 0$ and $\tau^\sharp_{\mathcal{D}}(t,a) = t$ if $a \geq t \geq 0$.

In summary, whenever we mention d, we attach to it either the unique evolution $t \mapsto \delta(t)$ governed by $\delta'(t) = -\varphi(\delta(t))$ starting from d at initial time 0 or the unique evolution $t \mapsto d(t) := \delta(t^\sharp - t)$ governed by $d'(t) = \varphi(d(t))$ and arriving at d at time $t^\sharp$, *without mentioning it explicitly.*

We associate with the Lagrangian, the causal map and the viability constraint function $\mathbf{k}$ the microsystem governing viable evolutions of the state:

Definition 13.4.1 ***[Microsystem]*** *We denote by*

$$F(d,x) \ := \ \{u \text{ such that } \mathbf{l}(d,x;u) \ < \ +\infty\} \tag{13.13}$$

the domain of the Lagrangian $\mathbf{l}$ *and by* $\mathcal{A}_{\mathbf{k}}(t^\sharp; d, x)$ *the set of evolutions* $x(\cdot)$ *governed by the system*

$$x'(t) \ \in \ F(d(t), x(t)) \tag{13.14}$$

"viable" in the sense that

$$\sup_{t \in [0,t^\sharp]} \mathbf{k}(d(t), x(t)) \ < \ +\infty$$

and arriving at x *at time* $t^\sharp$ *when* $d(t^\sharp) = d$.
When the function $\mathbf{k}$ *is associated with a viability environment* $d \rightsquigarrow K(d)$ *by its indicator (see Definition 18.6.1, p. 743)*

$$\mathbf{k}(d,x) \ := \ \psi_{K(d)}(x) = \psi_{\mathrm{Graph}(K)}(d,x)$$

then the viable evolutions are the ones satisfying

$$\forall\, t \in [0, t^\sharp], \ \ x(t) \ \in \ K(d(t))$$

13.4.2 The Variational Principle

13.4.2.1 Case of Boundary-Value Problems

We first assume (for simplicity of the formula) that the internal condition is actually a boundary condition: $\mathbf{c}(d,x) = +\infty$ whenever $d \in \mathrm{Int}(\mathcal{D})$.

The *valuation function* of the intertemporal optimization problem is defined by

$$U(d,x) := \inf_{x(\cdot)\in\mathcal{A}_{\mathbf{k}}(\tau^\sharp(d);d,x)} \left(\mathbf{c}(d(0),x(0)) + \int_0^{\tau^\sharp(d)} \mathbf{l}(d(t),x(t),x'(t))dt \right) \tag{13.15}$$

Theorem 13.4.2 *[The Viability Solution Solves the Variational Problem] The viability solution V defined by (13.12), p. 530:*

$$U(d,x) \;=\; V(d,x) \;:=\; \inf_{(d,x,y)\in \mathrm{Capt}_{(13.10)}(\mathcal{E}p(\mathbf{k}),\mathcal{E}p(\mathbf{c}))} y$$

is equal the valuation function U of the variational problem (13.15), p. 533.

13.4.2.2 General Case

In the general case, the statement of the intertemporal optimization problem is more intricate and requires further notations:

Theorem 13.4.3 *[The Viability Solution Solves the Variational Problem] We associate with the function* $\mathbf{c}$ *the functional* $\mathbf{J_c}$ *defined by*

$$\begin{cases} \mathbf{J_c}(t^\sharp;x(\cdot))(d,x) \\ := \mathbf{c}(d(0),x(0)) + \int_0^{t^\sharp} \mathbf{l}(d(\tau),x(\tau),x'(\tau))d\tau \end{cases}$$

with the function $\mathbf{k}$ *the functional* $I_{\mathbf{k}}$ *defined by*

$$\begin{cases} \mathbf{I_k}(t^\sharp;x(\cdot))(d,x) := \\ \sup_{s\in[0,t^\sharp]} \left(\mathbf{k}(d(s),x(s)) + \int_s^{t^\sharp} \mathbf{l}(d(\tau),x(\tau),x'(\tau))d\tau \right) \end{cases}$$

and with both functions $\mathbf{k}$ *and* $\mathbf{c}$ *the functionals defined by*

$$\begin{cases} \mathbf{L_{(k,c)}}(t^\sharp;x(\cdot))(d,x) \;:=\; \max(\mathbf{I_k}(t^\sharp;x(\cdot))(d,x),\mathbf{J_c}(t^\sharp;x(\cdot))(d,x)) \\ \mathbf{M_{(k,c)}}(t^\sharp;d,x) \;:=\; \inf_{x(\cdot)\in\mathcal{A}_{\mathbf{k}}(t^\sharp;d,x)} \mathbf{L_{(k,c)}}(t^\sharp;x(\cdot))(d,x) \end{cases}$$

Hence the viability solution V is equal the valuation function U of the variational problem defined by

$$\begin{cases} U(d,x) = \inf_{t^\sharp\in[0,\tau^\sharp(d)]} \mathbf{M}_{(\mathbf{k},\mathbf{c})}(t^\sharp;d,x) \\ = \inf_{t^\sharp\in[0,\tau^\sharp(d)]} \inf_{x(\cdot)\in\mathcal{A}_{\mathbf{k}}(t^\sharp;d,x)} \mathbf{L}_{(\mathbf{k},\mathbf{c})}(t^\sharp;x(\cdot))(d,x) \end{cases} \tag{13.16}$$

Proof. By Definition 2.10.2, p. 86 of viable-capture basins, to say that (d,x,y) belongs to the viable-capture basin $\text{Capt}_{(13.10)}(\mathcal{E}p(\mathbf{k}),\mathcal{E}p(\mathbf{c}))$ means that there exist some $t^\sharp \geq 0$ and a measurable function $v(\cdot) : [0,t^\sharp] \mapsto \text{Dom}(\mathbf{l})$ such that

$$t \in [0,t^\sharp] \mapsto (\delta(t),\xi(t),\eta(t)) \in \mathbb{R}^m \times \mathbb{R}^n \times \mathbb{R}^n$$

is a solution to (13.11)

$$\begin{cases} (i) \ \ \delta'(t) = -\psi(\delta(t)) \\ (ii) \ \ \xi'(t) = -v(t) \\ (iii) \ \eta'(t) = -\mathbf{l}(\delta(t),\xi(t);v(t)) \end{cases}$$

starting at x viable in the epigraph of $\mathbf{k}$ until time $t^\sharp \leq \tau^\sharp(d)$ when it belongs to the epigraph of the function $\mathbf{c}$, where

$$\begin{cases} (i) \ \ \xi(t) := x - \int_0^t v(\tau)d\tau \\ (ii) \ \eta(t) \leq \eta_0(t) := y - \int_0^t \mathbf{l}(\delta(\tau),\xi(\tau),v(\tau))d\tau \end{cases}$$

This implies that $t^\sharp \leq \tau^\sharp(d)$, that

$$\begin{cases} (i) \ \mathbf{c}\left(\delta(t^\sharp),\xi(t^\sharp)\right) \leq \eta(t^\sharp) \leq y - \int_0^{t^\sharp} \mathbf{l}(\delta(\tau),\xi(\tau),v(\tau))d\tau = \eta_0(t^\sharp) \\ (ii) \ \forall\, s \in [0,t^\sharp], \\ \quad \mathbf{k}(\delta(s),\xi(s)) \leq \eta(s) \leq y - \int_0^s \mathbf{l}(\delta(\tau),\xi(\tau),v(\tau))d\tau = \eta_0(s) \end{cases} \tag{13.17}$$

In the particular case of boundary conditions when $\mathbf{c}(d,x) = +\infty$ whenever $d \in \text{Int}(\mathcal{D})$, we obtain $t^\sharp = \tau^\sharp(d)$.
We introduce the set-valued map $\mathcal{F}$ defined by $\mathcal{F}$ defined by

$$\mathcal{F}(d,x) := \mathcal{E}p(\mathbf{I}(d,x;\cdot)) \cap [(c(\|x\|+\|d\|+1)B \times [-c(\|x\|+\|d\|+1),0])] \tag{13.18}$$

which has nonempty values.

Introducing the system

$$\begin{cases} (i) \ \ \delta'(t) = -\varphi(\delta(t)) \\ (ii) \ (\xi'(t),\eta'(t)) \in -\mathcal{F}(\delta(t),\xi(t)) \end{cases} \tag{13.19}$$

this implies that $\text{Capt}_{(13.10)}(\mathcal{E}p(\mathbf{k}),\mathcal{E}p(\mathbf{c})) \subset \text{Capt}_{(13.17)}(\mathcal{E}p(\mathbf{k}),\mathcal{E}p(\mathbf{c}))$. Since $\text{Graph}(\mathbf{l}) \subset \mathcal{E}p(\mathbf{l})$, the converse is true, so that equality

$$\text{Capt}_{(13.10)}(\mathcal{E}p(\mathbf{k}),\mathcal{E}p(\mathbf{c})) \;=\; \text{Capt}_{(13.17)}(\mathcal{E}p(\mathbf{k}),\mathcal{E}p(\mathbf{c})) \tag{13.20}$$

ensues.

Inequalities (13.17), p. 534 imply that

$$\sup_{s\in[0,t^\sharp]}\left(\mathbf{k}\left(\delta(s),\xi(s)\right)+\int_0^s \mathbf{l}(\delta(\tau),\xi(\tau),\upsilon(\tau))d\tau\right) \;\le\; y$$

The target condition implies that

$$\mathbf{c}\left(\delta(t^\sharp),\xi(t^\sharp)\right)+\int_0^{t^\sharp}\mathbf{l}(\delta(\tau),\xi(\tau),\upsilon(\tau))d\tau \;\le\; y$$

Let us set

$$\overleftarrow{\mathbf{J}}_{\mathbf{c}}(t^\sharp;\upsilon(\cdot))(d,x) := \; \mathbf{c}(\delta(t^\sharp),\xi(t^\sharp))+\int_0^{t^\sharp}\mathbf{l}(\delta(\tau),\xi(\tau),\upsilon(\tau))d\tau$$

and

$$\begin{cases}\overleftarrow{\mathbf{I}}_{\mathbf{k}}(t^\sharp;\upsilon(\cdot))(d,x) := \\ \sup\limits_{s\in[0,t^\sharp]}\left(\mathbf{k}(\delta(s),\xi(s))+\int_0^s \mathbf{l}(\delta(\tau),\xi(\tau),\upsilon(\tau))d\tau\right)\end{cases}$$

We posit

$$\begin{cases}\overleftarrow{\mathbf{L}}_{(\mathbf{k},\mathbf{c})}(t^\sharp;\upsilon(\cdot))(d,x) \\ := \; \max(\overleftarrow{\mathbf{I}}_{\mathbf{k}}(t^\sharp;x(\cdot),\upsilon(\cdot))(d,x),\, \overleftarrow{\mathbf{J}}_{\mathbf{c}}(t^\sharp;\upsilon(\cdot))(d,x))\end{cases}$$

We have proved that the viability and capturability conditions imply that there exist $t^\sharp \in [0,\tau^\sharp(d)]$ and $\upsilon(\cdot)$ such that

$$\overleftarrow{\mathbf{L}}_{(\mathbf{k},\mathbf{c})}(t^\sharp;\upsilon(\cdot))(d,x) \;\le\; y \tag{13.21}$$

Therefore, setting

$$U(d,x) \;:=\; \inf_{(t^\sharp;\upsilon(\cdot))}\overleftarrow{\mathbf{L}}_{(\mathbf{k},\mathbf{c})}(t^\sharp;\upsilon(\cdot))(d,x)$$

the viability solution $V(d,x) := \inf\limits_{(d,x,y)\in\text{Capt}_{(13.10)}(\mathcal{E}p(\mathbf{k}),\mathcal{E}p(\mathbf{c}))} y$ defined by (13.12), p. 530, satisfies inequality $U(d,x) \le V(d,x)$.

For proving the opposite inequality, take any $\varepsilon > 0$. Then there exist $\upsilon_\varepsilon(\cdot)$ and $t^\sharp_\varepsilon \in [0,\tau^\sharp(d)]$ such that

$$\overleftarrow{\mathbf{L}}_{(\mathbf{k},\mathbf{c})}(t^\sharp_\varepsilon;(\xi_\varepsilon(\cdot),\upsilon_\varepsilon(\cdot)))(d,x) \;\le\; U(d,x)+\varepsilon$$

Setting

$$\xi_\varepsilon(t) := x - \int_0^t \upsilon_\varepsilon(\tau)d\tau \text{ and}$$

$$\eta_\varepsilon(t) := U(d,x) - \int_0^t \mathbf{l}(\delta(\tau), \xi_\varepsilon(\tau), \upsilon_\varepsilon(\tau))d\tau - \varepsilon$$

we observe that $(\delta(\cdot), x_\varepsilon(\cdot), \eta_\varepsilon(\cdot))$ is a solution to (13.11), p. 530 starting at $(d, x, U(d,x) - \varepsilon)$, reaching the epigraph of $\mathbf{c}$ at time $t^\sharp_\varepsilon$ and viable in $\mathcal{E}p(\mathbf{k})$ on $[0, t^\sharp_\varepsilon]$. Therefore $(d, x, U(d,x) - \varepsilon)$ belongs to $\text{Capt}_{(13.10)}(\mathcal{E}p(\mathbf{k}), \mathcal{E}p(\mathbf{c}))$, and thus $U(d,x) - \varepsilon \geq V(d,x)$.

Letting ε converge to 0 implies that $U(d,x) \geq V(d,x)$ so that equality ensues.

Finally, setting $d(t) := \delta(t^\sharp - t)$, $x(t) := \xi(t^\sharp - t)$, $y(t) := \eta(t^\sharp - t)$, $u(t) := \upsilon(t^\sharp - t)$, $\mathbf{L}_{(\mathbf{k},\mathbf{c})}(t^\sharp; x(\cdot))(d,x) := \overleftarrow{\mathbf{L}}_{(\mathbf{k},\mathbf{c})}(t^\sharp; \upsilon(\cdot))(d,x)$, etc., we deduce that $x(\cdot) \in \mathcal{A}_{\mathbf{k}}(t^\sharp; d, x)$ is a solution arriving at x at time $t^\sharp$ and starting from $x(0) = \xi(t^\sharp)$ at time $t^\sharp$. Since $x'(t) = -\xi'(t^\sharp - t) = \upsilon(t^\sharp - t) = u(t)$, we infer that

$$\begin{cases} \mathbf{J}_{\mathbf{c}}(t^\sharp; x(\cdot))(d,x) \\ := \mathbf{c}(d(0), x(0)) + \displaystyle\int_0^{t^\sharp} \mathbf{l}(d(\tau), x(\tau), x'(\tau))d\tau \\ \mathbf{I}_{\mathbf{k}}(t^\sharp; x(\cdot))(d,x) := \\ \displaystyle\sup_{s\in[0,t^\sharp]} \left(\mathbf{k}(d(s), x(s)) + \int_{t^\sharp - s}^{t^\sharp} \mathbf{l}(d(\tau), x(\tau), x'(\tau))d\tau \right) \end{cases}$$

Hence we have proved that

$$V(d,x) = \inf_{t^\sharp \in [0, \tau^\sharp(d)]} \inf_{x(\cdot)\in\mathcal{A}_{\mathbf{k}}(t^\sharp; d,x)} \mathbf{L}_{(\mathbf{k},\mathbf{c})}(t^\sharp; x(\cdot))(d,x)$$

is the valuation function of the intertemporal optimization problem. □

Remark. Theorem 13.4.2, p. 533, follows from Theorem 13.4.3, p. 533 because assumption $\mathbf{c}(d,x) = +\infty$ whenever $d \in \text{Int}(\mathcal{D})$ implies that $t^\sharp = \tau^\sharp(d)$. □

Theorem 13.4.4 ***[Continuity Properties of the Viability Solution]*** *Assume that the Lagrangian is Marchaud. Then the viability solution* V *defined by (13.12), p. 530, is lower semicontinuous and its epigraph is equal to the viable-capture basin:*

$$\mathcal{E}p(V) = \text{Capt}_{(13.10)}(\mathcal{E}p(\mathbf{k}), \mathcal{E}p(\mathbf{c}))$$

Proof. Recalling inequality (13.8), p. 529 involved in the definition of Marchaud Lagrangian, we introduce the set-valued map $\mathcal{F}$ defined by (13.18), p. 534, *which has nonempty values* and the system (13.19), p. 534:

$$\begin{cases} (i)\ \delta'(t) = -\varphi(\delta(t)) \\ (ii)\ (\xi'(t), \eta'(t)) \in -\mathcal{F}(\delta(t), \xi(t)) \end{cases}$$

1. Inclusions $\text{Graph}(\mathbf{l}) \subset \mathcal{F} \subset \mathcal{E}p(\mathbf{l})$ imply

$$\begin{cases} \text{Capt}_{(13.10)}(\mathcal{E}p(\mathbf{k}), \mathcal{E}p(\mathbf{c})) \subset \text{Capt}_{(13.17)}(\mathcal{E}p(\mathbf{k}), \mathcal{E}p(\mathbf{c})) \\ \subset \text{Capt}_{(13.10)}(\mathcal{E}p(\mathbf{k}), \mathcal{E}p(\mathbf{c})) \end{cases}$$

   and equality (13.20), p. 535 implies that $\text{Capt}_{(13.10)}(\mathcal{E}p(\mathbf{k}), \mathcal{E}p(\mathbf{c})) = \text{Capt}_{(13.10)}(\mathcal{E}p(\mathbf{k}), \mathcal{E}p(\mathbf{c}))$.
   Hence, these three viable-capture basins coincide:

$$\begin{cases} \text{Capt}_{(13.10)}(\mathcal{E}p(\mathbf{k}), \mathcal{E}p(\mathbf{c})) = \text{Capt}_{(13.17)}(\mathcal{E}p(\mathbf{k}), \mathcal{E}p(\mathbf{c})) \\ = \text{Capt}_{(13.10)}(\mathcal{E}p(\mathbf{k}), \mathcal{E}p(\mathbf{c})) \end{cases} \tag{13.22}$$

2. The differential inclusion (13.19), p. 534 is Marchaud. Indeed, the graph of $(d, x) \rightsquigarrow \mathcal{E}p(\mathbf{l}(d, x; \cdot))$ is closed because it is the epigraph of the Lagrangian $\mathbf{l}$ is lower semicontinuous. Its values are convex since the Lagrangian is convex with respect to u. The set-valued map $\mathcal{F}$ being its intersection with $(d, x) \rightsquigarrow c(\|x\| + \|d\| + 1)B \times [-c(\|x\| + \|d\| + 1), 0]$ has linear growth. Consequently, the intersection $\mathcal{F}$ has a closed graph, convex values and linear growth, i.e., is a Marchaud set-valued map (see Definition 10.3.2, p. 384).
3. Hence Viability Theorem 2.15.5, p. 101 implies that the viable-capture basin $\text{Capt}_{(13.10)}(\mathcal{E}p(\mathbf{k}), \mathcal{E}p(\mathbf{c}))$ is closed. This implies in particular that $(d, x, V(d, x))$ belongs to this viable-capture basin, which, then, coincides with the epigraph of V. Being closed, the viability solution is lower semicontinuous. □

A voluminous literature is devoted to regularity theorems providing sufficient conditions for the viability solution to be continuous, Lipschitz, semi-concave, differentiable in such and such sense.

This study does not deal with the regularity properties of the viability solutions, but focuses on the existence of optimal evolutions and on microsystem regulating them.

13.5 Viability Implies Optimality

We now deduce from the viability theorems that there exists an optimal solution to the variational problem, actually, that all viable evolutions are optimal and satisfy the dynamic programming equations under viability constraints.

13.5.1 Optimal Evolutions

Theorem 13.5.1 *[Viable and Optimal evolutions] Assume that the Lagrangian is Marchaud. For any* $(d, x) \in \text{Dom}(V)$*, there exist* $t^\sharp \in [0, \tau^\sharp(d)]$ *such that* $d(t^\sharp) = d$ *and one evolution* $\overline{x}(\cdot) \in \mathcal{A}_{\mathbf{k}}(t^\sharp; d, x)$ *such that*

$$V(d, x) = \mathbf{L}_{(\mathbf{k},\mathbf{c})}(t^\sharp; \overline{x}(\cdot))(d, x)$$

achieves the minimum in the intertemporal optimization problem.

Proof. Consider *any* solution $t \mapsto (\delta(t), \xi(t), \eta(t))$ to the system (13.11) starting at $(d, x, V(d, x))$ *viable* in $\mathcal{E}p(\mathbf{k})$ until it reaches $\mathcal{E}p(\mathbf{c})$ at some finite time $t^\sharp$. At least one of such evolutions does exist since $(d, x, V(d, x))$ belongs to viable-capture basin $\text{Capt}_{(13.10)}(\mathcal{E}p(\mathbf{k}), \mathcal{E}p(\mathbf{c}))$ thanks to Theorem 13.4.4, p. 536.

It is associated with a control $v(\cdot)$ satisfying

$$\eta(t) \leq \eta_0(t) := V(d, x) - \int_0^t \mathbf{l}(\delta(\tau), \xi(\tau), v(\tau))d\tau$$

Inequality (13.21), p. 535, with $y = V(d, x)$, implies that

$$\overleftarrow{\mathbf{L}}_{(\mathbf{k},\mathbf{c})}(t^\sharp; v(\cdot))(d, x) \leq V(d, x)$$

and thus that $(t^\sharp; v(\cdot))$ is optimal.

Therefore, setting $d(t) := \delta(t^\sharp - t)$, $x(t) := \xi(t^\sharp - t)$, etc., we infer that there exist $t^\sharp$ and $(d(\cdot), x(\cdot)) \in \mathcal{A}_{\mathbf{k}}(t^\sharp; d, x)$ such that

$$V(d, x) = \overleftarrow{\mathbf{L}}_{(\mathbf{k},\mathbf{c})}(t^\sharp; v(\cdot))(d, x) = \mathbf{L}_{(\mathbf{k},\mathbf{c})}(t^\sharp; x(\cdot))(d, x)$$

achieves the minimum in the intertemporal optimization problem. □

Summary. Denoting by

$$\mathbb{T}_{(\mathbf{k},\mathbf{c})}(d, x) := \inf \left\{ t \in [0, \tau^\sharp(d)] \text{ such that } V(d, x) = \mathbf{M}_{(\mathbf{k},\mathbf{c})}(t; d, x) \right\}$$

the *initial time map* and by

$$\mathcal{O}_{\mathbf{k}}(t^{\sharp};d,x) \;:=\; \{x(\cdot) \text{ such that } \mathbf{L}_{(\mathbf{k},\mathbf{c})}(t^{\sharp};x(\cdot))(d,x) \;:=\; \mathbf{M}_{(\mathbf{k},\mathbf{c})}(t;d,x)\}$$

the *optimal map*, then the search of optimal evolutions splits into two steps:

1. Take $t^{\sharp} := \mathbb{T}_{(\mathbf{k},\mathbf{c})}(d,x)$.
2. Choose any evolution $\overline{x}(\cdot) \in \mathcal{O}_{\mathbf{k}}(t^{\sharp};d,x)$

For boundary condition functions $\mathbf{c}$ such that $\mathbf{c}(d,x) = +\infty$ whenever $d \in \mathrm{Int}(\mathcal{D})$, then $t^{\sharp} := \tau^{\sharp}(d)$. □

13.5.2 Dynamic Programming under Viability Constraints

Optimal evolutions satisfy the dynamic programming principle:

Theorem 13.5.2 ***[Dynamic Programming under Viability Constraints]*** *We assume that the Lagrangian* $\mathbf{l}$ *is Marchaud and that the function* $\mathbf{k}$ *is continuous in its domain. Consider an optimal evolution* $\overline{x}(\cdot) \in \mathcal{A}_{\mathbf{k}}(t^{\sharp};d,x)$. *Let* $\overline{\kappa} \in [0,t^{\sharp}]$ *be the first time when*

$$\mathbf{k}(d(\overline{\kappa}),\overline{x}(\overline{\kappa})) + \int_{\overline{\kappa}}^{t^{\sharp}} \mathbf{l}(d(\tau),\overline{x}(\tau),\overline{x}'(\tau))d\tau \;=\; V(d,x) \qquad (13.23)$$

Set $\kappa^{\sharp} \;:=\; \min(t^{\sharp},\overline{\kappa})$. *Then* $\overline{x}(\cdot)$ *satisfies the* dynamic programming equation*:*

$$\forall\, t \in [\kappa^{\sharp},t^{\sharp}],\;\; V(d(t),\overline{x}(t)) + \int_{t}^{t^{\sharp}} \mathbf{l}(d(\tau),\overline{x}(\tau),\overline{x}'(\tau))d\tau \;=\; V(d,x) \quad (13.24)$$

In particular, in the case without constraints $\mathbf{k}(d,x) = -\infty$, $\kappa^{\sharp} = 0$ *and the dynamic programming equation holds on the interval* $[0,t^{\sharp}]$.

Proof. By Theorem 2.15.2, p. 99, we know that the viable-capture basin $\mathrm{Capt}_{(13.10)}(\mathcal{E}p(\mathbf{k}),\mathcal{E}p(\mathbf{c}))$ of the epigraph of $\mathbf{c}$ under the auxiliary system (13.11), p. 530 is the unique bilateral fixed point

$$\mathrm{Capt}_{(13.10)}(\mathcal{E}p(V),\mathcal{E}p(\mathbf{c})) \;=\; \mathcal{E}p(V) \;=\; \mathrm{Capt}_{(13.10)}(\mathcal{E}p(\mathbf{k}),\mathcal{E}p(V))$$

1. Let $(d, x, V(d, x))$ belong to the viable-capture basin $\text{Capt}_{(13.10)}(\mathcal{E}p(V), \mathcal{E}p(\mathbf{c})) = \mathcal{E}p(V)$ of the epigraph of $\mathbf{c}$ under the auxiliary system (13.11), p. 530. There exist $t^\sharp \in [0, \tau^\sharp(d)]$ and $\overline{v}(\cdot)$ such that

$$t \mapsto (\delta(t), \overline{\xi}(t), \overline{v}(t))$$

is viable in the epigraph of V until it reaches the epigraph of $\mathbf{c}$ at time $t^\sharp$. Then the proof of Theorem 13.4.3, p. 533 implies that

$$\forall s \in [0, t^\sharp], \ V(\delta(s), \overline{\xi}(s)) + \int_0^s \mathbf{l}(\delta(\tau), \overline{\xi}(\tau), \overline{v}(\tau))d\tau \ \le \ V(d, x) \quad (13.25)$$

2. We shall deduce the opposite inequality from the second fixed point property $\mathcal{E}p(V) = \text{Capt}_{(13.10)}(\mathcal{E}p(\mathbf{k}), \mathcal{E}p(V))$. The assumption that $\mathcal{D}$ is a repeller under $-\varphi$ implies that $\mathcal{E}p(\mathbf{k})$ is also a repeller under structured characteristic system (13.11), p. 530, and thus $\mathcal{E}p(V) = \text{Viab}_{(13.10)}(\mathcal{E}p(\mathbf{k}), \mathcal{E}p(V))$. Hence

$$\complement(\text{Capt}_{(13.10)}(\mathcal{E}p(\mathbf{k}), \mathcal{E}p(V))) \ = \ \text{Abs}_{(13.10)}(\complement\mathcal{E}p(V), \complement\mathcal{E}p(\mathbf{k}))$$

We set

$$\complement\mathcal{E}p(V) \ := \ \{(d, x, y) \text{ such that } y \ < V(d, x)\} \ =: \overset{\circ}{\mathcal{H}yp} \ (V)$$

For any $\varepsilon > 0$, $(d, x, V(d, x) - \varepsilon)$ does not belong to $\mathcal{E}p(V) = \text{Capt}_{(13.10)}(\mathcal{E}p(\mathbf{k}), \mathcal{E}p(V))$, so that

$$(d, x, V(d, x) - \varepsilon) \ \in \ \text{Abs}_{(13.10)}(\overset{\circ}{\mathcal{H}yp} \ (V), \overset{\circ}{\mathcal{H}yp} \ (\mathbf{k}))$$

Therefore, for any $v(\cdot)$, there exists $\kappa_\varepsilon \le \tau^\sharp(d)$ such that $(\delta(\kappa_\varepsilon), \xi(\kappa_\varepsilon), \eta(\kappa_\varepsilon))$ reaches $\mathcal{E}p(\mathbf{k})$ and leaves it before eventually reaching $\mathcal{E}p(\mathbf{c})$. Hence κ_ε is defined by

$$\mathbf{k}(\delta(\kappa_\varepsilon), \xi(\kappa_\varepsilon)) + \int_0^{\kappa_\varepsilon} \mathbf{l}(\delta(\tau), \xi(\tau), v(\tau))d\tau \ = \ V(d, x) - \varepsilon$$

and, consequently,

$$\forall \ s \in [0, \kappa_\varepsilon], \ V(d, x) - \varepsilon \ \le \ V(\delta(s), \xi(s)) + \int_0^s \mathbf{l}(\delta(\tau), \xi(\tau), v(\tau))d\tau$$

Since $\kappa_\varepsilon \ \le \ \tau^\sharp(d) < +\infty$, a subsequence (again denoted by) κ_ε converges to some $\kappa \ \le \ \tau^\sharp(d)$ when $\varepsilon \to 0+$. The function $\mathbf{k}$ being continuous by assumption, we deduce that

$$\mathbf{k}(\delta(\kappa), \xi(\kappa)) + \int_0^{\kappa} \mathbf{l}(\delta(\tau), \xi(\tau), v(\tau))d\tau \ = \ V(d, x)$$

and that,

$$\forall\, s \in [0,\kappa],\ \ V(d,x) \ \leq\ V(\delta(s),\xi(s)) + \int_0^s \mathbf{l}(d(\tau),\xi(\tau),v(\tau))d\tau \quad (13.26)$$

This inequality holds in particular for the above viable evolution $(\delta(\cdot),\overline{\xi}(\cdot),\overline{\eta}(\cdot))$ on the interval $[0,t^\sharp]$ so that inequalities (13.25), p. 540 and (13.26), p. 541 imply equality

$$\forall\, s \in [0,\min(\kappa,t^\sharp)],\ \ V(d,x) \ =\ V(\delta(s),\overline{\xi}(s)) + \int_0^s \mathbf{l}(d(\tau),\overline{\xi}(\tau),\overline{v}(\tau))d\tau$$

ensues.

We derive the conclusion (13.24), p. 539 by setting $\overline{x}(t) := \overline{\xi}(t^\sharp - t)$ and $\overline{\kappa} := t^\sharp - \kappa$ satisfying (13.23), p. 539. □

13.6 Regulation of Optimal Evolutions

It is not enough to know the existence of optimal evolutions: the question arises whether we can compute it. For that purpose, we shall carve in the set-valued map $(d,x) \mapsto F(d,x)$ governing the evolution $x(\cdot)$ through differential inclusion $x'(t) \in F(d(t),x(t))$ a regulation map $(d,x) \mapsto R_V(d,x) \subset F(d,x)$ piloting optimal viable evolutions by differential inclusion $x'(t) \in R_V(d(t),x(t))$ until it reaches the terminal state x at optimal time $t^\sharp$.

The regulation map is characterized by the viability solution by a formula which uses the fact that the viability solution V is also the solution of the structured Hamilton–Jacobi inequality, which holds true for a Marchaud Lagrangian.

When V is any differentiable function, the regulation map associated with it is defined by:

$$R_V(d,x) \ := \left\{ \begin{array}{l} v \in F(d,x) \text{ such that} \\ \left\langle \frac{\partial V(d,x)}{\partial d}, \varphi(d) \right\rangle + \left\langle \frac{\partial V(d,x)}{\partial x}, v \right\rangle - \mathbf{l}(d,x;v) \ \geq\ 0 \end{array} \right\}$$

Remark: Lax–Oleinik Formula. Observe that if the Lagrangian $\mathbf{l}$ is Marchaud and if the function V satisfies the Hamilton–Jacobi inequality

$$\left\langle \frac{\partial V(d,x)}{\partial d}, \varphi(d) \right\rangle + \mathbf{l}\left(d,x;\frac{V(d,x)}{dx}\right) \ \geq\ 0$$

then the Legendre property (13.7), p. 529 of the Fenchel transform implies the *generalized Lax–Oleinik formula*

$$\partial \mathbf{l}\left(d, x; \frac{V(d,x)}{dx}\right) \in R_V(d,x)$$

Indeed, if we assume further that $R_V(d,x) = \{r(d,x)\}$ is a singleton and that the Lagrangian is differentiable with respect to u, this implies that

$$\frac{\partial V(d,x)}{\partial x} = \frac{d}{du}\mathbf{l}(d,x;r(d,x))$$

which is the *Lax–Oleinik formula*. General formulas (13.68), p. 561 relating the regulation map associated with the viability solution and its partial derivatives (or subdifferentials) with respect to state x are obtained under (much) stronger assumptions. They are consequences of the proof that the viability solution is the unique Barron–Jensen/Frankowska viscosity solution of the Hamilton–Jacobi. We postpone it to Sect. 13.10, p. 557 since we do not need this purely mathematical result for studying the regulation of optimal viable evolutions. □

When V is no longer differentiable, we consider its *epiderivative* $D_{\uparrow}^{\star\star}V(d,x)$ of V defined

$$\mathcal{E}p(D_{\uparrow}^{\star\star}V(d,x)) := T_{\mathcal{E}p(V)}^{\star\star}(d,x,V(d,x))$$

which is a directional derivative $(\delta, v) \mapsto D_{\uparrow}^{\star\star}V(d,x)(\delta,v) \in \mathbb{R} \cup \{-\infty\} \cup \{+\infty\}$ convex and lower semicontinuous instead of being linear (and continuous).

Definition 13.6.1 *[Regulation Map Associated with a Function] The regulation map R_V associated with a function V is defined by*

$$\begin{cases} R_V(d,x) := \\ \left\{v \in F(d,x) \text{ such that } D_{\uparrow}^{\star\star}V(d,x)(-\varphi(d),-v) + \mathbf{l}(d,x;u) \leq 0\right\} \end{cases} \quad (13.27)$$

We begin by proving that the regulation map associated with the viability solution has nonempty values:

Theorem 13.6.2 *[Regulation Map of the Viability Solution] If the Lagrangian $(d,p,v) \rightsquigarrow \mathbf{l}(d,x;v)$ is Marchaud, then the viability solution V defined by (13.12), p. 530 is the smallest lower semicontinuous function satisfying conditions (13.10), p. 530 and*

$$\inf_{v \in F(d,x)} \left(D_{\uparrow}^{\star\star} V(d,x)(-\varphi(d), -v) + \mathbf{l}(d,x;v) \right) \leq 0$$

is the contingent solution *(introduced by Hélène Frankowska) such that, whenever* $V(d,x) < \mathbf{c}(d,x)$*, the value* $R_V(d,x)$ *of the regulation map associated with the viability solution* V *is not empty.*

Proof. Since the Hamiltonian is Marchaud, so is the Lagrangian. Actually, the theorem remains true under the weaker assumption that the Lagrangian $\mathbf{l}$ is Marchaud. So is the set-valued map $\mathcal{F}$ defined by (13.18), p. 534 and, by (13.22), p. 537, the viable-capture basin $\mathcal{E}p(V) := \text{Capt}_{(13.10)}(\mathcal{E}p(\mathbf{k}), \mathcal{E}p(\mathbf{c}))$ is equal to $\text{Capt}_{(13.17)}(\mathcal{E}p(\mathbf{k}), \mathcal{E}p(\mathbf{c})) = \text{Capt}_{(13.10)}(\mathcal{E}p(\mathbf{k}), \mathcal{E}p(\mathbf{c}))$. It is then the largest closed subset between $\mathcal{E}p(\mathbf{c})$ and $\mathcal{E}p(\mathbf{k})$, locally viable in $\mathcal{E}p(V) \backslash \mathcal{E}p(\mathbf{c})$ thanks to Viability Theorem 11.4.6, p. 463, which also states that, whenever $(d,x,V(d,x)) \in \mathcal{E}p(V) \setminus \mathcal{E}p(\mathbf{c})$, i.e., whenever $V(d,x) < \mathbf{c}(d,x)$, there exists some $v \in \text{Dom}(\mathbf{l}(d,x;\cdot))$ such that

$$(-\varphi(d), -v, -\mathbf{l}(d,x;v)) \in T_{\mathcal{E}p(V)}(d,x,V(d,x))$$

Definition 2.15.6, p. 102 of the regulation map R for general differential inclusions and Definition 13.6.1, p. 542 imply that such v belongs to $R_V(d,x)$. □

Optimal evolutions do exist thanks to Theorem 13.4.3, p. 533. The question asked is how to regulate them. The first answer is provided by

Theorem 13.6.3 *[Regulation of Optimal Evolutions] If the Lagrangian* $(d,p,v) \rightsquigarrow \mathbf{l}(d,x;v)$ *is Marchaud, viable optimal evolutions* $x(\cdot) \in \mathcal{O}_{\mathbf{k}}(t^\sharp; d,x)$ *when* $t^\sharp \in \mathbb{T}_{(\mathbf{k},\mathbf{c})}(d,x)$ *is the optimal time are regulated by differential inclusion*

$$\forall\, t \in [0, t^\sharp[, \;\; x'(t) \in R_V(d(t), x(t))$$

and satisfy the terminal condition

$$d(t^\sharp) = d \text{ and } x(t^\sharp) = x$$

Proof. The proof of Theorem 13.5.1, p. 538 states that evolutions $t \mapsto (\delta(t), \xi(t), \eta(t))$ starting from $(d,x,V(d,x)) \in \mathcal{E}p(V)$ viable in $\mathcal{E}p(\mathbf{k})$ until they reach $\mathcal{E}p(\mathbf{c})$ at time $t^\sharp$ are optimal and regulated by

$$\forall\, t \in [0, t^\sharp], \;\; v(t) \in R(\delta(t), \xi(t))$$

thanks to Viability Theorem 11.4.6, p. 463. By setting $d(t) := \delta(t^\sharp - t)$, $x(t) := \xi(t^\sharp - t)$ and $x'(t) := v(t^\sharp - t)$, this is equivalent to saying that optimal viable evolutions $x(\cdot) \in \mathcal{O}_{\mathbf{k}}(t^\sharp; d, x)$ are regulated by the differential inclusion

$$x'(t) \in R_V(d(t), x(t))$$

and satisfy the terminal condition $d(t^\sharp) = d$ and $x(t^\sharp) = x$. □

We also know that

$$\mathbf{c}(d(0), x(0)) + \int_0^{t^\sharp} \mathbf{l}(d(\tau), x(\tau), x'(\tau))d\tau \leq V(d, x)$$

The initial causal variable $d(0) := \delta(t^\sharp)$ being known, the initial state $x(0)$ may be derived from the above formula which requires the knowledge of the evolution $x(\cdot)$ arriving at x and regulated by the regulation map R.

13.7 Aggregation of Hamiltonians

13.7.1 Aggregation

For simplicity, we consider problems without viability constraints. We introduce $j = 1, \ldots, J$ copies of the state space $Y := \mathbb{R}^n$ and the product space $X := \prod_{j=1}^{J} \mathbb{R}^n = \mathbb{R}^{Jn}$.

For each $j = 1, \ldots, n$, we introduce:

1. a lower semicontinuous convex Hamiltonian $\mathbf{l}_j^\star : \mathcal{D} \times Y \mapsto \mathbb{R} \cup \{+\infty\}$;
2. a function $\mathbf{c}_j : \mathcal{D} \times Y \mapsto \mathbb{R} \cup \{+\infty\}$

with which we associate the solutions to the *decentralized* Hamilton–Jacobi problems

$$\begin{cases} \left\langle \frac{\partial V_j(d, x_j)}{\partial d}, \varphi(d) \right\rangle + \mathbf{l}_j^\star \left(d, \frac{\partial V_j(d, x_j)}{\partial x_j} \right) = 0 \text{ on } \mathbb{R}^m \times \mathbb{R}^q \\ \text{satisfying } V_j(d, x_j) \leq \mathbf{c}_j(d, x_j) \end{cases} \quad (13.28)$$

We are now looking for the solution $W : \mathcal{D} \times Y \mapsto \mathbb{R} \cup \{+\infty\}$ to the *centralized* Hamilton–Jacobi problem

$$\begin{cases} \left\langle \dfrac{\partial W(d,y)}{\partial d}, \varphi(d) \right\rangle + \sum_{j=1}^{J} \mathbf{l}_j^\star \left(d, \dfrac{\partial W(d,y)}{\partial y} \right) = 0 \text{ on } \mathcal{D} \times \mathbb{R}^n \\ \text{satisfying } W(d,y) \le \mathbf{c}(d,y) \end{cases} \tag{13.29}$$

The link between the solution W of the centralized problem and solutions V_j to decentralized Hamilton–Jacobi problems is provided by the operation of inf-convolution (see Definition 18.8.1, p. 762): The *inf-convolution* $\bigstar_{j=1}^{J} \mathbf{v}_j : X \mapsto \mathbb{R} \cup \{+\infty\}$ of functions $\mathbf{v}_j$ is defined by

$$\mathcal{E}p(\bigstar_{j=1}^{J} \mathbf{v}_j) := \sum_{j=1}^{J} \mathcal{E}p(\mathbf{v}_j) \text{ (Minkowski sum of subsets)} \tag{13.30}$$

Lemma 18.8.3, p. 763 states that whenever

$$0 \in \operatorname{Int} \left(\{(p, \ldots, p)_{p \in Y^\star}\} + \prod_{j=1}^{J} \operatorname{Dom}(\mathbf{v}_j^\star) \right)$$

holds true, then there exist J elements $\overline{x}_j \in X_j$ such that

$$\sum_{j=1}^{J} \overline{x}_j = x \text{ and } \mathbf{v}(x) = \sum_{j=1}^{J} \mathbf{v}_j(\overline{x}_j) \tag{13.31}$$

In the case of two functions, we obtain

$$(\mathbf{u} \star \mathbf{v})(y) = \inf_{z} (\mathbf{u}(z) + \mathbf{v}(y - z))$$

from which the name of the operation is derived (when $\inf_y$ is replaced by $\int_y$ for the usual convolution in analysis).

We consider the inf-convolutions of the functions $\mathbf{c}_j$ and V_j:

$$\mathbf{c}(d,y) := \bigstar_{j=1}^{J} \mathbf{c}_j(y) \text{ and } W(d,y) := \bigstar_{j=1}^{J} V_j(y) \tag{13.32}$$

The natural question is to know whether, whenever the function $\mathbf{c}$ is the inf-convolution of the functions $\mathbf{c}_i$, the solution $W(d,y)$ of the centralized Hamilton–Jacobi problem (13.29), p. 545, is the inf-convolution of the solutions $V_j(d,x_j)$ of the decentralized Hamilton–Jacobi problem (13.29), p. 545.

The answer is positive under adequate assumptions.

More important is the second conclusion relating the regulation maps of the centralized and decentralized regulation maps $R_W(d,y)$ and $R_{V_j}(d,x_j)$. We shall prove that a centralized control v belongs to the centralized regulation map $R_W(d,y)$ at y if and only if there exist decentralized states x_j and controls u_j belonging to the regulation maps $R_{V_j}(d,x_j)$ at x_j satisfying

$$\sum_{j=1}^{J} x_j = y \text{ and } \sum_{j=1}^{J} u_j = v \tag{13.33}$$

Therefore, for any (d, y), an evolution of the centralized associated problem $y(\cdot) \in \mathcal{A}(t^\sharp(d); d, y)$ regulated by $y'(t) \in R_W(d(t), y(t))$ can be written in the form

$$y(t) = \sum_{j=1}^{J} x_j(t)$$

where the decentralized (optimal) evolutions $x_j(\cdot) \in \mathcal{A}_j(t^\sharp(d); d, x_j)$ regulated by $x'_j(t) \in R_{V_j}(d(t), x_j(t))$.

These results are quite intricate and it is both convenient and useful to cover other problems to deduce them from an abstract problem by setting

$$A : x = (x_1, \ldots, x_J) \in \mathbb{R}^q := \mathbb{R}^{Jn} \mapsto Ax := \sum_{j=1}^{J} x_j \in \mathbb{R}^n$$

and

$$\mathbf{m}^\star(d; q) = \sum_{j=1}^{J} \mathbf{l}_j^\star(d; q) \text{ and } \mathbf{m}(d; v) = \bigstar_{j=1}^{J} \mathbf{l}_j(d; v)$$

13.7.2 Composition of a Hamiltonian by a Linear Operator

Consider then the more general problem for arbitrary state space $\mathbb{R}^q$, with (much) higher dimension $q \geq n$ and a linear operator $A \in \mathcal{L}(\mathbb{R}^q, \mathbb{R}^n)$, the transpose of which is and its transpose $A^\star \in \mathcal{L}(\mathbb{R}^n, \mathbb{R}^q)$.

We shall assume throughout this section that the Hamiltonian $(d, p) \mapsto \mathbf{l}^\star(d; p)$ and the Lagrangian $(d, u) \mapsto \mathbf{l}(d; u)$ *do not depend upon the state variable* x.

Definition 13.7.1 *[Composition of a Hamiltonian by a Linear Operator] Let us consider a Hamiltonian* $(d, p) \mapsto \mathbf{l}^\star(d; p)$ *and linear operator* $A \in \mathcal{L}(\mathbb{R}^q, \mathbb{R}^n)$. *The composed Hamiltonian* $m^\star : (d, q) \in \mathbb{R}^m \times \mathbb{R}^n \mapsto m^\star(d, q) \in \mathbb{R} \cup \{+\infty\}$ *of* $\mathbf{l}^\star$ *by* $A^\star$ *is defined by*

$$m^\star(d, q) := \mathbf{l}^\star(d, A^\star q)$$

and we denote by $m(d, y) := \mathbf{l}^{\star\star}$ *its biconjugate.*

We also introduce two internal condition functions:

1. $\mathbf{c} : (d,x) \in \mathbb{R}^m \times \mathbb{R}^q \mapsto \mathbf{c}(d,x) \in \mathbb{R} \cup \{+\infty\}$
2. $\mathbf{b} : (d,y) \in \mathbb{R}^m \times \mathbb{R}^n \mapsto \mathbf{b}(d,y) \in \mathbb{R} \cup \{+\infty\}$

related by

$$\mathbf{b}(d,y) := \inf_{Ax=y} \mathbf{c}(d,x)$$

Theorem 18.2.5, p. 715 states that the assumptions: there exists a constant $c < +\infty$ such that

$$\begin{cases} (i) \ \forall (d,u) \in \mathrm{Dom}(\mathbf{l}), \forall \nu \in \mathbb{R}^n, \exists \mu \in \mathrm{Dom}(D_\uparrow \mathbf{l}(d,u)) \cap c\|\nu\| \text{ and } A\nu = \mu \\ (ii) \ \forall (d,x) \in \mathrm{Dom}(\mathbf{c}), \forall \nu \in \mathbb{R}^n, \exists \mu \in \mathrm{Dom}(D_\uparrow \mathbf{c}(d,x)) \cap c\|\nu\| \text{ and } A\nu = \mu \end{cases} \quad (13.34)$$

imply that the infimum is achieved in formulas

$$\begin{cases} (i) \ m(d,v) = \min_{Au=v} \mathbf{l}(d,u) \\ \quad \text{is the conjugate function of } m^\star(d,p) := \mathbf{l}^\star(d,A^\star q) \\ (ii) \ \mathbf{b}(d,y) = \min_{Ax=y} \mathbf{c}(d,x) \end{cases}$$

Theorem 13.7.2 ***[Link between Solutions of the Hamilton–Jacobi Equations]*** *We consider the two structured Hamilton–Jacobi problems*

$$\begin{cases} \left\langle \frac{\partial V(d,x)}{\partial d}, \varphi(d) \right\rangle + \mathbf{l}^\star\left(d, \frac{\partial V(d,x)}{\partial x}\right) = 0 \text{ on } \mathbb{R}^m \times \mathbb{R}^q \\ \text{satisfying } V(d,x) \le \mathbf{c}(d,x) \end{cases} \quad (13.35)$$

and

$$\begin{cases} \left\langle \frac{\partial W(d,y)}{\partial d}, \varphi(d) \right\rangle + \mathbf{l}^\star\left(d, A^\star \frac{\partial W(d,y)}{\partial y}\right) = 0 \text{ on } \mathbb{R}^m \times \mathbb{R}^n \\ \text{satisfying } W(d,y) \le \mathbf{b}(d,y) \end{cases} \quad (13.36)$$

Assume furthermore that the constraint qualification assumptions (13.34), p. 547 hold true. Then their viability solutions are related by formula

$$W(d,y) = \inf_{Ax=y} V(d,x) \quad (13.37)$$

Proof. Let us consider the operator $(\mathbf{1} \times A \times \mathbf{1}) : (d,x,\lambda) \in \mathcal{D} \times X \times \mathbb{R}_+ \mapsto (d,Ax,\lambda) \in \mathcal{D} \times Y \times \mathbb{R}_+$ and the two following characteristic systems

$$\begin{cases} (i) \ \delta'(t) \ = \ -\varphi(\delta(t)) \\ (ii) \ (\eta'(t), \zeta'(t)) \ \in \ -\mathcal{E}p(m(\delta(t);\cdot)) \ = \ -(\mathbb{I} \times A \times \mathbb{I})\mathcal{E}p(\mathbf{l}(\delta(t);\cdot)) \end{cases} \tag{13.38}$$

governing the evolution of $\eta(t) \in \mathbb{R}^n$ and

$$\begin{cases} (i) \ \delta'(t) \ = \ -\varphi(\delta(t)) \\ (ii) \ (\xi'(t), \zeta'(t)) \ \in \ -\mathcal{E}p(\mathbf{l}(\delta(t);\cdot)) \end{cases} \tag{13.39}$$

governing the evolution of $\xi(t) \in \mathbb{R}^q$. We introduce the viable-capture basins defining the epigraphs of the viability solutions

$$\begin{cases} (i) \ \text{Capt}_{(13.37)}(\mathcal{E}p(\mathbf{b})) \ = \ \mathcal{E}p(W) \text{ where } \mathcal{E}p(\mathbf{b}) \ = \ (\mathbb{I} \times A \times \mathbb{I})\mathcal{E}p(\mathbf{c}) \\ (ii) \ \text{Capt}_{(13.38)}(\mathcal{E}p(\mathbf{c})) \ = \ \mathcal{E}p(V) \end{cases} \tag{13.40}$$

We shall prove that

$$(\mathbb{I}\times A\times\mathbb{I})\text{Capt}_{(13.38)}(\mathcal{E}p(\mathbf{c}))=\text{Capt}_{(13.37)}((\mathbb{I}\times A\times\mathbb{I})\mathcal{E}p(\mathbf{c}))=\text{Capt}_{(13.37)}(\mathcal{E}p(\mathbf{b}))$$

1. *Proof of inequality* $W(d, Ax) \leq V(d, x)$. This inequality is always true. Take any $(d, x, V(d, x)) \in \text{Capt}_{(13.37)}(\mathcal{E}p(\mathbf{c}))$. There exist $t \mapsto \nu(t)$ and $t^\sharp \geq 0$ such that

$$\mathbf{c}\left(\delta(t^\sharp), x - \int_0^{t^\sharp} \nu(t)dt\right) + \int_0^{t^\sharp} \mathbf{l}(\delta(t);\nu(t))dt \ \leq \ V(d,x)$$

Since $\mathbf{b}(d, Ax) \leq \mathbf{c}(d, x)$ and $m(d, A\nu) \leq \mathbf{l}(d, \nu)$, we infer that, setting $\mu(t) := A\nu(t)$,

$$\mathbf{b}\left(\delta(t^\sharp), Ax - \int_0^{t^\sharp} \mu(t)dt\right) + \int_0^{t^\sharp} m(\delta(t);\mu(t))dt \ \leq \ V(d,x)$$

and thus, that $(d, Ax, V(d, x))$ belongs to $\text{Capt}_{(13.37)}(\mathcal{E}p(\mathbf{b})) \ = \ \mathcal{E}p(W)$. This implies that $W(d, Ax) \leq V(d, x)$ and that

$$(\mathbb{I} \times A \times \mathbb{I})\text{Capt}_{(13.38)}(\mathcal{E}p(\mathbf{c})) \ \subset \ \text{Capt}_{(13.37)}((\mathbb{I} \times A \times \mathbb{I})\mathcal{E}p(\mathbf{c}))$$

2. *Proof of inequality*: $\forall \ y, \ \exists \ x$ such that $Ax = y$ and $V(d, x) \leq W(d, y)$. Take any $(d, y, W(d, y)) \in \text{Capt}_{(13.38)}(\mathcal{E}p(\mathbf{b}))$. There exist an integrable function $t \mapsto \mu(t)$ and $t^\sharp \geq 0$ such that

$$\mathbf{b}\left(\delta(t^\sharp), x - \int_0^{t^\sharp} \mu(t)dt\right) + \int_0^{t^\sharp} m(\delta(t);\mu(t))dt \ \leq \ W(d,y)$$

Theorem 18.4.14, p. 734 and assumption (13.34)(i), p. 547 imply that the subset

$$\Phi(d,v) \;:=\; \{u \text{ such that } Au = v \text{ and } \mathbf{l}(d;u) \le m(d;v)\}$$

is not empty and the set-valued map Φ has a closed graph. The Measurable Selection Theorem (see for instance Theorem 8.1.13, p. 308 of *Set-Valued Analysis*, [27, Aubin & Frankowska]) implies that we can associate with the measurable function $\mu(\cdot)$ a measurable function $\nu(\cdot)$ such that, for almost all t, $\nu(t) \in \Phi(\delta(t), \mu(t))$. Furthermore, Theorem 18.4.14, p. 734 and assumption (13.34)(ii), p. 547 imply that we can associate with $y - \int_0^{t^\sharp} \mu(t)dt$ an element z such that $Az = y - \int_0^{t^\sharp} \mu(t)dt$ and $\mathbf{c}(z) = y - \int_0^{t^\sharp} \mu(t)dt$. Setting $x := z + \int_0^{t^\sharp} \nu(t)dt$, we have proved that

$$\begin{cases} \mathbf{c}\left(\delta(t^\sharp), x - \int_0^{t^\sharp} \nu(t)dt\right) + \int_0^{t^\sharp} \mathbf{l}(\delta(t);\nu(t))dt \\ = \mathbf{b}\left(\delta(t^\sharp), y - \int_0^{t^\sharp} \mu(t)dt\right) + \int_0^{t^\sharp} \mathbf{l}(\delta(t);\mu(t))dt \;\le\; W(d,y) \end{cases}$$

Therefore, $(d,y,W(d,y)) = (\mathbb{I} \times A \times \mathbb{I})(d,x,W(d,y))$ where $(d,x,W(d,y))$ belongs to $\text{Capt}_{(13.38)}(\mathcal{E}p(\mathbf{c})) \;=\; \mathcal{E}p(W)$. Consequently, $(d,y,W(d,y)) \in (\mathbb{I} \times A \times \mathbb{I})\mathcal{E}p(V)$, and thus, $V(d,x) \le W(d,y)$ and

$$\text{Capt}_{(13.37)}((\mathbb{I} \times A \times \mathbb{I})\mathcal{E}p(\mathbf{c})) \;\subset\; (\mathbb{I} \times A \times \mathbb{I})\text{Capt}_{(13.38)}(\mathcal{E}p(\mathbf{c}))$$

These two inequalities imply that $V(d,x) \le W(d,y) = W(d,Ax) \le V(d,x)$, so that

$$W(d,y) \;=\; \min_{Ax=y} V(d,x)$$

and $\text{Capt}_{(13.37)}(\mathbb{I} \times A \times \mathbb{I})(\mathcal{E}p(\mathbf{c})) = (\mathbb{I} \times A \times \mathbb{I})\text{Capt}_{(13.38)}(\mathcal{E}p(\mathbf{c}))$. □

We now compare the two regulation maps:

Theorem 13.7.3 ***[Links between the Regulation Maps]*** *We posit the assumptions of Theorem 13.7.2, p. 547. Consider the two regulation maps*

$$R_W(d,y) \;:=\; \left\{ \begin{array}{l} \mu \in \text{Dom}(m(d,y;\cdot)) \text{ such that} \\ \left\langle \frac{\partial W(d,y)}{\partial d}, \varphi(d) \right\rangle + \left\langle \frac{\partial W(d,y)}{\partial y}, \mu \right\rangle - m(d,y;\mu) \;\ge\; 0 \end{array} \right\}$$

and

$$R_V(d,x) \;:=\; \left\{ \begin{array}{l} \nu \in \text{Dom}(\mathbf{l}(d,x;\cdot)) \text{ such that} \\ \left\langle \frac{\partial V(d,x)}{\partial d}, \varphi(d) \right\rangle + \left\langle \frac{\partial V(d,x)}{\partial x}, \nu \right\rangle - \mathbf{l}(d,x;\nu) \;\ge\; 0 \end{array} \right\}$$

Inclusion $AR_V(d,x) \subset R_V(d,Ax)$ *always holds true. If we assume that the solution* V *satisfies*

$$\begin{cases} (i) \ \forall (d,u) \in \mathrm{Dom}(\mathbf{l}), \forall \nu \in \mathbb{R}^n, \exists \mu \in \mathrm{Dom}(D_\uparrow \mathbf{l}(d,u)) \cap c\|\nu\| \text{ and } A\nu = \mu \\ (ii) \ \forall (d,x) \in \mathrm{Dom}(V), \forall \nu \in \mathbb{R}^n, \exists \mu \in \mathrm{Dom}(D_\uparrow V(d,x)) \cap c\|\nu\| \text{ and } A\nu = \mu \end{cases} \tag{13.41}$$

equality

$$\exists\, x \text{ satisfying } Ax = y \text{ and } R_W(d,y) \;=\; AR_V(d;x) \tag{13.42}$$

ensues.

Proof. 1. *Proof of inclusion* $AR_V(d,x) \subset R_W(d,Ax)$. Let us consider $\nu \in R_V(d,x)$. This means that

$$(-\varphi(d), -\nu, -\mathbf{l}(d;\nu)) \;\in\; T_{\mathcal{E}p(V)}(d,x,V(d,x))$$

Therefore,

$$\begin{cases} (-\varphi(d), -A\nu, -\mathbf{l}(d;\nu)) \;=\; (\mathbb{I} \times A \times \mathbb{I})(-\varphi(d), -\nu, -\mathbf{l}(d;\nu)) \\ \in\; (\mathbb{I} \times A \times \mathbb{I}) T_{\mathcal{E}p(V)}(d,x,V(d,x)) \;\subset\; T_{(\mathbb{I} \times A \times \mathbb{I})\mathcal{E}p(V)}(d,Ax,V(d,x)) \\ =\; T_{\mathcal{E}p(W)}(d,y,V(d,x)) \end{cases}$$

Since by Proposition 6.1.4, p. 226 of *Set-Valued Analysis*, [27, Aubin & Frankowska], $T_{\mathcal{E}p(W)}(d,y,V(d,x)) = \mathrm{Dom}(D_\uparrow W(d,y)) \times \mathbb{R}$ if $W(d,y) < V(d,x)$ and $T_{\mathcal{E}p(W)}(d,x,V(d,x)) = T_{\mathcal{E}p(W)}(d,y,W(d,y))$ if $W(d,y) = V(d,x)$, we infer that $A\nu$ belongs to $R_W(d,Ax)$.

2. *Proof of inclusion* $\forall\, y$, $\exists\, x$ such that $Ax - y$ and $R_W(d,y) \subset AR_V(d,x)$. Let us consider $\mu \in R_W(d,y)$, i.e., satisfying

$$(-\varphi(d), -\mu, -m(d;\mu)) \;\in\; T_{\mathcal{E}p(W)}(d,y,W(d,y))$$

Theorem 18.4.14, p. 734 and assumption (13.41)(ii), p. 550 imply that there exist x such that $Ax = y$ and $V(d,x) = W(d,y)$.
By applying Theorem 18.4.14, p. 734 for the linear operator $\mathbb{I} \times A \times \mathbb{I}$ and the subset $\mathcal{E}p(V)$, we infer that assumption (13.41)(ii), p. 550 implies equality

$$(\mathbb{I} \times A \times \mathbb{I}) T_{\mathcal{E}p(V)}(d,x,V(d,x)) \;=\; T_{(\mathbb{I} \times A \times \mathbb{I})\mathcal{E}p(V)}(d,Ax,V(d,x)) \tag{13.43}$$

holds true. Then, there exist ν such that $A\nu = \mu$ and $\mathbf{l}(d,\nu) = m(d,\mu)$. Consequently,

$$\begin{cases} (-\varphi(d), -\mu, -m(d;\mu)) \;=\; (\mathbb{I} \times A \times \mathbb{I})(-\varphi(d), -\nu, -\mathbf{l}(d;\nu)) \\ \in\; (\mathbb{I} \times A \times \mathbb{I}) T_{\mathcal{E}p(W)}(d,y,W(d,y)) \;=\; (\mathbb{I} \times A \times \mathbb{I}) T_{\mathcal{E}p(W)}(d,x,V(d,x)) \end{cases}$$

Therefore,

$$(-\varphi(d), -\nu, -\mathbf{l}(d;\nu)) \in T_{\mathcal{E}p(V)}(d, x, V(d, x))$$

and we infer that ν belongs to $R_V(d, x)$. $\square$

13.8 Hamilton–Jacobi–Cournot Equations

The question whether we can obtain beforehand *the initial states* $x(0)$ *and the specific regulation maps* $\widetilde{R}(d, x(0), x)$ *(depending upon* $x(0)$*) driving optimal viable evolutions* $x(\cdot)$ arriving at terminal state x at optimal time $t^\sharp$ (see Definition 8.4.8, p. 288). This would avoid *computing initial states* $x(0)$ by solving the variational problem or, equivalently, avoid the regulation of *all* optimal evolutions from the terminal state through the regulation map R_V as in Theorem 13.6.3, p. 543.

For computing the formerly missing initial conditions, we just introduce an auxiliary parameter $\chi \in \mathbb{R}^n$ which plays the role of *candidate to be an initial condition.* We then extend internal and viability conditions $\widetilde{\mathbf{k}}$ and $\widetilde{\mathbf{c}}$ by setting

$$\begin{cases} (i) \ \widetilde{\mathbf{k}}(d, \chi, x) \ := \ \mathbf{k}(d, x) \\ (ii) \ \widetilde{\mathbf{c}}(d, x, x) \ := \ \mathbf{c}(d, x) \text{ and } \widetilde{\mathbf{c}}(d, \chi, x) := +\infty \ \ \chi \neq x \end{cases} \tag{13.44}$$

The answer is obtained by computing the viability solution $\widetilde{V}(d, \chi, x)$ to the Hamilton–Jacobi–Cournot partial differential equation

$$\left\langle \frac{\partial \widetilde{V}(d, \chi, x)}{\partial d}, \varphi(d) \right\rangle + \mathbf{l}^\star \left(d, x; \frac{\partial \widetilde{V}(d, \chi, x)}{\partial x} \right) \ = \ 0 \tag{13.45}$$

satisfying

$$\widetilde{\mathbf{k}}(d, \chi, x) \ \leq \ \widetilde{V}(d, \chi, x) \ \leq \ \widetilde{\mathbf{c}}(d, \chi, x) \tag{13.46}$$

We associate with the viability solution $\widetilde{V}$ of the Hamilton–Jacobi–Cournot equation:

1. the *Cournot regulation map* $R_{\widetilde{V}}$ defined by

$$\begin{cases} R_{\widetilde{V}}(d, \chi, x) \ = \\ \left\{ v \in F(d, x) \text{ such that } D_\uparrow^{\star\star}\widetilde{V}(d, \chi, x)(-\varphi(d), 0, -v) + \mathbf{l}(d, x) \ \leq \ 0 \right\} \end{cases}$$

2. the *Cournot map* $\mathbb{C}_{\widetilde{V}}$ defined by

$$\mathbb{C}_{\widetilde{V}}(d, x) \ := \ \left\{ \chi \text{ such that } \widetilde{V}(d, \chi, x) < \infty \right\} \tag{13.47}$$

Theorem 13.8.1 *[Regulation of Optimal Evolutions from Cournot Initial States] We posit the assumptions of Theorem 13.10.2, p. 558. For any initial state $\chi \in \mathbb{C}_{\widetilde{V}}(d,x)$ provided by the Cournot map, there exists at least one optimal viable optimal evolution $x(\cdot) \in \mathcal{O}_{\mathbf{k}}(t^\sharp; d, \chi, x)$ where $t^\sharp \in \mathbb{T}_{(\mathbf{k},\mathbf{c})}(d,x)$ starting from initial conditions $x(0) = \chi$, arriving at $x = x(t^\sharp)$ at optimal $t^\sharp$ when $d(t^\sharp) = d$.*

It is regulated by differential inclusion

$$\forall\, t \in [0, t^\sharp[, \quad x'(t) \;\in\; R_{\widetilde{V}}(d(t), \chi, x(t))$$

Proof. Let us consider the viability solution $(d,\chi,x) \mapsto \widetilde{V}(d,\chi,x)$ and the associated regulation map $R_{\widetilde{V}}$.

Its epigraph is the viable-capture basin of $\mathcal{E}p(\widetilde{c})$ viable in $\mathcal{E}p(\widetilde{k})$ under the characteristic system

$$\begin{cases} (i) \;\; \delta'(t) \;=\; -\varphi(\delta(t)) \\ (ii) \;\; \chi'(t) \;=\; 0 \\ (iii) \; (\xi'(t), \eta'(t)) \;\in\; -\mathcal{E}p(\mathbf{l}(\delta(t), \xi(t); \cdot)) \end{cases} \tag{13.48}$$

There exist at least some time $t^\sharp$ and one evolution $t \mapsto (\delta(t), \chi, \xi(t), \eta(t))$ starting from $(d, x, \chi, V(d,x)) \in \mathcal{E}p(\widetilde{V})$ viable in $\mathcal{E}p(\widetilde{\mathbf{k}})$ until it reaches $\mathcal{E}p(\widetilde{\mathbf{c}})$ at some time $t^\sharp$. The viability condition implies that $t^\sharp \leq \tau^\sharp(d)$ and that

$$\sup_{s \in [0,t^\sharp]} \left(\mathbf{k}\,(\delta(s), \xi(s)) + \int_0^s \mathbf{l}(\delta(\tau), \xi(\tau), v(\tau))d\tau \right) \;\leq\; \widetilde{V}(d,\chi,x)$$

so that

$$\overleftarrow{\mathbf{I}}_{\mathbf{k}}(t^\sharp; v(\cdot))(d,x) \;\leq\; \widetilde{V}(d,\chi,x)$$

and the target condition that

$$\widetilde{\mathbf{c}}\,(\delta(t^\sharp), \chi, \xi(t^\sharp)) + \int_0^{t^\sharp} \mathbf{l}(\delta(\tau), \xi(\tau), v(\tau))d\tau \;\leq\; \widetilde{V}(d,\chi,x) \;<\; +\infty$$

Since $\widetilde{\mathbf{c}}(d,\chi,x) := +\infty$ whenever $\chi \neq x$, we infer that $\chi = \xi(t^\sharp)$, that $\chi \in \mathbb{C}_{\widetilde{V}}(d,x)$ and that

$$\mathbf{c}(\delta(t^\sharp), \chi) + \int_0^{t^\sharp} \mathbf{l}(\delta(\tau), \xi(\tau), v(\tau))d\tau \;\leq\; \widetilde{V}(d,\chi,x) \;<\; +\infty$$

Setting

$$\begin{cases} \overleftarrow{\mathbf{J}}_{\mathbf{c}}(t^\sharp; v(\cdot))(d,\chi,x) \\ := \mathbf{c}(\delta(t^\sharp),\chi) + \int_0^{t^\sharp} \mathbf{l}(\delta(\tau),\xi(\tau),v(\tau))d\tau \text{ if } \xi(t^\sharp) = \chi \\ \text{and } +\infty \text{ otherwise} \end{cases}$$

we infer that

$$\overleftarrow{\mathbf{J}}_{\mathbf{c}}(t^\sharp; v(\cdot))(d,\chi,x) \leq \widetilde{V}(d,\chi,x)$$

and thus that, setting

$$\begin{cases} \overleftarrow{\mathbf{L}}_{(\mathbf{k},\mathbf{c})}(t^\sharp; v(\cdot))(d,\chi,x) \\ := \max(\overleftarrow{\mathbf{I}}_{\mathbf{k}}(t^\sharp; x(\cdot), v(\cdot))(d,x), \overleftarrow{\mathbf{J}}_{\mathbf{c}}(t^\sharp; v(\cdot))(d,\chi,x)) \end{cases}$$

we have proved that

$$\overleftarrow{\mathbf{L}}_{(\mathbf{k},\mathbf{c})}(t^\sharp; v(\cdot))(d,\chi,x) \leq \widetilde{V}(d,\chi,x)$$

The proof of Theorem 13.4.3, p. 533 implies that

$$\widetilde{V}(d,\chi,x) = \inf_{(t^\sharp; v(\cdot))} \overleftarrow{\mathbf{L}}_{(\mathbf{k},\mathbf{c})}(t^\sharp; v(\cdot))(d,\chi,x)$$

and that the viable evolution is regulated by

$$v(t) \in R_{\widetilde{V}}(\delta(t),\chi,\xi(t))$$

where, under assumptions of Theorem 13.10.2, p. 558, the regulation map $R_{\widetilde{V}}$ is equal to

$$R_{\widetilde{V}}(d,\chi,x) = \left\{ v \in F(d,x) \text{such that } D_{\uparrow}^{\star\star}\widetilde{V}(d,\chi,x)(-\varphi(d),0,-v) + \mathbf{l}(d,x) \leq 0 \right\}$$

By setting $d(t) := \delta(t^\sharp - t)$, $x(t) := \xi(t^\sharp - t)$ and $x'(t) := v(t^\sharp - t)$, this is equivalent to saying that an optimal viable evolution starting from $\chi \in \mathbb{C}_{\widetilde{V}}(d,x)$ is regulated by

$$v(t) \in R_{\widetilde{V}}(\delta(t),\chi,\xi(t))$$

and satisfy the terminal condition $d(t^\sharp) = d$ and $x(t^\sharp) = x$ and initial condition $x(0) = \chi$.

Hence, for each initial condition $\chi \in \mathbb{C}_{\widetilde{V}}(d,x)$, there exists an optimal viable evolution starting at χ and arriving at x in optimal time $t^\sharp$. □

Consequently, for any pair (d,x), if

1. $\mathbb{C}_{\widetilde{V}}(d,x) = \{\chi\}$ is a singleton, then there exists one optimal viable evolution starting at χ and regulated by $x'(t) \in R_{\widetilde{V}}(d(t),\chi,x(t))$ until it arrives at x at time $t^\sharp$,

2. $\mathbb{C}_{\widetilde{V}}(d,x) = \emptyset$, and no optimal evolution arrive at x,
3. $\mathbb{C}_{\widetilde{V}}(d,x)$ contains several initial states χ. From all $\chi \in \mathbb{C}_{\widetilde{V}}(d,x)$ start optimal viable evolutions regulated by $x'(t) \in R_{\widetilde{V}}(d(t),\chi,x(t))$ until they *collide* at x at time $t^{\sharp}$.

This is the property which motivates the terminology of Cournot map, since Antoine Cournot (1801–1877) suggested to capture one aspect of uncertainty or chance as the collision of several independent causal series.

13.9 Lax–Hopf Formula

Theorem 13.9.1 ***[Lax–Hopf Formula]*** *We assume that both φ and $\mathbf{l}$ do not depend upon d and x. Let us assume that the function $\mathbf{c}$ is lower semicontinuous and that the functions $\mathbf{k}$ and $\mathbf{l}$ are convex and lower semicontinuous. Then the viability solution V is equal to the* Lax–Hopf value function

$$V(d,x) \;=\; \max\left(\mathbf{k}(d,x),\; \inf_{t^{\sharp}\geq 0}\;\inf_{u\in \mathrm{Dom}(\mathbf{l})}\left(\mathbf{c}(d-t^{\sharp}\varphi, x-t^{\sharp}u)+t^{\sharp}\mathbf{l}(u)\right)\right) \tag{13.49}$$

which is the marginal function of a static minimization theorem.

The regulation map is the set of elements $u \in \mathrm{Dom}(\mathbf{l})$ minimizing this function:

$$\begin{cases} R_V(d,x) \;=\; \{u \in \mathrm{Dom}(\mathbf{l}) \text{ such that} \\ V(d,x) = \max\left(\mathbf{k}(d,x), \left(\mathbf{c}(d-\tau^{\sharp}(d)\varphi, x-\tau^{\sharp}(d)u)+\tau^{\sharp}(d)\mathbf{l}(u)\right)\right)\} \end{cases} \tag{13.50}$$

If the viability solution V and the Lagrangian $\mathbf{l}$ are differentiable and if $u := r(d,x) \in R_V(d,x)$ is the unique minimizer, Lax–Oleinik formula holds

$$\frac{\partial V}{\partial x} \;=\; \frac{d}{du}\mathbf{l}(d,x;r(d,x)) \tag{13.51}$$

For boundary value problems without constraints, the formula boils down to

$$V(d,x) \;=\; \inf_{u\in \mathrm{Dom}(\mathbf{l})}\left(\mathbf{c}(d-\tau^{\sharp}(d)\varphi, x-\tau^{\sharp}(d)u)+\tau^{\sharp}(d)\mathbf{l}(u)\right) \tag{13.52}$$

Proof. Since the epigraph of $\mathbf{k}$ is convex and since both the map φ and Lagrangian $\mathbf{l}$ do not depend on (d,x), differential inclusion

$$\begin{cases} (i) & \delta'(t) = -\varphi \\ (ii) & (\xi'(t), \eta'(t)) \in -\mathcal{E}p(\mathbf{l}(\cdot)) \end{cases} \tag{13.53}$$

can be written in the form

$$(\delta'(t), \xi'(t), \eta'(t)) \in \mathcal{G} := \{-\varphi\} \times -\mathcal{E}p(\mathbf{l}(\cdot))$$

where $\mathcal{G}$ is a constant closed convex subset. Lax–Hopf Formula (11.27), p. 469 of Theorem 11.5.4, p. 469 states that if the environment $\mathcal{E}p(\mathbf{k})$ is closed and convex, the target $C \subset K$ is closed and $\mathcal{G}$ is a closed convex subset, then the viable-capture basin enjoys the *Lax–Hopf formula*

$$\text{Capt}_{\mathcal{G}}(\mathcal{E}p(\mathbf{k}), \mathcal{E}p(\mathbf{c})) = \mathcal{E}p(\mathbf{k}) \cap (\mathcal{E}p(\mathbf{c}) - \mathbb{R}_+\mathcal{G})$$

Hence, the epigraph of the valuation function V, defined as the viable-capture basin $\text{Capt}_{(13.52)}(\mathcal{E}p(\mathbf{k}), \mathcal{E}p(\mathbf{c}))$ under the differential inclusion (13.53), is equal to

$$\begin{cases} \text{Capt}_{(13.52)}(\mathcal{E}p(\mathbf{k}), \mathcal{E}p(\mathbf{c})) = \text{Capt}_{\mathcal{G}}(\mathcal{E}p(\mathbf{k}), \mathcal{E}p(\mathbf{c})) \\ = \mathcal{E}p(\mathbf{k}) \cap \left(\mathcal{E}p(\mathbf{c}) - \bigcup_{\lambda \geq 0} \lambda(\{\varphi\} \times \mathcal{E}p(\mathbf{l}))\right) \end{cases} \tag{13.54}$$

Therefore (d, x, y) belongs the viable-capture basin if $\mathbf{k}(d, x) \leq y$ and if there exist $(\delta^\star, \xi^\star, \eta^\star) \in \mathcal{E}p(\mathbf{c})$, $t^\sharp > 0$ and $u \in \text{Dom}(\mathbf{l})$ such that

$$\mathbf{c}(d - t^\sharp, x - t^\sharp u) = \mathbf{c}(\delta^\star, \xi^\star) \leq \eta \leq y - t^\sharp \mathbf{l}(u)$$

This means that

$$\max\left(\mathbf{k}(d, x), \inf_{t^\sharp} \inf_{u} \left(\mathbf{c}(t - t^\sharp \varphi, x - t^\sharp u) + t^\sharp \mathbf{l}(u)\right)\right) = V(d, x)$$

which is the Lax–Hopf formula we were looking for. □

We associate with the *internal condition* function $(d, x) \mapsto \mathbf{c}(d, x) \in \mathbb{R} \cup \{+\infty\}$ the *domain map* $d \rightsquigarrow \mathbf{C}(d)$ defined by

$$\mathbf{C}(d) := \{x \text{ such that } \mathbf{c}(d, x) < +\infty\} \tag{13.55}$$

The question arises to know precisely the domains $\text{Dom}(V(t, \cdot))$ of the traffic profiles, i.e., the set of states x such that $V(t, x) < +\infty$: we shall prove that it is couched in terms of the domain map $\mathbf{C}$:

Theorem 13.9.2 ***[Domain of the Viability Solution]*** *We posit the assumptions of Theorem 13.9.1, p. 554. For any $d \in \mathcal{D}$, the domains of the function $V(d, \cdot)$ associated with the internal condition $\mathbf{c}$ are equal to*

$$\mathrm{Dom}(V(d,\cdot)) = \bigcup_{s\in[0,\tau^\sharp(d)],\ u\in\mathrm{Dom}(\mathbf{l})} (\mathbf{C}(d-s\varphi)+su) \tag{13.56}$$

If we assume furthermore that $\mathbf{C}$ *satisfies*

$$\forall\, s \in [0,\tau^\sharp(d)],\ \ \mathbf{C}(d-s\varphi) \subset \mathbf{C}(d-\tau^\sharp(d)\varphi)+(\tau^\sharp(d)-s)\mathrm{Dom}(\mathbf{l}) \tag{13.57}$$

then

$$\mathrm{Dom}(V(d,\cdot)) = \mathbf{C}(d-\tau^\sharp(d)\varphi)+\tau^\sharp(d)\mathrm{Dom}(\mathbf{l}) \tag{13.58}$$

Proof. For any $x \in \mathrm{Dom}(V(d,\cdot))$ and any $\varepsilon > 0$, there exist $s \in [0,\tau^\sharp(d)]$, $u \in \mathrm{Dom}(\mathbf{l})$ such that

$$\mathbf{c}(d-s\varphi, x-su)+s\mathbf{l}(u) \le V(d,x)+\varepsilon < +\infty$$

so that $x-su \in \mathbf{C}(d-s\varphi)$, and thus,

$$\mathrm{Dom}(V(d,\cdot)) \subset \mathbf{C}(d-s\varphi)+su \subset \bigcup_{s\in[0,\tau^\sharp(d)],\ u\in\mathrm{Dom}(\mathbf{l})} (\mathbf{C}(d-s\varphi)+su)$$

Conversely, let us take $x \in \bigcup_{s\in[0,\tau^\sharp(d)],\ u\in\mathrm{Dom}(\mathbf{l})} (\mathbf{C}(d-s\varphi)+su)$, and thus, take $s \in [0,\tau^\sharp(d)]$ and $u \in \mathrm{Dom}(\mathbf{l})$ such that $x \in \mathbf{C}(d-s\varphi)+su$. Therefore, $\mathbf{c}(d-s\varphi, x-su) < +\infty$ is finite and

$$V(d,x) \le \mathbf{c}(d-s\varphi, x-su)+s\mathbf{l}(u) < +\infty$$

so that $x \in \mathrm{Dom}(V)(d,\cdot)$.

As a particular case, for $s := \tau^\sharp(d)$

$$\mathbf{C}(d-\tau^\sharp(d)\varphi)+\tau^\sharp(d)\mathrm{Dom}(\mathbf{l}) \subset \mathrm{Dom}(V(t,\cdot))$$

Assume now that the tube $\mathbf{C}$ satisfies (14.22), p. 584. Take any $x \in \mathrm{Dom}(V(d,\cdot))$, with which we associate $s \in [0,\tau^\sharp(d)]$, $\xi \in \mathbf{C}(d-s\varphi)$ and $u \in \mathrm{Dom}(\mathbf{l})$ such that $x = \xi + su$ thanks to (14.21), p. 584. Therefore

$$\xi \in \mathbf{C}(d-s\varphi) \subset \mathbf{C}(d-\tau^\sharp(d)\varphi)+(\tau^\sharp(d)-s)\mathrm{Dom}(\mathbf{l})$$

so that there exists $v \in \mathrm{Dom}(\mathbf{l})$ such that $\xi \in \mathbf{C}(d-s\varphi)+(\tau^\sharp(d)-s)v$. Hence $x \in \mathbf{C}(d-s\varphi)+su+(\tau^\sharp(d)-s)v$. Since $\mathrm{Dom}(\mathbf{l})$ is convex and since u and v belong to it and $\tau^\sharp(d)-s \ge 0$, then $su+(\tau^\sharp(d)-s)v = \tau^\sharp(d)w$ where $w \in \mathrm{Dom}(\mathbf{l})$. Hence x belongs to $\mathbf{C}(d-s\varphi)+\tau^\sharp(d)\mathrm{Dom}(\mathbf{l})$ and thus,

$$\mathrm{Dom}(V(d,\cdot)) \subset \mathbf{C}(d - \tau^\sharp(d)\varphi) + \tau^\sharp(d)\mathrm{Dom}(\mathbf{l})$$

This completes the characterization of the domain of the traffic solution. □

13.10 Barron–Jensen/Frankowska Viscosity Solution

This (technical) section is devoted to the proof that the viability solution defined through a viable-capture basin is actually the unique solution to the structured Hamilton-Jacobi equation. This is done by translating the Frankowska property characterizing a viable-capture basin of a target at the level of differential inclusions in terms of gradients (or subdifferential) of the viability solutions.

For simplicity of the exposition, we begin with the assumption that the viability solution is differentiable, first in the case classical case without viability constraint, easier to formulate, next, under viability constraint, before dropping the differentiability assumption and proving that the viability solution is the Barron–Jensen/Frankowska Viscosity Solution (Theorem 13.10.3, p. 560).

Proposition 13.10.1 ***[Hamilton–Jacobi Equation without Viability Constraints]*** *We assume, for simplicity, that the viability solution is differentiable. We posit the following assumptions:*

1. *the Lagrangian is Marchaud,*
2. φ, $(d,x,v) \mapsto \mathbf{l}(d,x;v)$ *and the set-valued map* $(d,x) \rightsquigarrow F(d,x)$ *are Lipschitz.*

Then the viability solution is the unique function V solution to the Hamilton–Jacobi equation (13.9), p. 530

$$\text{if } V(d,x) < \mathbf{c}(d,x), \text{ then } \left\langle \frac{\partial V(d,x)}{\partial d}, \varphi(d) \right\rangle + \mathbf{l}^\star\left(d,x; \frac{\partial V(d,x)}{\partial x}\right) = 0$$

satisfying conditions $V(d,x) \leq \mathbf{c}(d,x)$.
The regulation map satisfies

$$\text{if } V(d,x) < \mathbf{c}(d,x), \text{ then } R_V(d,x) \subset \partial_p \mathbf{l}^\star\left(d,x; \frac{\partial V(d,x)}{\partial x}\right)$$

This is a consequence of the following theorem with state constraints:

Theorem 13.10.2 ***[Hamilton–Jacobi Equation with Viability Constraints]*** *We assume, for simplicity, that the viability solution is differentiable. We posit the following assumptions:*

1. *the Lagrangian is Marchaud,*
2. *φ, $(d,x,v) \mapsto \mathbf{l}(d,x;v)$ and the set-valued map $(d,x) \rightsquigarrow F(d,x)$ are Lipschitz*
3. *the viability constraint function* $\mathbf{k}$ *is differentiable and satisfies*

$$\left\langle \frac{\partial \mathbf{k}(d,x)}{\partial d}, \varphi(d) \right\rangle + \mathbf{l}^\star\left(d,x;\frac{\partial \mathbf{k}(d,x)}{\partial x}\right) \leq 0$$

Then the viability solution is the unique function V satisfying conditions (13.10), p. 530

$$\mathbf{k}(d,x) \leq V(d,x) \leq \mathbf{c}(d,x)$$

and solution to the Hamilton–Jacobi equation (13.9), p. 530 in the sense that

$$\begin{cases} (i) \text{ if } \mathbf{k}(d,x) < V(d,x) < \mathbf{c}(d,x), \text{ then} \\ \quad \left\langle \frac{\partial V(d,x)}{\partial d}, \varphi(d) \right\rangle + \mathbf{l}^\star\left(d,x;\frac{\partial V(d,x)}{\partial x}\right) = 0 \\ (ii) \text{ if } \mathbf{k}(d,x) = V(d,x) \leq \mathbf{c}(d,x), \text{ then} \\ \quad \left\langle \frac{\partial V(d,x)}{\partial d}, \varphi(d) \right\rangle + \mathbf{l}^\star\left(d,x;\frac{\partial V(d,x)}{\partial x}\right) \leq 0 \end{cases} \tag{13.59}$$

When $\mathbf{k}(d,x) < V(d,x)$, the regulation map satisfies

$$\text{if } \mathbf{k}(d,x) < V(d,x) < \mathbf{c}(d,x), \text{ then } R_V(d,x) \subset \partial_p \mathbf{l}^\star\left(d,x;\frac{\partial V(d,x)}{\partial x}\right) \tag{13.60}$$

This formula can be written

$$\text{if } \mathbf{k}(d,x) < V(d,x) < \mathbf{c}(d,x), \text{ then } \frac{\partial V(d,x)}{\partial x} \in \bigcap_{u \in R_V(d,x)} \partial_u \mathbf{l}\,(d,x;u) \tag{13.61}$$

and regarded as an extension of the Lax–Oleinik formula *to this general case.*

Proof. Actually, we shall derive the equality in formula (13.59), p. 558 from two inequalities, the first one valid when the Lagrangian is Marchaud, the second one under the Lipschitz conditions.

1. Since the Lagrangian is Marchaud, so is the set-valued map $\mathcal{F}$ defined by (13.18), p. 534 and, by (13.22), p. 537, the viable-capture basin

$\mathcal{E}p(V) := \text{Capt}_{(13.10)}(\mathcal{E}p(\mathbf{k}), \mathcal{E}p(\mathbf{c}))$ is equal to $\text{Capt}_{(13.17)}(\mathcal{E}p(\mathbf{k}), \mathcal{E}p(\mathbf{c})) = \text{Capt}_{(13.10)}(\mathcal{E}p(\mathbf{k}), \mathcal{E}p(\mathbf{c}))$.
It is then the largest closed subset between $\mathcal{E}p(\mathbf{c})$ and $\mathcal{E}p(\mathbf{k})$, locally viable in $\mathcal{E}p(V) \setminus \mathcal{E}p(\mathbf{c})$ thanks to the Viability Theorem 11.4.6, p. 463, which also states that, whenever $(d, x, V(d, x)) \in \mathcal{E}p(V) \setminus \mathcal{E}p(\mathbf{c})$, i.e., whenever $V(d, x) < \mathbf{c}(d, x)$, there exists some $v \in \text{Dom}(\mathbf{l})$ such that

$$(-\varphi(d), -v, -\mathbf{l}(d, x; v)) \in T_{\mathcal{E}p(V)}(d, x, V(d, x))$$

Recall that $N_{\mathcal{E}p(V)}(d, x, V(d, x)) := T^{\star}_{\mathcal{E}p(V)}(d, x, V(d, x))$ and that *the subdifferential* $\partial V(d, x)$ *is the set of* (p_d, p_x) *such that* $(p_d, p_x, -1) \in N_{\mathcal{E}p(V)}(d, x, V(d, x))$.
Therefore, we infer that

$$\forall\, v \in R_V(d, x),\ \forall\, (p_d, p_x) \in \partial V(d, x),\ \ 0 \leq \langle p_d, \varphi(d)\rangle + \langle p_x, v\rangle - \mathbf{l}(d, x; v) \tag{13.62}$$

and thus, that, by taking the supremum over $F(d, x) := \text{Dom}(\mathbf{l}(d, x; \cdot))$,

$$\forall\, (p_d, p_x) \in \partial V(d, x),\ \ 0 \ \leq\ \langle p_d, \varphi(d)\rangle + \mathbf{l}^{\star}(d, x; p_x) \tag{13.63}$$

2. Since $(d, x, v) \mapsto \mathbf{l}(d, x; v)$ and the set-valued map $(d, x) \rightsquigarrow F(d, x)$ are Lipschitz. it is easy to observe that the set-valued map $(d, x) \rightsquigarrow \mathcal{E}p(\mathbf{l}(d, x; \cdot))$ is Lipschitz. The viable-capture basin $\mathcal{E}p(V) := \text{Capt}_{(13.10)}(\mathcal{E}p(\mathbf{k}), \mathcal{E}p(\mathbf{c}))$ is the smallest closed subset between $\mathcal{E}p(\mathbf{c})$ and $\mathcal{E}p(\mathbf{k})$ backward invariant relatively to $\mathcal{E}p(\mathbf{k})$.
If we assume that $\mathcal{E}p(\mathbf{k})$ is itself backward invariant, then $\mathcal{E}p(V)$ is also backward invariant. It remains to translate properties of the Invariance Theorem 11.4.6, p. 463 in terms of subdifferentials.
Backward invariance of the epigraph $\mathcal{E}p(V)$ means that whenever $(d, x, V(d, x)) \in \mathcal{E}p(V)$, then

$$\forall v \in F(d, x), \forall (p_d, p_x) \in \partial V(d, x), (\varphi(d), v, \mathbf{l}(d, x; v)) \in T_{\mathcal{E}p(V)}(d, x, V(d, x))$$

This amounts to saying that

$$\forall\, v \in F(d, x),\ \forall\, (p_d, p_x) \in \partial V(d, x),\ \ \langle p_d, \varphi(d)\rangle +\ \langle p_x, u\rangle - \mathbf{l}(d, x; u)\ \leq\ 0$$

and thus, that

$$\forall\, (p_d, p_x) \in \partial V(d, x),\ \ \langle p_d, \varphi(d)\rangle + \mathbf{l}^{\star}(d, x; p_x)\ \leq\ 0 \tag{13.64}$$

In the same way, the epigraph of $\mathbf{k}$ is backward invariant if

$$\forall\, (q_d, q_x) \in \partial \mathbf{k}(d, x),\ \ \langle q_d, \varphi(d)\rangle + \mathbf{l}^{\star}(d, x; q_x)\ \leq\ 0$$

If $\mathbf{k}(d,x) < V(d,x) < \mathbf{c}(d,x)$, inequality (13.63), p. 559 and (13.64), p. 559 imply that the viability solution satisfies

$$\forall\, (p_d, p_x) \in \partial V(d,x),\;\; \langle p_d, \varphi(d)\rangle + \mathbf{l}^\star\,(d,x;p_x) \;=\; 0 \tag{13.65}$$

Taking into account that $\langle p_d, \varphi(d)\rangle = -\mathbf{l}^\star(d,x;p_x)$ in inequality (13.62), p. 559, we infer that

$$\forall\, v \in R_V(d,x),\; \forall\, (p_d,p_x) \in \partial V(d,x),\;\; 0 \;\leq\; \langle p_x, v\rangle - \mathbf{l}^\star\,(d,x;p_x) - \mathbf{l}(d,x;v)$$

The Legendre property (13.7), p. 529 of the Fenchel transform implies that this is equivalent to saying that $v \in \partial_p \mathbf{l}^\star(d,x;p_x)$ or that $p_x \in \partial \mathbf{l}_u(d,x;u)$. Consequently:

$$R_V(d,x) \;\subset\; \bigcap_{(p_d,p_x)\in \partial V(d,x)} \partial_p \mathbf{l}^\star(d,x;p_x)$$

or, equivalently,

$$\bigcup_{(p_d,p_x)\in \partial V(d,x)} p_x \;\subset\; \bigcap_{u\in R_V(d,x)} \partial_u \mathbf{l}(d,x;u)$$

Assuming that V is differentiable, we have proved the formulas (13.59), p. 558 and (13.60), p. 558 and, consequently, that the viability solution is the unique solution to the structured Hamilton–Jacobi equation. □

Under the assumptions of Theorem 13.10.2, p. 558, when the viability solution V is not differentiable, it is still the unique solution satisfying (13.65), p. 560:

$$\forall\, (p_d, p_x) \in \partial V(d,x),\;\; \langle p_d, \varphi(d)\rangle + \mathbf{l}^\star\,(d,x;p_x) \;=\; 0$$

Theorem 13.10.3 *[Barron–Jensen/Frankowska Viscosity Solution] We posit the following assumptions:*

1. *the Lagrangian is Marchaud,*
2. *φ, $(d,x,v) \mapsto \mathbf{l}(d,x;v)$ and the set-valued map $(d,x) \rightsquigarrow F(d,x)$ are Lipschitz*
3. *The viability constraint function $\mathbf{k}$ satisfies*

$$\forall\, (q_d, q_x) \in \partial \mathbf{k}(d,x),\;\; \langle q_d, \varphi(d)\rangle + \mathbf{l}^\star\,(d,x;q_x) \;\leq\; 0 \tag{13.66}$$

Then the viability solution is the unique function V satisfying conditions (13.10), p. 530

$$\mathbf{k}(d,x) \;\leq\; V(d,x) \;\leq\; \mathbf{c}(d,x)$$

and solution to the Hamilton–Jacobi equation (13.9), p. 530 in the sense that

$$\begin{cases} (i) \text{ if } \mathbf{k}(d,x) \; < \; V(d,x) \; < \; \mathbf{c}(d,x), \text{ then} \\ \quad \forall\,(p_d,p_x) \in \partial V(d,x), \;\; \langle p_d, \varphi(d)\rangle + \mathbf{l}^\star(d,x;p_x) \; = \; 0 \\ (ii) \text{ if } \mathbf{k}(d,x) \; = \; V(d,x) \; \leq \; \mathbf{c}(d,x), \text{ then} \\ \quad \forall\,(p_d,p_x) \in \partial V(d,x), \;\; \langle p_d, \varphi(d)\rangle + \mathbf{l}^\star(d,x;p_x) \; \leq \; 0 \end{cases} \tag{13.67}$$

which is the very definition of a Barron–Jensen/Frankowska viscosity solution for lower semicontinuous functions.
Furthermore, when $\mathbf{k}(d,x) < V(d,x)$*, the regulation map and the subdifferential of the viability solution with respect to* x *are related by*

$$\begin{cases} R_V(d,x) \;\subset\; \bigcap_{(p_d,p_x)\in\partial V(d,x)} \partial_p \mathbf{l}^\star(d,x;p_x) \\ \text{or, equivalently,} \\ \bigcup_{(p_d,p_x)\in\partial V(d,x)} \{p_x\} \;\subset\; \bigcap_{u\in R_V(d,x)} \partial_u \mathbf{l}(d,x;u) \end{cases} \tag{13.68}$$

Remark. The theorem holds true by dropping assumption (13.66), p. 560, but the conclusion translating the property that the epigraph of the viability solution is backward invariant relatively to the epigraph of $\mathbf{k}$ when $V(d,x) = \mathbf{k}(d,x)$ is quite technical and ugly:

$$\begin{cases} \text{if } \mathbf{k}(d,x) \; = \; V(d,x) \; \leq \; \mathbf{c}(d,x), \text{ then } \forall\,(p_d,p_x) \in \partial V(d,x), \\ \inf_{(q_d,q_x)\in\partial\mathbf{k}(d,x)}(\langle p_d - q_d, \varphi(d)\rangle + \mathbf{l}^\star(d,x;p_x - q_x)) \; \leq \; 0 \end{cases} \tag{13.69}$$

□

Chapter 14
Regulation of Traffic

14.1 Introduction

The advent of techniques to measure velocities of probe vehicles using GPS technology, for instance, complementing or replacing fixed sensing infrastructures such as density sensors of the road traffic sensors, motivates the revision of conceptual, mathematical algorithms and software based models used by the transportation engineering community. *Bruce Greenshields* used in 1933 photographic measurement methods for the first time to describe a *phenomenological law* described by a quadratic relation between vehicles and their density and flows, called the *fundamental diagram.*

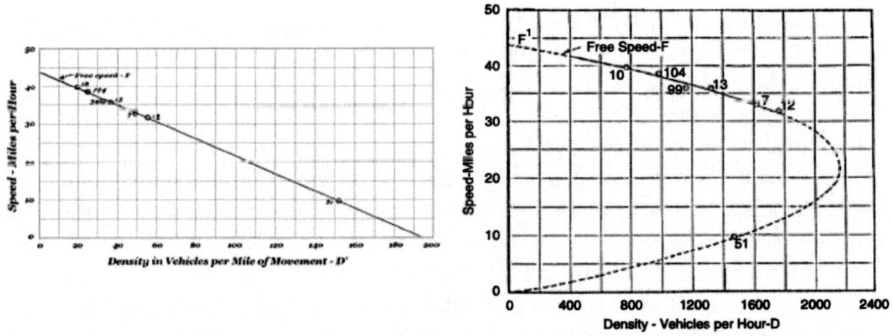

Fig. 14.1 Original 1933 Greenshields' Fundamental Diagrams. *The two first diagrams are the historical Greenshields' diagrams.*

Later, in the middle of the years 1950, Lighthill, Whitham and Richards, proposed a *partial differential equation* (conservation law) with concave flow function, the solution of which is the density of traffic at each time and at each position (see Chap. 16, p. 631).

J.-P. Aubin et al., *Viability Theory*, DOI 10.1007/978-3-642-16684-6_14,

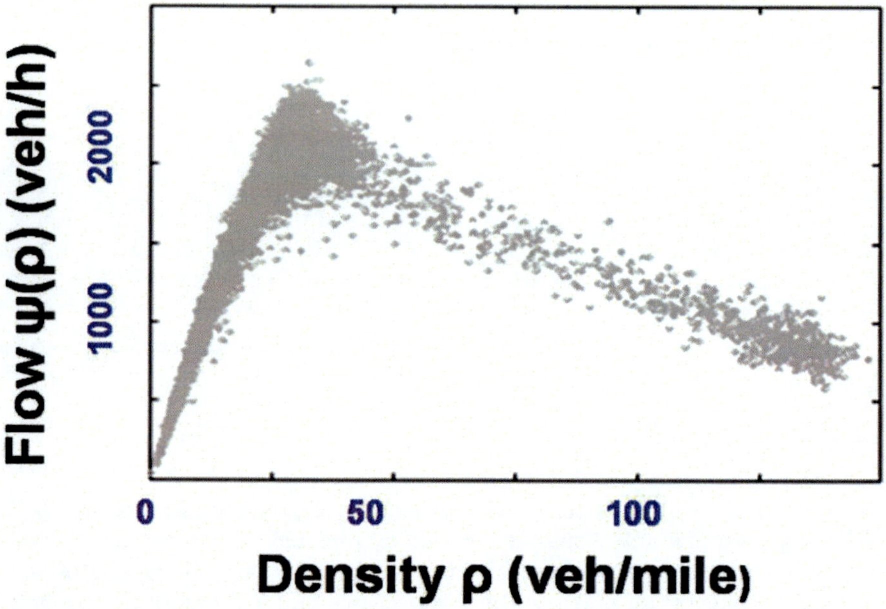

Fig. 14.2 Fundamental Diagram. *The fundamental diagram displays the physical relation between densities (abscissa) and the flow (ordinate). It increases from zero density to congestion density, and next decreases until jam density. The figure provides an empirical example of fundamental diagram.*

This attempt to characterize the behavior of congested traffic became the seminal model for numerous highway traffic flow studies available in the traffic engineering literature today.

Thirty five years later, *Gordon Newell* introduced the concept of "cumulative number" of vehicles passing at given position after a given time on a one-dimensional road, since density was the prevalent concept at the time. He acknowledged that *Karl Moskowitz*, an engineer from the California Department of Transportation who did not bother to publish, used this concept for some time to investigate properties of traffic.

It is thus convenient to call this specific Hamilton–Jacobi equation the "Moskowitz equation". Since data from the sensors can be written as Cauchy or Dirichlet boundary conditions, they were sufficient to determine a solution by classical methods.

Next, J.C. Luke and Newell discovered that these cumulative number functions are solutions to this Moskowitz equation and a *variational principle.* Daganzo, who took over these equations for studying them mathematically in 2004, wrote: *"Luke (1973) and Newell (1993) proposed the minimum operation as a way of selecting the unique and correct value at every point in space-time without proving it. It should be remembered in this respect that a "correct", i.e., physically meaningful, solution of the problem [...].* He also wrote in another paper that *"The least cost to reach a point is the vehicle number"*.

The "congestion variational principle" discovered by these authors states that *the solution to the Moskowitz equation is the value function of an optimal control problem minimizing an underlying (and hidden) congestion functional.* This suggests to regard this solution as a *congestion function* to stress this property. The minimal congestion evolutions are referred to as *"kinematic waves"* in the traffic engineering literature: we shall call them *minimal congestion evolutions* to underline their role. Also, we provide and characterize the *minimal congestion regulation map*, which is the set of (optimal) "controls" that minimal congestion evolutions must use to be "consistent" with the congestion function.

We give the name *"celerity"* to such a control and explain its role: this celerity is a form of "macroscopic velocity" attached to the road at each time and each position providing the optimal velocity that vehicles should follow for minimizing congestion, but not to a "microscopic velocity" attached to an arbitrary vehicle (the celerity is not the aggregated velocity commonly used in traffic flow engineering, which shares also a macroscopic property). The celerity is derived from the knowledge of the fundamental diagram and the traffic conditions. However, the celerity becomes the velocity of the optimal evolution (wave) of a vehicle at this time and at this position: using at this time and this position this celerity as the velocity, the minimal congestion regulation map provides the kinematic differential equation governing the evolution of this minimal congestion evolution. The celerity informs each driver to use the macroscopic celerity as the microscopic velocity allowing each driver. The celerity provided in a centralized way by the minimal congestion regulation map is used in a decentralized way as vehicle velocities to govern minimal congestion evolutions consistent with the minimal congestion function.

The issue became radically different when the traffic conditions are no longer classical boundary conditions, but are given along trajectories of probe vehicles in the interior of the domain: they are called "Lagrangian" conditions.

The Lagrangian data situation is even more complicated because the density measurements by fixed sensors have to be replaced by measurements of the velocities of vehicles provided by sensors *only on subsets of their trajectories.*

However, the other mathematical information obtained with the nature of the Lagrangian data involved is the Fenchel transform (see Definition 18.7.2, p.756) of the fundamental diagram. It is a convex decreasing function mapping maximum celerity to zero flow and null celerity to maximum flow. Its graph can be considered as a dual "celerity diagram", the fundamental diagram being regarded as a "density diagram".

The partial derivatives with respect to time and positions of the congestion function are interpreted respectively as the flow (or flux) of the traffic and its density. Hence the congestion function can be interpreted either as the Moskowitz–Newell cumulated number function, since it also measures the cumulated number of vehicles ahead of position x at time t, or the number of vehicles at position x before the time t, or the congestion of a given evolution on an interval, as well as many other interpretations, as we shall see in Lemma 14.2.2, p.567.

In this Chapter, we present both what could be called the *density paradigm* and the *celerity paradigm*.

The density paradigm is convenient whenever traffic is measured by densities, and naturally yields a congestion function as a solution to the Moskowitz partial differential equation, with initial and boundary conditions.

The celerity paradigm is mandatory whenever traffic is measured by celerities (traffic velocities), and naturally yields the congestion function as the value function of an optimal control problem on the evolutions and its velocities.

These two are equivalent, but these two dual approaches provide different sets of results. Viability theory and viability algorithms provide *at the same time*, and for free, *both the macroscopic or centralized congestion function and the minimal congestion regulation map*. Consequently, it dictates the optimal evolutions of each vehicle minimizing congestion.

In a nutshell, the viability approach provides the following results:

1. knowing the celerity diagram, calculated either through an empirical fundamental density diagram or directly through an empirical celerity diagram,
2. knowing the evolution of a finite number of probe vehicles,
3. the viability algorithms provide the minimal congestion regulation map, indicating to each vehicle, passing at given time through a given position, the velocity they must choose for minimizing congestion.

Knowing this minimal congestion regulation map, it is possible to reconstruct minimal congestion evolutions of vehicles, to compute their trajectories, to calculate their travel time, etc.

14.2 The Transportation Problem and its Viability Solution

We consider a single lane road where vehicles cannot pass each other. It is represented by an interval $[\underline{\gamma}, \overline{\gamma}]$. We can also choose $]-\infty, +\infty[$.

Definition 14.2.1 ***[Traffic Function, Density and Flux]*** *A differentiable* congestion function $V : (t, x) \mapsto V(t, x)$ *is a function where the first derivatives with respect to time and to position are respectively regarded as a flow (or a flow) and a density:*

$$\text{flow} = \frac{\partial V(t,x)}{\partial t} \text{ and density } = -\frac{\partial V(t,x)}{\partial x}$$

When the congestion function is not differentiable, one can still define generalized gradients $(p_t, p_x) \in \partial_- V(t,x)$ *representing pairs of flows and densities at* (t,x). *For any time* t, *the function* $V(t,\cdot) : x \mapsto V(t,x)$ *is called the* traffic profile *at time* t.

Thus the congestion function has different physical interpretations:

Lemma 14.2.2 ***[Physical Interpretations of the Congestion Function]*** *Knowing a congestion function* $(t,x) \mapsto V(t,x)$, *assumed to be differentiable for simplicity, we infer that*

$$\begin{cases} V(t,x) - V(t,\overline{\gamma}) := -\displaystyle\int_x^{\overline{\gamma}} \frac{\partial V(t,\xi)}{\partial x} d\xi \\ \text{is the incremental congestion on the interval } [x,\overline{\gamma}] \text{ at time } t \\ \text{representing the "cumulated vehicle count" after position } x\text{"} \\ V(t,x) - V(d,x) := \displaystyle\int_d^t \frac{\partial V(\tau,x)}{\partial t} d\tau \\ \text{is the incremental congestion on the interval } [d,t] \text{ at position } x \\ V(t,x) - V(d,\xi(d)) := \displaystyle\int_d^t \frac{dV(\tau,\xi(\tau))}{dt} d\tau \\ \text{is the incremental congestion on the interval } [d,t] \\ \text{along an evolution } \xi(\cdot) \text{ arriving at } x \text{ at time } t\text{: } \xi(t) = x \end{cases}$$

Congestion functions are decreasing functions of position and increasing functions of time, because, the shorter the time and the farther the distance, the smaller the number of cumulated vehicles, since they cannot pass each other.

Actually, we shall impose two types of conditions on the congestion function, a *phenomenological law* proposed by Greenshields, on one hand, and conditions on the congestion functions provided by several types of sensors (Eulerian for fixed sensors, Lagrangian for mobile ones). Instead of the conservation law (see Chap. 16, p.631) proposed by Lighthill, Whitham and Richards in the 1950s under the name of "LWR partial differential equation" providing the traffic density, we study here its Hamilton–Jacobi version providing the traffic congestion suggested by Moskowitz:

1. **The "Fundamental Diagram"** Lighthill, Whitham and Richards conservation law and Moskowitz's Hamilton–Jacobi equation *assume* that at (t,x), the *traffic density* $-\frac{\partial V(t,x)}{\partial x}$ and the *flow* $\frac{\partial V(t,x)}{\partial t}$ of the congestion function are related by the "*fundamental diagram*".

Definition 14.2.3 *[The Fundamental Diagram] The* fundamental density diagram *of transportation engineering is the graph* Graph($\mathbf{h}$) *of a density-flow function* $\mathbf{h}$ *associating with each density a flow. This graph is, in practice, empirically measured. The function* $\mathbf{h}$ *is the "Hamiltonian" governing the evolution of the congestion function through the Moskowitz partial differential equation*

$$\frac{\partial V(t,x)}{\partial t} = \mathbf{h}\left(-\frac{\partial V(t,x)}{\partial x}\right)$$

2. **Traffic Conditions**
Traffic conditions are prescribed by functions $\mathbf{c} : (t,x) \mapsto \mathbf{c}(t,x) \in \mathbb{R}_+ \cup \{+\infty\}$, with which we associate the *traffic domain map* $t \rightsquigarrow \mathbf{C}(t)$, which is the set-valued map defined by

$$\mathbf{C}(t) := \{x \text{ such that } \mathbf{c}(t,x) < +\infty\} \tag{14.1}$$

because the subset $\mathbf{C}(t)$ is the domain of the traffic profile $x \mapsto \mathbf{c}(t,x)$ (which is thus finite for all $x \in \mathbf{C}(t)$).
Traffic conditions require that

$$\forall\, x \in \mathbf{C}(t),\;\; V(t,x) \leq \mathbf{c}(t,x) \tag{14.2}$$

These general traffic conditions cover the following classical and less classical examples:

a. *Dirichlet boundary conditions* defined on the boundary of the domain. In this case, we set $\mathbf{c}(0,x) := \gamma_0(x)$ and $\mathbf{c}(t,x) = +\infty$ whenever $t > 0$. Hence $\mathbf{C}(0) = \text{Dom}(\gamma_0)$ and $\mathbf{C}(t) = \emptyset$,
b. *Eulerian conditions* measuring at each time the cumulative number of vehicles at *fixed* locations,
c. *Lagrangian "mobile" conditions* measuring few vehicle evolutions $t \mapsto \gamma_i(t)$ during some time intervals $]\underline{T}_i, \overline{T}_i]$ (by tracking probe vehicles). In the case of a finite number of Lagrangian conditions, $\mathbf{C}(t) = \bigcup_{i \text{ such that } \underline{T}_i \leq t \leq \overline{T}_i} \{\gamma_i(t)\}$ and the traffic condition $\mathbf{c}(t, \gamma_i(t))$ is defined on the graphs $(t, \gamma_i(t))_{t \geq 0}$ of the evolution of each vehicle i.

The traffic condition $\mathbf{c}$ is assumed to be decreasing with position and increasing with time. It can be used for labelling vehicles in a given position at a given time.
For initial conditions, the profile $x \mapsto \mathbf{c}(0,x)$, regarded as a cumulative number count after x, is a decreasing function which is usually for labelling vehicles and track their future.

For a fixed position x, the function $t \mapsto \mathbf{c}(t,x)$ measures the congestion of a vehicle at position x before time t provides a complementary label of a vehicle at position x departing at time t.

Definition 14.2.4 *[Canonical Labelling of a Probe Vehicle] Knowing*

a. the evolution $t \mapsto \gamma(t)$ of a probe vehicle on an interval $[d,T]$ where $t \geq d \geq 0$ and its velocity $t \mapsto \gamma'(t)$;
b. a traffic condition $(t,x) \mapsto \mathbf{c}(t,x)$ assumed to be decreasing with position and increasing with time,
c. the Lagrangian $\mathbf{l}$ defining the celerity diagram,

the traffic function $(t,x) \mapsto \mathbf{c}_{\gamma(\cdot)}(t,x)$ assigned to the evolution of the probe vehicle $\gamma(\cdot)$ on the interval $[d,T]$ by the formula

$$\begin{cases} \mathbf{c}_{\gamma(\cdot)}(t,\gamma(t)) := \mathbf{c}(d,\gamma(d)) + \max\left(0, \int_d^t \mathbf{l}(\gamma'(\tau)d\tau)\right) & \text{if } x = \gamma(t) \\ +\infty & \text{if } x \neq \gamma(t) \end{cases} \tag{14.3}$$

provides a Lagrangian traffic condition decreasing with position and increasing with time assigned to the probe vehicle.

This definition is in accordance with Daganzo's quotation *"The least cost to reach a point is the vehicle number"*.
We shall deduce from Theorem 14.4.2, p.581 that the viability solution (see Definition 14.4.1, p.580) that the congestion function $V_{\gamma(\cdot)}$ associated with the traffic condition $\mathbf{c}_{\gamma(\cdot)}$ satisfies the Lagrangian condition

$$\forall\, t \in [d,T], \;\; V_{\gamma(\cdot)}(t,\gamma(t)) \;=\; \mathbf{c}_{\gamma(\cdot)}(t,\gamma(t))$$

We observe that

a. if $T = d$, i.e., if the position $\gamma(d)$ of the probe vehicle is observed only at departure time d, then

$$\mathbf{c}_{\gamma(\cdot)}(d,\gamma(d)) \;=\; \mathbf{c}(d,\gamma(d))$$

b. If the probe vehicle is stationary for $t \geq d$, then

$$\mathbf{c}_{\gamma(\cdot)}(t,\gamma(d)) \;=\; \mathbf{c}(d,\gamma(d)) + \delta t$$

where $\delta = \mathbf{l}(0)$ is the maximal flow.

3. The Moskowitz Problem

Definition 14.2.5 *[The Moskowitz Problem] The complete model takes into account the two above requirements on the congestion function V:*

$$\begin{cases} (i)\ \forall\, t > 0, \forall\, x \notin \mathbf{C}(t), \ \mathbf{h}\left(-\frac{\partial V(t,x)}{\partial x}\right) = \frac{\partial V(t,x)}{\partial t} \\ (ii)\ \forall\, t > 0, \forall\, x \in \mathbf{C}(t), \ V(t,x) \le \mathbf{c}(t,x) \end{cases} \tag{14.4}$$

Even though some hard work for translating viability theorems in terms of partial differential equation is needed, the link between the concept of solution to the Moskowitz problem and capture basins allows us to prove the following properties of the congestion function V. We shall prove that the congestion function is a solution to the Moskowitz problem (14.4), p. 570 in a weak sense (viability solution) and "provides" the minimal congestion regulation map. Next, we shall prove that the congestion function V satisfies the following properties

- It is lower semicontinuous and is given by the Lax–Hopf formula (Theorem 14.4.3, p.583);
- It is the value function of a variational problem (Theorem 14.4.2, p.581);

In addition, the following properties hold:

- There exist minimal congestion evolutions satisfying the dynamic programming equations (Theorem 14.4.4, p.583);
- The domain of the congestion function V is characterized by an explicit formula (Theorem 14.4.5, p.584);
- The congestion function associated with a traffic condition which is the minimum of a finite family of traffic conditions is the minimum of the congestion functions associated with each traffic condition (Theorem 14.7.1, p.595);
- If the traffic condition is decreasing in position and increasing in time, so is the congestion function V. If the traffic evolutions involved in the Lagrangian conditions are increasing, so are the minimal congestion evolutions (Theorem 14.7.2, p.596).

For Cauchy and Lagrangian traffic conditions, the congestion function V can be expressed analytically and estimated below and above. Furthermore, the viability algorithms allow us to compute the congestion function and to regulate optimal evolutions.

14.3 Density and Celerity Flux Functions

In the Moskowitz framework, the Hamiltonian $\mathbf{h} : \mathbb{R} \mapsto \mathbb{R}$ is the concave density flow function the graph of which is called the "fundamental diagram". We shall classify them according to characteristic parameters $(\nu^\flat, \nu^\sharp, \omega, \delta)$ where

- at *null density* $p = 0$, the flow is equal to $\mathbf{h}(0) = 0$ and the celerity $\mathbf{h}'(0) := \nu^\flat \geq 0$, or, in the non differentiable case, $\nu^\flat \in \partial_+\mathbf{h}(0)$,
- at *jam density* $p = \omega$, the flow is again equal to 0 and the celerity $\mathbf{h}'(\omega) := -\nu^\sharp$ where $\nu^\sharp \geq 0$, or, in the non differentiable case, $-\nu^\sharp \in \partial_+\mathbf{h}(\omega)$,
- the *maximal flow* $\delta := \max_{p\in[0,\omega]} \mathbf{h}(p)$.

The interval $[\beta^\flat, \beta^\sharp]$ where $\mathbf{h}(p) = \delta$ of densities on which the flow functions reaches its maximum is called the *critical density interval.* If $\beta^\flat = \beta^\sharp$, then the common value is denoted by $\beta := \beta^\flat = \beta^\sharp$ and called the *critical density.*

Definition 14.3.1 ***[Flux Function]*** *We associate with scalars* $\nu^\flat \geq 0$, $\nu^\sharp \geq 0$, $\omega > 0$ and $\delta > 0$ *the class of* flow functions $\mathbf{h} : \mathbb{R} \mapsto \mathbb{R}$ *associated with those parameters which is any concave function* $\mathbf{h}$ *satisfying*

$$\begin{cases} \mathbf{h}(p) := \nu^\flat p & \text{if } p \in]-\infty, 0] \\ \mathbf{h}(p) \in [0, \delta] & \text{if } p \in [0, \omega] \\ \mathbf{h}(p) := \nu^\sharp(\omega - p) & \text{if } p \in [\omega, +\infty[\end{cases} \quad (14.5)$$

and

$$\begin{cases} (i)\ \mathbf{h}(0) = 0 \text{ and } \mathbf{h}'(0) = \nu^\flat \\ (ii)\ \mathbf{h}(\omega) = 0 \text{ and } \mathbf{h}'(\omega) = -\nu^\sharp \end{cases} \quad (14.6)$$

For solving the Moskowitz problem by viability techniques, we add to the concepts of density and flow the concept of "celerity". It has the dimension (in the physical sense) of a velocity, measuring, so to speak, the macroscopic velocity of the traffic. Here, the *density* and *celerity* are regarded as "dual variables". Their product, a *flow,* plays the role of *duality product,* as in mechanics, where position and velocity are dual variable the duality product of which is the power, or, in economics, commodity and price are dual variable the duality product of which is the value of the commodity.

Definition 14.3.2 ***[The Celerity Diagram]*** *The celerity diagram is the celerity-flow function* $\mathbf{l}$ *defined by*

$$\forall u, \quad \mathbf{l}(u) := \sup_{p} [\mathbf{h}(p) - \langle p, u \rangle] \tag{14.7}$$

Its graph Graph(**l**) *is called the fundamental* celerity diagram.

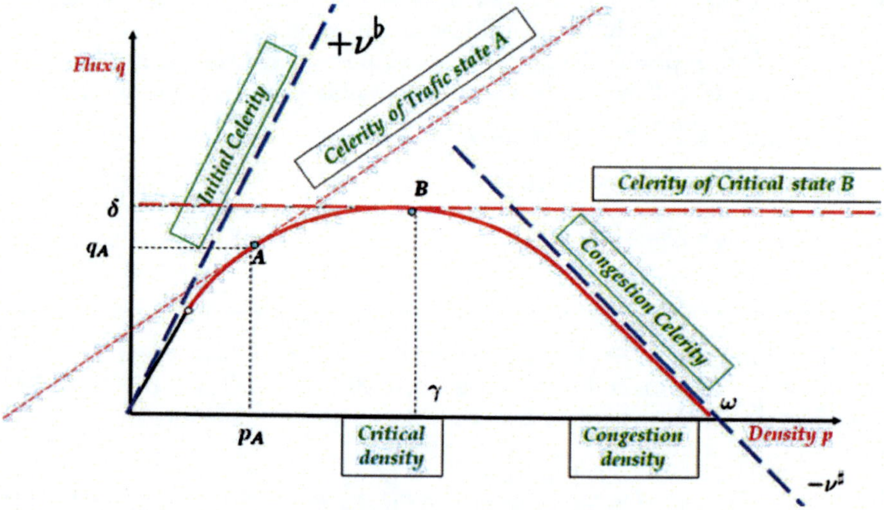

Fig. 14.3 Fundamental density Diagram.
Illustration of the concept of celerity, here, the derivative of the Hamiltonian, for densities at density equal to 0, to A, at critical density β and at jam density ω. The supremum δ of the Hamiltonian is the maximal flow.

We recall that the introduction of the celerity function is "mathematically" natural in the framework of duality theory in mechanics, in economics and in convex analysis (see Sect. 18.7, p. 755). The Fenchel Theorem 18.7.3, p.756, states that there exists a bijective correspondence between lower semicontinuous convex functions defined on a vector space and their conjugate functions defined on the dual. Here, there is a slight adaptation to perform, since the Hamiltonian **h** is concave and the Lagrangian **l** is convex. But the adaptation to this situation, tedious as it is, poses no problem: the following statements adapt the ones of Sect. 18.7, p. 755 to our specific case: conjugate functions are defined in Definition 18.7.2, p.756 and subdifferential in Definition 18.7.5, p.758. Fenchel Theorem 18.7.3, p.756 and Theorem 18.7.7, p.759 on the Legendre property of subdifferential of convex functions and their conjugate imply

Lemma 14.3.3 *[Celerity and Density Flux Functions] Assume that the Hamiltonian* **h** *is concave and upper semicontinuous. Then its*

Lagrangian $\mathbf{l}$ *defined by (14.7), p. 572*

$$\forall\, u,\ \mathbf{l}(u) := \sup_{p \in \mathrm{Dom}(\mathbf{h})} [\mathbf{h}(p) - \langle p, u\rangle]$$

is lower semicontinuous and convex. The Hamiltonian is related to the Lagrangian by the relation

$$\forall\, p,\ \mathbf{h}(p) = \inf_{u \in \mathrm{Dom}(\mathbf{l})} [\mathbf{l}(u) + \langle p, u\rangle] = -\mathbf{l}^\star(-p) \tag{14.8}$$

so that the Fenchel inequality

$$\forall\, u,\ \forall\, p,\ \langle p, u\rangle \leq \mathbf{l}(u) - \mathbf{h}(-p)$$

always holds true. Furthermore, $\overline{u}$ *achieves the minimum in the minimization problem* $\mathbf{h}(p) = \mathbf{l}(\overline{u}) + \langle p, \overline{u}\rangle$ *if and only if* $\overline{u} \in \partial_+\mathbf{h}(-p)$ *and* $\overline{p}$ *achieves the maximum in* $\mathbf{l}(u) := \mathbf{h}(\overline{p}) - \langle \overline{p}, u\rangle$ *if and only if* $\overline{p} \in \partial_-\mathbf{l}(u)$. *The three following statements are thus equivalent:*

$$\begin{cases} (i) & \langle \overline{p}, \overline{u}\rangle \geq \mathbf{l}(\overline{u}) - \mathbf{h}(-\overline{p}) \\ (ii) & \overline{u} \in \partial_+\mathbf{h}(-\overline{p}) \\ (iii) & \overline{p} \in \partial_-\mathbf{l}(\overline{u}) \end{cases} \tag{14.9}$$

Moreover, whenever $\mathbf{h}_i$ *is upper semicontinuous and concave (or* $\mathbf{l}_i$ *is lower semicontinuous and convex),* $i = 1,\ 2$, *then*

$$\mathbf{h}_1 \leq \mathbf{h}_2 \text{ if and only if } \mathbf{l}_1 \leq \mathbf{l}_2$$

Proof. This lemma is a consequence of Theorems 18.7.3, p.756 and 18.7.7, p.759. □

Therefore, since $\mathbf{h}(p) = -\mathbf{l}^\star(-p)$, the traffic solution V enjoys all the property of the solutions to Hamilton–Jacobi equations investigated in Chap. 13, p.523, when the causal variable is the time $t \in \mathbb{R}_+$.

Before "translating" them in this framework, we investigate the properties of these density and celerity flow functions.

Examples of Density and Celerity Flux Functions

Before proving the properties of the density and flow functions summarized in Proposition 14.3.5, p.578, we need to single out two main classes of examples of flow functions, one empirical, made of (non differentiable) trapezoidal density flow functions, the other, theoretical, the Greenshields quadratic one, less realistic but more commonly used because of its simple analytical properties.

1. **Trapezoidal Flux and Celerity Functions.**
In this example, we fix the following data: the *celerities* $\nu^{\flat} \geq 0$ and $\nu^{\sharp} \geq 0$, the *jam density* ω, and a *maximal flow* $0 \leq \delta \leq \overline{\delta}$ where $\overline{\delta} := \dfrac{\omega\nu^{\flat}\nu^{\sharp}}{(\nu^{\flat}+\nu^{\sharp})}$.

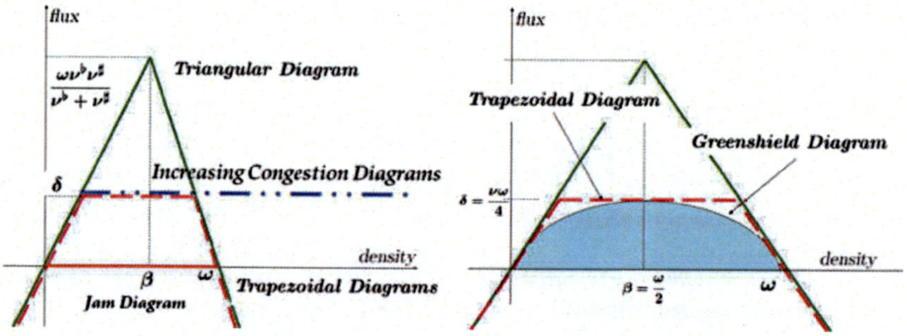

Fig. 14.4 Fundamental Trapezoidal and Greenshields Diagrams.
Left: This diagram displays the graphs of the trapezoidal, triangular, jam density and infinite congestion flow functions and other flow functions $\mathbf{h}_{(\alpha^{\sharp},\alpha^{\flat},\omega,\delta)}$ *between the jam and triangular celerity functions. Right: The Greenshields fundamental diagram and its "trapezoidal envelope" are displayed.*

Proposition 14.3.4 *[Trapezoidal Flux and Celerity Functions]*
Let us consider the trapezoidal density flow function $\mathbf{h}_{(\nu^{\flat},\nu^{\sharp},\omega,\delta)}$ *defined by*

$$\mathbf{h}_{(\nu^{\flat},\nu^{\sharp},\omega,\delta)}(p) := \begin{cases} \nu^{\flat}p & \textit{if } p \leq \beta^{\flat} \\ \delta & \textit{if } p \in [\beta^{\flat},\beta^{\sharp}] \\ \nu^{\sharp}(\omega-p) & \textit{if } p \geq \beta^{\sharp} \end{cases} \quad (14.10)$$

Its lower and upper critical densities are defined by

$$\beta^{\flat} := \frac{\delta}{\nu^{\flat}} \text{ and } \beta^{\sharp} := \frac{\nu^{\sharp}\omega-\delta}{\nu^{\sharp}}$$

The trapezoidal celerity flow function $\mathbf{l}_{(\nu^\flat,\nu^\sharp,\omega,\delta)}$ *associated with the* trapezoidal flow function *is equal to*

$$\mathbf{l}_{(\nu^\flat,\nu^\sharp,\omega,\delta)}(u) = \begin{cases} \frac{\delta}{\nu^\flat}(\nu^\flat - u) & \text{if } u \in [0,\nu^\flat] \\ \delta - \frac{\omega\nu^\sharp - \delta}{\nu^\sharp} u & \text{if } u \in [-\nu^\sharp, 0] \\ +\infty & \text{if } u \notin [-\nu^\sharp, \nu^\flat] \end{cases} \tag{14.11}$$

It is piecewise affine (affine on $[-\nu^\sharp, 0]$ *and* $[0, +\nu^\flat]$*) and satisfies* $\mathbf{l}(+\nu^\flat) = 0$, $\mathbf{l}(0) = \delta$ *and* $\mathbf{l}(-\nu^\sharp) = \omega\nu^\sharp$.

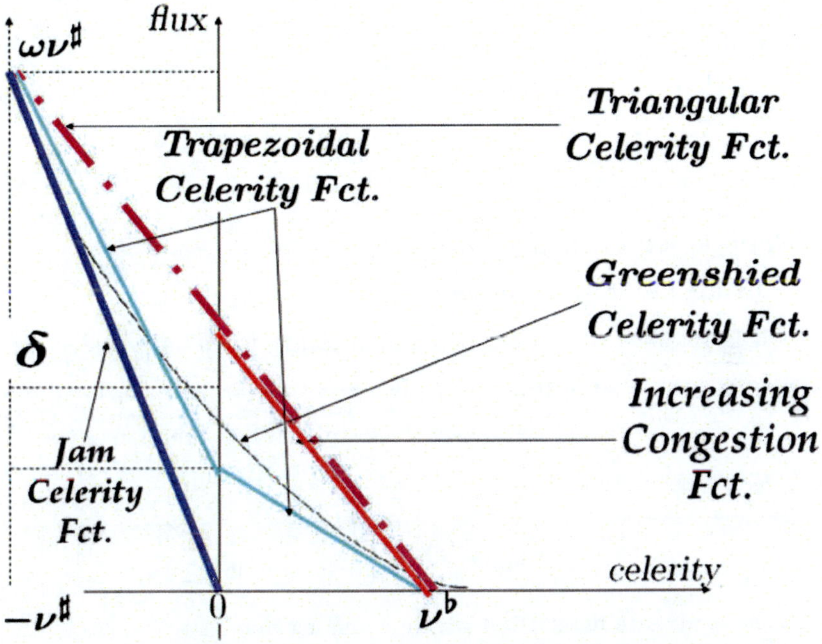

Fig. 14.5 Celerity Functions of Trapezoidal and Greenshields Diagrams. *This diagram displays the graphs of the trapezoidal, triangular, jam density and increasing triangular congestion flow functions* $\mathbf{l}_{(\nu^\flat,\nu^\sharp,\omega,\delta)}$. *Abscissas denote the celerity, ordinates the flow.*

Proof. For simplicity, we set $\mathbf{h}(p) := \mathbf{h}_{(\nu^\flat,\nu^\sharp,\omega,\delta)}(p)$ and $\mathbf{l}(u) := \mathbf{l}_{(\nu^\flat,\nu^\sharp,\omega,\delta)}(u)$. Remembering the values of the critical densities $\beta^\flat = \frac{\delta}{\nu^\flat}$ and $\beta^\sharp = \frac{\nu^\sharp\omega - \delta}{\nu^\sharp}$, the definition of the celerity function implies that

$$\begin{cases} \mathbf{l}(u) := \sup_p [\mathbf{h}(p) - \langle p, u\rangle] = \max\left(\sup_{p\leq\beta^\flat}\left[\langle p, \nu^\flat - u\rangle\right],\right. \\ \left.\left(\sup_{p\in[\beta^\flat,\beta^\sharp]}\left[\delta\,\langle p, u\rangle\right]\right), \left(\sup_{p\geq\beta^\sharp}\left[\langle p, \nu^\sharp + u\rangle\right] + \omega\nu^\sharp\right)\right) \end{cases}$$

Then, we infer that

a. the first term is infinite if $\nu^\flat - u < 0$ and equal to $\beta^\flat(\nu^\flat - u) = \frac{\delta}{\nu^\flat}(\nu^\flat - u)$ in the opposite case when $u \leq \nu^\flat$,
b. the second term is equal
   - if $u \geq 0$, to $\delta - u\beta^\flat = \frac{\delta}{\nu^\flat}(\nu^\flat - u)$
   - if $u \leq 0$, to $\delta^\sharp + u\beta^\sharp = \delta - \frac{\omega\nu^\sharp - \delta}{\nu^\sharp}u$
c. the third term is infinite if $u + \nu^\sharp < 0$ and equal to $\beta^\sharp(\nu^\sharp - u) + \omega\nu^\sharp = \delta - \frac{\omega\nu^\sharp - \delta}{\nu^\sharp}u$ in the opposite case when $u \geq -\nu^\sharp$.

We then deduce formula (14.11), p. 575. □

We single out several particular cases:

a. **Triangular Flux Function.**
   It is obtained when we choose the maximal flow $\overline{\delta} := \frac{\omega\nu^\flat\nu^\sharp}{(\nu^\flat + \nu^\sharp)}$. In this case, lower and upper critical densities collapse to the critical density $\overline{\beta} := \frac{\omega\nu^\sharp}{(\nu^\flat + \nu^\sharp)}$ and we recover the triangular flow function as the largest trapezoidal flow:

$$\mathbf{h}_{(\nu^\flat,\nu^\sharp,\omega,\overline{\delta})}(p) = \begin{cases} \nu^\flat p & \text{if } p \leq \overline{\beta} \\ \nu^\sharp(\omega - p) & \text{if } p \geq \overline{\beta} \end{cases}$$

   the conjugate function of which is defined by

$$\forall\, u \in [-\nu^\sharp, +\nu^\flat],\ \mathbf{l}_{(\nu^\flat,\nu^\sharp,\omega,\overline{\delta})}(u) = \frac{\omega\nu^\sharp}{(\nu^\flat + \nu^\sharp)}(\nu^\flat - u)$$

b. **Increasing Congestion Triangular Flux Functions.**
   They are obtained when $\omega := +\infty$, $\nu^\sharp = 0$, and thus defined by

$$\mathbf{h}_{(\nu^b,0,+\infty,\delta)}(p) := \begin{cases} \nu^b p \text{ if } p \in]-\infty, \dfrac{\delta}{\nu^b}[\\ \delta \quad \text{if } p \in [\dfrac{\delta}{\nu^b}, +\infty] \end{cases}$$

the celerity function of which is defined by

$$\forall\, u \in [0, \nu^b],\ \mathbf{l}_{(\nu^b,0,+\infty,\delta)}(u) = \frac{\delta}{\nu^b}(\nu^b - u)$$

c. **Symmetrical Trapezoidal Flux Functions.**
They are examples of the trapezoidal functions associated with a celerity ν:

$$\mathbf{h}_{(\nu,\nu,\omega,\delta)}(p) = \begin{cases} \nu p & \text{if } p \le \frac{\delta}{\nu} \\ \delta & \text{if } p \in [\frac{\delta}{\nu}, \omega - \frac{\delta}{\nu}] \\ \nu(\omega - p) & \text{if } p \ge \omega - \frac{\delta}{\nu} \end{cases} \tag{14.12}$$

the celerity function of which being equal to

$$\mathbf{l}_{(\nu,\nu,\omega,\delta)}(u) = \begin{cases} \dfrac{\delta}{\nu}(\nu - u) & \text{if } u \in [-0, \nu] \\ \delta - \dfrac{\omega\nu + \delta}{\nu} u & \text{if } u \in [-\nu, 0] \\ +\infty & \text{if } u \notin [-\nu, +\nu] \end{cases}$$

Hence $\mathbf{l}_{(\nu,\nu,\omega,\delta)}(\nu) = 0$, $\mathbf{l}_{(\nu,\nu,\omega,\delta)}(0) = \delta$ and $\mathbf{l}_{(\nu,\nu,\omega,\delta)}(-\nu) = \nu\omega$.

d. **Jam Flux Function.**
It is the other extreme case when we take $\delta := 0$, where all densities are jam densities, i.e., the lower and upper critical densities are the bounds of the density interval $\beta^b = 0$ and $\beta^\sharp = \omega$. We obtain *the jam flow function*:

$$\forall\, p \in [0, \omega],\ \mathbf{h}_{(\nu^b,\nu^\sharp,\omega,0)}(p) = 0$$

the celerity function of which is defined by

$$\forall\, u \in [-\nu^\sharp, +\nu^b],\ \mathbf{l}_{(\nu^b,\nu^\sharp,\omega,0)}(u) = \max(0, -\omega u)$$

The particular case $\delta := \dfrac{\omega\nu}{4}$ allows us to compare this particular symmetric trapezoidal function with the quadratic flow function introduced by *Greenshields*:

2. **Greenshields Flux Function.**
It is defined by

$$\mathbf{h}(p) = \begin{cases} \nu p & \text{if } p \le 0 \\ \dfrac{\nu}{\omega} p(\omega - p) & \text{if } p \in [0, \omega] \\ \nu(\omega - p) & \text{if } p \ge \omega \end{cases}$$

The Greenshields flow $\mathbf{h}$ reaches its maximum at critical density $\beta := \dfrac{\omega}{2}$ and is equal to $\delta := \frac{\nu\omega}{4}$. Its celerity function is equal to

$$\mathbf{l}(u) = \begin{cases} \frac{\omega}{4\nu}(\nu - u)^2 & \text{if } u \in [-\nu, +\nu] \\ +\infty & \text{if } u \notin [-\nu, +\nu] \end{cases}$$

Hence $\mathbf{l}(-\nu) = 0$, $\mathbf{l}(0) = \frac{\omega\nu}{4}$ and $\mathbf{l}(-\nu) = \nu\omega$ (compare with the corresponding trapezoidal function defined by (14.12), p. 577). □

The density and celerity flow functions $\mathbf{l}$ associated with the parameters $\nu^{\flat} \geq 0$, $\nu^{\sharp} \geq 0, \omega > 0$, $\delta > 0$ share common properties:

Proposition 14.3.5 *[Celerity Functions] Let $\mathbf{h}$ be a concave density flow function associated with the parameters $\nu^{\flat} > 0$, $\nu^{\sharp} \geq 0$, $\omega > 0$, $\delta \leq \overline{\delta}$. Then the associated convex celerity flow function $\mathbf{l}$ satisfies*

$$\begin{cases} 0 \leq \mathbf{l}(u) \leq \frac{\omega\nu^{\sharp}}{(\nu^{\flat} + \nu^{\sharp})}(\nu^{\flat} - u) & \text{if } u \in [0, \nu^{\flat}] \\ -\omega u \leq \mathbf{l}(u) \leq \frac{\omega\nu^{\sharp}}{(\nu^{\flat} + \nu^{\sharp})}(\nu^{\flat} - u) & \text{if } u \in [-\nu^{\sharp}, 0] \\ +\infty & \text{otherwise} \end{cases}$$

Therefore the function $u \mapsto \mathbf{l}(u)$ is decreasing (on its domain), satisfies

$$0 = \mathbf{l}(\nu^{\flat}) \leq \max(0, -\omega u) \leq \mathbf{l}(u) \leq \frac{\omega\nu^{\sharp}}{(\nu^{\flat} + \nu^{\sharp})}(\nu^{\flat} - u) \leq \mathbf{l}(-\nu^{\sharp}) = \omega\nu^{\sharp}$$

and the function $u \mapsto \frac{\mathbf{l}(u)}{|u|}$ is decreasing on the interval $[0, \nu^{\flat}[$, increasing on the interval $[-\nu^{\sharp}, 0[$, and, consequently, satisfies

$$\begin{cases} 0 = \frac{\mathbf{l}(\nu^{\flat})}{\nu^{\flat}} \leq \frac{\mathbf{l}(u)}{|u|} & \text{if } 0 < u \leq \nu^{\flat} \\ \omega = \frac{\mathbf{l}(-\nu^{\sharp})}{\nu^{\sharp}} \leq \frac{\mathbf{l}(u)}{|u|} & \text{if } -\nu^{\sharp} \leq u < 0 \end{cases}$$

Furthermore, $\mathbf{l}(0) = \sup_p \mathbf{h}(p) = \delta$ is the maximal flow *and its subdifferential $\partial\mathbf{l}(0)$ is its critical interval.*

Proof. Proposition 14.3.4, p. 574 implies that the lower and upper bounds on $\mathbf{h}$ satisfy

$$\mathbf{h}_{(\nu^{\flat}, \nu^{\sharp}, \omega, 0)}(p) \leq \mathbf{h}(p) \leq \mathbf{h}_{(\nu^{\flat}, \nu^{\sharp}, \omega, \overline{\delta})}(p)$$

We infer from the definition that

$$\max(0,-\omega u) \; = \; \mathbf{l}_{(\nu^\flat,\nu^\sharp,\omega,0)}(u) \; \leq \; \mathbf{l}(u) \; \leq \; \mathbf{l}_{(\nu^\flat,\nu^\sharp,\omega,\overline{\delta})}(u) \; = \; \frac{\omega\nu^\sharp}{(\nu^\flat+\nu^\sharp)}(\nu^\flat-u)$$

by (14.11), p. 575 with $\delta := 0$ and $\delta := \overline{\delta}$ respectively.

Since $\mathbf{h}(0) = 0$ and $\nu^\flat = \mathbf{h}'(0)$ (or $\nu^\flat \in \partial_+\mathbf{h}(0)$), we deduce from Lemma 14.3.3, p.572 that for all u, $\langle 0,\nu^\flat\rangle = \mathbf{h}(0) - \mathbf{l}(\nu^\flat) \geq \mathbf{h}(0) - \mathbf{l}(u)$, which boils down to $0 = \mathbf{l}(\nu^\flat) = \inf_u \mathbf{l}(u)$. In the case when $\omega > 0$ and $\nu^\sharp > 0$ are strictly positive and finite, conditions $\mathbf{h}(\omega) = 0$ and $-\nu^\sharp = \mathbf{h}'(\omega)$ (or $-\nu^\sharp \in \partial_+\mathbf{h}(\omega)$) imply that

$$\langle -\nu^\sharp,\omega\rangle \; = \; \mathbf{h}(\omega)-\mathbf{l}(-\nu^\sharp) \text{ and } \forall\, u \in [-\nu^\sharp,\nu^\flat], \;\; \langle \omega,u\rangle \geq \mathbf{h}(\omega)-\mathbf{l}(u) = -\mathbf{l}(u)$$

which can be written

$$\nu^\sharp\omega \; = \; \mathbf{l}(-\nu^\sharp) \text{ and } \forall\, u \in [-\nu^\sharp,\nu^\flat], \;\; -\omega u \; \leq \; \mathbf{l}(u)$$

Since $0 = \mathbf{l}(\nu^\flat) \leq \mathbf{l}(u)$ and since $\mathbf{l}$ is convex, we deduce that $\mathbf{l}$ is decreasing: take any $u \in [-\nu^\sharp,\nu^\flat]$ and $w := \alpha u + (1-\alpha)\nu^\flat \in [u,\nu^\flat]$ for $\alpha \in [0,1]$. Therefore $\mathbf{l}(w) \leq \alpha\mathbf{l}(u) + (1-\alpha)\mathbf{l}(\nu^\flat) = \alpha\mathbf{l}(u) \leq \mathbf{l}(u)$. Hence, if $w \geq u$, $\mathbf{l}(w) \leq \mathbf{l}(u)$.

Since $\mathbf{l}$ is positive, then, for any $u \in [0,\nu^\flat[$, we observe that $0 = \dfrac{\mathbf{l}(\nu^\flat)}{\nu^\flat} \leq \dfrac{\mathbf{l}(u)}{|u|}$. On the other hand, for any $u \in]-\nu^\sharp,0]$, we deduce from inequalities $-\omega u \; \leq \; \mathbf{l}(u)$ that, by dividing by $|u| > 0$, $\omega = \dfrac{\mathbf{l}(-\nu^\sharp)}{\nu^\sharp} \leq \dfrac{\mathbf{l}(u)}{|u|}$ since $u \leq 0$.

It is actually easy to check that the function $u \mapsto \dfrac{\mathbf{l}(u)}{|u|}$ is decreasing on the interval $[0,\nu^\flat[$ and increasing on the interval $[-\nu^\sharp,0[$. □

14.4 The Viability Traffic Function

The celerity function plays the role of a *Lagrangian*, as the fundamental diagram played the role of an Hamiltonian, in the sense that we associate with partial differential equation (14.4)(i), p. 570 and the celerity function $\mathbf{l}$ the *characteristic system*

$$\begin{cases} (i) \;\; \tau'(t) \; = \; -1 \\ (ii) \;\; (x'(t),y'(t)) \; \in \; -\mathcal{E}p(\mathbf{l}) \end{cases} \tag{14.13}$$

controlled by celerities $u(\cdot)$.

We take for environment $K := \mathbb{R}_+ \times [\underline{\gamma},\overline{\gamma}] \times \mathbb{R}$ and for target $C := \mathcal{E}p(\mathbf{c})$ for finite roads or $K := \mathbb{R}_+ \times \mathbb{R}_+ \times \mathbb{R}$.

Definition 14.4.1 ***[Congestion Function as Viability Solution to the Moskowitz Problem]*** *Let us consider the epigraph* $\mathcal{E}p(\mathbf{c})$ *of the traffic condition* $\mathbf{c}$*. The* congestion function *(associated with the traffic condition* $\mathbf{c}$*)* V *(or* $V_{\mathbf{c}}$ *when the reference to the traffic condition is useful) to the Moskowitz problem (14.4), p. 570 defined by the following formula*

$$V(T,x) := \inf_{(T,x,y)\in \mathrm{Capt}_{(14.13)}(\mathbb{R}_+\times\mathbb{R}\times\mathbb{R},\mathcal{E}p(\mathbf{c}))} y \qquad (14.14)$$

is called the congestion function.

We shall recall that the congestion function, when it is differentiable, is a solution to the Moskowitz equation satisfying the traffic conditions. Otherwise, when it is not differentiable, but only lower semicontinuous, we can give a meaning to a solution as a solution in the Barron–Jensen/Frankowska sense, using for that purpose subdifferential of lower semicontinuous functions defined in non-smooth analysis. This is not important for two reasons: all other properties of congestion functions that are proven in this Chapter are derived directly from the properties of capture basins without using the concept of derivatives, usual or generalized. Derivatives are used only in the last section, for constructing the minimal congestion regulation map providing the kinematic differential inclusion governing the evolutions minimizing the congestion.

Making the above definition explicit, one can prove that the viability congestion function satisfies the *congestion variational principle* for the Moskowitz problem (14.4), p. 570.

We regard $d \in [0,T]$ as a *departure time*, the associated *travel time* being equal to $s := T-d$. We consider the family $\mathcal{A}(d,T;x)$ of absolutely continuous traffic evolutions $\xi(\cdot)$ starting at departure time d at $\xi(d) \in \mathbf{C}(d)$ and arriving at time T at x. We assign to such a traffic evolution two "traffic values":

- the *departure traffic value* $\mathbf{c}(d,\xi(d))$ at the state $\xi(d)$ at departure time d,
- the cumulated *celerity traffic value* $\int_d^T \mathbf{l}(\xi'(\tau))d\tau$ on the celerity $\xi'(\cdot)$ of the evolution $\xi(\cdot) \in \mathcal{A}(d,T;x)$ on the interval $[d,T]$.

We associate with each departure time $d \in [0,T]$ the minimal *travel traffic value* over the traffic evolutions $\xi(\cdot) \in \mathcal{A}(d,T;x)$ defined on the travel interval $[d,T]$, defined by

$$J(d,T;x) \;:=\; \inf_{\xi(\cdot)\in\mathcal{A}(d,T;x)}\left(\int_d^T \mathbf{l}(\xi'(\tau))d\tau + \mathbf{c}(d,\xi(d))\right)$$

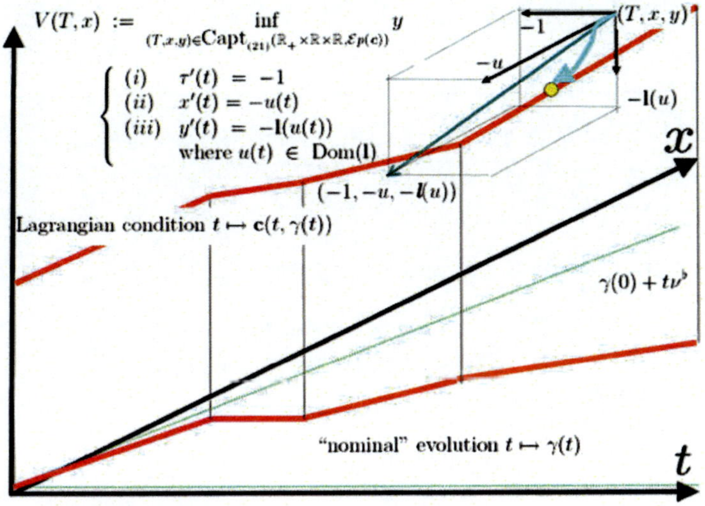

Fig. 14.6 Viability Solution.

The figure illustrates Definition 14.4.1, p.580 of the viability solution for a Lagrangian condition $t \mapsto \mathbf{c}(t,\gamma(t))$ defined on a nominal $t \mapsto \gamma(t)$ in the case when $\nu^\sharp = 0$. Its epigraph is the capture basin of the epigraph of the Lagrangian condition under system (14.13), p. 579, which is mentioned in the figure. The point $(T,x,y) + (-1,-u,-\mathbf{l}(u))$ is represented as well as the element $\left(T-t^\star, x-\int_0^{t^\star} u(\tau)d\tau, y-\int_0^{t^\star}\mathbf{l}(u(\tau))d\tau\right) \in \mathcal{E}p(\mathbf{c})$ where an evolution reaches the epigraph of the Lagrangian condition. By taking $y := V(T,x)$ and changing the direction of the arrow, we obtain an evolution starting from the Lagrangian condition and minimizing the cumulated congestion. The domain $t \rightsquigarrow [\gamma(0), \gamma(0)+t\nu^\sharp]$ of the viability solution is represented.

Theorem 14.4.2 ***[Congestion Variational Principle]*** *The viability congestion function $V(t,x)$ to the traffic problem (14.4), p. 570 minimizes the travel traffic value with respect to departure time d and evolutions $\xi(\cdot) \in \mathcal{A}(d,T;x)$:*

$$\begin{cases} V(T,x) \;=\; \inf_{d\in[0,T]} J(d,T;x) \\ \;=\; \inf_{d\in[0,T]} \inf_{\xi(\cdot)\in\mathcal{A}(d,T;x)} \left(\int_d^T \mathbf{l}(\xi'(\tau))d\tau + \mathbf{c}(d,\xi(d))\right) \end{cases} \quad (14.15)$$

Alternatively, one can formulate this variational principle in terms of the celerity $u(\cdot)$ instead of traffic evolutions $\xi(\cdot) \in \mathcal{A}(d,T;x)$: $V(T,x)$ =

$$\inf_{u(\cdot)\in L^1(0,T;\mathrm{Dom}(\mathbf{l})),\ t^\star\in[0,T]} \left(\int_0^{t^\star} \mathbf{l}(u(\tau))d\tau + \mathbf{c}\left(T - t^\star, x - \int_0^{t^\star} u(\tau)d\tau\right)\right) \tag{14.16}$$

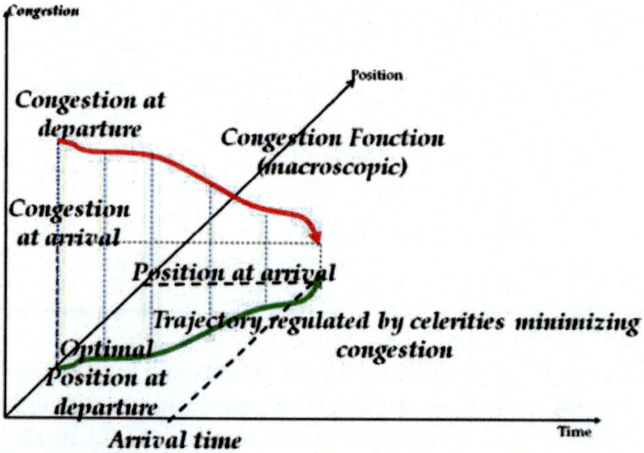

Fig. 14.7 Optimal Evolutions.

The figure displays an optimal evolution minimizing the intertemporal optimization arriving at a given position at a given time. The 3D curve represents the graph of the traffic (or congestion) function $V(t,x)$. Knowing the arrival time and position, at departure time, the regulation rule of the microsystem provides the initial position and the velocities of the optimal evolution, and thus, the departure position.

Proof. This is a consequence of Theorem 13.4.2, p.533 when $\mathbf{h}(p) := -\mathbf{l}^\star(-p)$ and when the structuring variable is the time $d = -t \in \mathbb{R}_+$. □

For proving that there exists a minimal congestion evolution achieving the minimum in the congestion function, we need to prove that the congestion function is lower semicontinuous, i.e., that the capture basin $\mathrm{Capt}_{(14.13)}(\mathbb{R}_+ \times \mathbb{R} \times \mathbb{R}, \mathcal{E}p(\mathbf{c}))$ characterizing it is closed. This will be easier to prove thanks to the Lax–Hop formula, which also provides a simpler formula of the congestion function.

The independence of the Hamiltonian $\mathbf{h}(p)$ on (t,x) and the convexity of the celerity function imply the *Lax–Hopf formula* for partial differential equations.

Theorem 14.4.3 *[The Lax–Hopf Formula for the Viability Traffic Function] Assume that the Hamiltonian* **h** *is a flow function defined in Definition 14.3.1, p.571. Then the viability congestion function is lower semicontinuous and is given by the following formula:*

$$V(T,x) = \inf_{d\in[0,T]} \inf_{\xi\in\mathbf{C}(d)} \left[\mathbf{c}(d,\xi) + (T-d)\mathbf{l}\left(\frac{x-\xi}{T-d}\right)\right] \tag{14.17}$$

which can be written in the following form:

$$V(T,x) = \inf_{s\in[0,T]} \inf_{u\in\text{Dom}(\mathbf{l})} [\mathbf{c}(T-s, x-su) + s\mathbf{l}(u)] \tag{14.18}$$

Proof. This is a consequence of Theorem 13.9.1, p.554 when $\mathbf{h}(p) := -\mathbf{l}^\star(-p)$ and when the structuring variable is the time $d = -t \in \mathbb{R}_+$. □

The traffic solution satisfies the dynamic programming equation:

Theorem 14.4.4 *[The Dynamic Programming Equation] Assume that the Hamiltonian* **h** *is a flow function defined in Definition 14.3.1, p.571. There exists an optimal departure date $d^\star$ reaching the minimum in*

$$V(T,x) = J(d^\star,T;x) = \inf_{d\in[0,T]} J(d,T;x)$$

and a minimal congestion evolution $\xi^\star(\cdot) := \xi^\star_{(T,x)}(\cdot) \in \mathcal{A}(d^\star,T;x)(\cdot)$ achieving the minimum

$$V(T,x) = J(d^\star,T;x) = \mathbf{c}(d^\star,\xi^\star(d^\star)) + \int_{d^\star}^T \mathbf{l}(\xi^{\star\prime}(\tau)d\tau \tag{14.19}$$

Such optimal date and evolution satisfy the dynamic optimality property

$$\forall\, s\in[d^\star,T],\ V(T,x) = V(s,\xi^\star(s)) + \int_s^T \mathbf{l}(\xi^{\star\prime}(\tau)d\tau \tag{14.20}$$

Furthermore, on minimal congestion evolutions $\xi^\star(\cdot)$,

$$\text{the function } t : [d^\star,T] \mapsto V(t,\xi^\star(t)) \text{ is increasing}$$

Proof. This is a consequence of Theorem 13.5.2, p.539. □

The question arises to know precisely the domains $\text{Dom}(V(t,\cdot))$ of the traffic profiles, i.e., the set of states x such that $V(t,x) < +\infty$: we shall prove that it is couched in terms of the set-valued map $\mathbf{C}$:

Theorem 14.4.5 ***[Domain of the Viability Traffic Function]*** *Assume that the Hamiltonian* $\mathbf{h}$ *is a flow function defined in Definition 14.3.1, p.571. For any* $t \geq 0$*, the domains of the traffic profiles* $V(t,\cdot)$ *associated with the traffic condition* $\mathbf{c}$ *are equal to*

$$\text{Dom}(V(t,\cdot)) \;=\; \bigcup_{s\in[0,t],\ u\in\text{Dom}(\mathbf{l})} (\mathbf{C}(t-s)+su) \tag{14.21}$$

If we assume furthermore that $\mathbf{C}$ *satisfies*

$$\forall\, t \geq 0,\;\; \mathbf{C}(t) \subset \mathbf{C}(0)+t\text{Dom}(\mathbf{l}) \tag{14.22}$$

then

$$\text{Dom}(V(t,\cdot)) \;=\; \mathbf{C}(0)+t\text{Dom}(\mathbf{l}) \tag{14.23}$$

Proof. This statement follows from Theorem 13.9.2, p.555. □

14.5 Analytical Properties of the Viability Traffic Functions

We derive from properties of the density flow functions associated with parameters $\nu^\flat \geq 0,\; \nu^\sharp \geq 0, \omega > 0,\; \delta > 0$ exposed in Proposition 14.3.5, p. 578 the following statements:

- lower and upper estimates of the viability congestion function to the Moskowitz problem (14.4), p. 570,
- the domains of the traffic profiles $x \mapsto V(t,x)$, which are proven to be equal to $\mathbf{C}(0)+t[-\nu^\sharp,\nu^\flat]$ whenever we assume that for every $t \geq 0$, $\mathbf{C}(t) \subset \mathbf{C}(0)+t[-\nu^\sharp,\nu^\flat]$

expressed only in terms of the traffic condition $\mathbf{c}$, its associated set-valued map $\mathbf{C}$ and the four characteristic parameters $\nu^\flat \geq 0,\; \nu^\sharp \geq 0, \omega > 0$ and $\delta{>}0$. They are consequently valid for any density flow function $\mathbf{h}$ associated those parameters. The Lax–Hopf formula provides a simpler formula of the congestion function, but we may need more information about the upper and lower bounds of the traffic conditions, which are more precise forms of "Theorems of Maximum" in the partial differential equation literature.

Proposition 14.5.1 *[Estimates of the Moskowitz Function] Let us associate with the function* $\mathbf{c}$ *and parameters* $\nu^\flat \geq 0$, $\nu^\sharp \geq 0, \omega > 0$ *and* $\delta > 0$ *the following subset*

$$\mathbf{E}(t,x) := \left\{(s,u) \in [0,t] \times [-\nu^\sharp, +\nu^\flat] \text{ such that } x \in \mathbf{C}(t-s) + su\right\} \tag{14.24}$$

Hence the viability congestion function to Moskowitz traffic problem (14.4), p. 570 is given by

$$V(t,x) := \inf_{(s,u)\in\mathbf{E}(t,x)} [\mathbf{c}(t-s, x-su) + s\mathbf{l}(u)] \tag{14.25}$$

Let us set

$$\underline{\mathbf{c}}(t,x) := \inf_{(s,u)\in\mathbf{E}(t,x)} (\mathbf{c}(t-s, x-su) + \max(0, -s\omega u))$$

and

$$\overline{\mathbf{c}}(t,x) := \inf_{(s,u)\in\mathbf{E}(t,x)} \left(\mathbf{c}(t-s, x-su) + s\frac{\omega\nu^\sharp}{(\nu^\flat + \nu^\sharp)}(\nu^\flat - u)\right)$$

Then, the viability congestion function V *satisfies the estimates*

$$\underline{\mathbf{c}}(t,x) \leq V(t,x) \leq \overline{\mathbf{c}}(t,x) \tag{14.26}$$

Proof. Since the celerity flow function $\mathbf{l}$ satisfies

$$\max(0, -\omega u) \leq \mathbf{l}(u) \leq \frac{\omega\nu^\sharp}{(\nu^\flat + \nu^\sharp)}(u - \nu^\flat)$$

thanks to Proposition 14.3.5, p. 578, then inequality $\underline{\mathbf{c}}(t,x) \leq V(t,x) \leq \overline{\mathbf{c}}(t,x)$ is straightforward from Lax–Hopf formula. □

Remark. Observe that

1. the lower estimate

$$\inf_{(s,u)\in\mathbf{E}(t,x)} \mathbf{c}(t-s, x-su) \leq \underline{\mathbf{c}}(t,x)$$

   holds,
2. the lower estimate $\underline{\mathbf{c}}$ is the solution to the Moskowitz problem (14.4), p. 570 when the Hamiltonian is *the jam flow function* $\mathbf{h}$ defined by:

$$\forall\, p \in [0,\omega],\;\; \mathbf{h}_{(\nu^\flat,\nu^\sharp,\omega,0)}(p) = 0$$

the conjugate function of which is defined by

$$\forall\, u \in [-\nu^\sharp, +\nu^\flat],\ \mathbf{l}_{(\nu^\flat,\nu^\sharp,\omega,0)}(u) \;=\; \max(0, -\omega u)$$

3. the upper estimate $\overline{\mathbf{c}}$ is the solution to the Moskowitz problem (14.4), p. 570 when the Hamiltonian $\mathbf{h}$ is the triangular density flow function

$$\mathbf{h}_{(\nu^\flat,\nu^\sharp,\omega,\overline{\delta})}(p) \;=\begin{cases} \nu^\flat p & \text{if } p \leq \overline{\beta} \\ \nu^\sharp(\omega - p) & \text{if } p \geq \overline{\beta} \end{cases}$$

the conjugate function of which is the celerity flow function defined by

$$\forall\, u \in [-\nu^\sharp, +\nu^\flat],\ \mathbf{l}_{(\nu^\flat,\nu^\sharp,\omega,\overline{\delta})}(u) \;=\; \frac{\omega\nu^\sharp}{(\nu^\flat + \nu^\sharp)}(u\nu^\flat - u)$$

4. by taking the values 0, $-\nu^\sharp$ and $\nu^\flat$, we obtain the further estimate

$$\overline{\mathbf{c}}(t,x) \;\leq \min \begin{bmatrix} \inf_{(s,\nu^\flat)\in\mathbf{E}(t,x)} \mathbf{c}(t-s, x - s\nu^\flat), \\ \inf_{(s,0)\in\mathbf{E}(t,x)} \mathbf{c}(t-s,x) + s\dfrac{\omega\nu^\sharp\nu^\flat}{(\nu^\flat+\nu^\sharp)} \quad , \\ \inf_{(s,-\nu^\sharp)\in\mathbf{E}(t,x)} (\mathbf{c}(t-s, x + s\nu^\sharp) + s\omega\nu^\sharp) \end{bmatrix} \square$$

14.5.1 Cauchy Initial Conditions

The *Cauchy initial condition* requires that at initial time $t = 0$, the initial traffic profile $\gamma_0(\cdot) : x \mapsto \gamma_0(x) \in \mathbb{R} \cup \{+\infty\}$ is given. We thus require that

$$\forall\, x \in \mathbb{R},\ V(0,x) \;\leq\; \gamma_0(x)$$

(actually, we shall prove that in this case, these inequalities are equalities). We extend the Cauchy traffic condition by setting $\mathbf{c}(0,x) := \gamma_0(x)$ and $\mathbf{c}(t,x) = +\infty$ whenever $t > 0$. In this case, $\mathbf{C}(0) = \mathrm{Dom}(\gamma_0)$ and $\mathbf{C}(t) = \emptyset$ otherwise.

Proposition 14.5.2 *[Cauchy Initial Conditions] The viability congestion function to the Moskowitz problem(14.4), p. 570 with Cauchy condition $\gamma_0(\cdot)$ satisfies initial condition*

$$\forall\, x \in \mathbb{R},\ V(0,x) \;=\; \gamma_0(x)$$

The domains of its traffic profiles are equal to

$$\forall\, t \geq 0,\ \mathrm{Dom}(V(t,\cdot)) \;=\; \mathrm{Dom}(\gamma_0) + t[-\nu^\sharp, \nu^\flat]$$

The viability congestion function is equal to

$$V(t,x) \;=\; \inf_{u\in[-\nu^\sharp,\nu^\flat]} (\gamma_0(x-tu)+t\mathbf{l}(u))$$

and satisfies the estimates

$$\begin{cases} \inf_{u\in[-\nu^\sharp,\nu^\flat]} (\gamma_0(x-tu)+t\max(0,-\omega u)) \;\leq\; V(t,x) \\ \leq\; \inf_{u\in[-\nu^\sharp,\nu^\flat]} \left(\gamma_0(x-tu)+t\frac{\omega\nu^\sharp}{(\nu^\flat+\nu^\sharp)}(\nu^\flat-u)\right) \end{cases} \tag{14.27}$$

Proof. The Cauchy condition is obviously a Moskowitz one because, 0 belonging to $[-\nu^\sharp,\nu^\flat]$, condition

$$\forall\, t>0,\;\; \mathbf{C}(t)=\emptyset \,\subset\, \mathbf{C}(0)+t[-\nu^\sharp,\nu^\flat]$$

hold. Then the domains of the associated traffic profiles are equal to

$$\forall\, t\geq 0,\;\; \mathrm{Dom}(V)(t,\cdot) \;=\; \mathbf{C}(0)+t[-\nu^\sharp,\nu^\flat] \;=\; \mathrm{Dom}(\gamma_0)+t[-\nu^\sharp,\nu^\flat]$$

thanks to Theorem 14.4.5, p. 584. The estimates follow from Proposition 14.5.1, p. 585. □

14.5.2 Lagrangian Traffic Conditions

Lagrangian traffic conditions are associated with:

1. the trajectory of an increasing "nominal" evolution $t\mapsto\gamma(t)$ of a given probe vehicle,
2. A Lagrangian condition $t\mapsto\mathbf{c}(t,\gamma(t))$ defined on this trajectory.

Eulerian conditions are particular cases of Lagrangian conditions whenever the traffic condition is constant ($\gamma(t)=\gamma$ for all $t\geq 0$).

Definition 14.5.3 ***[Lagrangian Traffic Conditions on Nominal Evolution]*** *We say that a traffic evolution $t\mapsto\gamma(t)$ is* consistent *with the celerities $\nu^\sharp$ and $\nu^\flat$ if it satisfies the "speed limit"*

$$\forall\, t\geq 0,\;\; \gamma'(t) \,\in\, [-\nu^\sharp,+\nu^\flat[\tag{14.28}$$

Lagrangian traffic conditions $\mathbf{c}(t,\gamma(t))$ are defined on (the graph of) the nominal trajectory. The viability congestion function to the Moskowitz *problem (14.4), p. 570 associated with any Lagrangian traffic condition requires that*

$$\forall t \geq 0, \;\; V(t,\gamma(t)) \;\leq\; \mathbf{c}(t,\gamma(t))$$

We begin by providing elementary properties of nominal evolutions:

Lemma 14.5.4 ***[Properties of Consistent Evolutions]*** *A nominal evolution is consistent with the celerities $\nu^\flat$ and $\nu^\sharp$ if and only if*

$$\forall t \geq 0, \;\; \forall s \in [0,t], \;\; -s\nu^\sharp \;\leq\; \gamma(t) - \gamma(t-s) \;\leq\; s\nu^\flat$$

In this case, the associated tube $\mathbf{C}(t) := \{\gamma(t)\}$ is a Moskowitz tube, *so that the domain of the viability congestion function is equal to*

$$\forall t \geq 0, \;\; \mathrm{Dom}(V(t,\cdot)) \;=\; [\gamma(0) - t\nu^\sharp, \gamma(0) + t\nu^\flat]$$

Proof. Indeed, the speed limit (14.28), p. 587 is obviously equivalent to

$$\forall t \geq 0, \;\; \forall s \in [0,t], \;\; -s\nu^\sharp \;\leq\; \gamma(t) - \gamma(t-s) \;\leq\; s\nu^\flat$$

from which we infer, taking $s = t$ in the above formula that

$$\forall t \geq 0, \;\; \gamma(0) - t\nu^\sharp \;\leq\; \gamma(t) \;\leq\; \gamma(0) + t\nu^\flat$$

i.e., that

$$\mathbf{C}(t) \;:=\; \{\gamma(t)\} \;\subset\; \gamma(0) + t[-\nu^\flat, \nu^\sharp] \;=\; \mathbf{C}(0) + t[-\nu^\flat, \nu^\sharp]. \qquad \square$$

The subset $\mathbf{E}(t,x)$ defined by (14.24), p. 585 is equal to

$$\mathbf{E}(t,x) \;:=\; \left\{ (s,u) \;\in\; [0,t] \times [-\nu^\sharp, +\nu^\flat] \text{ such that } u \;=\; \frac{x - \gamma(t-s)}{s} \right\}$$

Proposition 14.5.5 ***[Lagrangian Conditions]*** *Assume that the nominal evolution is consistent with the celerities $\nu^\flat$ and $\nu^\sharp$. Then the viability congestion function to partial differential equation (14.4), p. 570 associated with Lagrangian traffic condition $t \mapsto \mathbf{c}(t,\gamma(t))$ is defined by*

$$V(t,x) = \inf_{s\in[0,t]} \left[\mathbf{c}(t-s,\gamma(t-s)) + s\mathbf{l}\left(\frac{x-\gamma(t-s)}{s}\right)\right] \quad (14.29)$$

and satisfies the estimates

$$\begin{cases} \inf_{s\in[0,t]} \left(\mathbf{c}(t-s,\gamma(t-s)) + \max(0,\omega(\gamma(t-s)-x))\right) \le V(t,x) \\ \le \inf_{s\in[0,t]} \left(\mathbf{c}(t-s,\gamma(t-s)) + \frac{\omega\nu^\sharp}{(\nu^\flat+\nu^\sharp)}(s\nu^\flat + \gamma(t-s) - x)\right) \end{cases} \quad (14.30)$$

Proof. To say that $(s,u) \in \mathbf{E}(t,x)$ amounts to saying that $\mathbf{E}(t,x)$ is the graph of the map $s \mapsto \dfrac{x-\gamma(t-s)}{s}$. Since the Lagrangian traffic condition is defined only on the graph of the nominal evolution, we deduce that $\mathbf{c}(t-s,x-su)$ is finite if and only if $(s,u) \in \mathbf{E}(t,x)$, in which case $u := \dfrac{x-\gamma(t-s)}{s}$, so that general formula (14.25), p. 585 boils down to

$$V(t,x) = \inf_{s\in[0,t]} \left[\mathbf{c}(t-s,\gamma(t-s)) + s\mathbf{l}\left(\frac{x-\gamma(t-s)}{s}\right)\right]$$

Proposition 14.5.1, p. 585 implies the other statements. □

In order to make the general formula more operational, the question arises whether the *celerity map* $s \mapsto \dfrac{x-\gamma(t-s)}{s}$ from $[0,t]$ to $[-\nu^\sharp, +\nu^\flat]$ is surjective. Precisely,

Proposition 14.5.6 *[Surjectivity of the Celerity Map] Assume that the nominal evolution is consistent with the celerities $\nu^\flat$ and $\nu^\sharp$. Then the* celerity map $s \mapsto \dfrac{x-\gamma(t-s)}{s}$ *from $[0,t]$ to $[-\nu^\sharp, +\nu^\flat]$ is surjective:*

1. *when $x \in [\gamma(t), \gamma(0)+t\nu^\flat]$, for any $u \in \left[\dfrac{x-\gamma(0)}{t}, \nu^\flat\right]$, there exists $s \in [0,t]$ such that $u = \dfrac{x-\gamma(t-s)}{s}$. In particular, there always exists $s^\flat$ such that $\nu^\flat = \dfrac{x-\gamma(t-s^\flat)}{s^\flat}$ so that $V(t,x) \le \mathbf{c}(t-s^\flat,\gamma(t-s^\flat))$.*
2. *when $x \in [\gamma(0)-t\nu^\sharp, \gamma(t)]$, for any $u \in \left[-\nu^\sharp, \dfrac{x-\gamma(0)}{t}\right]$, there exists $s \in [0,t]$ such that $u = \dfrac{x-\gamma(t-s)}{s}$. In particular, there exists $s^\sharp$ such that $-\nu^\sharp = \dfrac{x-\gamma(t-s^\sharp)}{s^\sharp}$ so that $V(t,x) \le \mathbf{c}(t-s^\sharp,\gamma(t-s^\sharp)) + s^\sharp\omega\nu^\sharp$.*

Proof. We have to prove that for any $u \in [-\nu^\sharp, +\nu^\flat]$, the subsets

$$\mathbf{e}(t,x)^{-1}(u) \;:=\; \{s \in [0,t] \text{ such that } x - su = \gamma(t-s)\}$$

are not empty. We introduce the subsets

$$\mathbf{e}(t,x)^{-1}_{\leq}(u) \;:=\; \{s \in [0,t] \text{ such that } x - su \leq \gamma(t-s)\}$$

and

$$\mathbf{e}(t,x)^{-1}_{\geq}(u) \;:=\; \{s \in [0,t] \text{ such that } x - su \geq \gamma(t-s)\}$$

so that $\mathbf{e}(t,x)^{-1}(u) = \mathbf{e}(t,x)^{-1}_{\leq}(u) \cap \mathbf{e}(t,x)^{-1}_{\geq}(u)$. Since intervals are connected, this intersection is not empty whenever these subsets are closed, non empty and cover the interval. They are closed since the evolution $t \mapsto \gamma(t)$ is continuous, and they obviously cover the interval $[0,t]$. It remains to prove that they are not empty.

1. Case when $u \in \left[\frac{\gamma(t)-\gamma(0)}{t}, \nu^\flat\right]$ and $x \in [\gamma(t), \gamma(0)+tu]$. Since $x \geq \gamma(t)$, then $0 \in \mathbf{e}(t,x)^{-1}_{\geq}(u)$ because $x - 0u \geq \gamma(t-0)$, and since $x \leq \gamma(0)+tu$, $t \in \mathbf{e}(t,x)^{-1}_{\leq}(u)$. Therefore, there exists $s \in [0,t]$ such that $u = \frac{x-\gamma(t-s)}{s}$.
2. Case when $u \in \left[-\nu^\sharp, \frac{\gamma(t)-\gamma(0)}{t}\right]$ and $x \in [\gamma(0)+tu, \gamma(t)]$. Since $x \leq \gamma(t)$, then $0 \in \mathbf{e}(t,x)^{-1}_{\leq}(u)$ because $x - 0u \leq \gamma(t-0)$, and since $x \geq \gamma(0)+tu$, $t \in \mathbf{e}(t,x)^{-1}_{\geq}(u)$. There exists $s \in [0,t]$ such that $u = \frac{x-\gamma(t-s)}{s}$.

This completes the proof. □

14.5.3 Combined Traffic Conditions

We have listed particular traffic conditions, the Cauchy conditions, as well as the Dirichlet, Eulerian and Lagrangian conditions.

We now turn our attention to the combination of Cauchy and several Lagrangian conditions: the traffic conditions involve

- a Cauchy condition $\mathbf{c}_0$ such that $\mathbf{c}_0(0,x) = \gamma_0(x)$ and $\mathbf{c}_0(t,x) = +\infty$ whenever $t > 0$
- and/or Lagrangian conditions $\mathbf{c}_i(t,\gamma_i(t))$ satisfying $\mathbf{c}_i(t,x) = +\infty$ whenever $x \neq \gamma_i(t)$, $i \in I$.

We assume that these conditions satisfy

$$\forall\, i \in I,\;\; \gamma_i'(t) \;\in\; [-\nu^\sharp, +\nu^\flat] \tag{14.31}$$

We introduce the combined traffic condition

$$\mathbf{c}(t,x) \;:=\; \min\left[\gamma_0(x), \mathbf{c}_i(t,x)\right]$$

The domain $\mathbf{C}(t) := \mathrm{Dom}(\mathbf{c}(t,\cdot))$ of its profile is equal to

$$\mathbf{C}(t) \;=\; \begin{cases} \mathrm{Dom}(\gamma_0) & \text{if } t = 0 \\ \bigcup_{i \in I} \{\gamma_i(t)\} & \text{if } t > 0 \end{cases}$$

The Min Inf-Convolution Morphism Theorem 14.7.1, p.595 implies

Proposition 14.5.7 *[The Min-Morphism Property] Let us denote by*

- V_0 *the viability congestion function associated with the Cauchy condition* $\mathbf{c}_0$,
- V_i *the congestion function associated with the Lagrangian conditions* $\mathbf{c}_i$
- $V_{\mathbf{c}}$ *the congestion function associated with the combined condition* $\mathbf{c}$.

Then

$$V_{\mathbf{c}}(t,x) \;=\; \min[V_0(t,x), V_i(t,x)]$$

and is defined on $\mathrm{Dom}(V_{\mathbf{c}}(t,\cdot)) \;=\; \mathrm{Dom}(\gamma_0) + t[-\nu^\sharp, \nu^\flat]$. *It satisfies traffic condition*

$$V_{\mathbf{c}}(t,x) = \min\left[V_0(t,x), V_i(t,x)\right] \;\leq\; \mathbf{c}(t,x) \;:=\; \min\left[\gamma_0(x), \mathbf{c}_i(t,x)\right]$$

We deduce the following result:

Theorem 14.5.8 *[Departure Tube of Combined Traffic Conditions] Let us consider the departure tubes* $\mathbf{D}_i(t)$ *associated with the traffic conditions* $\mathbf{c}_i$ *and the departure tube* $\mathbf{D}(t)$ *of the combined traffic condition* $\mathbf{c}$. *Let us set* $I(t,x) := \{i \in I \text{ such that } \mathbf{c}_i(t,x) = \mathbf{c}(t,x)\}$. *Then*

$$\forall\, t \geq 0,\;\; \mathbf{D}(t) \subset \bigcap_{i \in I(t,x)} \mathbf{D}_i(t)$$

Proof. Let $x \in \mathbf{D}(t)$. Then there exists $i \in I(t,x)$ such that $\mathbf{c}_i(t,x) = \mathbf{c}(t,x)$ and j such that $V_j(t,x) = V(t,x)$. Therefore

$$V_i(t,x) \;\leq\; \mathbf{c}_i(t,x) \;=\; \mathbf{c}(t,x) = V_j(t,x) \;\leq\; V_i(t,x)$$

and thus, $V_i(t,x) = \mathbf{c}_i$, i.e., $x \in \mathbf{D}_i(t)$. $\square$

14.6 Decreasing Envelopes of Functions

Since *congestion functions must be increasing in time and decreasing in position for obvious "physical" reasons*, one must study under which conditions this property is true. Using the epigraphical approach, we have to provide an epigraphical characterization of decreasing functions.

This can be done in the framework of inf-convolution operators by specific functions (see Definition 18.8.1, p.762 and Lemmas 18.8.3, p.763 and 18.34, p.763).

There are many examples of "inf-convolution" operators $\mathbf{u} \mapsto \mathbf{v} \star \mathbf{u}$ by a function $\mathbf{v}$.

Examples: One interesting example is provided by the function $x \mapsto \mathbf{v}(x) := \lambda\|x\|$ because a function $\mathbf{u}$ is λ-Lipschitz if and only if $\mathbf{u} = \mathbf{v} \star \mathbf{u}$. The function $\mathbf{v} \star \mathbf{u}$ defined by

$$(\mathbf{v} \star \mathbf{u})(x) \ := \ \inf_y (\mathbf{u}(y) + \lambda\|x-y\|)$$

can be regarded as the λ-Lipschitz envelope of $\mathbf{u}$.

The classical Moreau–Yosida transform is the inf-convolution of a function with the quadratic function defined $\mathbf{v}(x) := \frac{1}{\lambda}\|x\|^2$:

$$(\mathbf{v} \star \mathbf{u})(x) \ := \ \inf_y \left(\mathbf{u}(y) + \frac{1}{\lambda}\|x-y\|^2\right)$$

It is used as a regularization procedure for approximating lower semicontinuous convex function by smooth ones.

We turn our attention to the example motivated by the monotonic properties of congestion functions. □

Let us consider a closed convex cone $P \subset X$. It defines an order relation by setting $y \geq x$ if and only if $y \in x + P$.

On the other hand, let us associate with the cone P its indicator function ψ_P defined by $\psi_P(x) = 0$ if $x \in P$ and $\psi_P(x) = +\infty$ otherwise. Therefore, $\mathcal{E}p(\psi_P) = P \times \mathbb{R}_+$. Consequently,

$$\mathcal{E}p(\psi_P \star \mathbf{u}) \ = \ \mathcal{E}p(\mathbf{u}) + P \times \mathbb{R}_+$$

Proposition 14.6.1 ***[Epigraphical Characterization of Decreasing Functions]*** *Let us consider an extended function* $\mathbf{u} : x \in X \mapsto \mathbf{u}(x) \in \mathbb{R} \cup$

$\{+\infty\}$ *and* $P \subset X$ *a closed convex cone. The two following statements are equivalent:*

- $\mathbf{u}$ *is* decreasing *in the sense that for any* $y \in x + P$*, then* $\mathbf{u}(y) \leq \mathbf{u}(x)$
- *The epigraph of the function* $\mathbf{u}$ *satisfies*

$$\mathcal{E}p(\mathbf{u}) + P \times \mathbb{R}_+ \;=\; \mathcal{E}p(\mathbf{u})$$

In this case, $\mathrm{Dom}(\mathbf{u}) = \mathrm{Dom}(\mathbf{u}) + P$ *(see Figs. 14.8, p.597 and 14.9, p.598).*

Proof. We observe first that inequality $\mathbf{u}(y) \leq \mathbf{u}(x)$ is equivalent to say that $(y, \mathbf{u}(x)) \in \mathcal{E}p(\mathbf{u})$.

1. *Assume that* $\mathbf{u}$ *is decreasing.* Since $P \times \mathbb{R}_+$ is a cone, inclusion $\mathcal{E}p(\mathbf{u}) \subset \mathcal{E}p(\mathbf{u}) + P \times \mathbb{R}_+$ is always true. For proving the converse inclusion, take any $y \in \mathrm{Dom}(\mathbf{u})$, $p \in P$, $\lambda \geq 0$ and set $x := y + p \geq y$. Hence

$$(x, \mathbf{u}(x)) + (p, \lambda) \;=\; (y, \mathbf{u}(x)) + (0, \lambda) \;\in\; \mathcal{E}p(\mathbf{u}) + \{0\} \times \mathbb{R}_+ \;=\; \mathcal{E}p(\mathbf{u})$$

   because $(y, \mathbf{u}(x)) \in \mathcal{E}p(\mathbf{u})$. Therefore, $\mathrm{Dom}(\mathbf{u}) + P \subset \mathrm{Dom}(\mathbf{u})$, and thus, is equal to it.
2. Conversely, assume that $\mathcal{E}p(\mathbf{u}) + P \times \mathbb{R}_+ \subset \mathcal{E}p(\mathbf{u})$ holds true, take any $p \in P$, $y := x + p \geq x$ and derive that $\mathbf{u}(y) \leq \mathbf{u}(x)$. Indeed,

$$(y, \mathbf{u}(x)) \;=\; (x, \mathbf{u}(x)) + (p, 0) \;\in\; \mathcal{E}p(\mathbf{u}) + P \times \mathbb{R}_+ \;\subset\; \mathcal{E}p(\mathbf{u})$$

   so that $(y, \mathbf{u}(x)) \in \mathcal{E}p(\mathbf{u})$, i.e., $\mathbf{u}(y) \leq \mathbf{u}(x)$. □

Definition 14.6.2 *[Decreasing Envelope of an Extended Function] Let us consider an extended function* $\mathbf{u} : x \in X \mapsto \mathbf{u}(x) \in \mathbb{R} \cup \{+\infty\}$ *and* $P \subset X$ *a closed convex cone. The function* $\mathbf{u}_{\searrow}$ *defined by*

$$\mathcal{E}p(\mathbf{u}_{\searrow}) \;:=\; \mathcal{E}p(\mathbf{u}) + P \times \mathbb{R}_+$$

is called the P-decreasing envelope *of the function* $\mathbf{u}$ *(or simply* decreasing envelope *if there is no ambiguity).*

In the same way, the two satements

- $\mathbf{u}$ is *increasing* is the sense that for any $y \in x + P$, then $\mathbf{u}(y) \geq \mathbf{u}(x)$
- The epigraph of the function $\mathbf{u}$ satisfies

$$\mathcal{E}p(\mathbf{u}) - P \times \mathbb{R}_+ \;=\; \mathcal{E}p(\mathbf{u})$$

are equivalent and the function $\mathbf{u}^{\nearrow}$ defined by

$$\mathcal{E}p(\mathbf{u}_{\nearrow}) := \mathcal{E}p(\mathbf{u}) - P \times \mathbb{R}_{+}$$

is called its *increasing envelope.*

Lemma 14.6.3 ***[Properties of Decreasing Envelopes]*** *The decreasing envelope* $\mathbf{u}_{\nearrow}$ *of* $\mathbf{u}$ *is larger than or equal to*

$$\inf_{p \in P} \mathbf{u}(x-p) \leq \mathbf{u}_{\searrow}(x) \text{ and } \mathrm{Dom}(\mathbf{u}_{\searrow}) \subset \mathrm{Dom}(\mathbf{u}) + P$$

Equality

$$\inf_{p \in P} \mathbf{u}(x-p) = \mathbf{u}_{\searrow}(x) \text{ and } \mathrm{Dom}(\mathbf{u}_{\searrow}) = \mathrm{Dom}(\mathbf{u}) + P \quad (14.32)$$

holds true whenever $\mathbf{u}$ *is lower semicontinuous and inf-compact (see Definition 18.6.2, p.744).*

For any family $\{\mathbf{u}_i\}_{i \in I}$ *of extended functions, the decreasing envelope of the infimum is the infimum of their decreasing envelopes:*

$$(\inf_{i \in I} \mathbf{u}_i)_{\searrow} = \inf_{i \in I} \mathbf{u}_{i \searrow} \quad (14.33)$$

Proof. To say that $(x, \mathbf{u}_{\searrow}(x))$ belongs to the epigraph of $\mathbf{u}_{\searrow}$ means that there exists $(y, \mathbf{u}(y)) \in \mathcal{E}p(\mathbf{u})$, $p \in P$ and $\lambda \geq 0$ such that $(x, \mathbf{u}_{\searrow}(x)) = (y, \mathbf{u}(y)) + (p, \lambda)$, i.e., such that $\mathbf{u}_{\searrow}(x) = \mathbf{u}(y) + \lambda = \mathbf{u}(x-p) + \lambda$, i.e., such that there exists $p \in P$ satisfying $\mathbf{u}(x-p) \leq \mathbf{u}_{\searrow}(x)$ and thus

$$\mathbf{w}(x) := \inf_{p \in P} \mathbf{u}(x-p) \leq \mathbf{u}_{\searrow}(x)$$

Therefore, if $x \in \mathrm{Dom}(\mathbf{u}_{\searrow})$, there exists some $p \in P$ such that $\mathbf{u}(x-p) < +\infty$, i.e., $x - p \in \mathrm{Dom}(\mathbf{u})$. Hence $\mathrm{Dom}(\mathbf{u}_{\searrow}) \subset \mathrm{Dom}(\mathbf{u}) + P$.

Conversely, there exists $\overline{p} \in P$ such that $\mathbf{u}(x - \overline{p}) = \mathbf{w}(x)$, since $\mathbf{u}$ is assumed to be lower semicontinuous and *inf-compact.* Consequently,

$$\begin{cases} (x, w(x)) = (x - \overline{p} + \overline{p}, \mathbf{u}(x - \overline{p})) \\ = (x - \overline{p}, \mathbf{u}(x - \overline{p})) + (\overline{p}, 0) \in \mathcal{E}p(\mathbf{u}) + P \times \mathbb{R}_{+} =: \mathcal{E}p(\mathbf{u}_{\searrow}) \end{cases}$$

so that $w(x) \geq \mathbf{u}_{\searrow}(x)$, and thus, $w(x) = \mathbf{u}_{\searrow}(x)$.

Let us consider a family of extended functions $\mathbf{u}_i$ and their infimum $u_I := \inf_{i \in I} \mathbf{u}_i$. Property (14.33), p. 594 follows from the fact that $\mathcal{E}p(\mathbf{u}_I) = \bigcup_{i \in I} \mathcal{E}p(\mathbf{u}_i)$ by Lemma 18.2.7, p.718 and that

$$\left(\bigcup_{i\in I}\mathcal{E}p(\mathbf{u}_i)\right)+P\times\mathbb{R}_+ \;=\; \bigcup_{i\in I}(\mathcal{E}p(\mathbf{u}_i)+(P\times\mathbb{R}_+))$$

by *distributivity property* (18.1), p. 714 of the Max-Plus algebra of subsets.
□

14.7 The Min-Inf Convolution Morphism Property and Decreasing Traffic Functions

Theorem 11.5.6, p.471 using the Lax–Hopf formula, once translated in terms of epigraphs, implies very useful properties on congestion functions. The epigraph of the minimum $\mathbf{u} := \min_{i\in I}\mathbf{u}_i$ being obviously the union $\mathcal{E}p(\mathbf{u}) = \bigcup_{i\in I}\mathcal{E}p(\mathbf{u}_i)$ of their epigraphs and the epigraph of their inf-convolution $\mathbf{u} := \bigstar_{i\in I}\mathbf{u}_i$ being the sum $\mathcal{E}p(\mathbf{u}) = \sum_{i\in I}\mathcal{E}p(\mathbf{u}_i)$ of their epigraphs, we obtain

Theorem 14.7.1 ***[The Min Inf-Convolution Morphism]*** *Assume that the Hamiltonian* $\mathbf{h}$ *is a flow function defined in Definition 14.3.1, p.571. Let us consider a finite family of traffic conditions* $\mathbf{c}_i$, $i \in I$. *We denote their viability congestion functions by* $V_{\mathbf{c}_i}$.

1. *The viability congestion function associated to the infimum* $\min_i(\mathbf{c}_i)$ *is the infimum of the viability congestion functions associated with traffic condition* $\mathbf{c}_i$*:*
$$V_{\min_i(\mathbf{c}_i)} \;=\; \min_i(V_{\mathbf{c}_i})$$
This allows us to study the contribution of each traffic condition whenever the traffic condition is their infimum.
2. *The viability congestion function associated with the inf-convolution* $\bigstar_i\mathbf{c}_i$ *of traffic conditions defined by*
$$\mathcal{E}p(\bigstar_i\mathbf{c}_i) \;:=\; \sum_i\mathcal{E}p(\mathbf{c}_i)$$
is the inf-convolution of the viability congestion functions associated with traffic condition $\mathbf{c}_i$*:*
$$V_{\bigstar_i(\mathbf{c}_i)} \;=\; \bigstar_i(V_{\mathbf{c}_i})$$

Proof. These formulas are consequences of Theorem 11.5.6, p.471 stating that the capture basin is a Max Plus morphism.

1. We recall that

$$\mathcal{E}p(\mathbf{c}_{\min_I}) = \bigcup_{i \in I} \mathcal{E}p(\mathbf{c}_i)$$

   Since the union of the epigraphs of functions $\mathbf{u}_i$ is the epigraph of their infimum, we deduce that if a finite number of traffic conditions $\mathbf{c}_i$ are given and denoting their viability congestion functions by $V_{\mathbf{c}_i}$, we infer that $V_{\min_i(\mathbf{c}_i)} = \min_i(V_{\mathbf{c}_i})$.
2. Since the sum of the epigraphs of functions $\mathbf{u}_i$ is the epigraph of their inf-convolution $\bigstar_i \mathbf{u}_i$ defined by

$$\bigstar_i \mathbf{u}_i(x) := \inf_{\sum_i x_i = x} \sum_i \mathbf{u}_i(x_i)$$

   we deduce that if a finite number of traffic conditions $\mathbf{c}_i$ are given and denoting their viability congestion functions by $V_{\mathbf{c}_i}$, we infer that $V_{\bigstar_i(\mathbf{c}_i)} = \bigstar_i(V_{\mathbf{c}_i})$.

This completes the proof. □

We deduce that if the traffic condition is increasing in time and decreasing in position, so is its associated traffic evolution.

Let us denote by $P := \mathbb{R}_- \times \mathbb{R}_+$ the order relation under which a traffic solution is decreasing: $(t, x) \succeq (s, y)$ if and only if $t \le s$ and $y \ge x$. Therefore, the congestion function V is decreasing along this preorder if and only if $\mathcal{E}p(V) = \mathcal{E}p(V) + P \times \mathbb{R}_+$, i.e., if $x_1 \le x_2$ and $t_1 \ge t_2$, then $V(t_2, x_2) \le V(t_1, x_1)$. Its *decreasing envelope* is defined by

$$V_\searrow(t, x) = \inf_{s \ge t,\ y \le x} V(s, y)$$

and its epigraph is equal to $\mathcal{E}p(V_\searrow) := \mathcal{E}p(V) + P \times \mathbb{R}_+$.

Theorem 14.7.2 ***[Monotonicity Property of the Traffic Function and its Optimal Viability Traffic Functions]*** *Assume that the Hamiltonian $\mathbf{h}$ is a flow function defined in Definition 14.3.1, p.571. Let $P := -\mathbb{R}_+ \times \mathbb{R}_+ \times \mathbb{R}_+$, $\mathbf{c} : t \in \mathbb{R}_+ \mapsto \mathbb{R} \cup \{+\infty\}$ be a lower semicontinuous traffic condition, $\mathbf{c}_\searrow$ its decreasing envelope, and $V_\mathbf{c}$ and $V_{\mathbf{c}_\searrow}$ the viability congestion functions associated with the traffic conditions $\mathbf{c}$ and $\mathbf{c}_\searrow$. Therefore, the viability congestion function $V_{\mathbf{c}_\searrow}$ associated with the P-decreasing envelope $\mathbf{c}_\searrow$ of the traffic condition $\mathbf{c}$ is the P-decreasing envelope $(V_\mathbf{c})_\searrow$ of the viability congestion function associated with the traffic condition $\mathbf{c}$:*

$$(V_{\mathbf{c}})_{\searrow} = V_{\mathbf{c}_{\searrow}} \tag{14.34}$$

Consequently, the viability congestion function $V_{\mathbf{c}_{\searrow}}$ associated with the P-decreasing envelope $\mathbf{c}_{\searrow}$ is decreasing in position and increasing in time. Furthermore, the minimal congestion evolutions $\xi^\star(\cdot)$ are increasing in time. Their velocities ${\xi^\star}'(t) \in \partial_+ \boldsymbol{h}(t, \xi^\star(t))$ are nonnegative, the associated densities $-\frac{\partial V}{\partial x}(t, \xi^\star(t))$ belong to the interval $[0, \beta^\sharp]$ and the flow $\frac{\partial V}{\partial t}(t, \xi^\star(t))$ range over the interval $[0, \delta]$. The velocities are equal to 0 on the critical interval $[\beta^\flat, \beta^\sharp]$.

Proof. This a consequence of the second statement of Theorem 14.7.1, p.595 by taking $\mathbf{c}_1 \equiv \mathbf{c}$ and $\mathbf{c}_2 := \psi_{\mathbb{R}_+ \times \mathbb{R}_+}$, the indicator of $\mathbb{R}_+ \times \mathbb{R}_+$. Then $\mathbf{c}_1 \equiv \mathbf{c}_2 = \mathbf{c}_{\searrow}$ and $V_{\mathbf{c}_{\searrow}} = V_{\mathbf{c}} \bigstar \mathbf{c}_2 = (V_{\mathbf{c}})_{\searrow}$. □

The question which arises deals with the construction of decreasing traffic condition in position.

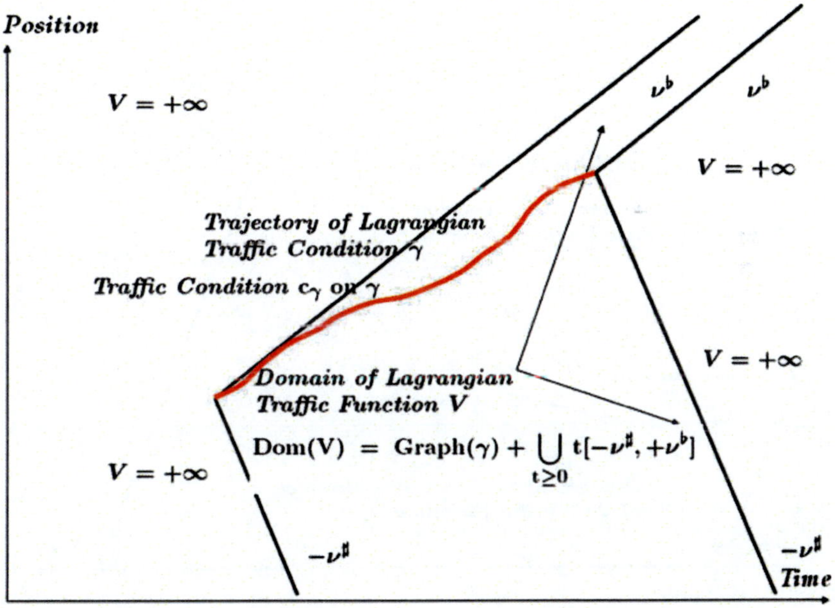

Fig. 14.8 Trajectory of a Traffic Condition and Domain of the Traffic Function. *This figure displays the trajectory of the traffic condition γ from time d to some time e. The domain of the congestion function is the Minkowski sum of the trajectory and the domain $[-\nu^\flat, \nu^\sharp]$, which is represented in this figure. The value of the congestion function is infinite outside of it. Note that it is not decreasing in position and increasing in time.*

Equation (14.32), p. 594 of Lemma 14.6.3, p.594 implies that the domain $\mathrm{Dom}(\mathbf{c}_{\searrow}) =: \mathrm{Graph}(\mathbf{C}_{\searrow})$ satisfies

$$\mathrm{Dom}(\mathbf{c}_{\searrow}) = \mathrm{Dom}(\mathbf{c}) + \mathbb{R}_{-} \times \mathbb{R}_{+}$$

When the domain of $\mathbf{c}$ is made of a single traffic condition $\gamma(\cdot) : t \in [\tau_\gamma^\flat, \tau_\gamma^\sharp[\mapsto \gamma(t)$, we infer that the domain of $\mathbf{c}$ can be regarded as the epigraph of the function $\gamma_{\nearrow}(\cdot)$

$$\mathcal{E}p(\gamma_{\nearrow}) := \mathcal{E}p(\gamma) + \mathbb{R}_{-} \times \mathbb{R}_{+}$$

Hence,

$$\gamma_{\nearrow}(t) := \inf_{s \geq t} \gamma(s)$$

Its domain is the interval $[0, \tau_\gamma^\sharp[$, it is equal to $\gamma(t) := \inf_{s \geq t} \gamma(s)$ on the interval $[0, \tau_\gamma^\flat]$, to $\gamma(t) := \inf_{s \geq t} \gamma(s)$ on the interval $[\tau_\gamma^\flat, \tau_\gamma^\sharp[$ and, if $\tau_\gamma^\sharp < +\infty$, to $+\infty$ whenever $t > \tau_\gamma^\sharp$.

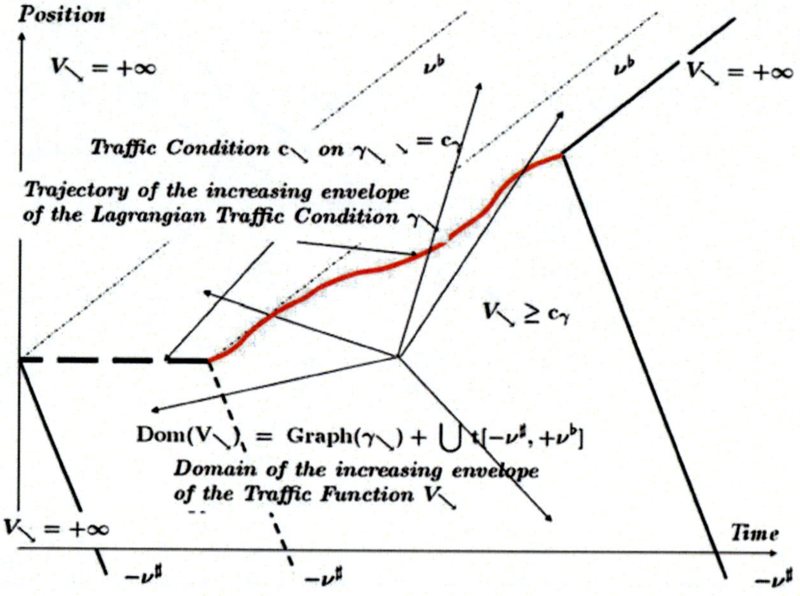

Fig. 14.9 Increasing Envelopes of a Traffic Condition and Domain of the Increasing Traffic Function.
We represent here the decreasing envelope of the trajectory described in Fig. 14.8, p.597: this envelope is obtained by "adding"a branch to the trajectory between the initial time 0 and the departure time d equal to the constant value $\mathbf{c}(d, \xi(d))$. The traffic solution obtained with this envelope is decreasing in position and increasing in time.

In the case when γ is increasing on its domain $[\tau_\gamma^\flat, \tau_\gamma^\sharp[$, extended to $+\infty$ elsewhere, we obtain the simple formula

$$\forall\, t \in [0, \tau_\gamma^\sharp[, \;\; \gamma_{\nearrow}(t) \;:=\; \max(\gamma(\tau_\gamma^\flat), \gamma(t))$$

14.8 The Traffic Regulation Map

For simplicity, we assume that the hamiltonian $\mathbf{h}$ and the traffic solutions are differentiable. The conclusions hold in the general case by weakening the concepts of derivatives (see Chap. 17, p. 681 on Hamilton–Jacobi–Bellman partial differential equations, of which the Moskowitz equation is a particular case), where these questions are solved in the general case.

Another reason to bypass this general study is that the viability algorithms for computing the congestion function and the minimal congestion regulation map do not use at all the mathematical consideration below.

Since the viable evolutions $(t, \xi(t), V(t, \xi(t)))$ are viable on the graph of the traffic map V, we derive that they satisfy the tangential condition $(1, \xi'(t), \mathbf{l}(\xi'(t))) \in T_{\text{Graph}(V)}(t, \xi(t), V(t, \xi(t)))$.

Hence, we define the regulation map in the following way:

Definition 14.8.1 ***[The Traffic Regulation Map]*** *The* minimal congestion regulation map *is defined by*

$$\begin{cases} R(t,x) \;:= \\ \left\{u \in \text{Dom}(\mathbf{l}) \mid (-1, -u, -\mathbf{l}(u)) \;\in\; T_{\text{Graph}(V)}(t, x, V(t,x))\right\} \end{cases} \quad (14.35)$$

Viability Theorem 11.3.4, p.455 implies the following consequence:

Theorem 14.8.2 ***[Regulation of Optimal Traffic Evolutions]*** *Assume that the Hamiltonian* $\mathbf{h}$ *is a flow function defined in Definition 14.3.1, p.571. Minimal congestion evolution* $\xi^\star(\cdot) \in \mathcal{A}(d^\star, T; x)$ *where* $V(d^\star, \xi^\star(d^\star)) = \mathbf{c}(d^\star, \xi^\star(d^\star))$ *are governed by differential inclusion*

$$\xi^{\star\prime}(t) \in R(t, \xi^\star(t)) \text{ starting at } \xi^\star(d^\star) \text{ at time } d^\star \tag{14.36}$$

Optimal evolutions satisfy

$$\frac{dV(t, \xi^\star(t))}{dt} = \mathbf{l}(\xi^{\star\prime}(t)) \geq 0$$

It would be enough to end the analysis of the congestion function here. However, the question arises to characterize further the regulation map in the case when the function V is differentiable. In this case, the tangent cone to the graph is the graph of the derivative:

$$T_{\text{Graph}(V)}(t, x, V(t,x)) = \text{Graph}(DV(t,x))$$

where $DV(t,x) : (\nu, u) \mapsto \frac{\partial V(t,x)}{\partial t}\nu + \left\langle \frac{\partial V(t,x)}{\partial t}, u \right\rangle$.

Proposition 14.8.3 ***[Traffic Regulation Map under Differentiability Assumptions]*** *Assume that the Hamiltonian and the congestion function are differentiable. Then*

$$R(t,x) := \frac{d\mathbf{h}}{dp}\left(-\frac{\partial V(t,x)}{\partial x}\right)$$

is the slope of the tangent to the Hamiltonian at density $-\frac{\partial V(t,x)}{\partial x}$.

Sketch of the proof. Indeed, to say that

$$(-1, -u, -\mathbf{l}(u)) \in T_{\text{Graph}(V)}(t, x, V(t,x)) = \text{Graph}(DV(t,x))$$

means that

$$\frac{\partial V(t,x)}{\partial t} - \left\langle \frac{\partial V(t,x)}{\partial x}, u \right\rangle = -\mathbf{l}(u)$$

Since V is the solution to the Moskowitz partial differential equation, this can be written

$$\mathbf{h}\left(-\frac{\partial V(t,x)}{\partial x}\right) - \left\langle \frac{\partial V(t,x)}{\partial x}, u \right\rangle = -\mathbf{l}(u)$$

By formula (14.9), p. 573 of Lemma 14.3.3, p.572, we infer that this is equivalent to say that $u = \frac{d\mathbf{h}}{dp}\left(-\frac{\partial V(t,x)}{\partial x}\right)$ or that $\frac{\partial V(t,x)}{\partial x} \in \partial \mathbf{l}(-u)$. □

14.9 From Density to Celerity: Shifting Paradigms

The congestion function being the value function of an optimal control problem, the role of density is played by celerities and the fundamental (density) diagram replaced by the celerity diagram. Consequently,

- The dynamic programming equation (14.19), p. 583 states that along a minimal congestion evolution $\xi(\cdot)$,

$$V(T,x) = \mathbf{c}(d^\star, \xi^\star(d^\star)) + \int_{d^\star}^{T} \mathbf{l}({\xi^\star}'(\tau)d\tau)$$

  This suggests to interpret $V_1(d;t,\xi(\cdot)) := \int_d^t \mathbf{l}(\xi'(\tau)d\tau)$ as the congestion function of a minimal congestion evolution $\xi(\cdot) \in \mathcal{A}(d;t,x)$ between d and t. The congestion variational principe states that at optimal departure date $d^\star$, given the "label" $\mathbf{c}(d^\star, \xi^\star)$ of the traffic condition on the departure graph, the congestion function at (t,x) is the sum of this label and of the minimal congestion along the evolution $\xi(\cdot) \in \mathcal{A}(d^\star;t,x)$.
  This would allow the computation of the congestion function and the minimal congestion regulation map, giving a complete answer to the management of traffic under minimal congestion directly from measures of celerities attached to the road and from measures of the velocities of the probe vehicles.
  Hence a meaningful physical interpretation of the congestion function, as it was advocated by Luke, Newell and Daganzo, is to regard it as a congestion function along an minimal congestion evolution.
- We also observed that the Hamiltonian $\mathbf{h}$ does not appear in the formula giving the congestion function, but only its Lagrangian $\mathbf{l}$. The motivation for introducing the Hamiltonian (and to derive the Lagrangian) was due to the fact that, since Greenshields (following vehicles with trucks) and, nowadays, with fixed sensors, the fundamental diagram can be empirically determined and traffic conditions were only Cauchy, Dirichlet or Eulerian boundary conditions.
- Also, the computation of the minimal congestion regulation map and its use to pilot the evolution of minimal congestion evolutions involve only the Lagrangian, and not the Hamiltonian.
  Hence the question arises to investigate whether the use of aerial recordings could provide empirical measures of the Lagrangian data.

- The use of probe vehicles could provide Lagrangian conditions on given trajectories by attaching to them their congestion labels (see Definition 14.2.4, p.569): knowing the velocity of a vehicle along its evolution, the Lagrangian condition **c** could measure the congestion along it

$$\mathbf{c}_{\gamma(\cdot)}(t,\gamma(t)) := \begin{cases} \mathbf{c}(d,\gamma(d)) & \text{if } t \in [0,d] \\ \mathbf{c}(d,\gamma(d)) + \int_d^t \mathbf{l}(\gamma'(\tau)d\tau) & \text{if } t \geq d \end{cases} \tag{14.37}$$

This would allow the computation of the congestion function and the minimal congestion regulation map, giving a complete answer to the management of traffic under minimal congestion directly from measures of celerities attached to the road and from measures of the velocities of the probe vehicles.

Chapter 15
Illustrations in Finance and Economics

15.1 Introduction

This chapter describes two problems motivated by financial mathematics (implicit evaluation of the volatility of portfolios) and of economic theory (bridging the gap between micro and macro economics). These are selected examples chosen for their intrinsic interest and for illustrating how viability concepts and theorems can be used to solve these questions. The focus of this chapter is not the place to expose and develop more examples.

The two questions of interest require the solution of first-order partial differential equations studied in more detail in Chaps. 13, p.523 and 17, p. 681. However, we made all efforts to make these sections self-contained, by assuming that the solutions are differentiable for sketching proofs which are provided in full detail in the previous chapters.

1. *Illustrations in Finance.* Section 15.2, p. 605 deals with one very specific, yet, important, problem in mathematical finance, the search for volatility. We focus our analysis on this topic because it does not require a detailed knowledge of financial mathematics. We tried to make its reading easier by minimizing the knowledge of financial mathematics and providing just the necessary definitions of this field. Mathematically, our investigation requires only some results on ***invariant absorption basins*** (see Definition 2.11.2, p.89) and convex analysis summarized in Sect. 18.7, p.755.
There are two basic approaches of the determination of the value of a portfolio confronted to financial constraints and objectives:

    (a) The *direct approach.* A "model", described by a partial differential equation the solution of which is the valuation function of the portfolio, *is given* and based on the knowledge of the *volatility* (stochastic or tychastic) of the uncertain evolution of the asset returns for providing a whole set of potential evolutions of prices (see Box 1, p.5);

J.-P. Aubin et al., *Viability Theory*, DOI 10.1007/978-3-642-16684-6_15,

(b) the *inverse approach.* Unfortunately, the volatility cannot be measured, so that the value functions given by these partial differential equations are not supported by *empirical values observed on the market.* Consequently, the question arises whether one can derive the knowledge of the volatility from empirical relations satisfied by portfolios (see Box 2, p.5): this question has been dealt with by many authors under the name of implicit or inverse volatility, mostly in the stochastic framework. We propose an answer to this problem in the tychastic framework by assuming known empirical relations between the value of the portfolio, the flow it generates and the prices of the shares, from which we deduce the (tychastic) volatility, allowing the manager to compute its portfolio at each instant from the knowledge of the actual asset return (or price).

Section 15.2, p. 605 attempts to uncover the links between those two approaches.

2. *Illustrations in Economics.*
Section 15.3, p. 620 deals with links between micro-economic and macroeconomic approaches of a dynamical economic problem.
Again, we choose the simplest case by assuming the existence of only one economic agent (a representative of many different agents) acting on the evolution of commodities (the case of n agents can be studied by using the tools of Sect. 13.7, p.544). The invisible hand (or a more visible central banker) is supposed to watch, regulate or control the evolution of some global "economic value" from which prices are derived.
In both microeconomic and macroeconomic cases, commodities are constrained to satisfy viability (scarcity) constraints, which makes the problem difficult to solve.
There are two approaches to this same problem:

(a) *From micro-economy to macro-economy.* By that, we mean that the behavior of the"micro-economic agent" is assumed to be known. The agent makes transactions (velocities of the evolution of commodities), based on his knowledge of

- the monetary rate or the global economy.
- a cost function allowing him to compute the cumulated cost of his transaction at each instant.

The behavior of the agent is the minimization of the (actualized) cumulated cost. The minimal value is regarded as the *"microeconomic" value*, depending on the current time and the commodity transaction at this time.

(b) *From macro-economy to micro-economy.* By that, we mean that the "macroeconomic agent" knows at each instant and for each commodity a *"macroeconomic" value* (a monetary value) of the economy, from which he derives the flows and the prices (which are the first derivatives, one with respect to time, the other one, the derivative with respect to commodity, with a minus sign). We assume that either the invisible hand observes a relation between the value, the flows and the prices (the price map), or a central banker imposes such a relation.

The question arises to know under which conditions the microeconomic and macro economic values are equal, and to extract the relations between the microeconomic and macroeconomic agents. Knowing the transactions and the interest rates of the microeconomic agent, the macro economic agent deduces the macroeconomic value and the price.
Knowing the macroeconomic economic value and the price, the microeconomic agent derives what is the interest rate of the economy and how his transactions depend on the commodity, the price and the interest rate, obeying an *emerging financial Walras law.*
The best has been made for this section to be self-contained. It uses some results on *viability kernels* (see Definition 2.10.2, p.86) and convex analysis summarized in Sect. 18.7, p. 755.

15.2 Uncovering Implicit Volatility in Macroscopic Portfolio Properties

Economic theory is dedicated to the analysis and the computation of supply and demand adjustment laws, among which the *Walras tâtonnement*, in the hope of explaining the mechanisms of price formation (see for instance *Dynamic economic theory: a viability approach* [22, Aubin]). In the last analysis, the choice of the prices is made by the invisible hand of the "Market", the new deity in which many economists and investors believe. Their worshippers may not realize that He may listen to their prayers, but that He is reacting to their actions in a carefully hidden way. Unfortunately, economic theory does not provide explicit or computable pricing mechanisms of assets and underlying, the commodities of the financial markets constituting portfolios.

In most financial scenarios, investors take into account their ignorance of the pricing mechanism. They assume instead that prices *evolve under uncertainty*, and that they can master this uncertainty. They still share the belief that the "Market knows best" how to regulate the prices, above all without human or political regulations. The question became to known how to master this uncertainty. For that, many of them trade the Adam Smith invisible hand against a Brownian movement, since it seems that this unfortunate hand is shaking like a particle on the surface of a liquid. It should then be enough to assume average returns and volatilities to be known for managing portfolios.

We accept the same attitude, but we exchange the Adam Smith invisible hand against tychastic uncertainty instead of a stochastic one on the formation of asset prices for deriving management rules of the portfolio.

Our analysis starts with the observation that we do know the variables on which uncertainty bears: the *future risky returns unknown at the investment date* when the initial investment must be computed. We also know all the potential (contingent) ways to manage a portfolio by determining the exposure (the risky component) of the portfolio. We know from experience that it is possible to restore viability when the liability is reached, by borrowing, instantly (through an impulse), an authorized loan to pay off the forthcoming ruin.

Hence, we choose here a tychastic approach to this problem instead of the stochastic one covered in so many articles and books. We just mention few facts on too well known stochastic uncertainty, we next proceed to comment the much less familiar aspect of uncertainty: tychastic uncertainty (see Sect. 10.10, p.433).

1. Stochastic uncertainty on the returns is described by a space Ω, a filtration $\mathcal{F}$, a Brownian process $W(t)$, a drift $\gamma(x)$ and a volatility $\sigma(x)$: $dR(t) = \gamma(x(t))dt + \sigma(x(t))dW(t)$. The set Ω is not described explicitly (one can always choose the space of all evolutions). Only the drift and volatility are assumed to be explicitly known.

    (a) The random events are not explicitly identified (the famous $\omega \in \Omega$ does not appear in the notations!).
    (b) Stochastic uncertainty does not study the "package of evolutions" (depending on $\omega \in \Omega$), but *functionals over this package*, such as the different moments and their statistical consequences (averages, variance, etc.) used as evaluation of risk.
    (c) Most properties are valid for *"almost all"* constant ω.

2. In this chapter, tyches are returns of the underlying on which the investor has no influence. The uncertainty is described by the tychastic map defined by

$$\mathcal{R}(t) := \left\{R \in \mathbb{R} \text{ such that } R \geq R^{\flat}(t)\right\}$$

    where $R^{\flat}(t)$ are the *lower bounds on returns* (instead of assuming that $dR(t) = \gamma(x(t))dt + \sigma(x(t))dW(t)$).

    (a) Tyches are identified (returns of the underlying, for example) which can then *be used* in dynamic management systems when they are actually observed and known at each date during the evolution.
    (b) For this reason, the results are computed in the worst case (*eradication of risk* instead of its *statistical evaluation*, as in robust control).
    (c) Required properties are valid for *"all" evolutions* of tyches $t \mapsto R(t) \in \mathcal{R}(t)$.

This section uses notations familiar in mathematical finance.

15.2.1 Managing a Portfolio under Unknown Volatility

For simplicity, we assume the return r_0 of a non risky asset known (a government bond, for instance), and we consider only one *risky asset*. The *price of one share* of this asset, called simply the *price*, is denoted by $S \in \mathbb{R}_+$. The velocity of a price evolution $S(t)$ is governed by its *return* $r(t) := \frac{S'(t)}{S(t)}$. Unfortunately, *the evolution of the return is unknown*: it is provided by an adequate law of supply and demand dictated by the invisible hand of a deity named "Market" who keeps its secret.

Definition 15.2.1 *[Portfolio Valuation] We denote by T the* exercise time, *which is the horizon at which a decision to buy or sell is exercised. The* portfolio valuation function $(T, S) \mapsto V(T, S)$ *associates with each exercise time T and initial price S (which is known) the* portfolio valuation $V(T, S)$*: it is the amount of an investment in capital at initial time needed to guarantee that its value at exercise time T under the unknown price $S(T)$ satisfies some prescribed constraints and objectives characterizing a "financial product". At any current (or spot) time $t \in [0, T]$, $V(T - t, S(t))$ is the value at time t of the portfolio in terms of the receding exercise time $T - t$ (also called time to maturity) and the present price $S(t)$.*

The simplest example of constraint is to ask the portfolio valuation to be positive and the value at exercise time to be larger or equal to a given amount. Financial experts have invented hundreds of them.

There are two basic approaches:

1. **From Micro to Macro: the Direct Approach.** In the direct approach, where a "model" of the uncertain evolution of the interest return is given, there are several categories of them, among which:
   (a) a *tychastic* one, based on the simple assumption that the returns, regarded as tyches (Definition 2.5.2, p.59), are ranging over an interval called the *tychastic domain*, the length of which is interpreted as a *tychastic volatility*;
   (b) a *stochastic* one, based on stochastic differential equations involving a coefficient regarded as a *volatility* (see Sect. 10.10, p.433).

   In both cases, stochastic and tychastic volatilities are unknown, since there is no available "volatilitometer".
   In both cases, it is proven that the portfolio valuation $(T, S) \mapsto V(T, S)$ is a solution to a partial differential equation, a linear second-order partial differential equation (among them, the Black and Scholes partial differential

equation) in the stochastic case, a nonlinear first-order Hamilton–Jacobi partial differential equation in the tychastic case.

2. **From Macro to Micro: the Inverse or Implicit Approach.** The fact is that the valuation functions given by these partial differential equations are not supported by *empirical values* $V(T, S)$ *observed on the market*, in particular due to the lack of knowledge of the stochastic or tychastic volatilities. Hence the question arises whether one can determine stochastic or tychastic volatilities from empirical observations, and thus, compute the value function. We investigate this problem in the tychastic case, using convex analysis and viability techniques on the basis of the following assumption: there exists an empirical relation between two sensitivity parameters explained below, the traduction of which is a first-order partial differential equation. The solution to this problem is then obtained through the *invariant absorption basin* of an adequate target (see Definition 2.11.2, p.89).

15.2.2 Implicit Portfolio Problem

Risk management indicators measuring the sensitivity of the *portfolio valuation*[1] $(T, S) \mapsto V(T, S)$ with respect to some parameters are currently used in mathematical finance because they have meaningful financial interpretations: they are called the *Greeks*, since some of them where denoted by Greek letters.

This macroscopic approach using only the exercise time T and the price S as variables on which the portfolio value depends, we use the following sensitivity parameters[2]:

Definition 15.2.2 *[The "Greeks"] We shall use the following sensitivity measures of a portfolio valuation:*

1. *The partial derivatives* $\Theta(T, S) := \dfrac{\partial V(T, S)}{\partial T}$ *with respect to the exercise time is the Greek* Theta.

[1] In short, we may use portfolio valuation instead of portfolio valuation function. This choice or terminology underlines the fact that it is not the value function in the sense of control theory, since the time T is a horizon and not the current time t, but a "portfolio valuation function": See Box 38, p.684.

[2] They require *a priori* the assumption that the portfolio value is sufficiently differentiable, as we shall do for the clarity of the exposition, but non-smooth analysis and the concept of viscosity solutions allow us to extend this definitions for continuous functions (see Chap. 17, p. 681 on Hamilton–Jacobi–Bellman partial differential equations).

2. *The partial derivatives* $\Delta(T,S) := \frac{\partial V(T,S)}{\partial S}$ *with respect to the price of the share is the Greek* Delta. *The* Delta *is regarded as the number of shares of the portfolio.*
3. *The product* $E(T,S) := -S\Delta(T,S) = -S\frac{\partial V(T,S)}{\partial S}$ *is the Greek* Epsilon, *also called the* exposition *of the portfolio. The exposition, being the risky value of the portfolio, allows the manager to deduce the number of shares when he knows the price of the asset (or, in some markets, to make the price by selling or buying a number of shares).*

Observe that the evolution of the portfolio value along time is governed by

$$\frac{dV(T-t,S(t))}{dt} = -r(t)E(T-t,S(t)) - \Theta(T-t,S(t)) \tag{15.1}$$

which involves the "Theta" Θ, the exposition E and the return $r(t)$. We shall assume three types of conditions on the portfolio value function $(T,S) \mapsto V(T,S)$ that a portfolio value function should satisfy:

We denote by r_0 the return of the riskiness asset.

1. **The Exposition Map and Equation**
   The macroscopic approach to the evaluation of a portfolio **assumes** that an empirical relation $\Theta = \mathbf{a}(E)$ between the *Theta* and the exposition *Epsilon* is available. The function $\mathbf{a} : E \mapsto \mathbf{a}(E)$ is *exposition map*. Therefore, knowing the exposition map $\mathbf{a}$, the value of the portfolio is a solution to the Hamilton–Jacobi partial differential equation

$$\frac{\partial V(T,S)}{\partial T} = \mathbf{a}\left(-S\frac{\partial V(T,S)}{\partial S}\right) - r_0 V(T,S)$$

2. **Value Requirement** described by an extended function $\mathbf{c} : (T,S) \mapsto \mathbf{c}(T,S) \in \mathbb{R} \cup \{+\infty\}$, requiring that

$$\forall\, T \geq 0,\ \forall\, S \geq 0,\ \ V(T,S) \leq \mathbf{c}(T,S)$$

   The standard example[3] is given by

$$\mathbf{c}(0,S) := \max(0, S - S_K) \text{ and } \forall\, T > 0,\ \ \mathbf{c}(T,S) := +\infty$$

   where S_K is regarded as a *striking price.*[4]
   In this case, condition $V(T,S) \leq \mathbf{c}(T,S)$ boils down to inequality $V(0,S) \leq \mathbf{c}(0,S)$.

[3] Example of portfolio replicating an *European options.*

[4] The striking price S_K is positive for portfolios replicating options and equal to 0 for standard portfolios with constraint and objective functions independent of the price S.

3. **Value Constraint** described by an extended function $\mathbf{k} : (T,S) \mapsto \mathbf{k}(T,S) \in \mathbb{R} \cup \{+\infty\}$, satisfying $\mathbf{k}(T,S) \leq \mathbf{c}(T,S)$, requiring that

$$\forall\, T \geq 0,\ \forall\, S \geq 0,\ \ \mathbf{k}(T,S) \ \leq \ V(T,S)$$

The function $\mathbf{k}$ is often called the *floor* (representing the flow of liabilities). The case "without constraints" is actually the case when $\mathbf{k}(T,S) = 0$, since the value of the portfolio must be positive. However, one can ask more stringent condition, such as

$$\mathbf{k}(T,S) \ := \ \max(0, Se^{\rho T} - S_K)$$

where ρ is a "guaranteed return" (the case when $\rho = 0$ boils down[5] to $\mathbf{k}(T,S) \ := \ \max(0, S - S_K)$).

This is not the place to develop more examples. For simplicity, however, we shall assume that $\forall\, T > 0,\ \ \mathbf{c}(T,S) := +\infty$.

Definition 15.2.3 *[The Implicit Problem of Portfolio Evaluation]* *Given the exposition map* $\mathbf{a} : E \mapsto \mathbf{a}(E)$, *the constraint function* $\mathbf{k}$ *and the objective function* $\mathbf{c}$, *the* implicit portfolio problem *is to find a portfolio value function* $(T,S) \mapsto V(T,S)$ *satisfying*

$$\begin{cases} \forall\, T \geq 0,\ \forall\, S \geq 0, \\ (i)\ \dfrac{\partial V(T,S)}{\partial T} \ = \mathbf{a}\left(-S\dfrac{\partial V(T,S)}{\partial S}\right) - r_0 V(T,S) \\ (ii)\ \mathbf{k}(T,S) \ \leq \ V(T,S) \ \leq \ \mathbf{c}(T,S) \end{cases} \tag{15.2}$$

If the exposition map $\mathbf{a}$ is convex and bounded below with derivatives lying in a compact interval A, we shall prove the following properties

1. There exists a solution to the implicit portfolio problem, called the viability portfolio value (Theorem 15.2.8, p.615).
2. There exists a *tychastic domain* A concealed in the exposition map $\mathbf{a}$ over which range the returns, the measure of it being the *tychastic volatility*.
3. There exists a utility function $\mathbf{u} : r \mapsto \mathbf{u}(r)$ such that the portfolio value function is also the supremum over the returns $r \in A$ of an intertemporal criterion involving the utility functions $\mathbf{u}$ and the constraint and objective functions $\mathbf{k}$ and $\mathbf{c}$.
4. The partial derivative of the portfolio value function with respect to prices (the Delta in finance terminology) provides the number of shares of the portfolio.

[5] For portfolios replicating *American options*.

5. The portfolio value $V(T,S)$ provides a *measure of risk* that is explained in Theorem 15.2.10, p.619: above $V(T,S)$, there exists a management rule of the portfolio such that, whatever the return in the tychastic domain, the value of the portfolio always satisfies the constraint and objective inequalities, and strictly below $V(T,S)$, whatever the management rule used, there exists at least on return function $r(\cdot)$ such that one of the constraints or objective inequalities is violated.

15.2.3 Emergence of the Tychastic Domain and the Cost Function

Expositions E and returns r are dual variables in the sense that their product has the dimension of a *flow*, because

$$rE = -\frac{\partial V(T,S)}{\partial S} Sr = -\frac{\partial V(T,S)}{\partial S}\frac{dS}{dt}$$

is the product of the velocity of S and $-\Delta(T,S)$, regarded as a number of shares of the risky asset. This is an invitation to involve the duality properties of convex analysis when the exposition map $\mathbf{a}$ is convex (see Sect. 18.7, p. 755).

Definition 15.2.4 ***[Exposition and Return Functions]***
We associate with the exposition map the utility function $\mathbf{u}$ *defined on returns* r *of the risky asset by*

$$\forall\, r,\ \ \mathbf{u}(r) \ := \ \inf_{E}\,[\mathbf{a}(E) + Er] \qquad (15.3)$$

The domain $\mathrm{Dom}(\mathbf{u})$ *of the utility function plays the role of the* tychastic domain *concealed in the exposition map* $\mathbf{a}$. *The measure of this domain is called the* implicit tychastic volatility.

The utility function is obviously an upper semicontinuous concave function measuring the utility of returns. The definition of a tychastic domain and volatility is justified by the forthcoming results.

The following statements adapt the ones of Sect. 18.7, p. 755 to our specific case: conjugate functions are defined in Definition 18.7.2, p.756 and subdifferential in Definition 18.7.5, p.758. Fenchel Theorem 18.7.3, p.756 and Theorem 18.7.7, p.759 on the Legendre property of subdifferential of convex functions and their conjugate imply

Lemma 15.2.5 ***[Utility Function and Exposition Map]*** *Assume that the exposition map* $\mathbf{a}$ *is convex and lower semicontinuous. Then its utility function* $\mathbf{u}$ *defined by (15.3), p. 611*

$$\forall\, r,\ \ \mathbf{u}(r) \ := \ \inf_{E \in \mathrm{Dom}(\mathbf{a})} [\mathbf{a}(E) + Er]$$

is upper semicontinuous and concave. The exposition map is related to the utility function by the relation

$$\forall\, E,\ \ \mathbf{a}(E) \ = \ \sup_{r \in \mathrm{Dom}(\mathbf{u})} [\mathbf{u}(r) - Er] \tag{15.4}$$

so that the Fenchel inequality

$$\forall\, E, \forall\, r,\ \ Er \ \geq \ \mathbf{u}(r) - \mathbf{a}(E)$$

always holds. Furthermore, $\overline{r}$ *achieves the maximum* $\mathbf{a}(E) \ = \ \mathbf{u}(\overline{r}) + Er$ *if and only if* $\overline{r} \in \partial_{-}\mathbf{a}(E)$ *and* $\overline{E}$ *achieves the minimum* $\mathbf{u}(r) \ := \ \mathbf{a}(\overline{E}) - \overline{E}r$ *if and only if* $E \in \partial_{+}E\mathbf{u}(\overline{r})$*. The three following statements are thus equivalent:*

$$\begin{cases} (i) \ \ \overline{E}\overline{r} \ = \ \mathbf{u}(\overline{r}) - \mathbf{a}(\overline{E}) \\ (ii) \ -\overline{r} \ \in \ \partial_{-}\mathbf{a}(\overline{E}) \\ (iii) \ \overline{E} \in \partial_{+}\mathbf{u}(\overline{r}) \end{cases} \tag{15.5}$$

The domain $\mathrm{Dom}(\mathbf{u})$ *is an interval which plays the role of a* concealed tychastic domain. *When it is bounded, it is denoted by* $[-\nu^{\flat}, +\nu^{\sharp}]$*:*

$$\mathrm{Dom}(\mathbf{u}) \ \subset \ [-\nu^{\flat}, +\nu^{\sharp}]$$

Hence the implicit tychastic versatility is bounded above by $\nu^{\sharp} - \nu^{\flat}$ *and* $\mathbf{u}(0) = \delta := \inf_E \mathbf{a}(E)$*. Whenever* $\mathbf{a}_i$ *is lower semicontinuous and (or* $\mathbf{u}_i$ *is upper semicontinuous and concave),* $i = 1,\ 2$*, then*

$$\mathbf{a}_1 \ \leq \ \mathbf{a}_2 \text{ if and only if } \mathbf{u}_1 \ \leq \ \mathbf{u}_2$$

15.2.4 The Viability Portfolio Value

We characterize the smallest solution to the implicit portfolio problem (15.2), p. 610, called the viability portfolio value, through an *invariant absorption basin* of an adequate target (see Definition 2.11.2, p.89) under a *characteristic tychastic system* involving the utility function $\mathbf{u}$ and its domain, the concealed tychastic domain:

$$\begin{cases} (i) & \tau'(t) = -1 \\ (ii) & S'(t) = r(t)S(t) \\ (iii) & y'(t) = r_0 y(t) - \mathbf{u}(r(t)) \\ & \text{where } r(t) \in \text{Dom}(\mathbf{u}) \end{cases} \tag{15.6}$$

controlled by returns $r(\cdot)$. We take for environment $K := \mathcal{E}p(\mathbf{k})$ and for target $C := \mathcal{E}p(\mathbf{c})$.

Definition 15.2.6 ***[Viability Portfolio Valuation]*** *Let us consider the epigraphs $\mathcal{E}p(\mathbf{k})$ and $\mathcal{E}p(\mathbf{c})$ of the constraint function* $\mathbf{k}$ *and the objective function* $\mathbf{c}$. *The* viability portfolio valuation V associated with the functions $\mathbf{k}$ *and* $\mathbf{c}$ *to problem (15.2), p. 610 is defined by the following formula*

$$V(T,x) := \inf_{(T,x,y)\in\text{Abs}_{(15.6)}(\mathcal{E}p(\mathbf{k}),\mathcal{E}p(\mathbf{c}))} y \tag{15.7}$$

We shall recall that the viability portfolio valuation function, when it is differentiable, is a solution to the exposition equation satisfying the constraint and objective conditions. Otherwise, when it is not differentiable, but only continuous, we can give a meaning of a solution in the viscosity sense, using for that purpose subdifferential and superdifferential of continuous functions defined in non-smooth analysis. It is not that important for two reasons: all other properties of viability portfolio valuation functions that are proven in Sect. 15.2, p. 605 are derived directly from the properties of absorption basins without using the concept of derivatives, usual or generalized. Derivatives are used only in the last section, for checking that the viability portfolio function is indeed a solution to the exposition partial differential equation.

We prove that the viability valuation of the portfolio satisfies the *portfolio variational principle.*

We set

$$R(t) := \int_0^t r(\tau)d\tau$$

We introduce:

- the *objective function* $e^{-r_0 t}\mathbf{c}(0, Se^{R(t)})$ on the state $Se^{R(t)}$ at initial time t,
- the cumulated *utility* $\int_0^t e^{-r_0\tau}\mathbf{u}(r(\tau))d\tau$ on the return $r(\cdot)$.

Theorem 15.2.7 ***[Portfolio Variational Principle]*** *We associate with the constraint and objective functions the following functionals on interest returns:*

$$\begin{cases} (i) & \mathbf{J_c}(T,S)(r(\cdot)) := \left(e^{-r_0 T}\mathbf{c}(0, Se^{R(T)}) + \int_0^T e^{-r_0\tau}\mathbf{u}(r(\tau))d\tau\right) \\ (ii) & \mathbf{I_k}(T,S)(r(\cdot)) := \\ & \sup_{t\in[0,T]} \left(e^{-r_0 t}\mathbf{k}(T-t, Se^{R(t)}) + \int_0^t e^{-r_0\tau}\mathbf{u}(r(\tau))d\tau\right) \\ (iii) & \mathbf{L_{(k,c)}}(T,S)(r(\cdot)) := \max(\mathbf{J_c}(T,S)(r(\cdot)), \mathbf{I_k}(T,S)(r(\cdot))) \end{cases} \tag{15.8}$$

The viability portfolio valuation $V(\mathbf{k},\mathbf{c})(t,x)$ to the implicit portfolio problem (15.2), p. 610 satisfies the portfolio variational principle:

$$V_{(\mathbf{k},\mathbf{c})}(T,S) = \sup_{r(\cdot)\in L^1(0,T;\mathrm{Dom}(\mathbf{u}))} \mathbf{L_{(k,c)}}(T,S)(r(\cdot)) \tag{15.9}$$

In the case when $\mathbf{k}(\cdot,S) = 0$, it takes the simpler form:

$$V_{\mathbf{c}}(T,S) = \sup_{r(\cdot)\in L^1(0,T;\mathrm{Dom}(\mathbf{u}))} \left(e^{-r_0 T}\mathbf{c}(0, Se^{R(T)}) + \int_0^T e^{-r_0\tau}\mathbf{u}(r(\tau))d\tau\right) \tag{15.10}$$

Proof. By definition of the absorption basin, to say that (T,S,y) belongs to the absorption basin $\mathrm{Abs}_{(15.6)}(\mathcal{E}p(\mathbf{k}), \mathcal{E}p(\mathbf{c}))$ means that for all measurable functions $r(\cdot) : [0,T] \mapsto \mathrm{Dom}(\mathbf{u})$, there exists some $t^\star \geq 0$ such that the solution

$$t \mapsto \left(T-t, Se^{R(t)}, e^{r_0 t}\left(y - \int_0^t e^{-r_0\tau}\mathbf{u}(r(\tau))d\tau\right)\right)$$

of the characteristic system (15.6), p. 613:

- reaches the target $\mathcal{E}p(\mathbf{c})$ at time $t^\star$,
- is viable in the environment $\mathcal{E}p(\mathbf{k})$ for all $t \in [0, t^\star]$.

The objective condition implies that

$$y - \int_0^{t^\star} e^{-r_0\tau}\mathbf{u}(r(\tau))d\tau \geq e^{-r_0 t^\star}\mathbf{c}\left(T-t^\star, Se^{R(t^\star)}\right)$$

Since we assumed that $\forall\, t > 0$, $\mathbf{c}(t,S) = +\infty$, we infer that $T - t^\star = 0$ and that

$$y - \int_0^T e^{-r_0\tau}\mathbf{u}(r(\tau))d\tau \geq e^{-r_0 T}\mathbf{c}\left(0, Se^{R(T)}\right)$$

i.e., that $\mathbf{J_c}(T,S)(r(\cdot)) \leq y$. On the other hand, since the solution is viable in $\mathcal{E}p(\mathbf{k})$ on $[0,t^\star] = [0,T]$ thanks to the constraint condition, we deduce that for any $t \in [0,T]$,

$$y - \int_0^t e^{-r_0\tau}\mathbf{u}(r(\tau))d\tau \;\geq\; e^{-r_0t}\mathbf{k}\left(T-t, Se^{R(t)}\right)$$

i.e., that $\mathbf{I_k}(T,S)(r(\cdot)) \leq y$. These two inequalities imply that $\mathbf{L_{(k,c)}}(T,S)(r(\cdot)) \leq y$, and thus, by taking the supremum $U(T,S) := \sup_{r(\cdot)} \mathbf{L_{(k,c)}}(T,S)(r(\cdot))$ over interest returns $r(\cdot)$, that $U(T,S) \leq y$. Taking the infimum over the (T,S,y) ranging over the absorption basin, we proved that $U(T,S) \leq V(T,S)$.

Assume now that $U(T,S) < V_{(\mathbf{k},\mathbf{c})}(T,S)$, i.e., that $(T,S,U(T,S))$ does not belong to the absorption basin $\text{Abs}_{(15.6)}(\mathcal{E}p(\mathbf{k}),\mathcal{E}p(\mathbf{c}))$. Then there exists an evolution $\left(T-t, Se^{R(t)}, e^{r_0t}\left(y - \int_0^t e^{-r_0\tau}\mathbf{u}(r(\tau))d\tau\right)\right)$ which is viable in the complement of $\mathcal{E}p(\mathbf{c})$ forever or until it leaves $\mathcal{E}p(\mathbf{k})$. Consequently, either

$$U(T,S) \;<\; e^{-r_0T}\mathbf{c}\left(0, Se^{R(T)}\right) + \int_0^T e^{-r_0\tau}\mathbf{u}(r(\tau))d\tau \;\leq\; U(T,S)$$

or there exists some $t^\star \in [0,T]$ such that

$$U(T,S) \;<\; e^{-r_0t^\star}\mathbf{k}\left(T-t^\star, Se^{R(t^\star)}\right) + \int_0^{t^\star} e^{-r_0\tau}\mathbf{u}(r(\tau))d\tau \;\leq\; U(T,S)$$

In both cases, we obtained the contradiction $U(T,S) < U(T,S)$. □

15.2.5 Managing the Portfolio

Theorem 15.2.8 *[The Viability Portfolio Valuation Solves the Implicit Portfolio Problem] Assume that the utility function* $\mathbf{u}$ *is concave and Lipschitz on its bounded domain. If the viability portfolio valuation function is differentiable, then it is the smallest solution to the implicit portfolio problem (15.2), p. 610:*

$$\begin{cases} \forall T \geq 0,\ \forall S \geq 0, \\ (i)\ \dfrac{\partial V(T,S)}{\partial T} \;=\; \mathbf{a}\left(-S\dfrac{\partial V(T,S)}{\partial S}\right) - r_0V(T,S) \\ (ii)\ \mathbf{k}(T,S) \;\leq\; V(T,S) \;\leq\; \mathbf{c}(T,S) \end{cases}$$

Furthermore, along an actual evolution of prices $S(t)$ *such that their returns* $r(t) := \frac{S'(t)}{S(t)} \in \text{Dom}(\mathbf{u})$ *range over the tychastic domain,* $-\frac{\partial V(T-t, S(t))}{\partial S} = -\Delta(T-t, S(t))$ *provides the number of shares of the portfolio.*

Proof. Recall first that when a function V is continuous, then $\complement \mathcal{E}p(V) = \mathcal{H}yp(V)$. Recall also the characterizations of the tangent cones to the epigraph and hypograph of V when it is differentiable:

$$\begin{cases} (i) \ T_{\mathcal{E}p(V)}(T, S, V(T,S)) \\ \quad = \{(\theta, \xi, \eta) \text{ such that } \frac{\partial V(T,S)}{\partial T}\theta + \frac{\partial V(T,S)}{\partial S}\xi - \eta \leq 0\} \\ (ii) \ T_{\mathcal{H}yp(V)}(T, S, V(T,S)) \\ \quad = \{(\theta, \xi, \eta) \text{ such that } \frac{\partial V(T,S)}{\partial T}\theta + \frac{\partial V(T,S)}{\partial S}\xi - \eta \geq 0\} \end{cases} \quad (15.11)$$

Since $\mathbb{R}_+ \times \mathbb{R} \times \mathbb{R}$ is a repeller under any system of the form $(-1, f(S,r), g(S,r))$, the epigraph of $\mathbf{k}$ is a repeller and the invariance kernel with target coincides with the absorption basin by Lemma 2.12.5, p.94. Inclusions

$$\mathcal{E}p(\mathbf{k}) \subset \mathcal{E}p(V) \subset \mathcal{E}p(\mathbf{c})$$

imply inequalities

$$\forall\, T,\ S \geq 0,\ \ \mathbf{k}(T,S) \leq V(T,S) \leq \mathbf{c}(T,S)$$

It remains to translate Proposition 11.4.10, p.464 on viscosity characterization of invariance kernels with targets for checking that the viability portfolio valuation function is a solution to the exposition partial differential equation (15.2)(i), p. 610: formula (11.22), p. 464 states that

$$\begin{cases} (i) \ \forall (T, S, V(T,S)) \in \mathcal{E}p(V) \backslash \mathcal{E}p(\mathbf{c}), \forall\, r \in \text{Dom}(\mathbf{u}) \\ \quad (-1, rS, r_0 V(T,S) - \mathbf{u}(r)) \in T_{\mathcal{E}p(V)}(T, S, V(T,S)) \\ (ii) \ \forall (T, S, V(T,S)) \in \text{Int}(\mathcal{E}p(\mathbf{k}) \cap \mathcal{H}yp(V)), \exists\ r \in \text{Dom}(\mathbf{u}) \text{ such that} \\ \quad (-1, rS, r_0 V(T,S) - \mathbf{u}(r)) \in T_{\mathcal{H}yp(V)}(T, S, V(T,S)) \end{cases}$$

In the first case, we infer that

$$\forall\, r \in \text{Dom}(\mathbf{u}),\ \ -\frac{\partial V}{\partial T} + \frac{\partial V}{\partial S} rS + \mathbf{u}(r) \leq r_0 V(T,S)$$

By taking the supremum with respect to $r \in \mathrm{Dom}(\mathbf{u})$, we obtain inequality $-\frac{\partial V}{\partial T} + \mathbf{a}\left(-S\frac{\partial V}{\partial S}\right) \leq r_0 V(T,S)$,

In the second case, we deduce that

$$\exists\ r \in \mathrm{Dom}(\mathbf{u}),\ -\frac{\partial V}{\partial T} - \left(-\frac{\partial V}{\partial S}S\right) r + \mathbf{u}(r) \ \geq\ r_0 V(T,S)$$

Indeed, the Fenchel inequality can be written in the form

$$rS\frac{\partial V}{\partial S} + \mathbf{u}(r) \ \leq\ \mathbf{a}\left(-S\frac{\partial V}{\partial S}\right)$$

Both inequalities imply that $-\frac{\partial V}{\partial T} + \mathbf{a}\left(-S\frac{\partial V}{\partial S}\right) \geq r_0 V(T,S)$.

This amounts to saying that the viability portfolio valuation is a solution to the implicit portfolio problem. □

We thus proved that when the utility function and the exposition map are related by

$$\mathbf{u}(r) := \inf_{E \in \mathrm{Dom}(\mathbf{a})} [\mathbf{a}(E) + Er] \text{ or } \mathbf{a}(E) = \sup_{r \in \mathrm{Dom}(\mathbf{u})} [\mathbf{u}(r) - Er]$$

the implicit portfolio value function coincides with the valuation function of the tychastic problem.

Knowing the portfolio value $V(T-t, S(t))$ at time t where the portfolio is given as an implicit portfolio valuation, the macroscopic approach allows the manager to derive its exposition $E(T-t, S(t))$ and the number of shares of the risky asset, but not utility of the unknown returns $r(t)$.

Knowing at each instant the utility function and the returns of the assets, the microscopic approach allows the manager to compute cumulated discounted utility of the known returns $r(t)$, but not the exposition $E(T-t, S(t))$.

The problem arises whether these two dual approaches are equivalent and allow the manager to derive the concealed returns from the known exposition (macroscopic approach) or to derive the unknown exposition from the observation of the return (microscopic approach).

Theorem 15.2.9 *[The Two Ways of Managing a Portfolio] Denote by $S(t)$ the effective price of the share at time t known to the manager. We obtain the inequality*

$$V(T,S) - \int_0^t e^{-r_0\tau}\mathbf{u}(r(\tau))d\tau \ \geq\ e^{-r_0 t}V(T-t, S(t))$$

Furthermore, the two strategies are equivalent:

- ***the macroscopic approach****: the manager knows the portfolio value* $V(T,S)$ *through the exposition map* $\mathbf{a}$ *and the exposition* $E(T-t, S(t))$ *at each time* t*. Then consistent returns* $r(t)$ *are given by the rule*

$$r(t) \in -\partial_+\mathbf{a}(E(T-t,S(t))) \tag{15.12}$$

so that the cumulated discounted utility $\int_0^t e^{-r_0\tau}\mathbf{u}(r(\tau))d\tau$ *can be computed.*

- ***the microscopic approach****: the manager does not have a direct access to the portfolio valuation function, but observes at each time* $t \in [0,T]$ *the return* $r(t)$*. He can also manage his portfolio by the management rule*

$$\forall\, t \in [0,T], \;\; E(T-t,S(t)) \in \partial_+\mathbf{u}(r(t))$$

These two decision rules are equivalent and, when they are used, we obtain the equality

$$V(T,S) - \int_0^t e^{-r_0\tau}\mathbf{u}(r(\tau))d\tau \; = \; e^{-r_0 t}V(T-t,S(t))$$

Proof. 1. We begin by proving the inequality. By Theorem 10.2.7, p.382, we know that

$$\text{Abs}_{(15.6)}(\mathcal{E}p(\mathbf{k}), \mathcal{E}p(\mathbf{c})) \; = \; \text{Abs}_{(15.6)}(\mathcal{E}p(\mathbf{k}), \mathcal{E}p(V))$$

This implies that for any $r(\cdot) \in \text{Dom}(\mathbf{u})$, the evolution

$$\left(T-t, Se^{R(t)}, e^{r_0 t}\left(V(T,S) - \int_0^t e^{-r_0\tau}\mathbf{u}(r(\tau))d\tau\right)\right)$$

is viable in $\mathcal{E}p(V)$ until it reaches the target $\mathcal{E}p(\mathbf{c})$ at time T. Therefore

$$\forall\, t \in [0,T], \;\; V(T,S) - \int_0^t e^{-r_0\tau}\mathbf{u}(r(\tau))d\tau \; \geq \; e^{-r_0 t}V(T-t,S(t))$$

2. Let us set

$$\begin{cases} y(t) := e^{r_0 t}\left(V(T,S) - \int_0^t e^{-r_0\tau}\mathbf{u}(r(\tau))d\tau\right) \\ z(t) := V(T-t,S(t)) \end{cases}$$

The two functions satisfy $y(0) = z(0) = V(T, S)$ and the above inequality means that $y(t) \geq z(t)$ on $[0, T]$.
Assume that we take the either one of the equivalent decision rules:

$$\begin{cases} (i) \;\; (E(T-t, S(t))) \in \partial_{+}\mathbf{u}(r(t)) \\ \quad\;\; \text{or, equivalently} \\ (ii) \; r(t) \in -\partial_{-}\mathbf{a}(E(T-t, S(t))) \end{cases}$$

For proving the equality $y(t) = z(t)$ on $[0, T]$, it is enough to check that they are the solutions of the same differential equation. It is easy to observe that $y'(t) = r_0 y(t) - \mathbf{u}(r(t))$ (starting at $V(T, S)$ at time 0).
On the other hand, let us consider the macroscopic solution $V(T, S)$ to the implicit portfolio problem, and in particular, partial differential equation (15.2), p. 610:

$$\frac{\partial V(T,S)}{\partial T} = \mathbf{a}\left(-S\frac{\partial V(T,S)}{\partial S}\right) - r_0 V(T,S)$$

and consider the exposition $E(T, S) := -S\dfrac{\partial V(T,S)}{\partial S}$.
The function $z(t) := V(T-t, S(t))$ starts at $V(T, S)$ and satisfies the differential equation

$$\begin{cases} z'(t) = -\dfrac{\partial V(T-t,S(t))}{\partial T} + \dfrac{\partial V(T-t,S(t))}{\partial S} S'(t) \\ = -\dfrac{\partial V(T-t,S(t))}{\partial T} - r(t)E(T-t,S(t)) \\ = -\mathbf{a}(E(T-t,S(t))) + r_0 V(T-t,S(t)) - r(t)E(T-t,S(t)) \end{cases}$$

i.e., $z'(t) = r_0 z(t) - \mathbf{a}(E(T-t, S(t))) - r(t)E(T-t, S(t))$.
By taking the management rule $(E(T-t, S(t))) \in \partial_{+}\mathbf{u}(r(t))$, or, equivalently, the return $r(t) \in -\partial_{-}\mathbf{a}(E(T-t, S(t)))$, we infer that $z'(t) = r_0 z(t) - \mathbf{u}(r(t))$ by Fenchel equality (15.17), p.624, characterizing subdifferentials. This implies that the two functions $y(\cdot)$ and $z(\cdot)$ are equal. □

We conclude by observing that the proof of Theorem 15.3.9, p.626 implies two properties which justify the use of the viability portfolio valuation as a measure of risk:

Theorem 15.2.10 ***[Measuring Risk with Viability Portfolio Valuations]*** *The viability portfolio valuation plays the role of a (non probabilistic) measure of risk of violating either the constraint inequality or the objective inequality:*

1. *if the investment exceeds the viability portfolio valuation $V(T,S)$: whatever the realization of an evolution of returns $r(t) \in \mathrm{Dom}(\mathbf{u})$ in the tychastic domain, the valuation of the portfolio managed by decision rule $\forall\, t \in [0,T],\;\; E(T-t, S(t)) \in \partial_+\mathbf{u}(r(t))$ always satisfies both the constraint and objective properties.*
2. *if the investment is strictly less than the viability portfolio valuation $V(T,S)$: whatever the management rule chosen, one of the constraint and objective properties is violated by the completion of at least one evolution of returns $r(t) \in \mathrm{Dom}(\mathbf{u})$.*

15.3 Bridging the Micro–Macro Economic Information Divide

15.3.1 Two Stories for the Same Problem

We throw away thousands of variables to keep only:

- the initial date is equal to 0, $t \geq 0$ denotes the current (evolving present) date and $s \geq t$ the (evolving) future dates;
- the *"economic value"* $m \geq 0$ or $y \geq 0$, to avoid using the polysemous word "money" overloaded with so many interpretations and functions;
- the *commodity* $x := (x_1, \ldots, x_h, \ldots, x_l) \in X := \mathbb{R}^l$ (regarded as a basket of goods), the commodity space, the components x_h of which denoting the quantity of units of good labelled h;
- the *price* $p := (p_1, \ldots, p_h, \ldots, p_l) \in X^\star := \mathbb{R}^{l\star} := \mathcal{L}(\mathbb{R}^l, \mathbb{R})$, a linear map associating with each commodity its value $\langle p, x\rangle := \sum_{h=1}^{l} p_h x_h$, where p_h denotes the value of the unit of commodity h (called the price of h in usual language);
- the *transaction* v or $x'(t)$ at time t of the evolving commodity $t \mapsto x(t)$;
- the *transaction value*$\langle p, x'(t)\rangle := \sum_{h=1}^{l} p_h x'_h(t)$.
- *interest rates* $r(t)$ with which we associate

$$R(t) \;:=\; \int_0^t r(\tau)d\tau$$

We assume here, for simplicity of notations and clarity of exposition, that there exists only one economic agent transforming commodities, the case of several agents being a more or less cumbersome conceptually straightforward task.

Definition 15.3.1 *[Scarcity and Viability Constraints] The scarcity (of available commodities) and viability (for the agent) are described by a time-dependent economic cost function* $\mathbf{k} : (t,x) \mapsto \mathbf{k}(t,x) \in \mathbb{R}_+ \cup \{+\infty\}$. *Setting*

$$K(t) := \{x \in X \text{ such that } \mathbf{k}(t,x) = 0\}$$

and

$$\widehat{K}(t) := \{x \in X \text{ such that } \mathbf{k}(t,x) < +\infty\}$$

the function $\mathbf{k}$ *can be interpreted as the cost of violating constraint* $x \in K(t)$ *by paying a penalty* $0 < \mathbf{k}(t,x) < +\infty$ *whenever* $x \in \widehat{K}(t) \setminus K(t)$ *and an infinite (or death) penalty when* $x \notin \widehat{K}(t)$.
In the classical case, $\mathbf{k}(t,x) = 0$ *whenever* $x \in K(t)$ *and* $\mathbf{k}(t,x) = +\infty$ *whenever* $x \notin K(t)$, *these two environments* $K(t)$ *and* $\widehat{K}(t)$ *coincide.*

1. **From Micro-Economy to Macro-Economy: the Direct Approach.** We denote by $\mathcal{O}(t,x)$ the subset of all (almost everywhere) differentiable commodity evolutions starting at departure time t with commodity x. We assume once and for all that the interest rates $r(\cdot)$ are integrable.
The micro–macro approach assumed known a transaction cost function:

Definition 15.3.2 *[Transaction Cost Function] A transaction cost* function $\mathbf{l} : (v,r) \in X \times \mathbb{R} \mapsto \mathbf{l}(v,r) \in \mathbb{R}_+ \cup \{+\infty\}$ *associates with any transaction* $v \in X$ *and any interest rate* r *its transaction cost* $\mathbf{l}(v,r)$.
We associate with the constraint function $\mathbf{k}$ *the* microeconomic value function

$$\begin{cases} V(t,x) := \inf_{x(\cdot)\in\mathcal{O}(t,x),\, r(\cdot)} \\ \sup_{s\geq t} e^{R(t)} \left(e^{-R(s)}\mathbf{k}(s,x(s)) + \int_t^s e^{-R(\tau)}\mathbf{l}(x'(\tau),r(\tau))d\tau \right) \end{cases} \tag{15.13}$$

which is the smallest value actualized at time t of the future largest sum of the cumulated transaction cost and constraint violation cost.
Denoting by $\mathcal{O}_{K(\cdot)}(t,x)$ *the subset of evolutions* $x(\cdot) \in \mathcal{O}(t,x)$ *viable in the tube* $K(\cdot)$ *in the sense that* $\forall\, s \geq t,\;\; x(s) \in K(s)$, *the microeconomic value function takes a simpler form when* $\mathbf{k}(t,x) = 0$ *whenever* $x \in K(t)$

and $+\infty$ *outside:*

$$V(t,x) := \inf_{x(\cdot)\in\mathcal{O}_{K(\cdot)}(t,x),\, r(\cdot)} e^{R(t)} \int_t^{+\infty} e^{-R(\tau)} \mathbf{l}(x'(\tau), r(\tau)) d\tau \quad (15.14)$$

The function $\mathbf{l}$ is assumed to be independent of time t and commodity x only for simplicity of the presentation.
The microeconomic approach ignores prices and deals only with transactions and interest rates.
Consequently, starting only with time-dependent constraint violation cost and transaction costs, we derived a microeconomic value function describing the behavior of the agent.

2. **From Macro-Economy to Micro-Economy: the Inverse Approach.** We introduce the *macroeconomic value function* $V : (t,x) \mapsto V(t,x)$ associating with time t and commodity x an economic value $V(t,x)$.
In the macroeconomic approach, we assume that the economic value function is indirectly accessible through its first-order derivatives:
   - the first-order partial derivative $y(t,x) := \frac{\partial V(t,x)}{\partial t}$, regarded as a *value flow* (or flow) for any fixed commodity x, the economic value being regarded as a stock of value flows,
   - the first-order partial derivative $\frac{\partial V(t,x)}{\partial x}$, related to the price by the relation $p(t,x) := -\frac{\partial V(t,x)}{\partial x}$

   The question arises to investigate whether the macroeconomic value function can be directly or empirically obtained through econometric measurements of a relation between economic value flows $y = \mathbf{h}(p,m)$ depending on prices p and macroeconomic values m.
   Knowing such relation $\mathbf{h}$, the economic value function can be recovered by solving the first-order Hamilton–Jacobi partial differential equation

$$\frac{\partial V(t,x)}{\partial t} = \mathbf{h}\left(-\frac{\partial V(t,x)}{\partial x}, V(t,x)\right) \quad (15.15)$$

   We also require that the economic value function satisfies the constraint violation cost

$$\forall\, t \geq 0 \text{ and } \forall\, x \in \widehat{K}(t),\;\; \mathbf{k}(t,x) \leq V(t,x)$$

Definition 15.3.3 *[The Price Map] The map* $\mathbf{h} : X^\star \times \mathbb{R} \mapsto \mathbb{R}$ *is regarded as the* price map *mapping prices* p *and macroeconomic values* *m into value flows* $\mathbf{h}(p,m)$. *The macroeconomic value function is defined as the solution to partial differential equation with constraints*

$$\begin{cases} \forall\, t \geq 0 \text{ and } x \in \widehat{K}(t), \\ (i) \ \dfrac{\partial V(t,x)}{\partial t} \ = \mathbf{h}\left(-\dfrac{\partial V(t,x)}{\partial x}, V(t,x)\right) \\ (ii)\ \mathbf{k}(t,x) \ \leq \ V(t,x) \end{cases} \tag{15.16}$$

The macroeconomic approach ignores transactions and deals only with prices.

3. **Bridging the Micro–Macro Economic Divide.** Both approaches aim at computing an economic value function V, the microeconomic one through the intertemporal minimization of the sum of constraint violation cost and cumulated transaction cost, the macroeconomic one by solving a partial differential equation relating prices to economic value flows satisfying a constraint requirement.
   The question arises to investigate under which conditions bearing on the transaction cost function $\mathbf{l}$ and on the price map $\mathbf{h}$ these two approaches provide the same economic value. We shall answer it by using tools of convex analysis and viability theory.

We assume that both the transaction cost function $\mathbf{l} : X \times \mathbb{R} \mapsto \mathbb{R} \cup \{+\infty\}$ and the transaction map $\mathbf{h} : X^\star \times \mathbb{R} \mapsto \mathbb{R} \cup \{+\infty\}$ are lower semicontinuous and convex. For relating them, we need to refer to Sect. 18.7, p. 755 for the Fenchel's Definition 18.7.3, p.756 of conjugate functions and Definition 18.7.5, p.758 for the definition of subdifferentials. Fenchel Theorem 18.7.3, p.756 and Theorem 18.7.7, p.759 on the Legendre property of subdifferential of convex functions and their conjugate adapted to our case imply:

Lemma 15.3.4 *[Conjugate Transaction Cost Function and Price Map] Assume that the transaction cost function* $\mathbf{l} : X \times \mathbb{R} \mapsto \mathbb{R} \cup \{+\infty\}$ *is convex and lower semicontinuous. Then the following conditions are equivalent:*

$$\begin{cases} (i) \ \ \mathbf{l}^\star(-p,y) \ = \ \sup_{v \in X,\ r \in \mathbb{R}} \left[ry - \langle p, v\rangle - \mathbf{l}(v,r)\right] \\ (ii) \ \mathbf{l}(v,r) \ = \ \sup_{p \in X^\star, y \in \mathbb{R}} \left[ry - \langle p, v\rangle - \mathbf{l}^\star(-p,y)\right] \\ (iii) \text{ the transaction cost function } \mathbf{l} \text{ is convex and lower semicontinuous} \\ (iv) \text{ the price map } \mathbf{l}^\star \text{ is convex and lower semicontinuous} \end{cases}$$

They satisfy the Fenchel inequality

$$\forall\, v \in X,\ p \in X^\star,\ r \in \mathbb{R},\ y \in \mathbb{R},\ \ ry - \langle p, v\rangle \ \leq \ \mathbf{l}(v,r) + \mathbf{l}^\star(-p,y)$$

The three following statements are thus equivalent:

$$\begin{cases} (i) & \overline{r}\overline{y} - \langle \overline{p}, \overline{v} \rangle \geq \mathbf{l}(\overline{v}, \overline{r}) + \mathbf{l}^\star(-\overline{p}, \overline{y}) \\ (ii) & (-\overline{p}, \overline{y}) \in \partial_- \mathbf{l}(\overline{v}, \overline{r}) \\ (iii) & (\overline{v}, \overline{r}) \in \partial_- \mathbf{l}^\star(-\overline{p}, \overline{y}) \end{cases} \quad (15.17)$$

We shall relate price maps and transaction cost function for bridging the micro–macro divide:

Definition 15.3.5 ***[Consistent Price Map and Transaction Cost Function]*** *A price map* $\mathbf{h}$ *is* consistent *with a transaction cost function* $\mathbf{l}$ *if* $\mathbf{h}(p, r) := \mathbf{l}^\star(p, r)$ *is the conjugate function of* $\mathbf{l}$.

15.3.2 The Viability Economic Value

We introduce a third concept of economic value, called the *viability economic value*, defined through the viability kernel under a *characteristic control system* involving the cost function $\mathbf{l}$:

$$\begin{cases} (i) & \tau'(t) = +1 \\ (ii) & x'(t) = v(t) \\ (iii) & y'(t) = r(t)y(t) - \mathbf{l}(v(t), r(t)) \\ & \text{where } (v(t), r(t)) \in \text{Dom}(\mathbf{l}) \end{cases} \quad (15.18)$$

controlled by transaction-rate pairs $(v(\cdot), r)$. We take for environment $K := \mathcal{E}p(\mathbf{k})$.

Definition 15.3.6 ***[Viability Economic Value]*** *Let us consider the epigraph* $\mathcal{E}p(\mathbf{k})$ *of the constraint function* $\mathbf{k}$. *The* viability economic value V *is defined by the following formula*

$$V(t, x) := \inf_{(t,x,y) \in \text{Viab}_{(15.18)}(\mathcal{E}p(\mathbf{k}))} y \quad (15.19)$$

Making the above definition explicit, one can prove that the viability economic value coincides with the microeconomic value through the *economic variational principle*.

Theorem 15.3.7 ***[Equality between Viability and Micro-Economic Values]*** *The viability economic value $V(t,x)$ to the economic value problem (15.16), p. 623 coincides with the microeconomic value:*

$$\begin{cases} V(t,x) := \inf_{x(\cdot)\in\mathcal{O}(t,x),\, r(\cdot)} \sup_{s\geq t} e^{R(t)} \\ \left(e^{-R(s)}\mathbf{k}(s,x(s)) + \int_t^s e^{-R(\tau)}\mathbf{l}(x'(\tau),r(\tau))d\tau\right) \end{cases} \tag{15.20}$$

Proof. Observe that a solution $y(\cdot)$ of equation (15.18)(iii), p. 624 starting at y is given by $y(r) = e^{R(r)}\left(y - \int_0^r e^{-R(\tau)}\mathbf{l}(v(\tau),r(\tau))d\tau\right)$.

By definition of viability kernels, to say that (t,x,y) belongs to the viability kernel $\mathrm{Viab}_{(15.18)}(\mathcal{E}p(\mathbf{k}))$ means that there exists an integrable function $(v(\cdot),r(\cdot)) \in \mathrm{Dom}(\mathbf{l})$ such that the solution

$$\left(t+r, \xi(r), \eta(r) := e^{\int_0^r \rho(\tau)d\tau}\left(-\int_0^r e^{-\int_0^\tau \rho(s)ds}\mathbf{l}(\nu(\tau),\rho(\tau))d\tau\right)\right) \in \mathcal{E}p(\mathbf{k})$$

is viable in the environment $\mathcal{E}p(\mathbf{k})$ for all $t \geq 0$. Therefore

$$e^{\int_0^r \rho(\tau)d\tau}\left(y - \int_0^r e^{-\int_0^\tau \rho(s)ds}\mathbf{l}(\nu(\tau),\rho(\tau))d\tau\right) \;\geq\; \mathbf{k}(t+r,\xi(r))$$

Setting $x(r) := \xi(r-t)$, $y(r) := \eta(r-t)$, $r(r) := \rho(r-t)$ and $v(r) := \nu(r-t)$, we observe that $x(\cdot) \subset \mathcal{O}(t,x)$, so that the above inequality reads

$$\forall\, s \geq t,\;\; e^{-R(t)}y - \int_t^s e^{-R(\tau)}\mathbf{l}(v(\tau),r(\tau))d\tau \;\geq\; e^{-R(s)}\mathbf{k}(s,\xi(s))$$

Taking the supremum over $s \geq t$ and the infimum over evolutions $x(\cdot) \in \mathcal{O}(t,x)$ and interest rates $r(\cdot)$ in the above inequality implies that

$$\begin{cases} U(t,x) := \inf_{x(\cdot)\in\mathcal{O}(t,x),\, r(\cdot)} \sup_{s\geq t} \\ e^{R(t)}\left(e^{-R(s)}\mathbf{k}(s,x(s)) + \int_t^s e^{-R(\tau)}\mathbf{l}(x'(\tau),r(\tau))d\tau\right) \;\leq\; y \end{cases}$$

Taking the infimum with respect to the y such that (t,x,y) belongs to the viability kernel, the definition of the viability economic value implies inequality $U(t,x) \leq V(t,x)$.

For proving the opposite equality, assume for a while that $U(t,x) < V(t,x)$ and choose any z such that $U(t,x) \leq z < V(t,x)$. Therefore, (t,x,z) does not belong to the viability kernel $\mathrm{Viab}_{(15.18)}(\mathcal{E}p(\mathbf{k}))$. Hence every evolution

$$(t+\tau, \xi(\tau), \zeta(\tau)) := (t+\tau, x(t+\tau), z(\tau+t))$$

satisfying $x(t) = \xi(0) = x$ and $z(t) = \zeta(0) = z$ leaves the environment $\mathcal{E}p(\mathbf{k})$ at a finite time $\tau^\sharp$:

$$z(\tau^\sharp + t) \; < \; \mathbf{k}(\tau^\sharp + t, x(t+\tau^\sharp))$$

Translating this inequality and setting $s^\sharp := t + \tau^\sharp$, we obtain

$$\begin{cases} U(t,x) \; \leq \; z \; < e^{R(t)} \left(e^{-R(s^\sharp)} \mathbf{k}(s^\sharp, x(s^\sharp)) + \int_t^{s^\sharp} e^{-R(\tau)} \mathbf{l}(x'(\tau), r(\tau)) d\tau \right) \\ \leq \; U(t,x) \end{cases}$$

Hence, we derived a contradiction, so that $U(t,x) = V(t,x)$. □

Definition 15.3.8 *[The Regulation Map] Assume that the price map* $\mathbf{h}$ *is convex and lower semicontinuous, satisfies* $\mathbf{h}(0) = 0$ *and satisfies the following inequalities: there exist* $\xi \in X$ *and constants* β *and* $x > 0$ *such that*

$$\langle p, \xi \rangle - \beta \; \leq \; \mathbf{h}(p,m) \; \leq \; c(\|p\| + |m| + 1)$$

Assume also for simplicity that viability economic value function is differentiable. The economic regulation map R *is defined by*

$$\begin{cases} R(t,x) \; := \; \{(v,r) \in \text{Dom}(\mathbf{l}) \text{ such that} \\ \dfrac{\partial V(t,x)}{\partial t} + \dfrac{\partial V(t,x)}{\partial x} v - rV(t,x) + \mathbf{l}(v,r) \; \leq \; 0 \} \end{cases} \tag{15.21}$$

We prove that the viability economic value function, when it is differentiable, is a solution to the Hamilton–Jacobi equation (15.22), p. 627, satisfying the constraint and objective conditions. Otherwise, when it is not differentiable, but only lower semicontinuous, we can give a meaning of a solution in the Barron-Jensen/Frankowska sense, using for that purpose subdifferential and superdifferential of lower semicontinuous functions defined in non-smooth analysis (see Chaps. 17, p. 681 and 18, p. 713).

Theorem 15.3.9 *[Equality between Viability and Macro-Economic Values] Assume that the transaction cost is a nonnegative, convex and lower semicontinuous. The viability economic value function* $(t,x) \mapsto V(t,x)$ *is the smallest lower semicontinuous function satisfying*

$$\begin{cases} \forall\, t \geq 0 \text{ and } x \in \widehat{K}(t), \\ (i)\ \dfrac{\partial V(t,x)}{\partial t} \leq \mathbf{h}\left(-\dfrac{\partial V(t,x)}{\partial x}, V(t,x)\right) \\ (ii)\ \mathbf{k}(t,x) \leq V(t,x) \end{cases} \tag{15.22}$$

and

$$\partial_{-}\mathbf{h}\left(-\frac{\partial V(t,x)}{\partial x}, V(t,x)\right) \subset R(t,x)$$

Furthermore, if the transaction cost function is Lipschitz, it is the unique solution satisfying

$$\begin{cases} \forall\, (t,x) \in \mathrm{Int}(\mathrm{Dom}(\mathbf{k})) \text{ such that } V(t,x) > \mathbf{k}(t,x), \\ \dfrac{\partial V(t,x)}{\partial t} = \mathbf{h}\left(-\dfrac{\partial V(t,x)}{\partial x}, V(t,x)\right) \\ \forall\, (t,x) \in \partial\mathrm{Dom}(\mathbf{k}) \text{ such that } V(t,x) = \mathbf{k}(t,x), \\ \dfrac{\partial V(t,x)}{\partial t} \geq \mathbf{h}\left(-\dfrac{\partial V(t,x)}{\partial x}, V(t,x)\right) \end{cases}$$

and

$$\partial_{-}\mathbf{h}\left(-\frac{\partial V(t,x)}{\partial x}, V(t,x)\right) = R(t,x)$$

Therefore, the viability economic value coincides with the macroeconomic value.

Proof. Inclusion

$$\mathcal{E}p(\mathbf{k}) \subset \mathcal{E}p(V)$$

implies inequalities

$$\forall\, t,\ x \geq 0,\ \ \mathbf{k}(t,x) \leq V(t,x)$$

Recall the characterization of the tangent cones to the epigraph when it is differentiable:

$$T_{\mathcal{E}p(V)}(t,x,V(t,x)) = \{(\theta,\xi,\eta) \text{ such that } \frac{\partial V(t,x)}{\partial t}\theta + \frac{\partial V(t,x)}{\partial x}\xi - \eta \leq 0\}$$

By Proposition 11.4.1, p.460, the epigraph $\mathcal{E}p(V)$ of the viability economic value is the smallest subset contained in $\mathcal{E}p(\mathbf{k})$ such that

$$\begin{cases} \forall\, (t,x,V(t,x)),\ \exists\, (v,r) \in \mathrm{Dom}(\mathbf{l}) \text{ such that} \\ (+1, v, rV(t,x) - \mathbf{l}(v,r)) \in T_{\mathcal{E}p(V)}(t,x,V(t,x)) \end{cases} \tag{15.23}$$

Assume for simplicity that the viability economic value function is differentiable. Then (v,r) belongs to $R(t,x)$ if and only if

$$\begin{cases} \forall\,(t,x,V(t,x)),\ \ \exists\,(v,r)\in \mathrm{Dom}(\mathbf{l}) \text{ such that} \\ \dfrac{\partial V(t,x)}{\partial t} + \dfrac{\partial V(t,x)}{\partial x}v - rV(t,x) + \mathbf{l}(v,r)\ \leq\ 0 \end{cases}$$

This means that, by (15.21), p. 626, for any $(t,x) \in \mathrm{Dom}(V)$, the value $R(t,x)$ of the regulation map is not empty.

By Lemma 15.3.4, p.623, this implies that

$$\begin{cases} \dfrac{\partial V(t,x)}{\partial t} \leq \displaystyle\sup_{(v,r)\in\mathrm{Dom}(\mathbf{l})} \left(rV(t,x) - \dfrac{\partial V(t,x)}{\partial x}v - \mathbf{l}(v,r)\right) \\ \leq \mathbf{h}\left(-\dfrac{\partial V(t,x)}{\partial x}, V(t,x)\right) \end{cases}$$

Hence $\dfrac{\partial V(t,x)}{\partial t} \leq \mathbf{h}\left(-\dfrac{\partial V(t,x)}{\partial x}, V(t,x)\right)$ and, furthermore, we observe that

$$\partial_{-}\mathbf{h}\left(-\frac{\partial V(t,x)}{\partial x}, V(t,x)\right) \subset R(t,x)$$

If we assume moreover that $\mathbf{l}$ is Lipschitz and that the constraint function is continuous in the interior of its domain, we infer from Theorem 11.4.5, p.463 that the epigraph enjoys the Frankowska property

$$\begin{cases} \forall\,(t,x)\in \mathrm{Int}(\mathrm{Dom}(\mathbf{k})) \text{ such that } V(t,x) > \mathbf{k}(t,x), \forall\,(v,r)\in\mathrm{Dom}(\mathbf{l}) \\ -\dfrac{\partial V(t,x)}{\partial t} - \dfrac{\partial V(t,x)}{\partial x}v + rV(t,x) - \mathbf{l}(v,r)\ \leq\ 0 \end{cases}$$

and that,

$$\begin{cases} \forall\,(t,x)\in \partial\mathrm{Dom}(\mathbf{k}) \text{ such that } V(t,x) = \mathbf{k}(t,x), \forall\,(v,r)\in\mathrm{Dom}(\mathbf{l}) \\ -\dfrac{\partial \mathbf{k}(t,x)}{\partial t} - \dfrac{\partial \mathbf{k}(t,x)}{\partial x}v + r\mathbf{k}(t,x) - \mathbf{l}(v,r)\ \leq\ 0 \end{cases}$$

We infer that

$$\begin{cases} \forall\ \ (t,x)\in \mathrm{Int}(\mathrm{Dom}(K)) \text{ such that } V(t,x) > \mathbf{k}(t,x) \\ -\dfrac{\partial V(t,x)}{\partial t} + \displaystyle\sup_{(v,r)\in\mathrm{Dom}(\mathbf{l})} \left(rV(t,x) - \dfrac{\partial V(t,x)}{\partial x}v - \mathbf{l}(v,r)\right) \\ = \mathbf{h}\left(\dfrac{\partial V(t,x)}{\partial x}, V(t,x)\right)\ \leq\ 0 \end{cases}$$

and that

$$\begin{cases} \forall\,(t,x) \in \partial\mathrm{Dom}(\mathbf{k}) \text{ such that } V(t,x) = \mathbf{k}(t,x), \\ -\dfrac{\partial \mathbf{k}(t,x)}{\partial t} + \sup_{(v,r)\in\mathrm{Dom}(\mathbf{l})} \left(r\mathbf{k}(t,x) - \dfrac{\partial \mathbf{k}(t,x)}{\partial x} v - \mathbf{l}(v,r) \right) \\ = \mathbf{h}\left(\dfrac{\partial \mathbf{k}(t,x)}{\partial x}, V(t,x) \right) \;\leq\; 0 \end{cases}$$

Therefore, in the interior of the domain of $\mathbf{k}$, $\mathbf{h}\left(-\frac{\partial V(t,x)}{\partial x}, V(t,x)\right) = \frac{\partial V(t,x)}{\partial t}$, and on the boundary, when $V(t,x) = \mathbf{k}(t,x)$, then $\mathbf{h}\left(-\frac{\partial V(t,x)}{\partial x}, V(t,x)\right) \leq \frac{\partial V(t,x)}{\partial t}$. Theorem 11.6.6, p.479 states that the viability economic value function is the unique solution to the partial differential equation (15.16)(i), p. 623, and furthermore, that

$$\forall\,(t,x) \in \mathrm{Int}(\mathrm{Dom}(\mathbf{k})),\;\; \partial_{-}\mathbf{h}\left(-\frac{\partial V(t,x)}{\partial x}, V(t,x)\right) \;=\; R(t,x)$$

This completes the proof of the theorem. □

Theorem 15.3.10 *[**The Macro-Micro and Micro–Macro Regulation Maps**] We introduce the two following regulation maps:*

1. ***From Macro to Micro:** the subdifferential map*

$$\partial\mathbf{h} : (p,y) \in X^{\star} \times \mathbb{R} \mapsto \partial\mathbf{h}(p,y) \subset X \times \mathbb{R}$$

associates with any price p and any macroeconomic value y the pairs $(x',r) \in \partial\mathbf{h}(p,y)$ made of a transaction x' and of a interest rate r at the microeconomic level. Knowing the economic value $V(t,x)$ and the price $p(t,x) := \frac{\partial V(t,x)}{\partial x}$, then the underlying economic agent knows how to determine his/her transaction $x'(t)$ and the interest rate allowing him/her to compute the transaction cost:

$$(x'(t), r(t)) \;\in\; \partial\mathbf{h}\left(p(t,x), V(t,x(t))\right)$$

2. ***From Micro to Macro:** the subdifferential map*

$$\partial\mathbf{l} : (v,r) \in X \times \mathbb{R} \mapsto \partial\mathbf{l}(v,r) \subset X \times \mathbb{R}$$

associates with any transaction v and any interest rate r the pairs $(p,y) \in \partial\mathbf{l}(v,r)$ made of a price p and of a macroeconomic value y. Knowing the transaction $x'(t)$ and the interest rate $r(t)$, the invisible hand (or the central banker) knows how to determine the economic value $V(t,x(t))$ and the price $p(t,x) := -\frac{\partial V(t,x(t))}{\partial x}$:

$$(p(t,x), V(t,x(t))) \in \partial \mathbf{l}(x'(t), r(t))$$

The emerging Walras law *states that the difference between the value of the transaction and the transaction cost is equal to the difference between the value flow generated by the transaction and the discounted macroeconomic value holds true:*

$$\langle p(t,x), x'(t)\rangle - \mathbf{l}(x'(t), r(t)) = \mathbf{h}(p(t,x), V(t,x)) - rV(t,x)$$

Proof. The Legendre property described in Lemma 15.17, p.624 states that these two regulation maps are inverse:

$$(x'(t), r(t)) \in \partial \mathbf{h}\,(p(t,x), V(t,x(t)))$$

if and only if

$$\left(-\frac{\partial V(t,x)}{\partial x}, V(t,x(t))\right) \in \partial \mathbf{l}(x'(t), r(t))$$

The Fenchel equality implies that, setting $p(t,x) := -\dfrac{\partial V(t,x)}{\partial x}$, regarded as the price,

$$\langle p(t,x), x'(t)\rangle - \mathbf{l}(x'(t), r(t)) = \mathbf{h}(p(t,x), V(t,x)) - rV(t,x)$$

which can be regarded as the *Walras law.* □

Chapter 16
Viability Solutions to Conservation Laws

16.1 Introduction

Several chapters (Chaps. 13, p.523, 14, p. 563, 17, p. 681, and Sects. 15, p. and 603) are devoted to first-order Hamilton–Jacobi–Bellman partial differential equations. This chapter presents a viability approach to another class of partial differential equations, *conservation laws.* We restrict our study to the Burgers equation (the canonical example of conservation laws) in Sect. 16.2, p.631 for illustrating this approach. We also include in this chapter the short Sect. 16.3, p.675 on a generalization of the Invariant Manifold Theorem to control problems which plays an important role in control theory. It is an addendum to Chap. 8 of the first edition of *Viability Theory* [18, Aubin] (1991) which is not repeated in this second edition.

16.2 Viability Solutions to Conservation Laws

We first introduce the Burgers equation:

Definition 16.2.1 ***[Burgers Partial Differential Equation]*** *The first-order partial differential equation*

$$\frac{\partial \mathbf{U}(t,x)}{\partial t} + \frac{\partial \mathbf{U}(t,x)}{\partial x}\mathbf{U}(t,x) = 0 \tag{16.1}$$

is called the Burgers equation.

This partial differential equation is often used in fluid mechanics as an approximate form of the one dimensional Navier-Stokes equation, and is

J.-P. Aubin et al., *Viability Theory*, DOI 10.1007/978-3-642-16684-6_16,

useful in dealing with nonstationary waves in a fluid. Being the simplest example of nonlinear hyperbolic conservation law, we restrict our study to it, although the viability approach allows us to treat more general cases.

We begin our study by imposing the traditional Cauchy/Dirichlet conditions (or initial/boundary value conditions)

$$\begin{cases} (i)\ \forall\, x \geq \xi, & \mathbf{U}(0,x) \;=\; \mathbf{U}_0(x) \text{ where } x \rightsquigarrow \mathbf{U}_0(x) \text{ is given} \\ (ii)\ \forall\, t \geq 0, & \mathbf{U}(t,\xi) \;=\; \Gamma_\xi(t) \text{ where } t \rightsquigarrow \Gamma_\xi(t) \text{ is given} \end{cases} \tag{16.2}$$

to the solutions to the Burgers equation on the environment defined by the time constraints $t \geq 0$ and space constraints $x \geq \xi$.

We add later in this chapter other conditions (additional intermediate Eulerian conditions, Lagrangian (mobile) conditions, etc.) on the solutions and assume that they satisfy other viability constraints. Section (16.2.10), p. 667 briefly presents the extension of the study to the controlled Burgers inclusion, when the 0 right hand side of the Burgers equation is replaced by a set-valued map, and when viability and Lagrangian conditions are described by general set-valued maps.

The specific feature of solutions to Burgers partial differential equation and other conservation laws is the *set-valued character* of their solutions. Problems without single-valued solutions, for instance when shocks are involved, are known since the work of Riemann. A considerable literature deals with many different ways to give a meaning to the solution to partial differential equation, involving by definition derivatives, whenever the solution is not necessarily differentiable, not even single-valued. Most of the approaches propose *distributions* as candidates for non differentiable solutions, since distributions are always differentiable, at the risk of loosing their pointwise character (see Sect. 18.9, p.765), despite the fact that the pointwise character could be a "physical requirement". In this chapter, we propose a novel approach suggesting set-valued maps instead of distributions as candidates for the concept of solution to partial differential equations such as the Burgers equation, since they are always differentiable too (see Sect. 18.5, p.738). Being novel, this other interpretation of set-valued solution requires more work and is very promising since it enables us to leverage other properties of viability theory. The viability approach to Burgers equation is still at its infancy and promises future breakthroughs. This is just the beginning of an unknown story.

We shall provide three solutions at once: the viability solution, the partial differential equation generalized solution and the tracking solution, and prove that they coincide (as in Sect. 13.2, p.524 in the case of Hamilton–Jacobi equations). We obtain in this way three equivalent interpretations of the same concept.

1. The *viability solution.*
We set $K := [\xi, +\infty[$. Constraints corresponding to Cauchy/Dirichlet problem are described by the set-valued map $\Psi : \mathbb{R}^2 \rightsquigarrow \mathbb{R}$ defined by

$$\Psi(t,x) := \begin{cases} \mathbb{R} \text{ if } x \geq \xi \text{ and } t \geq 0 \\ \emptyset \text{ if } x < \xi \text{ or } t < 0 \end{cases} \quad (16.3)$$

the graph of which is $\mathrm{Graph}(\Psi) := \mathbb{R}_+ \times K \times \mathbb{R}$.
We impose two conditions on the boundary $\partial(\mathrm{Graph}(\Psi)) := (\{0\} \times K \times \mathbb{R}) \cup (\mathbb{R}_+ \cup \{\xi\} \times \mathbb{R})$:

- The *Cauchy or initial condition* $\mathbf{U}_0 : X \rightsquigarrow \mathbf{U}_0(x) \subset \mathbb{R}$, that is extended for $t > 0$ by introducing the set-valued map $(t,x) \rightsquigarrow \mathbf{U}_0^+(t,x)$ defined by

$$\mathbf{U}_0^+(t,x) := \mathbf{U}_0(x) \text{ if } t = 0 \text{ and } \mathbf{U}_0^+(t,x) := \emptyset \text{ otherwise}$$

- The *Dirichlet or boundary condition* $t \mapsto \Gamma_\xi(t)$ at the lower bound ξ of $[\xi, +\infty[$. We extend this (set-valued) map by the set-valued map $\Gamma_\xi^+ : \mathbb{R}_+ \times X \rightsquigarrow X$ defined by:

$$\Gamma_\xi^+(t,x) := \begin{cases} \Gamma_\xi(t) \text{ if } x = \xi \\ \emptyset \quad\ \text{ if } x \neq \xi \end{cases}$$

We take these two conditions into account by introducing the set-valued map $\Phi \subset \Psi$ defined by

$$\Phi(t,x) := \mathbf{U}_0^+(t,x) \cup \Gamma_\xi(t,x) \text{ defined on } (\{0\} \times K) \cup (\mathbb{R}_+ \cup \{\xi\})$$

Definition 16.2.2 *[**Viability Solution to the Cauchy/Dirichlet Burgers Equation**] Let us introduce the "characteristic system"*

$$\begin{cases} (i) \quad \tau'(t) = -1 \\ (ii) \quad x'(t) = -y(t) \\ (iii) \; y'(t) = 0 \end{cases} \quad (16.4)$$

We shall say that the set-valued map $\mathbf{U} : \mathbb{R}_+ \times \mathbb{R}_+ \rightsquigarrow \mathbb{R}$ *defined by*

$$\mathrm{Graph}(\mathbf{U}) := \mathrm{Capt}_{(16.4)}(\mathrm{Graph}(\Psi), \mathrm{Graph}(\Phi)) \quad (16.5)$$

is the viability solution *to the Cauchy/Dirichlet Burgers problem.*

This definition should not look strange and artificial for the reader of this book by now familiar with the concept of capture basin (see Definition 2.10.2, p.86). For partial differential equation specialists, the characteristic system (16.4), p. 633 associated with the Burgers equations is quite classical, the solutions of which are called the *characteristics*, their favorite tools. In some sense, one can say that the viability solution defined through a capture basin "encapsulates" the method of characteristics which underlies the concept of capture basin under the characteristic system.

The graph of the viability solution being a capture basin, it inherits the properties of capture basins presented in this book. What remains to be done is to translate properties of capture basins in terms of set-valued maps and their derivatives. The only difficulty is due to the fact the notations become more and more intricate.

2. The viability solution is equal to *the unique set-valued solution with closed graph to the Burgers partial differential equation with Cauchy/Dirichlet conditions* in the Frankowska sense, when derivatives are the *graphical derivatives* of set-valued maps (coinciding with the usual derivative whenever the map is single-valued and differentiable). Graphical (set-valued) derivatives replace "weak derivatives" taken in spaces of distributions for giving a meaning to non differentiable solutions to the Burgers equation (see Sect. 18.9, p. 765 for a brief account of the differences between the two approaches, where "generalized derivatives" are either set-valued or distributions.) Non continuous single-valued maps which are solutions to the Burgers equation are not necessarily pointwise selections of the viability solution, but measures or distributions.
3. The viability solution coincides with the *tracking solution* to the Burgers problem, providing another property of the viability solution. In this "tracking approach", $T \geq 0$ denotes the (evolving) horizon and $t \in [0, T]$ the current time.

We are looking for a set-valued map $\mathbf{U} : \mathbb{R}_+ \times X \rightsquigarrow Y$ providing the velocities $y \in \mathbf{U}(T, x)$ such that there exists an evolution $y(\cdot)$ of velocities $y(t)$ satisfying

$$\begin{cases} (i) \;\; y(T) = y \\ (ii) \; \forall t \in [0, T], \; y(t) \in \mathbf{U}(t, x(t)) \end{cases}$$

Condition (ii) amounts to saying that the velocities "track" the state at each time.

The problem is to find such a set-valued map $\mathbf{U}$ when the evolution of velocities $y(t) := x'(t)$ is governed by a differential equation $y'(t) = g(t, x(t), y(t))$ or even, a differential inclusion $y'(t) \in G(t, x(t), y(t))$ or a control problem. This amounts to saying that the evolution of the states $x(t)$ is governed by a second-order differential equation $x''(t) = g(t, x(t), x'(t))$ or a second-order differential inclusion $x''(t) \in G(t, x(t), x'(t))$. The particular case when $g = 0$ provides the Burgers tracking problem.

For simplicity of the exposition, we take $K = X$ in this section.

Definition 16.2.3 *[**Burgers Tracking Problem**] Given the initial condition $\mathbf{U}_0$, a set-valued map $\mathbf{U} : \mathbb{R}_+ \times X \rightsquigarrow X$ is a solution to the*

Burgers tracking problem if it satisfies the Burgers tracking property*: any $y \in \mathbf{U}(T,x)$ satisfies*

$$x(T) = x \text{ and } \forall t \in [0,T], \;\; y \in \mathbf{U}(t, x(t)) \tag{16.6}$$

16.2.1 Viability Solution as Solution to Burgers Equation

Recall that $D^{\star\star}\mathbf{U}(t,x,y)$ denotes the convexified derivative of the set-valued map $\mathbf{U}$ at a point (t,x,y) of its graph (see Definition 18.5.5, p.740): it is defined by

$$T^{\star\star}_{\text{Graph}(F)}(x,y) \;=:\; \text{Graph}(D^{\star\star}F(x,y))$$

where $T^{\star\star}_K(x)$ is the closed convex hull to the tangent cone $T_K(x)$ to K at x.

Theorem 16.2.4 *[The Viability Solution is a Solution to Burgers Equation] Assume that the graphs of $\mathbf{U}_0$ and Γ_ξ are closed. The viability solution to the Burgers problem is the* **unique** *solution $\mathbf{U}$ with closed graph to the Burgers partial differential equation (16.1), p. 631, in the sense that*

$$\begin{cases} \quad \forall\, t > 0, \forall\, x > \xi, \forall\, y \in \mathbf{U}(t,x), \\ (i) \;\; 0 \in D^{\star\star}\mathbf{U}(t,x,y)(-1,-y) \\ (ii) \; 0 \in D^{\star\star}\mathbf{U}(t,x,y)(+1,+y) \end{cases} \tag{16.7}$$

satisfying the Cauchy/Dirichlet conditions

$$\begin{cases} (i) \;\; \forall\, x \geq \xi, \;\; \mathbf{U}(0,x) \;=\; \mathbf{U}_0(x) \\ (ii) \; \forall\, t \geq 0, \;\; \mathbf{U}(t,\xi) \cap \text{Int}(\mathbb{R}_+) \;\subset\; \Gamma_\xi(t) \;\subset\; \mathbf{U}(t,\xi) \end{cases} \tag{16.8}$$

and, when $x = \xi$, the further requirement

$$\forall\, t \geq 0, \; \forall\, y \in \mathbf{U}(t,\xi) \cap \mathbb{R}_+, \;\; 0 \;\in\; D^{\star\star}\mathbf{U}(t,\xi,y)(+1,+y) \tag{16.9}$$

Whenever the viability solution $\mathbf{U}$ is single-valued and differentiable at some point (t,x), setting then inclusions (16.7), p. 635 boil down to

$$\forall\, t > 0, \forall\, x > \xi, \;\; 0 \;=\; \frac{\partial \mathbf{U}(t,x)}{\partial t} + \frac{\partial \mathbf{U}(t,x)}{\partial x}\mathbf{U}(t,x) \tag{16.10}$$

Proof. Since the graph of the viability solution to the Cauchy/Dirichlet Burgers problem is the capture basin of the graph of Φ viable in the graph of Ψ and since the graph of Ψ is a repeller because equation (16.4)(i), p. 633 of the characteristic system implies that all solutions $(T-t, x(t), y(t))$ violate the constraint $T-t \geq 0$, Theorem 11.4.6, p.463, implies that the graph of the viability solution is the unique subset satisfying the inclusions

$$\text{Graph}(\Phi) \subset \text{Graph}(\mathbf{U}) \subset \text{Graph}(\Psi)$$

the tangential conditions

$$\forall\, t > 0,\ x > \xi,\ (-1, -y, 0) \in T^{\star\star}_{\text{Graph}(\mathbf{U})}(t, x, y) =: \text{Graph}(D^{\star\star}\mathbf{U}(t, x))$$

and

$$\begin{cases} (i)\ \forall\, t > 0,\ x > \xi,\ y \in \mathbf{U}(t, x), \textit{then} \\ \quad (+1, +y, 0) \in T^{\star\star}_{\text{Graph}(\mathbf{U})}(t, x, y) =: \text{Graph}(D^{\star\star}\mathbf{U}(t, x)) \\ (ii)\ \forall\, t > 0,\ y \in \mathbf{U}(t, \xi) \text{ such that } (+1, +y, 0) \in T^{\star\star}_{\text{Graph}(\Psi)}(t, \xi, y), \\ \quad \text{then } (+1, +y, 0) \in T^{\star\star}_{\text{Graph}(\mathbf{U})}(t, \xi, y) \end{cases}$$

This first inclusion means that $\mathbf{U}$ is the largest solution to inclusion (16.7)(i), p. 635, the second one that $\mathbf{U}$ is the smallest solution satisfying (16.7)(i), p. 635 and that whenever $x = \xi$

$$\forall\, y \geq 0,\ (+1, +y, 0) \in T^{\star\star}_{\text{Graph}(\mathbf{U})}(t, \xi, y)$$

i.e., that, at the boundary, only the weaker boundary requirement

$$\forall\, t \geq 0,\ \forall\, y \in \mathbf{U}(t, \xi) \cap \mathbb{R}_+,\ 0 \in D^{\star\star}\mathbf{U}(t, \xi, y)(+1, +y)$$

is needed.

Moreover, inclusion $\text{Graph}(\mathbf{U}) \subset \text{Graph}(\Psi)$ implies that viability constraints $\mathbf{U}(t, x) \subset \Psi(t, x)$ are satisfied and inclusions $\text{Graph}(\Phi) \subset \text{Graph}(\mathbf{U})$ that

$$\begin{cases} (i)\ \forall\, x \geq \xi,\ \mathbf{U}_0(x) \subset \mathbf{U}(0, x) \\ (ii)\ \forall\, t \geq 0,\ \Gamma_\xi(t) \subset \mathbf{U}(t, \xi) \end{cases}$$

By inclusions (11.18), p. 461 of Theorem 11.4.6, p.463, stating that

$$\text{Graph}(\mathbf{U}) \cap \text{Exit}_{(16.4)}(\text{Graph}(\Psi)) \subset \text{Graph}(\Phi)$$

we obtain the Cauchy/Dirichlet conditions. It is enough to observe that for any $t \geq 0$, $x \geq \xi$,

$$\begin{cases} (i)\ \forall\, y \in \mathbb{R},\ (0, x, y) \in \text{Exit}_{(16.4)}(\text{Graph}(\Psi)) \\ (ii)\ \forall\, y > 0,\ (t, \xi, y) \in \text{Exit}_{(16.4)}(\text{Graph}(\Psi)) \end{cases}$$

This is obvious for $t = 0$ since the first equation $\tau'(t) = -1$ of (16.4)(i), p. 633 implies that $(0, x, y)$ violates the constraint $t = 0$. For $x = \xi$, the second equation $x'(t) = -y(t) := y$ of (16.4)(i), p. 633 implies that $(0, x, y)$ violates the constraint $x = \xi$ whenever $y > 0$. Therefore for $t = 0$, $\mathbf{U}(0, x) \subset \Phi(0, x) = \mathbf{U}_0(x)$ and for $x = \xi$, $\mathbf{U}(t, \xi) \cap \overset{\circ}{\mathbb{R}}_+ \subset \Phi(t, \xi) = \Gamma_\xi(t) \subset \mathbf{U}(t, \xi)$. □

16.2.2 Viability and Tracking Solutions

Since the solutions starting from single-valued initial conditions may become set-valued, and since they can be regarded as initial conditions for future times, we are led to assume that initial conditions may be taken in the class of set-valued maps. Theorem 16.2.5 states that the viability solution is the unique solution (in the Barron–Jensen/Frankowska sense) to the Burgers tracking problem.

Theorem 16.2.5 ***[Viability Solution as the Unique Solution to the Burgers Tracking Problem]*** *The viability solution* $\mathbf{U}$ *(see Definition 16.2.2, p.633) is the* ***unique*** *solution* $\mathbf{V}$ *to the Cauchy Burgers tracking problem.*

Furthermore, $\mathbf{U}(t, x)$ *is the set of fixed points* $y \in \mathbf{U}_0(x - ty)$ *of the map* $y \rightsquigarrow \mathbf{U}_0(x - ty)$.

The viability solution satisfies the "maximum principle":

$$\forall (t, x) \in \mathbb{R}_+ \times X, \quad \sup_{y \in \mathbf{U}(t,x)} |y| \leq \sup_{x \in X} \sup_{y \in \mathbf{U}_0(x)} |y|$$

or, more precisely

$$\forall (t, x) \in \mathbb{R}_+ \times X, \ \mathbf{U}(t, x) \ \subset \ \mathrm{Im}(\mathbf{U}_0)$$

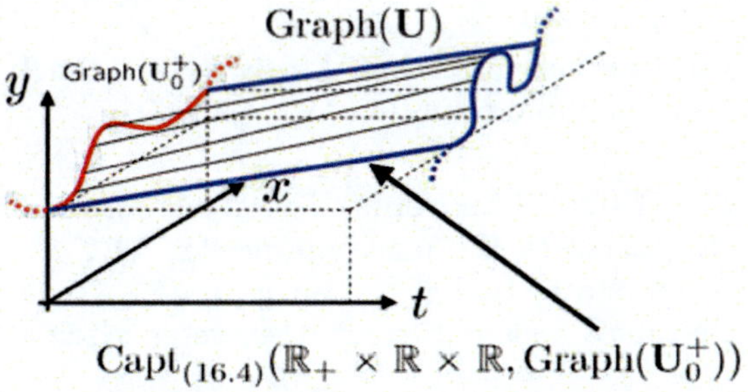

Fig. 16.1 Illustration of the definition of Graph(U).
$\mathrm{Graph}(\mathbf{U})$ *is the capture basin* $\mathrm{Capt}_{(16.4)}(\mathbb{R}_+ \times \mathbb{R} \times \mathbb{R}, \mathrm{Graph}(\mathbf{U}_0^+))$ *of the graph of the set-valued map* $\mathbf{U}_0^+$ *under the characteristic system (16.4).*

Proof. Even though this formula is a particular case of a general one (see Theorem 16.2.26), it is worth providing the direct proof of this very simple example.

Indeed, to say that (T, x, y) belongs to the capture basin $\text{Capt}_{(16.4)}(\mathbb{R}_+ \times \mathbb{R} \times \mathbb{R}, \text{Graph}(\mathbf{U}_0^+))$ amounts to saying that there exists a finite time $t^\star$ such that:

1. the value $(T - t^\star, x - yt^\star, y)$ of the solution to characteristic differential equation (16.4) at time $t^\star$ belongs to the graph of the set-valued map $\mathbf{U}_0^+$,
2. for all $t \in [0, t^\star]$, $(T - t, x - ty, y)$ belongs $\mathbb{R}_+ \times \mathbb{R} \times \mathbb{R}$.

The first condition means that $T - t^\star = 0$ and that $(x - yT, y)$ belongs to the graph of $\mathbf{U}_0$, i.e., that $y \in \mathbf{U}_0(x - yT)$. The second condition means that $t \in [0, T]$, $y(t) \in \mathbf{U}(t, x + (t - T)y)$, i.e., the tracking property on $[0, T]$. Actually, in the absence of viability constraints, it satisfies the tracking property

$$\forall t \geq 0, y \in \mathbf{V}(t, x + (t - T)y)$$

Furthermore, we have proved that $\mathbf{U}(T, x)$ is the set of fixed points of the map $y \rightsquigarrow \mathbf{U}_0(x - Ty)$.

When $T = 0$, we infer that $y \in \mathbf{U}_0(x)$, and thus, that $\mathbf{U}(0, x) \subset \mathbf{U}_0(x)$. By construction, $\mathbf{U}_0(x) \subset \mathbf{U}(0, x)$, so that the initial condition is satisfied.

Theorem 10.2.5 states that the graph of the viability solution is actually the unique graph of a set-valued map $\mathbf{V}$ between $\mathbf{U}_0^+$ and $\mathbb{R}_+ \times \mathbb{R} \times \mathbb{R}$ satisfying

$$\begin{cases} \text{Graph}(\mathbf{V}) = \text{Capt}_{(16.4)}(\text{Graph}(\mathbf{V}), \text{Graph}(\mathbf{U}_0^+)) \\ = \text{Capt}_{(16.4)}(\text{Graph}(\mathbb{R}_+ \times \mathbb{R} \times \mathbb{R}), \text{Graph}(\mathbf{V})) \end{cases}$$

The first condition means that for any $t \in [0, T]$, y belongs to $\mathbf{V}(T - t, x - yt)$. By the change of variable $s := T - t$, this means that for any $s \in [0, T]$, $y \in \mathbf{V}(s, x + (s - T)y)$.

The second condition means that for all $t \geq T$, $y \in \mathbf{V}(t, x + (t - T)y)$. If not, there would exist some $t^\sharp > T$ such that $(t^\sharp, x + (t^\sharp - T)y, y)$ does not belong to the graph of $\mathbf{V}$. Hence $(t^\sharp, x + (t^\sharp - T)y, y) \in \mathbb{R}_+ \times \mathbb{R} \times \mathbb{R} \setminus \text{Graph}(\mathbf{V})$ and, by construction, $(t^\sharp, x + (t^\sharp - T)y, y) \in \text{Capt}_{(16.4)}(\text{Graph}(\mathbb{R}_+ \times \mathbb{R} \times \mathbb{R}), \text{Graph}(\mathbf{V})) = \text{Graph}(\mathbf{V})$, a contradiction.

Hence these two conditions mean that $\mathbf{V}$ satisfies both the Burgers tracking property (16.6) and the initial condition $\mathbf{U}_0$. □

Remark: Fixed Point Theorems. The above formula requires fixed-point theorems to guarantee the non-emptiness of $\mathbf{U}(T, x)$. The Brouwer Fixed Point Theorem states that if the image $\text{Im}(\mathbf{U}_0) \subset [a, b]$ is bounded and if $\mathbf{U}_0$ is single-valued and continuous, then there exists a fixed point on the compact interval $[a, b]$. If $\mathbf{U}_0$ is set-valued map, if its graph is closed, if image $\mathrm{Im}(\mathbf{U}_0) \subset [a, b]$ is bounded and if the images are convex, the Kakutani Fixed Point Theorem, an extension of the Brouwer Fixed Point Theorem to set-valued maps, also guarantees the existence of a fixed point of the set-valued map $y \rightsquigarrow \mathbf{U}_0(x - Ty)$. The Banach–Picard Contraction Theorem implies existence and uniqueness of a fixed point of $y \rightsquigarrow \mathbf{U}_0(x - Ty)$ if $\mathbf{U}_0$ is single-valued and Lipschitz with constant λ and if $T < \frac{1}{\lambda}$ (because the map $y \mapsto \mathbf{U}_0(x - Ty)$ is Lipschitz with constant $T\lambda < 1$). $\square$

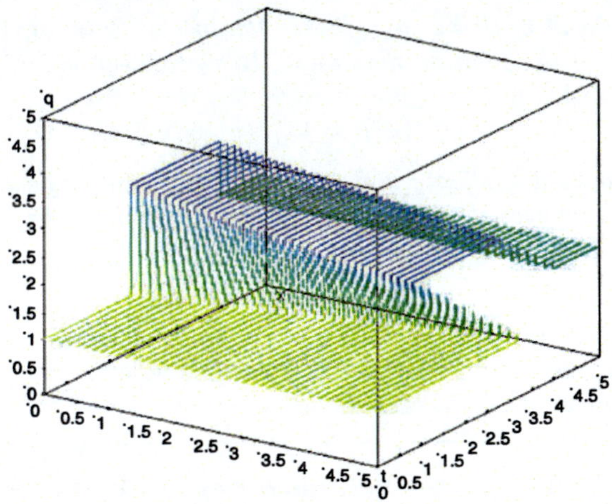

Fig. 16.2 Graph of the Viability Solution starting from a set-valued map initial condition.
The solution to the initial value problem to the Burgers partial differential equation $\frac{\partial \mathbf{U}(t,x)}{\partial t} + \frac{\partial \mathbf{U}(t,x)}{\partial x}\mathbf{U}(t,x) = 0$ *can be set-valued. Its graph is a viability kernel, which can be computed with the viability kernel algorithm.*

Consequently, even when the initial condition $\mathbf{U}_0$ is single-valued, the subset $\mathbf{U}(T, x)$ of fixed points may be set-valued for several values of T and x. The following cases do appear:

1. $\mathbf{U}(t, x) = \{y\}$ (singleton)
2. $\mathbf{U}(t, x) = \{y_1, \ldots, y_p\}$ (finite number of branches)
3. $\mathbf{U}(t, x) = S$ where S is an interval (shock)
4. $\mathbf{U}(t, x) = \emptyset$ (the solution ceases to exist)

(see Fig. 16.4 and Sect. 16.2.3 later).

These phenomena (shocks, several branches, singularity of solutions) are well known from the physicists' community. Paraphrasing Antony Jameson, "*shocks are not a* problem *of the solution, they are a* feature": shocks in fluids have been observed since the advent of supersonic testing facilities, and traffic jam shock waves in highways since urbanization triggered a demand exceeding highway capacity. One of the key questions which occupied scientists for decades has been to understand which of the numerous mathematical solutions represents the physical phenomena observed in nature.

Note that it is even possible to compute the inverse of the set-valued map $x \mapsto \mathbf{U}(t,x)$ knowing the inverse $y \rightsquigarrow \mathbf{U}_0^{-1}(y)$ of the initial condition $x \rightsquigarrow \mathbf{U}_0$ (which can be any set-valued map):

Proposition 16.2.6 ***[Inverse of the Viability Solution]*** *The inverse $y \mapsto \mathbf{U}^{-1}(t,y)$ of the solution map $t \mapsto \mathbf{U}(t,x)$ is given by*

$$\mathbf{U}^{-1}(t,y) = \mathbf{U}_0^{-1}(y) + ty$$

and thus, is defined explicitly and no longer in terms of fixed points. It is single-valued whenever the inverse of the initial condition is single-valued.

One of the first concerns with the Burgers tracking problem is to know under which initial conditions the solution is single-valued or the regions where the evolution is single-valued:

Proposition 16.2.7 ***[Single-Valuedness of the Viability Solution to Burgers Equation]*** *Assume that the initial condition $\mathbf{U}_0 : \mathbb{R} \rightsquigarrow \mathbb{R}$ is monotone (increasing) in the sense that there exists a constant $c \in \mathbb{R}$ (positive or negative) such that*

$$\forall (x_1,x_2),\ \forall y_1 \in \mathbf{U}_0(x_1), \forall\, y_2 \in \mathbf{U}_0(x_2),\ (y_1 - y_2)(x_1 - x_2) \geq\ c(x_1 - x_2)^2$$

Then the solution $\mathbf{U}(\cdot,\cdot)$ to the Burgers equation starting at $\mathbf{U}_0$ is single-valued whenever $t \geq 0$ if $c \geq 0$ and $0 \leq t < \frac{1}{|c|}$ if $c < 0$.

Proof. Since $\mathbf{U}(t,x)$ is the set of fixed points $y \in \mathbf{U}_0(x - yt)$, we deduce from the monotonicity condition on $\mathbf{U}_0$ that whenever $y_i \in \mathbf{U}(t,x)$, $i = 1,2$, then

$$\begin{cases} -t\|y_1 - y_2\|^2 = (y_1 - y_2)((x - y_1 t) - (x - y_2 t)) \\ \geq\ c\|(x - y_1 t) - (x - y_2 t)\| = ct^2\|y_1 - y_2\|^2 \end{cases}$$

so that

$$0\ \geq\ t(ct+1)\|y_1 - y_2\|^2$$

This implies that $y_1 = y_2$ whenever $t \geq 0$ and $ct + 1 > 0$. □

The viability solution benefits of all other properties of capture basins. For instance, the capture basin of a union of targets being (obviously) the union of the capture basins of each of the targets by Lemma 10.2.2, the map associating with any initial condition $\mathbf{U}_{0_i}(x)$ the solution $\mathbf{U}_i(t,x)$ is a *morphism*[1] with respect to the union (of set-valued maps):

Proposition 16.2.8 *[Morphism Property of the Viability Solution] Let $\mathbf{U}_i(t,x)$ denote the solution to the Burgers equation satisfying the initial condition $\mathbf{U}_i(0,x) = \mathbf{U}_{0_i}(x)$. Then*

$$\text{if } \mathbf{U}_0(x) := \bigcup_{i\in\mathbb{I}} \mathbf{U}_{0_i}(x), \text{ then } \mathbf{U}(t,x) = \bigcup_{i\in\mathbb{I}} \mathbf{U}_i(t,x)$$

In other words, one could say that the solution depends "unionly" on the initial conditions, instead of linearly. But this morphism property is as useful as the linearity property of solutions to linear systems.

We shall also derive from the stability properties of capture basins that the solution to the Burgers equations depends continuously on the initial conditions for an adequate concept of convergence: Since the initial conditions and the solutions may be set-valued maps, this stability property has a meaning when the convergence of the set-valued maps is defined by the convergence of their graphs (graphical convergence).

16.2.3 Piecewise Cauchy Conditions

Piecewise initial conditions are a standard case studies of the Burgers equation, as in the famous Riemann problem. Piecewise constant maps are of the form

$$\mathbf{U}_0(x) = \sum_{i=0}^{n} \beta_i \chi_{A_i}(x)$$

or piecewise linear maps of the form

$$\mathbf{U}_0(x) = \sum_{i=0}^{n} (\alpha_i x + \beta_i) \chi_{A_i}(x)$$

where the functions χ_{A_i} are the characteristic functions of the $n+1$ intervals A_i associated with a finite sequence $\delta_1 < \ldots < \delta_n$ by formulas

[1] The group structure $(+,0)$ of the vector space is replaced by the lattice structure $(\cup,\emptyset)$ on the subsets of the vector space, for which the maps associating an initial condition the solution of the semi-linear equation is a lattice-morphism.

$$\begin{cases} A_0 := [\xi, \delta_1[\\ A_i := [\delta_i, \delta_{i+1}[, \ i = 1, \ldots, n-1 \\ A_n := [\delta_n, +\infty[\end{cases}$$

These intervals form a *partition* of $K := [\xi, +\infty[$ and the initial condition is single-valued.

The Burgers equation being nonlinear, we cannot express the values of the solution $\mathbf{U}(t,x)$ as the sum of the values of solutions $\mathbf{U}_i(t,x)$ satisfying the initial conditions $(\alpha_i x + \beta_i)\chi_{A_i}(x)$.

However, we may use the remarkable morphism property stated in Proposition 16.2.8 to compute the solutions starting at the very same initial condition, but rewritten in the form of the amalgam (see Definition 18.3.14, p. 725) of affine functions $x \mapsto \alpha_i x + \beta_i$ on the partitions by subsets A_i

$$\mathbf{U}_0(x) = \bigcup_{i=0}^{n} (\alpha_i x + \beta_i) \Xi_{A_i}(x)$$

where the *mark* Ξ_A of A plays the set-valued role of a the characteristic functions of the interval A defined by

$$\Xi_A(x) := \Xi(A; x) \ := \begin{cases} 1 \text{ if } \ x \in A \\ \emptyset \text{ if } \ x \notin A \end{cases}$$

(see Definition 18.3.12, p. 724).

If $F : X \rightsquigarrow Y$ is a set-valued map, we denote by $F\Xi_A := F \cap \Xi_A : X \rightsquigarrow Y$ the set-valued map defined by

$$F(x)\Xi_A(x) := F(x)\Xi(A; x) \ := \begin{cases} F(x) \text{ if } \ x \in A \\ \emptyset \quad\ \ \text{ if } \ x \notin A \end{cases}$$

(See Definition 18.3.13, p. 725)

In particular, the *shock* at a point σ of magnitude $S \subset Y$ is described by $S\Xi_\sigma := S \cap \Xi_\sigma$ and associates with any x the subset S when $x = \sigma$ and the empty set otherwise (in other words, $S : x \rightsquigarrow S$ is regarded as a constant set-valued map).

Therefore, it is sufficient to compute the solutions $\mathbf{U}_i(t,x)$ to the Burgers equation starting at $(\alpha_i x + \beta_i)\Xi_{A_i}(x)$ or at shocks $S\Xi_\sigma$ to obtain the solution starting at $\mathbf{U}_0$.

Observe that whenever one can approximate an initial condition by piecewise constant (or even better, linear) set-valued maps, we shall be able to approximate the solution of the Burgers equation by solutions starting at these approximate solutions that can be explicitly computed.

We shall use this formula in simple cases:

Lemma 16.2.9 ***[Elementary Building Block]***

1. *If* $\mathbf{U}_0 := 0\Xi_A$*, then,* $\mathbf{U}(t,x) = 0\Xi(A; x)$.

2. *If* $\mathbf{U}_0 := \beta \Xi_A$*, then* $\mathbf{U}(t,x) = \beta\Xi(A+\beta t; x)$.
3. *If* $\mathbf{U}_0 := (\alpha x+\beta)\Xi_A(x)$*, then*
   - *If* $t \neq -\frac{1}{\alpha}$*, then* $\mathbf{U}(t,x) := \left(\frac{\alpha x+\beta}{1+\alpha t}\right) \Xi((1+\alpha t)A+\beta t; x)$
   - *If* $t = -\frac{1}{\alpha}$*, then there exists a shock of size* $\alpha A+\beta$ *at* $-\frac{\beta}{\alpha}$*:* $\mathbf{U}\left(-\frac{1}{\alpha}, x\right) := (\alpha A+\beta)\,\Xi\left(-\frac{\beta}{\alpha}; x\right)$.

   The location of the shock does not depend on A, but only on the coefficients α and β.
4. *If* $\mathbf{U}_0(x) = S\Xi_\sigma(x)$ *is a shock of size* S *at* $x = \sigma$*, then* $\mathbf{U}(t,x) = \left(\frac{x-\sigma}{t}\right) \Xi\,(tS+\sigma; x)$.

Now, combining these examples, we obtain

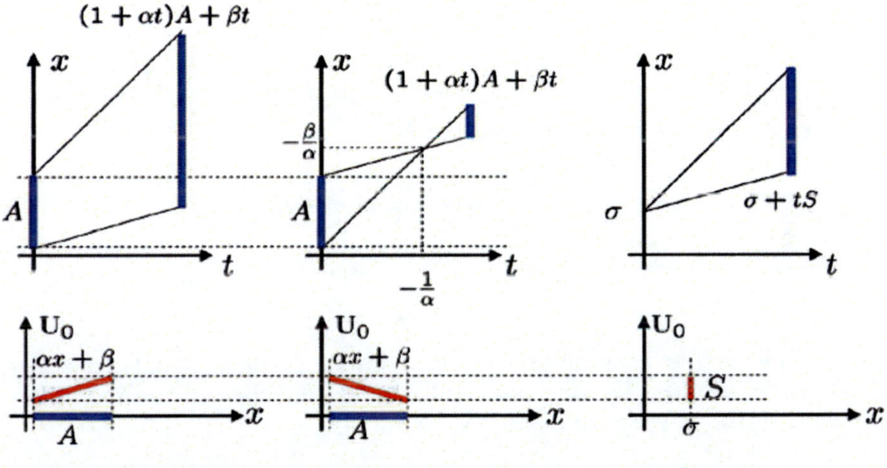

Fig. 16.3 Illustration of the construction of the building blocks solutions presented in Lemma 16.2.9.
Left: Case in which $\alpha > 0$, no shock appears. *Middle:* Case in which $\alpha < 0$, a shock appears at time $t = -\frac{1}{\alpha}$, at $x = -\frac{\beta}{\alpha}$. *Right:* Solution emanating from a shock initially located at $x = \sigma$, and of magnitude S.

Proposition 16.2.10 ***[Viability Solution to Burgers Equation for Piecewise Cauchy Conditions]*** *The viability solution to the Burgers problem satisfying the Cauchy condition*

$$\mathbf{U}(0,x) = \bigcup_{i \in I} (\alpha_i x + \beta_i) \Xi_{A_i}(x)$$

is equal to:

- *Case when* $t \neq -\frac{1}{\alpha_i}$ *for all* $i \in \mathbb{I}$,

$$\mathbf{U}(t,x) = \bigcup_{i \in \mathbb{I}} \left(\frac{\alpha_i x + \beta_i}{1 + \alpha_i t} \right) \Xi \left((1 + \alpha_i t) A_i + t\beta_i ; x \right) \tag{16.11}$$

The cardinal of the set $\mathbb{I}(t,x) := \{i \in \mathbb{I}$ *such that* $x \in (1 + \alpha_i t) A_i + t\beta_i\}$ *denotes the number of elements of* $\mathbf{U}(t,x)$ *and plays the role of a* "valuemeter", *i.e. it measures the degree of set-valuedness of the solution, which is the count of* $\mathbb{I}(t,x)$.

$$\mathbf{U}(t,x) = \left\{ \frac{\alpha_i x + \beta_i}{1 + \alpha_i t} \right\}_{i \in \mathbb{I}(t,x)}$$

- *Case when* $\alpha_i < 0$ *and* $t = -\frac{1}{\alpha_i}$ *for some* $i \in \mathbb{I}$*: we obtain shocks:*

$$\mathbf{U}\left(-\frac{1}{\alpha_i}, x\right) = (\alpha_i A_i + \beta_i) \, \Xi \left(-\frac{\beta_i}{\alpha_i} ; x \right)$$

at points $-\dfrac{\beta_i}{\alpha_i}$ *of size* $\alpha_i A_i + \beta_i$, *which plays the role of a* "valuemeter" *in case of shocks because we can write*

$$\mathbf{U}\left(-\frac{1}{\alpha_i}, -\frac{\beta_i}{\alpha_i}\right) = \alpha_i A_i + \beta_i$$

Proof. This is the consequence of formulas of Lemma 16.2.9 providing the viability solution for elementary Cauchy conditions and the morphism property stated in Proposition 16.2.8, p.641. □

Examples: We now provide examples of more specific (and classical) Cauchy conditions by applying Proposition 16.2.10 and the particular solutions already provided:

In particular, we obtain the following "building blocks" for constructing solutions to the Burgers equation with piecewise linear Cauchy conditions:

1. **O-Marks:**

$$\begin{cases} \text{if } \mathbf{U}_0 := 0\Xi_{]-\infty,0]} \text{ then } \mathbf{U}(t,x) = 0\Xi(]-\infty,0]\,;x) \\ \text{if } \mathbf{U}_0 := 0\Xi_{[0,\infty[} \text{ then } \mathbf{U}(t,x) = 0\Xi([0,+\infty[\,;x) \\ \text{if } \mathbf{U}_0 := 0\Xi_{]-\infty,1]} \text{ then } \mathbf{U}(t,x) = 0\Xi(]-\infty,1]\,;x) \end{cases}$$

2. **Marks:**

$$\begin{cases} \text{if } \mathbf{U}_0 := \Xi_{]-\infty,0]} \text{ then } \mathbf{U}(t,x) = \Xi(]-\infty,t]\,;x) \\ \text{if } \mathbf{U}_0 := \Xi_{[0,\infty[} \text{ then } \mathbf{U}(t,x) = \Xi([t,+\infty[\,;x) \\ \text{if } \mathbf{U}_0 := \Xi_{[1\infty[} \text{ then } \mathbf{U}(t,x) = \Xi([t+1,+\infty[\,;x) \\ \end{cases}$$

3. **Affine Maps**

- **Increasing Linear Maps**

$$\text{if } \mathbf{U}_0 := x\Xi_{[0,1]} \text{ then } \mathbf{U}(t,x) = \frac{x}{1+t}\Xi([0,1+t];x)$$

- **Decreasing Linear Maps and Emergence of Shocks**

$$\text{If } \mathbf{U}_0(x) = (1-x)\Xi_{[0,1]}, \text{ then } \mathbf{U}(t,x) = \begin{cases} \frac{1-x}{1-t} \text{ if } t \leq x < 1 \\ [0,1] \text{ if } x = 1 \,\&\, t = 1 \text{ (shock)} \\ \frac{x-1}{t-1} \text{ if } 1 < x \leq t \end{cases}$$

4. **Shocks**

$$\text{if } \mathbf{U}_0 := [0,1]\,\Xi_{\{1\}} \text{ then } \mathbf{U}(t,x) = \left(\frac{x-1}{t}\right)\Xi([1,1+t];x)$$

By combining them, we obtain

1. If the Cauchy condition $\mathbf{U}_0 := \xi_A = \Xi_A \cup 0\Xi_{\complement(A)}$ is the mark of A defined by

$$\mathbf{U}_0(x) := \begin{cases} 1 \text{ if } x \in A \\ 0 \text{ if } x \notin A \end{cases}$$

the solution $\mathbf{U}$ is given by

$$\mathbf{U}(t,x) = \Xi(A+t;x) \cup 0\Xi(\complement(A);x) = \begin{cases} 1 & \text{if } x \in A \cap (A+t) \\ 0 & \text{if } x \in \complement(A \cup (A+t)) = \complement A \backslash A+t \\ \{0,1\} & \text{if } x \in (A+t) \setminus A \\ \emptyset & \text{if } x \in A \setminus (A+t) \end{cases}$$

2. If the Cauchy condition is $\mathbf{U}_0 := \Xi_{]-\infty,0]} \cup 0\Xi_{[0,+\infty[}$ defined by

$$\mathbf{U}_0(x) := \begin{cases} 1 \text{ if } x \leq 0 \\ 0 \text{ if } x \geq 0 \end{cases}$$

(Riemann Problem), then the viability solution is given by the formula:

$$\mathbf{U}(t,x) = \Xi([-\infty,t]\,;x) \cup 0\Xi([0,+\infty[\,;x) = \begin{cases} 1 & \text{if } x \leq 0 \\ \{0,1\} & \text{if } 0 \leq x \leq t \\ 0 & \text{if } x \geq t \end{cases}$$

and the viability solution is set-valued: its images contains two points for $x \in [0,t]$. (see Fig. 16.4).

3. If the Cauchy condition is $\mathbf{U}_0 = 0\Xi_{]-\infty,0]} \cup \Xi_{[0,+\infty[}$ defined by

$$\mathbf{U}_0(x) := \begin{cases} 0 \text{ if } x \leq 0 \\ 1 \text{ if } x \geq 0 \end{cases}$$

then the viability solution becomes

$$\mathbf{U}(t,x) = \Xi(]-\infty,0];x) \cup \Xi([0,+\infty[;x) = \begin{cases} 0 \text{ if } x \leq 0 \\ \emptyset \text{ if } 0 \leq x \leq t \\ 1 \text{ if } x \geq t \end{cases}$$

and the viability solution is set-valued: its images are empty for $x \in [0,t]$ (see Fig. 16.4).

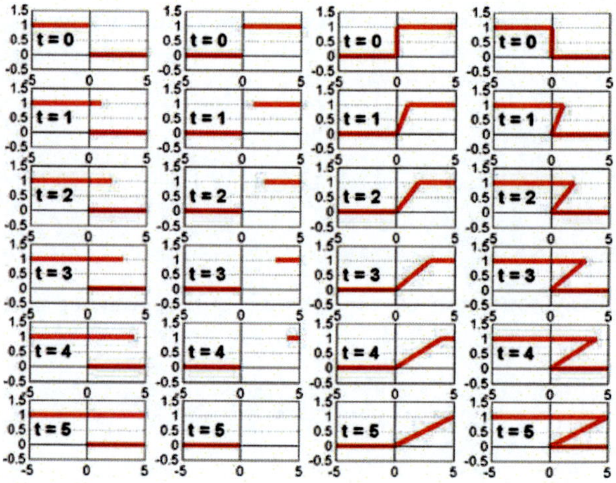

Fig. 16.4 Viability solution of the Burgers problem.
From left to right: ***Column 1:*** *Set-valued solution for the Riemann problem. The solution becomes instantaneously set-valued, due to a higher velocity of propagation of the $x \leq 0$ portion of the solution.* ***Column 2:*** *Set-valued solution of the opposite Cauchy condition. Note that the solution is set-valued, because on the interval $x \in [0,t]$, it is empty.* ***Column 3:*** *Single valued solution obtained from an initial set-valued condition, for the same Cauchy condition as for the middle column. This solution coincides with the usual entropy solution of Burgers equation.* ***Column 4:*** *Set-valued solution obtained from an initial set-valued condition, for the inverted Cauchy condition of Column 3. The Cauchy condition is set-valued and remains set-valued.*

4. If the Cauchy condition is $\mathbf{U}_0 := 0\Xi_{]-\infty,1]} \cup [0,1]\,\Xi_1 \cup \Xi_{[1,+\infty[}$ describing a shock at time 0 at $x = 1$, given by

$$\mathbf{U}_0(x) = \begin{cases} 0 & \text{if } x \le 1 \\ [0,1] & \text{if } x = 1 \\ 1 & \text{if } x \ge 1 \end{cases}$$

we obtain

$$\begin{aligned} \mathbf{U}(t,x) &= \Xi(]-\infty,1];x) \cup \left(\frac{x-1}{t}\right) \Xi([1,1+t];x) \cup \Xi([1+t,+\infty[;x) \\ &= \begin{cases} 0 & \text{if } x \le 1 \\ \frac{x-1}{t} & \text{if } 1 \le x \le 1+t \\ 1 & \text{if } x \ \ge \ 1+t \end{cases} \end{aligned}$$

The initial shock dissolves to an increasing linear function. This phenomenon is sometimes called *self similar expansion wave*, and is known in the field of fluid mechanics as *expansion fan* or *centered rarefaction*.

5. If the Cauchy condition is $\mathbf{U}_0 = 0\Xi_{]-\infty,0]} \cup x\Xi_{[0,1]} \cup \Xi_{[1,+\infty[}$, given by

$$\mathbf{U}_0(x) = \begin{cases} 0 \text{ if } x \le 0 \\ x \text{ if } x \in [0,1] \\ 1 \text{ if } x \ge 1 \end{cases}$$

then the solution $\mathbf{U}$ is given by

$$\begin{aligned} \mathbf{U}(t,x) &= \Xi(]-\infty,0];x) \cup \left(\frac{x}{1+t}\right) \Xi([0,1+t];x) \cup \Xi([1+t,+\infty[;x) \\ &= \begin{cases} 0 & \text{if } x \le 0 \\ \frac{x}{1+t} & \text{if } 0 \le x \le 1+t \\ 1 & \text{if } x \ \ge \ 1+t \end{cases} \end{aligned}$$

It is single-valued as a function of x for all $t \ge 0$ (see Fig. 16.4). The single valuedness of the viability solution is due to the increasing property of the initial value, a consequence of the general statement Proposition 16.2.7.

6. If the Cauchy condition is $\mathbf{U}_0 := \Xi_{]-\infty,0]} \cup (1-x)\Xi_{[0,1]} \cup 0\Xi_{[1,+\infty[}$ is given by

$$\mathbf{U}_0(x) = \begin{cases} 1 & \text{if } x \le 0 \\ 1-x & \text{if } x \in [0,1] \\ 0 & \text{if } x \ge 1 \end{cases}$$

Then one can check that the solution $\mathbf{U}$ is given by

$$\mathbf{U}(t,x) = \begin{cases} 1 & \text{if } x < \min(t,1) \\ \frac{1-x}{1-t} & \text{if } \min(t,1) \le x < 1 \text{ (single-valued)} \\ [0,1] & \text{if } x = 1 \;\&\; t = 1 \text{ (shock)} \\ \left\{0, 1, \frac{x-1}{t-1}\right\} & \text{if } 1 < x \le \max(t,1) \text{ (Z-shape)} \\ 0 & \text{if } x \ge \max(t,1) \end{cases}$$

As a function of x, the solution has a Z-shape after the shock that happened at $t = 1$ and $x = 1$. This is due to the fact that the Cauchy condition $\mathbf{U}_0$ is no longer increasing, and that it is single-valued on some finite time interval.

16.2.4 Dirichlet Boundary Conditions

Instead of taking $X := \mathbb{R}$, it might be useful to only consider $K := [\xi, +\infty[$. This could represent a half infinite shock tube (for which we are interested in the propagation of shocks emanating from an end and travelling far from the end).

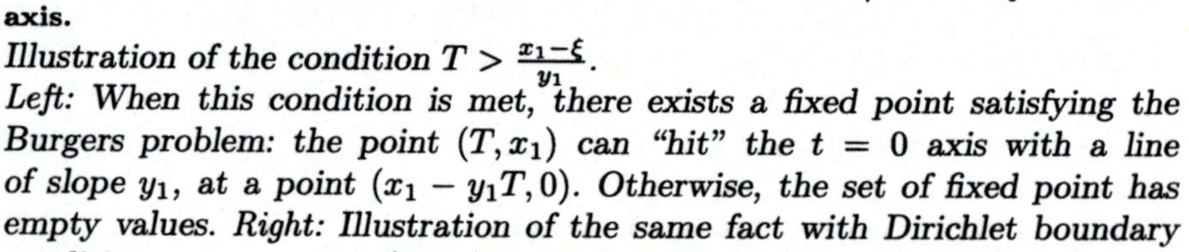

Fig. 16.5 Intersection of characteristic lines and initial/boundary condition axis.
Illustration of the condition $T > \frac{x_1-\xi}{y_1}$.
Left: When this condition is met, there exists a fixed point satisfying the Burgers problem: the point (T, x_1) *can "hit" the* $t = 0$ *axis with a line of slope* y_1*, at a point* $(x_1 - y_1 T, 0)$*. Otherwise, the set of fixed point has empty values. Right: Illustration of the same fact with Dirichlet boundary conditions: now a point* (T, x_2) *with corresponding* y_2 *"hits" the boundary of the domain at* $(\xi, T - \frac{x_2-\xi}{y_2})$*: the set of fixed points is not empty anymore because of the Dirichlet boundary conditions*

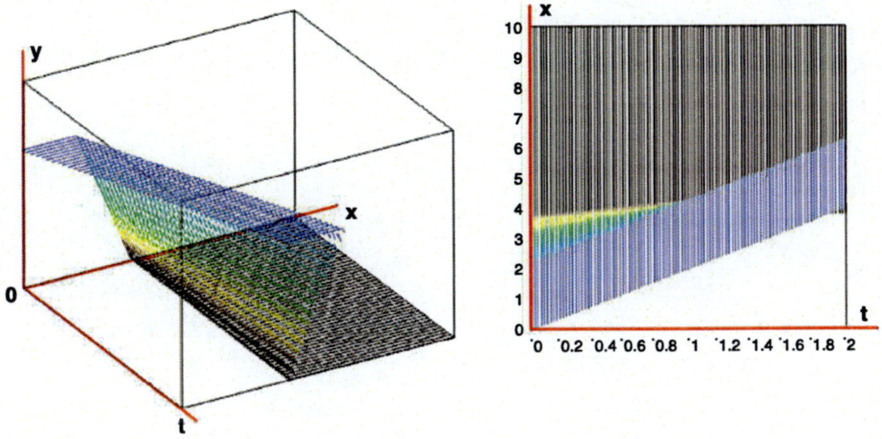

Fig. 16.6 Cauchy Problem with Constraints on Spatial Variables.
Left: The viability solution is the capture basin of the graph of the Cauchy condition $\mathbf{U}_0(x) := 2\Xi([0,1];x) \cup 2(2-x)\Xi([1,2];x) \cup 0\Xi([2,5];x)$. *It has empty values if* $T > \frac{x-\xi}{y}$. *Right: The domain of* U *to the Cauchy problem, which is no longer equal to* $\mathbb{R}_+ \times K$ *because of the presence of space constraints.*

Setting

$$\Psi_\xi(t,x) := \begin{cases} \mathbb{R} \text{ if } x \geq \xi \text{ and } t \geq 0 \\ \emptyset \text{ if } x < \xi \text{ or } t < 0 \end{cases}$$

the viability solution defined by

$$\mathrm{Graph}(\mathbf{U}) := \mathrm{Capt}_{(16.4)}(\mathrm{Graph}(\Psi_\xi), \mathrm{Graph}(\mathbf{U}_0^+))$$

is still the unique solution to the Cauchy problem for the Burgers problem, satisfying both the Burgers tracking property

$$\forall t \geq 0,\ \forall x \in [\xi, +\infty[,\ y \in \mathbf{V}(t, x + (t-T)y)$$

and the Cauchy condition $\mathbf{V}(0,x) = \mathbf{U}_0(x)$. The values $\mathbf{U}(T,x)$ are made of the fixed points

$$y \in \mathbf{U}_0(x - Ty) \cap \left]-\infty, \frac{x-\xi}{T}\right]$$

because $x - Ty$ must be larger than of equal to ξ. Such fixed points may no longer exist, and they never exist if $T > \frac{x-\xi}{y}$. See Figs. 16.5, p. 648 and 16.8, p. 651 for an illustration.

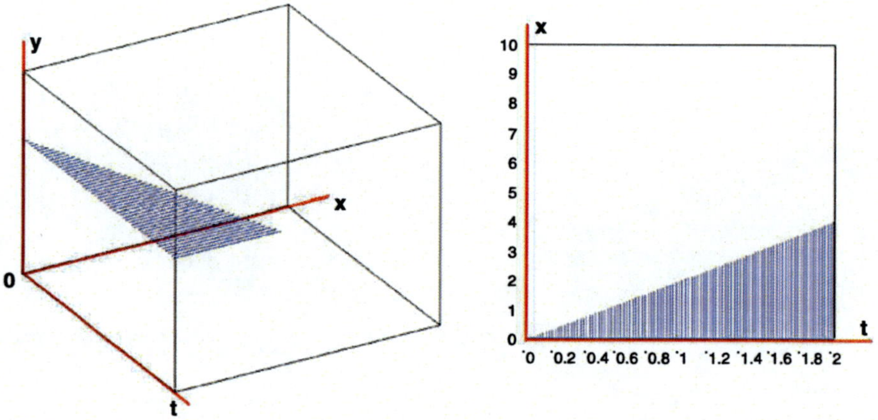

Fig. 16.7 Boundary-Value Problem in the case of Constraints on Spatial Variables.
The viability solution is the capture basin of the graph of the Dirichlet boundary condition $\varGamma_0(t) := 2\Xi([0,2];t)$. *It has empty values if* $T < \frac{x-\xi}{y}$.
Right: The domain of U *to the Cauchy problem, which is no longer equal to* $\mathbb{R}_+ \times K$ *because of the absence of Cauchy condition.*

For compensating for such empty values, we may "add" (in the union sense) to the initial data $\mathbf{U}_0$ other data, such as boundary-value data. Indeed, $x(t) \in K := [\xi, +\infty[$ at some time t come either from some initial state x at time 0 or from the lower bound ξ of $[\xi, +\infty[$ at a later time. We take this new fact into account by introducing *the sets* $\varGamma_\xi(t)$ *of velocities of states* $x(t)$ *arriving at time t at the lower bound* ξ *of* $[\xi, +\infty[$. We extend this (set-valued) map by the set-valued map $\varGamma_\xi : \mathbb{R}_+ \times X \rightsquigarrow X$ defined by:

$$\varGamma_\xi(t,x) := \begin{cases} \varGamma_\xi(t) & \text{if } x = \xi \\ \emptyset & \text{if } x \neq \xi \end{cases}$$

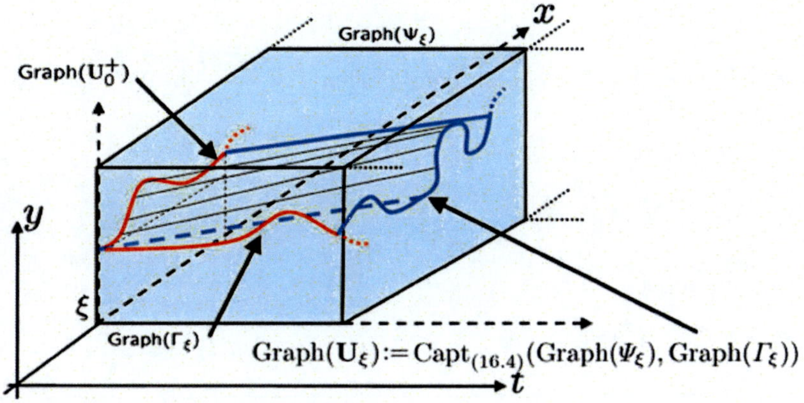

Fig. 16.8 Illustration of the definition of Graph(U).
Graph(**U**) *is the capture basin* Graph($\mathbf{U}_\xi$) := $\mathrm{Capt}_{(16.4)}$(Graph(Ψ_ξ), Graph(Γ_ξ)) *of the Cauchy condition by the augmented dynamics of the characteristic system (16.4).*

Definition 16.2.11 ***[Burgers Problem under Dirichlet Boundary Conditions]*** *Given the Dirichlet boundary condition* $\Gamma_\xi : \mathbb{R}_+ \times K \rightsquigarrow Y$*, a set-valued map* $\mathbf{V} : \mathbb{R}_+ \times K \rightsquigarrow Y$ *is a solution to the boundary-value problem for the Burgers problem if it satisfies*

1. *the* Burgers tracking property*:*

$$\forall t \geq \max\left(0, T - \frac{x-\xi}{y}\right),\ \forall x \in X,\ y \in \mathbf{U}\left(t, x + (t-T)y\right)$$

2. *the* Dirichlet boundary condition*:* $\mathbf{U}(t,\xi) := \Gamma_\xi(t)$,

We now prove the existence and the uniqueness of a solution to this problem:

Theorem 16.2.12 ***[Viability Solution of the Dirichlet Burgers Problem]*** *Assume that* $K := [\xi, +\infty]$*. The viability solution defined by*

$$\mathrm{Graph}(\mathbf{U}_\xi) := \mathrm{Capt}_{(16.4)}(\mathrm{Graph}(\Psi_\xi), \mathrm{Graph}(\Gamma_\xi)) \qquad (16.12)$$

is the ***unique*** *solution of the Burgers problem 16.2.11 with Dirichlet boundary conditions.*

Furthermore, $\mathbf{U}_\xi(T,x)$ *is the set of fixed point of the map*

$$y \rightsquigarrow \Gamma_\xi\left(T - \frac{x-\xi}{y}\right) \cap \left[\frac{x-\xi}{T}, +\infty\right[$$

where $T \geq \frac{x-\xi}{y}$ *(It is always empty when* $T < \frac{x-\xi}{y}$*). See Fig. 16.7 for an illustration.*

It satisfies the "maximum principle"

$$\forall (t,x) \in \mathbb{R}_+ \times K, \quad \sup_{y \in \mathbf{U}_\xi(t,x)} |y| \leq \sup_{t \in \mathbb{R}_+} \sup_{y \in \Gamma_\xi(t)} |y|$$

or, more precisely

$$\forall (t,x) \in \mathbb{R}_+ \times K, \ \mathbf{U}_\xi(t,x) \subset \mathrm{Im}(\Gamma_\xi)$$

Proof. The proof is analogous to the one of Theorem 16.2.5. To say that (T,x,y) belongs to the capture basin

$$\mathrm{Capt}_{(16.4)}(\mathrm{Graph}(\Psi_\xi), \mathrm{Graph}(\Gamma_\xi)) =: \mathrm{Graph}(\mathbf{U}_\xi)$$

amount to saying that there exists a finite time $t^\star(T,x,y)$ such that

1. the value $(T - t^\star(T,x,y), x - yt^\star(T,x,y), y)$ of the solution to characteristic differential equation (16.4) at time $t^\star(T,x,y)$ belongs to the graph of the set-valued map Γ_ξ
2. for all $t \in [0, t^\star(T,x,y)]$, $(T-t, x-ty, y)$ belongs to the graph of Ψ_ξ

The first condition means that $x - yt^\star(T,x,y) = \xi$ and that $(T - t^\star(T,x,y), \xi, y)$ belongs to the graph of Γ_ξ, i.e., that $y \in \Gamma_\xi(T - t^\star(T,x,y))$. It is sufficient to note that this amounts to saying that $t^\star(T,x,y) = \frac{x-\xi}{y}$ and $T \geq t^\star(T,x,y) = \frac{x-\xi}{y}$, or, equivalently, that $y \geq \frac{x-\xi}{T}$. In particular, we observe that $y \geq 0$.

The second condition means that for all $t \in [0, t^\star(T,x,y)]$, $x - yt \geq \xi$, i.e., that $t \leq t^\star(T,x,y) = \frac{x-\xi}{y}$.

Therefore, we have proved that $\mathbf{U}_\xi(T,x)$ is the set of fixed points of the set-valued map

$$y \rightsquigarrow \Gamma_\xi\left(T - \frac{x-\xi}{y}\right) \cap \left[\frac{x-\xi}{T}, +\infty\right[$$

Since $\mathrm{Graph}(\Gamma_\xi) \subset \mathrm{Graph}(\mathbf{U}_\xi)$, we know that $\Gamma_\xi(T) \subset \mathbf{U}_\xi(T,\xi)$. They are equal, because, if $y \in \mathbf{U}_\xi(T,\xi)$, then (T,ξ,y) belongs to the capture basin: Indeed, $(T - t^\star(T,\xi,y), \xi - t^\star(T,\xi,y)y, y) \in \mathrm{Graph}(\Gamma_\xi)$, and thus,

$\xi - t^\star(T,\xi,y)y \geq \xi$. Since $y > 0$, we infer that $t^\star(T,\xi,y) = 0$ so that $y \in \Gamma_\xi(T)$. Hence the Dirichlet boundary condition is satisfied.

Theorem 10.2.5 states that the graph of the viability solution is actually the unique graph of a set-valued map $\mathbf{V}$ between Γ_ξ and Ψ_ξ satisfying

$$\begin{cases} \text{Graph}(\mathbf{V}) = \text{Capt}_{(16.4)}(\text{Graph}(\mathbf{V}), \text{Graph}(\Gamma_\xi)) \\ \text{Capt}_{(16.4)}(\text{Graph}(\Psi_\xi), \text{Graph}(\mathbf{V})) \end{cases}$$

The first condition means that for any $t \in [0, t^\star(T,x,y)]$, y belongs to $\mathbf{V}(T-t, x-yt)$.

By the change of variable $s := T - t$, this means that for any $s \in [T - t^\star(T,x,y), T]$, $y \in \mathbf{V}(s, x + (s-T)y)$.

The second condition means that for all $t \geq T$, $y \in \mathbf{V}(t, x + (t-T)y)$. Indeed, for any $t \geq T$, we observe that $(t, x+(t-T)y, y)$ belongs to the graph of Ψ_ξ since $x + (t-T)y = \xi + \frac{t}{T}(x-\xi) \geq 0$. On the other hand, setting $t^\star := t - T \geq 0$, we see that $(T,x,y) = (t - t^\star, x + (t-T)y - t^\star y, y)$ belongs to the graph of $\mathbf{V}$. This implies that $(t, x+(t-T)y, y)$ belongs to the capture basin $\text{Capt}_{(16.4)}(\text{Graph}(\Psi_\xi), \text{Graph}(\mathbf{V}))$, equal to the graph of $\mathbf{V}$.

Hence these two conditions mean that $\mathbf{V}$ is a solution to the Burgers problem (16.2.11) with Dirichlet boundary conditions. □

16.2.5 Piecewise Dirichlet Conditions

Lemma 16.2.13 ***[Viability Solution when Dirichlet Boundary Conditions which are Marks]*** *Let us consider an interval $K := [\xi, +\infty[\subset X := \mathbb{R}$. Let us provide a Dirichlet boundary condition described by*

$$\Gamma_\xi(t) = \delta_\xi \Xi(\Delta_\xi; t)$$

where $\Delta_\xi \subset \mathbb{R}_+$ a time interval. If $T \geq \dfrac{x-\xi}{\delta_\xi}$, then the viability solution is equal to

$$\forall T \geq \frac{x-\xi}{\delta_\xi}, \ \mathbf{U}_\xi(T,x) = \delta_\xi \Xi\left(\Delta_\xi + \frac{x-\xi}{\delta_\xi}; t\right)$$

Proof. Theorem 16.2.12 implies that $\mathbf{U}_\xi(T,x)$ is the set of fixed points

$$y \in \delta_\xi \Xi\left(\Delta_\xi; T - \frac{x-\xi}{y}\right)$$

means that $y = \delta_\xi$, that $t^\star(T,x,y) := \frac{x-\xi}{\delta_\xi}$ and that $T - \dfrac{x-\xi}{\delta_\xi} \in \Delta_\xi$, i.e., $T \in \Delta_\xi + \frac{x-\xi}{\delta_\xi}$.

This means that $\delta_\xi \Xi\left(\Delta_\xi + \frac{x-\xi}{\delta_\xi}; t\right)$ is the set of fixed points $y \in \delta_\xi \Xi\left(\Delta_\xi; T - \frac{x-\xi}{\delta_\xi}; t\right)$. □

Therefore, as we did in the case of Cauchy condition, we can approximate a Dirichlet boundary condition Γ_ξ by a piecewise constant Dirichlet boundary condition:

Proposition 16.2.14 *[Viability Solution to Burgers Equation for Piecewise Constant Dirichlet Boundary Conditions]* *The viability solution to the Burgers tracking property satisfying the Dirichlet boundary condition*

$$\Gamma_\xi(t) = \bigcup_{j \in \mathbb{J}} \delta_j \Xi_{\Delta_j}(t)$$

where $\Delta_j \subset \mathbb{R}_+$ are time intervals, is equal to:

$$\mathbf{U}_\xi(t,x) = \bigcup_{j \in \mathbb{J}} \delta_j \Xi\left(\Delta_j + \frac{x-\xi}{\delta_j}; t\right)$$

which can be written in the form

$$\mathbf{U}(t,x) = \{\delta_j\}_{j \in \mathbb{J}(t,x)}$$

*where $\mathbb{J}(t,x) := \left\{ j \in \mathbb{J} \text{ such that } t \in \Delta_j + \frac{x-\xi}{\delta_j} \right\}$, the cardinal of which plays the role of a "*valuemeter*".*

16.2.6 Cauchy/Dirichlet Condition

Now, thanks to the Lemma 10.2.2 stating that the capture basin of a union of targets is the union of the capture basins, we can mix the initial and the Dirichlet boundary condition by taking the unions of the solution $\mathbf{U}_{\mathbf{U}_0}$ associated with the Cauchy condition $\mathbf{U}_0$ (which may have empty values) and of the solution $\mathbf{U}_\xi$ associated with the Dirichlet boundary condition Γ_ξ.

Let us consider the viability Solution (16.5), p. 633 to the Cauchy/Dirichlet Burgers equation defined by

$$\text{Graph}(\mathbf{U}) := \text{Capt}_{(16.4)}(\text{Graph}(\Psi_\xi), \text{Graph}(\mathbf{U}_0^+ \cup \Gamma_\xi))$$

Definition 16.2.15 ***[Viability Solution to the Cauchy/Dirichlet Burgers Tracking Problem]*** *A map* **U** *is a solution to the Cauchy/Dirichlet Burgers problem if it satisfies*

1. *the* Burgers tracking property*:*

$$\forall t \geq \max\left(0, T - \frac{x-\xi}{y}\right),\ \forall x \in X,\ y \in \mathbf{U}\left(t, x + (t-T)y\right)$$

2. *the* Cauchy condition $\mathbf{U}(0,x) := \mathbf{U}_0(x)$*,*
3. *the* Dirichlet boundary condition $\mathbf{U}(t,\xi) := \Gamma_\xi(t)$*,*

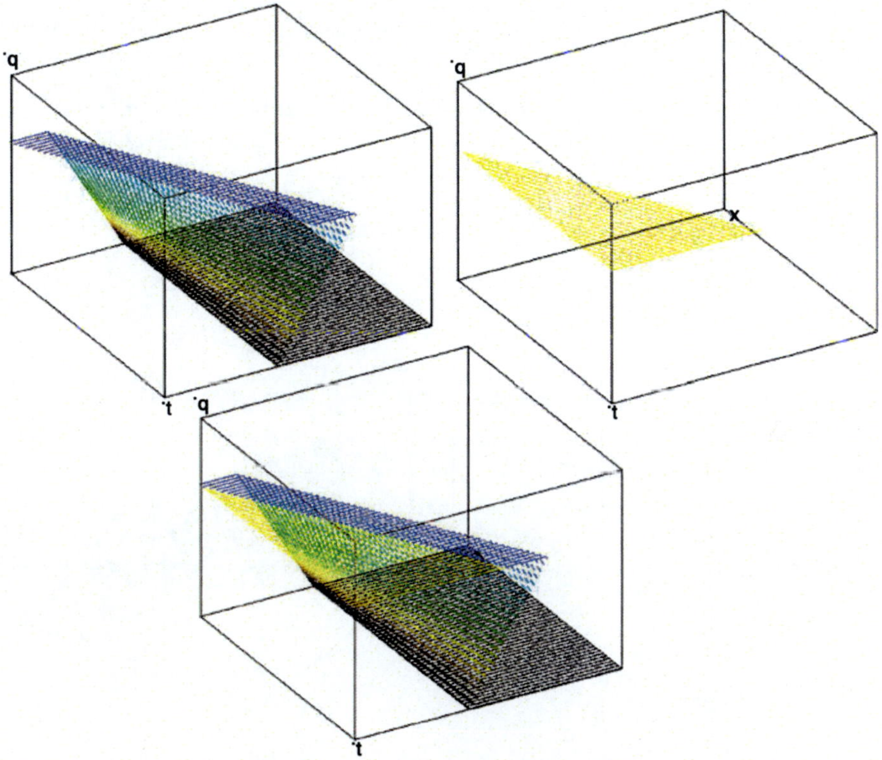

Fig. 16.9 Viability Kernel and Cauchy/Dirichlet Evolution.
*Thanks to the morphism property of the viability solution (the solution of a union of condition is the union of the solutions), the viability solution of the Cauchy/Dirichlet Burgers problem (*lower center*) is the union of the solution of the Cauchy problem (*upper left*) and of the solution to the boundary-value problem (*upper right*), as illustrated by this numerical example*

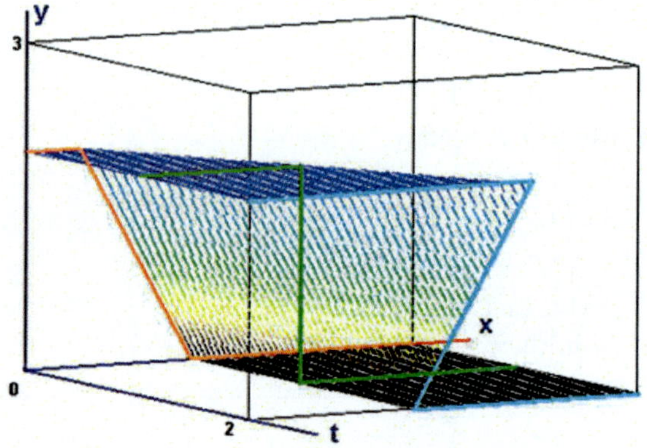

Fig. 16.10 Cauchy/Dirichlet Problem in the case of Constraints on Spatial Variables.
The viability solution is the capture basin of the union of the graph of the Cauchy condition $\mathbf{U}_0 := 2\Xi([0,1];x) \cup 2(2-x)\Xi([1,2];x) \cup 0\Xi([2,5];x)$ *and of the graph of the Dirichlet boundary condition* $\Gamma_0(t) := 2\Xi([0,2];t)$.

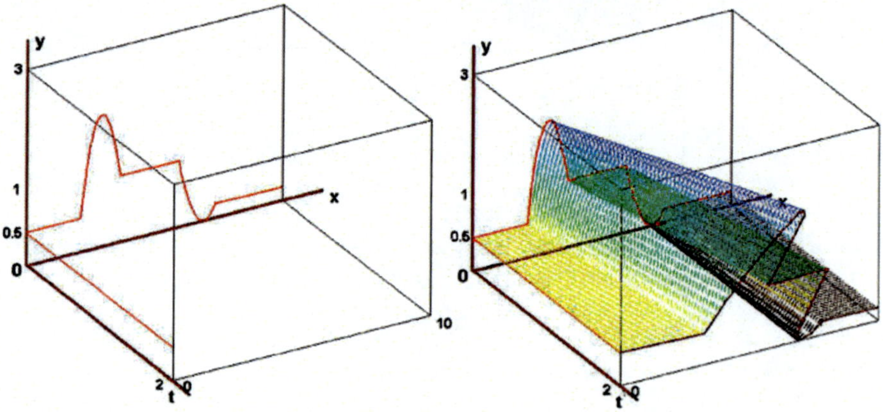

Fig. 16.11 Example of viability solution to the Burgers equation.
with initial data $\mathbf{U}_0(x) := \max(0.5, 2(1-(x-3)^2))\Xi([0,3];x) \cup \max(1, 2(1-(x-3)^2))\Xi([3,5];x) \cup \min(1,(x-7)^2)\Xi(5,7[];x) \cup \max(0.2,(x-7)^2)\Xi([7,10];x)$ *and Dirichlet boundary condition* $\Gamma_0(t) := 0.5\Xi([0,2];t)$.

Theorem 16.2.16 ***[Viability Solution for Cauchy/Dirichlet Conditions]*** *Assume that* $K := [\xi, +\infty]$ *and that* $\mathbf{U}_0(\xi) = \Gamma_\xi(0, \xi)$. *The viability solution* $\mathbf{U}$ *defined by (16.5) is the* ***unique*** *solution to the Cauchy/Dirichlet Burgers problem satisfying both the Burgers tracking property*

$$\forall t \geq \max\left(0, T - \frac{x - \xi}{y}\right), \ \forall x \in [\xi, +\infty[, \ y \in \mathbf{U}\,(t, x + (t - T)y)$$

and the initial and Dirichlet boundary conditions

$$\begin{cases} (i) \ \ \forall x \geq \xi, \ \mathbf{U}(0, x) = \mathbf{U}_0(x) \\ (ii) \ \forall t \geq 0, \ \mathbf{U}(t, \xi) = \Gamma_\xi(t) \end{cases}$$

It is the union $(t, x) \rightsquigarrow \mathbf{U}(t, x) := \mathbf{U}_{\mathbf{U}_0}(t, x) \cup \mathbf{U}_\xi(t, x)$ *of the viability solutions* $\mathbf{U}_{\mathbf{U}_0}(t, x)$ *associated with the initial datum* $\mathbf{U}_0$ *and the viability solution* $\mathbf{U}_\xi(t, x)$ *associated with the boundary datum* Γ_ξ.

Furthermore, $\mathbf{U}(T, x)$ *is the set of velocities* y *satisfying*

$$\begin{cases} y \in \mathbf{U}_0(x - Ty) & \text{if } T \leq \frac{x-\xi}{y} \\ y \in \Gamma_\xi\left(T - \frac{x-\xi}{y}\right) & \text{if } T \ \geq \ \frac{x-\xi}{y} \end{cases}$$

It satisfies the "maximum principle"

$$\forall (t, x) \in \mathbb{R}_+ \times K, \ \mathbf{U}_\xi(t, x) \ \subset \ \mathrm{Im}(\mathbf{U}_0) \cup \mathrm{Im}(\Gamma_\xi)$$

Proof. It is enough to observe that the graph of the $\mathbf{U}$ is the capture basin

$$\mathrm{Graph}(\mathbf{U}) := \mathrm{Capt}_{(16.4)}(\mathrm{Graph}(\Psi_\xi), \mathrm{Graph}(\mathbf{U}_0^+ \cup \Gamma_\xi)) \tag{16.13}$$

which is the union of the capture basins of the targets $\mathrm{Graph}(\mathbf{U}_0^+)$ and $\mathrm{Graph}(\Gamma_\xi)$. □

16.2.6.1 Example of Non-Strict Dirichlet Boundary Condition

Figure 16.10 provided an example of a "well-posed" intial/boundary value-problem. We may consider an intermediate situation between this well-posed case and the case of an initial value problem without Dirichlet boundary condition as in Fig. 16.6 by taking for Dirichlet boundary condition a non-strict set-valued map: See for instance Fig. 16.12.

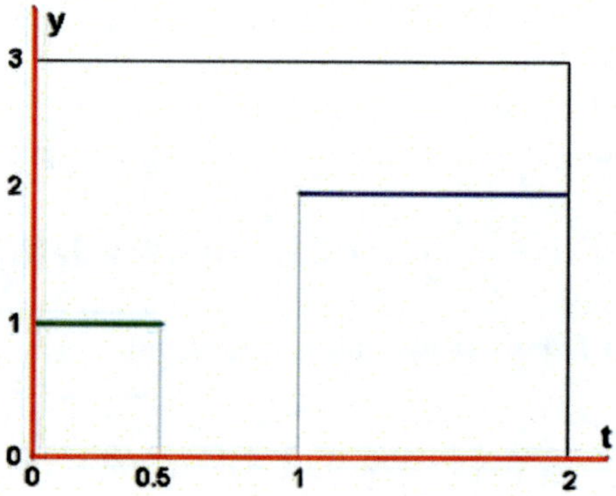

Fig. 16.12 Example of a Non-Strict Dirichlet Boundary Condition.
The Dirichlet boundary condition is defined by the map $\Gamma_0(t) := \Xi([0,0.5];t) \cup 1.5\Xi([1,2];t)$ *equal to* 1 *on* $[0,0.5]$, *to* 1.5 *on* $[1,2]$ *and to the empty set otherwise.*

The viability solution of such a non-strict Cauchy/Dirichlet problem may have non-empty values: See Fig. 16.13 for example.

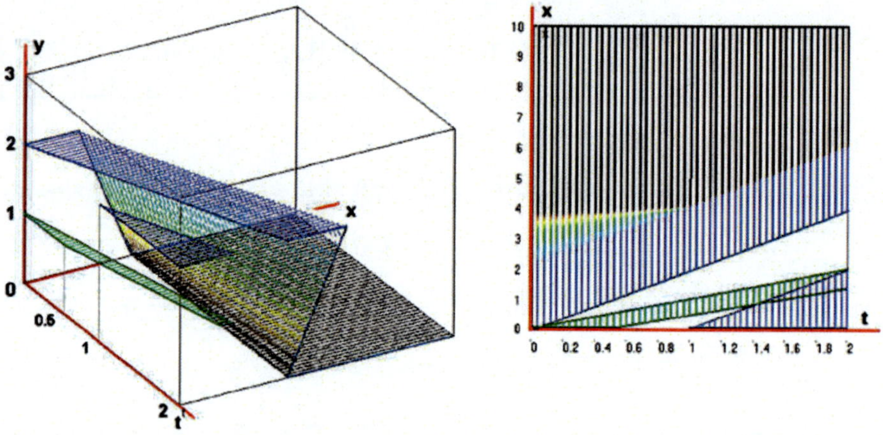

Fig. 16.13 Non-Strict Boundary-Value Problem in the case of constraints on spatial variables.
The viability solution is the capture basin of the union of graphs of the Cauchy condition $\mathbf{U}_0 := 2\Xi([0,1];x) \cup 2(2-x)\Xi([1,2];x) \cup 0\Xi([2,5];x)$ *and of the graph of the Dirichlet boundary condition defined by the map* $\Gamma_0(t) := \Xi([0,0.5];t) \cup 1.5\Xi([1,2];t)$. *It still has empty values in a subset of the area* $T < \dfrac{x-\xi}{y}$, *but has also non empty values thanks to the contribution of the Dirichlet boundary condition, either equal to a singleton or to a subset of two elements. Right: The domain of* U *is not equal to* $\mathbb{R}_+ \times K$.

16.2.7 Additional Eulerian Conditions

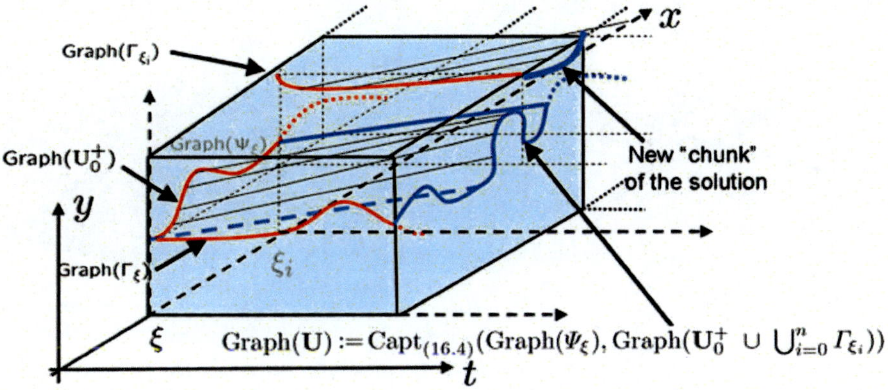

Fig. 16.14 Illustration of the definition of Graph(U).
Graph(**U**) *is the capture basin* Graph(**U**) $:=$ $\text{Capt}_{(16.4)}(\text{Graph}(\Psi_\xi),$ $\text{Graph}(\mathbf{U}_0^+ \cup \bigcup_{i=0}^n \Gamma_{\xi_i}))$ *of the initial and Eulerian boundary conditions by the augmented dynamics of the characteristic system (16.4). Note the additional chunk of intermediate data added at* ξ_i *in this Figure, and the corresponding added chunk of* **U**.

The morphism property shown earlier enables the construction of viability solutions obtained from Cauchy conditions "added at arbitrary locations along the x axis". Traditionally, in a partial differential equation setting, Cauchy conditions and boundary conditions are prescribed (so called Cauchy-Dirichlet problem). With the present method, we can add Eulerian conditions, which are Dirichlet conditions, not at the boundary, but in the interior of the domain. They constitute further conditions to take into account in the propagation of the solution. We denote by $\xi_i > \xi$ such locations. At each instant t, we add the Eulerian condition $\Gamma_{\xi_i}(t, \xi_i) := \Gamma_{\xi_i}(t)$. We extend this map by setting $\Gamma_{\xi_i}(t, x) = \emptyset$ whenever $x > \xi_i$.

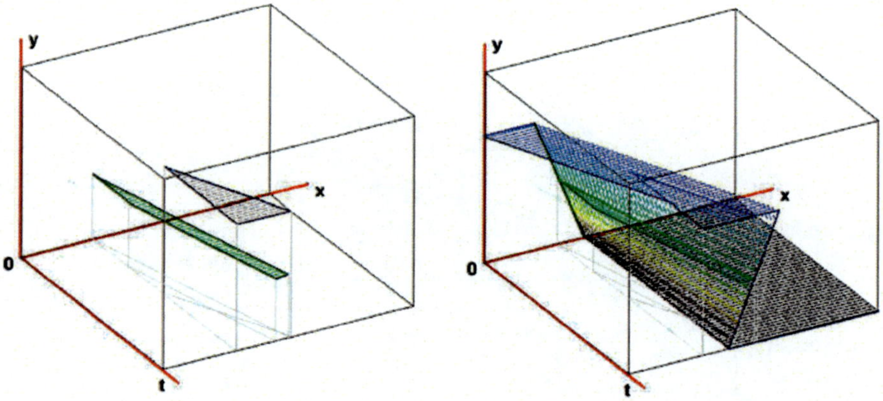

Fig. 16.15 Example of an Additional Intermediate Eulerian Condition.
$\Gamma_1(t) := \Xi([0,0.5];t) \cup 1.5\Xi([1,2];t)$ alone on the left and with Cauchy condition $\mathbf{U}_0 := 2\Xi([0,1];x) \cup 2(2-x)\Xi([1,2];x) \cup 0\Xi([2,5];x)$ on the right.

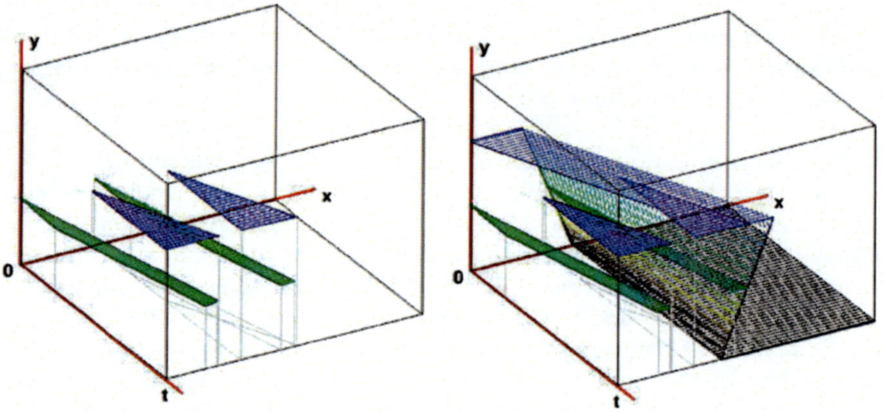

Fig. 16.16 Example of Additional Eulerian Conditions.
Union of a boundary condition $\Gamma_0(t) := \Xi([0,0.5];t) \cup 1.5\Xi([1,2];t)$ and of the additional intermediate condition $\Gamma_1(t) := \Xi([0,0.5];t) \cup 1.5\Xi([1,2];t)$ alone on the left and with Cauchy condition $\mathbf{U}_0 := 2\Xi([0,1];x) \cup 2(2-x)\Xi([1,2];x) \cup 0\Xi([2,5];x)$ on the right.

A slight modification of the proof of Theorem 16.2.12 implies the existence of an unique viability solution to the Burgers problem with an additional Eulerian conditions:

Lemma 16.2.17 ***[Viability Solution for Eulerian conditions]***
Assume that $K := [\xi, +\infty]$. The viability solution defined by

$$\text{Graph}(\mathbf{U}_{\xi_i}) := \text{Capt}_{(16.4)}(\text{Graph}(\Psi_\xi), \text{Graph}(\Gamma_{\xi_i})) \quad (16.14)$$

*is the **unique** solution satisfying both the Burgers tracking property*

$$\forall t \geq \max\left(0, T - \frac{x-\xi}{y}\right), \ \forall x \in X, \ y \in \mathbf{U}\left(t, x + (t-T)y\right)$$

and the additional intermediate condition at ξ_i

$$\forall t \geq 0 \ \mathbf{U}(t, \xi_i) = \Gamma_{\xi_i}(t, \xi_i)$$

Furthermore, $\mathbf{U}_{\xi_i}(T,x)$ is the set of fixed points of the map

$$y \rightsquigarrow \Gamma_{\xi_i}\left(T - \frac{x-\xi_i}{y}, \xi_i\right) \cap \left[\frac{x-\xi_i}{T}, +\infty\right[$$

where $T \geq \dfrac{x-\xi_i}{y}$.

Now, thanks again to the Lemma 10.2.2 stating that the capture basin of a union of targets is the union of the capture basins, we deduce that this problem still has a unique solution:

Theorem 16.2.18 *[**Viability Solution of for Eulerian Conditions**] Let us consider the sequence $\xi =: \xi_0 < \xi_1 < \cdots < x_n$. Given the Cauchy condition $\mathbf{U}_0$ and the additional Eulerian conditions Γ_{ξ_i}, $i = 0, 1, \ldots, n$, the union $\mathbf{U}$ of the viability solution $\mathbf{U}_{\mathbf{U}_0}$ associated with the Cauchy condition $\mathbf{U}_0$ and of the viability solutions $\mathbf{U}_{\xi_i}$ associated with the Eulerian conditions Γ_{ξ_i} is the **unique** solution to the Burgers problem satisfying: If $x \in [\xi_i, \xi_{i+1}[$, then $\mathbf{U}(T,x)$ is the set of velocities y such that:*

$$\begin{cases} y \in \mathbf{U}_0(x - Ty) & \text{if } T \in \left[0, \dfrac{x-\xi_i}{y}\right[\\ y \in \mathbf{U}_0(x-Ty) \cup \Gamma_{\xi_i}\left(T - \dfrac{x-\xi_i}{y}, \xi_i\right) & \text{if } T \in \left[\dfrac{x-\xi_i}{y}, \dfrac{x-\xi_{i-1}}{y}\right[\\ \vdots & \vdots \\ y \in \mathbf{U}_0(x-Ty) \cup \displaystyle\bigcup_{j=1}^{i} \Gamma_{\xi_j}\left(T - \dfrac{x-\xi_j}{y}, \xi_j\right) & \text{if } T \in \left[\dfrac{x-\xi_1}{y}, \dfrac{x-\xi_0}{y}\right[\\ y \in \displaystyle\bigcup_{j=0}^{i} \Gamma_{\xi_j}\left(T - \dfrac{x-\xi_j}{y}, \xi_j\right) & \text{if } T \geq \dfrac{x-\xi_0}{y} \end{cases}$$

Proof. It is enough to observe that the graph of the $\mathbf{U}$ is the capture basin

$$\text{Graph}(\mathbf{U}) := \text{Capt}_{(16.4)}\left(\text{Graph}(\Psi_{\xi}), \text{Graph}\left(\mathbf{U}_0^{+} \cup \bigcup_{i=0}^{n} \Gamma_{\xi_i}\right)\right) \quad (16.15)$$

which is the union of the capture basins of the targets $\text{Graph}(\mathbf{U}_0^{+})$ and $\text{Graph}(\Gamma_{\xi_i})$. □

16.2.8 Viability Constraints

It is quite useful to be able to impose constraints not only on state variables, as we did in the preceding sections, but also on the values $\mathbf{U}(t,x)$ of the Burgers problem. We thus require that the solution of the Cauchy/Dirichlet Burgers problem also satisfies the viability constraint

$$\forall t \geq 0,\ \forall x \in X,\ \mathbf{U}(t,x) \ \subset \ [a(t,x), b(t,x)]$$

remains in an interval $[a(t,x), b(t,x)]$ where $a : \mathbb{R}_+ \times \mathbb{R} \mapsto \mathbb{R}$ and $b : \mathbb{R}_+ \times \mathbb{R} \mapsto \mathbb{R}$ are two given functions satisfying $a(t,x) \leq b(t,x)$ for all $(t,x) \in \mathbb{R}_+ \times \mathbb{R}$.

This does not complicate the problem at all: It is enough to introduce the new set-valued map $\Psi_{\xi;(a,b)}$ defined by

$$\Psi_{\xi;(a,b)}(t,x) \ := \begin{cases} [a(t,x), b(t,x)] & \text{if } x \ \geq \ \xi \\ \emptyset & \text{if } x \ < \ \xi \end{cases}$$

Definition 16.2.19 *[**Viability Solution to the Cauchy/Dirichlet Burgers Problem Under Viability Constraints**] A set-valued map* $\mathbf{U}$ *is said to be a solution to the Cauchy/Dirichlet Burgers problem with viability constraints if it satisfies*

1. *the* Burgers tracking property*:* $\forall y \in \mathbf{U}(T,x)$, $\forall s \geq T$ *such that* $\forall t \in [T,s]$, $y \in \Psi_{\xi,(a,b)}(t, x+(t-T)y)$*, then*

$$\forall t \in \left[\max\left(0, T - \frac{x-\xi}{y}\right), s\right],\ \forall x \in X,\ y \in \mathbf{U}\left(t, x+(t-T)y\right)$$

2. *the* Cauchy condition $\mathbf{U}(0,x) := \mathbf{U}_0(x)$*,*
3. *the* boundary condition $\mathbf{U}(t,\xi) := \Gamma_{\xi}(t)$*,*
4. *the* viability constraints $\mathbf{U}(t,x) \ \subset \ [a(t,x), b(t,x)]$.

We shall say that the set-valued map $\mathbf{U} : \mathbb{R}_+ \times \mathbb{R}_+ \rightsquigarrow \mathbb{R}$ *defined by*

$$\text{Graph}(\mathbf{U}) := \text{Capt}_{(16.4)}(\text{Graph}(\Psi_{\xi;(a,b)}), \text{Graph}(\mathbf{U}_0^+ \cup \Gamma_\xi)) \qquad (16.16)$$

is the viability solution *to the Cauchy/Dirichlet Burgers Problem.*

The viability solution is still the unique solution to the Burgers problem with constraints:

Theorem 16.2.20 ***[Existence and Uniqueness of the Viability Solution Under Viability Constraint]*** *Let us set*

$$\begin{cases} a^\sharp(T,x,y) := \sup_{t\in\left[0,\min\left(T,\frac{x-\xi}{y}\right)\right]} a(T-t,x-ty) \\ b^\flat(T,x,y) := \inf_{t\in\left[0,\min\left(T,\frac{x-\xi}{y}\right)\right]} b(T-t,x-ty) \end{cases}$$

The viability solution $\mathbf{U}$ *is the* ***unique*** *solution* $\mathbf{V}$ *to the Cauchy Burgers problem under constraints* $\mathbf{U}(t,x) \subset [a(t,x),b(t,x)]$. *The value* $\mathbf{U}(T,x)$ *of this unique solution is the set of fixed points*

$$y \in \begin{cases} \mathbf{U}_0(x-Ty) \cap \left[a^\sharp(T,x,y), b^\flat(T,x,y)\right] & \text{if } T \le \dfrac{x-\xi}{y} \\ \Gamma_\xi\left(T-\frac{x-\xi}{y}\right) \cap \left[a^\sharp(T,x,y), b^\flat(T,x,y)\right] & \text{if } T \ge \dfrac{x-\xi}{y} \end{cases}$$

Proof. The proof is analogous to the proofs of Theorems 16.2.5 and 16.2.12 and is a consequence of the general Theorem 16.2.26: We mention only the modification.

To say that (T,x,y) belongs to the capture basin

$$\text{Capt}_{(16.4)}(\text{Graph}(\Psi_{\xi;(a,b)}), \text{Graph}(\mathbf{U}_0^+ \cup \Gamma_\xi)) =: \text{Graph}(\mathbf{U})$$

amounts to saying that there exists a finite time $t^\star(T,x,y)$ such that

1. the value $(T-t^\star(T,x,y), x-yt^\star(T,x,y), y)$ of the solution to characteristic differential equation (16.4) at time $t^\star(T,x,y) \in [0,T]$ belongs to the graph of the set-valued map $\mathbf{U}_0^+ \cup \Gamma_\xi$,
2. for all $t \in [0,t^\star(T,x,y)]$, $(T-t, x-ty, y)$ belongs to the graph of $\Psi_{\xi;(a,b)}$

The first condition means that $t^\star(T,x,y) = \min\left(T, \frac{x-\xi}{y}\right)$ and thus, that $y \in \mathbf{U}_0(x-Ty)$ if $T \le \frac{x-\xi}{y}$ and $y \in \Gamma_\xi\left(T-\frac{x-\xi}{y}\right)$ otherwise.

The second condition means that for all $t \in [0,t^\star(T,x,y)] = \left[0, \min\left(T,\frac{x-\xi}{y}\right)\right]$, $a(T-t,x-ty) \le y \le b(T-t,x-ty)$, or, equivalently, that

$$\sup_{t\in\left[0,\min\left(T,\frac{x-\xi}{y}\right)\right]} a(T-t,x-ty) \leq y \leq \inf_{t\in\left[0,\min\left(T,\frac{x-\xi}{y}\right)\right]} b(T-t,x-ty)$$

The rest of the proof remains the same. □

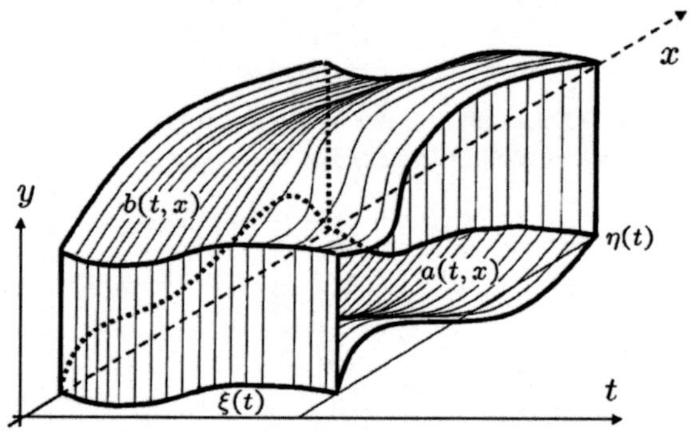

Fig. 16.17 Dirichlet Boundary Conditions, Viability Constraints for the Burgers Problem.
Illustration of the set of constraints for the Burgers problem with mobile boundary conditions and viability constraints. The mobile boundary are represented by the vertical surfaces $\xi(t)$ and $\eta(t)$. The constraints $a(t,x)$ and $b(t,x)$ are also shown. The viability solution of the Burgers problem is now contained within this set.

16.2.9 Lagrangian Conditions

Lagrangian conditions are "mobile value conditions" defined not only for fixed positions ξ_i, but on mobile trajectories of evolutions $t \mapsto \xi(t)$. This is in contrast of "Eulerian conditions" which are constant (static) conditions. For simplicity, we analyze this question for two space and Lagrangian conditions only: we require instead that the evolution of $x(t)$ must satisfy at each instant inequalities

$$\xi(t) \leq x(t) \leq \eta(t)$$

We thus introduce two Lagrangian (set-valued) maps

$$\Gamma_\xi(t) := \Gamma_\xi(t,\xi(t)) \;\; \& \;\; \Gamma_\eta(t) := \Gamma_\eta(t,\eta(t))$$

which we extend to set-valued maps $\Gamma_\xi : \mathbb{R}_+ \times X \rightsquigarrow X$ and $\Gamma_\eta : \mathbb{R}_+ \times X \rightsquigarrow X$ satisfying

$$\Gamma_\xi(t,x) := \emptyset \text{ if } x \neq \xi(t) \ \&\ \Gamma_\eta(t,x) := \emptyset \text{ if } x \neq \eta(t)$$

We also require that the solution satisfies the viability constraint

$$\forall t \geq 0,\ \forall x \in X,\ \mathbf{U}(t,x) \subset [a(t,x), b(t,x)]$$

For investigating this new case, we introduce the new set-valued map $\Psi_{(\xi,\eta;a,b)}$

$$\Psi_{(\xi,\eta;a,b)}(t,x) := \begin{cases} [a(t,x), b(t,x)] & \text{if } x \in [\xi(t), \eta(t)] \\ \emptyset & \text{if } x \notin [\xi(t), \eta(t)] \end{cases}$$

Definition 16.2.21 ***[Viability Solution to the for Lagrangian Conditions under Viability Constraints]*** *Let us denote by $t_\xi(T,x,y)$ the smallest fixed-point of the map $t \mapsto \frac{x-\xi(T-t)}{y}$, by $t_\eta(T,x,y)$ the smallest fixed-point of the map $t \mapsto \frac{x-\eta(T-t)}{y}$. A set-valued map $\mathbf{U}$ is said to be a solution to the Cauchy Burgers problem with two Lagrangian conditions if it satisfies*

1. *the* Burgers tracking property*: $\forall y \in \mathbf{U}(T,x)$, $\forall s \geq T$ such that $\forall t \in [\max(0, T - t_\xi(T,x,y), T - t_\eta(T,x,y)), s]$, then*

$$\forall x \in X,\ y \in \mathbf{U}(t, x + (t-T)y)$$

2. *the* Cauchy condition $\mathbf{U}(0,x) := \mathbf{U}_0(x)$,
3. *the* Lagrangian conditions $\mathbf{U}(t,\xi(t)) := \Gamma_\xi(t)$ *and* $\mathbf{U}(t,\eta(t)) := \Gamma_\eta(t)$,
4. *the* viability constraints $\mathbf{U}(t,x) \subset [a(t,x), b(t,x)]$.

We shall say that the set-valued map $\mathbf{U} : \mathbb{R}_+ \times \mathbb{R}_+ \rightsquigarrow \mathbb{R}$ defined by

$$\text{Graph}(\mathbf{U}) := \text{Capt}_{(16.4)}(\text{Graph}(\Psi_{(\xi,\eta;a,b)}), \text{Graph}(\mathbf{U}_0^+ \cup \Gamma_\xi \cup \Gamma_\eta)) \quad (16.17)$$

is the viability solution *to the Cauchy Burgers problem with Lagrangian conditions.*

The viability solution is still the unique solution to the Burgers problem with constraints:

Theorem 16.2.22 ***[Viability Solution for Lagrangian Conditions]*** *Let us denote by $t_\xi(T,x,y)$ the smallest fixed-point of the map*

$t \mapsto \dfrac{x-\xi(T-t)}{y}$, *by* $t_\eta(T,x,y)$ *the smallest fixed-point of the map* $t \mapsto \dfrac{x-\eta(T-t)}{y}$. *We now set*

$$\begin{cases} a^\sharp(T,x,y) := \sup_{t\in[0,\min(T,t_\xi(T,x,y),t_\eta(T,x,y))]} a(T-t,x-ty) \\ b^\flat(T,x,y) := \inf_{t\in[0,\min(T,t_\xi(T,x,y),t_\eta(T,x,y))]} b(T-t,x-ty) \end{cases}$$

The viability solution $\mathbf{U}$ *is the* ***unique*** *solution to the Cauchy Burgers problem for Lagrangian conditions and viability constraints* $\mathbf{U}(t,x) \subset [a(t,x),b(t,x)]$. *The value* $\mathbf{U}(T,x)$ *of this unique solution is the set of fixed points*

$$y \in \begin{cases} \mathbf{U}_0(x-Ty) \cap \left[a^\sharp(T,x,y), b^\flat(T,x,y)\right] \\ \text{if } T = \min\left(T, t_\xi(T,x,y), t_\eta(T,x,y)\right) \\ \\ \Gamma_\xi\left(T - t_\xi(T,x,y)\right) \cap \left[a^\sharp(T,x,y), b^\flat(T,x,y)\right] \\ \text{if } t_\xi(T,x,y) = \min\left(T, t_\xi(T,x,y), t_\eta(T,x,y)\right) \\ \\ \Gamma_\eta\left(T - t_\eta(T,x,y)\right) \cap \left[a^\sharp(T,x,y), b^\flat(T,x,y)\right] \\ \text{if } t_\eta(T,x,y) = \min\left(T, t_\xi(T,x,y), t_\eta(T,x,y)\right) \end{cases}$$

Proof. The proof is analogous to the proofs of Theorem 16.2.20: We mention only the modification.

To say that (T,x,y) belongs to the capture basin

$$\mathrm{Capt}_{(16.4)}(\mathrm{Graph}(\Psi_{(\xi,\eta;a,b)}), \mathrm{Graph}(\mathbf{U}_0^+ \cup \Gamma_\xi \cup \Gamma_\eta)) =: \mathrm{Graph}(\mathbf{U})$$

amount to saying that there exists a finite time $t^\star(T,x,y)$ such that

1. the value $(T-t^\star(T,x,y), x-yt^\star(T,x,y), y)$ of the solution to characteristic differential equation (16.4) at time $t^\star(T,x,y) \in [0,T]$ belongs to the graph of the set-valued map $\mathbf{U}_0^+ \cup \Gamma_\xi \cup \Gamma_\eta$,
2. for all $t \in [0, t^\star(T,x,y)]$, $(T-t, x-ty, y)$ belongs to the graph of $\Psi_{(\xi,\eta;a,b)}$

The first condition means that $t^\star(T,x,y) = \min\left(T, t_\xi(T,x,y), t_\eta(T,x,y)\right)$ and thus, that $y \in \mathbf{U}_0(x-Ty)$ if $T = t_\xi(T,x,y)$, that $y \in \Gamma_\xi\left(T - t_\xi(T,x,y)\right)$ if $t_\xi(T,x,y) = \min\left(T, t_\xi(T,x,y), t_\eta(T,x,y)\right)$ and that $y \in \Gamma_\eta\left(T - t_\xi(T,x,y)\right)$ if $t_\eta(T,x,y) = \min\left(T, t_\xi(T,x,y), t_\eta(T,x,y)\right)$.

The second condition means that for all $t \in [0, t^\star(T,x,y)] = [0, \min\left(T, t_\xi(T,x,y), t_\eta(T,x,y)\right)]$, $a(T-t, x-ty) \le y \le b(T-t, x-ty)$, or, equivalently, that

$$\begin{cases} \sup_{t\in[0,\min(T,t_\xi(T,x,y),t_\eta(T,x,y))]} a(T-t,x-ty) \le y \\ \le \inf_{t\in[0,\min(T,t_\xi(T,x,y),t_\eta(T,x,y))]} b(T-t,x-ty) \end{cases}$$

The rest of the proof remains the same. □

16.2.10 Regulating the Burgers Controlled Problem

As mentioned before, it is restrictive to assume that the velocities remain constant, i.e., that their acceleration is equal to 0. Even prescribing a right hand side $x''(t) = g(t, x(t), x'(t))$ for defining the acceleration may not be sufficient for studying control problems, since we might have to regulate (control, pilot) the evolution of the states $x(t)$.

Therefore, instead of taking the velocities constant, we may leave the choice of the accelerations $x''(\cdot)$ "open" in some closed convex set $G(t, x(t), x'(t))$, i.e., regard the acceleration $x''(t) =: u(t)$ as a control ranging in the subset $G(t, x(t), x'(t))$ depending upon the time, the state and the velocity. This allows us to regulate (or to pilot, to control) the velocity of the states by introducing controls.

Hence, we have to define tracking problems for the second-order differential inclusion $x''(t) \in G(t, x(t), x'(t))$.

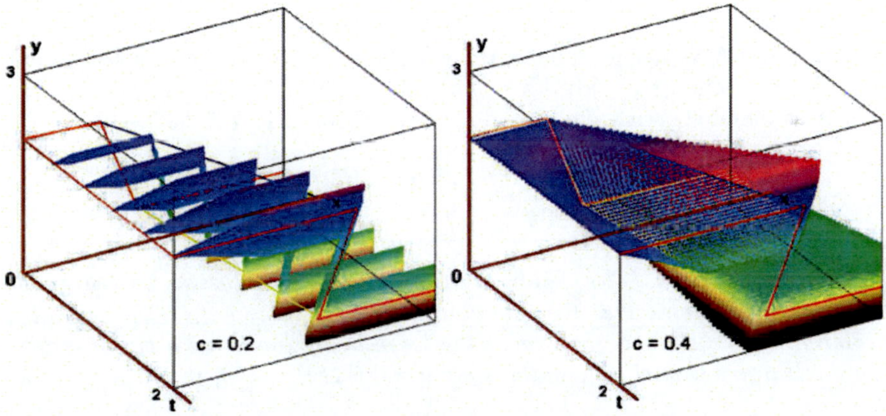

Fig. 16.18 Viability Solution to the Controlled Burgers Equation. $\frac{\partial \mathbf{U}(t,x)}{\partial t} + \frac{\partial \mathbf{U}(t,x)}{\partial x}\mathbf{U}(t,x) = u$ *where* $u \in [-c, +c]$ *for 2 values of* $c = 0.2$ *&* 0.4 *with Cauchy condition equal to* $\mathbf{U}_0 := 2\Xi([0,1];x) \cup 2(2-x)\Xi([1,2];x) \cup 0\Xi([2,5];x)$ *and boundary condition equal to* $\Gamma_0(t) := \Xi([0,0.5];t) \cup 1.5\Xi([1,2];t)$. *It still has the familiar Z-shape, but with "thick" values. They are "split" for* $c = 0.2$ *for the sake of clarity.*

We define a viability solution to such a tracking problem in the following way:

Definition 16.2.23 ***[Viability Solution to a Controlled Problem]*** *Let us consider two set-valued maps Ψ and $\Phi \subset \Psi$ from $\mathbb{R}_+ \times \mathbb{R}$ to $\mathbb{R}$ and a second-order differential inclusion $x''(t) \in G(t, x(t), x'(t))$.*

Introduce the "characteristic system of a differential inclusion"

$$\begin{cases} (i) \ \tau'(t) = -1 \\ (ii) \ x'(t) = -y(t) \\ (ii) \ y'(t) \in -G(\tau(t), x(t), y(t)) \end{cases} \quad (16.18)$$

The set-valued map $\mathbf{U}_{(\Psi,\Phi)} : \mathbb{R}_+ \times K \rightsquigarrow Y$ defined by

$$\text{Graph}(\mathbf{U}_{(\Psi,\Phi)}) := \text{Capt}_{(16.18)}(\text{Graph}(\Psi), \text{Graph}(\Phi)) \quad (16.19)$$

is called the viability solution to the controlled Burgers problem.

16.2.10.1 Viability Solution to Burgers Partial Differential Inclusions

Recall that $D^{\star\star}\mathbf{U}(t, x, y)$ denotes the convexified derivative of the set-valued map $\mathbf{U}$ at a point (t, x, y) of its graph (see Definition 18.5.5, p.740).

Let us associate with the set-valued map Ψ the map $\Psi^{\Rightarrow}$ defined by

$$\text{Graph}(\Psi^{\Rightarrow}) \ := \ \text{Exit}_{(16.18)}(\text{Graph}(\Psi))$$

Theorem 16.2.24 ***[The Viability Solution is a Solution to Burgers Inclusion]*** *Assume that the set-valued map G is Marchaud. The viability solution to the Burgers problem is the largest solution $\mathbf{U}$ with closed graph to the Burgers partial differential equation (16.18), p. 668, in the sense that*

$$\forall \, y \in \mathbf{U}(t, x) \setminus \Phi(t, x), \ \ D^{\star\star}\mathbf{U}(t, x, y)(-1, -y) \cap G(t, x, y) \ \neq \ \emptyset \quad (16.20)$$

satisfying

$$\Phi(t, x) \ \subset \ \mathbf{U}(t, x) \ \subset \ \Psi(t, x)$$

It satisfies moreover

$$\mathbf{U}(t, x) \cap \Psi^{\Rightarrow}(t, x) \ \subset \ \Phi(t, x) \ \subset \mathbf{U}(t, x) \quad (16.21)$$

The regulation map $R_{\mathbf{U}}$ is then equal to

$$\forall\, y \in \mathbf{U}(t,x) \setminus \Phi(t,x), \;\; R_{\mathbf{U}}(t,x,y) \;=\; D^{\star\star}\mathbf{U}(t,x,y)(-1,-y) \cap G(t,x,y) \tag{16.22}$$

*If we assume moreover that the set-valued map G is Lipschitz, then the viability solution is the **unique** solution to*

$$\forall\, y \in \mathbf{U}(t,x,y) \cap D^{\star\star}\Psi(t,x,y)(+1,+y), \;\; G(t,x,y) \subset D^{\star\star}\mathbf{U}(t,x,y)(+1,+y) \tag{16.23}$$

Remark: "Boundary Conditions". Inclusion $\mathbf{U}(t,x) \cap \Psi^{\Rightarrow}(t,x) \subset \Phi(t,x)$ (see inclusions (11.18), p. 461 of Theorem 11.4.1, p.460) is the "mother of all conditions" concealed in the set-valued map Φ, as we saw in the proof of Theorem 16.2.4, p.635 for the Cauchy/Burgers condition. □

Remark: Required Properties of the Constraint Map Ψ. To be applied in concrete examples of constraints depicted by the set-valued map Ψ, Theorem 16.2.24, p.668 requires the following information on Ψ:

- the knowledge or the exit set of the graph of Ψ (for characterizing further conditions relating $\mathbf{U}(t,x)$ and $\Phi(t,x) \subset \mathbf{U}(t,x)$);
- the knowledge of the derivative $D^{\star\star}\Psi(t,x,y)$ of the set-valued map Ψ (for obtaining uniqueness of the solution in the class of set-valued maps with closed graphs) instead of knowing only that it is largest solution satisfying (16.22), p. 669. □

Proof. (Theorem 16.2.24, p.668). It is exactly the same than the proof of Theorem 16.2.4, p.635. We thus sketch it.

Theorem 11.4.1, p.460 states that the graph of the viability solution $\mathbf{U}$ allows us to compute the regulation map $R_{\mathbf{U}}$. It associates with any triple (t,x,y) the set of controls $v \in G(t,x,y)$ such that

$$(-1,-y,-v) \;\in\; T^{\star\star}_{\mathrm{Graph}(\mathbf{U})}(t,x,y) \;=:\; \mathrm{Graph}(D^{\star\star}\mathbf{U}(t,x,y))$$

which can be written in the form $v \in D^{\star\star}\mathbf{U}(t,x,y)(-1,-y) \cap G(t,x,y)$.

The viability solution is thus the largest closed set-valued map between Φ and Ψ satisfying $D^{\star\star}\mathbf{U}(t,x,y)(-1,-y) \cap G(t,x,y) \neq \emptyset$ and

$$\Phi(t,x) \;\subset\; \mathbf{U}(t,x) \;\subset\; \Psi(t,x)$$

Furthermore

$$\forall\, y \in \mathbf{U}(t,x,y) \setminus \Phi(t,x), \quad R_{\mathbf{U}}(t,x,y) \;\neq\; \emptyset$$

and is equal to $D^{\star\star}\mathbf{U}(t,x,y)(-1,-y) \cap G(t,x,y)$.

Formula (10.12), p. 400 of Theorem 10.5.2, p.400 implies that

$$\mathrm{Graph}(\mathbf{U}) \cap \mathrm{Exit}_{(16.18)}(\mathrm{Graph}(\Psi)) \;\subset\; \mathrm{Graph}(\Phi)$$

which means that

$$\mathbf{U}(t,x) \cap \Psi^{\Rightarrow}(t,x) \subset \Phi(t,x)$$

If furthermore G is Lipschitz, Theorem 11.4.6, p.463 implies that the viability solution is the **unique** closed graph set-valued map satisfying

$$\begin{cases} (i) \ \forall\, y \in \mathbf{U}(t,x,y) \setminus \Phi(t,x), \ R_{\mathbf{U}}(t,x,y) = D^{\star\star}\mathbf{U}(t,x,y)(-1,-y) \\ (ii) \ \forall y \in \mathbf{U}(t,x,y) \cap D^{\star\star}\Psi(t,x,y)(+1,+y), G(t,x,y) \subset D^{\star\star}\mathbf{U}(t,x,y)(+1,+y) \end{cases}$$

This completes the proof. □

16.2.10.2 Viability Solution to Controlled Burgers Tracking Problem

Definition 16.2.25 *[Controlled Burgers Tracking Property] Let us consider two set-valued maps Ψ and $\Phi \subset \Psi$ from $\mathbb{R}_+ \times \mathbb{R}$ to $\mathbb{R}$ and a second-order differential inclusion $x''(t) \in G(t,x(t),x'(t))$. We shall say that a set-valued map $\mathbf{V} : \mathbb{R}_+ \times \mathbb{R} \rightsquigarrow \mathbb{R}$ satisfies the* tracking property *if for every $y \in \mathbf{V}(T,x)$,*

1. ***there exists** a solution $x(\cdot)$ to $x''(t) \in G(t,x(t),x'(t))$ and a time $s^\star \in [0,T]$ such that*

$$\begin{cases} (i) \ \ x(T) = x \text{ and } x'(T) = y \\ (ii) \ \ x'(s^\star) \in \Phi(s^\star, x(s^\star)) \\ (iii) \ \forall t \in [s^\star, T], \ x'(t) \in \mathbf{V}(t,x(t)) \end{cases}$$

2. ***all solutions** $x(\cdot)$ to $x''(t) \in G(t,x(t),x'(t))$ such that*

$$\begin{cases} (i) \ x(T) = x \text{ and } x'(T) = y \\ (ii) \text{ for all } t \in [T,s], \ x'(t) \in \Psi(t,x(t)) \end{cases}$$

satisfy

$$\text{for all } t \in [T,s], \ x'(t) \in \mathbf{V}(t,x(t))$$

The viability solution is still the unique solution to the Burgers tracking problem with constraints:

Theorem 16.2.26 *[Viability Solution to a general Burgers Problem] Let us consider two set-valued maps $\Psi : \mathbb{R}_+ \times X \rightsquigarrow X$ and $\Phi : \mathbb{R}_+ \times X \rightsquigarrow X$ contained in Ψ. Assume that for all $y \in \Psi(T,x)$ and for all $t \geq T$, $y \in \Psi(t, x+(t-T)y)$. The viability solution $\mathbf{U}_{(\Psi,\Phi)} : \mathbb{R}_+ \times \mathbb{R}_+ \rightsquigarrow \mathbb{R}$*

*is the **unique** solution to the Burgers problem associated with* (Ψ, Φ). *Furthermore,* $\mathbf{U}(T,x)$ *is the set of fixed points*

$$y \in \Phi(T - t^\star(T,x,y), x - t^\star(T,x,y)y) \cap \bigcap_{t \in [0,t^\star(T,x,y)]} \Psi(T-t, x-ty)$$

where $t^\star(T,x,y)$ *is the first instant* t *when* $(T-t, x-ty, y)$ *reaches the graph of* Φ.

Proof. To say that (T,x,y) belongs to the capture basin

$$\mathrm{Capt}_{(16.4)}(\mathrm{Graph}(\Psi), \mathrm{Graph}(\Phi)) =: \ \mathrm{Graph}(\mathbf{U})$$

amounts to saying that there exists a finite time $t^\star(T,x,y)$ such that

1. the value $(T-t^\star(T,x,y), x-yt^\star(T,x,y), y)$ of the solution to characteristic differential equation (16.4) at time $t^\star(T,x,y) \in [0,T]$ belongs to the graph of the set-valued map Φ.
2. for all $t \in [0, t^\star(T,x,y)]$, $(T-t, x-ty, y)$ belongs to the graph of Ψ.

The first condition means that $y \in \Phi(T - t^\star(T,x,y))$.

The second condition means that for all $t \in [0, t^\star(T,x,y)]$, $y \in \Psi(T-t, x-ty)$.

Therefore, we have proved that $\mathbf{U}(T,x)$ is the set of fixed points of the set-valued map

$$y \rightsquigarrow \Phi(T - t^\star(T,x,y), x - t^\star(T,x,y)y) \cap \bigcap_{t \in [0,t^\star(T,x,y)]} \Psi(T-t, x-ty)$$

Theorem 10.2.5 states that the graph of the viability solution is actually the unique graph of a set-valued map $\mathbf{V}$ between Φ and Ψ satisfying

$$\begin{cases} \mathrm{Graph}(\mathbf{V}) \ = \ \mathrm{Capt}_{(16.4)}(\mathrm{Graph}(\mathbf{V}), \mathrm{Graph}(\Phi)) \\ = \ \mathrm{Capt}_{(16.4)}(\mathrm{Graph}(\Psi), \mathrm{Graph}(\mathbf{V})) \end{cases}$$

The first condition means that for any $t \in [0, t^\star(T,x,y)]$, y belongs to $\mathbf{V}(T-t, x-yt)$. By the change of variable $s := T-t$, this means that for any $s \in [T - t^\star(T,x,y), T]$, $y \in \mathbf{V}(s, x+(s-T)y)$.

Let $s > T$ such that $\forall t \in [T,s]$, $y \in \Psi(t, x+(t-T)y)$. Then the second condition implies that

$$\forall t \in [T,s], \ y \in \mathbf{V}\left(t, x+(t-T)y\right)$$

Indeed, it is sufficient to check that $(t, x + (t - T)y, y)$ belongs to the capture basin $\mathrm{Capt}_{(16.4)}(\mathrm{Graph}(\Psi), \mathrm{Graph}(\mathbf{V}))$ of the graph of $\mathbf{V}$. This is the case because, for the finite time $t^\star := t - T \geq 0$,

$$(t - t^\star, x + (t - T)y - t^\star y, y) = (T, x, y) \in \mathrm{Graph}(\mathbf{V})$$

Therefore $(t, x + (t - T)y, y)$ belongs to $\mathrm{Capt}_{(16.4)}(\mathrm{Graph}(\Psi)), \mathrm{Graph}(\mathbf{V}) = \mathrm{Graph}(\mathbf{V})$ and thus

$$y \in \mathbf{V}(t, x + (t - T)y)$$

Hence these two conditions mean that $\mathbf{V}$ is a solution to the Burgers problem associated with the data (Φ, Ψ). □

16.2.10.3 Example

We exploit this result in the case of Cauchy condition $\mathbf{U}_0$ without boundary constraints and by restricting our analysis to the case when $G(t, x, y) \equiv [-1, +1]$ for saving notations. We obtain the following formulas:

Proposition 16.2.27 *[Example: Case of Cauchy Conditions]*
Let us consider the case when

1. *the right hand side* $G(t, x, y) \equiv [-1, +1]$ *is the constant unit interval* $[-1, +1]$,
2. *there is no constraint:* $\Psi(t, x) \equiv \mathbb{R}$,
3. *the set-valued map* $\mathbf{U}_0^+$ *is associated to an initial datum* $\mathbf{U}_0 : \mathbb{R} \rightsquigarrow \mathbb{R}$.

Then $y \in \mathbf{U}(T, x)$ *if and only if there exists an open loop control* $t \mapsto u(t) \in [-1, +1]$ *such that*

$$y \in \mathbf{U}_0\left(x - Ty + \int_0^T (T - s)u(s)ds\right) + \int_0^T u(s)ds$$

Furthermore, $\mathbf{U}$ *is the unique set-valued map* $\mathbf{V}$ *satisfying the Cauchy condition* $\mathbf{V}(0, x) = \mathbf{U}_0(x)$ *such that* $y \in \mathbf{V}(T, x)$ *if and only if*

1. ***there exists*** *an open loop control* $t \mapsto u(t) \in [-1, +1]$ *such that,* $\forall t \in [0, T]$,

$$y - \int_t^T u(T - \tau)d\tau \in \mathbf{V}\left(t, x + (t - T)y + \int_t^T (\tau - s)u(T - \tau)d\tau\right)$$

2. ***for all*** *open loop controls* $t \mapsto v(t) \in [-1,+1]$, $\forall t \geq T$,

$$y - \int_t^T v(T-\tau)d\tau \in \mathbf{V}\left(t, x + (t-T)y + \int_t^T (\tau - t)v(T-\tau)d\tau\right)$$

Proof. Indeed, to say that (T, x, y) belongs to

$$\text{Graph}(\mathbf{U}_{(\Psi,\mathbf{U}_0^+)}) := \text{Capt}_{(16.18)}(\text{Graph}(\Psi), \text{Graph}(\mathbf{U}_0^+))$$

amounts to saying that there exists a backward evolution $(x(\cdot), y(\cdot))$ to system (16.18) starting at (x, y) such that $(x(T), y(T))$ belongs to the graph of $\mathbf{U}_0$. Such solution can be written

$$\begin{cases} y(t) = y - \int_0^t u(s)ds \\ x(t) = x - \int_0^t y(s)ds = x - ty + \int_0^t (t-s)u(s)ds \end{cases}$$

where $t \mapsto u(t) \in [-1,+1]$ is an open loop control and since

$$\int_0^t \int_0^s u(\tau)dsd\tau = \int_0^t u(\tau)d\tau \left(\int_\tau^t ds\right) = \int_0^t (t-\tau)u(\tau)d\tau$$

Taking $t = T$, we obtain

$$y \in \mathbf{U}_0\left(x - Ty + \int_0^T (T-s)u(s)ds\right) + \int_0^T u(s)ds$$

Theorem 10.2.5 states that the graph of the viability solution is actually the unique graph of a set-valued map $\mathbf{V}$ between $\mathbf{U}_0^+$ and $\mathbb{R}_+ \times \mathbb{R} \times \mathbb{R}$ satisfying

$$\begin{aligned} \text{Graph}(\mathbf{V}) &= \text{Capt}_{(16.18)}(\text{Graph}(\mathbf{V}), \{0\} \times \text{Graph}(\mathbf{U}_0)) \\ &= \text{Capt}_{(16.18)}(\text{Graph}(\mathbb{R}_+ \times \mathbb{R} \times \mathbb{R}), \text{Graph}(\mathbf{V})) \end{aligned}$$

The first condition means that for any $t \in [0, T]$, $y(t)$ belongs to $\mathbf{V}(T - t, x(t))$. This means that

$$\forall t \in [0,T],\ y - \int_0^t u(s)ds \in \mathbf{V}\left(T-t, x - ty + \int_0^t (t-s)u(s)ds\right)$$

By the change of variable $s := T - t$ and $v(s) := u(T-s)$, this means that for any $s \in [0, T]$,

$$\forall s \in [0,T],\ y - \int_s^T v(\tau)d\tau \in \mathbf{V}\left(s, x + (s-T)y + \int_s^T (\tau - s)v(\tau)d\tau\right)$$

The second condition amounts to saying that for all $t \geq T$ and for all open loop controls $t \mapsto u(t) \in [-1,+1]$,

$$\forall s \geq T,\ y - \int_s^T v(\tau)d\tau \in \mathbf{V}\left(s, x + (s-T)y + \int_s^T (\tau - s)v(\tau)d\tau\right)$$

If not, there would exist some $t^\sharp > T$ such that

$$y - \int_{t^\sharp}^T u(T-\tau)d\tau \in \mathbf{V}\left(t^\sharp, x + (t^\sharp - T)y + \int_{t^\sharp}^T (\tau - s)u(T-\tau)d\tau\right)$$

belongs to $\mathbb{R}_+ \times \mathbb{R} \times \mathbb{R} \setminus \text{Graph}(\mathbf{V})$. But, by construction,

$$y - \int_{t^\sharp}^T u(T-\tau)d\tau \in \mathbf{V}\left(t^\sharp, x + (t^\sharp - T)y + \int_{t^\sharp}^T (\tau - s)u(T-\tau)d\tau\right)$$

belongs to $\text{Capt}_{(16.18)}(\text{Graph}(\mathbb{R}_+ \times \mathbb{R} \times \mathbb{R}), \text{Graph}(\mathbf{V})) = \text{Graph}(\mathbf{V})$, a contradiction. □

Proposition 16.2.28 *[Example: Case of Marks] Let us consider the case when*

1. *the right hand side* $G(t,x,y) \equiv [-1,+1]$ *is the constant unit interval* $[-1,+1]$,
2. *there is no constraint:* $\forall\, t \geq 0,\ \forall\, x \in \mathbb{R},\ \Psi(t,x) \equiv \mathbb{R}$,
3. *the set-valued map* $\mathbf{U}_0^+$ *is associated to an initial datum* $\mathbf{U}_0 : \mathbb{R} \rightsquigarrow \mathbb{R}$,
4. $\mathbf{U}_0(x) := \beta\Xi(A;x)$.

Then the solution is given by $\mathbf{U}(T,x) =$

$$\left\{\left(\beta + \int_0^T u(s)ds\right)\Xi\left(A + T\beta + \int_0^T su(s)ds; x\right)\right\}_{u(\cdot)\in L^1(0,T;[-1,+1])}$$

16.3 The Invariant Manifold Theorem

The idea behind the *Invariant Manifold* approach is the reduction of a system

$$\begin{cases} (i)\ x'(t) = f(x(t)) \\ (ii)\ y'(t) = g(x(t), y(t)) \end{cases}$$

to one of them by looking for a map $\mathbf{U} : X \mapsto Y$ tracking the solution $x(\cdot)$ of the first system by $y(t) := \mathbf{U}(x(t))$, solution to the second system. This means that

$$\forall\, t \geq 0,\ \ (x(t), y(t)) \ \in \ \text{Graph}(\mathbf{U})$$

or, in other words, that the graph of the map $\mathbf{U}$ is invariant under the above system. Hence the name of invariant (or center) manifold for denoting the graph of this map.

Jacques Hadamard was the first to prove the existence of an invariant manifold for systems

$$\begin{cases} (i)\ x'(t) = f(x(t)) \\ (ii)\ y'(t) = -My(t) + g(x(t)) \end{cases}$$

showing that the invariant manifold was the graph of the solution to the partial differential equation

$$\forall x,\ 0 \ = \ \frac{d\mathbf{U}(x)}{dx} f(x) - g(x, \mathbf{U}(x))$$

It is given by the explicit analytical formula

$$\mathbf{U}(x) := - \int_0^{+\infty} e^{-Mt} g(\vartheta_f(t, x))dt$$

where $\vartheta_f(t, x)$ is the flow (or reachable map) generated by f (see Definition 8.4.1, p.284), whenever the largest eigenvalue

$$\lambda \ := \ \inf_{\|x\|=1} \langle Mx, x \rangle$$

of the matrix M is large enough for the above integral to exist.

The situation is more complex for nonlinear problems, a manifold of (local) invariant manifold theorems were proved.

In Sect. 8.3 of the first edition of *Viability Theory* [18, Aubin] (1991), the invariant manifold theorem was extended to control systems

$$\begin{cases} (i)\ x'(t) = f(x(t), u(t)) \\ (ii)\ y'(t) = -My(t) + g(x(t), u(t)) \\ \quad\quad \text{where } u(t) \in \mathbf{U}(x(t)) \end{cases}$$

It was proved that under growth conditions involving the largest eigenvalue of M, there exists a unique solution to the partial differential inclusion

$$\forall x, \forall y \in \mathbf{U}(x),\ \exists\, u \in \mathbf{U}(x) \mid 0 \in \frac{d\mathbf{U}(x)}{dx} f(x,u) + My - g(x,u)$$

For further details and results, see Chap. 8 of the first edition of *Viability Theory* [18, Aubin] (1991).

We only present here a theorem due to Hélène Frankowska extending the Hadamard Invariant Manilfold Theorem to nonlinear control systems

$$\begin{cases} (i)\ \ x'(t) = f(x(t),u(t)) \\ (ii)\ y'(t) = g(x(t),y(t),u(t)) \\ \quad\ \ \text{where } u(t) \in \mathbf{U}(x(t)) \end{cases} \tag{16.24}$$

subjected to adequate growth conditions.

Theorem 16.3.1 [Strictness of the Viability Solution under Growth Conditions] *Assume that the dynamics of the controlled system*

$$\begin{cases} (i)\ \ x'(t) = f(x(t),u(t)) \\ (ii)\ y'(t) = g(x(t),y(t),u(t)) \\ \quad\ \ \text{where } u(t) \in \mathbf{U}(x(t)) \end{cases} \tag{16.25}$$

are Marchaud maps satisfying the two following growth conditions:

$$\begin{cases} (i)\ \ \sup_{x \in X} \sup_{u \in \mathbf{U}(x)} \|f(x,u)\| \le c(\|x\|+1) \\ (ii)\ \sup_{(x,y) \in X \times Y} \sup_{u \in \mathbf{U}(x)} \frac{\langle g(x,y,u), y\rangle}{\|y\|} \le -\lambda\|y\| + d(\|x\|+1)^{\alpha} \end{cases} \tag{16.26}$$

If $\lambda > c\alpha$, then there exists a solution $\mathbf{U} : X \rightsquigarrow Y$ with nonempty values *to the tracking problem* satisfying the growth condition

$$\forall x \in X,\ \|\mathbf{U}(x)\| \le \frac{d}{\lambda - c\alpha}(\|x\|+1)^{\alpha} \tag{16.27}$$

the graph of which is both backward viable and invariant *under (16.25). It the is largest closed set-valued map satisfying growth conditions (16.27), p.676 and solution to the partial differential inclusion*

$$\forall x, \forall y \in \mathbf{U}(x),\ \exists\, u \in \mathbf{U}(x) \mid 0 \in \frac{d\mathbf{U}(x)}{dx} f(x,u) - g(x,y,u)$$

This statement is a consequence of the following result:

Theorem 16.3.2 *[***Non-Emptiness of the Values of the Viability Solution under Constraints***]*

Let $\Psi : X \rightsquigarrow Y$ be a closed set-valued map. Assume that the dynamics of the controlled system (16.25) are Marchaud maps satisfying growth conditions

$$\begin{cases} (i) & \sup_{(x,y)\in \mathrm{Graph}(\Psi)} \sup_{u\in \mathbf{U}(x)} \|f(x,u)\| \leq c(\|x\|+1) \\ (ii) & \sup_{(x,y)\in \mathrm{Graph}(\Psi)} \sup_{u\in \mathbf{U}(x)} \frac{\langle g(x,y,u), y\rangle}{\|y\|} \leq -\lambda\|y\| + d(\|x\|+1)^{\alpha} \end{cases}$$

and that

1. *the graph* Graph(Ψ) *of Ψ is invariant under (16.25),*
2. *the domain* Dom(Ψ) *of Ψ is backward viable under (16.25),*
3. *the growth of Ψ is polynomial:*

$$\forall x \in \mathrm{Dom}(\Psi),\ \|\Psi(x)\| \leq b(\|x\|+1)^{\beta}$$

Let $\mathbf{U} : X \rightsquigarrow Y$ be the viability solution to the tracking problem under (16.25) defined by

$$\mathrm{Graph}(\mathbf{U}) := \mathrm{Viab}_{(16.29)}(\mathrm{Graph}(\Psi)) \tag{16.28}$$

where

$$\begin{cases} (i)\ x'(t) = -f(x(t),u(t)) \\ (ii)\ y'(t) = -g(x(t),y(t),u(t)) \\ \quad\ \text{where } u(t) \in \mathbf{U}(x(t)) \end{cases} \tag{16.29}$$

If $\lambda > c\max(\alpha,\beta)$, then Dom(**U**) = Dom($\Psi$)*: for every $x \in$* Dom(Ψ), **U**$(x) \neq \emptyset$*. Furthermore, the graph of* **U** *is both* ***backward viable and invariant*** *under (16.25).*

Proof. Nothing guarantees that the values of the viability solution **U** are not empty. We shall derive non emptiness of the values **U**(x) for every $x \in$ Dom(Ψ) satisfying the assumptions of the theorem.

Let $x_0 \in$ Dom(Ψ) and $x(\cdot)$ be a solution starting from x_0 to differential inclusion $x'(t) = -f(x(t),u(t))$ where $u(t) \in \mathbf{U}(x(t))$, viable in the domain of Ψ (which exists since the system is Marchaud and Dom(Ψ) backward viable). The growth condition on f implies that

$$\|x(t)\| \leq (\|x_0\|+1)e^{ct} - 1$$

Let T_n be a sequence converging to ∞, $y_n \in \Psi(x(T_n))$ be chosen in $\Psi(x(T_n))$ and $y_n(\cdot)$ be the solutions to

$$y_n'(t) = -g(x(t), y_n(t), u(t)) \ \& \ y_n(T_n) = y_n$$

Setting $\overrightarrow{x}_n(t) := x(T_n - t)$ and $\overrightarrow{y}_n(t) := y_n(T_n - t)$, we see that the pair $(\overrightarrow{x}_n(\cdot), \overrightarrow{y}_n(\cdot))$ is a solution to the control system (16.25) starting from $(x(T_n), y_n) \in \text{Graph}(\Psi)$. Since this graph is assumed to be backward invariant, we infer that for any $t \geq 0$, $\overrightarrow{y}_n(t) \in \Psi(\overrightarrow{x}_n(t))$, i.e., that

$$\forall T_n \geq t, \ y_n(t) = \overrightarrow{y}_n(T_n - t) \in \Psi(\overrightarrow{x}_n(T_n - t)) = \Psi(x(t))$$

The growth condition on g implies that

$$\frac{d}{dt}(e^{\lambda t}\|\overrightarrow{y}_n(t)\|) = e^{\lambda t}\left(\left\langle \overrightarrow{y}_n'(t), \frac{\overrightarrow{y}_n(t)}{\|\overrightarrow{y}_n(t)\|}\right\rangle + \lambda\|\overrightarrow{y}_n(t)\|\right) \leq e^{\lambda t} d(\|x(T_n - t)\| + 1)^\alpha$$

so that, integrating from 0 to $T_n - t \geq 0$, we obtain

$$e^{\lambda(T_n - t)}\|\overrightarrow{y}_n(T_n - t)\| \leq \|\overrightarrow{y}_n(0)\| + d\int_0^{T_n - t} e^{\lambda\tau}(\|x(T_n - \tau)\| + 1)^\alpha d\tau$$

Since $\overrightarrow{y}_n(0) = y_n \in \Psi(x(T_n))$, we infer that

$$\|\overrightarrow{y}_n(0)\| \leq b(\|x(T_n)\| + 1)^\beta \leq b(\|x_0\| + 1)^\beta e^{c\beta T_n}$$

On the other hand,

$$\|x(T_n - \tau)\| + 1 \leq e^{c(T_n - t - \tau)}(\|x(t)\| + 1)$$

so that

$$\|\overrightarrow{y}_n(T_n - t)\| \leq e^{-(\lambda - c\beta)T_n} e^{\lambda t} b(\|x_0\| + 1)^\beta + d(\|x(t)\| + 1)^\alpha \int_0^{T_n - t} e^{-(\lambda - c\alpha)(T_n - t - \tau)} d\tau$$

and thus, that

$$\|\overrightarrow{y}_n(T_n - t)\| \leq e^{-(\lambda - c\beta)T_n} e^{\lambda t} b(\|x_0\| + 1)^\beta + \frac{d(\|x(t)\| + 1)^\alpha}{\lambda - c}\left(1 - e^{-(\lambda - c\alpha)(T_n - t)}\right)$$

or, for every $T_n \geq t$,

$$\|y_n(t)\| \leq e^{-(\lambda - c\beta)T_n} e^{\lambda t} b(\|x_0\| + 1)^\beta + \frac{d(\|x(t)\| + 1)^\alpha}{\lambda - c\alpha}\left(1 - e^{-(\lambda - c)(T_n - t)}\right)$$

In particular, for $t = 0$, we obtain inequality

$$\|y_n(0)\| \leq e^{-(\lambda - c\beta)T_n} b(\|x_0\| + 1)^\beta + \frac{d(\|x_0\| + 1)^\alpha}{\lambda - c\alpha}\left(1 - e^{-(\lambda - c\alpha)T_n}\right)$$

Since $\lambda > c\max(\alpha, \beta)$, we infer that $y_n(0)$ is bounded, and thus, that a subsequence (again denoted by) converges to some $y_0^\star$.

Therefore, control system (16.29) being Marchaud, a subsequence (again denoted by) $(x(\cdot), y_n(\cdot))$ converges to a solution $(x(\cdot), y(\cdot))$ to control system (16.29) starting at $(x_0, y_0^\star)$ and satisfying for every $t \geq 0$, $y(t) \in \Psi(x(t))$. This proves that $(x_0, y_0^\star)$ belongs to the viability kernel of the graph of Ψ under control system (16.29), i.e., to the graph of the viability solution $\mathbf{U}$. Hence we have proved that $\mathbf{U}(\cdot)$ is not empty. □

When the set-valued map Ψ is not imposed, we can associate with the growth conditions an appropriate set-valued map Ψ that satisfies the assumptions of Theorem 16.3.2:

Proof. (Theorem 16.3.1, p.676). We apply Theorem 16.3.2 to the set-valued map $\Psi : X \rightsquigarrow Y$ defined by

$$\Psi(x) := \left\{ y \in Y \mid \|y\| \leq \frac{d}{\lambda - c\alpha}(\|x\| + 1)^\alpha \right\}$$

We have to prove that the graph of Ψ is invariant, i.e., that its complement is backward invariant under thanks to Theorem 10.5.7. □

If not, there would exist $y_0 \notin \Psi(x_0)$ and a solution $(x(\cdot), y(\cdot))$ to the system (16.29) starting from (x_0, y_0) reaching the graph of Ψ at time t^*. Since $y'(t) = -g(x(t), y(t), u(t))$, we deduce that

$$\frac{d}{dt}(e^{-\lambda t}\|y(t)\|) = e^{-\lambda t}\left(\left\langle y'(t), \frac{y(t)}{\|y(t)\|}\right\rangle - \lambda\|y(t)\|\right) \;\geq\; -e^{-\lambda t} d(\|x(t)\|+1)^\alpha$$

Integrating from 0 to t this inequality, we obtain

$$e^{-\lambda t}\|y(t)\| - \|y_0\| \;\geq\; -d\int_0^t e^{-\lambda\tau}(\|x(\tau)\| + 1)^\alpha d\tau$$

and thus

$$\begin{cases} \|y_0\| \leq e^{-\lambda t}\|y(t)\| + d(\|x_0\| + 1)^\alpha \displaystyle\int_0^t e^{-(\lambda - c\alpha)\tau} d\tau \\ \leq \; e^{-\lambda t}\|y(t)\| + \dfrac{d}{\lambda - c\alpha}(\|x_0\| + 1)^\alpha \left(1 - e^{-(\lambda - c\alpha)t}\right) \end{cases}$$

Therefore, since $(x(t^*), y(t^*))$ belongs to $\mathrm{Graph}(\Psi)$, we can use the estimate $y(t^*) \leq \frac{d}{\lambda - c\alpha}(\|x(t^*)\| + 1)^\alpha \leq \frac{d}{\lambda - c\alpha}(\|x_0\| + 1)^\alpha e^{c\alpha t}$ and obtain:

$$\|y_0\| \leq \frac{d}{\lambda - c\alpha}(\|x_0\| + 1)^\alpha \;<\; \|y_0\|$$

that is the contradiction we were looking for.

Furthermore, the tangential conditions imply that the set-valued map $\mathbf{U}$ is the largest closed set-valued map satisfying growth conditions (16.27), p.676 and solution to the partial differential inclusion

$$\forall x, \forall y \in \mathbf{U}(x), \ \exists\, u \in \mathbf{U}(x) \mid 0 \in \frac{d\mathbf{U}(x)}{dx} f(x,u) - g(x,y,u) \ \square$$

Chapter 17
Viability Solutions to Hamilton–Jacobi–Bellman Equations

17.1 Introduction

We summarized the main results of the Hamilton–Jacobi–Bellman strategy to study intertemporal optimization in the "optimal control survival kit", Sect. 4.11, p. 168. Chapters 4, p. 125 and 14, p. 563 and Sects. 15.2, p. 605 and 15.3, p. 620 provided many examples of value functions of a series of optimization problems over state-control pairs solutions to control systems which were characterized in terms of viability kernels and capture basins of auxiliary systems, which we referred to as ***viability episolutions***.

The purpose of this chapter is to study classes of intertemporal optimization problems summarized in Sect. 17.6, p.708 covering these particular cases and numerous other ones.

This allows us to use the Viability and Invariance Theorems 11.3.4, p. 455 and 11.3.7, p. 457 to derive that theses value functions are "generalized" solutions to Hamilton–Jacobi–Bellman partial differential equation. This is what *Hélène Frankowska* discovered and uncovered at the end of the 1980s. This chapter summarizes her results by answering formally these questions. Unfortunately, these results become more and more technical to take into account the lack of differentiability of solutions to partial differential equations, due to the respect of viability constraints. The question arises to give a mathematical sense to the derivative of non-differentiable functions, known under the generic name of *weak derivatives*, *graphical* and *epigraphical derivatives*, *subdifferentials*, *superdifferentials* and *generalized gradients*, an example of permanent revolutions of earlier definitions of the concept of derivative which started with *Pierre de Fermat*, summarized in Sect. 18.9, p. 765.

However, the ideas behind these concepts are very simple: we just *translate* in terms of value functions and partial differential equation the previous Viability and Invariance Theorems and the Frankowska property (11.19, p. 462), as well as other results gathered in this book.

J.-P. Aubin et al., *Viability Theory*, DOI 10.1007/978-3-642-16684-6_17,

But translation is not always easy, as at least one of the French speaking authors of this book knows too well. This will be the case here, since we shall have to pass from tangent and normal cones to generalized derivatives (actually, epiderivatives) and generalized gradients of a lower semicontinuous extended function, which is not differentiable in the usual sense. This requires the formalism of nonsmooth analysis summarized in Chap. 18, p. 713.

17.2 Viability Solutions to Hamilton–Jacobi–Bellman Problems

Hence, in this chapter, we shall prove that:

1. value functions of optimal control problems are solutions to associated Hamilton–Jacobi–Bellman partial differential equations,
2. solutions to Hamilton–Jacobi–Bellman partial differential equations of a certain class are value functions of an underlying optimal control problem, playing the role of representation formulas.

The first problem is an issue in control theory, the second one is a topic of partial differential equation analysis, using their specific tool box. These two problems are actually two faces of a viability problem (actually, a capturability one) of an auxiliary control system we are about to describe, the stone allowing us to "kill" the two problems, i.e., *to solve a class of optimal control problems and a class of Hamilton–Jacobi–Bellman partial differential equations.*

For this purpose, we will place all these problems under the same umbrella, which takes the form of the general auxiliary systems described by four functions:

37 ***[The Gang of Four Functions.]*** *The auxiliary system and hence, the class of optimal control problems and Hamilton–Jacobi–Bellman partial differential equations, is described by the control system*

$$(\mathcal{S}): \quad x'(t) \;=\; f(x(t), u(t)) \text{ where } u(t) \in U(x(t)) \tag{17.1}$$

where we assume that

$$\forall\, x \in K, \;\; U(x) \;\neq\; \emptyset$$

and the four extended functions:

1. *the* Lagrangian $\mathbf{l} : X \times \mathcal{U} \mapsto \mathbb{R}_+$*, assigning to each state-control pair* (x,u) *its transient cost* $\mathbf{l}(x,u)$*,*
2. *the* discount rate $\mathbf{m} : X \times \mathcal{U} \mapsto \mathbb{R}_+$*, assigning to each state-control pair* (x,u) *its discount rate* $\mathbf{m}(x,u)$ *which may depend upon the state* x *and controlled by* u*,*

3. *the* constraint function $\mathbf{k} : \mathbb{R}_+ \times X \mapsto \mathbb{R} \cup \{+\infty\}$
4. $\mathbf{c} : \mathbb{R}_+ \times X \mapsto \mathbb{R} \cup \{+\infty\}$, *the* target function, *regarded as a* spot cost *in control theory or a target condition (Cauchy function, Dirichlet boundary condition, etc.) in partial differential equations.*

The main example is the case $\mathbf{k}(t,x) := \psi_K(x)$ where ψ_K is the *indicator* of K (see Definition 18.6.1, p. 743), which is a cost function defined by $\psi_K(x) := 0$ when $x \in K$ and $\psi_K(x) := +\infty$ otherwise. It is used to enforce that the state variable x belongs to K. The constraint $u \in U(x)$ can also be represented by taking for function $\mathbf{k}(t,x) := \psi_{\text{Graph}(U)}(x,u)$ the indicator of its graph. The absence of viability constraints is obtained by taking $\mathbf{k}(t,x) \equiv -\infty$.

Hence the auxiliary viability/capturability problem is made of

1. the auxiliary control system (17.2)

$$\begin{cases} (i) & \tau'(t) = -1 \\ (ii) & x'(t) = f(x(t), u(t)) \\ (iii) & y'(t) = \mathbf{m}(x(t), u(t))y(t) - \mathbf{l}(x(t), u(t)) \\ & \text{where } u(t) \in U(x(t)) \end{cases} \tag{17.2}$$

2. the environment $\mathcal{K} := \mathcal{E}p(\mathbf{k})$ and the target $\mathcal{C} := \mathcal{E}p(\mathbf{c})$.

Definition 17.2.1 ***[Viability Solutions of Hamilton–Jacobi–Bellman]*** *Let us consider the auxiliary control system (17.2), p. 683 and the extended functions* $\mathbf{k}$ *and* $\mathbf{c}$*. The four* viability solutions *to the Hamilton–Jacobi–Bellman problem are defined in terms of capture basins and absorption basins of epigraphs and hypographs defined respectively by*

$$\begin{cases} W^{\inf}_{(\mathbf{k},\mathbf{c})}(T,x) & := \inf_{(T,x,y) \in \mathrm{Capt}_{(17.2)}(\mathcal{E}p(\mathbf{k}), \mathcal{E}p(\mathbf{c}))} y \\ W^{\sup}_{(\mathbf{k}_\downarrow,\mathbf{c}_\downarrow)} & := \sup_{(T,x,y) \in \mathrm{Capt}_{(17.2)}(\mathcal{H}yp(\mathbf{k}_\downarrow), \mathcal{H}yp(\mathbf{c}_\downarrow))} y \\ W^{\sup}_{(\mathbf{k},\mathbf{c})}(T,x) & := \inf_{(T,x,y) \in \mathrm{Abs}_{(17.2)}(\mathcal{E}p(\mathbf{k}), \mathcal{E}p(\mathbf{c}))} y \\ W^{\inf}_{(\mathbf{k}_\downarrow,\mathbf{c}_\downarrow)}(T,x) & := \sup_{(T,x,y) \in \mathrm{Abs}_{(17.2)}(\mathcal{H}yp(\mathbf{k}_\downarrow), \mathcal{H}yp(\mathbf{c}_\downarrow))} y \end{cases} \tag{17.3}$$

Viability Solutions $W^{\inf}_{(\mathbf{k},\mathbf{c})}$ *and* $W^{\sup}_{(\mathbf{k},\mathbf{c})}$ *are called* episolutions *and viability solutions* $W^{\inf}_{(\mathbf{k}_\downarrow,\mathbf{c}_\downarrow)}$ *and* $W^{\sup}_{(\mathbf{k}_\downarrow,\mathbf{c}_\downarrow)}$ *are called* hyposolutions.

Each of those viability solutions is

1. the value function of an optimization problem of a functional depending on the four functions $(\mathbf{l}, \mathbf{m}, \mathbf{k}, \mathbf{c})$ (see Definition 37, p. 682), which are described in Theorem 17.6.1, p.708, at the end of this chapter;
2. the "generalized" solution to

$$\begin{cases} (i) \quad \forall\, t \geq 0, x \in \Omega^V_{(\mathbf{k},\mathbf{c})}(t), \ \dfrac{\partial V(t,x)}{\partial t} + \mathbf{h}\left(x, V(t,x), \dfrac{\partial V(t,x)}{\partial x}\right) = 0 \\ (ii) \ \forall\,(t,x) \in \mathbb{R}_+ \times X, \ \ V(t,x) \leq \mathbf{c}(t,x) \\ (iii) \ \forall\,(t,x) \in \mathbb{R}_+ \times X, \ \ \mathbf{k}(t,x) \leq V(t,x) \end{cases} \tag{17.4}$$

where

$$\Omega^V_{(\mathbf{k},\mathbf{c})}(t) := \{x \in \in X \text{ such that } \mathbf{k}(t,x) \leq V(t,x) < \mathbf{c}(t,x)\}$$

and
$\mathbf{h} : X \times X \times \mathbb{R} \mapsto \mathbb{R} \cup \{+\infty\}$ is the *Hamiltonian* defined in terms of the four functions of the gang.
Condition (17.4)(ii), p. 684 subsumes *Cauchy conditions, Dirichlet conditions* or *Lagrangian conditions*, according to the choice of the target function $\mathbf{c}$. For instance, taking $\mathbf{c}(t,x) := +\infty$ whenever $t > 0$ and $\mathbf{c}(0,x) := U_0(x)$, it will be easy to show that condition (17.4)(ii), p. 684 is the *Cauchy condition* $V(0,x) = U_0(x)$. Inequality (17.4)(iii), p. 684 describes viability constraints.

Since the proofs are very similar (exchanging epigraphs and hypographs one hand, capture basins and absorption basins on the other one), we shall study in detail the viability solution

$$W^{\inf}_{(\mathbf{k},\mathbf{c})}(T,x) := \inf_{(T,x,y) \in \mathrm{Capt}_{(17.2)}(\mathcal{E}p(\mathbf{k}),\mathcal{E}p(\mathbf{c}))} y \tag{17.5}$$

38 [Why Using Backward Time] The use of backward time governed by differential equation $\tau'(t) = -1$ in the first component of the auxiliary system is justified by the following property: the environment $\mathcal{K} := \mathcal{E}p(\mathbf{k})$ is a *backward invariant repeller* under the auxiliary system (17.2), p. 683. The second reason will be clarified in Definition 17.3.3, p. 688 of the *valuation function* $V(T,x)$: the time involved in the solution $V(T,x)$ is no longer the current time $t \in [0,T]$, but the *horizon* T before which the current time t belongs. Hence the slight modification of the classical concept of *value function* $V(t,x)$ depending upon the current time $t \in [0,T]$.

17.3 Valuation Functions and Viability Solutions

17.3.1 A Class of Intertemporal Functionals

We introduce a set-valued map $U : X \rightsquigarrow \mathcal{U}$ and positive extended functions $\mathbf{k}$ (constraint function) and $\mathbf{c}$ (objective function) satisfying

$$\forall (t,x) \in \mathbb{R}_+ \times X,\ 0 \ \leq\ \mathbf{k}(t,x) \ \leq\ \mathbf{c}(t,x) \ \leq\ +\infty$$

We set

$$\widehat{\mathbf{m}}(t) := \int_0^t \mathbf{m}(x(\tau), u(\tau))d\tau$$

without mentioning explicitly the dependence on $(x(\cdot), u(\cdot))$ for simplifying notations.

We associate with any extended function $\mathbf{v} : \mathbb{R}_+ \times X \times \mathcal{U} \mapsto \mathbb{R}_+ \cup \{+\infty\}$, any $t \in [0,T]$ and any $(x(\cdot), u(\cdot)) \in \mathcal{S}(x)$ the functional

$$\begin{cases} \mathbf{J}_{\mathbf{v}}(t; (x(\cdot), u(\cdot)))(T,x) \\ := \ e^{-\widehat{\mathbf{m}}(t)}\mathbf{v}(T-t, x(t), u(t)) + \displaystyle\int_0^t e^{-\widehat{\mathbf{m}}(\tau)}\mathbf{l}(x(\tau), u(\tau))d\tau \end{cases}$$

The function $\mathbf{c} : \mathbb{R}_+ \times C \mapsto \mathbb{R} \cup \{+\infty\}$ is used for imposing target conditions. The function $\mathbf{k}$ is used for imposing *viability constraints*.

We define the maximal cost up to the current time t by:

$$\mathbf{I}_{\mathbf{k}}(t; (x(\cdot), u(\cdot)))(T,x) := \sup_{s \in [0,t]} \mathbf{J}_{\mathbf{k}}(s; (x(\cdot), u(\cdot)))(T,x)$$

We next integrate this cumulated cost together with the cost $\mathbf{J}_{\mathbf{c}}(t; (x(s), u(s)))(T,x)$ associated with the target function $\mathbf{c}$ by introducing the new cost functions

$$\mathbf{L}_{(\mathbf{k},\mathbf{c})}(t; (x(\cdot),u(\cdot)))(T,x) := \max(\mathbf{I}_{\mathbf{k}}(t; x(\cdot),u(\cdot))(T,x), \mathbf{J}_{\mathbf{c}}(t; (x(\cdot), u(\cdot)))(T,x))$$

Finally, we set

$$V_{(\mathbf{k},\mathbf{c})}(x(\cdot), u(\cdot))(T,x) \ := \ \inf_{t \in [0,T]} \mathbf{L}_{(\mathbf{k},\mathbf{c})}(t; (x(\cdot), u(\cdot)))(T,x)$$

This is the class of criteria that we shall minimize in this chapter.

Note that if $\mathbf{k}_1 \leq \mathbf{k}_2$ and $\mathbf{c}_1 \leq \mathbf{c}_2$, then $V_{(\mathbf{k}_1,\mathbf{c}_1)} \leq V_{(\mathbf{k}_2,\mathbf{c}_2)}$.

Example 1. For that purpose, we denote by $\mathbf{0}$ the function defined by

$$\mathbf{0}(t,x) = \begin{cases} 0 & \text{if } t \geq 0, \\ +\infty & t < 0 \end{cases}$$

and by $\mathbf{u}_\infty$ the function defined by

$$\mathbf{u}_\infty(t,x) := \begin{cases} \mathbf{u}(x) \text{ if } t = 0 \\ +\infty \text{ if } t \neq 0 \end{cases} \tag{17.6}$$

Proposition 17.3.1 *[Three Examples of Intertemporal Functionals]* *Assume that* $\mathbf{l} \geq 0$. *Then the functional*

$$(x(\cdot), u(\cdot)) \mapsto V_{(\mathbf{k},\mathbf{c})}(x(\cdot), u(\cdot))(T,x)$$

takes the following forms for the various choices of pairs $(\mathbf{k}, \mathbf{c})$ *associated with a given* time-independent function $\mathbf{u}$*:*

$$\begin{cases} (i)\ V_{(\mathbf{0},\mathbf{u}_\infty)}((x(\cdot),u(\cdot)))(T,x) = \quad \mathbf{J_u}(T;(x(\cdot),u(\cdot))(x)) \\ (ii)\ V_{(\mathbf{0},\mathbf{u})}((x(\cdot),u(\cdot)))(T,x) \ = \inf_{t\in[0,T]} \mathbf{J_u}(t;(x(\cdot),u(\cdot)))(x) \\ (iii)\ V_{(\mathbf{u},\mathbf{u}_\infty)}((x(\cdot),u(\cdot)))(T,x) = \sup_{t\in[0,T]} \mathbf{J_u}(t;(x(\cdot),u(\cdot)))(x) \end{cases}$$

Proof. 1. Case when $\mathbf{k} = 0$ and $\mathbf{l} \geq 0$: We obtain

$$\mathbf{L}_{(\mathbf{0},\mathbf{c})}(t;(x(\cdot),u(\cdot)))(T,x) = \mathbf{J_c}(t;(x(\cdot),u(\cdot)))(T,x) \tag{17.7}$$

Indeed, we observe that taking $\mathbf{k} = \mathbf{0}$, then

$$\mathbf{J_0}(t;(x(\cdot),u(\cdot)))(T,x) = \int_0^t e^{\int_0^\tau \mathbf{m}(x(s),u(s))ds}\mathbf{l}(x(\tau),u(\tau))d\tau$$

so that,

$$\mathbf{I_0}(t;(x(\cdot),u(\cdot)))(T,x) = \sup_{s\in[0,t]} \int_0^s e^{\int_0^\tau \mathbf{m}(x(\sigma),u(\sigma))d\sigma}\mathbf{l}(x(\tau),u(\tau))d\tau$$

If $\mathbf{l} \geq 0$, we infer that

$$\mathbf{I_0}(t;(x(\cdot),u(\cdot)))(T,x) = \int_0^t e^{\int_0^\tau \mathbf{m}(x(s),u(s))ds}\mathbf{l}(x(\tau),u(\tau))d\tau$$

Therefore, for any positive cost function $\mathbf{c}$, we have $\mathbf{L}_{(\mathbf{0},\mathbf{c})} = \mathbf{J_c}$. Taking $\mathbf{c} := \mathbf{u}_\infty$, we obtain

$$\mathbf{J}_{\mathbf{u}_\infty}(t;(x(\cdot),u(\cdot)))(T,x) := \begin{cases} \mathbf{J_u}(T;(x(\cdot),u(\cdot)))(x) \text{ if } t = T \\ +\infty \qquad\qquad\qquad\qquad\quad \text{if } t \in [0,T[\end{cases}$$

since $\mathbf{c}(t,x) := \mathbf{u}_\infty(t,x)$ that takes infinite values for $t > 0$.

Therefore,

a. when $\mathbf{k}(t,x) := \mathbf{0}(t,x)$ and $\mathbf{c}(t,x) := \mathbf{u}_\infty(t,x)$, the two above remarks imply that

$$V_{(\mathbf{0},\mathbf{u}_\infty)}((x(\cdot),u(\cdot)))(T,x) = \mathbf{J}_\mathbf{u}(T;(x(\cdot),u(\cdot)))(x)$$

b. when $\mathbf{k}(t,x) := \mathbf{0}(t,x)$ and $\mathbf{c}(t,x) := \mathbf{u}(x)$, we obtain

$$V_{(\mathbf{0},\mathbf{u})}((x(\cdot),u(\cdot)))(T,x) = \inf_{t\in[0,T]} \mathbf{J}_\mathbf{u}(t;(x(\cdot),u(\cdot)))(x)$$

2. Case when $\mathbf{c} := \mathbf{k}_\infty$ where

$$\mathbf{k}_\infty(t,x) := \begin{cases} \mathbf{k}(0,x) & \text{if } t=0 \\ +\infty & \text{if } t>0 \end{cases}$$

We prove that

$$\mathbf{L}_{(\mathbf{k},\mathbf{k}_\infty)}(t;(x(\cdot),u(\cdot)))(T,x) := \begin{cases} \mathbf{I}_\mathbf{k}(T;(x(\cdot),u(\cdot)))(T,x) & \text{if } t=T \\ +\infty & \text{if } t\in[0,T[\end{cases} \tag{17.8}$$

so that

$$\begin{aligned} V_{(\mathbf{k},\mathbf{k}_\infty)}(x(\cdot),u(\cdot))(T,x) &= \mathbf{I}_\mathbf{k}(T;x(\cdot),u(\cdot))(T;(x(\cdot),u(\cdot)))(T,x) \\ &= \sup_{t\in[0,T]} \mathbf{J}_\mathbf{k}(t;x(\cdot),u(\cdot))(T,x) \end{aligned} \tag{17.9}$$

Indeed, we see that $\mathbf{J}_{\mathbf{k}_\infty}(t;(x(\cdot),u(\cdot)))(T,x) = +\infty$ if $t < T$ and $\mathbf{J}_{\mathbf{k}_\infty}(T;(x(\cdot),u(\cdot)))(T,x) = \mathbf{J}_\mathbf{k}(T;(x(\cdot),u(\cdot)))(T,x)$.
In particular, when $\mathbf{k}(t,x) := \mathbf{u}(x)$ and $\mathbf{c}(t,x) := \mathbf{u}_\infty(t,x)$, we obtain

$$V_{(\mathbf{u},\mathbf{u}_\infty)}((x(\cdot),u(\cdot)))(T,x) = \sup_{t\in[0,T]} \mathbf{J}_\mathbf{u}(t;(x(\cdot),u(\cdot)))(x) \quad \square$$

17.3.2 Valuation Functions and Viability Solutions

In summary, we have introduced the following functionals:

Definition 17.3.2 *[A Class of Functionals to minimize] Given two functions* $\mathbf{k} : \mathbb{R}_+ \times X \times \mathcal{U} \mapsto \mathbb{R}\cup\{+\infty\}$ *and* $\mathbf{c} : \mathbb{R}_+ \times X \mapsto \mathbb{R}\cup\{+\infty\}$, *we* *set*

$$\begin{cases} \mathbf{J_c}(t;(x(\cdot),u(\cdot)))(T,x) \\ := e^{-\widehat{\mathbf{m}}(t)}\mathbf{c}(T-t,x(t)) + \int_0^t e^{-\widehat{\mathbf{m}}(\tau)}\mathbf{l}(x(\tau),u(\tau))d\tau \\ \mathbf{I_k}(t;(x(\cdot),u(\cdot)))(T,x) := \\ \sup_{s\in[0,t]}\left(e^{-\widehat{\mathbf{m}}(s)}\mathbf{k}(T-s,x(s)) + \int_0^s e^{-\widehat{\mathbf{m}}(\tau)}\mathbf{l}(x(\tau),u(\tau))d\tau\right) \\ \mathbf{L_{(k,c)}}(t;(x(\cdot),u(\cdot)))(T,x) \\ := \max(\mathbf{I_k}(t;x(\cdot),u(\cdot))(T,x), \mathbf{J_c}(t;(x(\cdot),u(\cdot)))(T,x)) \\ V_{(\mathbf{k},\mathbf{c})}(x(\cdot),u(\cdot))(T,x) := \inf_{t\in[0,T]} \mathbf{L_{(k,c)}}(t;(x(\cdot),u(\cdot)))(T,x) \end{cases} \quad (17.10)$$

We shall minimize the functional $V_{(\mathbf{k},\mathbf{c})}(x(\cdot),u(\cdot))(T,x)$ over the state-control pairs solutions to a control system:

Definition 17.3.3 *[The Valuation Function] Let us consider the control system*

$$x'(t) = f(x(t),u(t)) \text{ where } u(t) \in U(x(t)) \quad (17.11)$$

In this section, we denote by $\mathcal{S} := \mathcal{S}_{(17.11)}$ the evolutionary system generated by control system (17.11): it associates with any initial state x the set $\mathcal{S}(x) := \mathcal{S}_{(17.11)}(x)$ of state-control pairs $(x(\cdot),u(\cdot))$ governed by control system (17.11) starting from the initial state x at time 0, instead of the set of evolutions of the states $x(\cdot)$.

The valuation function $V^{\inf}_{(\mathbf{k},\mathbf{c})}$ *for the minimization of the intertemporal control problem is defined by*

$$V^{\inf}_{(\mathbf{k},\mathbf{c})}(T,x) := \inf_{(x(\cdot),u(\cdot))\in\mathcal{S}(x)} \inf_{t\in[0,T]} \mathbf{L_{(k,c)}}(t;x(\cdot),u(\cdot))(T,x) \quad (17.12)$$

When viability constraints are absent, the valuation function boils down to the simpler expression

$$\begin{cases} V^{\inf}_{\mathbf{c}}(T,x) := \\ \inf_{(x(\cdot),u(\cdot))\in\mathcal{S}(x)} \inf_{t\in[0,T]} \left(e^{-\widehat{\mathbf{m}}(t)}\mathbf{c}(T-t,x(t)) + \int_0^t e^{-\widehat{\mathbf{m}}(\tau)}\mathbf{l}(x(\tau),u(\tau))d\tau\right) \end{cases} \quad (17.13)$$

We use here the term *valuation function* instead of the classical *value function* used in control theory, because, in our case, the valuation function depends upon the horizon T whereas the value function depends of the current time $t \in [0, T]$ (see Comment 38, p. 684). We do not present this study in this volume of the book.

The first result links viability episolutions to valuation functions:

Theorem 17.3.4 *[Viability Characterization of Valuation Functions] Let us assume that the extended functions* $\mathbf{k}$ *and* $\mathbf{c}$ *are nontrivial. Then the valuation function* $V^{\inf}_{(\mathbf{k},\mathbf{c})}$ *and the viability episolution* $W^{\inf}_{(\mathbf{k},\mathbf{c})}$ *coincide:*

$$W^{\inf}_{(\mathbf{k},\mathbf{c})}(T,x) \;=\; V^{\inf}_{(\mathbf{k},\mathbf{c})}(T,x) \;:=\; \inf_{(T,x,y)\in \mathrm{Capt}_{(17.2)}(\mathcal{E}p(\mathbf{k}),\mathcal{E}p(\mathbf{c}))} y \qquad (17.14)$$

Proof. Let us set $\mathcal{K} := \mathcal{E}p(\mathbf{k})$ and $\mathcal{C} := \mathcal{E}p(\mathbf{c})$.

We begin by observing that a solution $(\tau(\cdot), x(\cdot), y(\cdot))$ to control system (17.2) starting from (T, x, y) is given by $t \mapsto (T - t, x(t), e^{\widehat{\mathbf{m}}(t)}(y - z(t)))$ where

$$z(t) \;:=\; \int_0^t e^{-\widehat{\mathbf{m}}(\tau)}\mathbf{l}(x(\tau), u(\tau))d\tau$$

because solutions to (17.2)(iii) are given explicitly by the formula

$$y(t) \;:=\; e^{\widehat{\mathbf{m}}(t)}\left(y - \int_0^t e^{-\widehat{\mathbf{m}}(\tau)}\mathbf{l}(x(\tau), u(\tau))d\tau\right)$$

Therefore, to say that (T, x, y) belongs to $\mathrm{Capt}_{(17.2)}(\mathcal{E}p(\mathbf{k}), \mathcal{E}p(\mathbf{c}))$ amounts to saying that there exists a solution $(\tau(\cdot), x(\cdot), y(\cdot))$ to the characteristic control system (17.2) starting from (T, x, y) and $t^\star \geq 0$ such that $(t - t^\star, x(t^\star), y(t^\star))$ belongs to the target $\mathcal{E}p(\mathbf{c})$, i.e., such that

$$\mathbf{c}\,(T - t^\star, x(t^\star)) \;\leq\; y(t^\star) := e^{\widehat{\mathbf{m}}(t^\star)}(y - z(t^\star))$$

and it is viable in $\mathcal{E}p(\mathbf{k})$. It implies that $t^\star \in [0, T]$.

Recalling the definition of the functional $\mathbf{J_c}$, this his can be written in the form

$$\mathbf{J_c}(t^\star; (x(\cdot), u(\cdot))) \;=\; e^{-\widehat{\mathbf{m}}(t^\star)}\mathbf{c}\,(T - t^\star, x(t^\star)) + z(t^\star) \;\leq\; y$$

On the other hand, since $(T - t, x(t), y(t))$ is a solution to the auxiliary system (17.2) viable in the epigraph of $\mathbf{k}$, we know that $\forall\, s \in [0, t^\star]$, $(T - s, x(s), y(s))$ satisfies $\mathbf{k}(t - s, x(s)) \leq y(s)$, i.e.,

$$\forall s \in [0, t^\star],\ \mathbf{k}(t-s, x(s)) \le e^{-\widehat{\mathbf{m}}(s)}(y - z(s))$$

which can be written in the form

$$\mathbf{I}_{\mathbf{k}}(t; x(\cdot), u(\cdot))(T, x) = \sup_{s\in[0,t^\star]} \left(e^{-\widehat{\mathbf{m}}(s)}\mathbf{k}(T-s, x(s)) + z(s)\right) \le y$$

This implies that

$$\mathbf{L}_{(\mathbf{k},\mathbf{c})}(t^\star; (x(\cdot), u(\cdot))) \le y$$

and thus, by taking the infimum over $t^\star \in [0, T]$, we obtain

$$V_{(\mathbf{k},\mathbf{c})}(x(\cdot), u(\cdot))(T, x) \le y \qquad (17.15)$$

and next, by taking the infimum over the pairs $(x(\cdot), u(\cdot)) \in \mathcal{S}(x)$, that

$$V^{\inf}_{(\mathbf{k},\mathbf{c})}(T, x) \le y$$

Therefore, taking the infimum over $y \in \mathbb{R}$ such that $(T, x, y) \in \mathrm{Capt}_{(17.2)}(\mathcal{E}p(\mathbf{k}), \mathcal{E}p(\mathbf{c}))$, we have proved that $V^{\inf}_{(\mathbf{k},\mathbf{c})}(T, x) \le W^{\inf}_{(\mathbf{k},\mathbf{c})}(T, x)$.

For proving the converse inequality, we associate with every $\varepsilon > 0$ a state-control pair $(x_\varepsilon(\cdot), u_\varepsilon(\cdot)) \in \mathcal{S}(x)$ and a time $t_\varepsilon \in [0, t]$ such that,

$$\mathbf{L}_{(\mathbf{k},\mathbf{c})}(t; (x_\varepsilon(\cdot), u_\varepsilon(\cdot)))(T, x) \le V^{\inf}_{(\mathbf{k},\mathbf{c})}(T, x) + \varepsilon$$

Therefore, setting

$$y_\varepsilon(t) := e^{\widehat{\mathbf{m}}(t)}(V^{\inf}_{(\mathbf{k},\mathbf{c})}(T, x) + \varepsilon - z_\varepsilon(t))$$

where

$$z_\varepsilon(t) := \int_0^t e^{-\widehat{\mathbf{m}}(\tau)}\mathbf{l}(x_\varepsilon(\tau), u_\varepsilon(\tau))d\tau$$

We observe that the function $s \mapsto (t-s, x_\varepsilon(s), y_\varepsilon(s))$ starts from $(t, x, V^{\inf}_{(\mathbf{k},\mathbf{c})}(T, x) + \varepsilon)$, is a solution to the auxiliary micro–macro control system (17.2), satisfying $\mathbf{k}(t-s, x_\varepsilon(s), u_\varepsilon(s)) \le y_\varepsilon(s)$ for $s \le t_\varepsilon$ and reaching the target $\mathcal{C} := \mathcal{E}p(\mathbf{c})$ at time t_ε.

This implies that $(T, x, V^{\inf}_{(\mathbf{k},\mathbf{c})}(T, x) + \varepsilon)$ belongs to the capture basin $\mathrm{Capt}_{(17.2)}(\mathbb{R}_+ \times K \times \mathbb{R}, \mathcal{E}p(\mathbf{c}))$, and thus, that $W^{\inf}_{(\mathbf{k},\mathbf{c})}(T, x) \le V^{\inf}_{(\mathbf{k},\mathbf{c})}(T, x) + \varepsilon$. Letting ε converge to 0 provides the converse inequality. □

Comment 38, p. 684 mentioned that backward time was used for parameterizing the time T as the horizon instead of the current time t. This motivates the fact than instead of considering evolutions $x(\cdot) \in \overleftarrow{\mathcal{S}}(x) := \mathcal{S}_{(17.11)}(x)$ starting from x at time 0, we could use its *backward shift evolution* that we shall denote by $\xi(t) := \overset{\vee}{\kappa}(T)(x(\cdot))(t) := x(T-t)$ *arriving* at x at horizon time T (see Definition 8.2.2, p.276).

Definition 17.3.5 ***[Evolution Arriving to a State at Horizon Time]*** *Let $d \in [0,T]$ be regarded as a* departure time. *We denote by $\mathcal{A}(d,T;x) := \mathcal{A}_{(17.11)}(d,T;x)$ the set of evolutions $\xi(\cdot)$ governed by control system (17.11), p. 688 defined on the interval $[d,T]$ and arriving at time T at x.*

Hence, a simple change of variables $d = T - t$ allows us to reformulate the valuation function as the minimal value of the intertemporal value function on evolutions $\xi \in \mathcal{A}(d,T;x)$ We set

$$\widehat{\mathbf{m}}(t) := \int_t^T \mathbf{m}(\xi(\tau), u(\tau))d(\tau)$$

without mentioning explicitly the dependence on $(\xi(\cdot), u(\cdot))$ for simplifying notations.

We associate with any extended function $\mathbf{v} : \mathbb{R}_+ \times X \times \mathcal{U} \mapsto \mathbb{R}_+ \cup \{+\infty\}$, any $t \geq 0$ and any $(x(\cdot), u(\cdot)) \in \mathcal{S}(x)$ the functional

$$\begin{cases} \mathbf{J_v}(d; (\xi(\cdot), u(\cdot)))(T,x) \\ := e^{-\widehat{\mathbf{m}}(d)}\mathbf{v}(t, \xi(t), u(t)) + \displaystyle\int_d^T e^{-\widehat{\mathbf{m}}(\tau)}\mathbf{l}(\xi(\tau), u(\tau))d\tau \end{cases}$$

$$\begin{cases} \mathbf{J_c}(d; (\xi(\cdot), u(\cdot)))(T,x) \\ := e^{-\widehat{\mathbf{m}}(t)}\mathbf{c}(d, \xi(d)) + \displaystyle\int_d^T e^{-\widehat{\mathbf{m}}(\tau)}\mathbf{l}(\xi(\tau), u(\tau))d\tau \\ \mathbf{I_k}(d; (\xi(\cdot), u(\cdot)))(T,x) := \\ \displaystyle\sup_{s\in[d,T]} \left(e^{-\widehat{\mathbf{m}}(s)}\mathbf{k}(s, \xi(s)) + \int_s^T e^{-\widehat{\mathbf{m}}(\tau)}\mathbf{l}(\xi(\tau), u(\tau))d\tau \right) \\ \mathbf{L_{(k,c)}}(d; (\xi(\cdot), u(\cdot)))(T,x) \\ := \max(\mathbf{I_k}(d; \xi(\cdot), u(\cdot))(T,x), \mathbf{J_c}(d; (\xi(\cdot), u(\cdot)))(T,x)) \end{cases} \tag{17.16}$$

$$V^{\text{inf}}_{(\mathbf{k},\mathbf{c})}(T,x) := \inf_{d\in[0,T]} \inf_{(\xi(\cdot),u(\cdot))\in\mathcal{A}(d,T;x)} \mathbf{L_{(k,c)}}(d; \xi(\cdot), u(\cdot))(T,x) \tag{17.17}$$

17.3.3 Existence of Optimal Evolution

We shall prove in this section the valuation functional is lower semicontinuous and achieves its minimum:

Theorem 17.3.6 *[Existence of Optimal Evolutions] Let us assume that:*

1. the control system $x'(t) = f(x(t), u(t))$ where $u(t) \in U(x(t))$ is Marchaud,
2. the function $u \mapsto \mathbf{m}(x,u)$ is concave and the function $u \mapsto \mathbf{l}(x,u)$ and $(t,x,u) \mapsto \mathbf{k}(t,x)$ are convex,
3. the functions $(x,u) \mapsto \mathbf{m}(x,u)$ is upper semicontinuous and $(x,u) \mapsto \mathbf{l}(x,u)$, $(t,x) \mapsto \mathbf{c}(t,x)$ and $(t,x) \mapsto \mathbf{k}(t,x)$ are lower semicontinuous,
4. the functions $\mathbf{l}$ and $\mathbf{m}$ are positive and the suprema $m^\sharp := \sup_{(x,u)\in \text{Graph}(U)} \mathbf{m}(x,u)$ and $l^\sharp := \sup_{(x,u)\in \text{Graph}(U)} \mathbf{l}(x,u)$ are finite.

(17.18)

The viability episolution is lower semicontinuous and its epigraph

$$\mathcal{E}p(V^{\text{inf}}_{(\mathbf{k},\mathbf{c})}) = \text{Capt}_{(17.2)}(\mathcal{K}, \mathcal{C})$$

is equal to its capture basin $\text{Capt}_{(17.2)}(\mathcal{K}, \mathcal{C})$ of the epigraph of $\mathbf{c}$ under the auxiliary system (17.2), p. 683. Furthermore,

1. there exists at least one optimal evolution minimizing the valuation functional $V^{\text{inf}}_{(\mathbf{k},\mathbf{c})}$;
2. optimal evolutions $x(\cdot)$ are the components of the evolutions $t \mapsto (T - t, x(t), y(t))$ governed by auxiliary system (17.2), p. 683 starting at $(T, x, V^{\text{inf}}_{(\mathbf{k},\mathbf{c})}(T,x))$ and viable until the first time $t^\star$ when

$$e^{-\widehat{\mathbf{m}}(t^\star)}\mathbf{c}(T - t^\star, x(t^\star)) + \int_0^{t^\star} e^{-\widehat{\mathbf{m}}(\tau)}\mathbf{l}(x(\tau), u(\tau))d\tau = V^{\text{inf}}_{(\mathbf{k},\mathbf{c})}(T,x)$$

This result can be derived from Theorem 11.5.8, p.472. However, for deriving it directly from viability theorems, we have to check that the auxiliary control system (17.2) is Marchaud. It is the actually differential inclusion

$$(\tau'(t), x'(t), y'(t)) \in F_0(\tau(t), x(t), y(t))$$

where

$$F_0(\tau, x, y) := \{(-1, f(x,u), \mathbf{m}(x,u)y - \mathbf{l}(x,u))\}_{u\in U(x)} \tag{17.19}$$

The values of this set-valued map are not necessarily convex or closed if we assume only that the function $\mathbf{m}$ is upper semicontinuous and concave in u and $\mathbf{l}$ is lower semicontinuous and convex in u.

But, by subtracting positive controls on the third component of the set valued map, we obtain the set-valued map

$$F_{\infty}(\tau, x, y) := \{(-1, f(x,u), \mathbf{m}(x,u)y - \mathbf{l}(x,u) - \pi)\}_{u \in U(x),\, \pi \geq 0} \quad (17.20)$$

and the associated differential inclusion

$$(\tau'(t), x'(t), y'(t)) \in F_{\infty}(\tau(t), x(t), y(t))$$

We shall prove that this the graph of this set-valued map is closed and that its values are convex, but it is not bounded.

In order to obtain a Marchaud map, we set a finite bound

$$0 \leq \pi \leq \alpha(x, u, y) := \mathbf{m}(x,u)y - \mathbf{l}(x,u) + m^{\sharp}y + l^{\sharp}$$

on the new control π. We set

$$F(\tau, x, y) := \{(-1, f(x,u), \mathbf{m}(x,u)y - \mathbf{l}(x,u) - \pi)\}_{(u \in U(x), \pi \in [0, \alpha(x,u,y)])} \quad (17.21)$$

and differential inclusion

$$(\tau'(t), x'(t), y'(t)) \in F(\tau(t), x(t), y(t))$$

We shall prove that the set-valued map F is Marchaud and that the capture basins of the target $\mathcal{E}p(\mathbf{c})$ under these three differential inclusions are equal.

Lemma 17.3.7 *[Marchaud Auxiliary Systems] We posit the assumptions (17.18), p. 692 of Theorem 17.3.6, p. 692. Then the set-valued map F is Marchaud and its growth is linear:*

$$\|F(\tau, x, y)\| \leq \max(1, c(\|x\| + 1), m^{\sharp}y + l^{\sharp})$$

Remark. Actually, we shall prove that the growth of F_0 is linear, that F is Marchaud and the graph of F_{∞} is closed and that its values are convex. □

Proof. We shall check successively that:

1. *The growth of F is linear:*

$$\|F(\tau, x, y)\| \leq \max(1, c(\|x\| + 1), m^{\sharp}y + l^{\sharp})$$

Indeed, the first component of (τ, x, y) is bounded by 1, the second one by $c(\|x\|+1)$ since the control system is assumed to be Marchaud and we observe that the functions $\mathbf{l}$ and $\mathbf{m}$ being strictly positive by assumption, that $-(m^\sharp y + l^\sharp) \leq \mathbf{m}(x,u)y - \mathbf{l}(x,u) - \pi \leq m^\sharp y \leq m^\sharp y + l^\sharp$.

2. *The values $F_\infty(\tau, x, y)$ of the set-valued map F_∞ are convex*: Indeed, for convex weight $\lambda_i \geq 0$ such that $\sum \lambda_i = 1$, we can write

$$\begin{cases} \sum \lambda_i(-1, f(x,u_i), \mathbf{m}(x,u_i)y - \mathbf{l}(x,u_i) - \pi_i) \\ = (-1, f(x,\overline{u}), \mathbf{m}(x,\overline{u})y - \mathbf{l}(x,\overline{u}) - \overline{\pi}) \end{cases}$$

where $\overline{u} := \sum \lambda_i u_i$ and

$$\overline{\pi} := \sum \lambda_i \pi_i + \sum \lambda_i (\mathbf{l}(x,u_i) - \mathbf{m}(x,u_i)y) + \mathbf{m}(x,\overline{u})y - \mathbf{l}(x,\overline{u})$$

Hence $\overline{\pi}$ because $u \mapsto \mathbf{m}(x,u)y - \mathbf{l}(x,u)$ is concave by assumption and it is easy to check that $\overline{\pi} \leq \mathbf{m}(x,\overline{u})y - \mathbf{l}(x,\overline{u}) + m^\sharp y + l^\sharp$.

3. *The graph of the set-valued map F is closed:* Indeed, let us consider a sequence of elements $((\tau_n, x_n, y_n), (-1, f(x,u_n), \lambda_n))$ where $u_n \in U(x_n)$, belonging to the graph of F converging to $((\tau, x, y), (-1, v, \lambda))$ where $\lambda_n := \mathbf{m}(x_n, u_n)y - \mathbf{l}(x_n, u_n) - \pi_n$ and where $\pi_n \geq 0$. Since the set-valued map U is Marchaud and since x_n converges to x, a subsequence (again denoted by) of u_n converges to some $u \in U(x)$. Since f is continuous, we infer that $v = f(x,u)$. Since the function $(\tau, x, y, u) \mapsto (\mathbf{m}(x,u)y - \mathbf{l}(x,u))$ is upper semicontinuous by assumption, $\mathbf{m}(x_n, u_n)y_n - \mathbf{l}(x_n, u_n)$ converges to some $\mu \leq \mathbf{m}(x,u)y - \mathbf{l}(x,u)$. Let us set $\varpi := \mathbf{m}(x,u)y - \mathbf{l}(x,u) - \mu \geq 0$. Therefore π_n converges to $\mu - \lambda = \mathbf{m}(x,u)y - \mathbf{l}(x,u) - \varpi - \lambda$. In other words, setting $\overline{\pi} := \pi + \varpi$, we find that $\lambda = \mathbf{m}(x,u)y - \mathbf{l}(x,u) - \overline{\pi}$ where it is easy to check that $0 \leq \overline{\pi} \leq \mathbf{m}(x,u)y - \mathbf{l}(x,u) + m^\sharp y + l^\sharp$. Hence

$$((\tau, x, y), (-1, v, \lambda)) = ((\tau, x, y), (-1, f(x,u), \mathbf{m}(x,u)y - \mathbf{l}(x,u) - \overline{\pi}))$$

belongs to the graph of F. This implies it is closed.

Hence, we have proved that the graph of the set-valued map F is Marchaud. □

We now check that we can replace the set-valued map F_0 by the Marchaud map F or the map F_∞ without changing the capture basin, and thus, the viability episolution:

Lemma 17.3.8 *[**Equality between Capture Basins**] The capture basins of the epigraph of the function* $\mathbf{c}$ *by systems (17.2), (17.21) and (17.20) coincide:*

$$\text{Capt}_{(17.2)}(\mathcal{K}, \mathcal{E}p(\mathbf{c})) = \text{Capt}_{(21)}(\mathcal{K}, \mathcal{C}) = \text{Capt}_{(17.20)}(\mathcal{K}, \mathcal{C})$$

Furthermore,

$$\text{Capt}_{(21)}(\mathcal{K}, \mathcal{C}) = \text{Capt}_{(21)}(\mathcal{K}, \mathcal{C}) - \{0\} \times \{0\} \times \mathbb{R}_+$$

Proof. Inclusions

$$\text{Capt}_{(17.2)}(\mathcal{K}, \mathcal{E}p(\mathbf{c})) \subset \text{Capt}_{(21)}(\mathcal{K}, \mathcal{C}) \subset \text{Capt}_{(17.20)}(\mathcal{K}, \mathcal{C})$$

are obvious.

For proving that

$$\text{Capt}_{(17.20)}(\mathcal{K}, \mathcal{C}) \subset \text{Capt}_{(17.2)}(\mathcal{K}, \mathcal{C})$$

let us consider an element $(T, x, y) \in \text{Capt}_{(17.20)}(\mathbb{R}_+ \times K \times \mathbb{R}, \mathcal{E}p(\mathbf{c}))$ and show that it belongs to $\text{Capt}_{(17.2)}(\mathbb{R}_+ \times K \times \mathbb{R}, \mathcal{E}p(\mathbf{c}))$. We begin by observing that a solution $(\tau(\cdot), x(\cdot), y(\cdot))$ to the control system (17.20) starting from (T, x, y) is given by $t \mapsto (T - t, x(t), e^{\widehat{\mathbf{m}}(t)}(y - z(t) - \varpi(t)))$ where

$$z(t) := \int_0^t e^{-\widehat{\mathbf{m}}(\tau)} \mathbf{l}(x(\tau), u(\tau)) d\tau \text{ and } \varpi(t) := \int_0^t e^{-\widehat{\mathbf{m}}(\tau)} \pi(\tau) d\tau \geq 0$$

Therefore, the proof of Theorem 17.3.4, p. 689 where we replace $z(t)$ by $z(t) + \varpi(t)$ states that it implies that there exists $t^\star \in [0, t]$ such that

$$\mathbf{c}(T - t^\star, x(t^\star)) \leq y(t^\star) \leq e^{\widehat{\mathbf{m}}(t^\star)}(y - z(t^\star) - \varpi(t^\star)) \leq e^{\widehat{\mathbf{m}}(t^\star)}(y - z(t^\star))$$

and that

$$\forall s \in [0, t^\star], \ \mathbf{k}(t - s, x(s)) \leq e^{-\widehat{\mathbf{m}}(s)}(y - z(s) - \varpi(t^\star)) \leq e^{-\widehat{\mathbf{m}}(s)}(y - z(s))$$

This means that $(T, x, y) \in \text{Capt}_{(17.2)}(\mathbb{R}_+ \times K \times \mathbb{R}, \mathcal{E}p(\mathbf{c}))$, so that, the three capture basins coincide.

We also observe that whenever $(T, x, y) \in \text{Capt}_{(17.20)}(\mathcal{K}, \mathcal{C})$ and $y \leq z$, inequalities

$$\begin{cases} \mathbf{c}(T - t^\star, x(t^\star)) \leq y(t^\star) \leq e^{\widehat{\mathbf{m}}(t^\star)}(y - z(t^\star) - \varpi(t^\star)) \\ \leq e^{\widehat{\mathbf{m}}(t^\star)}(y - z(t^\star)) \leq e^{\widehat{\mathbf{m}}(t^\star)}(z - z(t^\star)) \end{cases}$$

and that

$$\begin{cases} \forall s \in [0, t^\star], \ \mathbf{k}(t - s, x(s)) \leq e^{-\widehat{\mathbf{m}}(s)}(y - z(s) - \varpi(t^\star)) \\ \leq e^{-\widehat{\mathbf{m}}(s)}(y - z(s)) \leq e^{-\widehat{\mathbf{m}}(s)}(z - z(s)) \end{cases}$$

imply that for any $z \geq y$, $(T,x,z) \in \mathrm{Capt}_{(17.20)}(\mathbb{R}_+ \times K \times \mathbb{R}, \mathcal{E}p(\mathbf{c}))$. The proof is completed. □

Proof. (Theorem 17.3.6, p. 692). Since F is a Marchaud control system, the capture basin of a target is closed whenever the target $\mathcal{C}$ and the environment $\mathcal{K}$ are closed and the complement of the target in the environment is a closed backward repeller: This is the case thanks to Comment 38, p. 684. Since we have proved that

$$\mathrm{Capt}_{(17.2)}(\mathcal{K},\mathcal{C}) = \mathrm{Capt}_{(17.2)}(\mathcal{K},\mathcal{C}) - \{0\} \times \{0\} \times \mathbb{R}_+$$

we infer that $\mathrm{Capt}_{(17.2)}(\mathcal{K},\mathcal{C})$ is an epigraph, and thus, the epigraph of the viability episolution. Since it is closed because the auxiliary system is Marchaud, the valuation functional is lower semicontinuous, and, in particular, $y := V^{\mathrm{inf}}_{(\mathbf{k},\mathbf{c})}(T,x)$, we infer that by taking $y := V^{\mathrm{inf}}_{(\mathbf{k},\mathbf{c})}(T,x)$, the triple $(T,x,V^{\mathrm{inf}}_{(\mathbf{k},\mathbf{c})}(T,x))$ belongs to the capture basin $\mathrm{Capt}_{(17.2)}(\mathcal{K},\mathcal{C})$ of the epigraph of $\mathbf{c}$ under the auxiliary system (17.2), p. 683. In the proof of Theorem 17.3.4, p. 689, we have proved that any evolution $t \mapsto (T-t, x(t), y(t))$ starting from $(T,x,V^{\mathrm{inf}}_{(\mathbf{k},\mathbf{c})}(T,x))$ satisfies inequality (17.15), p. 690 stating that

$$V_{(\mathbf{k},\mathbf{c})}(x(\cdot),u(\cdot))(T,x) \leq V^{\mathrm{inf}}_{(\mathbf{k},\mathbf{c})}(T,x)$$

Since

$$V^{\mathrm{inf}}_{(\mathbf{k},\mathbf{c})}(T,x) := \inf_{(x(\cdot),u(\cdot))\in\mathcal{S}(x)} V_{(\mathbf{k},\mathbf{c})}(x(\cdot),u(\cdot))(T,x)$$

by definition (17.12), p. 688 of the valuation functional, we infer that the state-control pair is optimal. □

17.3.4 Dynamic Programming Equation

We did not yet exploit the bilateral fixed point characterization of capture basins, which happens to imply the celebrated Isaac–Bellman *dynamic programming equation.*

Theorem 17.3.9 ***[Dynamic Programming Equation]*** *We posit the assumptions of Theorem 17.3.6, p. 692. Optimal evolutions $x(\cdot)$ satisfy* dynamic programming equation

$$e^{-\widehat{\mathbf{m}}(s)} V^{\mathrm{inf}}_{(\mathbf{k},\mathbf{c})}(T-s, x(s)) + \int_0^s e^{-\widehat{\mathbf{m}}(\tau)} \mathbf{l}(x(\tau),u(\tau))d\tau = V^{\mathrm{inf}}_{(\mathbf{k},\mathbf{c})}(T,x) \tag{17.22}$$

for all $s \in [0, t^\star]$ *where* $t^\star$ *is the first time when*

$$e^{-\widehat{\mathbf{m}}(t^\star)}\mathbf{c}(T-t^\star, x(t^\star)) + \int_0^{t^\star} e^{-\widehat{\mathbf{m}}(\tau)}\mathbf{l}(x(\tau), u(\tau))d\tau \; = \; V^{\inf}_{(\mathbf{k},\mathbf{c})}(T,x)$$

Proof. By Theorem 2.15.5, p. 101, we know that the capture basin $\text{Capt}_{(17.2)}(\mathcal{K}, \mathcal{C})$ of the epigraph of $\mathbf{c}$ under the auxiliary system (17.2), p. 683 is the bilateral fixed point

$$\text{Capt}_{(17.2)}(\mathcal{E}p(V^{\inf}_{(\mathbf{k},\mathbf{c})}), \mathcal{C}) \; = \; \mathcal{E}p(V^{\inf}_{(\mathbf{k},\mathbf{c})}) \; = \; \text{Capt}_{(17.2)}(\mathcal{K}, \mathcal{E}p(V^{\inf}_{(\mathbf{k},\mathbf{c})}))$$

Let $(T, x, V^{\inf}_{(\mathbf{k},\mathbf{c})}(T,x))$ belong to the capture basin $\text{Capt}_{(17.2)}(\mathcal{E}p(V^{\inf}_{(\mathbf{k},\mathbf{c})}), \mathcal{C})$ of the epigraph of $\mathbf{c}$ under the auxiliary system (17.2), p. 683. The proof of Theorem 17.3.4, p. 689 implies that there exist a state-control pair $(x(\cdot), u(\cdot)) \in \mathcal{S}(x)$ and $t^\star$ such that,

$$e^{-\widehat{\mathbf{m}}(t^\star)}\mathbf{c}(T-t^\star, x(t^\star)) + \int_0^{t^\star} e^{-\widehat{\mathbf{m}}(\tau)}\mathbf{l}(x(\tau), u(\tau))d\tau \; \leq \; V^{\inf}_{(\mathbf{k},\mathbf{c})}(T,x)$$

and such that, for every $s \in [0, t^\star]$,

$$e^{-\widehat{\mathbf{m}}(s)}V^{\inf}_{(\mathbf{k},\mathbf{c})}(T-s, x(s)) + \int_0^{s} e^{-\widehat{\mathbf{m}}(\tau)}\mathbf{l}(x(\tau), u(\tau))d\tau \; \leq \; V^{\inf}_{(\mathbf{k},\mathbf{c})}(T,x)$$

We shall deduce the opposite inequality from the second fixed point property, implying actually that

$$\begin{cases} V^{\inf}_{(\mathbf{k},\mathbf{c})}(T,x) \; \leq \; \inf_{(x(\cdot),u(\cdot))\in\mathcal{S}(x)} \inf_{s\in[0,t]} \\ \left(e^{-\widehat{\mathbf{m}}(s)}V^{\inf}_{(\mathbf{k},\mathbf{c})}(T-s, x(s)) + \int_0^{s} e^{-\widehat{\mathbf{m}}(\tau)}\mathbf{l}(x(\tau), u(\tau))d\tau \right) \end{cases} \tag{17.23}$$

Indeed, for any $z < V^{\inf}_{(\mathbf{k},\mathbf{c})}(T,x)$, the triple (T,x,z) does not belong to $\mathcal{E}p(V^{\inf}_{(\mathbf{k},\mathbf{c})}) \; = \; \text{Capt}_{(17.2)}(\mathcal{K}, \mathcal{E}p(V^{\inf}_{(\mathbf{k},\mathbf{c})}))$. This means that, by Lemma 2.12.2, p. 92,

$$(T,x,z) \; \in \; \text{Inv}_{(17.2)}(\overset{\circ}{\mathcal{H}yp}\,(V^{\inf}_{(\mathbf{k},\mathbf{c})}), \complement\mathcal{K})$$

because

$$\complement\mathcal{E}p(V^{\inf}_{(\mathbf{k},\mathbf{c})}) \; := \; \{(T,x,y) \text{ such that } y \; < \; V^{\inf}_{(\mathbf{k},\mathbf{c})}(T,x)\} \; =: \overset{\circ}{\mathcal{H}yp}\,(V^{\inf}_{(\mathbf{k},\mathbf{c})})$$

Therefore, for any $(x(\cdot), u(\cdot)) \in \mathcal{S}(x)$ and for any $s \in [0, T]$,

$$e^{-\widehat{\mathbf{m}}(t)}\left(z - \int_0^s e^{-\widehat{\mathbf{m}}(\tau)}\mathbf{l}(x(\tau), u(\tau))d\tau\right) < V^{\inf}_{(\mathbf{k},\mathbf{c})}(T, x)$$

and thus,

$$z < e^{-\widehat{\mathbf{m}}(s)}V^{\inf}_{(\mathbf{k},\mathbf{c})}(T - s, x(s)) + \int_0^s e^{-\widehat{\mathbf{m}}(\tau)}\mathbf{l}(x(\tau), u(\tau))d\tau$$

Letting z converge to $V^{\inf}_{(\mathbf{k},\mathbf{c})}(T, x)$, we obtain

$$V^{\inf}_{(\mathbf{k},\mathbf{c})}(T, x) \leq e^{-\widehat{\mathbf{m}}(s)}V^{\inf}_{(\mathbf{k},\mathbf{c})}(T - s, x(s)) + \int_0^s e^{-\widehat{\mathbf{m}}(\tau)}\mathbf{l}(x(\tau), u(\tau))d\tau$$

In particular, taking for evolution the optimal one and $t := t^\star$, we obtain equation (17.22), p. 696.

On the other hand, taking the infimum over state-control pairs $(x(\cdot), u(\cdot)) \in \mathcal{S}(x)$ and $t \in [0, T]$, we obtain (17.23), p. 697. □

17.4 Solutions to Hamilton–Jacobi Equation and Viability Solutions

17.4.1 Contingent Solution to the Hamilton–Jacobi Equation

We set:

$$\Omega^V_{(\mathbf{k},\mathbf{c})}(t) := \{x \in X \text{ such that } \mathbf{k}(t, x) \leq V(t, x) < \mathbf{c}(t, x)\} \quad (17.24)$$

Recall that the *convexified epiderivative* $D^{\star\star}_{\uparrow}\mathbf{u}(x)$ of the extended function $\mathbf{u}$ at $x \in \text{Dom}(\mathbf{u})$ is defined by

$$\mathcal{E}p(D^{\star\star}_{\uparrow}\mathbf{u}(x)) := T^{\star\star}_{\mathcal{E}p(\mathbf{u})}(x, \mathbf{u}(x))$$

where $T^{\star\star}_K(x)$ is the closed convex hull to the tangent cone $T_K(x)$ to K at x (see Definition 18.6.12, p. 749). The convexified epiderivative $u \mapsto D^{\star\star}_{\uparrow}\mathbf{u}(x)(u)$ is a lower semicontinuous convex function.

Let us associate with the extended function $\mathbf{k}$ the extended function $\mathbf{k}^{\Rightarrow}$ defined by

$$\mathcal{E}p(\mathbf{k}^{\Rightarrow}) := \text{Exit}_{(17.2)}(\mathcal{E}p(\mathbf{k}))$$

Theorem 17.4.1 ***[Contingent Frankowska Solution]*** *We posit the assumptions (17.18), p. 692 of Lemma 17.3.7, p. 693. The viability episolution V (see Definition 17.2.1, p. 683) is the* ***smallest*** *lower semicontinuous solution greater than or equal to* $\mathbf{c}$ *and satisfying:* $\forall\, t \geq 0, \forall\, x \in \Omega^V_{(\mathbf{k},\mathbf{c})}(t)$,

$$\inf_{u \in U(x)} \left(D_\uparrow^{\star\star} V(t,x)(-1, f(x,u)) + \mathbf{l}(x,u) - \mathbf{m}(x,u)V(t,x) \right) \leq 0 \quad (17.25)$$

the constraint and target $\mathbf{k}(t,x) \leq V(t,x)$ *and the target conditions*

$$\mathbf{k}(t,x) \leq V(t,x) \leq \mathbf{c}(t,x) \quad (17.26)$$

Furthermore, it satisfies

$$V(t,x) \leq \mathbf{c}(t,x) \leq \max(V(t,x), \mathbf{k}^{\Rightarrow}(t,x))$$

If the system is Lipschitz, then V is the ***largest*** *lower semicontinuous solution smaller than or equal to* $\mathbf{c}$ *satisfying:* $\forall\, t \geq 0, \forall\, x \in X$,

$$\sup_{u \in U(x)} \left(D_\uparrow^{\star\star} V(t,x)(1, -f(x,u)) + \mathbf{m}(x,u)V(t,x) - \mathbf{l}(x,u) \right) \leq 0 \quad (17.27)$$

We need the Lemma 18.6.18, p. 753 on tangent cones to epigraphs for proving Theorem 17.4.1:

Proof. It is divided in two parts:

1. First, we observe that $\mathcal{E}p(V) \backslash \mathcal{E}p(\mathbf{c})$ is the set of (t,x,y) such that $V(t,x) \leq y < \mathbf{c}(t,x)$ and is non empty if and only if $x \in \Omega^V_{(\mathbf{k},\mathbf{c})}(t)$. It is a repeller since $\mathcal{K}$ is a repeller by Comment 38, p. 684.

   Since the auxiliary system is Marchaud by Lemma 17.3.7, p. 693, the first statement of Theorem 11.4.6, p. 463 states that the capture basin is the largest closed subset between the epigraph of $\mathbf{c}$ and $\mathbb{R}_+ \times X \times \mathbb{R}$ such that $\forall t \geq 0, \forall x \in \Omega^V_{(\mathbf{k},\mathbf{c})}(t), \forall\, y$ such that $V(t,x) \leq y < \mathbf{c}(t,x)$,

   $$\exists u \in U(x) \text{ such that } (-1, f(x,u), \mathbf{m}(x,u)y - \mathbf{l}(x,u)) \in T^{\star\star}_{\mathcal{E}p(V)}(t,x,y)$$

   If $y = V(t,x)$, then

   $$T^{\star\star}_{\mathcal{E}p(V)}(t,x,V(t,x)) =: \mathcal{E}p(D_\uparrow^{\star\star} V(t,x))$$

   so that we infer that there exists $u \in U(x)$

$$D_{\uparrow}^{\star\star}V(t,x)(-1,f(x,u)) \le \mathbf{m}(x,u)y - \mathbf{l}(x,u)$$

from which inequality (17.25)

$$\inf_{u\in U(x)} (D_{\uparrow}V(t,x)(-1,f(x,u)) + \mathbf{l}(x,u) - \mathbf{m}(x,u)V(t,x)) \le 0$$

ensues.

Conversely, since $D_{\uparrow}^{\star\star}V(t,x)(-1,\cdot)$ is lower semicontinuous and $U(x)$ is compact, inequality (17.25) implies the existence of $u \in U(x)$ such that

$$(-1,f(x,u),\mathbf{m}(x,u)y - \mathbf{l}(x,u)) \in T_{\mathcal{E}p(V)}^{\star\star}(t,x,V(t,x))$$

When $y > V(t,x)$, then Lemma 18.6.18, p. 753 implies that

$$(-1,f(x,u),\mathbf{m}(x,u)y - \mathbf{l}(x,u)) \in T_{\mathcal{E}p(V)}^{\star\star}(t,x,y)$$

because $(-1,f(x,u))$ belongs to the domain of $D_{\uparrow}^{\star\star}V(t,x)$.

2. For proving inequality $\mathbf{c}(t,x) \le \max(V(t,x),\mathbf{k}^{\Rightarrow}(t,x))$, we deduce from formula (11.18), p. 461 of Theorem 11.4.1, p.460 that

$$\mathcal{E}p(V)\cap \text{Exit}_{(17.2)}(\mathcal{E}p(\mathbf{k})) \subset \mathcal{E}p(\mathbf{c})$$

which means that

$$\mathcal{E}p(\max(V,\mathbf{k}^{\Rightarrow})) = \mathcal{E}p(V)\cap\mathcal{E}p(\mathbf{k}^{\Rightarrow}) \subset \mathcal{E}p(\mathbf{c})$$

3. Since the environment $\mathcal{K}$ is backward invariant thanks to Comment 38, p. 684, then the second statement of Theorem 11.4.6, p. 463 implies that the capture basin is the smallest closed subset containing the epigraph of $\mathbf{c}$ such that $\mathcal{E}p(V)$ is backward invariant, and thus satisfy: $\forall\ (t,x,y) \in \mathcal{E}p(V)$, $\forall\ u \in U(x)$,

$$(-1,f(x,u),\mathbf{m}(x,u)y - \mathbf{l}(x,u)) \in T_{\mathcal{E}p(V)}^{\star\star}(t,x,y)$$

If $y = V(t,x)$, then so that we infer that for all $u \in U(x)$

$$D_{\uparrow}^{\star\star}V(t,x)(1,-f(x,u)) \le \mathbf{l}(x,u) - \mathbf{m}(x,u)y$$

from which we derive inequality (17.27). Conversely, since for all $u \in U(x)$, $(1,-f(x,u))$ belongs to the domain of $D_{\uparrow}V(t,x)$, we derive that

$$(1,-f(x,u),\mathbf{l}(x,u) - \mathbf{m}(x,u)y) \in T_{\mathcal{E}p(V)}^{\star\star}(t,x,y)$$

holds whenever $y \ge V(t,x)$ thanks to Lemma 18.6.18, p. 753. □

17.4.2 Barron–Jensen/Frankowska Solution to the Hamilton–Jacobi Equation

Instead of characterizing capture basins in terms of tangent cones and translating them in terms of contingent Frankowska episolutions, we translate them in the equivalent formulation of Barron–Jensen/Frankowska solutions, a weaker concept of viscosity solutions requiring only the lower semicontinuity of the solution instead of its continuity. For simplicity of the exposition at this stage, we still involve the domain $K \subset X$ by assuming that $\mathbf{k}(t,x) := +\infty$ whenever $x \notin K$.

We introduce the Hamiltonian

$$\mathbf{h}(t,x,y,p) := \sup_{u \in U(x)} (\mathbf{m}(x,u)y - \mathbf{l}(x,u) - \langle p, f(x,u)\rangle)$$

Definition 17.4.2 *[**Barron–Jensen/Frankowska Solutions**] The Barron–Jensen/Frankowska solution V is a lower semicontinuous function satisfying inequalities*

$$\mathbf{k}(t,x) \le V(t,x) \le \mathbf{c}(t,x)$$

and

$$\begin{cases} (i) \ \forall t > 0, \ \forall x \in \Omega^V_{(\mathbf{k},\mathbf{c})}(t), \ \forall (p_t,p_x) \in \partial V(t,x), \\ \quad p_t + \mathbf{h}(t,x,V(t,x),p_x) \ \ge \ 0 \\ (ii) \ \forall t > 0, \ \forall x \in \Omega^V_{(\mathbf{k},\mathbf{c})}(t), \\ \quad \forall (p_t,p_x) \in (\mathrm{Dom}(D^{\star\star}_{\uparrow} V(t,x)))^-, \ \ p_t + \sigma(-f(x,U(x)),p_x) \ \ge \ 0 \end{cases} \tag{17.28}$$

The main theorem of this chapter is to prove that the viability episolution (see Definition 17.2.1, p. 683) is the unique Barron–Jensen/Frankowska solution:

Theorem 17.4.3 *[**Barron–Jensen/Frankowska Solution**] We posit the assumptions (17.18), p. 692 of Lemma 17.3.7, p. 693. The viability episolution V is the **smallest** lower semicontinuous solution greater than or equal to* $\mathbf{c}$ *and satisfying*

$$\begin{cases} (i) \ \forall t > 0, \ \forall x \in \Omega^V_{(\mathbf{k},\mathbf{c})}(t), \ \forall (p_t, p_x) \in \partial V(t,x), \\ \quad p_t + \mathbf{h}(t,x,V(t,x),p_x) \ \geq \ 0 \\ (ii) \ \forall t > 0, \ \forall x \in \Omega^V_{(\mathbf{k},\mathbf{c})}(t), \\ \quad \forall (p_t,p_x) \in (\mathrm{Dom}(D^{\star\star}_{\uparrow} V(t,x)))^-, \ \ p_t + \sigma(-f(x,U(x)),p_x) \ \geq \ 0 \end{cases} \tag{17.29}$$

the constraint functions $\mathbf{k}(t,x) \leq V(t,x)$ *and the target conditions* $V(t,x) \leq \mathbf{c}(t,x)$.

If the system is Lipschitz, then V *is the* ***largest*** *lower semicontinuous solution smaller than or equal to* $\mathbf{c}$ *satisfying*

$$\begin{cases} (i) \ \forall t \geq 0, \ \forall x \in X, \ \forall (p_t, p_x) \in \partial V(t,x), \\ \quad p_t + \mathbf{h}(t,x,V(t,x),p_x) \ \leq \ 0 \\ (ii) \ \forall t \geq 0, \ \forall x \in \Omega^V_{(\mathbf{k},\mathbf{c})}(t), \ \forall (p_t,p_x) \in (\mathrm{Dom}(D^{\star\star}_{\uparrow} V(t,x)))^-, \\ \quad p_t + \sigma(-f(x,U(x)),p_x) \ \leq \ 0 \end{cases} \tag{17.30}$$

and thus, the ***unique*** *lower semicontinuous Barron–Jensen/Frankowska solution, which satisfies also inequalities*

$$V(t,x) \ \leq \ \mathbf{c}(t,x) \ \leq \ \max(V(t,x), \mathbf{k}^{\Rightarrow}(t,x))$$

Observe that under the Lipschitz assumptions, the viability episolution satisfies

$$\begin{cases} \quad \forall t > 0, \ \forall x \in \Omega^V_{(\mathbf{k},\mathbf{c})}(t), \\ (i) \ \forall (p_t,p_x) \in \partial V(t,x), \ \ p_t + \mathbf{h}(t,x,V(t,x),p_x) \ = \ 0 \\ (ii) \ \forall (p_t,p_x) \in (\mathrm{Dom}(D^{\star\star}_{\uparrow} V(t,x)))^-, \ \ p_t + \sigma(-f(x,U(x)),p_x) \ = \ 0 \end{cases} \tag{17.31}$$

Proof. As the proof of Theorem 17.4.1, p. 699, it is divided in two parts:

1. Since the auxiliary system is Marchaud by Lemma 17.3.7, p. 693, the first statement of Theorem 11.6.7, p. 480 states that $\forall t \ \geq \ 0, \forall x \in \Omega^V_{(\mathbf{k},\mathbf{c})}(t), \forall \ y | V(t,x) \ \leq \ y \ < \ \mathbf{c}(t,x)$, there exists $u \in U(x)$ such that $\forall \ (p_t,p_x) \in \partial V(t,x)$

$$\begin{cases} \langle (p_t,p_x,-\lambda),(-1,f(x,u),\mathbf{m}(x,u)y - \mathbf{l}(x,u)) \rangle \\ = -p_t + \langle p_x, f(x,u) \rangle + \lambda(\mathbf{l}(x,u) - \mathbf{m}(x,u)y) \ \leq \ 0 \end{cases} \tag{17.32}$$

By Lemma 18.6.18, if $y = V(t,x)$, $(p_t,p_x,-\lambda) \in N_{\mathcal{E}p(V)}(t,x,y)$ means that either $\lambda > 0$, and that, taking $\lambda = 1$, $(p_t,p_x) \in \partial V(t,x)$ or that $\lambda = 0$, and that $(p_t,p_x) \in (\mathrm{Dom}(D^{\star\star}_{\uparrow} V(t,x)))^-$. If $y > V(t,x)$, $(p_t,p_x,-\lambda) \in N_{\mathcal{E}p(V)}(t,x,y)$ also means that $\lambda = 0$, and that $(p_t,p_x) \in (\mathrm{Dom}(D^{\star\star}_{\uparrow} V(t,x)))^-$.

Consequently, condition (17.32), p. 702 can be written in the following form:

- Case when $y = V(t,x)$ and $\lambda = 1$:

$$\begin{cases} \forall t > 0, \forall x \in \Omega^V_{(\mathbf{k},\mathbf{c})}(t),\ \forall\,(p_t,p_x) \in \partial V(t,x),\ \text{then} \\ -p_t \ \leq\ \langle -p_x, f(x,u)\rangle - \mathbf{l}(x,u) + \mathbf{m}(x,u)y \ \leq\ \mathbf{h}(t,x,y,p_x) \end{cases}$$

  Hence

$$\forall\,(p_t,p_x) \in \partial V(t,x),\ \ p_t + \mathbf{h}(t,x,V(t,x),p_x) \ \geq\ 0$$

- Case when $y \geq V(t,x)$ and $\lambda = 0$:

$$\begin{cases} \forall t > 0, \forall x \in \Omega^V_{(\mathbf{k},\mathbf{c})}(t),\ \forall\,(p_t,p_x) \in (\mathrm{Dom}(D^{\star\star}_{\uparrow}V(t,x)))^{-},\ \text{then} \\ -p_t - \sup_{u\in U(x)} \langle p_x, -f(x,u)\rangle \\ =\ -p_t - \sigma(-f(x,U(x)),p_x) \ \leq\ 0 \end{cases}$$

  (Recall that this condition disappears whenever the viability episolution V is epidifferentiable, and, in particular, when the episolution is Lipschitz).

2. Since the environment is backward invariant and the auxiliary system, is Lipschitz, the second statement of Theorem 11.6.7, p. 480 states that $\forall\, u \in U(x)$,

$$\begin{cases} \forall\,(p_t,p_x,-\lambda) \in N_{\mathcal{E}p(V)}(t,x,y), \\ \langle (p_t,p_x,-\lambda),(1,-f(x,u),\mathbf{l}(x,u)-\mathbf{m}(x,u)y)\rangle \\ =\ p_t + \langle -p_x, f(x,u)\rangle + \lambda(\mathbf{m}(x,u)y - \mathbf{l}(x,u)) \ \leq\ 0 \end{cases} \tag{17.33}$$

This implies that $\lambda \geq 0$.
Consequently, condition (17.33), p. 703 can be written in the following form:

- Case when $y = V(t,x)$ and $\lambda = 1$:

$$\begin{cases} \forall t > 0, \forall x \in X,\ \forall\,(p_t,p_x) \in \partial V(t,x),\ \text{then} \\ p_t + \sup_{x\in U(x)}(\langle -p_x, f(x,u)\rangle + \mathbf{m}(x,u)y - \mathbf{l}(x,u)) \\ =\ p_t + \mathbf{h}(t,x,y,p_x) \ \leq\ 0 \end{cases}$$

- Case when $y \geq V(t,x)$ and $\lambda = 0$:

$$\begin{cases} \forall t > 0, \forall x \in X,\ \forall\,(p_t,p_x) \in (\mathrm{Dom}(D^{\star\star}_{\uparrow}V(t,x)))^{-},\ \text{then} \\ p_t + \sup_{u\in U(x)} \langle p_x, -f(x,u)\rangle \ =\ p_t + \sigma(-f(x,U(x)),p_x) \ \leq\ 0 \end{cases} \qquad \square$$

17.4.3 The Regulation Map of Optimal Evolutions

The *associated regulation map* R for regulating viable evolutions of the auxiliary system is defined by: $\forall\, t > 0,\ x \in \Omega^V_{(\mathbf{k},\mathbf{c})}(t)$,

$$\widehat{R}(t,x,y) \;:=\; \{u \mid (-1, f(x,u), \mathbf{m}(x,u))y - \mathbf{l}(x,u) \;\in\; T^{\star\star}_{\mathcal{E}p(V)}(t,x,V(t,x))\}$$

(see Definition 2.15.6, p. 102). The partial derivative $\partial_x V(t,x)$ with respect to x is the subset of elements p_x such that there exists p_t satisfying $(p_t, p_x) \in \partial V(t,x)$.

> **Definition 17.4.4** *[The Regulation Map] The* regulation map *is the set-valued map*
> $$(t,x) \rightsquigarrow R(t,x) \;:=\; \widehat{R}(t,x,V(t,x))$$

17.5 Other Intertemporal Optimization Problems

17.5.1 Maximization Problems

Replacing epigraphs of functions $\mathbf{k}$ and $\mathbf{c}$ by hypographs of functions $\mathbf{k}_\downarrow$ and $\mathbf{c}_\downarrow$ provides the maximization over the state-control pairs of the functional $V^\flat_{(\mathbf{k}_\downarrow,\mathbf{c}_\downarrow)}(x(\cdot),u(\cdot))(T,x)$ we are about to define and to characterize in terms of capture basin. Hence, this value functional exhibits all the properties of the capture basin, that we leave as an exercise to translate, as we did for the valuation functional $V_{(\mathbf{k},\mathbf{c})}(x(\cdot),u(\cdot))(T,x)$ (that we could have written in the form $V^\flat_{(\mathbf{k}_\uparrow,\mathbf{c}_\uparrow)}(x(\cdot),u(\cdot))(T,x)$ to stress the symmetry between these two examples). We just provide the proof of the viability characterization in terms of capture basins.

> **Definition 17.5.1** *[A Class of Functionals to Maximize] Given two functions* $\mathbf{k}_\downarrow : \mathbb{R}_+ \times X \times \mathcal{U} \mapsto \mathbb{R} \cup \{-\infty\}$ *and* $\mathbf{c}_\downarrow : \mathbb{R}_+ \times X \mapsto \mathbb{R} \cup \{-\infty\}$,

we set

$$
\begin{cases}
\mathbf{J}_{\mathbf{c}_\downarrow}(t;(x(\cdot),u(\cdot)))(T,x) \\
:= e^{-\widehat{\mathbf{m}}(t)}\mathbf{c}_\downarrow(T-t,x(t)) + \int_0^t e^{-\widehat{\mathbf{m}}(\tau)}\mathbf{l}(x(\tau),u(\tau))d\tau \\
\mathbf{I}^\flat_{\mathbf{k}_\downarrow}(t;(x(\cdot),u(\cdot)))(T,x) := \\
\inf_{s\in[0,t]}\left(e^{-\widehat{\mathbf{m}}(s)}\mathbf{k}_\downarrow(T-s,x(s)) + \int_0^s e^{-\widehat{\mathbf{m}}(\tau)}\mathbf{l}(x(\tau),u(\tau))d\tau\right) \\
\mathbf{L}^\flat_{(\mathbf{k}_\downarrow,\mathbf{c})}(t;(x(\cdot),u(\cdot)))(T,x) \\
:= \min(\mathbf{I}_{\mathbf{k}_\downarrow}(t;x(\cdot),u(\cdot))(T,x), \mathbf{J}_{\mathbf{c}_\downarrow}(t;(x(\cdot),u(\cdot)))(T,x)) \\
V^\flat_{(\mathbf{k}_\downarrow,\mathbf{c}_\downarrow)}(x(\cdot),u(\cdot))(T,x) := \sup_{t\in[0,T]} \mathbf{L}_{(\mathbf{k}_\downarrow,\mathbf{c}_\downarrow)}(t;(x(\cdot),u(\cdot)))(T,x)
\end{cases}
\tag{17.34}
$$

We associate the valuation functional

$$
V^{\text{sup}}_{(\mathbf{k}_\downarrow,\mathbf{c}_\downarrow)}(T,x) := \sup_{(x(\cdot),u(\cdot))\in\mathcal{S}(x)} V^\flat_{(\mathbf{k}_\downarrow,\mathbf{c}_\downarrow)}(x(\cdot),u(\cdot))(T,x) \tag{17.35}
$$

Let us consider the new auxiliary control system

$$
\begin{cases}
(i) \quad \tau'(t) = -1 \\
(ii) \quad x'(t) = f(x(t),u(t)) \\
(iii) \quad y'(t) = \mathbf{m}(x(t))y(t) - \mathbf{l}(x(t),u(t)) \\
\qquad \text{where } u(t) \in U(x(t))
\end{cases}
\tag{17.36}
$$

Hence

$$
V^{\text{sup}}_{(\mathbf{k}_\downarrow,\mathbf{c}_\downarrow)}(T,x) := \sup_{(T,x,y)\in\text{Capt}_{(17.36)}(\mathcal{E}p(\mathbf{k}),\mathcal{H}yp(\mathbf{c}_\downarrow))} y
$$

17.5.2 Two Other Classes of Optimization Problems

Until now, we have characterized the valuation functionals $V_{(\mathbf{k},\mathbf{c})}$ and $V^{\text{sup}}_{(\mathbf{k}_\downarrow,\mathbf{c}_\downarrow)}$ by capture basins of the epigraph and the hypograph of the function $\mathbf{k}$, which inherited the properties of capture basins, that we have entirely made explicit in the case of the valuation functional $V_{(\mathbf{k},\mathbf{c})}$ and left as an exercise for the valuation functional $V^{\text{sup}}_{(\mathbf{k}_\downarrow,\mathbf{c}_\downarrow)}$.

In this section, we shall introduce two more valuation functionals, $V^{\text{sup}}_{(\mathbf{k},\mathbf{c})}$ and $V^{\text{inf}}_{(\mathbf{k}_\downarrow,\mathbf{c}_\downarrow)}$, which are characterized by absorption basins, which inherit the properties of absorption basins, the translation of which in terms of solutions of Hamilton–Jacobi–Bellman partial differential equation we leave as exercises.

We consider the functionals defined by (17.10), p. 688:

$$\begin{cases} \mathbf{J_c}(t;(x(\cdot),u(\cdot)))(T,x) \\ := e^{-\widehat{\mathbf{m}}(t)}\mathbf{c}(T-t,x(t)) + \int_0^t e^{-\widehat{\mathbf{m}}(\tau)}\mathbf{l}(x(\tau),u(\tau))d\tau \\ \\ \mathbf{I_k}(t;(x(\cdot),u(\cdot)))(T,x) := \\ \sup_{s\in[0,t]} \left(e^{-\widehat{\mathbf{m}}(s)}\mathbf{k}(T-s,x(s)) + \int_0^s e^{-\widehat{\mathbf{m}}(\tau)}\mathbf{l}(x(\tau),u(\tau))d\tau \right) \\ \\ \mathbf{L_{(k,c)}}(t;(x(\cdot),u(\cdot)))(T,x) \\ := \max(\mathbf{I_k}(t;x(\cdot),u(\cdot))(T,x), \mathbf{J_c}(t;(x(\cdot),u(\cdot)))(T,x)) \\ \\ V_{(\mathbf{k},\mathbf{c})}(x(\cdot),u(\cdot))(T,x) := \inf_{t\in[0,T]} \mathbf{L_{(k,c)}}(t;(x(\cdot),u(\cdot)))(T,x) \end{cases}$$

and the valuation functional

$$V^{\mathrm{sup}}_{(\mathbf{k},\mathbf{c})}(T,x) := \sup_{(x(\cdot),u(\cdot))\in\mathcal{S}(x)} V_{(\mathbf{k},\mathbf{c})}(x(\cdot),u(\cdot))(T,x) \tag{17.37}$$

Let us consider the auxiliary system (17.2), p. 683:

$$\begin{cases} (i) \;\; \tau'(t) = -1 \\ (ii) \;\; x'(t) = f(x(t),u(t)) \\ (iii) \;\; y'(t) = \mathbf{m}(x(t))y(t) - \mathbf{l}(x(t),u(t)) \\ \quad\quad \text{where } u(t)\in U(x(t)), \text{ and } \mathbf{k}(t,x(t),u(t)) \le y(t) \end{cases}$$

Theorem 17.5.2 ***[Viability Characterization of Valuation Functions]*** *The valuation function* $V^{\mathrm{sup}}_{(\mathbf{k},\mathbf{c})}$ *satisfies*

$$V^{\mathrm{sup}}_{(\mathbf{k},\mathbf{c})}(T,x) := \inf_{(T,x,y)\in \mathrm{Abs}_{(17.2)}(\mathcal{E}p(\mathbf{k}),\mathcal{E}p(\mathbf{c}))} y$$

Proof. Inequality

$$V^{\mathrm{sup}}_{(\mathbf{k},\mathbf{c})}(T,x) \le \inf_{(T,x,y)\in \mathrm{Abs}_{(17.2)}(\mathcal{E}p(\mathbf{k}),\mathcal{E}p(\mathbf{c}))} y$$

is straightforward. It is proven in the same way than the first inequality in the proof of Theorem 17.3.4, p. 689 (where, at the end, we replace the infimum inf by the supremum sup).

For proving the opposite inequality, we take any $z < U(T,x) := \inf_{(T,x,y)\in \text{Abs}_{(17.2)}(\mathcal{E}p(\mathbf{k}),\mathcal{E}p(\mathbf{c}))} y$ which we let converge to $U(T,x)$. Since the triple (T,x,z) does not belong to $\text{Abs}_{(17.2)}(\mathcal{K},\mathcal{E}p(\mathbf{c}))$, Lemma 2.12.2, p. 92 implies that

$$(T,x,z) \in \text{Capt}_{(17.2)}(\overset{\circ}{\mathcal{H}yp}(\mathbf{c}), \complement\mathcal{K})$$

because

$$\complement\mathcal{E}p(\mathbf{c}) := \{(T,x,y) \text{ such that } y < \mathbf{c}(T,x) =: \overset{\circ}{\mathcal{H}yp}(\mathbf{c})\}$$

Therefore, there exists $(x(\cdot),u(\cdot)) \in \mathcal{S}(x)$ such that for any $t \in [0,T]$, such that

$$e^{-\widehat{\mathbf{m}}(t)}\left(z - \int_0^t e^{-\widehat{\mathbf{m}}(\tau)}\mathbf{l}(x(\tau),u(\tau))d\tau\right) < \mathbf{c}(T-t,x(t))$$

and thus,

$$z \le e^{-\widehat{\mathbf{m}}(t)}\mathbf{c}(T-t,x(t)) + \int_0^t e^{-\widehat{\mathbf{m}}(\tau)}\mathbf{l}(x(\tau),u(\tau))d\tau$$

Letting z converge to $U(T,x)$, we obtain

$$\begin{cases} U(T,x) \le \inf\limits_{s\in[0,T]} \left(e^{-\widehat{\mathbf{m}}(s)}\mathbf{c}(T-st,x(s)) + \int_0^s e^{-\widehat{\mathbf{m}}(\tau)}\mathbf{l}(x(\tau),u(\tau))d\tau\right) \\ \le \sup\limits_{(x(\cdot),u(\cdot))\in\mathcal{S}(x)} \inf\limits_{s\in[0,T]} \left(e^{-\widehat{\mathbf{m}}(s)}\mathbf{c}(T-st,x(s)) + \int_0^s e^{-\widehat{\mathbf{m}}(\tau)}\mathbf{l}(x(\tau),u(\tau))d\tau\right) \\ = \sup\limits_{(x(\cdot),u(\cdot))\in\mathcal{S}(x)} \inf\limits_{t\in[0,T]} \mathbf{J_c}(t;(x(\cdot),u(\cdot)))(T,x) \\ \le \sup_{(x(\cdot),u(\cdot))\in\mathcal{S}(x)} V_{(\mathbf{k},\mathbf{c})}(x(\cdot),u(\cdot))(T,x) = V^{\sup}_{(\mathbf{k},\mathbf{c})}(T,x) \end{cases}$$

Hence we have proved the viability characterization of $V^{\sup}_{(\mathbf{k},\mathbf{c})}(T,x)$ by absorption basins. □

Replacing epigraphs by hypographs, we minimize valuation functionals $(x(\cdot),u(\cdot)) \mapsto V^{\flat}_{(\mathbf{k}_\downarrow,\mathbf{c}_\downarrow)}(x(\cdot),u(\cdot))(T,x)$:

$$V^{\inf}_{(\mathbf{k}_\downarrow,\mathbf{c}_\downarrow)}(T,x) := \inf_{(x(\cdot),u(\cdot))\in\mathcal{S}(x)} \sup_{t\in[0,T]} L_{(\mathbf{k}_\downarrow,\mathbf{c}_\downarrow)}(t;x(\cdot),u(\cdot))(T,x) \quad (17.38)$$

17.6 Summary: Valuation Functions of Four Classes of Intertemporal Optimization Problems

In conclusion, we have studied four classes of optimization problems associated with:

1. the *Lagrangian* $\mathbf{l} : X \times \mathcal{U} \mapsto \mathbb{R}_+$, assigning to each state-control pair (x, u) its transient cost $\mathbf{l}(x, u)$,
2. the *discount rate* $\mathbf{m} : X \times \mathcal{U} \mapsto \mathbb{R}_+$, assigning to each state-control pair (x, u) its discount rate $\mathbf{m}(x, u)$ which may depend upon the state x and controlled by u,
3. the *constraint function* $\mathbf{k} : \mathbb{R}_+ \times X \mapsto \mathbb{R} \cup \{+\infty\}$
4. $\mathbf{c} : \mathbb{R}_+ \times X \mapsto \mathbb{R} \cup \{+\infty\}$, the *target function*, regarded as a *spot cost* in control theory or a target condition (Cauchy function, Dirichlet boundary condition, etc.) in partial differential equations.

Recall that we set

$$\begin{cases} \mathbf{J_c}(t; (x(\cdot), u(\cdot)))(T, x) \\ := e^{-\widehat{\mathbf{m}}(t)}\mathbf{c}(T-t, x(t)) + \displaystyle\int_0^t e^{-\widehat{\mathbf{m}}(\tau)}\mathbf{l}(x(\tau), u(\tau))d\tau \\ \mathbf{I_k}(t; (x(\cdot), u(\cdot)))(T, x) := \\ \displaystyle\sup_{s\in[0,t]} \left(e^{-\widehat{\mathbf{m}}(s)}\mathbf{k}(T-s, x(s)) + \int_0^s e^{-\widehat{\mathbf{m}}(\tau)}\mathbf{l}(x(\tau), u(\tau))d\tau \right) \\ \mathbf{L_{(k,c)}}(t; (x(\cdot), u(\cdot)))(T, x) \\ := \max(\mathbf{I_k}(t; x(\cdot), u(\cdot))(T, x), \mathbf{J_c}(t; (x(\cdot), u(\cdot)))(T, x)) \end{cases}$$

and

$$\begin{cases} \mathbf{J_{c_\downarrow}}(t; (x(\cdot), u(\cdot)))(T, x) \\ := e^{-\widehat{\mathbf{m}}(t)}\mathbf{c}_\downarrow(T-t, x(t)) + \displaystyle\int_0^t e^{-\widehat{\mathbf{m}}(\tau)}\mathbf{l}(x(\tau), u(\tau))d\tau \\ \mathbf{I}^\flat_{\mathbf{k}_\downarrow}(t; (x(\cdot), u(\cdot)))(T, x) := \\ \displaystyle\inf_{s\in[0,t]} \left(e^{-\widehat{\mathbf{m}}(s)}\mathbf{k}_\downarrow(T-s, x(s)) + \int_0^s e^{-\widehat{\mathbf{m}}(\tau)}\mathbf{l}(x(\tau), u(\tau))d\tau \right) \\ \mathbf{L}^\flat_{(\mathbf{k}_\downarrow,\mathbf{c})}(t; (x(\cdot), u(\cdot)))(T, x) \\ := \min(\mathbf{I}_{\mathbf{k}_\downarrow}(t; x(\cdot), u(\cdot))(T, x), \mathbf{J}_{\mathbf{c}_\downarrow}(t; (x(\cdot), u(\cdot)))(T, x)) \end{cases}$$

Theorem 17.6.1 *[Valuation Functions of Four classes of Intertemporal Optimization Problems] Under assumptions of Theorems 17.3.4, p.689 and 17.5.2, p.706, the four viability episolutions* *and hyposolutions defined in Definition 17.2.1, p.683 are equal to the four following valuation functions of intertemporal optimization problems:*

$$\begin{cases} V^{\inf}_{(\mathbf{k},\mathbf{c})}(T,x) & := \inf_{(x(\cdot),u(\cdot))\in\mathcal{S}(x)} \inf_{t\in[0,T]} \mathbf{L}_{(\mathbf{k},\mathbf{c})}(t;(x(\cdot),u(\cdot)))(T,x) \\ & = \inf_{(T,x,y)\in\mathrm{Capt}_{(17.2)}(\mathcal{E}p(\mathbf{k}),\mathcal{E}p(\mathbf{c}))} y \\ V^{\sup}_{(\mathbf{k}_\downarrow,\mathbf{c}_\downarrow)} & := \sup_{(x(\cdot),u(\cdot))\in\mathcal{S}(x)} \sup_{t\in[0,T]} \mathbf{L}_{(\mathbf{k}_\downarrow,\mathbf{c}_\downarrow)}(t;(x(\cdot),u(\cdot)))(T,x) \\ & = \sup_{(T,x,y)\in\mathrm{Capt}_{(17.2)}(\mathcal{H}yp(\mathbf{k}_\downarrow),\mathcal{H}yp(\mathbf{c}_\downarrow))} y \\ V^{\sup}_{(\mathbf{k},\mathbf{c})}(T,x) & := \sup_{(x(\cdot),u(\cdot))\in\mathcal{S}(x)} \inf_{t\in[0,T]} \mathbf{L}_{(\mathbf{k},\mathbf{c})}(t;(x(\cdot),u(\cdot)))(T,x) \\ & = \inf_{(T,x,y)\in\mathrm{Abs}_{(17.2)}(\mathcal{E}p(\mathbf{k}),\mathcal{E}p(\mathbf{c}))} y \\ V^{\inf}_{(\mathbf{k}_\downarrow,\mathbf{c}_\downarrow)}(T,x) & := \inf_{(x(\cdot),u(\cdot))\in\mathcal{S}(x)} \sup_{t\in[0,T]} \mathbf{L}_{(\mathbf{k}_\downarrow,\mathbf{c}_\downarrow)}(t;(x(\cdot),u(\cdot)))(T,x) \\ & = \sup_{(T,x,y)\in\mathrm{Abs}_{(17.2)}(\mathcal{H}yp(\mathbf{k}_\downarrow),\mathcal{H}yp(\mathbf{c}_\downarrow))} y \end{cases} \tag{17.39}$$

Consequently, the valuation functions of these four classes of intertemporal optimization problems inherit the properties of capture and absorption basins proved in this book and "translated" in this chapter.

Part IV
Appendices

Chapter 18
Set-Valued Analysis at a Glance

18.1 Introduction

The purpose of this chapter is to present a short introduction to the main concepts of Set-Valued Analysis used throughout this book, and regarded here as a "Toolbox" for viability theory, optimal control, differential games and their applications to mathematical economics and finance.

There is no need to justify here the introduction of set-valued maps which goes back, in mathematical economics, for instance, to the celebrated Gérard Debreu's *Theory of Value* [76, Debreu]. However, a differential calculus of these set-valued maps has been designed and efficiently exploited in the 1980s actually, for the very purpose of differentiating regulation maps in viability theory and control, in order to give a meaning to heavy solutions which is what is summarized in these notes. More details can be found in the monographs on convex analysis, to *Optima and Equilibria*, [19, Aubin] for optimization and convex analysis, to *Set-Valued Analysis*, [27, Aubin & Frankowska], for instance, and to the exhaustive book *Variational Analysis*, [178, Rockafellar & Wets], among many other ones on these topics. One can find in [23, Aubin] another approach to a differential calculus of set-valued maps, called the *mutational calculus*, which is also appropriate to define differential equations (called mutational equations) governing the evolution of subsets. We also summarize some results in convex analysis used in Chaps. 15, p. 603 and 14, p. 563.

18.2 Notations

In this book, we use the definition of *positive*, *negative*, *increasing* and *decreasing* in the sense of *positive or null*, *negative or null*, *increasing or constant* and *decreasing or constant*. Otherwise, we add the adjective *strictly*.

J.-P. Aubin et al., *Viability Theory*, DOI 10.1007/978-3-642-16684-6_18,

The reason why we do not use the words *non negative*, *non strictly positive*, *non decreasing* and *non increasing* is that, if a function is (strictly) positive, a non (strictly) positive function is a function which is negative or null for at least one point, and not for all points!

Definition 18.2.1 ***[Notations: Spaces and Hyperspaces]*** *Usually, (finite dimensional vector) spaces are denoted by X, Y, Z or $\mathcal{U}$, $\mathbf{X}$, etc. when needed. Infinite dimensional spaces are usually denoted by script capital letters, such as $\mathcal{C}(0,+\infty,X)$. Notation $\mathcal{P}(X)$ (or 2^X) denotes the* hyperspace *(or* power space*) of subsets $A \subset X$.*

For a set $K \subset X$, we set

$$\complement K := K^c := X \backslash K = \text{ the complement of } K$$

and

$$K \backslash B := K \cap \complement B \text{ the complement of } B \text{ in } K$$

We denote by $\overline{K}$ or $\mathrm{cl}(K)$ the *closure* of K, by $\overset{\circ}{K}$ or $\mathrm{Int}(K)$ its *interior*, by

$$\widehat{K} := X \backslash \mathrm{Int}(K) = \overline{\complement K}$$

the complement of the interior of K.

Definition 18.2.2 ***[Topological Regularity]*** *A subset K is said to be* topological regular *if it is the closure of its interior: $K = \overline{\mathrm{Int}(K)}$.*

When X is vector space, and if $A \subset X$ and $B \subset X$, we set

$$A+B := \bigcup_{b \in B}(A+b) \text{ and } A \ominus B := \bigcap_{b \in B}(A-b)$$

the *sum and the Minkowski difference* of the subset A and B.

Note that the *distributivity property*

$$\left(\bigcup_{i \in I} A_i\right) + B = \bigcup_{i \in I}(A_i + B) \tag{18.1}$$

The *unit ball* of X is denoted by B (or B_X if the space must be mentioned). We denote by

$$d_K(x) := d(x,K) := \inf_{y \in K} \|x-y\|$$

the *distance from x to K*, where we set $d(x,\emptyset) := +\infty$. The *ball of radius $r > 0$ around K in X* is denoted by

$$B_X(K,r) := \{x \in X \mid d(x,K) \leq r\} = \overline{K + rB_X}$$

When there is no ambiguity, we set

$$B(K,r) := B_X(K,r)$$

The balls $B(K,r)$ are neighborhoods of K. When K is compact, each neighborhood of K contains such a ball around K.

When X is a Banach space (or any topological vector space), we denote by $X^\star := \mathcal{L}(X,\mathbb{R})$ the (topological) *dual*, i.e., the space of *continuous linear* functions $p : x \mapsto p(x) =: \langle p, x\rangle \in \mathbb{R}$ and by $X^{\star\star}$ its *bidual*. Recall than when $X := \mathbb{R}^n$ is a finite dimensional vector space, *all the norms are equivalent and all linear operators from a finite dimensional vector space to another are continuous.*

Definition 18.2.3 *[Support Functions and Barrier Cones] Let us consider a subset $K \subset \mathbb{R}^n$. Its* support function *is defined by*

$$\sigma(K,p) \;:=\; \sup_{x\in K} \langle p, x\rangle$$

The barrier cone *of K is defined by $K^\triangledown := \{p \in X^\star \mid \sigma_K(p) < +\infty\} =$* Dom$(\sigma_K)$ *and its* polar cone *by $K^\star := \{p \in X^\star \mid \sigma_K(p) \leq 0\}$.*
If $B := \{x$ such that $\|x\| \leq 1\}$, the dual norm is defined by $\|p\|_\star := \sigma(B,p)$.

We recall one (among many equivalent) statement of the *Separation Theorem*:

Theorem 18.2.4 *[Separation Theorem] Let us consider a subset $K \subset \mathbb{R}^n$. Then its closed convex hull $\overline{\text{co}}(K)$ is defined by an infinite number of linear inequality constraints:*

$$\overline{\text{co}}(K) \;=\; \{x \text{ such that}, \;\forall\, p \in X^\star, \;\; \langle p, x\rangle \;\leq\; \sigma(K,p)\}$$

We quote a simple but important closedness criterion:

Theorem 18.2.5 *[Closed Image Theorem] Let $A \in \mathcal{L}(X_1, X_2)$. If $K \subset X_1$ is closed, condition*

$$\text{Im}(A^\star) + K^\nabla = X_1^\star \quad (18.2)$$

implies that $A(K)$ is closed and that the subsets $K \cap A^{-1}(x_2)$ are compact.

Proof. Assume that $Ax_n := y_n$ converges to y. Since any $p \in X_1^\star = A^\star q + r$ where $q \in X_2^\star$ and $r \in K^\nabla = \text{Dom}(\sigma_K)$, we infer that

$$\begin{cases} \langle p, x_n \rangle \leq \langle q, Ax_n \rangle + \langle r, x_n \rangle \\ \leq \langle q, y_n \rangle + \sigma_K(r) < +\infty \end{cases}$$

Hence x_n is bounded, and thus relatively compact, and converges to some x, so that $A(K)$ is closed. □

Notations for maps and functions follow as possible the following rules:

39 ***[Notations: Maps and Functions]*** *Set-valued maps are usually denoted by F, G, H, U, etc., and their single valued counterparts by f, g, h, u. Tubes are denoted by $\mathbf{K} : t \rightsquigarrow K(t)$. Numerical and extended functions are defined by* **a**, **b**, **c**, **k**, **l**, **u**, **v**, *etc. In order to be consistent with traditional notations in control theory, we use the notations V, W for value and valuation functions.*

The basic notation for an evolutionary system is the set-valued map $\mathcal{S} : X \rightsquigarrow \mathcal{C}(0, +\infty; X)$. We keep the same notation for systems $\mathcal{S} : X \rightsquigarrow \mathcal{C}(-\infty, +\infty; X)$ governing evolutions from $\mathbb{R}$ to X.

When the evolutionary system is generated by a system

$$x'(t) = f(x(t), u(t)) \text{ where } u(t) \in U(x(t))$$

and when we need to use explicitly the controls, we set

40 ***[Notations: Evolutionary Systems]*** *We set:*

- *in Chap. 4, p. 125,*

$$\begin{cases} \mathcal{P}(x) := \{(x(\cdot), u(\cdot)) \text{ such that } x(0) = x \\ \text{and } x'(t) = f(x(t), u(t)) \text{ where } u(t) \in U(x(t)) \} \end{cases}$$

- *in Chap. 6, p. 199,*

$$\mathcal{P}(x, u) := \{(x(\cdot), u(\cdot)) \text{ such that } x(\cdot) \in \mathcal{S}(x) \text{ and } u(0) = u\}$$

- *in Chap. 8, p. 273,*

$$\begin{cases} \mathcal{S}(y,z) := \{x(\cdot) \in \mathcal{S}(y) \text{ such that } \exists\, t^\star < +\infty \text{ satisfying } x(t^\star) = z\} \\ \mathcal{P}(y,z) := \{(x(\cdot),u(\cdot)) \text{ such that } x(\cdot) \in \mathcal{S}(y,z)\} \end{cases}$$

- *in Sect. 8.10, p. 309, for time dependent dynamics,*

$$\mathcal{S}(T,x) := \{x(\cdot) \text{ such that } x(T) = x\}$$

so that $\mathcal{S}(0,x) = \mathcal{S}(x)$.

As a rule,

- *whenever we mention that the evolutions are viable in a subset* K *(or a tube* $\mathbf{K} : t \rightsquigarrow K(t)$*), we use the superscript* ^{K}*: for example,* $\mathcal{S}^K$, $\mathcal{P}^K$, *etc.*
- *whenever we split an evolution* $x(\cdot)$ *in forward* $\overrightarrow{x}(\cdot)$ *and backward* $\overleftarrow{x}(\cdot)$ *components, we use the same symbols for every concept associated with the forward–backward decomposition: for instance,* $\overleftarrow{\mathcal{S}}$, $\overleftarrow{\mathcal{P}}$, $\overleftarrow{\text{Viab}}_{\mathcal{S}}(K,C) = \text{Viab}_{\overleftarrow{\mathcal{S}}}(K,C)$, *etc.*

We recall the definition of infimum and supremum and some of their elementary properties:

Definition 18.2.6 *[Infima] Let* $A \subset \mathbb{R}$ *be a subset of* $\mathbb{R}$. *Then* $\inf(A) \in \mathbb{R} \cup \{-\infty\}$ *is the element of* $\overline{A}$ *satisfying*

$$\begin{cases} (i)\ \forall\, x \in A,\ \inf(A) \le x \\ (ii)\ \forall\, \lambda > \inf(A),\ \exists\, x_\lambda \in A \text{ such that } x_\lambda \le \lambda \end{cases} \tag{18.3}$$

We set by convention

$$\inf(\emptyset) := +\infty \text{ and } \inf(X) := -\infty$$

If we know in advance that $\inf(A) > -\infty$, *then, by taking* $\varepsilon := \lambda - \inf(A)$, *condition (18.2.6), p. 717 boils down to*

$$\begin{cases} (i)\ \forall\, x \in A,\ \inf(A) \le x \\ (ii)\ \forall\, \varepsilon > 0,\ \exists\, x_\varepsilon \in A \text{ such that } x_\varepsilon \le \inf(A) + \varepsilon \end{cases}$$

If $\mathbf{v} : X \mapsto \mathbb{R} \cup \{+\infty\}$ *is a nontrivial extended function, the infimum* $\inf_{x \in K} \mathbf{v}(x) := \inf(\mathbf{v}(K)) \in \mathbb{R} \cup \{-\infty\}$ *is the infimum of the image* $\mathbf{v}(K) \subset \mathbb{R} \cup \{+\infty\}$.

We recall the following formulas on infimum of functions on unions of sets:

Lemma 18.2.7 *[Infima on Unions] Let* $f : X \mapsto \mathbb{R} \cup \{+\infty\}$. *Then*

$$\inf_{x \in \bigcup_{i=1}^{n} A_i} f(x) = \inf_{i=1,\ldots,n} \inf_{x \in A_i} f(x)$$

Proof. Let us set $A := \bigcup_{i=1}^{n} A_i$. Since $A_i \subset A$, then $\inf_{x \in A} f(x) \le \inf_{x \in A_i} f(x)$, so that $\inf_{x \in A} f(x) \le \inf_{i=1,\ldots,n} \inf_{x \in A_i} f(x)$.

Conversely, for any $\varepsilon > 0$, there exist $i \in \{1, \ldots, n\}$ and $x_\varepsilon \in A_i$ such that $f(x_\varepsilon) \le \inf_{x \in A} f(x) + \varepsilon$. Therefore $\inf_{i=1,\ldots,n} \inf_{x_i \in A_i} f(x_i) \le f(x_\varepsilon) \le \inf_{x \in A} f(x) + \varepsilon$, so that, by letting ε converge to 0, we infer that $\inf_{i=1,\ldots,n} \inf_{x \in A_i} f(x) \le \inf_{x \in A} f(x)$. □

Let us point out the following

Lemma 18.2.8 *[Infimum of Unions and Intersections] Consider a family* $A_i \subset \mathbb{R}_+$ *of subsets of* $\mathbb{R}_+$. *Then*

$$\inf \left(\bigcup_{i \in I} (A_i + \mathbb{R}_+) \right) = \inf_{i \in I} (\inf A_i)$$

and

$$\inf \left(\bigcap_{i \in I} (A_i + \mathbb{R}_+) \right) = \sup_{i \in I} (\inf A_i)$$

Proof. The first statement is obvious. Let us prove the second one, set $a := \inf \left(\bigcap_{i \in I} (A_i + \mathbb{R}_+) \right)$ and $b := \sup_{i \in I} \inf A_i$. For proving that $b \le a$, let $\varepsilon > 0$ be chosen. By definition of a, for any $i \in I$, there exists $a_i \in A_i$ such that $a_i \le a + \varepsilon$. In other words, for any $i \in I$, $\inf A_i \le a + \varepsilon$, i.e., $b := \sup_{i \in I} \inf A_i \le a + \varepsilon$.

Conversely, for establishing that $b \ge a$, by definition of b, for any $i \in I$, there exists $a_i \in A_i$ such that $a_i \le b + \varepsilon$, i.e., such that $b + \varepsilon \in A_i + \mathbb{R}_+$. Therefore $b + \varepsilon$ belongs to $\bigcap_{i \in I} (A_i + \mathbb{R}_+)$. Hence $b + \varepsilon \ge a$ and thus, $a = b$. □

18.3 Set-Valued Maps

Set-valued maps and their graphs have been introduced in the very first Box 3, p. 12 of this book. We provide a more elaborate definition:

Definition 18.3.1 *[Set-Valued Map] A* set-valued *map* $F : X \rightsquigarrow Y$ *associates with any* $x \in X$ *a subset* $F(x) \subset Y$ *(which may be the empty set* $\emptyset$*). The symbol "*$\rightsquigarrow$*" denotes set-valued maps whereas the classical symbol "*$\mapsto$*" denotes single-valued maps.*
The graph Graph(F) *of a set-valued map* F *is the set of pairs* $(x, y) \in X \times Y$ *satisfying* $y \in F(x)$. *Its* domain Dom(F) *is the subset of elements* $x \in X$ *such that* $F(x)$ *is not empty and its* image $\text{Im}(F) = \bigcup_{x \in X} F(x)$ *is the union of the values* $F(x)$ *of* F *when* x *ranges over* X. *The* inverse F^{-1} *of* F *is the set-valued map from* Y *to* X *defined by*

$$x \in F^{-1}(y) \iff y \in F(x) \iff (x, y) \in \text{Graph}(F)$$

We add few other concepts and state some more statements of properties, referring to [27, Aubin & Frankowska] or [174, Rockafellar & Wets] for additional results, their proofs and a bibliography. We add few other definitions and state some more properties.

Definition 18.3.2 *[Direct Images and Focuses] Let* $F : X \rightsquigarrow Y$ *be a set-valued map. Let* $\mathcal{P}(X)$ *and* $\mathcal{P}(Y)$ *denote the* hyperspaces *of subsets* $A \subset X$ *and* $B \subset Y$. *There are three ways to extend* $F : X \rightsquigarrow Y$ *into (single-valued) maps from* $\mathcal{P}(X)$ *to* $\mathcal{P}(Y)$:

1. *through* extension *(or* (canonical) extension*) (again denoted by)* $F : \mathcal{P}(X) \mapsto \mathcal{P}(Y)$ *of* F *defined by:*

$$\forall\, A \in \mathcal{P}(X),\ \ F(A) := \bigcup_{x \in A} F(x) = \{y \mid A \cap F^{-1}(y) \neq \emptyset\} \in \mathcal{P}(Y)$$

where $F(A)$ *is called the* image *of* A *under* F,

2. *through the* focus $\widehat{F} : \mathcal{P}(X) \mapsto \mathcal{P}(Y)$ *of* F *defined by:*

$$\forall\, A \in \mathcal{P}(X),\ \ \widehat{F}(A) := \bigcap_{x \in A} F(x) = \{y \mid A \subset F^{-1}(y)\} \in \mathcal{P}(Y)$$

where $\widehat{F}(A)$ *i is called the* focus *of* A *by* F.

Their inverse are defined by

Definition 18.3.3 *[Inverse Images, Focuses and Cores] Let $F : X \rightsquigarrow Y$ be a set-valued map, $\mathcal{P}(X)$ and $\mathcal{P}(Y)$ their* hyperspaces.

1. *The* canonical inverse extension $F^{-1} : \mathcal{P}(Y) \mapsto \mathcal{P}(X)$ *of F^{-1} is defined by:*

$$\forall\, B \in \mathcal{P}(Y),\ \ F^{-1}(B) := \bigcup_{y \in B} F^{-1}(y) = \{x \mid B \cap F(x) \neq \emptyset\} \in \mathcal{P}(Y)$$

where $F^{-1}(B)$ is called the inverse image *of B under F.*

2. *The* inverse focus $\widehat{F}^{-1} : \mathcal{P}(Y) \mapsto \mathcal{P}(X)$ *of F is defined by:*

$$\forall\, B \in \mathcal{P}(Y),\ \ \widehat{F}^{-1}(B) := \bigcap_{y \in B} F^{-1}(y) = \{x \mid B \subset F(x)\} \in \mathcal{P}(X)$$

where $\widehat{F}^{-1}(B)$ is called the inverse focus *of B by F.*

3. *The* core *(or* (inverse) core*) $F^{\ominus 1} : \mathcal{P}(Y) \mapsto \mathcal{P}(X)$ of F is defined by*

$$\forall\, B \in \mathcal{P}(Y),\ \ F^{\ominus 1}(B) := \{x \mid F(x) \subset B\}$$

where $F^{\ominus 1}(B)$ is called the core *of B by F.*

We observe that

$$\forall y \in \mathrm{Im}(F),\ \widehat{F}^{-1}(\{y\}) = F^{\ominus 1}(\{y\}) = F^{-1}(y)$$

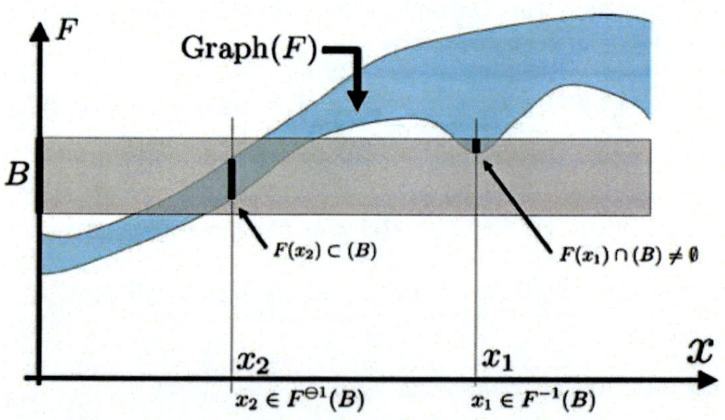

Fig. 18.1 Inverse Image and Core.
Illustration of the core and the inverse of a set valued map.

The following "partition property" of cores plays an important role:

Lemma 18.3.4 *[The Partition Lemma] Let us consider a* partition *of $K = B_1 \cup_\emptyset B_2$ by two disjoint subsets $B_i \subset Y$. This means that $K = B_1 \cup B_2$ and that $B_1 \cap B_2 = \emptyset$.*

If $F : X \rightsquigarrow Y$ is a set-valued map, then

$$F^{\ominus 1}(K) \;=\; F^{-1}(B_1) \cup F^{\ominus 1}(B_2) \text{ and } F^{-1}(B_1) \cap F^{\ominus 1}(B_2) \;=\; \emptyset$$

is a partition of the core of K by the inverse image of B_1 and the core of B_2. It implies formula

$$F(\complement A) = \complement\left((F^{-1})^{\ominus 1}(A)\right) \;\&\; F^{-1}(\complement B) = \complement F^{\ominus 1}(B)$$

Proof. To say that $x \in F^{\ominus 1}(K)$ means that $F(x) \subset K = B_1 \cup B_2$. Since $B_1 \cap B_2 = \emptyset$, this can happen only if $F(x) \cap B_1 \neq \emptyset$ or if $F(x) \subset B_2$, but not at the same time. □

We review some other obvious properties:

Lemma 18.3.5 *[Monotonicity] The three extensions satisfy the following properties:*

1. *The images and inverse images are increasing: If $A_1 \subset A_2$, then $F(A_1) \subset F(A_2)$ and if $B_1 \subset B_2$, then $F{-}1(B_1) \subset F^{-1}(B_2)$ and transform unions into unions:*

$$F\left(\bigcup_{i\in I} A_i\right) = \bigcup_{i\in I} F(A_i) \;\&\; F^{-1}\left(\bigcup_{i\in I} B_i\right) = \bigcup_{i\in I} F^{-1}(B_i)$$

2. *The focus and inverse focus are decreasing: If $A_1 \subset A_2$, then $\widehat{F}(A_2) \subset \widehat{F}(A_1)$ and if $B_1 \subset B_2$, then $\widehat{F^{-1}}(B_2) \subset \widehat{F^{-1}}(B_1)$ and transform unions into intersections:*

$$\widehat{F}\left(\bigcup_{i\in I} A_i\right) = \bigcap_{i\in I} \widehat{F}(A_i) \;\&\; \widehat{F^{-1}}\left(\bigcup_{i\in I} B_i\right) = \bigcap_{i\in I} \widehat{F^{-1}}(B_i)$$

3. *The core is increasing: if $B_1 \subset B_2$, then $F^{\ominus 1}(B_1) \subset F^{\ominus 1}(B_1)$ and transforms intersections into intersections:*

$$F^{\ominus 1}\left(\bigcap_{i\in I} B_i\right) \;=\; \bigcap_{i\in I} F^{\ominus 1}(B_i)$$

18.3.1 Graphical Approach

We summarize here several conventions used all over this book:

Definition 18.3.6 *[Graphical Operations on Set-Valued Maps] Let $\mathcal{P}$ be a property of a subset (for instance, closed, convex, etc.) Since we do emphasize the interpretation of a set-valued map as a graph (instead of a map from a set to another one), we shall say as a general rule that a* set-valued map satisfies property $\mathcal{P}$ if and only if its graph satisfies it.

If $\blacktriangle$ denotes an operation mapping sets to sets (complement, closure, etc.), we denote by $\blacktriangle(F)$ the set-valued map defined "graphically" from $F : X \rightsquigarrow Y$ by

$$\text{Graph}(\blacktriangle(F)) := \blacktriangle(\text{Graph}(F))$$

The set-valued map $\complement F : X \rightsquigarrow Y$ is defined by $(\complement F)(x) := \complement F(x)$ because

$$\text{Graph}(\complement(F)) := \complement(\text{Graph}(F))$$

For instance:

Definition 18.3.7 *[Closed and/or Convex Maps] A set-valued map F is said to be* closed *if its graph is closed,* convex *if its graph is convex and a* process *if its graph is a cone.* Closed convex processes *are set-valued maps the graphs of which are closed convex cones, the set-valued analogues of continuous linear operator, with which they share a large number of properties.*

Another graphical property is the concept of selection or extension:

Definition 18.3.8 *[Selection] Let E and F be two set-valued maps from X to Y. We say that E is a* selection *of F, or that $E \subset F$ is* contained *in F, or that F is an* extension *of E , if the graph of E is contained in the graph of F:*

$$E \subset F \iff \text{Graph}(E) \subset \text{Graph}(F)$$

or, equivalently, if for any $x \in X$, $E(x) \subset F(x)$, or, in the same way, if for any $y \in Y$, $E^{-1}(y) \subset F^{-1}(y)$.

Another important graphical operation is the composition of set-valued maps:

Lemma 18.3.9 *[Composition of Maps] Let us consider set-valued maps $\Phi : X \rightsquigarrow \overline{X}$ and $\Psi : Y \rightsquigarrow \overline{Y}$. We associate with a set-valued map $F : X \rightsquigarrow Y$ the set-valued map*

$$(\Phi \boxtimes \Psi)(F) : \overline{X} \rightsquigarrow \overline{Y}$$

"graphically" defined by

$$\mathrm{Graph}((\Phi \boxtimes \Psi)(F)) = (\Phi \times \Psi)(\mathrm{Graph}(F))$$

It is equal to the composition product

$$(\Phi \boxtimes \Psi)(F) \;=\; \Psi \circ F \circ \Phi^{-1} : x \rightsquigarrow \Psi(F(\Phi^{-1}(x)))$$

of the set-valued maps Φ^{-1}, F and Ψ.

We also obtain the following graphical characterization of focus and inverse focus:

Lemma 18.3.10 *[Graphical Characterization of Focus] The following statements are equivalent:*

$$\begin{cases} (i) \;\; A \subset \widehat{F}^{-1}(B) \\ (ii) \;\; B \subset \widehat{F}(A) \\ (iii) \; A \times B \subset \mathrm{Graph}(F) \end{cases} \tag{18.4}$$

Graphical operations have to be distinguished from the "setwise" (as we say "pointwise") operations on the values (and not, the graphs) of the set-valued maps:

Definition 18.3.11 *[Operations on Values of Set-Valued Maps] If the images of a set-valued map F are closed, convex, bounded, compact, and so on, we say that F is* closed-valued, convex-valued, bounded-valued, compact-valued, *and so on. When $\bigstar : Y \times Y \mapsto Y$ denotes an operation on subsets, we use the same notation for the operation on set-valued maps, which is defined by*

$$F_1 \bigstar F_2 : x \rightsquigarrow F_1(x) \bigstar F_2(x)$$

We define in that way $F_1 \cap F_2$, $F_1 \cup F_2$, $F_1 + F_2$ (in vector spaces), etc.

18.3.2 Amalgams of Set-Valued Maps

Amalgams of feedbacks (see Definition 10.8.1, p. 423) and concatenations of evolutions (see Definition 2.8.1, p. 69) provide examples of amalgams of set-valued maps.

Definition 18.3.12 *[Characteristic Set-Valued Maps or Marks] We associate with any pair of subsets $A \subset X$ and $B \subset Y$ the set-valued map $\Xi_A^B : X \rightsquigarrow Y$ the graph $\text{Graph}(\Xi_A^B) := A \times B$ of which is the product of subsets A and B, and thus defined by:*

$$\Xi_A^B(x) := \begin{cases} B & \text{if } x \in A \\ \emptyset & \text{if } x \notin A \end{cases} \tag{18.5}$$

The set-valued map $\Xi_A^B : X \rightsquigarrow Y$ is called the mark *of the product $A \times B$.*

When $B := Y$, we set $\Xi_A := \Xi_A^Y$, so that $\Xi_A(x) = Y$ if $x \in A$ and $\Xi_A(x) = \emptyset$ if $x \notin A$ (they play the role of *characteristic functions* χ_A of the subset A, assigning to every x the value 1 if its belongs to A and of 0 otherwise). This is the reason why these set-valued maps Ξ_A^B can be called of *marks* of the product $A \times B$. If the notation of A is too long, we set $\Xi^B(A; x) := \Xi_A^B(x)$.

Obviously,

$$(\Xi_A^B)^{-1} = \Xi_B^A$$

and

$$(\Phi \boxtimes \Psi)\Xi_A^B = \Xi_{\Phi(A)}^{\Psi(B)}$$

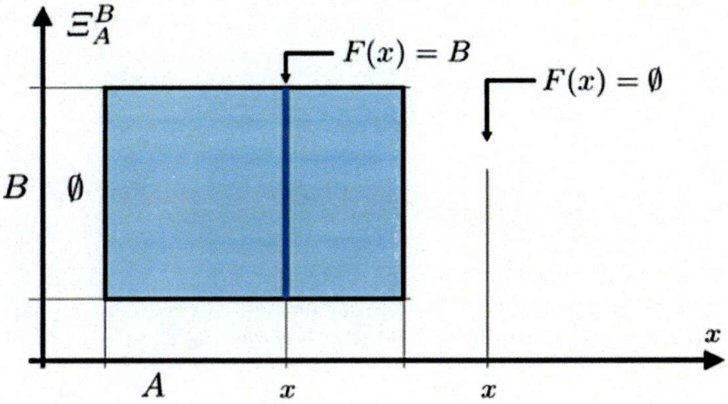

Fig. 18.2 Graph of the set-valued map $F := \Xi_A^B$.

Definition 18.3.13 *[Graphical Restriction] When $F : X \rightsquigarrow Y$ is a set-valued map from X to Y, the intersection $F \cap \Xi_A^B$ is called the* graphical restriction *of F to $A \times B$ since* $\text{Graph}(F \cap \Xi_A^B) = \text{Graph}(F) \cap A \times B$*:*

$$(F \cap \Xi_A^B)(x) := \begin{cases} F(x) \cap B & \text{if } x \in A \\ \emptyset & \text{if } x \notin A \end{cases}$$

If $B := Y$, we also use the classical notation $F|_A := F \cap \Xi_A$.

Therefore $(F \cap \Xi_A^B)^{-1} = F^{-1} \cap \Xi_B^A$ is equal to

$$(F \cap \Xi_A^B)^{-1}(y) := \begin{cases} F^{-1}(y) \cap A & \text{if } y \in B \\ \emptyset & \text{if } y \notin B \end{cases}$$

Definition 18.3.14 *[Amalgam of Set-Valued Maps] Let us consider a family $\{F_i\}_{i \in I}$ of set-valued maps $F_i : X \rightsquigarrow Y$ and a family $\{A_i \times B_i\}_{i \in I}$ of subsets $A_i \times B_i \subset X \times Y$.* The amalgam of set-valued maps $\{F_i\}_{i \in I}$ over the family $\{A_i \times B_i\}_{i \in I}$ *is the set-valued map* $\bigcup_{i \in I} F_i \cap \Xi_{A_i}^{B_i}$ *defined by*

$$\left(\bigcup_{i \in I} F_i \cap \Xi_{A_i}^{B_i} \right)(x) = \bigcup_{i \in I(x)} F_i(x) \cap B_i \text{ where } I(x) := \{i \in I \text{ such that } x \in A_i\}$$

We note that

$$\text{Graph}\left(\bigcup_{i\in I} F_i \cap \Xi_{A_i}^{B_i}\right) = \bigcup_{i\in I}(\text{Graph}(F_i) \cap (A_i \times B_i))$$

18.3.2.1 Amalgams of Single-Valued Functions

When Y is a vector space, when $B_i := Y$ for all $i \in I$, when the subsets $A_i \subset X$ form a *partition* of X and when the maps $f_i := F_i$ are *single-valued*, we recover the simple functions of integration theory and of its many applications:

$$\bigcup_{i\in I}(f_i \cap \Xi_{A_i})(x) = \sum_{i\in I} \chi_{A_i}(x) f_i(x)$$

where χ_{A_i} denotes the characteristic function of the subset A_i.

If the subsets A_i do not form a partition, the above formula does not make sense any longer: if all the subsets $I(x) := \{i \in I \text{ such that } x \in A_i\}$ are finite, for example, $\bigcup_{i\in I}(f_i \cap \Xi_{A_i})$ is a set-valued map associating with any x the finite set $\bigcup_{i\in I(x)} f_i(x)$ whereas $\sum_{i\in I}\chi_{A_i}(x) f_i$ is a function associating with any x the number $\sum_{i\in I(x)} f_i(x)$ of elements of this set.

Remark.
By using *characteristic maps (marks)* Ξ_A^Y instead of *characteristic functions*, the addition of numbers of elements of a set is replaced by the union of the subsets and the neutral element 0 by the empty set $\emptyset$. The extension by 0 of functions is replaced by the extension of a set-valued map by empty set. In this case, this extension of a single-valued map is a set-valued map because the empty set is a set. It also replaces the value $+\infty$ used to extend an *extended function* in convex and nonsmooth analysis. □

18.3.2.2 Concatenations of Evolutions

Concatenations of evolutions are specific amalgams of evolutions $x_i(\cdot) \in \mathcal{C}(-\infty, +\infty; X)$.

Recall that the translation $\kappa(T)x(\cdot)$ of an evolution $x(\cdot)$ is defined by $(\kappa(T)x(\cdot))(t) := x(t-T)$. This is a translation to the right if T is positive and to the left if T is negative.

Consider a sequence of evolutions $x_i(\cdot) : \mathbb{R}_+ \mapsto X$ and a strictly increasing sequence of instants t_i, $i \in \mathbb{N}$. We set $t_{-1} := -\infty$. If the sequence t_i, $i = 0, \ldots, N$ is finite, we set $t_{N+1} := +\infty$. The duration $\tau_i := t_{i+1} - t_i$, $i \geq 0$ is called *cadence*. In the case of a finite sequence. The last cadence $\tau_N = +\infty$ is infinite. If the family of cadences is given, we recover the sequence of instants by formula $t_{i+1} := t_i + \tau_i$. An infinite sequence of instants t_i (or of cadences τ_i) is a *Zeno sequence* if it converges to a finite number T.

Definition 18.3.15 *[Concatenation of Evolutions] The amalgam*

$$x_0(\cdot)\Diamond_{t_1}x_1(\cdot)\cdots\Diamond_{t_i}x_i(\cdot)\cdots := \bigcup_{i\geq 0}(\kappa(t_i)x_i(\cdot))\cap \Xi_{[t_i,t_{i+1}]} \tag{18.6}$$

is called the concatenation of evolutions $x_i(\cdot)$ *over the increasing sequence* t_i. *It is given explicitly by is defined by*

$$\forall\, i\geq 0,\ \ t\in]t_i,t_{i+1}],\ \ x(t)\ :=\ x_i(t-t_i)$$

In particular, the concatenation

$$(x(\cdot)\diamond_T y(\cdot))(t)\ := \begin{cases} x(t) & \text{if } t\in[0,T] \\ y(t-T) & \text{if } t\geq T\end{cases}$$

of two evolutions is equal to

$$x(\cdot)\diamond_T y(\cdot) = x(\cdot)\cap \Xi_{[0,T[}\cup \kappa(T)y(\cdot)\cap \Xi_{]T,+\infty]}$$

(See Definition 2.8.1, p. 69).

We observe that when $A\subset\mathbb{R}$ is an interval, then

$$\kappa(T)\Xi_A = \Xi_{T+A}$$

and thus, that

$$\kappa(T)\left(\bigcup_{i\in I}x_i(\cdot)\cap \Xi_{]t_i,t_{i+1}]}(\cdot)\right) = \left(\bigcup_{i\in I}\kappa(T)x_i(\cdot)\cap \Xi_{]t_i+T,t_{i+1}+T]}(\cdot)\right)$$

18.4 Limits of Sets

18.4.1 Upper and Lower Limits

Limits of sets have been introduced by *Paul Painlevé* in 1902 before the formalization of metric spaces by *Fréchet* in 1906, and thus, without the concept of topology. They have been popularized by *Kuratowski* in his famous book *Topologie* and thus, often called *Kuratowski lower and upper limits* of sequences of sets. They are defined without the concept of a topology on the hyperspace $\mathcal{P}(E)$ derived from a topology on the underlying set E.

Definition 18.4.1 ***[Limits of Sets]*** *Let $(K_n)_{n\in\mathbf{N}}$ be a sequence of subsets of a metric space E. We say that the subset*

$$\mathrm{Limsup}_{n\to\infty}K_n := \left\{x \in E \mid \liminf_{n\to\infty} d(x, K_n) = 0\right\}$$

is the Painlevé–Kuratowski upper limit *of the sequence K_n and that the subset*

$$\mathrm{Liminf}_{n\to\infty}K_n \;:=\; \{x \in E \mid \lim_{n\to\infty} d(x, K_n) = 0\}$$

is its Painlevé–Kuratowski lower limit. *A subset K is said to be the* Painlevé–Kuratowski limit *or the* set limit *of the sequence K_n if*

$$K = \mathrm{Liminf}_{n\to+\infty}K_n = \mathrm{Limsup}_{n\to+\infty}K_n =: \mathrm{Lim}_{n\to+\infty}K_n$$

Lower and upper limits are obviously *closed.* We also see at once that

$$\mathrm{Liminf}_{n\to+\infty}K_n \subset \mathrm{Limsup}_{n\to+\infty}K_n$$

and that the upper limits and lower limits of the subsets K_n and of their closures $\overline{K_n}$ do coincide, since $d(x, K_n) = d(x, \overline{K_n})$.

Any decreasing sequence of subsets K_n has a limit, which is the intersection of their closures:

$$\text{if } K_n \subset K_m \text{ when } n \geq m, \text{ then } \mathrm{Lim}_{n\to+\infty}K_n = \bigcap_{n\geq 0} \overline{K_n}$$

An upper limit may be empty (no subsequence of elements $x_n \in K_n$ has a cluster point).

Concerning sequences of singleta $\{x_n\}$, the set limit, when it exists, is either empty (the sequence of elements x_n is not converging), or is a singleton made of the limit of the sequence.

It is easy to check that:

Proposition 18.4.2 ***[Limits of Subsets]*** *If $(K_n)_{n\in\mathbf{N}}$ is a sequence of subsets of a metric space, then* $\mathrm{Liminf}_{n\to\infty}K_n$ *is the* set of limits of sequences $x_n \in K_n$ *and* $\mathrm{Limsup}_{n\to\infty}K_n$ *is the set* of cluster points of sequences $x_n \in K_n$, *i.e., of limits of subsequences $x_{n'} \in K_{n'}$.*

In other words, upper limits are "thick" cluster points and lower limits are "thick" limits.

18.4.2 Upper Semicompact and Upper and Lower Semi-Continuous Maps

Definition 18.4.3 ***[Upper and Lower Semicontinous Maps]*** *Let us consider a set-valued map $F : X \rightsquigarrow Y$.*

- Lower semicontinuous *at $x \in \mathrm{Dom}(F)$ if and only if for any $y \in F(x)$ and for any sequence of elements $x_n \in \mathrm{Dom}(F)$ converging to x, there exists a sequence of elements $y_n \in F(x_n)$ converging to y. It is said to be* lower semicontinuous *if it is lower semicontinuous at every point $x \in \mathrm{Dom}(F)$.*
- Upper semicompact *at x if for every sequence $x_n \in \mathrm{Dom}(F)$ converging to x and for every sequence $y_n \in F(x_n)$, there exists a subsequence y_{n_p} converging to some $y \in F(x)$. It is said to be* upper semicompact *if it is upper semicompact at every point $x \in \mathrm{Dom}(F)$.*
- Upper semicontinuous *at x if for every $\varepsilon > 0$ there exists an $\eta(\varepsilon, x) \leq \varepsilon$ such that*

$$\forall y \in B(x, \eta(\varepsilon, y)),\ F(y) \subset F(x) + \varepsilon B$$

Examples The set-valued map F_1 defined by

$$F_1(x) := \begin{cases} [-1,+1] & \text{if } x \neq 0 \\ \{0\} & \text{if } x = 0 \end{cases}$$

is lower semicontinuous at zero but not upper semicompact at zero.

The set-valued map $F_2 : \mathbb{R} \rightsquigarrow \mathbb{R}$ defined by

$$F_2(x) := \begin{cases} \{0\} & \text{if } x \neq 0 \\ [-1,+1] & \text{if } x = 0 \end{cases}$$

is upper semicompact at zero but not lower semicontinuous at zero. □

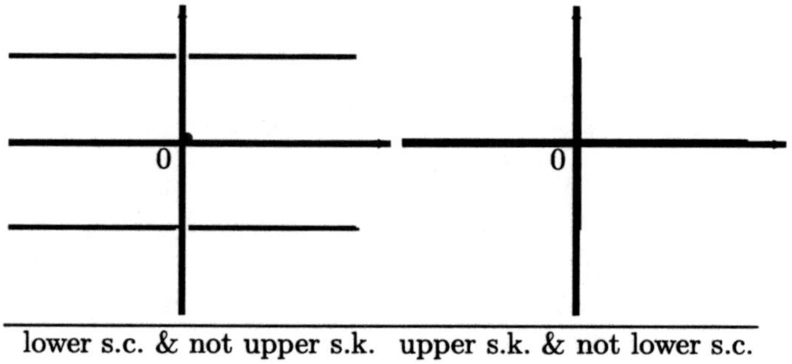

Fig. 18.3 Semicontinuous and Noncontinuous Maps.
"lower s.c. stands for lower semicontinuous and "upper s.k." for upper semicompact.

The connections between semi-continuity of set-valued maps and set limits are given by

Proposition 18.4.4 ***[Semicontinuity and Set Limits]*** *A point* (x, y) *belongs to the* closure of the graph *of a set-valued map* $F : X \rightsquigarrow Y$ *if and only if*

$$y \in \text{Limsup}_{x' \to x} F(x')$$

and F *is* lower semicontinuous *at* $x \in \text{Dom}(F)$ *if and only if*

$$F(x) \subset \text{Liminf}_{x' \to x} F(x')$$

Any upper semicompact map is closed (its graph is closed) and the converse statement is true if the images of the set-valued map remain in a compact subset.

Terminological Remarks. This proposition led several authors to call upper semicontinuous maps the ones which are closed in our terminology. Since upper semicontinuity is polysemous, *Terry Rockafellar* proposed to call *outer semicontinuous* at some point x the set-valued maps satisfying

$$\text{Limsup}_{x' \to x} F(x') \subset F(x)$$

in such a way that a set-valued map is outer semicontinuous if and only if is closed. *Inner semicontinuity* is synonymous of lower semicontinuity. □

Lemma 18.4.5 ***[Truncation of a Set-Valued Map]*** *Let* $F : X \rightsquigarrow Y$ *be a* closed *set-valued map and* $r : X \mapsto \mathbb{R}$ *be a bounded upper semicontinuous function. If the* dimension of Y is finite, *then the set-valued map* $F_r : X \rightsquigarrow Y$

defined by

$$F_r(x) := F(x) \cap r(x)B \tag{18.7}$$

is upper semicompact.

The following stability statements are of current use:

Definition 18.4.6 *[Marginal Functions] Consider a set-valued map* $F : X \rightsquigarrow Y$ *and a function* $\mathbf{u} : \text{Graph}(F) \mapsto \mathbb{R}$. *We associate with them the* marginal function $\mathbf{v}^\sharp : X \mapsto \mathbb{R}$ *defined by*

$$\mathbf{v}^\sharp(x) := \sup_{y \in F(x)} \mathbf{u}(x, y)$$

We deduce the continuity properties of the marginal maps.

Theorem 18.4.7 *[Maximum Theorem] Let metric spaces* X, Y, *a set-valued map* $F : X \rightsquigarrow Y$ *and a function* $\mathbf{u} : \text{Graph}(F) \mapsto \mathbb{R}$ *be given.*

1. *If* $\mathbf{u}$ *is lower semicontinuous and* F *is lower semicontinuous, so is the marginal function* $\mathbf{v}^\sharp$.
2. *If* $\mathbf{u}$ *is upper semicontinuous and* F *is upper semicontinuous, so is the marginal function* $\mathbf{v}^\sharp$.

Proof. See the proof of Theorem 1.4.16 of *Set-Valued Analysis*, [27, Aubin & Frankowska]. □

If a set-valued map F is lower semicontinuous (*resp.* upper semicompact), then the function $(x, y) \mapsto d(y, F(x))$ is upper semicontinuous (*resp.* lower semicontinuous).

18.4.3 Tangent Cones

Tangential characterizations

$$\begin{cases} (i)\ \forall\, x \in K \setminus C, & F(x) \cap T_K^{\star\star}(x) \neq \emptyset \\ (ii)\ \forall\, x \in K \setminus C, & F(x) \subset T_K^{\star\star}(x) \end{cases}$$

of viability and invariance properties studied in Chap. 11, p. 437 involve tangent cones:

Definition 18.4.8 *[Tangent and Convexified Tangent Cones] Let $K \subset X$ be a subset of a normed vector space X and $x \in K$. Since the tangent cone $T_K(x)$ is the set of elements v such that there exists a sequence of elements $h_n > 0$ converging to 0 and a sequence of $v_n \in X$ converging to v satisfying*

$$\forall n \geq 0, \ x + h_n v_n \in K$$

we deduce that the tangent cone $T_K(x)$ is the upper limit of the subsets $(K - x)/h$ *(regarded as "set differential quotients")*

$$T_K(x) := \text{Limsup}_{h \to 0+} \frac{K - x}{h}$$

We denote by

$$T_K^{\star\star}(x) \ := \ \overline{co}(T_K(x))$$

its closed convex hull, called the convexified tangent cone.

Therefore $T_K(x)$ is always a closed cone of "tangent directions". It is a vector subspace whenever K is a smooth manifold.

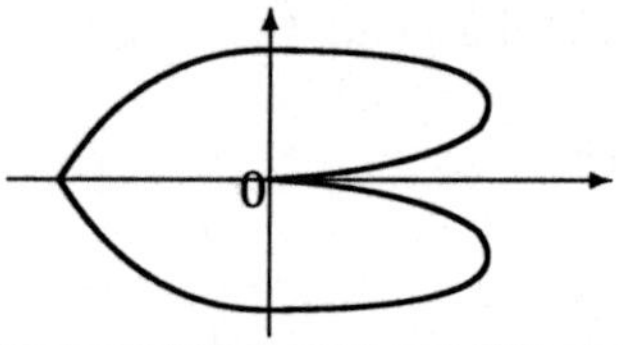

Subset K such that $T_K(0) = X$.

Fig. 18.4 Contingent cone at a boundary point may be the entire space.

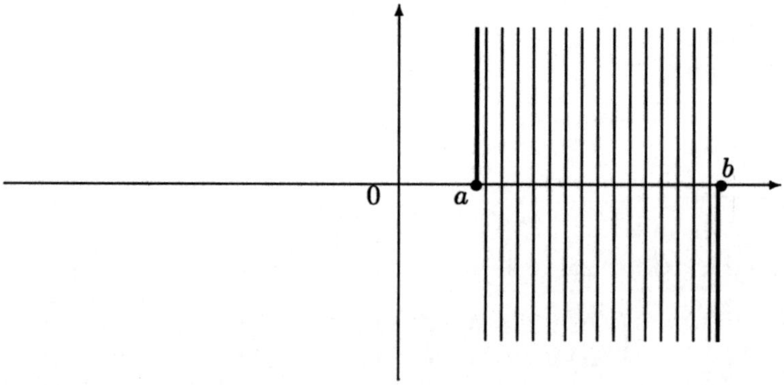

Fig. 18.5 The Graph of $T_{[a,b]}(\cdot)$

The tangent cone $T_{[a,b]}(x) = \mathbb{R}$ *if* $a < x < b$, $T_{[a,b]}(a) = \mathbb{R}_+$ *and* $T_{[a,b]}(b) = \mathbb{R}_-$. *The graph of the tangent cone is locally compact around* $(a, 0)$ *and* $(b, 0)$.

Definition 18.4.9 *[**Sleek Subsets**] A subset K is said to be sleek at* $x \in K$ *if* $T_K(\cdot)$ *is lower semicontinuous at this point.*

Tangent cones to sleek subsets are convex:

Theorem 18.4.10 *[**Convexity of Tangent Cones to Convex and Sleek Subsets**] If K is sleek at x, then the tangent cone* $T_K(x)$ *is a closed convex cone.*

Proof. See the proof of Theorem 4.1.10 of *Set-Valued Analysis*, [27, Aubin & Frankowska]. □

For convex subsets K, the tangent cone coincides with the closed cone spanned by $K - x$:

Proposition 18.4.11 *[**Tangent Cones to Convex Subsets**] Let us assume that K is convex. Then K is sleek and the tangent cone* $T_K(x)$ *to K at x is equal*

$$T_K(x) = \overline{\bigcup_{h>0} \frac{K - x}{h}}$$

Terminological Remarks. As mentioned in Chap. 11, p. 437, tangent cones have long been called *contingent cones* or *Bouligand tangent cones.* The two decades 1960–1980s saw the eclosion of many other definitions bearing the names of as many authors, including the *Clarke tangent cone*, which is always convex (which explains its popularity), to the price of often being trivial (i.e., reduced to $\{0\}$). It happens that the fundamental theorems (Viability Theorem 11.3.4, p. 455 and the Inverse Function Theorem 9.7.1, p. 366 among them) do not use Clarke tangent cones, but simply the contingent ones. Theorem 18.4.10, p. 733 stating that the tangent cones to sleek subsets of finite dimensional vector space coincide with Clarke tangent cones as well as Theorem 11.2.7, p. 447 do not require the convexity of the tangent cones. □

18.4.4 Polar Cones

In Convex Analysis, vector subspaces are replaced by cones (*where subtraction is forbidden*) and orthogonal subspaces by "polar cones":

Definition 18.4.12 *[Polar Cones] We associate with any subset $P \subset X$ of a finite dimensional vector space X its* polar cone *$P^\star \subset X^\star$ defined by*

$$p \in P^\star \textit{ if and only if } \forall v \in P,\ \langle p, v\rangle \leq 0$$

The polar cone is usually denoted $P^\star := P^-$ to underline the choice of the inequality. When P is a vector subspace, then $P^\star = P^\perp$ is the orthogonal space of P.

The Bipolar Theorem characterizes the closed convex cone spanned by a subset:

Theorem 18.4.13 *[Bipolar Theorem] The polar $P^\star$ of a subset $P \subset X$ is a closed convex cone and its* bipolar cone *is the cone $P^{\star\star}$. The* bipolar cone *$P^{\star\star} = \overline{co}(P)$ coincides with the closed convex cone spanned by a subset P. Consequently, P is a closed convex cone if and only if $P^{\star\star} = P$.*

See the proof of Theorem 2.4.3 of *Set-Valued Analysis*, [27, Aubin & Frankowska]. □

We recall the characterization of the tangent cone to the image by a linear operator:

Theorem 18.4.14 *[Tangent Cone to Image] Let $A \in \mathcal{L}(X, Y)$ be a linear operator, $K \subset X$ and $x_0 \in K$. Then*

$$AT_K(x) \subset T_{A(K)}(Ax)$$

If we assume that for some $x_0 \in K$, there exist constants $c > 0$ and $\alpha \in [0, 1[$ such that, for any $x \in K \cap B(x_0, \eta)$,

$$\forall v \in Y,\ \exists u \in T_K(x) \cap \|u\|B \text{ such that } \|Au - v\| \leq \alpha\|u\| \quad (18.8)$$

then equality

$$AT_K(x) = T_{A(K)}(Ax)$$

holds true.

Proof. The first inclusion is obvious and for proving the converse, we take $v \in T_{A(K)}(Ax)$. Then, there exist $h_n > 0$ converging to 0 and v_n converging to v such that $Ax + h_n v_n \in A(K)$. Then Constrained Inverse Function Theorem 3.4.5, p. 96 of *Set-Valued Analysis*, [27, Aubin & Frankowska] states that assumption (18.8), p. 734 implies that there exist constants $\gamma > 0$ and $\lambda > 0$ such that, for any $(x, Ax) \in (K \times A(K)) \cap B((x_0, Ax_0), \gamma)$,

$$\exists\, \xi \in K \text{ such that } A\xi \ = \ Ax + h_n v_n \text{ and } \|\xi - x\| \ \leq \ \lambda h_n \|v_n\|$$

Setting $u_n := \dfrac{\xi - x}{h_n}$, we infer that $Au_n = v_n$ and that $\|u_n\| \leq \lambda\|v_n\| \leq \|v\|+1$ for n large enough. Since the dimension of X is finite, this implies that there exists a subsequence (again denoted by) u_n converging to some u belonging to $T_K(x) \cap \lambda\|v\|B$ satisfying $Au = v$. □

18.4.5 Projectors

Definition 18.4.15 *[Projector] Let $K \subset X$ be a closed subset. The subset $\Pi_K(y)$ denotes the set of best approximations of y by elements of K:*

$$\Pi_K(y) := \{x \in K \textit{such that } \|y - x\| = \inf_{v \in K} \|y - v\|\}$$

We shall say that the set-valued map Π_K is the projector (of best approximation) *onto K.*

When the subset K is closed and convex, the projector is a single-valued map:

Theorem 18.4.16 *[The Projection Theorem] Let $K \subset X$ be a closed convex set of a finite dimensional vector space X supplied with a norm derived from the Euclidian scalar product $\langle \cdot, \cdot \rangle$ for instance. Then the projector Π_K is single-valued and is characterized by*

$$x = \Pi_K(y) \textit{ if and only if } x \in K \text{ and } \forall z \in K,\ \langle x - y, x - z\rangle \leq 0$$

The Moreau Theorem extends to the closed convex cones the Projection Theorem:

Theorem 18.4.17 *[The Moreau Projection Theorem] Let X be a finite dimensional vector space (a Hilbert space for a scalar product, identified with its dual), $P \subset X$ be a closed convex cone.*

Let us consider the projector Π_P onto a closed convex cone P and the projector $\Pi_{P^\star}$ onto its polar cone $P^\star$. Then the two following properties are equivalent:

$$\forall x \in X, \begin{cases} (i) & x = v + p \text{ where } v \in P, \ p \in P^\star \ \& \ \langle p, v \rangle = 0 \\ (ii) & v = \Pi_P(x) \text{ and } p = \Pi_{P^\star}(x) \end{cases} \tag{18.9}$$

Furthermore, the projectors satisfy

$$\forall x \in X, \ \|\Pi_P(x)\| \leq \|x\| \ \& \ \forall p \in X^\star, \ \|\Pi_{P^\star}(p)\|_\star \leq \|p\|_\star$$

We refer to *Neural networks and qualitative physics* [21, Aubin] for the definition and properties of orthogonal right inverses of a surjective linear operator:

Proposition 18.4.18 *[Orthogonal Right Inverse] Let us consider a surjective linear operator $A \in \mathcal{L}(X, Y)$ (for any $y \in Y$, the problem $Ax = y$ has at least a solution). We may select the solution $\overline{x}$ with minimal norm, i.e., a solution to the minimization problem with linear equality constraints*

$$A\overline{x} = y \ \& \ \|\overline{x}\| = \min_{Ax=y} \|x\|$$

The solution to this problem is given by the formula

$$\overline{x} = A^\star (AA^\star)^{-1} y$$

The operator $A^+ := A^\star(AA^\star)^{-1}$ is called the orthogonal right inverse of A.

We can supply the space Y with the final scalar product

$$((y_1, y_2))^A := \langle A^+ y_1, A^+ y_2 \rangle$$

and its associated final norm

$$\|y\|^A := \inf_{Ax=y} \|x\|$$

The orthogonal right inverse allows us to provide explicit formulas for quadratic minimization problems:

Proposition 18.4.19 ***[Orthogonal Right Inverse of a Subset]*** *Assume that $A \in \mathcal{L}(X,Y)$ is surjective and that $M \subset Y$ is closed and convex. Denote by Π_M^A the orthogonal projector on M when Y is supplied with the scalar product $((y_1, y_2))^A$. Let $u \in X$ and $v \in Y$ be given. Then the unique solution $\overline{x}$ to the minimization problem*

$$\inf_{Ax \in M+v} \|x - u\|$$

is equal to

$$\overline{x} = u - A^+(1 - \Pi_M^A)(Au - v)$$

When $M \subset Y$ is a closed convex cone, the solution can also be written in the form

$$\overline{x} = u - A^\star \Pi_{M^\star}^{A^\star}(AA^\star)^{-1}(Au - v)$$

For a proof, see Chap. 2 of *Neural networks and qualitative physics: a viability approach*, [21, Aubin].

18.4.6 Normals

For smooth subsets, the tangent space is a vector space and its orthogonal space is the normal space. For convex cones, vector subspaces are replaced by cones and orthogonal subspaces by "polar cones":

Definition 18.4.20 ***[Normal Cones]*** *We denote by*

$$P^\star \;:=\; \{p \in X^\star \text{ such that } \forall\, x \in P,\;\; \langle p, x\rangle \;\leq\; 0\}$$

the polar cone of P. The polar cone $N_K(x) := (T_K(x))^\star$ is called the normal cone *to K at x. The convexified tangent cone is defined by $T_K^{\star\star} = N_K^\star(x)$*

Terminological Remarks. The normal cone is also called the *Bouligand normal cone*, or the *contingent normal cone*, or also, the *sub-normal cone* and more recently, the *regular normal cone* by *Terry Rockafellar* and *Wets*. In this book, only (regular) normals are used, so that we shall drop the adjective "regular". □

Therefore, the tangential conditions characterizing viability and invariance properties are equivalent to the following normal conditions:

$$\begin{cases} (i)\;\; \forall\, x \in K \setminus C, \;\; \forall\, p \in N_K(x), \;\; \sigma(F(x), -p) \;\geq\; 0 \\ (ii)\; \forall\, x \in K \setminus C, \;\; \forall\, p \in N_K(x), \;\; \sigma(F(x), p) \;\leq\; 0 \end{cases}$$

of viability and invariance properties studied in Chap. 11, p. 437 involving tangent cones.

18.5 Graphical Analysis

18.5.1 Graphical Convergence of Maps

Pointwise, compact and uniform convergence of sequences of set-valued maps are ill adapted, whereas, being characterized by their graphs, convergence of graphs may, and indeed, does, provide convenient results:

Definition 18.5.1 *[Graphical Convergence] Let us consider a sequence of set-valued maps $F_n : X \rightsquigarrow Y$. The set-valued map $F^\sharp := \text{Lim}^\sharp{}_{n\to+\infty} F_n$ from X to Y defined by*

$$\text{Graph}(\text{Lim}^\sharp{}_{n\to+\infty} F_n) := \text{Limsup}_{n\to\infty}\text{Graph}(F_n)$$

is called the (graphical) upper limit *of the set-valued maps F_n.*

Even for single-valued maps, this is a weaker convergence than the pointwise convergence:

Proposition 18.5.2 *[Graphical and Pointwise Convergence]*

1. *If $f_n : X \mapsto Y$ converges pointwise to f, then, for every $x \in X$, $f(x) \in f^\sharp(x)$. If the sequence $\{f_n\}_{n\geq 0}$ is equicontinuous, then $f^\sharp(x) = \{f(x)\}$.*
2. *Let $\Omega \subset \mathbb{R}^n$ be an open subset. If a sequence $f_n \in L^p(\Omega)$ converges to f in $L^p(\Omega)$, then*

$$\text{for almost all } x \in \Omega,\ f(x) \in f^\sharp(x)$$

3. *If a sequence $f_n \in L^p(\Omega)$ converges* weakly *to f in $L^p(\Omega)$, then*

$$\text{for almost all } x \in \Omega,\ f(x) \in \overline{\text{co}} f^\sharp(x)$$

18.5.2 Derivatives of Set-Valued Maps

Let $F : X \rightsquigarrow Y$ be a set-valued map. Graphical derivatives of set-valued maps have been introduced at the end of the 1970s in the framework of viability theory (see for instance *Applied nonlinear analysis*, [26, Aubin & Ekeland], and *Set-valued analysis*, [27, Aubin & Frankowska]).

We introduce the *differential quotients*

$$u \rightsquigarrow \nabla_h F(x,y)(u) := \frac{F(x+hu)-y}{h}$$

of a set-valued map $F : X \rightsquigarrow Y$ at $(x, y) \in \text{Graph}(F)$.

Following *Gottfried Leibniz*, we propose the following definition of derivative of a set-valued map as a limit of its differential quotients, but for the graphical limit:

Definition 18.5.3 ***[Derivative of a Set-Valued Map]*** *The* (contingent graphical) derivative $DF(x,y)$ *of* F *at* $(x,y) \in \text{Graph}(F)$ *is the graphical upper limit of differential quotients:*

$$DF(x,y) := \text{Lim}^{\sharp}{}_{h \to 0+} \nabla_h F(x,y) \tag{18.10}$$

A set-valued map F *is said to be* differentiable *at* (x,y) *if* $\text{Dom}(DF(x,y)) = X$.

In other words, v belongs to $DF(x,y)(u)$ if and only if there exist sequences $h_n \to 0^+$, $u_n \to u$ and $v_n \to v$ such that $\forall n \geq 0, \quad y + h_n v_n \in F(x + h_n u_n)$.

It is easy to check that *a Lipschitz set-valued map is differentiable.*

In particular, if $f : X \mapsto Y$ is a single valued function, we set $Df(x) = Df(x, f(x))$.

On the other hand, *Pierre de Fermat* introduced, before *Gottfried Leibniz*, the concept of derivatives, stating that the graph of the differential $f'(x)$ of f at x is the tangent space to the graph of f at the point (x, y) where $y = f(x)$. We deduce the fundamental formula on the graph of the contingent derivative stating their equivalence:

Theorem 18.5.4 ***[Fermat Formulation of Derivatives]*** *The graph of the contingent derivative of a set-valued map is the tangent cone to its graph: for all* $(x,y) \in \text{Graph}(F)$,

$$\text{Graph}(DF(x,y)) = T_{\text{Graph}(F)}(x,y) \tag{18.11}$$

Proof. Indeed, we know that the tangent cone

$$T_{\mathrm{Graph}(F)}(x,y) = \mathrm{Limsup}_{h\to 0+} \frac{\mathrm{Graph}(F) - (x,y)}{h}$$

is the upper limit of the differential quotients $\frac{\mathrm{Graph}(F)-(x,y)}{h}$ when $h \to 0+$. It is enough to observe that

$$\mathrm{Graph}(\nabla_h F(x,y)) = \frac{\mathrm{Graph}(F) - (x,y)}{h}$$

and to take the upper limit to conclude. □

Definition 18.5.5 *[**Convexified Derivative**] The* convexified derivative *$D^{\star\star}F(x,y)$ of the set-valued map F at (x,y) is defined by*

$$T^{\star\star}_{\mathrm{Graph}(F)}(x,y) \ =: \ \mathrm{Graph}(D^{\star\star}F(x,y))$$

where $T^{\star\star}_K(x)$ is the closed convex hull to the tangent cone $T_K(x)$ to K at x.

The co-derivative *$D^\star F(x,y) : Y^\star \rightsquigarrow X^\star$ defined by*

$$\forall\, q \in Y^\star, \ \ p \in D^\star F(x,y)(q) \textit{ if and only if } (p,-q) \in N_{\mathrm{Graph}(F)}(x,y)$$

We can easily compute the derivative of the inverse of a set-valued map F (or even of a non injective single-valued map): *The contingent derivative of the inverse of a set-valued map F is the inverse of the contingent derivative*:

$$D(F^{-1})(y,x) \ = \ DF(x,y)^{-1}$$

If K is a subset of X and f is a single-valued map which is Fréchet differentiable around a point $x \in K$, then *the contingent derivative of the restriction of f to K is the restriction of the derivative to the tangent cone*:

$$D(f|_K)(x) = D(f|_K)(x, f(x)) = f'(x)|_{T_K(x)}$$

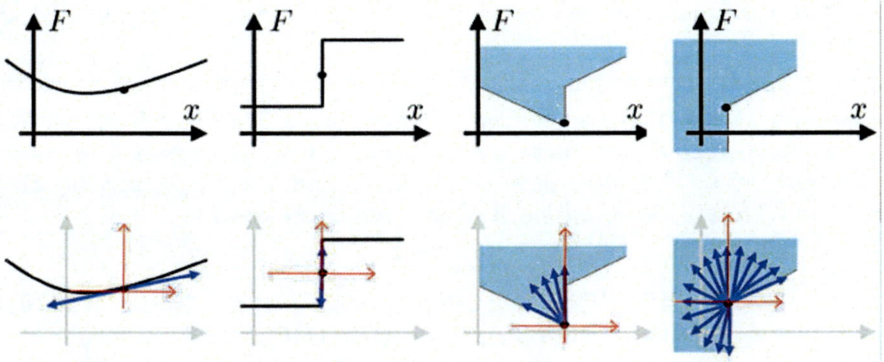

Fig. 18.6 Examples of Contingent Derivatives.
The upper row displays the point in the graph where the contingent derivative is taken. The lower row shows the graph of the corresponding derivative at this point.

Observe that

$$\mathrm{Dom}(DF(x,y)) \subset T_{\mathrm{Dom}(F)}(x)$$

and that, whenever $y \in \mathrm{Int}(K)$, then

$$DF(x,y)(u) := Y \text{ whenever } u \in T_{\mathrm{Dom}(F)}(x) \tag{18.12}$$

18.5.3 Derivative of Numerical Set-Valued Maps

Let us consider the case when $Y := \mathbb{R}$ and $F : X \rightsquigarrow \mathbb{R}$. Therefore, $DF(x,y)$ and $D^{\star\star}F(x,y)$ are maps from X to $\mathbb{R}$, the graphs of which are respectively a closed cone and a closed convex cone.

Definition 18.5.6 ***[Subdifferential of Numerical Map]*** *Let $F : X \rightsquigarrow \mathbb{R}$. The graph of the convexified derivative $D^{\star\star}F(x,y)$ being a closed convex cone, the images $D^{\star\star}F(x,y)(u) \subset \mathbb{R}$ are closed intervals denoted by*

$$D^{\star\star}F(x,y) := [d_{\uparrow}(F(x,y))(u), d_{\downarrow}(F(x,y))(u)] \tag{18.13}$$

The subdifferential $\partial_{-}F(x,y)$ and superdifferential $\partial_{+}F(x,y)$ of the numerical set-valued map F are respectively defined by

$$\begin{cases} (i)\ \partial_{-}F(x,y) := \{p \in X^{\star} \text{ such that } \forall\, u,\ \langle p,u\rangle \le d_{\uparrow}(F(x,y))(u)\} \\ (ii)\ \partial_{+}F(x,y) := \{p \in X^{\star} \text{ such that } \forall\, u,\ d_{\downarrow}(F(x,y))(u) \le \langle p,u\rangle\} \end{cases} \tag{18.14}$$

With these notations, we can compute the co-derivative of the numerical map F:

Lemma 18.5.7 *[Co-Derivative of Numerical Map] Let $F : X \rightsquigarrow \mathbb{R}$. The graph of $D^\star F(x,y)$ being a closed convex cone, it is enough to know the value of $D^\star F(x,y)(\lambda)$ at $\lambda := +1$, $\lambda = 0$ and $\lambda = -1$ for deriving the values $D^\star F(x,y)(\lambda)$ of all $\lambda \in \mathbb{R}$. These values are equal to*

$$\left\{\begin{array}{lcl} (i) \; D^\star(F(x,y))(+1) & = & \partial_- F(x,y) \\ (ii) \; D^\star(F(x,y))(-1) & = & -\partial_+ F(x,y) \\ (iii) \; D^\star(F(x,y))(0) & = & (\mathrm{Dom}(DF(x,y)))^\star \end{array}\right. \tag{18.15}$$

Note that $D^\star(F(x,y))(0) := \{0\}$ whenever F is differentiable at (x,y). Furthermore, for any $y \in \mathrm{Int}(F(x))$, the co-derivative $D^\star F(x,y)(\lambda) = \emptyset$ are empty for $\lambda \neq 0$.

Proof. To say that $p \in D^\star F(x,y)(+1)$ means that for any $u \in X$, $\forall\, v \in [d_\uparrow(F(x,y))(u), d_\downarrow(F(x,y))(u)]$, $\langle p,u\rangle \le (+1)v$. Therefore, $p \in D^\star F(x,y)(+1)$ if and only if $\forall\, u \in X$, $\langle p,u\rangle \le d_\downarrow(F(x,y))(u)$. In the same way, one can check that $p \in D^\star F(x,y)(-1)$ if and only if $\forall\, u \in X$, $d_\downarrow(F(x,y))(u)(x,y)(u) \le \langle -p,u\rangle$. Finally, $p \in D^\star F(x,y)(0)$ means simply that $\forall\, u \in X$, $\langle p,u\rangle \le 0$, i.e., that $p \in (\mathrm{Dom}(DF(x,y)))^\star$, the polar cone of the domain of the derivative of F. Consequently, if F is *differentiable* in the sense that when $\mathrm{Dom}(DF(x,y)) = X$, then $D^\star F(x,y)(0) = \{0\}$.

Property (18.12), p. 741 implies that whenever $y \in \mathrm{Int}(F(x))$ and $u \in T_{\mathrm{Dom}(F)}(x)$, then $d_\downarrow(F(x,y))(u) = -\infty$ and $d_\uparrow(F(x,y))(u) = +\infty$. Hence, the co-derivative $D^\star F(x,y)(\lambda) = \emptyset$ are empty for $\lambda \neq 0$. □

18.6 Epigraphical Analysis

18.6.1 Extended Functions

Extended functions have been introduced in Definition 4.2.1, p. 131.

Since the order relation on the real numbers is involved in the definition of the Lyapunov property (as well as in minimization problems and other dynamical inequalities), we no longer characterize a real-valued function by its graph, but rather by its *epigraph*

$$\mathcal{E}p(\mathbf{v}) := \{(x,\lambda) \in X \times \mathbb{R} \mid \mathbf{v}(x) \le \lambda\}$$

(see Definition 4.2.2, p. 131).

A function is said to be *nontrivial* if its domain is not empty. Any function $\mathbf{v}$ defined on a subset $K \subset X$ can be regarded as the extended function $\mathbf{v}_K$ equal to $\mathbf{v}$ on K and to $+\infty$ outside of K, whose domain is K.

The graph of a real-valued (finite) function is then the intersection of its epigraph and its hypograph.

Definition 18.6.1 ***[Indicator of a Set]*** Indicators ψ_K of subsets K *are cost functions defined by*

$$\psi_K(x) := 0 \text{ if } x \in K \text{ and } +\infty \text{ if not}$$

which characterize subsets (as characteristic functions $\chi_K(x) := e^{-\psi_K(x)}$ *do for other purposes) provide important examples of extended functions.*

Since

$$\mathcal{E}p(\psi_K) = K \times \mathbb{R}_+$$

we deduce that the indicator ψ_K is lower semicontinuous if and only if K is closed and that ψ_K is convex if and only if K is convex. One can regard the sum $\mathbf{v} + \psi_K$ as the restriction of $\mathbf{v}$ to K.

Remark: Toll sets. The indicator ψ_K can be regarded as a *membership cost* to K: it costs nothing to belong to K, and $+\infty$ to step outside of K. In the same context, positive extended functions $\mathbf{v} : X \mapsto [0, +\infty]$ can be regarded as some kind of *fuzzy sets*, called *toll sets*, since their membership cost is not only 0 or $+\infty$, but can also be any positive cost. The set of x such that $\mathbf{v}(x) = 0$ is called the *core of the toll set* $\mathbf{v}$, the set of x such that $\mathbf{v}(x) = +\infty$ the *complement of the toll set* $\mathbf{v}$ and the set of x such that $0 < \mathbf{v}(x) < +\infty$ the *boundary of the toll set* $\mathbf{v}$. Definition 15.3.1, p.621 of constraint function in economics provides can be reformulated by saying the constrained tube is a "toll tube". □

We also remark that some properties of a function are actually properties of their epigraphs and λ*-lower section* or λ*-lower level set*

$$\mathbf{L}^{\leq}_{\mathbf{v}}(\lambda) \;:=\; \{x \in K \text{ such that } \mathbf{v}(x) \leq \lambda\}$$

(see Definition 10.9.5, p.429). For instance, *an extended function* $\mathbf{v}$ *is convex (resp. positively homogeneous) if and only if its epigraph is convex (resp. a cone).*

Definition 18.6.2 *[Lower Semicontinuous and inf-Compact Extended Functions] An extended function* $\mathbf{v}$ *is said to be* lower semicontinuous *if its lower level-set* $\mathbf{L}^{\leq}_{\mathbf{v}}(\lambda)$ *are closed for all* $\lambda \in \mathbb{R}$*. It is called* inf-compact *if furthermore there exists* $\lambda \in \mathbb{R}$ *such that its lower level-set* $\mathbf{L}^{\leq}_{\mathbf{v}}(\lambda)$ *is nonempty and compact.*

Lemma 18.6.3 *[Lower Semicontinuous Functions] Consider an extended function* $\mathbf{v} : X \mapsto \mathbb{R} \cup \{\pm\infty\}$*. The conditions are equivalent:*

1. *The epigraph of* $\mathbf{v}$ *is closed*
2. *For all* $\lambda \in \mathbb{R}$*, the level-sets* $\mathbf{L}^{\leq}_{\mathbf{v}}(\lambda)$ *of* $\mathbf{v}$ *are closed*
3. *The function* $\mathbf{v}$ *satisfies:*

$$\forall x \in X, \ \mathbf{v}(x) \ \leq \ \liminf_{y \to x} \mathbf{v}(y)$$

Any lower semicontinuous and inf-compact extended function achieves it infimum at some $\overline{x} \in \mathrm{Dom}(\mathbf{v})$.

Proof. Assume that the epigraph of $\mathbf{v}$ is closed and pick $x \in X$. There exists a sequence of elements x_n converging to x such that

$$\lim_{n \to +\infty} \mathbf{v}(x_n) = \liminf_{x' \to x} \mathbf{v}(x')$$

Hence, for any $\lambda > \liminf_{x' \to x} \mathbf{v}(x')$, there exist N such that, for all $n \geq N$, $\mathbf{v}(x_n) \leq \lambda$, i.e., such that $(x_n, \lambda) \in \mathcal{E}p(\mathbf{v})$. By taking the limit, we infer that $\mathbf{v}(x) \leq \lambda$, and thus, that $\mathbf{v}(x) \leq \liminf_{x' \to x} \mathbf{v}(x')$. The converse statements are obvious.

Since at least one level set $\mathbf{L}^{\leq}_{\mathbf{v}}(\lambda_0)$ is nonempty and compact, the decreasing family of closed level-sets $\mathbf{L}^{\leq}_{\mathbf{v}}(\lambda)$ when $\inf_{x \in X} \mathbf{v}(x) < \lambda \leq \overline{\lambda}$ in the compact subset $\mathbf{L}^{\leq}_{\mathbf{v}}(\lambda_0)$ has a nonempty (compact) intersection $\bigcap_{\inf_{x \in X} \mathbf{v}(x) < \lambda \leq \overline{\lambda}} \mathbf{L}^{\leq}_{\mathbf{v}}(\lambda)$. This intersection is equal to the set of elements achieving the infimum of $\mathbf{v}$. □

Let us point out the following

Lemma 18.6.4 ***[Interior of an Epigraph]*** *A pair* $(x,\lambda) \in \text{Int}(\mathcal{E}p(\mathbf{v}))$ *belongs to the interior of the epigraph of* $\mathbf{v} : X \mapsto \mathbb{R} \cup \{+\infty\}$ *if and only if*

$$x \in \text{Int}(K) \text{ and } \limsup_{y \to x} \mathbf{v}(y) \; < \; \lambda \qquad (18.16)$$

Proof. Indeed, to say that $(x,\lambda) \in \text{Int}(\mathcal{E}p(\mathbf{v}))$ amounts to saying that there exist a neighborhood $N(x)$ of x and $\varepsilon > 0$ such that $(y,\mu) \in (\mathcal{E}p(\mathbf{v}))$ for all $y \in N(x)$ and $\mu \geq \lambda - \varepsilon$. This is equivalent to saying that $x \in \text{Int}(\mathcal{E}p(\mathbf{v}))$ and that

$$\limsup_{y \to x} \mathbf{v}(y) \; \leq \; \sup_{y \in N(x)} \mathbf{v}(y) \; \leq \; \lambda - \varepsilon \; < \; \lambda$$

Consequently, if $\mathbf{v}$ is upper semicontinuous in the interior of its domain, then the interior of its epigraph $\text{Int}(\mathcal{E}p(\mathbf{v}))$ is the set of pairs (x,λ) such that $\mathbf{v}(x) < \lambda$. □

We have defined the *epilevel function* of a tube $\mathbf{K} : \mathbb{R} \rightsquigarrow X$ in Definition 10.9.6, p. 430, defined by (10.26), p. 430:

$$\Lambda^{\uparrow}_{\mathbf{K}}(x) \; := \; \inf \{\lambda \text{ such that } x \in \mathbf{K}(\lambda)\} \; = \; \inf_{(\lambda,x) \in \text{Graph}(\mathbf{K})} \lambda$$

where $\text{Graph}(\mathbf{K})$ is a subset of $\mathbb{R}_+ \times X$. We adopt an analogous definition for subset of $X \times \mathbb{R}$:

Definition 18.6.5 ***[Epi-envelopes]*** *A subset* $\mathcal{M} \subset X \times \mathbb{R}$ *is an* epigraph *if*

$$\mathcal{M} + \{0\} \times \mathbb{R}_+ \; = \; \mathcal{M}$$

We associate with a subset $\mathcal{M} \subset X \times \mathbb{R}_+$ *its* epi-envelope $\mathbf{v}_{\mathcal{M}} : X \mapsto \mathbb{R}_+ \cup \{+\infty\}$ *defined by*

$$\mathbf{v}_{\mathcal{M}}(x) \; := \; \inf_{(x,w) \in \mathcal{M}} w$$

and its epi-closure *defined by*

$$\mathbf{v}_{\overline{\mathcal{M}}}(x) \; := \; \inf_{(x,w) \in \overline{\mathcal{M}}} w$$

If $\mathcal{M} := \mathcal{E}p(\mathbf{v})$*, its epi-envelope*

$$\widehat{\mathbf{v}}(x) := \mathbf{v}_{\mathcal{E}p(\mathbf{v})}(x) = \inf_{w \geq \mathbf{v}(x)} w$$

is called the epi-envelope *of* **v** *and its epi-closure the* epi-closure *of* **v**.

We observe that

$$\mathcal{M} + \{0\} \times \mathbb{R}_+ \subset \mathcal{E}p(\mathbf{v}_{\mathcal{M}}) \subset \overline{\mathcal{M} + \{0\} \times \mathbb{R}_+}$$

Lemma 18.6.6 ***[Epigraphs of epi-envelopes]*** *Let $\mathcal{M} \subset X \times \mathbb{R}_+$ be a closed subset. Then its epi-closure $\mathbf{v}_{\mathcal{M}}$ is lower semicontinuous and*

$$\mathcal{E}p(\mathbf{v}_{\mathcal{M}}) = \mathcal{M} + \{0\} \times \mathbb{R}_+$$

Proof. For proving that the epi-closure $\mathbf{v}_{\mathcal{M}}$ is lower semicontinuous whenever $\mathcal{M}$ is closed, let (x_n, λ_n) be a sequence of $\mathcal{E}p(\mathbf{v}_{\mathcal{M}})$ converging to (x, λ). We know that there exists $\alpha_n \leq \lambda_n + \frac{1}{n}$ such that (x_n, α_n) belongs to $\mathcal{M}$. Since $0 \leq \alpha_n \leq \lambda_n + \frac{1}{n}$, a subsequence (again denoted by) α_n converges to some $\alpha \leq \lambda$, and thus, (x, α) belongs to $\mathcal{M}$ and consequently, (x, λ) belongs to $\mathcal{E}p(\mathbf{v}_{\mathcal{M}})$. □

Let us consider a family of extended functions $\mathbf{v}_{i \in I}$ and their pointwise supremum $\sup_{i \in I} \mathbf{v}_i$ defined by

$$\forall x \in X, \ (\sup_{i \in I} \mathbf{v}_i)(x) := \sup_{i \in I}(\mathbf{v}_i(x))$$

and their pointwise infimum $\inf_{i \in I} \mathbf{v}_i$ defined by

$$\forall x \in X, \ (\inf_{i \in I} \mathbf{v}_i)(x) := \inf_{i \in I}(\mathbf{v}_i(x))$$

We first observe that

Lemma 18.6.7 ***[Pointwise suprema and infima]*** *The epigraph of the pointwise supremum $\mathbf{v} := \sup_{i \in I} \mathbf{v}_i$ of a family of functions $\mathbf{v}_i$ is the intersection of their epigraphs:*

$$\mathcal{E}p(\sup_{i \in I} \mathbf{v}_i) = \bigcap_{i \in I} \mathcal{E}p(\mathbf{v}_i)$$

and thus,

$$\mathbf{v}_{\bigcap_{i \in I} \mathcal{E}p(\mathbf{v}_i)} = \sup_{i \in I} \mathbf{v}_{\mathcal{M}_i}$$

Let $\mathcal{M}_{i\in I}$ be a family of subsets $\mathcal{M}_i \subset X\times\mathbb{R}$. Then the epilevel function of the union of the $\mathcal{M}_i$ is the infimum of the epilevel functions of the set $\mathcal{M}_i$:

$$\mathbf{v}_{\bigcup_{i\in I}\mathcal{M}_i} = \inf_{i\in I}\mathbf{v}_{\mathcal{M}_i}$$

Proof. The first statement is obvious. For proving the second one, we observe

$$\begin{cases} \mathbf{v}_{\bigcup_{i\in I}\mathcal{M}_i}(x) = \inf_{(x,y)\in\bigcup_{i\in I}\mathcal{M}_i} y \\ = \inf_{i\in I}\inf_{(x,y)\in\mathcal{M}_i} y = \inf_{i\in I}\mathbf{v}_{\mathcal{M}_i}(x) \end{cases} \quad \square$$

18.6.2 Epidifferential Calculus

Definition 18.6.8 ***[Epilimits]*** *The epigraph of the* lower epilimit $\lim^\sharp_{\uparrow n\to+\infty}\mathbf{u}_n$ *of a sequence of extended functions* $\mathbf{u}_n : X \mapsto \mathbb{R}\cup\{+\infty\}$ *is the upper limit of the epigraphs:*

$$\mathcal{E}p(\lim^\sharp_{\uparrow n\to+\infty}\mathbf{u}_n) := \text{Limsup}_{n\to+\infty}\mathcal{E}p(\mathbf{u}_n)$$

The function $\lim^\flat_{\uparrow n\to+\infty}\mathbf{u}_n$ *whose epigraph is the lower limit of the epigraphs of the functions* $\mathbf{u}_n$

$$\mathcal{E}p(\lim^\flat_{\uparrow n\to+\infty}\mathbf{u}_n) := \text{Liminf}_{n\to+\infty}\mathcal{E}p(\mathbf{u}_n)$$

is the upper epilimit of the functions $\mathbf{u}_n$

One can check that

$$\left(\lim^\sharp_{\uparrow n\to+\infty}\mathbf{u}_n\right)(x_0) = \liminf_{n\to\infty, x\to x_0}\mathbf{u}_n(x)$$

When $\mathbf{u}$ is an extended function, we associate with it its epigraph and the tangent cones to this epigraph. This leads to the concept of epiderivatives of extended functions.

Definition 18.6.9 ***[Epiderivatives]*** *Let* $\mathbf{u} : X \mapsto \mathbb{R}\cup\{\pm\infty\}$ *be a nontrivial extended function and* x *belong to its domain.*

We associate with it the differential quotients

$$u \rightsquigarrow \nabla_h \mathbf{u}(x)(u) := \frac{\mathbf{u}(x+hu) - \mathbf{u}(x)}{h}$$

The epiderivative $D_\uparrow \mathbf{u}(x)$ of $\mathbf{u}$ at $x \in \mathrm{Dom}(\mathbf{u})$ is the lower epilimit of its differential quotients:

$$D_\uparrow \mathbf{u}(x) := \mathrm{lim}^\sharp_{\uparrow h \to 0+} \nabla_h \mathbf{u}(x) \qquad (18.17)$$

We shall say that the function $\mathbf{u}$ is contingently epidifferentiable *at x if for any $u \in X$, $D_\uparrow \mathbf{u}(x)(u) > -\infty$ (or, equivalently, if $D_\uparrow \mathbf{u}(x)(0) = 0$).*

The alternative definition is a generalization of Fermat's concept of derivative:

Theorem 18.6.10 ***[Fermat Formulation of Epiderivatives]*** *Let $\mathbf{u} : X \mapsto \mathbb{R} \cup \{\pm\infty\}$ be a nontrivial extended function and x belong to its domain. The epigraph of the epiderivative $D_\uparrow \mathbf{u}(\cdot)$ is equal to the tangent cone to the epigraph of $\mathbf{u}$ at $(x, \mathbf{u}(x))$ is*

$$\mathcal{E}p(D_\uparrow \mathbf{u}(x)) = T_{\mathcal{E}p(\mathbf{u})}(x, \mathbf{u}(x)) \qquad (18.18)$$

Proof. The first statement is obvious. For proving the second one, we recall that the tangent cone

$$T_{\mathcal{E}p(\mathbf{u})}(x, \mathbf{u}(x)) = \mathrm{Limsup}_{h \to 0+} \frac{\mathcal{E}p(\mathbf{u}) - (x, \mathbf{u}(x))}{h}$$

is the upper limit of the differential quotients $\frac{\mathcal{E}p(\mathbf{u}) - (x, \mathbf{u}(x))}{h}$ when $h \to 0+$. It is enough to observe that

$$\mathcal{E}p(\nabla_h \mathbf{u}(x)) = \frac{\mathcal{E}p(\mathbf{u}) - (x, \mathbf{u}(x))}{h}$$

to conclude that

$$\mathcal{E}p(D_\uparrow \mathbf{u}(x)) := T_{\mathcal{E}p(\mathbf{u})}(x, \mathbf{u}(x)) \quad \square$$

Consequently, *the epigraph of the epiderivative at x is a closed cone. It is then lower semicontinuous and positively homogeneous whenever $\mathbf{u}$ is contingently epidifferentiable at x.*

We observe that the epiderivative of the indicator function ψ_K at $x \in K$ is the indicator of the tangent cone to K at x:

$$D_\uparrow \psi_K(x) = \psi_{T_K(x)}$$

making precise the intuition stating that the tangent cone $T_K(x)$ plays the role of a "derivative of a set", as the limit of differential quotients $\frac{K-x}{h}$ of sets.

Let us mention the following pointwise translation of the definition of epiderivatives:

Lemma 18.6.11 ***[Pointwise Characterization of Epiderivatives]*** *The epiderivative $D_\uparrow \mathbf{u}(x)$ is equal to*

$$\forall u \in X,\ D_\uparrow \mathbf{u}(x)(u) \ = \ \liminf_{h \to 0+, u' \to u} \frac{\mathbf{u}(x+hu') - \mathbf{u}(x)}{h}$$

Proof. See the proof of Proposition 6.1.3 of *Set-Valued Analysis*, [27, Aubin & Frankowska]. □

As for derivatives of set-valued maps, since viability and invariance tangential conditions actually involve the closed convex hull of the tangent cone, we shall need the concept of convexified epiderivative:

Definition 18.6.12 ***[Convexified Epiderivative]*** *The* convexified epiderivative $D_\uparrow^{\star\star}\mathbf{u}(x)$ *of the extended function* $\mathbf{u}$ *at* $x \in \mathrm{Dom}(u)$ *is defined by*

$$\mathcal{E}p(D_\uparrow^{\star\star}\mathbf{u}(x)) \ := \ T_{\mathcal{E}p(\mathbf{u})}^{\star\star}(x, \mathbf{u}(x))$$

where $T_K^{\star\star}(x)$ is the closed convex hull to the tangent cone $T_K(x)$ to K at x.

Under adequate assumptions, *the epiderivative of the restriction of* $\mathbf{u}$ *to K is the restriction of the epiderivative to the tangent cone*:

Lemma 18.6.13 ***[Epiderivatives of a Restriction]*** *Let $K \subset X$ be a closed subset and $\mathbf{u} : X \mapsto \mathbb{R} \cup \{+\infty\}$ be an extended function. We denote by $\mathbf{u}|_K := f + \psi_K$ the restriction to $\mathbf{u}$ at K. Inequality*

$$D_\uparrow \mathbf{u}(x)|_{T_K}(x) \;\leq\; D_\uparrow \mathbf{u}|_K(x)$$

always holds true. It is an equality when **u** *is differentiable from the right.*

Proof. Indeed, let $x \in K \cap \mathrm{Dom}(\mathbf{u})$. If u belongs to $T_K(x)$, there exist $h_n \to 0+$ and $x_n := x + h_n u_n \in K$ such that

$$D_\uparrow \mathbf{u}(x)(u) \leq \liminf_{n \to +\infty} \frac{\mathbf{u}(x_n) - \mathbf{u}(x)}{h_n} = \liminf_{n \to +\infty} \frac{\mathbf{u}|_K(x_n) - \mathbf{u}|_K(x)}{h_n}$$

which implies the inequality. If $\mathbf{u}$ is differentiable from the right, the differential quotient converges to the common value $D_\uparrow \mathbf{u}(x) = D_\uparrow \mathbf{u}|_K(x) = D_\downarrow \mathbf{u}|_K(x)$. □

For locally Lipschitz functions, the epiderivatives are finite:

Proposition 18.6.14 *[Derivatives of Locally Lipschitz Functions] If* $\mathbf{u} : X \mapsto \mathbb{R} \cup \{+\infty\}$ *is Lipschitz around* $x \in \mathrm{Int}(\mathrm{Dom}(\mathbf{u}))$*, then the epiderivative* $D_\uparrow \mathbf{u}(x)$ *is Lipschitz: there exists* $\lambda > 0$ *such that*

$$\forall u \in X,\; D_\uparrow \mathbf{u}(x)(u) = \liminf_{h \to 0+} \frac{\mathbf{u}(x + hu) - \mathbf{u}(x)}{h} \leq \lambda \|u\|$$

Proof. Since $\mathbf{u}$ is Lipschitz on some ball $B(x, \eta)$, the above inequality follows immediately from

$$\forall u \in \eta B,\; \frac{\mathbf{u}(x + hu) - \mathbf{u}(x)}{h} \leq \frac{\mathbf{u}(x + hu') - \mathbf{u}(x)}{h} + \lambda(\|u\| + \|u' - u\|)$$

by taking the liminf when $h \to 0+$ and $u' \to u$. □

The Ekeland Theorems 3.3.1, p.94, and 6.1.12, p.226, of *Set-valued Analysis*, [27, Aubin & Frankowska] provide an approximate minimizer of a bounded from below lower semicontinuous function in a given neighborhood of a point. This localization property is very useful and explains the importance of this result which has been extensively used since its discovery in 1974.

Theorem 18.6.15 *[The Ekeland Variational Principle] Let*

$$\mathbf{v} : X \mapsto \mathbf{R} \cup \{+\infty\}$$

be a lower semicontinuous nontrivial extended bounded from below function defined on a complete metric space X. *Let* $x_0 \in \mathrm{Dom}(\mathbf{v})$ *and* $\varepsilon > 0$ *be fixed. Then there exists* $x_\varepsilon \in X$, *a solution to*

$$\begin{cases} (i) \;\; \mathbf{v}(x_\varepsilon) + \varepsilon d(x_0, x_\varepsilon) \;\leq\; \mathbf{v}(x_0) \\ \\ (ii) \; \forall\, x \neq x_\varepsilon, \;\; \mathbf{v}(x_\varepsilon) \;<\; \mathbf{v}(x) + \varepsilon d(x, x_\varepsilon) \end{cases} \tag{18.19}$$

When X *is a finite dimensional vector space, it can be written in the form: there exists* $x_\varepsilon \in X$ *satisfying*

$$\begin{cases} (i) \;\; \mathbf{v}(x_\varepsilon) + \varepsilon \|x_\varepsilon - x_0\| \;\leq\; \mathbf{v}(x_0) \\ \\ (ii) \; \forall\, u \;\in\; X, \;\; 0 \;\leq\; D_\uparrow \mathbf{v}(x_\varepsilon)(u) + \varepsilon \|u\| \end{cases} \tag{18.20}$$

18.6.3 Generalized Gradients

Definition 18.6.16 *[Subgradients] Let* $\mathbf{u} \,:\, X \;\mapsto\; \mathbb{R} \cup \{+\infty\}$ *be a nontrivial extended function. The continuous linear functionals* $p \in X^\star$ *satisfying*

$$\forall v \in X, \; \langle p, v \rangle \leq D_\uparrow \mathbf{u}(x)(v)$$

are called the (regular) subgradients *of* $\mathbf{u}$ *at* x, *which constitute the (possibly empty) closed convex subset*

$$\partial_- \mathbf{u}(x) := \{p \in X^\star \mid \forall v \in X, \;\; \langle p, v \rangle \;\leq\; D_\uparrow \mathbf{u}(x)(v)\}$$

called the (regular) subdifferential *of* $\mathbf{u}$ *at* x_0.

Naturally, when $\mathbf{u}$ is Fréchet differentiable at x, then

$$D_\uparrow \mathbf{u}(x)(v) = \langle f'(x), v \rangle$$

so that *the subdifferential* $\partial_- \mathbf{u}(x)$ *is reduced to the gradient* $\mathbf{u}'(x)$.

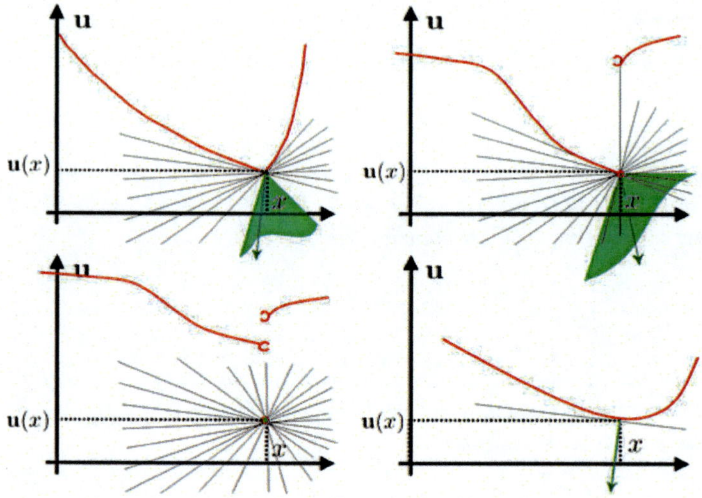

Fig. 18.7 Subgradients.
Illustration of subgradient for different class of functions, continuous, discontinuous, upper or lower semicontinuous.

We observe that

$$\partial_{-}\mathbf{u}(x) + N_K(x) \subset \partial(\mathbf{u}|_K)(x)$$

If $\mathbf{u}$ is differentiable at a point $x \in K$, then the *subdifferential of the restriction is the sum of the gradient and the normal cone*:

$$\partial_{-}(\mathbf{u}|_K)(x) = \mathbf{u}'(x) + N_K(x)$$

We also note that the subdifferential of the indicator of a subset is the normal cone:

$$\partial_{-}\psi_K(x) = N_K(x)$$

and that

$$\begin{cases} (i) \;\; (p,-1) \in N_{\mathcal{E}p(\mathbf{u})}(x,\mathbf{u}(x)) \text{ if and only if } p \in \partial_{-}\mathbf{u}(x) \\ (ii) \; (p,0) \in N_{\mathcal{E}p(\mathbf{u})}(x,\mathbf{u}(x)) \text{ if and only if } p \in \text{Dom}(D_{\uparrow}\mathbf{u}(x))^{-} \end{cases}$$

We also deduce that

$$N_{\mathcal{E}p(\mathbf{u})}(x,\mathbf{u}(x)) = \{\lambda(q,-1)\}_{q\in\partial_{-}\mathbf{u}(x),\,\lambda>0} \bigcup \{(q,0)\}_{q\in\text{Dom}(D_{\uparrow}\mathbf{u}(x))^{-}}$$

The subset $\text{Dom}(D_{\uparrow}\mathbf{u}(x))^{-} = \{0\}$ whenever the domain of the epiderivative $D_{\uparrow}\mathbf{u}(x)$ is dense in X. This happens when $\mathbf{u}$ is locally Lipschitz and when the dimension of X is finite:

Proposition 18.6.17 ***[Viscosity Characterization of Subdifferentials]*** *Let X be a finite dimensional vector space, $\mathbf{u} : X \mapsto \mathbb{R} \cup \{\pm\infty\}$ be a nontrivial extended function and $x_0 \in \mathrm{Dom}(\mathbf{u})$. Then the subdifferential $\partial_{-}\mathbf{u}(x)$ is the set of elements $p \in X^{\star}$ satisfying*

$$\liminf_{x \to x_0} \frac{\mathbf{u}(x) - \mathbf{u}(x_0) - \langle p, x - x_0 \rangle}{\|x - x_0\|} \geq 0 \tag{18.21}$$

Proof. See the proof of Proposition 6.4.8 of *Set-Valued Analysis*, [27, Aubin & Frankowska]. □

The equivalent formulation (18.21) of the concept of subdifferential has been introduced by Crandall & P.-L. Lions for defining *viscosity solutions* to Hamilton–Jacobi equations.

18.6.4 Tangent and Normal Cones to Epigraphs

We shall need the following technical Lemma 18.6.18 to prove that viability solutions of Hamilton–Jacobi–Bellman equations are actually contingent Frankowska solutions (see Theorem 17.4.1, p. 699) and Barron–Jensen/Frankowska solutions (see Theorem 17.4.3, p. 701):

Lemma 18.6.18 ***[Tangent and Normal Cones to Epigraphs]*** *Assume that $\mathbf{v} : X \mapsto \mathbb{R}_{+} \cup \{+\infty\}$ is a nontrivial extended function and that its convexified epiderivative (see Definition 18.6.12, p.749) $D_{\uparrow}^{\star\star}\mathbf{v}(x)(u)$ is finite.*

1. *if $w = \mathbf{v}(x)$, then, for any $\mu \geq D_{\uparrow}^{\star\star}\mathbf{v}(x)(u)$, the pair $(u, \mu) \in T_{\mathcal{E}p(\mathbf{v})}^{\star\star}(x, \mathbf{v}(x))$*
2. *if $w > \mathbf{v}(x)$, then, for any $\mu \in \mathbb{R}$, the pair (u, μ) belongs to the convexified tangent cone $T_{\mathcal{E}p(\mathbf{v})}^{\star\star}(x, w)$ to the epigraph of $\mathbf{v}$ at (x, w).*

Consequently, a pair (p, λ) belongs to the normal cone $N_{\mathcal{E}p(\mathbf{v})}(x, w)$ to the epigraph of $\mathbf{v}$ at (x, w) if and only if

1. *if $w = \mathbf{v}(x)$, then either*
   - *$\lambda = 0$ and thus $p \in (\mathrm{Dom}(D_{\uparrow}^{\star\star}\mathbf{v}(x)))^{-}$,*
   - *or $\lambda < 0$ and $\frac{p}{|\lambda|} \in \partial_{-}\mathbf{v}(x)$*

2. *if* $w > \mathbf{v}(x)$, *then* $\lambda = 0$ *and thus* $p \in (\mathrm{Dom}(D_{\uparrow}^{\star\star}\mathbf{v}(x)))^-$. *In particular, if the domain of* $D_{\uparrow}\mathbf{v}(x)$*is dense in* X, *then* $p = 0$. *This is the case whenever* $\mathbf{v}$ *is Lipschitz around* x.

Proof. We successively prove the following points:

1. Let (u, μ) belong to $T_{\mathcal{E}p(\mathbf{v})}(x, \mathbf{v}(x))$. Then we know that there exist sequences $h_n > 0$ converging to 0, u_n converging to u and μ_n converging to μ such that $(x + h_n u_n, \mathbf{v}(x) + h_n\mu_n)$ belongs to $\mathcal{E}p(\mathbf{v})$. Therefore, for $w > \mathbf{v}(x)$ and $\mu \in \mathbb{R}$ and h_n small enough,

$$(x+h_n u_n, w+h_n\mu) = (x+h_n u_n, \mathbf{v}(x)+h_n\lambda_n)+(0, w-\mathbf{v}(x)+h_n(\mu-\mu_n))$$

belongs to the epigraph of $\mathbf{v}$ since $w - \mathbf{v}(x) > 0$. This implies that the pair $(u, \mu) \in T_{\mathcal{E}p(\mathbf{v})}(x, w)$, and thus, belongs to the convexified tangent cone $T_{\mathcal{E}p(\mathbf{v})}^{\star\star}(x, w)$.
2. Let us consider now a pair (p, λ) belonging to the normal cone $N_{\mathcal{E}p(\mathbf{v})}(x, w) := (T_{\mathcal{E}p(\mathbf{v})}(x, w))^-$ to the epigraph of $\mathbf{v}$ at (x, w): Therefore,

$$\forall\, (u, \mu) \in T_{\mathcal{E}p(\mathbf{v})}^{\star\star}(x, w),\ \ \langle (u, \mu), (p, \lambda)\rangle\ = \langle p, u\rangle + \lambda\mu\ \leq\ 0$$

   a. Examine first the case when $w = \mathbf{v}(x)$, for which $(u, \mu) \in T_{\mathcal{E}p(\mathbf{v})}^{\star\star}(x, \mathbf{v}(x))$ if and only if $u \in \mathrm{Dom}(D_{\uparrow}^{\star\star}\mathbf{v}(x))$ and $\mu \geq D_{\uparrow}^{\star\star}\mathbf{v}(x)(u)$. If $\lambda > 0$, we obtain a contradiction because, when $\mu \to +\infty$, $\langle p, u\rangle + \lambda\mu \to +\infty$. Hence
      - either $\lambda < 0$, and thus, dividing by $|\lambda|$ and taking $\mu := D_{\uparrow}^{\star\star}\mathbf{v}(x)(dp)$, we obtain

$$\forall\, u \in \mathrm{Dom}(D_{\uparrow}^{\star\star}\mathbf{v}(x)),\ \left\langle \frac{p}{|\lambda|}, u\right\rangle - D_{\uparrow}^{\star\star}\mathbf{v}(x)(u)\ \leq\ 0$$

        which means that $\dfrac{p}{|\lambda|} \in \partial_-\mathbf{v}(x)$
      - or $\lambda = 0$ and we obtain

$$\forall\, u \in \mathrm{Dom}(D_{\uparrow}^{\star\star}\mathbf{v}(x)),\ \langle p, u\rangle\ \leq\ 0$$

        which means that $p \in (\mathrm{Dom}(D_{\uparrow}^{\star\star}\mathbf{v}(x)))^-$ by definition of the polar cone.

   b. When $w > \mathbf{v}(x)$, inequalities

$$\forall\, (u, \mu) \in T_{\mathcal{E}p(\mathbf{v})}^{\star\star}(x, w),\ \ \langle (u, \mu), (p, \lambda)\rangle\ = \langle p, u\rangle + \lambda\mu\ \leq\ 0$$

   imply that $\lambda = 0$. Indeed, otherwise, $\lambda\mu$ converges to $+\infty$ when $\mu \to +\infty$ when $\lambda > 0$ and when $\mu \to -\infty$ when $\lambda < 0$ since μ is allowed to range over $\mathbb{R}$. Therefore $p \in (\mathrm{Dom}(D_\uparrow^{\star\star}\mathbf{v}(x)))^-$ because, since $\lambda = 0$, the above inequalities imply that $\langle p, u\rangle \le 0$ for all $u \in \mathrm{Dom}(D_\uparrow^{\star\star}\mathbf{v}(x))$.

If the domain of $D_\uparrow^{\star\star}\mathbf{v}(x)$ is dense in X, then its polar cone $(\mathrm{Dom}D_\uparrow\mathbf{v}(x))^-$ is $\{0\}$. Therefore, $p = 0$ since it belongs to this set. □

18.7 Convex Analysis: Moreau–Rockafellar Subdifferentials

The first consequence of convexity assumptions is the *Minimax Theorem*. Consider any function $\varphi : K \times L \mapsto \mathbb{R}$. It is obvious that

$$\sup_{y\in L}\inf_{x\in K}\varphi(x,y) \le \inf_{x\in K}\sup_{y\in L}\varphi(x,y)$$

The question arises whether this inequality becomes an inequality and under which conditions. *Von Neumann* proved that convexity assumptions were sufficient:

Theorem 18.7.1 *[**Minimax Theorem**] Let us consider two convex subsets $K \subset X$ and $L \subset Y$ of finite dimensional vector spaces X and Y and a function $\varphi : K \times L \mapsto \mathbb{R}$ such that*

$$\begin{cases} (i)\ \forall\, y \in L, & x \mapsto \varphi(x,y) \text{ is convex} \\ (ii)\ \forall\, x \in K, & y \mapsto \varphi(x,y) \text{ is concave} \end{cases}$$

1. *If K is compact and, for all $y \in L$, $x \mapsto \varphi(x,y)$ is lower semicontinuous, then*

$$\sup_{y\in L}\inf_{x\in K}\varphi(x,y) = \inf_{x\in K}\sup_{y\in L}\varphi(x,y) \tag{18.22}$$

and there exists $\overline{x} \in L$ such that the "lop-sided minsup equality"

$$\sup_{y\in L}\varphi(\overline{x},y) = \sup_{y\in L}\varphi(\overline{x},y) = \inf_{x\in K}\sup_{y\in L}\varphi(x,y) = \sup_{y\in L}\inf_{x\in K}\varphi(x,y) \tag{18.23}$$

holds true

2. *If furthermore, L is compact and, for all $x \in K$, $y \mapsto \varphi(x,y)$ is upper semicontinuous, then there exist $\overline{y} \in L$ such that $\forall\,(x,y) \in K \times L$,*

$$\varphi(\overline{x},y) \le \inf_{x\in K}\sup_{y\in L}\varphi(x,y) = \sup_{y\in L}\inf_{x\in K}\varphi(x,y) \le \varphi(x,\overline{y}) \tag{18.24}$$

Proof. See the proof of Theorems 8.1 and 8.2 of *Optima and Equilibria*, [19, Aubin]. □

When $\mathbf{u}$ is a lower semicontinuous convex function, the generalized gradient coincides with the subdifferential introduced by *Jean Jacques Moreau* and *Terry Rockafellar* for convex functions in the early 1960s. This follows from the *Werner Fenchel* characterization of lower semicontinuous convex functions in terms of conjugate functions:

Definition 18.7.2 *[Conjugate Functions] Consider a nontrivial extended function* $\mathbf{u} : X \mapsto \mathbb{R} \cup \{+\infty\}$*. Its* Fenchel conjugate function *is the extended function* $\mathbf{u}^\star : X^\star \mapsto \mathbb{R} \cup \{+\infty\}$ *defined by*

$$\forall\, p \in X^\star, \quad \mathbf{u}^\star(p) \ := \ \sup_{x \in X} \left(\langle p, x\rangle - \mathbf{u}(x)\right) \tag{18.25}$$

Therefore,

$$\forall\, p \in X^\star, \ \forall\, x \in X, \ \langle p, x\rangle \ \leq \ \mathbf{u}(x) + \mathbf{u}^\star(p) \ \text{Fenchel Inequality} \tag{18.26}$$

The bi-conjugate function *function* $\mathbf{u} : X \mapsto \mathbb{R} \cup \{+\infty\}$ *is the extended function* $\mathbf{u}^{\star\star} : X \mapsto \mathbb{R} \cup \{+\infty\}$ *defined by*

$$\forall\, x \in X, \quad \mathbf{u}^{\star\star}(x) \ := \ \sup_{p \in X^\star} \left(\langle p, x\rangle - \mathbf{u}^\star(p)\right)$$

We observe that $\forall\, x \in X, \ \ \mathbf{u}(x) \leq \mathbf{u}^{\star\star}(x)$. The equality holds if and only $\mathbf{u}$ is nontrivial, convex and lower semicontinuous. This was derived from the Separation Theorem by *Fenchel*:

Theorem 18.7.3 *[The Fenchel Theorem] A nontrivial function* $\mathbf{u} : X \mapsto \mathbb{R} \cup \{+\infty\}$ *is convex and lower semicontinuous if and only if the function* $\mathbf{u} = \mathbf{u}^{\star\star}$ *is equal to its biconjugate.*

Proof. See the proof of Theorem 3.1 of *Optima and Equilibria*, [19, Aubin]. □

Let $\beta : x \mapsto \beta(x)$ be a lower semicontinuous convex function defined everywhere ($\mathrm{Dom}(\beta) = X$) and ψ_F the indicator of F defined by: $\psi_F(x) = 0$ if $x \in F$ and to $+\infty$ outside of F.

Remark: Restriction of a Function. The *restriction* $\mathbf{u}(\cdot)$ of $\beta(\cdot)$ to F can be written in the form

$$\mathbf{u}(x) := \beta(x) + \psi_F(x)$$

Denoting $\sigma_F(p) := \sup_{x\in F} \langle p, x\rangle$ the support function of F, we can associate with any p at least one $\overline{q}$ such that

$$\mathbf{u}^\star(p) := \inf_q (\beta^\star(q) + \sigma_F(p-q)) = \beta^\star(\overline{q}) + \sigma_F(p-\overline{q})$$

- If $\beta(x) := 0$, then
$$\mathbf{u}^\star(p) := \sigma_F(p)$$
- If $\beta(x) := \langle \pi, x\rangle + \gamma$, then $\beta^\star(\pi) = -\gamma$ and for $p \neq \pi$, $\beta^\star(p) = +\infty$. Therefore
$$\mathbf{u}^\star(p) := \sigma_F(p-\pi) - \gamma$$
- If $\beta(x) := \rho\,\|c\|\,x + \gamma$, then $\beta^\star(p) = \psi_{\rho B_\star} - \gamma$. Therefore
$$\mathbf{u}^\star(p) := \inf_{\|q\|_\star \leq \rho} \sigma_F(p-q) - \gamma \;\square$$

Recall that two lower semicontinuous convex functions satisfy $\mathbf{u}_1 \leq \mathbf{u}_2$ if and only if $\mathbf{u}_2^\star \leq \mathbf{u}_1^\star$. We deduce from the above examples the following dual characterization of estimates of $\mathbf{u}$ and $\mathbf{u}^\star$:

Lemma 18.7.4 ***[Dual Estimates]*** *Let* $\mathbf{u}$ *be a lower semicontinuous convex function. We introduce two subsets* $F \subset G$ *and two finite lower semicontinuous convex functions* $\alpha \leq \beta$. *The following conditions are equivalent:*

$$\forall\, p,\ \inf_q (\beta^\star(q) + \sigma_F(p-q)) \leq \mathbf{u}^\star(p) \leq \inf_q (\alpha^\star(q) + \sigma_G(p-q)) \quad (18.27)$$

and

$$\begin{cases} \overline{\mathrm{co}}(F) \subset \mathrm{Dom}(\mathbf{u}) \subset \overline{\mathrm{co}}(G) \\ \forall\, x \in \overline{\mathrm{co}}(F),\ \mathbf{u}(x) \leq \beta(x) \\ \forall\, x \in \overline{\mathrm{co}}(G),\ \alpha(x) \leq \mathbf{u}(x) \end{cases} \quad (18.28)$$

In particular, the two following conditions are equivalent:

$$\forall\, p,\ \sigma_{\mathrm{Dom}(\mathbf{u})}(p) - c \leq \mathbf{u}^\star(p) \leq c\|p\|_\star + \alpha \quad (18.29)$$

and

$$\begin{cases} \mathrm{Dom}(\mathbf{u}) \subset cB \text{ and is closed} \\ \forall\, x \in \mathrm{Dom}(\mathbf{u}),\ -\alpha \leq \mathbf{u}(x) \leq c \end{cases} \quad (18.30)$$

Proof.

1. To say that $\forall\, p,\ \inf_q (\beta^\star(q) + \sigma_F(p-q)) \leq \mathbf{u}^\star(p)$ amounts to saying that $\mathbf{u}(x) \leq \psi_{\overline{\mathrm{co}}(F)}(x) + \beta(x)$. This means that for all $x \in \overline{\mathrm{co}}(F)$, $\mathbf{u}(x) \leq \beta(x)$ and thus, since β is assumed to be finite, that $\overline{\mathrm{co}}(F) \subset \mathrm{Dom}(\mathbf{u})$.

2. To say that $\forall\, p,\ \ \mathbf{u}^\star(p) \le \inf_q (\alpha^\star(q) + \sigma_G(p-q))$ amounts to saying that $\psi_{\overline{\mathrm{co}}(G)}(x)+\alpha(x) \le \mathbf{u}(x)$. This means that for all $x \in \mathrm{Dom}(\mathbf{u})$, $\alpha(x) \le \mathbf{u}(x)$ and thus, since α is assumed to be finite, that $\mathrm{Dom}(\mathbf{u}) \subset \overline{\mathrm{co}}(G)$.

The second statement is deduced by taking $F := \mathrm{Dom}(\mathbf{u})$, $G := cB$, $\beta(\cdot) \equiv c$ and $\alpha(\cdot) \equiv \alpha$. □

For lower semicontinuous convex functions, subgradients coincide with Moreau–Rockafellar sub-subdifferentials:

Definition 18.7.5 ***[Subdifferential of a Convex Function]*** *Consider a nontrivial function* $\mathbf{u} : X \mapsto \mathbb{R} \cup \{+\infty\}$ *and* $x \in \mathrm{Dom}(\mathbf{u})$*. The closed convex subset* $\partial\mathbf{u}(x)$ *defined by*

$$\partial\mathbf{u}(x) = \{p \in X \text{ such that } \langle p, x\rangle \ = \ \mathbf{u}(x) + \mathbf{u}^\star(p)\} \ \text{(Fenchel equality)}$$

(which may be empty) is called the Moreau–Rockafellar subdifferential *of* $\mathbf{u}$ *at* x*. We say that* $\mathbf{u}$ *is* subdifferentiable at x *if* $\partial\mathbf{u}(x) \neq \emptyset$*.*

For convex functions, this concept coincides with the general one.

Proposition 18.7.6 ***[Epiderivative of a Convex Function]*** *Let* $\mathbf{u} : X \mapsto \mathbb{R}_+$ *be a nontrivial extended convex function. Then*

1. the epiderivative is equal to

$$D_\uparrow\mathbf{u}(x)(u) = \liminf_{u' \to u} \left(\inf_{h>0} \frac{\mathbf{u}(x+hu') - \mathbf{u}(x)}{h} \right)$$

2. the subdifferential $\partial_-\mathbf{u}(x)$ *coincides with Moreau–Rockafellar subdifferential* $\partial\mathbf{u}(x)$*.*

Proof. For proving the first statement, take $0 < h_1 \le h_2$,

$$\mathcal{E}p(\nabla_{h_2}\mathbf{u}(x)) = \frac{\mathcal{E}p(\mathbf{u}) - (x, \mathbf{u}(x))}{h_2} \subset \frac{\mathcal{E}p(\mathbf{u}) - (x, \mathbf{u}(x))}{h_1} = \mathcal{E}p(\nabla_{h_1}\mathbf{u}(x))$$

i.e.,

$$\forall u \in X,\ \nabla_{h_1}\mathbf{u}(x)(u) \le \nabla_{h_2}\mathbf{u}(x)(u)$$

Therefore,

$$\begin{cases} \forall u \in X,\ D\mathbf{u}(x)(u) \ := \ \lim\limits_{h \to 0+} \dfrac{\mathbf{u}(x+hu) - \mathbf{u}(x)}{h} \\ = \ \inf\limits_{h>0} \dfrac{\mathbf{u}(x+hu) - \mathbf{u}(x)}{h} \ \le \ \mathbf{u}(x+v) - \mathbf{u}(x) \end{cases} \tag{18.31}$$

and this function $D\mathbf{u}(x)$ is convex with respect to u. Since the epigraph of $D\mathbf{u}(x)$ is the increasing union of the epigraphs of the differential quotients $\nabla_h \mathbf{u}(x)$, we infer that

$$D_{\uparrow}\mathbf{u}(x)(u) := \liminf_{u' \to u} D\mathbf{u}(x)(u')$$

For proving the second statement, take $p \in X^{\star}$ satisfying the Fenchel equality:

$$\partial \mathbf{u}(x) \;=\; \{p \in X \text{ such that } \langle p, x \rangle \;=\; \mathbf{u}(x) + \mathbf{u}^{\star}(p)\}$$

For any $v \in X$, Fenchel inequality

$$\partial \mathbf{u}(x+hv) \;\leq\; \{p \in X \text{ such that } \langle p, x \rangle \;\leq\; \mathbf{u}(x+hv) + \mathbf{u}^{\star}(p)\}$$

and Fenchel equality imply that

$$\forall\, v \in X,\; \forall\, h > 0,\;\; \langle p.v \rangle \;\leq\; \frac{\mathbf{u}(x+hv) - \mathbf{u}(x)}{h}$$

We thus infer that

$$\forall\, v \in X,\;\; \langle p, v \rangle \;\leq\; D_{\uparrow}\mathbf{u}(x)(v)$$

and thus, that p belongs to $\partial_{-}\mathbf{u}(x)$.

Conversely, take $p \in \partial_{-}\mathbf{u}(x)$, $x \in X$ and consider any $v \in X$. By (18.31), p. 758, we infer that

$$\langle p, x+v \rangle - \langle p, x \rangle \;:=\; \langle p, v \rangle \;\leq\; D_{\uparrow}\mathbf{u}(x)(v) \;\leq\; \mathbf{u}(x+v) - \mathbf{u}(x)$$

and thus, that

$$\mathbf{u}^{\star}(p) \;=\; \sup_{v \in X}(\langle p, x+v \rangle - \mathbf{u}(x+v)) \;\leq\; \langle p, x \rangle - \mathbf{u}(x) \;\leq\; \mathbf{u}^{\star}(p)$$

Hence $\langle p, x \rangle = \mathbf{u}(x) + \mathbf{u}^{\star}(p)$. □

Therefore, we deduce at once from the Fenchel Theorem the crucial *Legendre property* of subdifferentials stating that the inverse of the set-valued map $x \rightsquigarrow \partial \mathbf{u}(x)$ is the set-valued map $p \rightsquigarrow \partial \mathbf{u}^{\star}(p)$:

Theorem 18.7.7 *[Legendre Property of Subdifferentials] Let* $\mathbf{u} : X \mapsto \mathbb{R} \cup \{+\infty\}$ *be a nontrivial lower semicontinuous convex function. Then the three following conditions are equivalent:*

$$\begin{cases} (i) & \langle p, x \rangle \;=\; \mathbf{u}(x) + \mathbf{u}^{\star}(p) \\ (ii) & p \;\in\; \partial_{-}\mathbf{u}(x) \\ (iii) & x \;\in\; \partial_{-}\mathbf{u}^{\star}(p) \end{cases} \qquad (18.32)$$

In other words

$$(\partial_{-}\mathbf{u}(\cdot))^{-1} = \partial_{-}\mathbf{u}^{\star}(\cdot)$$

Let us also mention the following simple – but useful – remark:

Proposition 18.7.8 *[Example of the use of the Fermat Rule] Assume that* $\mathbf{u} := \mathbf{v} + \mathbf{w}$ *is the sum of a differentiable function* $\mathbf{v}$ *and a convex function* $\mathbf{w}$. *If* $\overline{x}$ *minimizes* $\mathbf{u}$, *then*

$$-\mathbf{v}'(\overline{x}) \in \partial\mathbf{w}(\overline{x})$$

Proof. Indeed, for $h > 0$ small enough, $\overline{x} + h(y - \overline{x}) = (1 - h)\overline{x} + hy$ so that

$$0 \leq \frac{\mathbf{u}(\overline{x} + h(y - \overline{x})) - \mathbf{u}(\overline{x})}{h} \leq \frac{\mathbf{u}(\overline{x} + h(y - \overline{x})) - \mathbf{u}(\overline{x})}{h} + \mathbf{w}(y) - \mathbf{w}(\overline{x})$$

thanks to the convexity of $\mathbf{w}$. Letting h converge to 0 yields

$$0 \leq \langle \mathbf{v}'(\overline{x}), y - \overline{x} \rangle + \mathbf{w}(y) - \mathbf{w}(\overline{x})$$

so that $-\mathbf{v}'(\overline{x})$ belongs to $\partial\mathbf{w}(\overline{x})$. □

We recall the following important property of convex functions defined on finite dimensional vector spaces:

Theorem 18.7.9 *[Continuity of a Convex Function] An extended convex function* $\mathbf{u}$ *defined on a finite dimensional vector-space is locally Lipschitz and subdifferentiable on the interior of its domain. Therefore, when* x *belongs to the interior of the domain of* $\mathbf{u}$, *there exists a constant* λ_x *such that*

$$\forall u \in X,\ D_{\uparrow}\mathbf{u}(x)(u) = \inf_{h>0} \frac{\mathbf{u}(x + hu) - \mathbf{u}(x)}{h} \leq \lambda_x \|u\|$$

Proof. See the proof of Theorem 4.1 of *Optima and Equilibria*, [19, Aubin]. □

We shall need to adapt convex analysis to concave analysis, associating with a convex function $\mathbf{v}$ the function $\mathbf{l}$ defined by

$$\mathbf{l}(x) := -\mathbf{v}(-x)$$

$$\mathbf{l}^{\boxtimes}(p) := \inf_{x \in \mathrm{Dom}(\mathbf{l})} [\langle p, x \rangle - \mathbf{l}(x)] = -\mathbf{v}^{\star}(-p)$$

The Fenchel Theorem 18.7.3, p.756 states that $\mathbf{l} = \mathbf{l}^{\boxtimes\boxtimes}$ if and only if $\mathbf{l}$ is concave, upper semicontinuous, and non trivial (i.e. $\mathrm{Dom}(\mathbf{l}) := \{x \mid \mathbf{v}(x) > -\infty\} \neq \emptyset$).

The epigraph $\mathcal{E}p(\mathbf{v})$ of an extended function $\mathbf{v}$ is the set of pairs $(x, \lambda) \in X \times \mathbb{R}$ such that $\mathbf{v}(x) \leq \lambda$ and the hypograph $\mathcal{H}yp(\mathbf{l})$ of a function $\mathbf{l}$ is the set of pairs $(x, \mu) \in X \times \mathbb{R}$ such that $\mu \leq \mathbf{l}(x)$. Note that the hypograph of $\mathbf{l}$ is related to the epigraph of $\mathbf{v}$ by the relation

$$(x, \lambda) \in \mathcal{H}yp(\mathbf{l}) \text{ if and only if } (-x, -\lambda) \in \mathcal{E}p(\mathbf{v})$$

An extended function is lower semicontinuous if and only if its epigraph is closed and upper semicontinuous if and only if its hypograph is closed.

Definition 18.7.10 ***[Hypoderivatives and Superdifferentials]*** *The hypoderivative $D_{\downarrow}\mathbf{l}(x)$ and the epiderivative $D_{\uparrow}^{\star\star}\mathbf{v}(x)$ are related to the tangent cones of the hypograph of $\mathbf{l}$ and epigraph of $\mathbf{v}$ by the relations*

$$\mathcal{H}yp(D_{\downarrow}\mathbf{l}(x)) := T_{\mathcal{H}yp(\mathbf{l})}(x, \mathbf{l}(x)) \text{ and } \mathcal{E}p(D_{\uparrow}^{\star\star}\mathbf{v}(x)) := T_{\mathcal{E}p(\mathbf{v})}(x, \mathbf{v}(x))$$

The superdifferential $\partial_{+}\mathbf{l}(x)$ of the concave function $\mathbf{l}$ at p is defined by

$$u \in \partial_{+}\mathbf{l}(x) \text{ if } \forall v \in X, \ \langle u, v \rangle \geq D_{\downarrow}\mathbf{l}(x)(v)$$

and the subdifferential $\partial_{-}\mathbf{v}(x)$ is defined by

$$u \in \partial_{-}\mathbf{v}(x) \text{ if } \forall v \in X, \ \langle u, v \rangle \leq D_{\uparrow}^{\star\star}\mathbf{v}(x)(v)$$

We infer that

$$\forall v \in X, \ D_{\downarrow}\mathbf{l}(x)(v) = -D_{\uparrow}\mathbf{v}(x)(v)$$

and that

$$p \in \partial_{+}\mathbf{l}(x) \text{ if and only if } p \in -\partial_{-}\mathbf{v}(x)$$

The superdifferential $\partial_{+}\mathbf{l}(x)$ and the subdifferential $\partial_{-}\mathbf{v}(x)$ are related to the normal cones of the hypograph of $\mathbf{l}$ and epigraph of $\mathbf{v}$ by the relations

$$u \in \partial_{+}\mathbf{l}(x) \text{ if and only if } (-u, 1) \in N_{\mathcal{H}yp(\mathbf{l})}(x, \mathbf{l}(x))$$

and

$$u \in \partial_{-}\mathbf{v}(x) \text{ if and only if } (p, -1) \in N_{\mathcal{E}p(\mathbf{v})}(x, \mathbf{v}(x))$$

Recall the Legendre inversion formula:

$$u \in -\partial_{+}\mathbf{l}(x) \text{ if and only if } x \in \partial_{-}\mathbf{v}^{\star}(p)$$

18.8 Weighted Inf-Convolution

The weighted inf-convolution of functions $\mathbf{v}_j$ is defined through the weighted aggregation of their epigraphs:

Definition 18.8.1 ***[Weighted Inf-Convolution]*** *Let us consider J lower semicontinuous $\mathbf{v}_j : X_j \mapsto \mathbb{R} \cup \{+\infty\}$.*

The weighted inf-convolution $\bigstar_{j=1}^{J} A_j \mathbf{v}_j : X \mapsto \mathbb{R} \cup \{+\infty\}$ *of functions $\mathbf{v}_j$ and operators $A_j \in \mathcal{L}(X_j, X)$ is defined by*

$$\mathcal{E}p(\bigstar_{j=1}^{J} A_j \mathbf{v}_j) := \sum_{j=1}^{J} (A_j \times 1)\mathcal{E}p(\mathbf{v}_j) \tag{18.33}$$

where 1 *denotes the identity operator of* $\mathbb{R}$

If the functions $\mathbf{v}_j$ are convex, so is their weighted inf-convolution $\bigstar_{j=1}^{J} A_j \mathbf{v}_j$ because the weighted sum of images of convex epigraphs by linear operators is convex. Since the sum of closed subsets is not necessarily closed, we shall need some assumption for showing that the weighted inf-convolution $\bigstar_{j=1}^{J} A_j \mathbf{v}_j$ of lower semicontinuous functions $\mathbf{v}_j$ is lower semicontinuous. This is the purpose of the next two lemmas:

Lemma 18.8.2 ***[Epigraphs of Weighted Inf-Convolutions]*** *Let us set*

$$\mathbf{v}(x) := \inf_{\sum_{j=1}^{J} A_j x_j = x} \sum_{j=1}^{J} \mathbf{v}_j(x_j)$$

Then

$$\sum_{j=1}^{J} (A_j \times 1)\mathcal{E}p(\mathbf{v}_j) \subset \mathcal{E}p(\mathbf{v}) \subset \overline{\sum_{j=1}^{J} (A_j \times 1)\mathcal{E}p(\mathbf{v}_j)}$$

Proof. We observe first that $\sum_{j=1}^{J} (A_j \times 1)\mathcal{E}p(\mathbf{v}_j) \subset \mathcal{E}p(\mathbf{v})$.

Indeed, take any $(x,y) := \sum_{j=1}^J (A_j x_j, y_j) \in \sum_{j=1}^J (A_j \times 1)\mathcal{E}p(\mathbf{v}_j)$. This means that for any $j = 1,\dots,J$, $\mathbf{v}_j(x_j) \le y_j$, that $\sum_{j=1}^J A_j x_j = x$ and that $\sum_{j=1}^J y_j = y$. Hence

$$\mathbf{v}(x) \le \sum_{j=1}^J \mathbf{v}_j(x_j) \le \sum_{j=1}^J y_j = y$$

Second, let us check that $\mathcal{E}p(\mathbf{v}) \subset \overline{\sum_{j=1}^J (A_j \times 1)\mathcal{E}p(\mathbf{v}_j)}$. By definition of the infimum, we can associate with every $\varepsilon > 0$ elements x_{j_ε} such that $\sum_{j=1}^J \mathbf{v}_j(x_{j_\varepsilon}) \le \mathbf{v}(x) + \varepsilon$. This implies that $(x, \mathbf{v}(x)+\varepsilon) \in \sum_{j=1}^J (A_j \times 1)\mathcal{E}p(\mathbf{v}_j)$. It is enough to let ε converge to 0. □

Lemma 18.8.3 *[Lower Semicontinuity of Weighted Inf-Convolutions] Setting*

$$D := \left\{(A_j^\star p)_{j=1,\dots,J}\right\}_{p\in X^\star} \subset \prod_{j=1}^J X_j^\star$$

the weighted inf-convolution $\bigstar_{j=1}^J A_j \mathbf{v}_j$ *of lower semicontinuous* $\mathbf{v}_j$ *functions is lower semicontinuous whenever assumption*

$$0 \in \text{Int}\left(D + \prod_{j=1}^J \text{Dom}(\mathbf{v}_j^\star)\right) \tag{18.34}$$

holds true. Furthermore, if the functions $\mathbf{v}_j$ *are inf-compact, so is their weighted inf-convolution* $\bigstar_{j=1}^J A_j \mathbf{v}_j$ *and there exist J elements* $\overline{x}_j \in X_j$ *such that*

$$\sum_{j=1}^J A_j \overline{x}_j = x \text{ and } \mathbf{v}(x) = \sum_{j=1}^J \mathbf{v}_j(\overline{x}_j)$$

Proof. Let us consider a sequence

$$(x_n, y_n) := \sum_{j=1}^J (A_j x_{j_n}, y_{j_n}) \in \sum_{j=1}^J (A_j \times 1)\mathcal{E}p(\mathbf{v}_j)$$

converging to (x, y) and prove that $(x, y) := \sum_{j=1}^{J}(A_j x_j, y_j) \in \sum_{j=1}^{J}(A_j \times 1)\mathcal{E}p(\mathbf{v}_j)$. Therefore, for any $j = 1, \ldots, J$, $\mathbf{v}_j(x_{j_n}) \leq y_{j_n}$, that $\sum_{j=1}^{J} A_j x_{j_n} = x_n$ converges to x and that $\sum_{j=1}^{J} y_{j_n} = y_n$ converges to y.

Assumption (18.34), p. 763 implies that there exists $\eta > 0$ such that

$$\eta \overrightarrow{B}_\star \in D + \prod_{j=1}^{J} \mathrm{Dom}(\mathbf{v}_j^\star)$$

We deduce that the sequence $\overrightarrow{x_n} = (x_{1_n}, \ldots x_{j_n}, \ldots x_{J_n})$ is bounded. Indeed, for any $\overrightarrow{p} \in \eta B_\star$, there exist $q \in X^\star$ and $r_j \in \mathrm{Dom}(\mathbf{v}_j^\star)$ such that

$$\forall\, j = 1, \ldots J, \ p_j \ = \ A_j^\star q + r_j$$

and thus, that

$$\langle \overrightarrow{p}, \overrightarrow{x_n} \rangle \ = \ \sum_{j=1}^{J} \langle p_j, x_{j_n} \rangle \ = \ \left\langle q, \sum_{j=1}^{J} A_j x_{j_n} \right\rangle + \sum_{j=1}^{J} \langle r_j, x_{j_n} \rangle$$

Observing that

$$\langle r_j, x_{j_n} \rangle \ \leq \ \mathbf{v}_j(x_{j_n}) + \mathbf{v}_j^\star(r_j) \leq y_{j_n} + \mathbf{v}_j^\star(r_j)$$

we infer that

$$\langle \overrightarrow{p}, \overrightarrow{x_n} \rangle \ \leq \ \langle q, x_n \rangle + y_n + \sum_{j=1}^{J} \mathbf{v}_j^\star(r_j) \ < \ +\infty$$

since the sequences x_n and y_n are bounded, being convergent. This implies that the sequence $\overrightarrow{x_n}$ is bounded. Therefore, a subsequence (again denoted by) $\overrightarrow{x_n}$ converges to some $\overrightarrow{x}$ such that $\sum_{j=1}^{J} A_j x_j = x$ and such that $0 \leq \mathbf{v}_j(x_{j_n}) \leq y_{j_n}$. Since the sequence y_n is bounded, the subsequences y_{j_n} are also bounded and converge to some y_i such that $\sum_{j=1}^{J} y_j = x$.

Since $\mathbf{v}_j(x_{j_n}) \leq y_{j_n}$ and since $\mathbf{v}_j(\cdot)$ is lower semicontinuous, we infer that $\mathbf{v}_j(x_j) \leq y_j$, i.e., that $(x_j, y_j) \in \mathcal{E}p(\mathbf{v}_j)$. Consequently,

$$(x, y) \ = \ \sum_{j=1}^{J}(A_j x_j, y_j) \ \in \ \sum_{j=1}^{J}(A_j \times 1)\mathcal{E}p(\mathbf{v}_j)$$

which is then closed. □

Lemma 18.8.4 *[Subdifferential of Weighted Inf-Convolutions]* *Assume that condition (18.34), p. 763 hods true and that the functions* $\mathbf{v}_j$ *are lower semicontinuous, inf-compact and convex. Then there exist* J *elements* x_j *such that*

$$\sum_{j=1}^{J} A_j x_j \; = \; x$$

and that the following conditions are equivalent:

$$\begin{cases} (i) \quad p \in \partial \left(\bigstar_{j=1}^{J} A_j \mathbf{v}_j \right)(x) \\ (ii) \; \forall\, j = 1, \ldots J, \;\; A_j^\star p \in \partial \mathbf{v}_j(x_j) \\ (iii) \; \forall\, j = 1, \ldots J, \;\; x_j \in \partial \mathbf{v}_j^\star(A_j^\star p) \end{cases} \tag{18.35}$$

In other words,

$$\partial \left(\bigstar_{j=1}^{J} A_j \mathbf{v}_j \right)(x) \; = \; \bigcup_{\sum_{j=1}^{J} A_j x_j = x} \; \bigcap_{j=1}^{J} {A_j^\star}^{-1} \partial \mathbf{v}_j(x_j) \tag{18.36}$$

Proof. Conditions (18.35)(ii) and (iii) are equivalent and equivalent to inequalities

$$\forall\, j = 1, \ldots, J, \;\; 0 \; = \; \langle A_j^\star p_j, x_j \rangle - \mathbf{v}_j(x_j) - \mathbf{v}_j^\star(A_j^\star p)$$

Summing then for $j = 1$ to J, we obtain

$$\forall\, j = 1, \ldots, J, \;\; 0 \; = \; \sum_{j=1}^{J} \langle A_j^\star p_j, x_j \rangle - \sum_{j=1}^{J} \mathbf{v}_j(x_j) - \mathbf{v}_j^\star(A_j^\star p)$$

The converse is true since, for all $j = 1$ to J, $0 \leq \langle A_j^\star p_j, x_j \rangle - \mathbf{v}_j(x_j) - \mathbf{v}_j^\star(A_j^\star p)$. □

18.9 The Graal of the Ultimate Derivative

After *Pierre de Fermat*, *Isaac Newton* and *Gottfried Leibniz* three centuries ago, *Augustin–Louis Cauchy* formalized the concept of derivative a little more than one century ago, and, for that purpose, defined rigorously the concept of limit. Since then, there was a consensus on the formalization of derivatives as limits of difference quotients for the *pointwise convergence*. It was so strong that the concept of derivative became a permanent reality, protected from any dissident view. However, some mathematicians motivated by applications are no longer free to choose the assumptions and the rules of the game. Should the nondifferentiable functions popping up in so many fields be deprived forever from the benefits of some properties of the derivatives? This is quite natural, though, because each problem demands its own amount of properties that the derivative should enjoy, i.e., its own degree of regularity. Without going too far by always requiring minimal assumptions, some problems could not be solved by sticking to the richest structure. The right balance between generality and readability is naturally a subjective choice.

Mathematicians of this period still asked many properties for the derivatives of functionals and were not ready to give away linearity.

These definitions were too restrictive, so that they were weakened in several ways, and led to a ménagerie of concepts: strong or weak *Fréchet* and *Gâteaux derivatives*, *Dini directional semiderivatives* or derivatives from the right, to quote a few.

However, this was not enough, the topologies used to define the limits of difference quotients were still too strong for allowing more maps to retain some kind of differentiability. But weakening the topologies allows us to get more limits at the price of obtaining these limits outside the set of single-valued maps. This was even worse than loosing the linearity of the directional derivatives.

However, *Serge Sobolev* and *Laurent Schwartz* did dare to introduce weak derivatives and *distributions* in the 1940s for obtaining solutions to partial differential equations, *Jean Jacques Moreau* and *Terry Rockafellar* to define *set-valued subdifferential* of convex functions to implement the Fermat rule in optimization at the beginning of the 1960s, the 1980s witnessing the eclosion of *graphical derivatives* of set-valued maps and set-valued analysis for dealing, for instance, with control systems and differential games, the 1990s the appearance of *mutations of set-valued maps* for grasping new kind of differential equations – called mutational equations – governing the evolution of sets or, more generally, elements of a metric space. This process of differentiating "less and less differentiable maps", so to speak, continues its random course to unknown shores.

The strong requirement of pointwise convergence of differential quotients can be weakened in (at least) two ways, each way sacrificing different groups of properties of the usual derivatives:

- *Fix the direction* v and take the limit of the function $x \mapsto \nabla_h f(x)(v)$ in the weaker sense of distributions (see for instance [16, Aubin]). The limit $D_v f$ may then be a *distribution*, and no longer a single-valued map. However, it coincides with the usual limit when f is Gâteaux differentiable. Moreover, one can define difference quotients of distributions, take their limit, and thus, *differentiate distributions.*

  Distributions are no longer functions or maps defined on $\mathbf{R}^n$, so *they loose the pointwise character of functions* and maps, but retain the *linearity* of the operator $f \mapsto D_v f$, mandatory for using the theory of linear operators for solving partial differential equations.

- *Fix the direction* x and take the limit of the function $v \mapsto \nabla_h f(x)(v)$ in the weaker sense of *graphical convergence* and *epigraphical convergence* (Definitions 18.5.1, p. 738 and 18.6.8, p. 747): the limit $Df(x)$ may then be a set-valued map, and *no longer a single-value map*. However, it coincides with the usual limit when f is Gâteaux differentiable. Moreover, one can define difference quotients of set-valued maps, take their limit, and thus, differentiate set-valued maps. These graphical derivatives and epiderivatives keep the pointwise character of functions and maps, mandatory for implementing the Fermat Rule, proving inverse function theorems under constraints (see Theorem 9.7.1, p. 366) and solving first-order systems of partial differential equations, for instance, but *loose the linearity* of the map $f \mapsto Df(x)$.

In both cases, the approaches are similar: They use (different) concepts of convergence *weaker than the pointwise convergence* for increasing the possibility for the difference-quotients to converge. But the price to pay is the loss of some properties by passing to these weaker limits (the pointwise character for distributional derivatives, the linearity of the differential operator for graphical derivatives).

For first-order systems of partial differential equation, either conservation laws (Chap. 16, p. 631) or Hamilton–Jacobi–Bellman equations (Chap. 17, p. 681), graphical and epigraphical derivatives happened to be quite appropriate, whereas distributional derivatives are quite efficient for second-order partial differential equations.

Chapter 19
Convergence and Viability Theorems

19.1 Introduction

This chapter is mainly devoted to the proof of a series of Viability Theorems leading to Theorem 11.3.4, p.455. This is done by approximating the differential inclusion by the simple explicit finite-difference scheme (the Euler method). In this setting, being a discrete scheme, the viability characterization of an environment is trivial (see Theorem 2.9.3, p.72).

When the right hand side F of the differential inclusion is Marchaud, the problem then boils down to prove that:

- the sequence of approximate evolutions converges;
- the limit is a solution to the differential inclusion.

As for differential equations, it is classical to prove the uniform convergence of the approximated evolutions (from the Ascoli Compactness Theorem). For differential equations, the convergence of derivatives follows under continuity assumptions, and the pointwise convergence of both the evolutions and their derivatives allow us to derive that the limit is a solution of the differential equation.

This strategy does not work as well for differential inclusions since the evolutions no longer determine in a unique way their velocities. Hence, we need to prove their convergence, under assumptions as weak as possible (Marchaud differential inclusions). This follows from the deep Convergence Theorem 19.2.4, p.772 based on the fundamental theorems of functional analysis and proved in Sect. 19.2, p.770. This is why its proof has been relegated to the end of this book. This Convergence Theorem implies also the convergence of discrete viability kernels with target to the continuous ones.

Once Convergence Theorem 19.2.4, p.772 proved, we devote Sect. 19.3, p.774, to the approximation of viability kernels and regulation map under time-discretizations of the environment, the target et the control system.

J.-P. Aubin et al., *Viability Theory*, DOI 10.1007/978-3-642-16684-6_19,

These theorems allow us to prove several Viability Theorems under more and more restrictive assumptions in Sect. 19.4, p.781.

19.2 The Fundamental Convergence Theorems

We introduced maps F under which the (global) Viability Theorem 11.3.4, p.455 is true. But other (local) Viability Theorems are true for a larger class of set-valued maps:

Definition 19.2.1 ***[Zaremba Maps]*** *A set-valued map $F : X \rightsquigarrow Y$ is called a* Zaremba map *if*

$$\forall x \in X, \begin{cases} (i) \;\; F(x) \text{ is convex and compact} \\ (ii) \; p \in Y^{\star}, \quad x \mapsto \sigma(F(x), p) \text{ is upper semicontinuous} \end{cases} \quad (19.1)$$

where $\sigma(F(x), p) := \sup_{y \in F(x) \langle p,y \rangle}$ is the support function of $F(x)$.

We shall need the following property of Marchaud maps:

Lemma 19.2.2 ***[Marchaud and Zaremba Maps]*** *Every Marchaud set-valued map is a Zaremba one.*

Proof. Proving that for each $p \in Y^{\star}$, the support function is upper semicontinuous amounts to proving that for any $\lambda < +\infty$, the subset

$$K(p, \lambda) \; := \; \{x \;\text{ such that } \lambda \; \leq \; \sigma(F(x), p)\}$$

is closed. Consider thus a sequence $x_n \in K(p, \lambda)$ converging to some $x \in X$. Since F is Marchaud, the subsets $F(x_n)$ are compact, so that there exists $y_n \in F(x_n)$ satisfying

$$\lambda \; \leq \; \sigma(F(x_n), p) \; = \; \langle p, y_n \rangle \; \leq \; c(\|x_n\| + 1)\|p\| \; \leq \; \gamma \|p\|$$

Since y_n belongs to the compact ball of radius γ, a subsequence (again denoted by) y_n converges to some y. Since (x_n, y_n) belongs to the graph $\text{Graph}(F)$, which is closed, we infer that $(x, y) \in \text{Graph}(F)$. On the other hand, we infer that

$$\lambda \; \leq \; \langle p, y \rangle \; \leq \; \sigma(F(x), p)$$

so that x belongs to $K(p,\lambda)$, which is, then, closed. Hence F is a Zaremba map. □

Recall that the *Lipschitz norm* $\|F\|_\Lambda$ of a map $F : x \rightsquigarrow Y$ is the smallest Lipschitz constants of F (see Definition 10.3.5, p.385).

Theorem 19.2.3 *[Convergence Theorem with* **a priori** *Estimates]* *Let $F : X \rightsquigarrow Y$ be a Zaremba map. Consider sequences of Lipschitz evolutions $x_n(\cdot) \in \mathrm{Lip}(0,T;X)$ and of integrable evolutions $y_m(\cdot) \in L^1(0,T;X)$ "approximating" the graph of F in the sense that for all $\varepsilon > 0$ there exists N such that, for all $n \geq N$,*

$$\text{for almost all } t \in [0,T], \ (x_n(t), y_n(t)) \in \mathrm{Graph}(F) + \varepsilon(B \times B)$$

Assume also that the sequences $x_n(\cdot)$ and $y_n(\cdot)$ satisfy the following a priori *estimates*

$$\begin{cases} (i) \ \sup_{n\geq 0} \|x_n\|_\Lambda \leq \alpha \\ (ii) \ \sup_{n\geq 0} \|y_n(t)\| \leq \alpha \end{cases} \tag{19.2}$$

Then subsequences (again denoted by) $x_n(\cdot)$ and $y_n(\cdot)$ satisfying:

- *$x_n(\cdot)$ converges uniformly on compact intervals to an evolution $x(\cdot)$,*
- *$y_n(\cdot)$ converges weakly to an evolution $y(\cdot)$ in $L^1(0,T;X)$,*

and their limits satisfy

$$\text{for almost all } t \in [0,T], \ y(t) \in F(x(t))$$

Proof. Consider sequences $x_n(\cdot)$ and $y_n(\cdot)$ of approximate solutions $x_n(\cdot)$ approximating the graph of F and satisfying the *a priori* estimates (19.2), p.771. They imply that for all $t \in [0,T]$, the sequence $x_n(t)$ remains in a bounded set and that the sequence $x_n(\cdot)$ is *equicontinuous*, because their Lipschitz constants are bounded by α. We then deduce from Ascoli's Theorem that the sequence remains in a compact subset of the Banach space $\mathcal{C}(0,T;X)$, and thus, that a subsequence (again denoted) $x_n(\cdot)$ converges uniformly to some function $x(\cdot)$.

On the other hand, the sequence $y_n(\cdot)$ being bounded in $L^\infty(0,T;X)$, the dual of the Banach space $L^1(0,T;X)$, it is weakly relatively compact thanks to Alaoglu's Theorem.[1] The Banach space $L^\infty(0,T;X)$ is contained in $L^1(0,T;X)$ with a stronger topology.[2] The identity map being continuous for the norm topologies, is still continuous for the weak topologies. Hence the

[1] Alaoglu's Theorem states that any *bounded subset of the dual of a Banach space is weakly compact*.

[2] Since the Lebesgue measure on $[0,T]$ is finite, we know that

sequence $y_n(\cdot)$ is weakly relatively compact in $L^1(0,T;X)$ and a subsequence (again denoted) $y_n(\cdot)$ converges weakly to some function $v(\cdot)$ belonging to $L^1(0,T;X)$.

In summary, we have proved that

$$\begin{cases} (i)\ x_n(\cdot) \text{ converges uniformly to } x(\cdot) \\ (ii)\ y_n(t) \text{ converges weakly to } y(\cdot) \text{ in } L^1(0,T;X) \end{cases}$$

□

The rest of the proof is the consequence of the following:

Theorem 19.2.4 *[Convergence Theorem] Let $F : X \rightsquigarrow Y$ be a Zaremba map. Consider sequences of Lipschitz evolutions $x_n(\cdot) \in$ Lip$(0,T;X)$ and of integrable evolutions $y_m(\cdot) \in L^1(0,T;X)$ "approximating" the graph of F in the sense that for all $\varepsilon > 0$ there exists N such that, for all $n \geq N$,*

$$\text{for almost all } t \in [0,T], \ (x_n(t), y_n(t)) \in \text{Graph}(F) + \varepsilon(B \times B) \quad (19.3)$$

Assume also that sequences $x_n(\cdot)$ and $y_n(\cdot)$ converge in the following sense:

$$\begin{cases} (i)\ x_n(\cdot) \text{ converges uniformly to } x(\cdot) \\ (ii)\ y_n(t) \text{ converges weakly to } y(\cdot) \text{ in } L^1(0,T;X) \end{cases}$$

Then their limits satisfy

$$\text{for almost all } t \in [0,T], \ y(t) \in F(x(t))$$

Proof. Recall that in a Banach space, the closure (for the normed topology) of a set coincides with its weak closure (for the weakened topology[3]).

$$L^\infty(0,T;X) \subset L^1(0,T;X)$$

with a stronger topology. The weak topology $\sigma(L^\infty(0,T;X), L^1(0,T;X))$ (weak-star topology) is stronger than the weakened topology $\sigma(L^1(0,T;X), L^\infty(0,T;X))$ since the canonical injection is continuous. Indeed, we observe that the seminorms of the weakened topology on $L^1(0,T;X)$, defined by finite sets of functions of $L^\infty(0,T;X)$, are seminorms for the weak-star topology on $L^\infty(0,T;X)$, since they are defined by finite sets of functions of $L^1(0,T;X)$.

[3] By definition of the weakened topology, the continuous linear functionals and the weakly continuous linear functionals coincide. Therefore, the closed half-spaces and weakly closed half-spaces are the same. The Hahn–Banach Separation Theorem, which holds in Hausdorff locally convex topological vector spaces, states that closed convex subsets are the intersection of the closed half-spaces containing them. Since the weakened topology is locally convex, we then deduce that closed convex subsets and weakly closed convex subsets do coincide. This result is known as *Mazur's theorem*.

We apply this result: for every m, the function $y(\cdot)$ belongs to the weak closure of the convex hull $\text{co}(\{y_p(\cdot)\}_{p\geq m}.)$ It coincides with the (strong) closure of $\text{co}(\{y_p(\cdot)\}_{p\geq m})$. Hence we can choose functions

$$v_m(\cdot) := \sum_{p=m}^{\infty} a_m^p y_p(\cdot) \in \text{co}(\{y_p(\cdot)\}_{p\geq m})$$

(where the coefficients a_m^p are positive or equal to 0 but for a finite number of them, and where $\sum_{p=m}^{\infty} a_m^p = 1$) *which converge strongly* to $y(\cdot)$ in $L^1(0,T;X)$. This implies that the sequence $v_m(\cdot)$ converges strongly to the function $y(\cdot)$ in $L^1(0,T;X)$.

Thus, there exists another subsequence (again denoted by) $v_m(\cdot)$ such that[4]

$$\text{for almost all } t \in [0,T], \;\; v_m(t) \text{ converges to } y(t)$$

Let $t \in [0,T]$ such that $x_m(t)$ converges to $x(t)$ in X and $v_m(t)$ converges to $y(t)$ in X. Let $p \in X^\star$ be such that $\sigma(F(x(t)),p) < +\infty$ and let us choose $\lambda > \sigma(F(x(t)),p)$. Since F is a Zaremba map, $x \mapsto \sigma(F(x),p)$ is upper semicontinuous. Hence there exists a neighborhood $\mathcal{V}$ of 0 in X such that

$$\forall\, u \in x(t) + \mathcal{V}, \quad \text{then } \sigma(F(u),p) \;\leq\; \lambda \tag{19.4}$$

Let N_1 be an integer such that

$$\forall\, q \geq N_1, \quad x_q \in x(t) + \frac{1}{2}\mathcal{V}$$

Let $\eta > 0$ be given. Assumption (19.3), p.772, of the theorem implies the existence of N_2 and of elements (u_q, v_q) of the graph of F such that

---

[4] *Strong convergence of a sequence in Lebesgue spaces L^p implies that some subsequence converges almost everywhere.* Let us consider indeed a sequence of functions f_n converging *strongly* to a function f in L^p. We can associate with it a subsequence f_{n_k} satisfying

$$\|f_{n_k} - f\|_{L^p} \;\leq\; 2^{-k}; \quad \cdots < n_k < n_{k+1} < \cdots$$

Therefore, the series of integrals

$$\sum_{k=1}^{\infty} \int \|f_{n_k}(t) - f(t)\|_X^p dt < +\infty$$

is convergent. The Monotone Convergence Theorem implies that the series

$$\sum_{k=1}^{\infty} \|f_{n_k}(t) - f(t)\|_X^p$$

converges almost everywhere. For every t where this series converges, we infer that the general term converges to 0.

$$\forall\, q \geq N_2,\ \ u_q \in x_q(t) + \frac{1}{2}\mathcal{V},\ \ \|y_q(t) - v_q\| \ \leq\ \eta$$

Therefore u_q belongs to $x(t) + \mathcal{V}$ and we deduce from (19.4) that

$$\begin{cases} \langle p, y_q(t)\rangle \ \leq\ \langle p, v_q\rangle + \eta\|p\|_\star \\ \leq\ \sigma(F(u_q), p) + \eta\|p\|_\star \\ \leq\ \lambda + \eta\|p\|_\star \end{cases}$$

Let us fix $N \ \geq\ \max(N_1, N_2)$, multiply the above inequalities by the nonnegative a_m^q and add them up from $q = 1$ to ∞. We obtain:

$$\langle p, v_m(t)\rangle \ \leq\ \lambda + \eta\|p\|_\star$$

By letting m go to infinity, it follows that

$$\langle p, y(t)\rangle \ \leq\ \lambda + \eta\|p\|_\star$$

Letting now λ converge to $\sigma(F(x(t)), p)$ and η to 0, we obtain:

$$\langle p, y(t)\rangle \ \leq\ \sigma(F(x(t)), p)$$

Since this inequality is automatically satisfied for those p such that

$$\sigma(F(x(t)), p) \ =\ +\infty$$

it thus holds true for every $p \in X^\star$. Hence, the images $F(x)$ being closed and convex, the Separation Theorem implies that $y(t)$ belongs to $F(x(t))$. The Convergence Theorem ensues. □

19.3 Upper Convergence of Viability Kernels and Regulation Maps

19.3.1 Prolongation of Discrete Time Evolutions

For mapping sequences to functions and functions to sequences, we introduce the prolongation operators $\mathbf{p}_h^0$ and $\mathbf{p}_h$ and the restriction operator $\mathbf{r}^h$ defined in the following way (see for instance *Approximation of Elliptic Boundary-Value Problems*, [15, Aubin]):

The prolongation operators $\mathbf{p}_h^0$ and $\mathbf{p}_h$ map any discrete time evolution $\overrightarrow{x}_h := (x_0^h, \ldots, x_n^h, \ldots)$ to respectively the step function $\mathbf{p}_h^0\overrightarrow{x}^h \in L^1(0, \infty; X)$ and the piecewise linear function $\mathbf{p}_h\overrightarrow{x}^h \ \in\ \mathcal{C}(0, \infty; X)$ interpolating this sequence at the nodes nh:

$$\begin{cases} \forall\, n \geq 0, \ \ \forall\, t \in [nh, (n+1)h[, \\ (i) \ \ \mathbf{p}_h^0 \overrightarrow{x}^h(t) \ := \ x_n^h \\ (ii) \ \ \mathbf{p}_h \overrightarrow{x}^h(t) \ := \ x_n^h + \nabla^h \overrightarrow{x}_n^h(t - nh) \text{ where } \nabla^h \overrightarrow{x}_n^h \ := \ \dfrac{x_{n+1}^h - x_n^h}{h} \\ (iii) \ \dfrac{d}{dt} \mathbf{p}_h \overrightarrow{x}^h(t) \ = \ \mathbf{p}_h^0 \nabla^h \overrightarrow{x}^h(t) \end{cases} \tag{19.5}$$

We set

$$\begin{cases} (i) \ \ \|\overrightarrow{x}^h\|_\infty \ := \ \sup_{j \geq 0} \|x_j^h\| \\ (ii) \ \|\mathbf{p}_h \overrightarrow{x}^h\|_\infty \ := \ \sup_{t \geq 0} \|\mathbf{p}_h^0 \overrightarrow{x}^h(t)\| \end{cases}$$

and we observe that

$$\begin{cases} (i) \ \ \|\mathbf{p}_h \overrightarrow{x}^h\|_\infty \ \leq \ \|\mathbf{p}_h^0 \overrightarrow{x}^h\|_\infty \\ (ii) \ \|\mathbf{p}_h \overrightarrow{x}^h - \mathbf{p}_h^0 \overrightarrow{x}^h\|_\infty \ \leq \ h \|\nabla^h \overrightarrow{x}^h\|_\infty \end{cases} \tag{19.6}$$

The restriction operator $\mathbf{r}_h$ maps any continuous function $x(\cdot) \in \mathcal{C}(0, \infty; X)$ the sequence $\mathbf{r}_h x(\cdot)$ defined by

$$\forall\, j \geq 0, \ \ \mathbf{r}_h x(\cdot)_j \ := \ x(hj)$$

We observe that any continuous function $x(\cdot)$ can be approximated by the functions $(\mathbf{p}_h \mathbf{r}_h x)(\cdot)$.

Definition 19.3.1 *[Time Discretization of Sets and Systems] Let $K \subset X$ be any subset. We shall say that subsets $K^h \subset X$ are discretizations of the subset K if for all positive ε, there exists $h_\varepsilon > 0$ such that for all $h \in]0, h_\varepsilon]$, $K^h \subset K + \varepsilon B$.*
Consider a control system (f, U). Discrete control systems (f_h, U_h) defined by single-valued maps $f_h : X \times \mathcal{U} \rightsquigarrow X$ and set-valued maps $U_h : X \rightsquigarrow \mathcal{U}$ are discretizations of the control systems if for all positive ε, there exists $h_\varepsilon > 0$ such that for all $h \in]0, h_\varepsilon]$,

$$\begin{cases} \text{Graph}\,(f_h) \ \subset \ \text{Graph}(f) + \varepsilon(B_X \times B_{\mathcal{U}} \times B_X) \\ \text{Graph}\,(U_h) \ \subset \ \text{Graph}(U) + \varepsilon(B_X \times B_{\mathcal{U}}) \end{cases} \tag{19.7}$$

A discrete time evolution $\overrightarrow{x}^h := (x_0^h, \ldots, x_n^h, \ldots)$ is regulated by the system (f_h, U_h) if, for all $n \geq 0$,

$$\begin{cases} (i) \ \ \nabla^h \overrightarrow{x}_n^h \ := \ f_h(\overrightarrow{x}_n^h, \overrightarrow{u}_n^h) \ (\text{or } \overrightarrow{x}_{n+1}^h \ = \ \overrightarrow{x}_n^h + h f_h(\overrightarrow{x}_n^h, \overrightarrow{u}_n^h)) \\ (ii) \ \overrightarrow{u}_n^h \ \in \ U_h(\overrightarrow{x}_n^h) \end{cases} \tag{19.8}$$

19.3.2 Upper Convergence

These important theorems have a long history, to which participated Pierre Cardaliaguet and Marc Quincampoix during the 1990s for justifying the convergence of viability kernels obtained by the viability algorithms.

We begin by studying the conditions under which the prolongations of viable sequences converge to a viable evolution:

Lemma 19.3.2 *[**Convergence of Viable Discrete Evolutions**] Let $\mathcal{V}(K^h, C^h)$ be the set of sequences $\overrightarrow{x}^h$ viable in K^h forever or until a date N when it reaches the target C^h and $\mathcal{V}(K, C)$ the set of evolutions $x(\cdot)$ viable in K forever or until a time T when it reaches the target T (see Definition 2.2.3, p.49) where $T := Nh$. Assume that:*

1. *the environments K^h and targets C^h are discretizations of the environment K and the target C;*
2. *sequences $\overrightarrow{x}^h \in \mathcal{V}(K^h, C^h)$ satisfy* a priori *estimates*

$$\|\overrightarrow{x}^h\|_\infty \leq \beta \text{ and } |\nabla^h \overrightarrow{x}^h\|_\infty \leq \beta$$

Then there exists a subsequence (again denoted by) $\mathbf{p}_h \overrightarrow{x}^h(\cdot)$ converging uniformly over compact intervals to a limit $x(\cdot)$ which belongs to $\mathcal{V}(K, C)$. In other words,

$$\text{Limsup}_{h \to 0+} \mathcal{V}(K^h, C^h) \subset \mathcal{V}(K, C)$$

Proof. Let us consider a sequence $\overrightarrow{x}^h \in \mathcal{V}^h(K^h, C^h)$. Therefore, $d(\mathbf{p}_h^0 \overrightarrow{x}^h(s), K^h) \leq h \sup_{j \leq N} \|x_j^h\|$ and $d(\mathbf{p}_h^0 \overrightarrow{x}_h^0(Nh), C^h) \leq h\|x_N^h\|$. Since $K^h \subset K + \alpha h B$ and that $C^h \subset C + \alpha h B$, we infer that

$$\mathbf{p}_h^0 \overrightarrow{x}^h(\cdot) \in \mathcal{V}(K + h(\alpha + \|x^h\|_\infty), C + h(\alpha + \|x^h\|_\infty))$$

Assumptions on the sequence $\overrightarrow{x}^h$ imply that the sequences $\mathbf{p}^h(\cdot)$ and $\frac{d}{dt}\mathbf{p}^h(\cdot)$ are bounded. Hence, the Ascoli compactness Theorem implies that the prolongations $\mathbf{p}^h \overrightarrow{x}^h(\cdot)$ remain in a compact subset. Therefore a subsequence (again denoted by) converges to a limit $x(\cdot)$ uniformly on compact intervals. It belongs to $\mathcal{V}(K+h(\alpha+\beta)B, C+h(\alpha+\beta)B)$. For any fixed h, Lemma 10.3.9, p.388 implies that this set is closed (for the compact convergence). Hence, the limit $x(\cdot)$ belongs to $\mathcal{V}(K+h(\alpha+\beta)B, C+h(\alpha+\beta)B)$. Since this holds true for any h, we infer that $x(\cdot)$ belongs to $\mathcal{V}(K, C)$. □

Viability kernel and regulation maps under approximate control systems (f_h, U_h) converge to the viability kernel and the regulation of the control system (f, U) in the following sense:

Theorem 19.3.3 ***[Upper Convergence of Viability Kernels and Regulation Maps]*** *Let us consider a Marchaud control system* (f, U) *and a sequence of time discretizations* (f_h, U_h) *and time discretizations* K^h *and* C^h. *Assume that evolutions* $(\overrightarrow{x}^h, \overrightarrow{u}^h)$ *regulated by the discrete control systems (19.8), p.775 satisfy the* a priori *estimates*

$$\exists\, \beta > 0 \text{ such that } \sup_h \max(\|\overrightarrow{x}^h\|_\infty, \|\overrightarrow{u}^h\|_\infty, \|\nabla^h \overrightarrow{x}^h\|_\infty) \le \beta \qquad (19.9)$$

Then the upper limits of the viability kernels and the graphs of the regulation maps under the discrete systems (f_h, U_h) *are contained in the viability kernel and the graph of the regulation map under the control system* (f, U)*:*

$$\begin{cases} (i) \ \operatorname{Limsup}_{h \to 0+} \operatorname{Viab}_{(f_h, U_h)}(K^h, C^h) \subset \operatorname{Viab}_{(f,U)}(K, C) \\ (ii) \ \operatorname{Limsup}_{h \to 0+} \operatorname{Graph}(R_{K^h}) \subset \operatorname{Graph}(R_K) \end{cases} \qquad (19.10)$$

Proof. Let us introduce the set-valued map $H : X \rightsquigarrow X \times X$ defined by

$$H(x) := \{(u, v) \text{ such that } u \in U(x) \text{ and } v := f(x, u)\}$$

We observe that if the system (f, U) is Marchaud (see Definition 11.3.3, p.454), so is the set-valued map H.

Consider the following discretizations (f_h, U_h) of the control system (f, U). Let us take any $x_0 = \lim_{h \to 0} x_0^h$ where $x_0^h \in \operatorname{Viab}_{(f_h, U_h)}(K^h, C^h)$. Then there exists a discrete evolution $\overrightarrow{x}^h := (x_0^h, \ldots, x_n^h, \ldots)$ to the system (19.8), p.775 viable in K_h forever or until it reaches the target C^h.

Since the prolongations $\mathbf{p}_h^0 \overrightarrow{x}^h(\cdot)$, $\mathbf{p}_h^0 \overrightarrow{u}^h(\cdot)$ and $\mathbf{p}_h^0 \nabla^h \overrightarrow{x}^h(\cdot)$ are step functions, we infer that for all t,

$$\begin{cases} (\mathbf{p}_h^0 \overrightarrow{x}^h(t), \mathbf{p}_h^0 \overrightarrow{u}^h(t), \mathbf{p}_h^0 \nabla^h \overrightarrow{x}^h(t)) \in \operatorname{Graph}(f_h) \\ \subset \operatorname{Graph}(f) + \varepsilon(B_X \times B_{\mathcal{U}} \times B_X) \\ (\mathbf{p}_h^0 \overrightarrow{x}^h(t), \mathbf{p}_h^0 \overrightarrow{u}^h(t)) \in \operatorname{Graph}(U_h) \subset \operatorname{Graph}(U) + \varepsilon(B_X \times B_{\mathcal{U}}) \end{cases} \qquad (19.11)$$

This can be written

$$(\mathbf{p}_h^0 \overrightarrow{x}^h(t), \mathbf{p}_h^0 \overrightarrow{u}^h(t), \mathbf{p}_h^0 \nabla^h \overrightarrow{x}^h(t)) \in \operatorname{Graph}(H) + \varepsilon(B_X \times B_{\mathcal{U}} \times B_X)$$

On the other hand, *a priori* estimates (19.9), p.777 satisfied by the solutions to the discrete time control system (19.8), p.775 imply that

$$\max \left(\|\mathbf{p}_h^0 \overrightarrow{x}^h\|_\infty, \|\mathbf{p}_h^0 \overrightarrow{u}^h\|_\infty, \left\| \frac{d}{dt}\mathbf{p}_h \overrightarrow{x}^h \right\|_\infty \right) \leq \beta$$

Convergence Theorem 19.2.4, with $F : H$, $x_h(\cdot) := \mathbf{p}_h \overrightarrow{x}^h(\cdot)$ and $y_h(\cdot) := (\mathbf{p}_h^0 \overrightarrow{u}^h(\cdot), \mathbf{p}_h^0 \nabla^h \overrightarrow{x}^h(\cdot))$, implies that there exist subsequences (again denoted by) $\mathbf{p}^h \overrightarrow{x}^h(\cdot)$ converging uniformly on compact intervals to a function $x(\cdot)$ and $(\mathbf{p}_h^0 \overrightarrow{u}^h(\cdot), \mathbf{p}_h^0 \nabla^h \overrightarrow{x}^h(\cdot))$ converging weakly to $y(\cdot) := (u(\cdot), v(\cdot))$ in $L^1(0,T;X)$ which satisfy, for almost all t, $(u(t), v(t)) \in H(x(t))$. Since $\frac{d}{dt}\mathbf{p}_h \overrightarrow{x}^h = \mathbf{p}_h^0 \nabla^h \overrightarrow{x}^h$ and since $|\mathbf{p}_h \overrightarrow{x}^h - \mathbf{p}_h^0 \overrightarrow{x}^h\|_\infty \leq h\|\nabla^h \overrightarrow{x}^h\|_\infty \leq h\beta$ thanks to (19.6), p.775, we infer that $v(\cdot) = x'(\cdot)$ in $L^1(0,T;X)$ satisfy, for almost all t, $(u(t), x'(t)) \in H(x(t))$, i.e., are governed by the control system

$$x'(t) = f(x(t), u(t)) \text{ where } u(t) \in U(x(t))$$

Let us denote by

$$R_{K^h}(\overrightarrow{x}_n^h) := \left\{ \overrightarrow{u}_n^h \in U_h(\overrightarrow{x}_n^h) \text{ such that } \overrightarrow{x}_n^h + h f_h(\overrightarrow{x}_n^h, \overrightarrow{u}_n^h) \in K^h \right\}$$

the regulation map governing the discrete time evolutions $\overrightarrow{x}^h \in \mathcal{V}(K^h, C^h)$ viable in K^h forever or until they reach C^h in finite time. Then Lemma 19.3.2, p.776 implies that $x(\cdot)$ belongs to $\mathcal{V}(K, C)$.

This implies that x_0 belongs to the viability kernel $\text{Viab}_{(f,U)}(K, C)$. Since the limit $x(\cdot)$ is viable in K until it reaches C, then, for almost all t, $f(x(t), u(t)) = x'(t) \in T_K(x(t))$, In other words, for almost all t, $u(t) \in R_K(x(t))$.

This means that

$$\text{Limsup}_{h \to 0+} \text{Viab}_{(f_h, U_h)}(K^h, C^h) \subset \text{Viab}_{(f,U)}(K, C)$$

and

$$\text{Limsup}_{h \to 0+} \text{Graph}(R_{K^h}) \subset \text{Graph}(R_K)$$

This prove the upper convergence of viability kernels with targets and of the graphs of the regulation maps. □

Theorem 19.3.4 ***[Restrictions of Solutions]*** *Let us assume that $F : X \rightsquigarrow X$ is Lipschitz and bounded by $\|F\|_\infty := \sup_{x \in X} \|F(x)\| < +\infty$. We set*

$$\alpha := \frac{\|F\|_\Lambda \|F\|_\infty}{2}$$

Setting

$$G_h^\alpha := \mathbf{1} + hF + \alpha h^2 B$$

the restriction operator $\mathbf{r}_h$ *maps the solution map* $\mathcal{S}_F$ *into the solution map* $\overrightarrow{\mathcal{S}}_{G_h^\alpha}$*:*

$$\mathbf{r}_h\left(\mathcal{S}_F(x_0)\right) \subset \overrightarrow{\mathcal{S}}_{G_h^\alpha}(x_0) \tag{19.12}$$

Assume furthermore that $\frac{G_h(x)-x}{h} \subset F(x)+\varepsilon B$*. The map* $\mathbf{p}_h$ *satisfies*

$$\begin{cases} \mathbf{p}_h(\overrightarrow{\mathcal{S}}_{G_h}(x_0^h)) \subset \mathcal{S}_F(x_0) \\ +e^{\|F\|_\Lambda t}\|x0 - x_0^h\| + h(\|F\|_\infty + \varepsilon)(e^{\|F\|_\Lambda t}-1) \end{cases}$$

Proof. Take any solution $x(\cdot) \in \mathcal{S}_F(x_0)$ to the differential inclusion $x' \in F(x)$, which satisfies $\forall\, t \geq s \geq 0$,

$$x(t) - x(s) \in \int_s^t F(x(\tau))d\tau$$

Since F is bounded, we deduce that

$$\|x(t) - x(s)\| \leq (t-s)\|F\|_\infty \tag{19.13}$$

On the other hand, since F is Lipschitz,

$$F(x(\tau)) \subset F(x(s)) + \|F\|_\Lambda \|x(\tau) - x(s)\| B,$$

and consequently

$$x(t) - x(s) \in (t-s)\left[F(x(s)) + \|F\|_\Lambda \left(\int_s^t \|x(\tau)-x(s)\|d\tau\right) B\right] \tag{19.14}$$

Hence:

$$\forall\, t \geq s \geq 0,\ \ x(t) - x(s) \in (t-s)F(x(s)) + \frac{\|F\|_\Lambda \|F\|_\infty}{2}(t-s)^2 B$$

and thus, for $j = 0, \ldots, N-1$,

$$x((j+1)h) \in x(jh) + hF(x(jh)) + \alpha h^2 B =: G_h^\alpha(x(jh))$$

The sequence $\mathbf{r}_h x$ is then a solution to the discrete system G_h^α.

Consider now a solution $\overrightarrow{x}^h$ starting at x_0^h to the discrete system G_h and its associated piecewise linear interpolation $\mathbf{p}_h \overrightarrow{x}^h$. By the Filippov Theorem 11.3.9, we know that there exists a solution $x_h(\cdot) \in \mathcal{S}_F(x_0)$ satisfying inequality

$$\begin{cases} \|x_h(t) - \mathbf{p}_h \overrightarrow{x}^h(t)\| \\ \leq e^{\|F\|_\Lambda t}\left(\|x_0^h - x_0\| + \int_0^t d(\mathbf{p}_h \overrightarrow{x}^h)'(s), F(\mathbf{p}_h \overrightarrow{x}^h(s))e^{-\|F\|_\Lambda s} ds\right) \end{cases}$$

But, on each interval $[jh,(j+1)h[$,

$$(\mathbf{p}_h \overrightarrow{x}^h)'(s) = v_j^h := \frac{x_{j+1}^h - x_j^h}{h} \ \& \ \mathbf{p}_h \overrightarrow{x}^h(s) = x_j^h + (s-jh)v_j^h$$

Since $v_j^h \in F(y_j^h(jh)) + \varepsilon B$ and since F is Lipschitz, we deduce that

$$\begin{cases} d((\mathbf{p}_h \overrightarrow{x}^h)'(s), F(\mathbf{p}_h \overrightarrow{x}^h(s)) \\ \leq \|F\|_\Lambda(\|y_j^h - x_j^h - (s-jh)v_j^h)\| \\ \leq h\|F\|_\Lambda((\|F\|_\infty + \varepsilon)) \end{cases}$$

Therefore

$$\begin{cases} e^{\|F\|_\Lambda t} \int_0^t d((\mathbf{p}_h \overrightarrow{x}^h)'(s), F(\mathbf{p}_h \overrightarrow{x}^h(s))e^{-\|F\|_\Lambda s} \\ \leq h e^{\|F\|_\Lambda t}\|F\|_\Lambda(\|F\|_\infty + \varepsilon)\int_0^t e^{-\|F\|_\Lambda} ds \\ = h(\|F\|_\infty + \varepsilon)(e^{\|F\|_\Lambda t} - 1) \end{cases}$$

We thus infer that

$$\|x_h(t) - \mathbf{p}_h \overrightarrow{x}^h(t)\| \leq e^{\|F\|_\Lambda t}\|x_0^h - x_0\| + (\varepsilon + h(\|F\|_\infty + \varepsilon))(e^{\|F\|_\Lambda t} - 1) \ \square$$

Theorem 19.3.5 ***[Convergence of Viability Kernels]*** *Assume f that F is Lipschitz, bounded and satisfy*

$$1 + hF + \frac{\|F\|_\infty \|F\|_\Lambda}{2} h^2 B \subset G_h$$

Then

$$\text{Viab}_F(K,C) \subset \text{Liminf}_{h \to 0+} \text{Viab}_{G_h}(K,C)$$

Proof. Let x_0 belong to the viability kernel $\text{Viab}_F(K,C)$ of a closed subset K with closed target C. Since F is Lipschitz and bounded, there exists a solution $x(\cdot)$ to the differential inclusion $x' \in F(x)$ viable in K forever or until it reaches C. Theorem 19.3.4 implies that its image $\mathbf{r}_h x$ is a solution to the discrete system G_h, which is then viable in K until it reaches C. Hence $(\mathbf{r}^h x(\cdot))(0)$ belongs to $\text{Viab}_{G_h}(K,C)$. $\square$

Remark. By taking constant environments and targets, we infer that

$$\text{Viab}_F(K,C) \subset \text{Liminf}_{h \to 0+} \text{Viab}_{G_h}(K,C)$$

The limit of the viability kernels under G_h is thus equal to the viability kernel under F. $\square$

19.4 Proofs of Viability Theorems

We gather in this section the proof of the Viability Theorem as well as a series of more abstract ones needed to derive it.

We begin by proving the necessary conditions. Next, we prove a general Viability Theorem valid in a relatively compact environment K (this means that each element of $x_0 \in K$ has a compact neighborhood $B_K(x_0, r) := K \cap (x_0 + rB)$). In a finite dimensional vector space, open subsets, closed subsets, intersections of open and closed subsets are relatively compact. When the environment is open, we obtain the generalization to differential inclusions of the Peano Theorem on the existence of a solution to a differential equation with continuous right hand side. When the environment is closed, we derive the local Viability Theorem, when solutions are allowed to blow up in finite time. Additional linear growth condition excludes this possibility for obtaining the Viability Theorem 11.3.4, p.455 of the survival kit 2.15, p. 98, that we did use in most parts of the book.

19.4.1 Necessary Condition for Viability

Proposition 19.4.1 *[**Necessary Condition**] Let us assume that F is a Zaremba map. Let us consider a solution $x(\cdot)$ to differential inclusion (19.17) starting at x_0 and satisfying*

$$\forall\, T > 0,\ \exists\, t \in]0, T] \text{ such that } x(t) \in K \tag{19.15}$$

(Naturally, viable solutions do satisfy this property.) Then

$$F(x_0) \cap T_K(x_0) \neq \emptyset$$

Proof. By assumption (19.15), there exists a sequence $t_n \to 0+$ such that $x(t_n) \in K$. Since F is Zaremba, the support functions $\sigma(F(x(\cdot)), p)$ are upper semicontinuous at x_0. We thus can associate with any $p \in X^\star$ and $\varepsilon > 0$ an $\eta_p > 0$ such that

$$\forall\, \tau \in [0, \eta_p],\ < p, x'(\tau) > \leq \sigma(F(x(\tau)), p) \leq \sigma(F(x_0), p) + \varepsilon \|p\|_\star$$

By integrating this inequality from 0 to t_n, setting $v_n := \dfrac{x(t_n) - x_0}{t_n}$ and dividing by $t_n > 0$, we obtain for n larger than some N_p

$$\forall\, p \in X^\star,\ \forall\, n \geq N_p,\ \langle p, v_n \rangle \leq \sigma(F(x_0), p) + \varepsilon \|p\|_\star$$

Therefore, v_n lies in a bounded subset of a finite dimensional vector space, so that a subsequence (again denoted) v_n converges to some $v \in X$ satisfying

$$\forall\, p \in X^\star, \quad < p, v > \leq \sigma(F(x_0), p) + \varepsilon \|p\|_\star$$

By letting ε converge to 0, we deduce that v belongs to the closed convex hull of $F(x_0)$.

On the other hand, since for any n, $x(t_n) = x_0 + t_n v_n$ belongs to K, we infer that v belongs to the contingent cone $T_K(x_0)$ since

$$\begin{cases} \liminf_{n\to\infty} d_K(x_0 + hv)/h \\ \leq \lim_{n\to\infty} \|x_0 + t_n v - x(t_n)\|/t_n \;=\; \lim_{n\to\infty} \|v_n - v\| \;=\; 0 \end{cases}$$

The intersection $F(x_0) \cap T_K(x_0)$ is then nonempty, so that the necessary condition ensues. □

19.4.2 The General Viability Theorem

Theorem 19.4.2 *[General Viability Theorem] Let $F : X \rightsquigarrow X$ be a Zaremba map. Assume that K is relatively compact and satisfy tangential condition (11.5), p. 447:*

$$\forall x \in K \cap \overset{\circ}{B}(x_0, r), \; F(x) \cap T_K^{\star\star}(x) \neq \emptyset \qquad (19.16)$$

Then for any initial state $x_0 \in K$, there exist a strictly positive $T > 0$ and a solution on $[0, T[$ to differential inclusion

$$\text{for almost all } t \in [0, T], \; x'(t) \in F(x(t)) \qquad (19.17)$$

starting from x_0, viable in K.

Proof. We construct approximate solutions by modifying Euler's method to take into account the viability constraints, then deduce from available estimates that a subsequence of these solutions converges in some sense to a limit, and finally, check that this limit is a viable solution to differential inclusion (19.17).

1. **Construction of Approximate Solutions**
   By assumption, there exists $r > 0$ such that $B_K(x_0, r) := K \cap (x_0 + rB)$ is compact. We set

$$C := F(B_K(x_0, r)) + B, \quad T := r/\|C\|$$

and observe that C is compact.

2. Theorem 11.2.7, p.447 implies that for all $\varepsilon > 0$, and for any "graphical approximation" F_ε (in the sense that $\text{Graph}(F_\varepsilon) \subset \text{Graph}(F) + \varepsilon(B \times B)$) of F, then

$$\begin{cases} \exists\, \eta(\varepsilon) > 0 \text{ such that } \forall x \in B_K(x_0, r),\ \forall h \in [0, \eta(\varepsilon)]\,, \\ (x + hF(x)) \cap B_K(x_0, r) \neq \emptyset \end{cases} \tag{19.18}$$

Setting $G_h(x) := (x + hF(x)) \cap B_K(x_0, r)$, we see that this discretization of the set-valued map F governs viable discrete evolutions $\overrightarrow{x}^h$ starting from $X_0 \in K$ defined by

$$x^h_{j+1} \in (x^h_j + hF(x^h_j)) \cap B_K(x_0, r)$$

viable in $B_K(x_0, r)$ for $j = 0, \ldots, J^h$ where J^h is the smallest integer largest than $T := \dfrac{r}{\|C\|}$. Indeed, inequality

$$\|x^h_j - x_0\| \leq \sum_{i=0}^{i=j-1} \|x^h_{i+1} - x^h_i\| \leq \sum_{i=0}^{i=J^h-1} h\,\|v^h_j\| \leq J^h\|C\|$$

implies that the discrete evolution is viable in K and in the ball $B(x_0, r)$ on the interval $[0, T]$. It satisfies the *a priori* estimates

$$\|\overrightarrow{x}^h\|_\infty \leq \beta \text{ and } |\nabla^h \overrightarrow{x}^h\|_\infty \leq \beta \tag{19.19}$$

3. They imply *a priori* estimates (19.9), p.777 of the Convergence Theorem 19.3.3, p.777. It implies that the limit of a converging subsequence is a solution to the differential inclusion, viable in K thanks to Lemma 19.3.2, p.776. □

19.4.3 The Local Viability Theorem

Recall that by Definition 4.3.3, p.135, the minimal time functional

$$\varpi_{(K,C)}(x(\cdot)) := \inf\{t \geq 0 \mid x(t) \in C \ \&\ \forall s \in [0, t],\ x(s) \in K\,\}$$

is infinite when it is viable in $K \setminus C$ and finite when it captures the target C.

Theorem 19.4.3 *[Local Viability Theorem] Let $K \subset X$ and $C \subset K$ be two closed subsets. Assume that the right hand side F of inclusion $x' \in F(x)$ is Zaremba and satisfies tangential condition (11.5), p. 447:*

$$\forall x \in K \setminus C,\ F(x) \cap T_K^{\star\star}(x) \neq \emptyset \tag{19.20}$$

Then for any initial state $x_0 \in K \setminus C$, there exists a strictly positive $T^\star > 0$ and a solution on $[0, T^\star[$ to differential inclusion $x'(t) \in F(x(t))$ on the interval $[0, T^\star[$, starting from x_0, viable in K forever or until it reaches C where $T^\star$ satisfies

$$\begin{cases} \text{either } T^\star = \varpi_{(K,C)}(x(\cdot)) \\ \quad \text{or} \quad T^\star < \varpi_{(K,C)}(x(\cdot)) \quad \text{and} \quad \limsup_{t \to T^\star -} \|x(t)\| = \infty \\ \qquad \text{“blows up” before reaching the target } C \end{cases} \tag{19.21}$$

Proof. 1. The subset $K \setminus C$ is locally compact since it is the intersection closed subset and an open subset and since the dimension of X is finite. If x_0 belongs to $K \setminus C$, there exists $\rho > 0$ such that $(K \setminus C) \cap B(x_0, \rho)$ is compact. By Theorem 19.4.2, p.782, there exists an evolution $x(\cdot)$ governed by differential inclusion $x' \in F(x)$ viable in $(K \setminus C) \cap B(x_0, \rho)$ on an interval $[0, S[$ for some positive S.

2. We claim that starting from any x_0, the evolution can be extended by an evolution defined on a maximal $[0, T^\star[$ which is viable in $K \setminus C$ either forever ($\varpi_{(K,C)}(x(\cdot)) = +\infty$) or until it reaches C in finite time $\varpi_{(K,C)}(x(\cdot))$. We introduce the set of pairs $\{(T, x(\cdot))\}$ where $x(\cdot)$ is an evolution governed by $x' \in F(x)$ viable in $K \setminus C$ forever or reaches C in finite time. We define the order relation $\prec$ by

$$(T, x(\cdot)) \prec (S, y(\cdot)) \text{ if and only if } T \leq S \ \&\ \forall\, t \in [0, T[,\ \ x(t) = y(t)$$

Since every totally ordered subset has obviously an upper bound, Zorn's Lemma implies that any solution $y(\cdot)$ defined on some interval $[0, S[$ can be extended to an evolution $x(\cdot)$ defined on a maximal interval $[0, T^\star[$.

3. We shall deduce from the maximality of $(T^\star, x(\cdot))$ that if $T^\star$ is finite, we cannot have

$$\gamma := \limsup_{t \to T^\star -} \|x(t)\| < \varpi_{(K,C)}(x(\cdot))$$

Indeed, if $\gamma < +\infty$, there would exist a constant $\eta \in]0, T^\star[$ such that

$$\forall\, t \in [T^\star - \eta, T^\star[,\ \ \|x(t)\| \leq \gamma + 1$$

Since F is upper semicontinuous with compact images on the compact subset $K \cap (\gamma + 1)B$, we infer that

$$\forall\, t \in [T^\star - \eta, T^\star[,\quad x'(t) \in F(K \cap (\gamma + 1)B), \quad \text{which is compact}$$

and thus bounded by a constant ρ. Therefore, for all $\tau, \sigma \in [T^\star - \eta, T^\star[$, we obtain:

$$\|x(\tau) - x(\sigma)\| \leq \int_{\sigma}^{\tau} \|x'(s)ds\| \leq \rho|\tau - \sigma|$$

Hence the Cauchy criterion implies that $x(t)$ has a limit from the left when $t \to T^{\star}$. We denote by $x(T^{\star})$ this limit, which belongs to K because it is closed. Let $T_k < T^{\star}$ converge to $T^{\star}$. Equalities

$$x(T_k) = x_0 + \int_0^{T_k} x'(\tau)d\tau$$

and Lebesgue's Theorem imply that by letting $k \to \infty$, we obtain:

$$x(T^{\star}) = x_0 + \int_0^{T^{\star}} x'(\tau)d\tau$$

(a) either $x(T^{\star}) \in K \backslash C$, and we can find a viable solution starting at $x(T^{\star})$ on some interval $[T^{\star}, S[$ where $S > T^{\star}$ thanks to Theorem 19.4.2, p.782, a contradiction to the maximality of $(T^{\star}, x(\cdot))$;
(b) or $x(T^{\star}) \in C$ and thus, $T^{\star} \geq \varpi_{(K,C)}(x(\cdot))$, a contradiction of $T^{\star} < \varpi_{(K,C)}(x(\cdot))$. □

19.4.4 Proof of the Viability Theorem 11.3.4

We have gathered all the tools needed to prove this Theorem.

Linear growth of F or other conditions imply an *a priori* estimate of the form $\|x(t)\| \leq \varphi(t)$ where $\forall\, t \geq 0$, $\varphi(t) < +\infty$, so that the second condition $T^{\star} < +\infty$ and $\limsup_{t \to T^{\star}-} \|x(t)\| = \infty$ of (19.21), p.784 being excluded, there exits a solution which exists for all positive t and viable in K without blowing up in finite time.

Since the growth of F is linear,

$$\exists\, c \geq 0, \quad \text{such that } \forall\, x \in \text{Dom}(F), \quad \|F(x)\| \leq c(\|x\| + 1)$$

Therefore, any solution to differential inclusion (19.17) satisfies the estimate:

$$\|x'(t)\| \leq c(\|x(t)\| + 1)$$

The function $t \to \|x(t)\|$ being locally Lipschitz, it is almost everywhere differentiable. Therefore, for any t where $x(t)$ is different from 0 and differentiable, we have

$$\frac{d}{dt}\|x(t)\| = \left\langle \frac{x(t)}{\|x(t)\|}, x'(t) \right\rangle \leq \|x'(t)\|$$

These two inequalities imply the estimates:

$$\|x(t)\| \leq (\|x_0\| + 1)e^{ct} \ \& \ \|x'(t)\| \leq c(\|x_0\| + 1)e^{ct} \tag{19.22}$$

Hence, for any $T^\star > 0$, we infer that

$$\limsup_{t \to T\star-} \|x(t)\| < +\infty$$

Theorem 19.4.3 implies that we can extend the solution on the interval $[0, \varpi_{(K,C)}(x(\cdot))[$. □

Remark. One can find an extension of Viability Theorem to the case of time dependent differential inclusions when the dependence on time is measurable (see Sect. 11.7, p. 395, of *Viability Theory*, [18, Aubin]. We only reproduce the statement of the viability theorem on measurable time dependence by *Peter Tallos* and refer to the literature for subsequent results. □

Theorem 19.4.4 ***[Measurable Time Dependence]*** *Let X be a finite dimensional vector-space X and $F : \mathbb{R}_+ \times K \rightsquigarrow X$ be a nontrivial set-valued map satisfying*

$$\begin{cases} i) \ \forall x \in K, \ t \rightsquigarrow F(t,x) \text{ is measurable} \\ ii) \ \forall t \geq 0, \ x \rightsquigarrow F(t,x) \\ \quad \text{is upper semicontinuous with compact convex values} \\ iii) \ \exists\, c(\cdot) \in L^1(0,\infty; \mathbf{R}_+) \ \textit{such that} \ \|F(t,x)\| \leq c(t)(\|x\|+1) \end{cases}$$

If the tangential condition

$$\text{for almost all } t \geq 0, \ \ F(t,x) \cap T_K(x) \neq \emptyset$$

holds, the environment K is viable under F: for any initial stage $x_0 \in K$, there exists a solution to the differential inclusion $x'(t) \in F(t, x(t))$ starting at x_0 which is viable in K.

19.4.5 Pointwise Characterization of Differential Inclusions

A solution $x(\cdot)$ to a Marchaud differential inclusion being Lipschitz, is almost everywhere differentiable (in the usual sense), but also graphically differentiable in the sense that its (contingent graphical) derivative of the graphical derivative of $x(\cdot)$. Theorem 18.5.4, p.739 states that the graphical derivative $Dx(t) := Dx(t, x(t))$ is defined by

$$\text{Graph}(Dx(t)) := T_{\text{Graph}(x)}(t, x)$$

has always non empty values, in particular, for the time direction 1:

$$\forall\, t \geq 0,\ \ \forall\, u \in X,\ \ Dx(t)(1) \ = \ \text{Limsup}_{h \to 0+} \left\{ \frac{x(t+h) - x(t)}{h} \right\} \neq \emptyset$$

We show that we can trade the "almost all t" in the definition of differential inclusions using classical derivatives by the "all t" by using graphical derivatives.

Lemma 19.4.5 ***[Pointwise Characterization of Solutions to Differential Inclusions]*** *Assume that the right hand side F of the differential inclusion is Marchaud. The two following statements are equivalent for characterizing a solution $x(\cdot)$ to the differential inclusion "$x' \in F(x)$":*

$$\begin{cases} (i) \ \ \textbf{for almost all} \ \ t \geq 0, \ \ x'(t) \ \in \ F(x(t)) \\ (ii) \ \textbf{for all} \ \ t \geq 0, \ \ Dx(t)(1) \cap F(x(t)) \ \neq \ \emptyset \end{cases} \tag{19.23}$$

Proof. Let us consider a solution $x(\cdot)$ to differential inclusion

$$\text{for almost all } \ t \in [0, T], \ \ x'(t) \ \subset \ F(x(t))$$

in the usual sense. By definition, the graph

$$\text{Graph}(x) \ := \ \{(t, x(t))\}_{t \geq 0} \ \subset \ \mathbb{R}_+ \times X$$

of the solution is viable under the auxiliary differential inclusion

$$\begin{cases} (i) \ \ \tau'(t) \ = \ 1 \\ (ii) \ x'(t) \ \in \ F(x(t)) \end{cases}$$

The Viability Theorem 11.3.4, p. 455 states that, for any $t \geq 0$, there exists $u \in F(x)$ such that

$$(1, u) \ \in \ (\{1\} \times F(x)) \cap T_{\text{Graph}(x)}(t, x)$$

and conversely.

This means that for every $t \geq 0$, there exists $u \in F(x)$ such that $u \in Dx(t)(1)$. □

References

1. Aamo, O.M., Krstic, M.: Flow Control by Feedback: Stabilization and Mixing. Springer, UK (2002)
2. Abraham, R., Marsden, J.E.: Foundations of Mechanics (AMS Chelsea Publishing), 2nd edn. American Mathematical Society, Providence (1978)
3. Alart, P., Maisonneuve, O., Rockafellar, R.T.: Nonsmooth Mechanics and Analysis: Theoretical and Numerical Advances, Springer, New York (2005)
4. Alur, R., Arzen, K.-E., Baillieul, J., Hanzinger, T.A., Hristu-Varsakelis, D., Levine, W.S.: Handbook of networked and embedded control systems (Control Engineering), Birkhauser, Boston (2005)
5. Ambrosio, L.: Corso introduttivo alla teoria geometrica della misura ed alle superfici minime, Scuola Normale di Pisa (1997)
6. Ambrosio, L.: Lecture Notes on geometric evolution problems, distance functions and viscosity solutions, Università di Pavia (1997)
7. Antsaklis, P.J., Michel, A.N.: A linear systems primer, Birkhauser, Boston (2010)
8. Ariyu, K.B., Krstic, M.: Real-Time Optimization by Extremum-Seeking Control. Wiley, New York (2003)
9. Astrom, K.J., Murray, R.M.: Feedback systems: an introduction for scientists and engineers. Princeton University Press, Princeton (2008)
10. Attouch, H.: Variational convergence for functions and operators, Research Notes in Mathematics. Pitman, London (1984)
11. Aliprantis, C.D.: Problems in Equilibrium Theory. Springer, New York (1996)
12. Amann, H.: Ordinary differential equations, Walter de Gruyter, Berlin (1990)
13. Ambrosio, L.: Lecture Notes on geometric evolution problems, distance functions and viscosity solutions, Università di Pavia (1997)
14. Arbib, M. (ed.): The handbook of brain theory and neural networks, Bradford Books and MIT, Cambridge, Massachusetts (1995)
15. Aubin, J.-P.: Approximation of Elliptic Boundary-Value Problems. Wiley-Interscience, New York (1972)
16. Aubin, J.-P.: Applied Functional Analysis, 2nd edn. 1999, Wiley-Interscience, New York (1977,1989)
17. Aubin, J.-P.: Mathematical Methods of Game and Economic Theory. Birkhäuser, Boston (1979,1982)
18. Aubin, J.-P.: Viability Theory. Birkhäuser, Boston (1991)
19. Aubin, J.-P.: Optima and Equilibria. Springer, Berlin (1993)
20. Aubin, J.-P.: Initiation à l'analyse appliquée. Masson, Paris (1994)
21. Aubin, J.-P.: Neural Networks and Qualitative Physics: A Viability Approach. Cambridge University Press, Cambridge (1996)

J.-P. Aubin et al., *Viability Theory*, DOI 10.1007/978-3-642-16684-6,

22. Aubin, J.-P.: Dynamic Economic Theory: A Viability Approach. Springer, Berlin (1997)
23. Aubin, J.-P.: Mutational and Morphological Analysis: Tools for Shape Regulation and Morphogenesis. Birkhäuser, Boston (2000)
24. Aubin, J.-P.: La mort du devin, l'émergence du démiurge. Essai sur la contingence, la viabilité et l'inertie des systèmes, Éditions Beauchesne (2010); English version: The Demise of the Seer, the Rise of the Demiurge. Essay on contingency, viability and inertia of systems (in preparation)
25. Aubin, J.-P., Cellina, A.: Differential inclusions, Set-Valued Maps and Viability Theory, Springer-Verlag, Grundlehren der math. Wiss. # 264 (1984)
26. Aubin, J.-P., Ekeland, I.: Applied Nonlinear Analysis. Wiley-Interscience, New York (1984)
27. Aubin, J.-P., Frankowska, H.: Set-Valued Analysis. Birkhäuser, Boston (1990)
28. Bacelli, F., Cohen, G., Olsder, G., Quadrat, J.-P.: Synchronisation and Linearity. Wiley, New York (1992)
29. Balasko, Y.: Foundations of the Theory of General Equilibrium. Academic, San Diego (1988)
30. Bardi, M., Capuzzo Dolcetta, I.: Optimal Control and Viscosity Solutions to Hamilton-Jacobi-Bellman Equations. Birkhäuser, Boston (1998)
31. Banach, S.: Théorie des opérations linéaires. Monographie Matematyczne, Panstowowe Wyd Nauk (1932)
32. Baras, J.S., Mirelli, V.: Recent Advances in Stochastic Calculus (Progress in Automation and Information Systems). Springer, Berlin (1990)
33. Basar, T., Bernhard, P.: H^∞-Optimal Control and Related Minimax Design Problems. A Dynamic Game Approach. Birkhäuser, Boston (1991)
34. Ben-Akiva, M., Lerman, S.: Discrete Choice Analysis: Theory and Application to Travel Demand (transportation studies). MIT, Cambridge (1985)
35. Beer, G.: Topologies on Closed and Closed Convex Sets. Kluwer, Dordrecht (1993)
36. Bensoussan, A., Lions, J.L.: Contrôle impulsionnel et inéquations quasi-variationnelles. Dunod, Paris (1982)
37. Bensoussan, A.: Stochatic Control by Functional Analysis Methods. North-Holland, Amsterdam (1982)
38. Berge, C.: Espaces topologiques et fonctions multivoques. Dunod, Paris (1959)
39. Bernhard, P.: Commande optimale, décentralisation et jeux dynamiques. Dunod, Paris (1976)
40. Bezout, E., Feron, E.: General Theory of Algebraic Equations. Princeton University Press, Princeton (2006)
41. Bouligand, G.: Introduction à la géométrie infinitésimale directe. Gauthier-Villars, Paris (1932)
42. Bonnans, J.-F. Gilbert, J.C., Lemarechal, C., Sagastizábal, C.A.: Numerical optimization: Theoretical and Practical Aspects (Universitext), 2nd edn. Springer, Berlin (2006)
43. Bonnard, B., Faubourg, L., Trelat, E.: Mécanique céleste et contrôle des véhicules spatiaux (Mathématiques et Applications), French edition. Springer, Berlin (2005)
44. Bonneuil, N.: Introduction à la modélisation démographique. Armand Colin, Paris (1997)
45. Borrelli, F.: Constrained Optimal Control of Linear and Hybrid Systems. Springer, Berlin (2003)
46. Bourbaki, N.: Fonctions d'une variable réelle. Théorie élémentaire. Masson, Paris (1982)
47. Bourbaki, N.: Eléments d'histoire des mathématiques. Masson, Paris (1984)
48. Bourbaki, N.: Topologie générale, Masson, Paris (1990)
49. Boyd, S., El Ghaoui, L., Feron, E., Balakrishnan, V.: Linear matrix inequalities in system and control theory (studies in applied and numerical mathematics), Society for Industrial Mathematics, Philadelphia (2007)

50. Bressan, A., Piccoli, B.: Introduction to mathematical theory of control (Applied mathematics), American Institute of Mathematical Sciences, Springfield (2007)
51. Brézis, H.: Opérateurs maximaux monotones et semi-groupes de contraction dans les espaces de Hilbert. North-Holland, Amsterdam-London (1973)
52. Bruning, E., Blanchard, P.: Calculus of Variations in Mathematical Physics. Springer, Berlin (1991)
53. Bullo, F., Cortes, J., Martinez, S.: Distributed Control of Robotic Networks: A Mathematical Approach to Motion Coordination Algorithms, Princeton series in applied mathematics. Princeton University Press, Princeton (2009)
54. Carathéodory, C.: Calculus of Variations and Partial Differential Equations of the First Order, 3 edn. American Mathematical Society, Providence (1935). Reprinted in 1999
55. Carlson, D.A., Haurie, A.: Infinite Horizon Optimal Control: Theory and Applications Lecture Notes in Economics and Mathematical Systems. Springer, Berlin (1987)
56. Carr, J.: Applications of Centre Manifold Theory. Springer, New York (1981)
57. Cassandras, C.G., Lygeros, J.: Stochastic Hybrid Systems (Automation and Control Engineering). CRC, Boca Raton (2006)
58. Cassandras, C.G., Lafortune, S.: Introduction to Discrete Event Systems, 2nd edn. 2009. Springer, New York (2007)
59. Castaing, C.H., Valadier, M.: Convex Analysis and Measurable Multifunctions. Springer, Berlin (1977)
60. Cea, J.: Optimisation : théorie et algorithmes. Dunod, Paris (1971)
61. Chow, S.N., Hale, J.K.: Methods of Bifuration Theory. Springer, New York (1982)
62. Christofides, P.D.: Nonlinear and Robust Control of PDE Systems. Birkhauser, New York (2001)
63. Clarke, F.H.: Optimization and Nonsmooth Analysis. Wiley-Interscience, New York (1983)
64. Coiffet, Ph.: La robotique. Hermès, Paris (1986)
65. Colonius, F.: Optimal Periodic Control. Lecture Notes in MAthematics, vol. 1313. Springer, New York (1988)
66. Culler, D., Singh, J.P., Gupta, A.: Parallel computer architecture: A hardware/software approach, The Morgan Kaufmann Series in Computer Architecture and Design. Morgan Kaufmann, San Francisco (1998)
67. Cronin, J.: Differential Equations. Introduction and Qualitative Theory. Dekker, New York (1994)
68. Daganzo, C.F.: Fundamentals of Transportation and Traffic Operations. Pergamon, Oxford (1977)
69. Daganzo, C.F.: Logistics Systems Analysis, 3rd edn. Springer, Berlin (1999)
70. Da Prato, G.: Introduction to Differential Stochastic Equations. Appunti dei corsi, Scuola Normale di Pisa (1995)
71. Da Prato, G.: Stochastic Evolution Equations by Semi-Group Methods. Centre de Recerca Matematica, Barselona (1997)
72. Da Prato, G., Zabczyk, J.: Stochastic Evolution Equations in Infinite Dimensions Cambridge University Press, Cambridge (1992)
73. Da Prato, G., Zabczyk, J.: Ergodicity for Infinite Dimensional Systems. Cambridge University Press, Cambridge (1996)
74. Dautray, R., Lions, J.-L.: Analyse mathématique et calcul numérique pour les Sciences et Techniques. Masson, Paris (1985)
75. Day, R.H.: Complex Economic Dynamics, vol. I, An Introduction to Dynamical Systems and Market Mechanims. MIT, Cambrigde (1994)
76. Debreu, G.: Theory of Value. Wiley, New York (1959)
77. De Lara, M., Doyen, L.: Sustainable Management of Natural Resources: Mathematical Models And Methods. Springer, Berlin (2008)
78. Delchamps, D.: State Space and Input-Output Linear Systems. Springer, New York (1987)

79. Delfour, M., Zolesio, J.-P.: Shapes and Geometries: Analysis, Differential Calculus and Optimization. Society for Industrial Mathematics, Philadelphia (1987)
80. Devaney, R.L.: An Introduction to Chaotic Dynamical Systems. Addison-Wesley, Redwood City, CA (1989)
81. Dontchev, A., Rockafellar, R.T.: Implicit Functions and Solution Mappings. Springer, Berlin (2009)
82. Dieudonne, J.: History of Functional Analysis. Mathematical Studies # 49, North-Holland, Amsterdam (1981)
83. Dini, U.: Lezioni di analisi inifinitesimale. Nistri (1877)
84. Dini, U.: Fundamenti per la teoria delle funzioni di variabili reali. Nistri (1878)
85. Dordan, O.: Analyse Qualitative. Masson, Paris (1995)
86. Dubois, D., Prade, H.: Fuzzy Sets and Systems – Theory and Applications. Academic, New York (1980)
87. Dubois, D., Prade, H.: Théorie des possibilités. Masson, Paris (1988)
88. Dolicni, P.J., Canudas-de-Wit, C., Bechart, H.: Dry Clutch Control for Automative Applications. Advances in Industrial Control. Springer, Berlin (2010)
89. Dullerud, G.E.: Control of Uncertain Sampled-Data Systems. Birkhauser, Boston (1995)
90. Dullerud, G.E., Paganini, F.: A Course in Robust Control Theory: A Convex Approach. Springer, Berlin (2010)
91. Edelstein-Keshel, L.: Mathematical Models in Biology. Random House, New York (1987)
92. Ekeland, I.: Théorie des jeux et applications économiques. PUF, Paris (1974)
93. Ekeland, I.: Eléments d'économie mathématique. Hermann, Paris (1979)
94. Ekeland, I.: La calcul, l'imprévu. Seuil, Paris (1984)
95. Ekeland, I.: Au hasard. Seuil, Paris (1991)
96. Ekeland, I.: Le Chaos. Flammarion, Paris (1995)
97. Egerstedt, M., Martin, C.: Control Theoretic Splines: Optimal Control, Statistics, and Path Planning. Princeton Series in Applied Mathematics. Princeton University Press, Princeton (2009)
98. Evans, C.: Partial Differential Equations. AMS, Providence (1998)
99. Farkas, M.: Periodic Motions. Springer, Berlin (1994)
100. Fudenberg, D., Tirole, J.: Game Theory. MIT, Cambridge (1991)
101. Françoise, J.-P.: Oscillations en biologie. Analyse qualitative et modèles. Springer, New York (2005)
102. Franklin, G.F., Powell, J.D., Emami-Naeini, A.: Feeback Control of Dynamic Systems, 6th edn. Prentice Hall, New Jersey (2009)
103. Frankowska, H.: Value function in optimal control. In: Agrachev, A.A. (ed.) Mathematical Control Theory. Lecture Notes, vol. 8. ICTP, http://users.ictp.it/ pub off/lectures/lns008/Frankowska/Frankowska02.pdf (2002)
104. Frankowska, H.: Control under state constraints. In: Proceedeings of the International Congress of Mathematicians (ICM 2010), Hyderabad (2010)
105. Filippov, A.F.: Differential Equations with Discontinuous Right-Hand Side. Nauka and Kluwer, Boston (1985)
106. Fleming, W.H., Rishel, R.W.: Deterministic and Stochastic Optimal Control. Springer, Berlin (1975)
107. Garavello, M., Piccoli, B.: Traffic Flow on Networks (Applied Mathematics). American Institute of Mathematical Sciences, Providence (2006)
108. Genocchi, A., Peano, G.: Calcolo differenziale e principi di calcolo integrale. Bocca (1884)
109. Glass, L., Mackey, M.C.: From Clocks to Chaos. The Rhytms of Life. Princeton University Press, Princeton (1988)
110. Gruene, L.: Asymptotic Behavior of Dynamical and Control Systems Under Perturbation and Discretization. Springer, Berlin (2002)

111. Gulick, D.: Encounters with Chaos. McGraw-Hill, New York (1992)
112. Hadamard, J.: Essai sur la psychologie de l'invention dans le domaine mathématique. Jacques Gabay (1993)
113. Haurie, A., Viguier, L.: The Coupling of Climate and Economic Dynamics: Essays on Integrated Assessment. Springer, New York (2010)
114. Hebb, D.: The Organization of Behavior. Wiley, New York (1949)
115. Heijmans, H.: Morphological Image Operators. Academic, Boston (1994)
116. Hersh, R.: What are Mathematics Today ? Oxford University Press, Oxford (1997)
117. Hespanha, J.P.: Linear Systems Theory. Princeton University Press, New Jersey (2009)
118. Hildenbrand, W., Kirman, A.P.: Equilibrium Analysis. North-Holland, Netherlands (1998)
119. Hiriart-Urruty, J.-B. Lemarechal, C.: Convex Analysis and Minimization Algorithms, (2 volumes). Springer, Berlin (1994)
120. Hirsch, M.W., Pugh, C. Shub, M.: Invariant Manifolds. Lecture Notes in Mathematics. Springer, New York (1977)
121. Hsieh, H., Balarin, F., Sangiovanni-Vincentelli, A.L.: Synchronous Equivalence: Formal Methods for Embedded Systems. Springer, Berlin (2001)
122. Hofbauer, J., Sigmund, K.: Evolutionary Games and Population Dynamics. Cambridge University Press, Cambridge (1988)
123. Iannelli, M.: Mathematical Theory of Age-Structured Population Dynamics. Giardini, Pisa (1995)
124. Iordache, M., Ansaklis, P.J.: Supervisory Control of Concurrent Systems: A Petri Net Structural Approach (Systems and Control: Foundations and Applications). Birkhauser, Boston (2006)
125. Irem, Histoire des problèmes, histoire des mathématiques, Édition Marketin, Ellipses, París (1993)
126. Isaacs, R.: Differential Games. Wiley, New York (1965)
127. Isidori, A.: Nonlinear Control Systems. Springer, Berlin (1989)
128. Jordan, M.I.: Learning in Graphical Models (Adaptive Computation and Machine Learning). MIT, Cambridge (1998)
129. Khalil, H.K.: Nonlinear Systems, 3rd edn. Prentice Hall, New Jersey (2001)
130. Keyfitz, N.: Applied Mathematical Demography. Wiley Intersecience, New York (1977)
131. Kohonen, T.: Self-Organization and Associative Memory. Series in Information Sciences, vol. 8. Springer, Berlin (1984)
132. Kovaleskaia: L'aventure d'une mathématicienne. Belin, Paris (1993)
133. Krasovski, N.N., Subbotin, A.I.: Positional Differential Games. Springer, Nauka, Moscow (1974)
134. Kryszewski, W.: Homotopy Properties of Set-Valued Mappings. Univ. Nicolas Copernic, Torun (1997)
135. Krstic, M., Kanellakopoulos, I., Kokotovic, P.V.: Nonlinear and Adaptive Control Design. Adaptive and Learning Systems for Signal Processing, Communications and Control Series. Wiley-Interscience, New York (1995)
136. Krstic, M. Smyshlyaev, A.: Boundary Control of PDEs: A Course on Backstepping Designs (Advances in Design and Control). Society for industrial and applied mathematics, Philadelphia, PA (2008)
137. Kumar, P.R., Varaiya, P.: Stochastic Systems: Estimation, Identification and Adaptive Control, Prentice-Hall Information and System Sciences Series. Prentice Hall, New Jersey (1986)
138. Kuratowski, K.: Topologie, vols. 1 and 2, 4th edn. corrected, Panstowowe Wyd Nauk, (Academic, New York, 1966) (1958)
139. Lasota, A., Mackey, M.C.: Probabilistic Properties of Deterministic Systems. Cambridge University Press, New York (1985)

140. Lee, E.A. Varaiya, P.: Structure and Interpretation of Signals and Systems. Addison Wesley, Reading (2002)
141. Lee, I., Leung, J.Y.-T., Son, S.H.: Handbook of Real-time and Embedded Systems (Chapman and Hall/CRC Computer and Information Science Series), CRC, Boca Raton (2007)
142. Leitmann, G.: The Calculus of Variations and Optimal Control. Plenum, New York (1981)
143. Liberzon, D.: Switching in Systems and Control (Systems and Control: Foundations and Applications). Birkhauser, Boston (2003)
144. Lions, J.-L.: Contrôle optimal de systèmes gouvernés par des equations aux dérivées partielles. Dunod, Paris (1968)
145. Lions, J.-L.: Optimal Control of Systems Described by Partial Differential Equations. Springer, Berlin (1971)
146. Lions, J.-L.: Sur quelques questions d'analyse, de mécanique et de contrôle optimal, Presses de l'Université de Montréal (1976)
147. Lions, P.-L.: Generalized Solutions of Hamilton-Jacobi Equations. Pitman, London (1982)
148. Litrico, X., Fromion, V.: Modeling and Control of Hydrosystems. Springer, Berlin (2009)
149. Loulou, R., Waaub, J.-P., Zaccour, G.: Energy and Environment. Springer, New York (2010)
150. Lorenz, T.: Mutational Analysis. A Joint Framework for Cauchy Problems in and Beyond Vector Spaces. Series: Lecture Notes in Mathematics, vol. 1996. Springer, Berlin (2010)
151. Mahmassani, H.S.: In Perpetuial Motion: Travel Behavior Research Opportunities and Application Challenges. Pergamon/Elsevier Science, Amsterdam (2002)
152. Marro, G.: Fondamenti di teoria dei sistemi, Patron Editore (1975)
153. Matheron, G.: Random Sets and Integral Geometry. Wiley, New York (1975)
154. Mawhin, J.: Analyse Fondements techniques évolution. De Boeck Université (1992)
155. Mawhin, J., Willems, M.: Critical Point and Hamiltonian Systems. Springer, Berlin (1993)
156. Mesbahi, M., Egerstedt, M.: Graph Theoretic Methods in Multiagent Networks. Princeton University Press, Princeton (2010)
157. Mettler, B.: Identification Modeling and Characteristics of Miniature Rotorcraft. Springer, New York (2010)
158. Meyer, Y.: Ondelettes et opérateurs. Tome 1. Ondelettes. Hermann, Paris (1990)
159. Moin, P.: Fundamentals of Engineering Numerical Analysis. Cambridge University Press, Cambridge (2010)
160. Moulin, H.: Théorie des jeux pour l'économie et la politique. Hermann, Paris (1980)
161. Murray, J.: Biomathematics. Springer, Berlin (1990)
162. Murray, R.M., Li, Z., Sastry, S.S.: A mathematical Introduction to Robotic Manipulation. CRC, Boca Raton (1994)
163. Monod, J.: Chance and Necessity. Vintage, New York (1971)
164. Murray, J.: Theoretical Biology. Springer, Berlin (1990)
165. Nagumo, M., Yamaguti, M., Nirenberg, L., Mizohata, S., Sibuya, Y.: Collected Papers. Springer, New York (1993)
166. Najman, L., Talbot, H.: Mathematical Morphology. Springer, Berlin (2010)
167. Nonnenmacher, T.F., Losa, G.A, Weibel, E.R.: Fractals in Biology and Medecine. Birkhauser, Boston (1994)
168. Nordon, D.: Les mathématiques pures n'existent pas. Actes Sud (1993)
169. Oden, J.T.: Qualitative Methods in Nonlinear Mechanics. Prentice Hall, New Jersey (1985)
170. Patterson, D.A., Hennessy, J.L.: Computer Organizationa and Design: The Hardware/Software Interface. The Morgan Kaufmann Series in Computer Architecture and Design. Morgan Kaufmann, Boston, MA (2008)

171. Pflug, G.C.: Optimization of Stochastic Models. Klüwer, Dordrecht (1996)
172. Pironneau, O.: Méthodes des éléments finis pour les fluides. Masson, Paris (1988)
173. Rockafellar, R.T.: Convex Analysis. Princeton University Press, Princeton (1970)
174. Rockafellar, R.T., Wets, R.: Variational Analysis. Springer, Berlin (1997)
175. Roseau, M.: Equations Différentielles. Masson, Paris (1976)
176. Rouché, N., Mawhin, J.: Equations Différentielles (Tome 1 : Théorie générale). Masson, Paris (1973)
177. Rouché, N., Mawhin, J.: Equations différentielles ordinaires (Tome 2 : stabilité et solutions périodiques). Masson, Paris (1973)
178. Rockafellar, R.T., Wets, R.: Variational Analysis. Springer, Berlin (1997)
179. Rudin, W.: Analyse réelle et complexe. Masson, Paris (1992)
180. Rus, T., Rus, D.: Execution Support Environment, System Software and Software Systems, Volume 2. Worlds Scientific Publishing Company Inc, River Edge, NJ, USA (1994)
181. Russell, S., Norvig, P.: Artifical Intelligence: A Modern Approach, 3rd edn. Prentice Hall, New Jersey (2009)
182. Saari, D.G.: Geometry of Voting. Springer, New York (1994)
183. Saari, D.G.: Collisions, Rings, and Other Newtonian N-Body Problems. American Mathematical Society, Providence, RI (2005)
184. Saari, D.G.: Disposing Dictators, Demystifying Voting Paradoxes: Social Choice Analysis. Cambridge University Press, Cambridge (2008)
185. Saligrama, V.: Networked Sensing Information and Control. Springer, New York (2008)
186. Sastry, S.: Nonlinear Systems: Analysis, Stability and Control. Springer, New York (2010)
187. Schmitt, M., Mattioli, J.: Morphologie Mathématique. Masson, Paris (1991)
188. Schwartz, L.: Méthodes mathématiques pour la physique. Hermann, Paris (1969)
189. Schwartz, L.: Les tenseurs. Hermann, Paris (1975)
190. Schwartz, L.: Théorie des Distributions. Hermann, Paris (1978)
191. Schwartz, L.: Topologie Générale et Analyse Fonctionelle. Hermann, Paris (1980)
192. Schwartz, L.: Cours d'Analyse (tomes I and II), Hermann (1981)
193. Schwartz, L.: Un mathématicien aux prises avec le siècle. Odile Jacob, Paris (1997)
194. Segel, L.A.: Modeling Dynamic Phenomena in Molecular and Cellular Biology. Cambridge University Press, Cambridge (1984)
195. Serra, J.: Image Analysis and Mathematical Morphology. Academic, London (1982)
196. Sigmund, K.: Games of Life. Oxford University Press, Oxford (1993)
197. Smoller, J.: Shocks Waves and Reaction-Diffusion Equations. Springer, Berlin (1983)
198. Smyshlyaev, A., Krstic, M.: Adaptive Control of Parabolic PDEs. Princeton University Press, Princeton (2010)
199. Sokokowski, J., Zolesio, J.-P.: Introduction to Shape Optimization. Springer, Berlin (1992)
200. Speyer, J.L., Jacobson, D.H.: Primer on Optimal Control Theory. Advances in Design and Control. Society for Industrial and Applied Mathematics, Philadelphia, PA (2010)
201. Stroock, D.W., Varadhan, S.R.S.: Multidimensional Diffusion Processes. Springer, Berlin (1979)
202. Tabuada, P.: Verification and Control of Hybrid Systems: A Symbolic Approach. Springer, New York (2009)
203. Temam, R.: Navier-Stokes Equations. North-Holland, Amsterdam (1984)
204. Temam, R.: Infinite Dimensional Dynamical Systems in Mechanics and Physics. Springer, Berlin (1988)
205. Tilman, D., Kareiva, P.: The Role of Space in Population Dynamics and Interspecific Interactions. Pinceton University Press, Pinceton (1998)
206. Treves, F.: Topological Vector Spaces, Distributions and Kernels. Academic, New York (1967)

207. Vasquez, R., Krstic, M.: Control of Turbulent and Magnetohydrodynamic Channel Flows: Boundary Stabilization and State Estimation (Systems and Control: Foundations and Applications). Birkhauser, Boston (2007)
208. Von Neumann, J., Morgenstern, O.: Theory of Games and Economic Behaviour. Princeton University Press, Princeton (1944)
209. Walrand, J., Varaiya, P.P.: High-Performance Commication Networks. The Morgan Kaufmann Series in Networking. Morgan Kaufmann, Boston, MA (1996)
210. Webb, G.: Theory of Nonlinear Age Dependent Population Dynamics. Marcell Dekker, New York (1985)
211. Weidlich, W., Haag, G.: Concepts and Models of a Quantitative Sociology: The Dynamics of Interacting Populations. Springer, Berlin (1983)
212. Visintin, A.: Differential Models of Hysteresis. Springer, New York (1994)
213. Wiggins, S.: Global Bifurcations and Chaos: Analytical Methods. Springer, New York (1988)
214. Wiggins, S.: Introduction to Nonlinear Systems and Chaos. (1990)
215. Wigner, E.P.: The Unreasonable Effectiveness of Mathematics in the Natural Sciences. Commun. Pure Appl. Math. **13**, 1–14 (1960)
216. Wonham, W.M.: Linear Multivariable Control. A Geometric Approach. Springer, New York (1985)
217. Young, L.C.: Lectures on the Calculus of Variations and Optimal Control Theory. Saunders, Philadelphia (1969)
218. Young, L.C.: Mathematicians and Their Times. Mathematical Studies # 48, North-Holland, New York (1981)
219. Ygnace, J.-L.: Les systèmes de transport intelligents, La Documentation Française (2000)
220. Yoshizawa, T.: Stability Theory by Lyapunov's Second Method. The mathematical society of Japan (1966)
221. Zabczyk, J.: Mathematical Control Theory: An Introduction. Birkhäuser, Boston (1992)
222. Zabczyk, J.: Chance and Decision: Stochastic Control in Discrete Time. Quaderni, Scuola Normale di Pisa (1996)
223. Zaccour, G.: Decision and Control in Management Science: Essays in Honor of Alain Haurie. Springer, Berlin (2010)
224. Zeidler, E.: Nonlinear Functional Analysis and Its Applications. Springer, New York (1984)

Index